A

GOLD

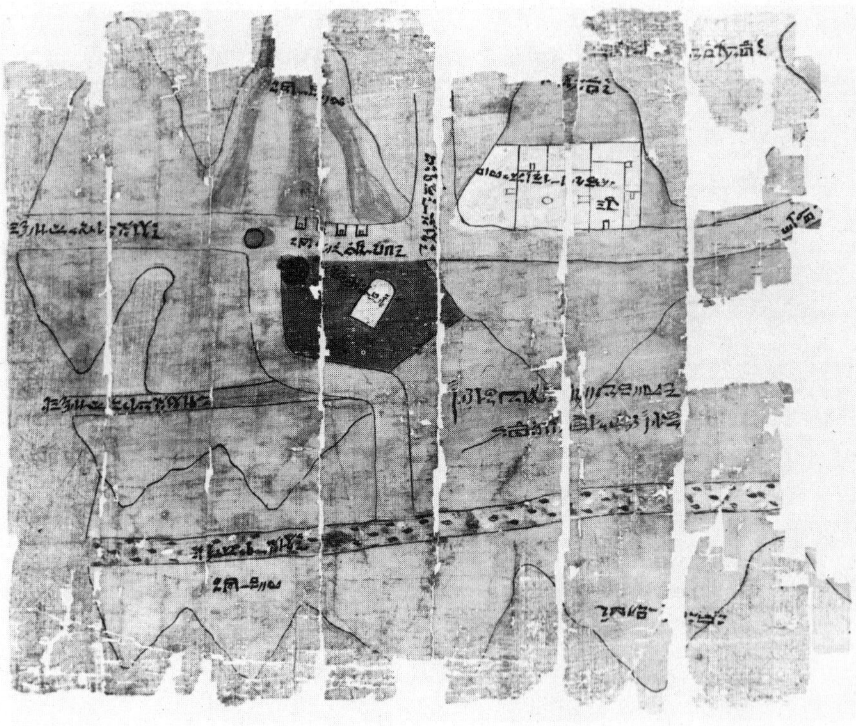

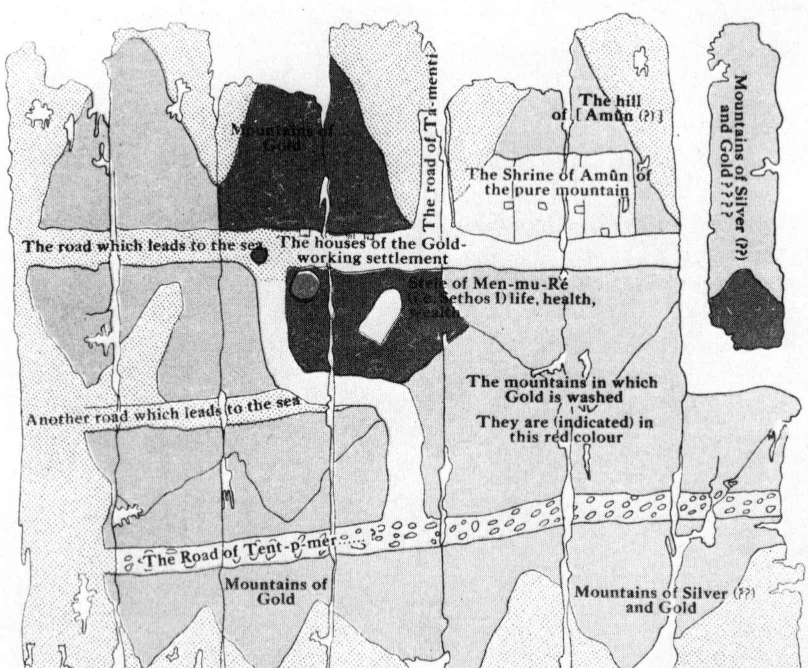

Top: The original Turin Papyrus of the Egyptian gold mine (circa 1320 B.C.), the oldest geological map extant. (Photo courtesy Il Soprintendente per le antichita egizie, Prof. S. Curto, Turino). *Bottom:* Interpretation of the script on the Turin Papyrus of the Egyptian gold mine (after Ball, 1942; see chapter 3).

GOLD

History and Genesis of Deposits

Robert W. Boyle
Geological Survey of Canada
Ottawa, Ontario, Canada

Sponsored by
SOCIETY OF ECONOMIC GEOLOGISTS
and
SOCIETY OF ECONOMIC GEOLOGISTS FOUNDATION

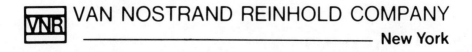

VNR VAN NOSTRAND REINHOLD COMPANY
New York

Library of Congress Catalog Card Number: 86-15675
ISBN: 0-442-21162-7

Manufactured in the United States of America.

Published by Van Nostrand Reinhold Company Inc.
115 Fifth Avenue
New York, New York 10003

Van Nostrand Reinhold Company Limited
Molly Millars Lane
Wokingham, Berkshire RG11 2PY, England

Van Nostrand Reinhold
480 La Trobe Street
Melbourne, Victoria 3000, Australia

Macmillan of Canada
Division of Canada Publishing Corporation
164 Commander Boulevard
Agincourt, Ontario MIS 3C7, Canada

15 14 13 12 11 10 9 8 7 6 5 4 3 2 1

Library of Congress Cataloging in Publication Data
Boyle, R. W. (Robert William), 1920-
Gold : history and genesis of deposits.

Includes bibliographies and index.
1. Gold ores. 2. Gold. I. Title.
TN420.B63 1987 553.4'1 86-15675
ISBN 0-442-21162-7

CONTENTS

CHAPTER 6: GOLD DURING THE TRANSITION TO MODERN SCIENTIFIC VIEWS 65

Theories of Origin of Gold and Other Mineral Deposits / 66
References and Selected Bibliography / 76

CHAPTER 7: GOLD IN THE MODERN ERA 79

Outline of Theories of Origin of Mineral Deposits in the Modern Era / 81
References and Selected Bibliography / 82

CHAPTER 8: GEOCHEMISTRY OF GOLD 85

Geochemistry of Gold / 88
8-1 BISCHOF, K. GUSTAV: Gediegenes Gold / 89
Lehrbuch der chemischen und physikalischen Geologie, 3 vols., Adolph Marcus, Bonn, 1855, pp. 2050-2057
8-2 CLARKE, FRANK W.: Gold / 97
The data of geochemistry, *U.S. Geol. Survey Bull. 770,* 1924, pp. 656-663
8-3 FERSMAN, A. E.: Gold (Au-At. Wt. 79) / 105
Translated from *Geochemistry,* vol. 4, Leningrad, 1939, pp. 262-271
8-4 GOLDSCHMIDT, V. M.: Gold / 118
Geochemistry, A. Muir, ed., Clarendon Press, Oxford, 1954, pp. 197-205
8-5 RANKAMA, K., and TH. G. SAHAMA: Silver, Gold / 128
Geochemistry, The University of Chicago Press, Chicago, 1950, pp. 702-707
8-6 PETROVSKAYA, N. V.: An Outline of the Geochemistry of Gold / 135
Translated from *Native Gold,* Idz. "Nauka", Moscow, 1973, pp. 8-20
Specialized Papers on the Geochemistry of Gold / 151
8-7 LENHER, V.: The Transportation and Deposition of Gold in Nature / 153
Econ. Geology **7:**744-750 (1912)
8-8 SMITH, F. G.: The Alkali Sulphide Theory of Gold Deposition / 169
Econ. Geology **38:**561-590 (1943)
8-9 LAKIN, W. H., G. C. Curtain, A. E. Hubert, H. T. Shacklette, and K. G. Doxtader: Geochemistry of Gold in the Weathering Cycle / 178
U.S. Geol. Survey Bull. 1330, 1974, pp. 71-75
8-10 FREISE, F. W.: The Transportation of Gold by Organic Underground Solutions / 182
Econ. Geology **26:**421-431 (1931)
General Summary / 194
References and Selected Bibliography / 194

CHAPTER 9: THE ORIGIN OF EPIGENETIC GOLD DEPOSITS— THE ORE-MAGMA THEORY 197

The Ore-Magma Theory of the Origin of Auriferous Veins / 197
9-1A SPURR, J. E.: The Origin of Ore Magmas or Solutions: Veindikes / 205
The Ore Magmas, McGraw-Hill, New York, 1923, pp. 1-3
9-1B SPURR, J. E.: The Origin of Fissure Veins / 207
The Ore Magmas, McGraw-Hill, New York, 1923, pp. 707-709
9-2 FARMIN, R.: Host-Rock Inflation by Veins and Dikes at Grass Valley, California / 210
Econ. Geology **36:**143, 163-174 (1941)

14-2 HORWOOD, C. B.: Concluding Remarks / 342
The Gold Deposits of the Rand, Charles Griffin & Co., London, 1917, pp. 369-393
14-3 GRATON, L. C.: Hydrothermal Origin of the Rand Gold Deposits / 367
Econ. Geology **25** (suppl. to no. 3):182-185 (1930)
14-4 MELLOR, E. T.: The Conglomerates of the Witwatersrand / 373
Inst. Mining and Metallurgy Trans. **25:**261-291 (1916)
14-5 LIEBENBERG, W. R.: The Occurrence and Origin of Gold and Radioactive Minerals in the Witwatersrand System, the Dominion Reef, the Ventersdorp Contact Reef and the Black Reef / 404
Geol. Soc. South Africa Trans. **58:**218-222 (1955)
14-6 PRETORIUS, D. A.: The Depositional Environment of the Witwatersrand Goldfields: A Chronological Review of the Speculations and Observations / 409
Minerals Sci. Eng. **7**(1):18-20, 23-47 (1975)
14-7 HALLBAUER, D. K., and K. T. VAN WARMELO: Fossilized Plants in Thucholite from Precambrian Rocks of the Witwatersrand, South Africa / 439
Precambrian Res. **1:**199-212 (1974)
Other Deposits / 453
References / 453

CHAPTER 15: GOLD DEPOSITS—PLACERS 455

Eluvial Placers / 456
Alluvial Placers / 462
15-1 TYRRELL, J. B.: The Law of the Pay-Streak in Placer Deposits / 465
Inst. Mining and Metallurgy Trans. **21:**593-613 (1912)
15-2 LINDGREN, W.: Gold of the Tertiary Gravels / 489
The Tertiary gravels of the Sierra Nevada of California, *U.S. Geol. Survey Prof. Paper 73,* 1911, pp. 65-76
15-3 MACKAY, B. R.: Economic Geology: Placer Gold / 504
Beauceville Map-area, Quebec, *Canada Geol. Survey Mem. 127,* 1921, pp. 63-69
15-4 SHILO, N. A., and Yu. V. SHUMILOV: Mechanisms of Behaviour of Gold during Placer Formation Processes in the North-East of the USSR / 514
25th Internat. Geol. Congress, Australia, Abstracts, vol. 1, 1976, p. 224
15-5 ZHELNIN, S. G., and Yu. A. TRAVIN: Comparative Features of Placer Geology in the Modern Structure of the North-Eastern U.S.S.R. / 516
25th Internat. Geol. Congress, Australia, Abstracts, vol. 1, 1976, pp. 229-230
Special Problems of Placer Gold / 518
15-6 MERTIE, J. B., JR.: Placer Gold in Alaska / 520
Washington Acad. Sci. Jour. **30:**(3):114-124 (1940)
15-7 CHENEY, E. S., and T. C. PATTON: Origin of the Bedrock Values of Placer Deposits / 532
Econ. Geology **62**(6):852-853 (1967)
15-8 GUNN, C. B.: Origin of the Bedrock Values of Placer Deposits / 534
Econ. Geology **63**(1):86 (1968)
15-9 TUCK, R.: Origin of the Bedrock Values of Placer Deposits / 535
Econ. Geology **63**(2):191-193 (1968)
15-10 KROOK, L.: Origin of the Bedrock Values of Placer Deposits / 537
Econ. Geology **63**(7):844-846 (1968)
References / 542

CHAPTER 16: OXIDATION AND SECONDARY ENRICHMENT OF GOLD DEPOSITS **545**

16-1 EMMONS, W. H.: Gold / 548
The Enrichment of Ore Deposits, *U.S. Geol. Survey Bull. 625,* 1917, pp. 305-324
16-2 SMIRNOV, S. S.: Gold / 569
The Zone of Oxidation of Sulphidic Mineral Deposits, Izd. Akad. Nauk USSR, Moscow, 1951, pp. 198-207
16-3 ZVYAGINTSEV, O: Review: Behavior of Gold in the Zone of Oxidation of Auriferous Sulphide Deposits / 577
Geochemistry **6:**683-685 (1959)
16-4 LESURE, F. G.: Residual Enrichment and Supergene Transport of Gold, Calhoun Mine, Lumpkin County, Georgia / 581
Econ. Geology **66:**178 (1971)
References / 582

CHAPTER 17: GOLD DEPOSITS-SPECIAL TOPICS **583**

Associated Minerals and Elements in Auriferous Deposits / 583
Au/Ag Ratios and Fineness of Native Gold in Auriferous Deposits / 584
Wall Rock Alteration Effects in Auriferous Deposits / 587
Structural Environment of Deposition of Epigenetic Gold Deposits / 596
References / 598
17-1 WHITE, W. H.: The Mechanism and Environment of Gold Deposition in Veins / 599
Econ. Geology **38:**512-532 (1943)
17-2 EBBUTT, F.: Relationships of Minor Structures to Gold Deposition in Canada / 619
Structural Geology of Canadian Ore Deposits, Canadian Institute of Mining and Metallurgy, Montreal, 1948, pp. 64-69, 73, 77

CHAPTER 18: THE ECONOMICS OF GOLD AND GOLD MINING **627**

History of the Economics of Gold to the End of the Middle Ages / 628
Economics of Gold during the Transition to Modern Times / 643
Economics of Gold and Gold Mining during the Modern Era / 646
Uses of Gold / 655
The Future of Gold / 657
Conclusions / 660
References and Selected Bibliography / 661

Author Citation Index / 663
Subject Index / 671

FOREWORD

GOLD: History and Genesis of Deposits is the product of an effort by the Society of Economic Geologists to publish materials that will expand knowledge concerning timely, specific topics important to the study of economic geology and to economic geologists. A volume on gold was selected for a general review-type publication because of the importance of the gold mining industry in the 1980s. The officers and council of the Society of Economic Geologists authorized the preparation of this book on gold in 1981, and Dr. Robert W. Boyle was selected as its author.

Dr. Boyle has extensive experience in the study of gold deposits. He has an international reputation and a broad interest and understanding of the gold mining industry, the origin of gold deposits, and the history of gold as a metal and ore from prehistoric times to recent. Dr. Boyle uses important publications on gold deposits as source materials to document the various pathways of geological thought over time to introduce the reader to modern concepts. The book contains a wealth of information concerning gold.

The Society of Economic Geologists hopes the book will appeal to a broad spectrum of readers wishing to develop a better understanding of gold and gold geology. The book is written to present gold, which has long been viewed as a symbol of wealth and power, as a mineral commodity. Knowledge of the geology of gold has changed and developed over an extended history but the long grassroots-type interest in it has changed little. Information concerning both interests is to be found in the book, giving the reader the opportunity to follow the historic development of an understanding of the occurrences and sources of gold.

In the 1960s the Society considered and rejected the idea of publishing a series of single-topic, review benchmark-type volumes. The idea was revived in the 1980s with this volume on gold to be the first of a series. This idea is now in abeyance and is not likely to be revived. The reception of this volume may hold the future of other commodity-oriented, review-type books prepared under the guidance of the Society of Economic Geologists.

This volume owes much in its concept to E. L. Ohle and H. L. James; in carrying it from manuscript to publication to Brian J. Skinner, Editor of *Economic Geology* and to Charles S. Hutchinson, Jr., of Van Nostrand Reinhold Company; for moving the book through its long development phase to the past chairmen of the SEG Publications Committee—Robert O. Rye, Eric S. Cheney, and particularly John F. Slack for overseeing the review process; and to Roger P. Ashley, Richard W. Hutchinson, and J. James Eidel for technical reviews.

SOCIETY OF ECONOMIC GEOLOGISTS

PREFACE

The general philosophy underlying this volume on gold can be simply stated as the collection, sifting, and analysis of landmark papers on the history and genesis of the deposits of the most noble of metals. This volume seems necessary today since the literature on the geology and geochemistry of gold is so vast, so scattered, and in the case of many classic and historic papers, quite inaccessible to many economic geologists, particularly those who have little or no access to major libraries because of their isolation in mining camps, located in many cases far from urban centers. This book is designed to alleviate this problem, at least in part, although it should be emphasized that no single volume on gold can ever hope to include all of the significant conceptual papers on the history and genesis of the deposits of this precious metal.

The book opens with a general historical introduction on gold followed by a background chapter on the geochemistry of the element and the types of auriferous deposits. This background sets the stage for chapters containing the classic papers on gold from the world literature, either in their entirety, in significant extracts, or in abstract form. The papers chosen to cover the subjects of the volume comprise principally papers from the last century to relatively recent contributions. Each paper or group of papers is accompanied by an appropriate introduction and concluding commentary. Where the papers are in a language other than English an attempt has been made to provide translations for those of more recent date.

The introductions and commentaries that accompany each paper or group of papers in the volume are intended to set the papers into historical perspective and to emphasize their importance in the growth of ideas relating to the history of gold and the genesis of its deposits. Sufficient references are given at the end of each chapter to enable the interested reader to pursue these subjects at greater length.

The history of gold is a long one, going back to the dawn of civilization, at least some 5,000 years. The theories postulated to explain the origin of the various types of gold deposits have, likewise, had a long life. To set the development of the varied theories of origin of auriferous deposits into a historical context is particularly difficult. Often at any one point in time, two or more theories have been current, whereas at other times only one theory has generally prevailed. In addition, some theories of the origin of epigenetic and placer gold deposits have been cyclic, appearing from time to time during the history of economic geology. Despite these problems I think it is possible to define a certain historical sequence of theories respecting epigenetic gold deposits of certain general types. As followed in this book these are an early period when an igneous magmatic theory prevailed, a period when magmatic hydrothermal theories predominated, a short period marked by granitization theories, and a recent period in which exhalite theories have been postulated. Secretion theories have had a cyclic history. Introduced first by Agricola during the Renaissance, they waned shortly thereafter to reappear in the latter part of the nineteenth century as lateral secretion *(sensu stricto)*; forgotten for nearly 50 years they have again appeared in a modern guise as metamorphic and hydrothermal secretion theories *(sensu lato)*. The enigmatic auriferous quartz-pebble conglomerates, the largest known gold deposits, have been explained in many ways, the predominating theories of their origin being strictly placer, modified placer, and hydrothermal. Likewise, the origin of the gold dust and nuggets in placers have from time to time had advocates of a purely mechanical process of concentration, an essentially chemical mode of formation, and a combined chemical-mechanical mechanism of accumulation.

Numerous classifications of gold deposits have been suggested since the inception of economic geology as a science. Most of these, twenty or more in number, are based mainly on genetic principles, are therefore speculative and often more deviative than useful, points emphasized by J. M. Maclaren many years ago in his great monograph on gold deposits. Genetic classifications are unsatisfactory for gold deposits because gold has such a diversity of deposits, originating from many physical and chemical processes, that each deposit type is a genetic study in itself. I was confronted with this situation in writing my monograph on gold in 1979; my answer to the problem was to frame a strictly taxonomic classification of auriferous deposits based on their geological setting and geochemical character. This classification, modified slightly to accommodate recent work, appears in Chapter 1. The reader may find this classification somewhat of a departure from most textbooks on economic geology, but it seems to me that this type of taxonomic classification is more valuable than one based on speculative genetic theories. By stating the geological and geochemical facts clearly and objectively, it should be possible in most cases to discern the origin of particular types of gold deposits, and more importantly, to predict the environment where one may prospect for similar deposits.

The scientific and economic literature on gold is vast. That available for the period prior to the fall of the Roman Empire in the west in 476 A.D. is, however, meager, although originally the number of references to gold was evidently large, judging from the writings of Pliny the Elder in his *Historia naturalis*. That much of this early literature was lost is probably due to the edict of the Emperor Diocletian, passed about 290 A.D., ordering the destruction of all works on alchemical subjects, and especially on gold and silver, throughout the Roman Empire—a measure taken to prohibit the makers of gold and silver to amass riches which might have permitted them to organize revolts against the empire. During Medieval times the literature on gold was again restricted mainly because of the general decline in gold mining due to wars, brigandage, and political unrest. With the advent of more stable times and especially after the invention of the printing press circa 1445, the literature on gold has increased consistently to the present day. These vagaries in the availability of the literature on gold have caused some difficulty in the selection of landmark papers. Nevertheless in selecting papers, abstracts, and quotations concerning the geology, geochemistry, theories of origin, and economics of gold deposits, I have striven to maintain a proper historical balance in accordance with the literature available through the ages.

The landmark papers on gold in this book provide well-defined points in the progress of the science of economic geology, especially in the study of the development of the theories of origin of economic deposits, not only of gold but also of the various elements with which this precious metal is associated. Such a study is not merely a subject of antiquarian interest. On the contrary, an adequate knowledge of the work and theories of the past should be an indispensible part of the training of any economic geologist who aspires to understand the processes that lead to concentrations of the elements. Often by looking back at the landmark papers one can mark out a clearer path ahead in this understanding.

Acknowledgments

In the selection of the classic papers that constitute "Landmarks" in the history and theories of origin of gold deposits I have corresponded with and sought advice from many geologists and geochemists throughout the world who are familiar with the literature on gold. In this respect I wish to thank collectively all those who responded to my queries about "Landmark Papers", and in particular I wish to thank many individuals who through the years have supplied me with reprints and other information on gold, especially Frank Ebbutt of Toronto; H. S. Bostock, J. F. Henderson, and I. R. Jonasson of the Geological Survey of Canada; A. S. Radtke, formerly of the United States Geological Survey; J.-J. Bache, BRGM, Orleans, France; M. Ziauddin, Geological Survey of India; D. A. Pretorius, University of the Witwatersrand, South Africa; and N. A. Shilo, N. V. Petrovskaya, I. S. Rozhkov, and N. V. Roslyakova and N. A. Roslyakov, of the Academy of the Sciences, USSR.

Translations of the classical papers in Russian were done by C. de Leuchtenberg during the first stages of preparation of this book and later by members of the Translation Bureau, Multilingual Services Division, Department of the Secretary of State, Ottawa. The drafting of a number of figures was carried out in the Geological Survey of Canada under the supervision of R. F. Daugherty. The library staff of the Geological Survey of Canada, especially A. E. Bourgeois and D. E. Tedford, are thanked for their inestimable help in identifying sources and obtaining copies of most of the classical papers reproduced or commented upon in this volume. I also wish to thank the Librarians of the Ramakrishna Institute, Calcutta, for their assistance while researching ancient science, mining, and metallurgy in India. The typing of the manuscript was carried out in the Geological Survey of Canada word processing center under the supervision of Miss Janet Legere, to whom I owe a sincere vote of thanks for a job well done.

Drs. R. P. Ashley and R. W. Hutchinson have read the text and offered a number of improvements for which the writer is especially grateful.

Finally, I wish to express my gratitude to my wife, Marguerite, whose help in the field in documenting many of the features of the classical gold fields and in assisting in the referencing, reading, and correcting of the manuscript has been more valuable than fine gold itself.

ROBERT W. BOYLE

CONTENTS BY AUTHOR

GOLD

Introduction

Long has man travelled in the realms of gold.

Gold has been called the first folly of man, the whore of civilization, a barbarous relic, the savior of civilization, and a host of other fanciful, if at times derogatory, epithets. Nevertheless, man has retained a curious fascination for the metal for more than 5,000 years, undoubtedly promoted by its great natural beauty and universal durability.

Gold has a long history, and the most noble of metals is many things to different persons. To the geochemist it is a rare metal, the geochemistry of which is intricate and complex; to the mining engineer and metallurgist it presents a challenge of extraction from the earth and from its ores; to the artist, goldsmith, and jeweler it is a metal of superb and everlasting beauty; to the industrial artist it is a metal with unique properties useful in electronics and many other artifices of man; to the numismatist it is a coinage metal with a long and interesting history; and finally to the economist, a valuable standard against which wealth is measured and an imperishable medium for balancing international accounts.

The term *gold* is said by the scholars and philologists to come from the Sanskrit *jvalita,* derived from *jval* 'to shine.' Our word *gold* derives from the Anglo-Saxon *gold,* a word apparently corrupted from the Teutonic *gulth* 'glowing or shining metal.' The Latin term for gold, *aurum,* and the earlier Sabine *ausum* are said to be words of early Italian origin related to *aurora* meaning 'glowing dawn.' Another version has it that the Latin word *aurum* derives from the Hebrew *aor* meaning 'light.' The Latin term is preserved in the chemical symbol for gold, Au, and in the terminology of its salts, *aurous* and *auric.*

In the literature of nearly all cultures the word *gold* and its derivatives appear more often than any other word tinging metaphor and simile and providing innumerable parables, analogies, and proverbs. We have Golden Ages, the Golden Apple, the Golden Ass, the Golden Bough, the Gold Coast, Gold Cups, Golden Bulls, the Golden Calf, the Golden Gate, the Golden Gloves, the

Golden Rule, and the Gold Standard, to name only a few metaphors. In simile we have such commonplace sayings as heart of gold, good as gold, and so on. Perhaps the two best-known proverbs involving gold are "All that glistens is not gold" and "Gold is where you find it."

The Egyptians used the most perfect of planar geometric figures, the circle, as the symbol for gold, the most perfect and noblest of the metals. The alchemists associated gold with the sun *(Sol)* or with the Greek sun-god *(Apollo)* and represented it by the symbol of perfection, the circle with a dot at the center, or by the circle with a crown of rays to represent the king or Apollo of metals. To the early Hindu philosophers gold was the "mineral light"; to the early Western philosophers the metal was the image of solar light and hence of the divine intelligence of the universe.

The desire for gold has markedly influenced man's history: the cry "gold" has lured men across oceans and continents, over the highest mountain ranges, into Arctic tundras and scorching deserts, and through nearly impenetrable jungles. Its gleam prompted the expeditions and conquests of Jason of Thessaly, Cyrus and Darius of Persia, Alexander of Macedon, Caesar of Rome, Columbus of Genoa, Vasco da Gama and Amerigo Vespucci of Portugal, Cortez and Pizzarro of Spain, Raleigh of England, and many others throughout history. Gold has carried the torch of civilization to the remotest regions of the world; unfortunately the *auri sacra fames** has also wrought terrible acts of slavery, war, and bitter contention upon mankind. So it is also with most other materials of this earth.

To make gold from baser metals was a major preoccupation of the alchemists (as were also their ceaseless efforts to discover the elixir of life and the fountain of youth). The fruits of their labors gave us the rudiments of modern chemistry.

Gold has a widespread occurrence in practically every country of the world and has influenced the exploration and settlement of most. In Africa, Europe, and Asia ancient gold mines are known in Egypt, Spain, France, Great Britain, Yugoslavia, Romania, Greece, Turkey, Saudi Arabia, Iran, India, China, Japan, and the U.S.S.R. Ancient placers have yielded gold from the rivers Tagus, Guadalquivir, Tiber, Po, Rhone, Rhine, Hebrus (Maritsa), Nile, Zambezi, Niger, Senegal, Pactolus (in ancient Lydia), Oxus (Amu Darya that flows through the golden land of Samarkand), Indus, Ganges, Lena, Aldan, Amur, Yangtze, and a multitude of others. The artisans of the earliest civilizations of Anatolia (Çatal Hüyük), Mesopotamia (Sumer), and the Indus Valley (Harappa and Mohenjo-Daro) worked in gold obtained from many sites in the Caucasus and Middle Asia, the Middle East, and the Indian Peninsula. The Egyptians mined gold extensively in eastern Egypt and Sudan (Nubia) as far back as 4,000 years ago. It was from them that the Persians, Greeks, and Romans in turn learned the techniques of gold prospecting, mining, and metallurgy. The Greeks and Romans mined gold ores from the extensive metalliferous regions of their empires. Pliny the Elder (A.D. 23-79) in his *Historia naturalis,* written in the early years of our era, repeatedly mentions the mining and metallurgy of gold,

---

**Quid non mortalia pectora cogis,*
Auri sacra fames!

Accursed thirst for gold!
What dost thou not compel mortals
to do.

Virgil, *Aeneid*

and during the Renaissance Agricola referred to it, as had many others before him during the Middle Ages.

Compared with the gold placers and mines of the Old World, those in parts of the New may be as ancient, although it would appear that the aborigines of North and South America placed little emphasis on gold beyond its use in ornaments, jewelery, sacrificial knives, and the like. Columbus of Genoa found the natives of Hispaniola (Haiti) in possession of gold nuggets in 1492, a fact that excited the Spaniards to later pursue their conquests of Mexico and South America, where in 1550 they found their Eldorado in the fabulous placer deposits of the Chocó in Colombia. In Brazil gold was sought by the Portuguese during the last half of the sixteenth century, but the deposits found were small and mined only sporadically during the seventeenth century. In 1693 economic deposits of gold were discovered in Minas Gerais, and for a century thereafter this state was one of the world's major sources of the precious metal. One of these deposits, the famous Morro Velho, has been mined by underground workings for almost a century and a half and is still productive.

Although silver has been the most important precious metal of Mexico, gold has also been won from many of the silver-gold deposits, mostly of Tertiary age. The bedrock deposits of the great silver-gold vein system of the Veta Madre at Guanajuato were found in 1550 and exploited almost immediately thereafter. El Oro, one of the premier gold districts but now largely exhausted, was discovered in 1521, developed extensively by 1530, and mined intermittently for nearly 400 years thereafter, producing more than 5 million oz. of gold.

Since the beginning of the nineteenth century prospecting for gold has ranged widely over Canada and the United States, resulting in many great placer gold rushes, first to California in 1848, then to British Columbia in 1857, and later to the Klondike in Yukon in 1896 and Nome in Alaska in 1899. After the exhaustion or near-exhaustion of many of the placers, attention turned to bedrock deposits during the last half of the nineteenth century and the first half of the present century. The Mother Lode and Grass Valley in California and the famous Comstock Lode in Nevada were discovered and developed in the 1850s. The gold telluride deposits of Cripple Creek in Colorado were located in 1892, and by 1905 the Tonopah and Goldfield districts in Nevada were under development. The Homestake mine, at Lead, South Dakota, was found in 1876 and brought into production soon thereafter.

In eastern Canada lode gold was first worked in Nova Scotia in the late 1850s, followed in 1866 by the first discovery of lode gold in the Canadian Shield near Madoc, Hastings County, Ontario. After the discovery of the native silver deposits at Cobalt, Ontario, in 1903, prospectors ranged widely over the Precambrian areas of Ontario, Quebec, Manitoba, Saskatchewan, and Northwest Territories. In Ontario and Quebec, Abitibi and Larder Lake were discovered in 1906; Porcupine, 1909; Swastika, 1910; Kirkland Lake, 1911; Matachewan, 1916; Rouyn (Noranda), 1924; and Red Lake, 1925. In Manitoba the Rice Lake district was discovered in 1911, and in Northwest Territories, the deposits in the sediments of the Yellowknife area were discovered in 1933 and those in the greenstones in 1935. The most recent discoveries in the Canadian Shield are the large auriferous orebodies in the Hemlo area, northwestern Ontario, originally indicated in 1869 and extensively developed in the early 1980s.

In western Canada the bedrock gold deposits in British Columbia first attracted attention in 1863 during the first great placer gold rushes to the province. Little work was done, however, on most of the discoveries and many were forgotten. The area in the vicinity of the Cariboo Gold Quartz and Island Mountain mines in the Barkerville district was prospected in 1860, and some mining was done in 1876 and a few years thereafter at these mine sites; large scale mining, however, did not commence until 1933 and 1934 respectively at the two mines. In 1897 the Cadwallader gold belt in the Bridge River district, containing the Bralorne and Pioneer deposits, was prospected, but it was not until 1928 that the Pioneer mine was brought into production, followed in 1932 by the Bralorne mine. Rossland in West Kootenay District was located in 1889 and brought into production in 1890, the Premier mine in Stewart District in 1918, and the Zeballos gold belt on the west coast of Vancouver Island was discovered and developed in 1934.

The discovery of payable placer gold in Australasia was made first in 1851 near Bathurst, New South Wales, Australia, by Edward Hammond Hargraves. There followed then the discoveries of large eluvial placer and bedrock gold deposits in Australia at Bendigo and Ballarat (1851), Gympie (1867), Charters Towers (1870), Beaconsfield in Tasmania (1876), Mount Morgan (1886), Kimberley (1886), Coolgardie (1892), Kalgoorlie (1893), and Tennant Creek (1932). In Papua New Guinea the rich gold placers of the Morobe field were discovered by the Spaniards in 1528 but brought into extensive production only in 1926.

There are few references to gold in the journals of the early explorers of New Zealand, and it was left mainly to the prospectors who had followed the great gold rushes to California and later to Australia to establish the presence of gold in commercial amounts (Salmon, 1963). Alluvial gold was first discovered in 1852 at Coromandel (Hauraki Goldfield) North Island in Driving Creek by Charles Ring; later at Collingwood in Nelson, South Island, in 1857; at Gabriels Gully in Otago by Gabriel Read in 1861; and in a number of sites along the western coastline of the South Island in 1864. Lode gold in the Hauraki Goldfield, the main centers being Waihi, Thames, Karangahake, and Coromandel, was first exploited in the 1860s; the quartz deposits in Otago, South Island, were first worked in the 1870s; and the productive quartz lodes at Reefton produced their first gold in the early 1870s.

Gold and gold tellurides are widely distributed in the Fijian Islands, and the presence of these minerals was evidently known to the early Fijians. Baron A. B. de Este is said to have discovered gold in the Tavua area in 1872, but some sixty years were to pass before any serious prospecting was carried out. In 1932 B. Borthwick and J. Sinclair discovered payable gold on Vunisina Creek, a small tributary of the Nasivi River. Further investigation of this prospect, associated with the TertiaryVatukoula volcanic caldera, led to the development of a number of mines, of which the famous Emperor mine has been in continuous production since about 1935 (Fraser, 1954).

Elsewhere in the Pacific region gold had been sought and mined long before the Christian era. The gold-silver mines at Radjang Lebong, Indonesia, are said to have been exploited in a desultory manner by the Hindus as early as 900 B.C. and more systematically by others since 1700. The deposits at Bau, Sarawak, likewise have a long history but have received detailed attention only since 1820. In the Philippine Archipelago gold was known to the Chinese and Hindus from the beginning of our era; the Spaniards found gold there as early as 1524, but extensive exploitation of the deposits took place only after Spain ceded the Philippines to the United States in 1898.

The Chinese have mined gold for millenia, the first indication of this pursuit being in the artifacts of the Shang dynasty (1765 B.C.). More recently considerable gold mining has been carried out in Archean, Proterozoic, and younger terranes in Shandong, Yunnan, Kansu, Szechuan, and other provinces. Likewise in Korea gold mining is an old technology going back to at least the beginning of the Christian Era in mining districts such as those of Unsan, Nurupi, Sak Ju, and Sen Sen. Similarly in Japan the search for gold was in progress long before the Christian era, judging from archaeological evidence; the methods of searching for and mining the metal were probably introduced from Korea. The Sado mine on Sado Island in the Japan Sea, the largest gold and silver producer in Japan, was discovered in 1542 and has been in production intermittently since that time.

India has long been the site of gold mining, first from placers and then in more modern times from the oxidized and primary zones of a variety of auriferous deposits. Pliny, writing at the beginning of our era in his *Historia naturalis,* mentions the gold of India, and the land of Ophir mentioned in I Kings 10:11 in the Old Testament can, according to some authorities, be equated with India. It is certain that gold placers and the rich oxidized zones of auriferous deposits were worked in India long before the Christian era, as evidenced by archaeological data and written records. According to Allchin (1962) large scale mining in India began with the Mauryan colonization of the Deccan about the end of the fourth century B.C. The discovery of the Kolar field would seem to date from the beginning of the Christian era, probably coeval with that of the Hutti field to the north. The modern mining of the famous Champion Lode in the Kolar field, rediscovered in 1873, began about 1880 and has continued since that date.

The Soviet Union has long been a legendary source of gold. The land of Colchis drained by the river Phasis (the modern Rioni) in Georgian SSR is reputed to have provided great quantities of gold. Similarly, the Persians are said to have obtained much gold from the Scythians, a polyglot group of tribes that inhabited the region north of the Black Sea, and from various Iranian and other tribes who inhabited the Ural-Uzbek-Altai region. The golden road to Samarkand was known centuries before Christ. With time the monopoly of gold mining became the sole preserve of the Imperial Czars, who pursued extensive placer and lode mining first in the Urals, beginning about 1774, and later in many parts of Siberia, especially in the Altai region where alluvial deposits were exploited as early as 1820. In 1829 the placer deposits of the Lena were first exploited and in 1840 those of the Yenisei Ridge came into production; the placers in the drainage system of the Amur were apparently first worked around 1867, and those in the Far Eastern maritime area appear to have been first exploited around 1870 or earlier. In recent years many more gold districts and deposits, both placer and bedrock (e.g., Muruntau, Uzbek. SSR), have been developed and brought into production, making the Soviet Union the second largest gold producer in the world.

Gold from West Africa found its way into Europe as early as the tenth century and probably before. Most of this gold came by Sahara caravan to

Barbary and thence to Europe, the original sources being the kingdoms of Ghana, Mali, and Songhai. It is said that much of this gold came from a region known as Wangara (probably the basin of the Falémé, a tributary of the Senegal River and noted for its placers), but considering the aurificity of West Africa it seems likely that the gold had a much more widespread source. In any event, one of the motives of the Portuguese voyages inspired by Henry the Navigator was to ascertain and exploit the west African gold. The Portuguese were soon followed by a host of English, French, Dutch, Danish, and Spanish entrepreneurs. It is thought that annually more than a quarter of a million ounces of gold reached Europe during the fifteenth and sixteenth centuries from African sources. Based mainly on native workings, numerous gold deposits, both bedrock and placer, were rediscovered during the latter part of the nineteenth century throughout Senegal, Guinea, Sierra Leone, Ghana, Nigeria, and the other nations of the Gold Coast. The Precambrian auriferous Tarkwa conglomerates of Ghana were developed in a modern way during the period 1876-1882 by Pierre Bonnat, the father of modern gold mining in the Gold Coast (Junner, 1935). In 1895, Ashanti Goldfields Corporation began work in the Obuasi district of Ghana, developing the Ashanti and other mines, which have produced the largest proportion of gold since 1900 in the countries of the Gold Coast. All of these deposits are of Precambrian age.

Natives worked gold deposits in Zimbabwe perhaps as far back as the beginning of our era. Long lines of ancient workings in the oxidized zones followed the strike of many Precambrian (Archean) deposits that were later developed by modern mining methods at the beginning of the present century. The Gaika mine was developed in 1894, the Globe and Phoenix in 1895-1900, the Eldorado in 1905, the Antelope in 1908, the Cam and Motor in 1909, the Shamva in 1909-1910, and numerous others during the period 1895-1911.

In South Africa an event in 1834 has influenced the history of gold ever since: Carel Kruger discovered gold on the Witwatersrand, or White Waters Ridge, while on a hunting expedition north of the Vaal River. However, little attention was paid to the find because the goldfields of Barberton and the DeKaap Valley were the chief focus of gold prospecting through 1885. In 1886 George Harrison, an Australian golddigger, and George Walker, an Englishman, discovered payable gold reef on the Witwatersrand. This discovery soon led to the development of the great reefs that constitute the largest gold deposits known and that made South Africa the foremost gold producer in the world. During a century of mining, some 4,000 million metric tons of ore have been treated from the Witwatersrand deposits, resulting in the recovery of 37 million kg of gold (1,200 million oz). The story of the discovery of the remarkable deposits of the Witwatersrand, of the growth of Johannesburg, and of the men who struggled for control of not only the gold reefs but of what is now South Africa is told admirably by Crisp (1974).

It is estimated that the total amount of gold won from the earth to the end of 1985 is about 3.85 billion (3.85×10^9) troy oz (120×10^9 g). Of this amount 2% was produced prior to 1492, 8% during the period 1492-1800, 20% during the interval 1801-1900, and 70% from 1901-1985 (all figures being rough estimates). In volume the total amount of gold won from the earth would occupy 6,300 m^3 or an 18.5 m cube, a small volume of metal indeed to have so influenced the toil, trials, tribulations, and destiny of man for 5,000 years.

The current annual world production of gold is about 1,338 metric tons (Table 18-2). Some 50% of this production is derived from quartz-pebble conglomerate deposits, some 20% from nonlithified eluvial and alluvial placers, and the remainder from the various vein and disseminated deposits.

The general literature on gold reaches back some 5,000 years, almost to the birth of writing. The most ancient accounting tablets of the Sumerians of Mesopotamia, scribed about 3100 B.C., mention the metal as do also the pictographics, phonographics (word-syllabic systems), and hieroglyphics on the tablets and papyri of the most ancient Hittite, Elamite, Egyptian, Cretan, Indian (Harappan), and Chinese civilizations.

Initial mention of gold in a geological context appeared about 1320 B.C. on the famous *Carte des mines d'or,* a Rameside papyrus and fragments depicting a gold mining region in Egypt (Frontispiece). Since that time the geological and geochemical literature on gold has multiplied prodigiously, so that today there are perhaps more references to gold in the literature of the earth sciences than for any other element in the periodic table of the elements.

Gold has played a unique and prominent part in the history of the theory of mineral deposits, especially in the theories advanced through the ages to explain the origin of veins and placers. For this reason, and to provide a background, I have included an opening chapter on the types and geochemistry of auriferous deposits; I have also included in later chapters pertinent details on the various theories of the origin of mineral deposits and discussed briefly the philosophical, social, and scientific milieu within which these various theories developed.

REFERENCES AND SELECTED BIBLIOGRAPHY

Acosta, J. de, 1940. *Historia natural y moral de las Indias,* Fondo de Cultura Economica, Mexico City, 638p. Also *The Natural and Moral History of the Indies,* Hakluyt Society, no. 60-61, 2 vols. 1880.

Adams, F. D., 1938. *The Birth and Development of the Geological Sciences,* Williams & Wilkins, Co., Baltimore, 506p.

Agricola, G., 1556. *De re metallica,* Basle. H. C. and L. H. Hoover, Transl., *Mining Mag., London.* 1912, 637p.

Aitchison, L., 1960. *A History of Metals,* 2 vols., MacDonald & Evans, London, 647p.

Allchin, F. R., 1962. Upon the antiquity and methods of gold mining in ancient India, *Jour. Econ. Social History of the Orient* **5** (pt. 2, 197):195-211.

Andrée, J., 1922. *Bergbau in der Vorzeit,* C. Kabische, Leipzig, 72p.

Barba, A. A., 1640. *El Arte de los Metales,* Madrid. (Trans. by R. E. Douglass and E. P. Mathewson as *The Art of the Metals,* John Wiley & Sons, New York, 1923, 288p.)

Berkner, L., 1964. *The Scientific Age,* Yale Univ. Press, New Haven, 137p.

Bernal, J. D., 1954. *Science in History,* Watts & Co., London, 967p.

Biringuccio, V., 1540. *Pirotechnia,* Venice. C. S. Smith and M. T. Gnudi, Trans., M.I.T. Press, Cambridge, Mass., 1959, 477p.).

Blainey, G., 1969. *The Rush that Never Ended, A History of Australian Mining,* Melbourne Univ. Press, Melbourne, 389p.

Boas, M., 1970. *The Scientific Renaissance, 1450-1630,* Fontana-Collins, London, 350p.

Boyle, R. W., 1979. The geochemistry of gold and its deposits, *Canada Geol. Survey Bull. 280,* 584p.

Bromehead, C. E. N., 1940. The evidence for ancient mining, *Geog. Jour.* **96:**101-120.

Buranelli, V., 1979. *Gold, An Illustrated History,* Hammond, Maplewood, N.J., 224p.

Caley, E. R., 1964. *Analysis of Ancient Metals,* Macmillan, New York, 176p.
Caley, E. R., and J. F. C. Richards, 1956. *Theophrastus on Stones,* Ohio State Univ., Columbus, Ohio, 238p.
Cambridge Ancient History, 1982-, 12 vol., Cambridge Univ. Press, Cambridge.
Cambridge Medieval History, 1966-, 9 vol., Cambridge Univ. Press, Cambridge.
Canadian Institute of Mining and Metallurgy, 1948, 1957. *Structural Geology of Canadian Ore Deposits,* 2 vols., Canadian Inst. Min. Metall., Montreal.
Cartwright, A. P., 1962. *The Gold Miners,* Purnell & Sons, Johannesburg, 340p.
Chang, Kwang-Chih, 1978. *The Archaeology of Ancient China,* 3rd ed., Yale Univ. Press, New Haven, 537p.
Cline, W., 1937. *Mining and Metallurgy in Negro Africa,* George Banta, Menasha, Wis., 155p.
Crisp, R., 1974. *The Outlanders,* Granada Publishing, Mayflower Books, Frogmore, St. Albans, Herts., England.
Crook, T., 1933. *History of the Theory of Ore Deposits,* Thos. Murby & Co., London, 163p.
Cumenage, E., and F. Robellaz, 1898. *L'or dans la nature,* P. Vicq-Dunod et Cie, Editeurs, Paris, 106p.
Curle, J. H., 1905. *The Gold Mines of the World,* 3rd ed., George Routledge & Sons, London, 308p.
Davies, O., 1935. *Roman Mines in Europe,* Clarendon Press, Oxford, 291p.
Del Mar, A. 1902. *A History of the Precious Metals,* 2nd ed., Cambridge Encyclopedia Company, New York, 478p.
Dunn, E. J., 1929. *Geology of Gold,* Charles Griffin & Co., London, 303p.
Emmons, W. H., 1937. *Gold Deposits of the World,* McGraw-Hill, New York, 562p.
Forbes, R. J., 1964. Gold, in *Studies in Ancient Technology,* E. J. Brill, Leiden, Holland, pp. 151-192.
Foster, R. P., ed., 1984. *Gold '82: The Geology, Geochemistry and Genesis of Gold Deposits,* A. A. Balkema, Rotterdam, 753p.
Fraser, J. A., 1954. *Gold Dish and Kava Bowl,* J. M. Dent & Sons, London, 262p.
Friedensburg, F., 1953. *Die Metallischen Rohstoffe,* heft 3, *Gold,* F. Enke Verlag, Stuttgart, 234p.
Gardiner, A., 1961. *Egypt of the Pharaohs,* Clarendon Press, Oxford, 461p.
Gmelin, 1950-1954. *Gmelins Handbuch der anorganischen Chemie,* System-nummer 62, *Gold,* pts. 1 and 2, Weinheim/Bergstrasse, Verlag Chemie, GMBH, 406p.
Govett, M. H., and M. R. Harrowell, 1982. *Gold: World Supply and Demand,* Australian Mineral Economics Pty., Sydney, Australia, 455p.
Green, T., 1973. *The World of Gold Today,* Walker & Company, New York, 287p.
Hanula, M. R., R. M. Longo, L. F. Jones, J. A. Miller, and L. C. Brown, eds., 1982. *The Discoverers,* Pitt Pub. Co., Toronto, 317p.
Hawkes, J., 1973. *The First Great Civilizations,* Knopf, New York, 465p.
Hawkes, J., and L. Woolley, 1963. *Prehistory and the Beginnings of Civilization,* Harper & Row, New York, 873p.
Healy, J. F., 1978. *Mining and Metallurgy in the Greek and Roman World,* Thames & Hudson, London, 316p.
Hobson, B., 1971. *Historic Gold Coins of the World,* Doubleday, Garden City, N.Y., 192p.
Holmyard, E. J., 1957. *Alchemy,* Penguin Books, Toronto, 281p.
International Geological Congress, 1930. *The Gold Resources of the World,* Bureau of the Congress, Session 15, Pretoria, South Africa, 457p.
International Gold Corporation, 1968- , *Gold Bulletin,* International Gold Corporation, Marshalltown, South Africa.
Isserow, S., and H. Zahnd, 1943. Chemical knowledge in the Old Testament, *Jour. Chem. Ed.* **20:**327-335.
Jacob, W., 1831. *Historical Inquiry into the Production and Consumption of the Precious Metals,* 2 vols., London.
Jensen, M. L., and A. M. Bateman, 1979. *Economic Mineral Deposits,* 3rd ed., John Wiley & Sons, New York, 593p.
Junner, N. R., 1935. Gold in the Gold Coast, *Gold Coast Geol. Survey Mem. 4,* 76p.

Keesing, N., ed., 1967. *History of the Australian Gold Rushes,* Angus & Robertson, Melbourne, 412p.

Keynes, J. M., 1930. *A Treatise on Money,* 2 vols., Harcourt, Brace & Co. New York.

Levey, M., 1959. *Chemistry and Chemical Technology in Ancient Mesopotamia,* Elsevier, Amsterdam, 242p.

Li, C., 1948. *The Chemical Arts of Old China,* Jour. Chem. Ed., Easton, Pa., 215p.

Lindgren, W., 1933. *Mineral Deposits,* 4th ed., McGraw-Hill, New York, 930p.

Lucas, A., 1948. *Ancient Egyptian Materials and Industries,* 3rd ed., Longmans, London, 580p.

Maclaren, J. M., 1908. *Gold, Its Geological Occurrence and Geographical Distribution,* Mining Jour., London, 687p.

Mellor, J. W., 1923. *A Comprehensive Treatise on Inorganic and Theoretical Chemistry,* vol. 3, pp. 491-618, Longmans, Green & Co., London.

Morgan, E. V., 1965. *A History of Money,* Pelican Books, Baltimore, Md., 237p.

Morrell, W. P., 1940. *The Gold Rushes,* Adam & Charles Black, London, 427p.

Napier, J., 1879. *Manufacturing Arts in Ancient Times,* Alexander Gardner, Paisley Scotland, 367p.

Needham, J., 1954-1983. *Science and Civilization in China,* 6 vols., Cambridge Univ. Press, Cambridge.

Park, C. F., and R. A. MacDiarmid, 1970. *Ore Deposits,* W. H. Freeman & Co., San Francisco, 522p.

Partington, J. R., 1935. *Origins and Development of Applied Chemistry,* Longmans, Green & Co., London, 597p. Reprint Arno Press, New York, 1975.

Partington, J. R., 1960. *A Short History of Chemistry,* Harper Torchbooks, 415p.

Partington, J. R., 1961. *A History of Chemistry,* 4 vols., Macmillan, London.

Patterson, C. C., 1971. Native copper, silver, and gold accessible to early metallurgists, *Am. Antiquity* **36**(3):286-321.

Paul, W., 1970. *Mining Lore,* Morris Printing Co., Portland, Oreg., 940p.

Petrovskaya, N. V., 1973. *Native Gold* (in Russian), Izd. "Nauka", Moscow, 347p.

Pledge, H. T., 1966. *Science since 1500,* H.M. Stationery Office, London, 357p.

Pliny the Elder (Gaius Plinius Secundus), A.D. 77. *Historia Naturalis,* 39 Books, Rome.

Proust, G. P., 1920. *L'or—prospection, gisement, extraction,* Gauthier-Villars et Cie, Editeurs, Paris, 319p.

Quiring, H., 1948. *Geschichte des Goldes,* F. Enke Verlag, Stuttgart, 318p.

Raguin, E., 1961. *Géologie des gîtes minéraux,* Masson & Cie, Editeurs, Paris, 686p.

Ray, P. C., 1904, 1925. *History of Hindu Chemistry,* 2 vols., Chuckervarty & Chatterjee, Calcutta. Rev. ed. in one vol., as *History of Chemistry in Ancient and Medieval India,* Indian Chemical Society, Calcutta, 1956.

Rickard, T. A., 1932. *Man and Metals,* 2 vols., McGraw-Hill, New York, 1068p.

Ridge, J. D., ed., 1968. *Ore Deposits of the United States, 1933-1967,* 2 vols., Am. Inst. Min. Metall. Petrol. Eng., New York.

Rose, T. K., and W. A. C. Newman, 1937. *The Metallurgy of Gold,* 7th ed., Charles Griffin & Co., London, 561p.

Rosenfeld, A., 1965. *The Inorganic Raw Materials of Antiquity,* Frederick A. Praeger, New York, 245p.

Rosenthal, E., 1970. *Gold! Gold! Gold!,* Macmillan, London, 372p.

Routhier, P., 1963. *Les gisements métallifères,* t. I and II, Masson et Cie, Paris.

Salmon, J. H. M., 1963. *A History of Goldmining in New Zealand,* R. E. Owen, Government Printer, Wellington, 309p.

Sagui, C. L., 1930. Economic geology and allied sciences in ancient times, *Econ. Geology* **25:**65-86.

Sarton, G., 1945-1948. *Introduction to the History of Science,* 3 vols., Carnegie Inst. Washington, Williams & Wilkins Co., Baltimore, 2155p.

Schneiderhöhn, H., 1955. *Erzlagerstätten,* Gustav Fischer Verlag, Stuttgart, 375p.

Shepherd, R., 1980. *Prehistoric Mining and Allied Industries,* Academic Press, London, 272p.

Shilo, N. A., 1981. *Fundamentals of the Study of Placers,* "Nauka", Moscow, 383p.

Singer, C., E. J. Holmyard, A. R. Hall, and T. I. Williams, 1954-1958. *A History of Technology,* 5 vols., Clarendon Press, Oxford.

Smirnov, V. I., 1976. *Geology of Mineral Deposits,* Mir Publishers, Moscow, 520p.
Sprague de Camp, L., 1963. *The Ancient Engineers,* Ballantine Books, Toronto, 450p.
Street, A., and W. Alexander, 1951. *Metals in the Service of Man,* Penguin Books, Harmondsworth, Middlesex, U.K., 237p.
Summers, R., 1969. Ancient mining in Rhodesia, *Natl. Mus. Rhodesia Mem. 3,* Salisbury, 236p.
Taylor, F. S., 1949. *The Alchemists (Founders of Modern Chemistry),* Henry Schuman, New York, 246p.
Thorndike, L., 1923-1958. *A History of Magic and Experimental Science,* 8 vols., Columbia Univ. Press, New York.
Tylecote, R. F., 1962. *Metallurgy in Archaeology,* Edward Arnold, London, 368p.
Tylecote, R. F., 1976. *A History of Metallurgy,* The Metals Society, London, 182p.
Vilar, P., 1975. *A History of Gold and Money, 1450-1920,* New Left Books, London, 360p.
Weeks, M. E., and H. M. Leicester, 1968. *Discovery of the Elements,* 7th ed., Jour. Chem. Ed., Easton, Pa., 896p.
Wertime, T. A., 1973. The beginnings of metallurgy: A new look, *Science* **182:**875-887.
Wilson, A. J., 1979. *The Pick and the Pen,* Mining Jour. Books, London, 308p.
Williams, G. J., 1974. Economic geology of New Zealand, *Australasian Inst. Min. Metall. Monograph ser. 4,* 2nd ed., 490p.
Zahnd, H., and D. Gillis, 1946. Chemical knowledge in the New Testament, *Jour. Chem. Ed.* **23:**90-97, 128-134.

CHAPTER

1

General Geochemistry of Gold and Types of Auriferous Deposits

Eldorado!

This chapter deals only briefly with the general geochemistry of gold and with the geological and geochemical setting of its varied types of deposits. The text is intended only as a background for understanding the chapters that follow. Details should be sought in the various monographs and papers recorded in the selected bibliographies listed at the end of this and later chapters.

Gold is the most noble of metals, and its geochemistry is conditioned principally by this fact. Compared with other elements in the periodic table the terrestrial abundance of gold (0.005 ppm), is low compared with copper (50 ppm) and silver (0.07 ppm) the accompanying two elements in group IB, and approximately equal to that of platinum (0.005 ppm), the adjacent element in group VIII.

Two general types of auriferous deposits are recognized—lode (vein) deposits and placers. The enigmatic quartz-pebble conglomerate deposits, the largest known auriferous concentrations on earth, have generally been classified as modified paleo-placers, but some geologists have considered them to be of hydrothermal origin and akin to lode deposits.

The quartz-pebble conglomerate deposits supply 50% or more of the world's annual gold production. The remaining half is provided by the other types of auriferous deposits, including the vein and disseminated types, eluvial and alluvial placers, and the various byproduct sources such as polymetallic veins, lodes, massive sulfide bodies, and stockworks.

GENERAL GEOCHEMISTRY OF GOLD

Gold is a member of group IB of the periodic table, which includes copper, silver, and gold. In its chemical reactions gold resembles silver in some respects, but its chemical character is markedly more noble. The principal oxidation states of gold are + 1 (aurous) and + 3 (auric). These states are unknown as aquo-ions in solutions, the element being present mainly in complexes of the type $[Au(CN)_2]^-$, $[AuCl_2]^-$, $[Au(OH)_4]^-$, $[AuCl_4]^-$, and $[AuS]^-$. There is only one naturally occurring isotope of gold: ^{197}Au.

In nature, gold occurs predominantly in the native state or as a major constituent of various alloys containing mainly silver, copper, or platinum metals. Several gold and gold-silver tellurides are known, of which the most common are sylvanite, calaverite, petzite, krennerite, and nagyagite. The antimonide, aurostibite, $AuSb_2$, occurs in some auriferous deposits, and there is also an argentiferous gold selenide, fischesserite, Ag_3AuSe_2, an argentiferous gold sulfide, uytenbogaardtite, Ag_3AuS_2, and a bismuthide, maldonite, Au_2Bi, which is fairly well differentiated. The principal ore minerals of gold are the native metal, aurostibite, and the various tellurides.

The abundance of gold in the upper lithosphere is about 0.005 ppm and the Au/Ag ratio is about 0.07. The average gold content of igneous-type rocks in parts per million is ultramafic, 0.004; gabbro-basalt, 0.007; diorite-andesite, 0.005; and granite-rhyolite, 0.003. The average gold content of sedimentary rocks in parts per million is sandstone and conglomerate, 0.03; normal shale, 0.004; and limestone, 0.003. Certain graphitic shales, sulfide schists, phosphorites, and some types of sandstones and conglomerates may contain up to 2 ppm Au or more.

The average gold content of soils is 0.005 ppm, and the average for natural fresh waters is 0.00003 ppm. Sea and ocean waters contain an average of 0.000012 ppm Au. Gold is a trace constituent of many plants and animals. Some coals are slightly enriched in gold, with 0.05 to 0.1 ppm Au in the ash.

AURIFEROUS DEPOSITS

Gold is won from deposits mined essentially for the metal and as a byproduct of the mining and treatment of nickel, copper, zinc, lead, and silver ores. Nine principal types of deposits, exploited mainly for gold, are listed subsequently. The classification of these deposits is that which I suggested in my monograph on gold (Boyle, 1979), revised to include more recent data; it is based essentially on the general morphology and chemical constitution of a deposit-type and on its geological and geochemical setting, particularly the nature of its host rocks. Because of the great diversity of auriferous deposits it is thought that this manner of classification is as factual and as objective as can be devised, and that it is relatively independent of speculative genetic theories. Numerous other classifications have been suggested and are discussed in the papers and monographs listed in the selected bibliographies at the end of this and later chapters. Many of these classifications, particularly those of epigenetic gold deposits, are based on magmatic or hydrothermal concepts and are largely speculative. Concerning classifications of gold deposits one should take heed of the admonition by Maclaren (1908, 42).

> Auriferous veins or deposits may be of any form, may occur in any rock, and may have received their gold from various sources. Particular classifications based on obviously adventitious characters, as similarity of form of deposit, or identity of matrix or of associated minerals, can therefore serve no useful purpose, either scientific or

> economic. Such classifications have been current for many years. Some have certainly been suggestive, but the majority have helped the miner and prospector not a whit, and have proved a source of confusion and embarrassment to the student.

For purposes of this book the following types of auriferous deposits are distinguished:

1. Auriferous porphyry dykes, sills, and stocks; auriferous pegmatites, coarse-grained granitic bodies, aplites, and albitites

The indigenous gold content of these granitic rocks is invariably low, of the order of 0.003 ppm. Certain albitites and quartz-feldspar porphyry dykes and stocks with indigenous pyrite and pyrrhotite may contain up to 0.10 ppm Au and 1 ppm Ag, principally in the sulfide minerals. Porphyritic, aplitic, and granitic bodies of this type are common in Precambrian, mainly Archean, terranes and in younger rocks throughout the world. Most are of an intrusive nature, probably the anatectic products of deep-seated granitization. None so far are known to be of economic value, although many are probably the sources of the gold, silver, and other metals secondarily concentrated in the fractures, faults, and shear zones in the porphyry and albitic bodies themselves and in their nearby host rocks (see type 7.1).

2. Carbonatites and carbonatite-related bodies

These bodies are extremely complex magmatic and hydrothermal assemblages of rocks in which four stages can usually be recognized: (1) an ultramafic followed by an alkalic magmatic phase, (2) a magmatic dyke phase, (3) a (magmatic?) carbonatite phase, and (4) a late hydrothermal phase. The third and fourth phases are marked by extensive replacement processes in some complexes. Most carbonatites are zoned, often in a ring pattern. The late hydrothermal stage is commonly marked by sulfide mineralization that occupies late fractured and faulted parts of the rocks of all zones. In some carbonatites, however, sulfides, mainly pyrite, pyrrhotite, chalcopyrite and molybdenite, are widely distributed as (indigenous) disseminations in the fenites and rocks of all zones of the complexes; in the great Palabora deposit in South Africa the chalcopyrite and bornite occur in a disseminated (indigenous) form in a number of rock types, but the main concentrations are in the fractured transgressive carbonate (sövite) complex and in transgressive carbonate veinlets that cut several rock types.

Carbonatites and carbonatite-related bodies (e.g., carbonate-barite-fluorite-sulfide veins, dykes, and disseminations) are characterized by a distinctive suite of elements that includes Na, K, Fe, Ba, Sr, rare earths, Ti, Zr, Hf, Nb, Ta, U, Th, Cu, Zn, P, S, F, and more rarely Li, Be, and Pb. Most of the rocks comprising carbonatites are low in gold and silver (0.005 ppm Au and 0.1 ppm Ag). The late stage carbonate-sulfide mineralization, however, commonly contains slightly enriched amounts of both gold and silver. The silver is present mainly in galena, tetrahedrite, and other such minerals; the gold is invariably native and occurs in association with pyrite, pyrrhotite, molybdenite, chalcopyrite, and other copper sulfides.

Few if any carbonatites are enriched enough in gold and silver to constitute economic orebodies. However, the fact that the sulfide phases exhibit enrichments in the two precious metals suggests that deposits of this type should be considered as possible gold deposits.

The recently discovered Cu-U-Au deposits at Olympic Dam (Roxby Downs) in South Australia and the auriferous quartz-carbonate counterparts of the rare-earth carbonate deposits at Mountain Pass, San Bernardino County, California, may be carbonatite-related bodies.

3. Auriferous skarn-type deposits

Gold is a frequent constituent of skarn deposits, in which it is commonly more abundant than the literature would indicate. Most skarn deposits yield gold as a byproduct of copper and lead-zinc mining, but many of these deposits are greatly enriched in gold and silver and are mined essentially for the two precious metals.

The general features of skarn deposits are well known and need not be described in detail here. Most of the deposits occur in highly metamorphosed terranes, particularly those containing carbonate rocks or carbonate-bearing pelites, and in which there has been much granitization and injection of granitic rocks. Some deposits occur near the contacts of granitic bodies and have long been called *contact metamorphic;* others are developed in favorable reactive beds or zones some distance from granitic contacts. The deposits contain a characteristic suite of early-developed Ca-Mg-Fe silicate and oxide minerals and a lower-temperature, generally later, suite of silicate, carbonate, sulfide, and arsenide minerals. The gold minerals include native gold and various tellurides. Most of the skarn deposits worked essentially for gold contain much pyrite and/or arsenopyrite.

The elements most frequently enriched with gold in skarn deposits are Fe, S, Cu, Ag, Zn, Pb, Mo, As, Bi, and Te. There is commonly a positive correlation between Au and Cu in some skarn deposits. Tungsten is a common trace element in gold-bearing skarn deposits. The element belongs to the early phase of skarnification, whereas gold tends to be precipitated late in the mineralization processes. The two elements may, therefore, be negatively correlated, for it is common to find skarn deposits that are rich in tungsten (scheelite) but practically devoid of gold and vice versa. The Au/Ag ratio of the auriferous skarn-type ores is variable but is commonly greater than 1.

Auriferous skarn deposits occur at widespread points in the Canadian Cordillera, particularly in the Hedley district of British Columbia, where the Nickel Plate and French mines worked arsenopyrite-pyrite orebodies in skarn developed in Triassic limestone and limy argillites. In the Canadian Shield, auriferous skarn deposits occur mainly in the Grenville Province, examples being the lead-zinc-silver-gold ores of the Tetreault mine near Quebec and the New Calumet mine northwest of Ottawa. Elsewhere in the world auriferous skarn deposits have a widespread distribution, especially in belts of carbonate rocks invaded by diorites, monzonites, granodiorites, and granites. Here belong the auriferous skarn deposits at Cable in Montana, the La Luz and Rosita mines in Nicaragua, a number of mines in the Altai-Sayan of USSR, and the skarn deposits of Bau, Sarawak, and the Suian district of Korea.

4. Gold-silver and silver-gold veins, stockworks, lodes, mineralized pipes and irregular silicified bodies in fractures, faults, shear zones, sheeted zones and breccia zones essentially in volcanic terranes

Representatives of this type of deposit are widespread throughout the folded and relatively flat-lying volcanic terranes of the earth. The deposits occur in rocks of all ages, but the largest number occur in those of Precambrian and Tertiary age.

The favorable host rocks are commonly basalts, andesites, latites, trachytes, and rhyolites. In Precambrian rocks, such assemblages are usually referred to as *greenstones*. Many deposits of Precambrian age occur in tuffs, agglomerates, and sediments interbedded with the volcanic flows, particularly in banded iron-formations. In the older terranes, the rocks are generally regionally metamorphosed and have the characteristic regional metamorphic facies outward from igneous or granitized centers. The younger rocks generally show the effects of chloritization, carbonatization, hydration, and pyritization (propylitization) over broad zones, but locally some of the andesites and rhyolites may be relatively fresh.

In the older rocks, the deposits are veins, lodes, stockworks, pipes, and irregular mineralized masses generally in extensive fracture and shear-zone systems. Some occur in drag folds. The deposits in the younger rocks are usually confined to fissures, fractures, faults, and brecciated zones that cut the volcanic rocks of calderas and generally have a limited horizontal and vertical extent. Others, however, are associated with fracture and fault systems that extend for many kilometers across volcanic sequences and their associated intrusive granitoids. A few deposits in young volcanic terranes occur in or near the throats of extinct (or present day) hot springs and/or in siliceous hot spring aprons.

The mineralization of these particular deposits is characterized essentially by quartz, carbonate minerals, pyrite, arsenopyrite, base-metal sulfide minerals, and a variety of sulfosalt minerals. The principal gold minerals are the native metal and various tellurides; aurostibite occurs in some deposits. Characteristic types of wall rock alteration are generally developed adjacent to and in the vicinity of nearly all deposits in this class. In the old Precambrian rocks, the most common types of alteration are chloritization, carbonatization, sericitization, pyritization, arsenopyritization, and silicification. In the younger rocks, propylitization (chloritization and pyritization) is especially characteristic, and there may also be a development of adularization, silicification, kaolinization, sericitization, and more rarely alunitization.

The elements commonly concentrated in this class of deposits include Cu, Ag, Zn, Cd, Hg, B, Tl, Pb, As, Sb, Bi, V, Se, Te, S, Mo, W, Mn, Fe, CO_2 and SiO_2; less commonly Ba, Sr, U, Th, Sn, Cr, Co, Ni, and F. Hg and Sb are particularly characteristic of the younger deposits. The Au/Ag ratio of the ores is generally greater than 1 in most Precambrian and in some younger deposits; in many Tertiary deposits the ratio is less than 1.

Deposits of this type are widespread in the Precambrian greenstone belts of the world; examples include Yellowknife, Northwest Territories, Canada; Red Lake and Timmins, Ontario, Canada; Kolar goldfield, India; Kalgoorlie goldfield, western Australia; and the Cam and Motor, Dalny, and other similar mines in Zimbabwe. Younger representatives are the Mother Lode system of California (Mesozoic); Comstock Lode, Nevada (Tertiary); Goldfield, Nevada (Tertiary); Cripple Creek, Colorado (Tertiary); Săcărîmb (Nagyág), Romania (Tertiary); Coromandel gold belt, New Zealand (Tertiary); Emperor mine, Fiji (Tertiary); Lebong and other auriferous districts, Indonesia (Tertiary); Lepanto mine, Philippines (Tertiary); Kasuga mine, Japan (Tertiary), and the Belaya Gora and other similar deposits in the far eastern USSR (Tertiary).

5. Auriferous veins, lodes, sheeted zones, and saddle reefs in faults, fractures, bedding-plane discontinuities and shears, drag folds, crushed zones, and openings on anticlines essentially in sedimentary terranes; also replacement tabular and irregular bodies developed near faults and fractures in chemically favorable beds

These deposits are widespread throughout the world and have produced a large amount of gold and silver; they are often referred to as "Bendigo type". The deposits are developed predominantly in sequences of shale, sandstone, and graywacke dominantly of marine origin. Such sequences are invariably folded, generally in a complex manner, metamorphosed, granitized, and invaded by granitic rocks, forming extensive areas of slate, argillite, quartzite, graywacke, and their metamorphic equivalents. Near the granitic bodies, various types (kyanite, andalusite, cordierite) of quartz-mica schists and hornfels are developed and grade imperceptibly into relatively unmetamorphosed slates, argillites, quartzites and graywackes marked by the development of sericite, chlorite and other low-grade metamorphic minerals. Most of the gold deposits are developed in the lower-grade facies. A few economic deposits occur in the granitic batholiths and stocks that invade the graywacke-slate sequences.

The principal gangue mineral in these deposits is quartz; feldspar, mica, chlorite, and minerals such as rutile are subordinate. Among the metallic minerals, pyrite and arsenopyrite are most common, but galena, chalcopyrite, sphalerite, and pyrrhotite also occur. Molybdenite, bismuth minerals, and tungsten minerals are local. Stibnite occurs in abundance in a few deposits, but is relatively rare in most deposits. Acanthite, tetrahedrite-tennantite, and other sulfosalts are not common in these deposits. Carbonate minerals, mainly calcite and ankerite, are common but not abundant. The valuable ore minerals are native gold, generally low in silver, auriferous pyrite, and auriferous arsenopyrite. Telluride minerals are relatively rare, and aurostibite is an uncommon mineral in these deposits.

A few deposits in this category are tabular or irregular replacement (disseminated) bodies developed in carbonate rocks or calcareous argillites and shales. The principal minerals in these deposits are quartz, fluorite, pyrrhotite, pyrite, arsenopyrite, sphalerite, galena, chalcopyrite, and stibnite.

As a general rule, wall rock alteration associated with these deposits is minimal, and the quartz veins, saddle reefs, and irregular masses are frozen against the slate, argillite or graywacke wall rocks. In places, thin zones of mild chloritization, sericitization, and carbonatization are present. Some veins are marked by thick black zones (up to 15 cm wide) of tourmalinized rock. Disseminated pyrite and arsenopyrite are common in the wall rocks of most of these deposits. This pyrite and arsenopyrite is usually auriferous.

The elements exhibiting a high frequency of occurrence in this type of gold deposit include Cu, Ag, Mg, Ca, Zn, Cd, (Hg), B, (In), (Tl), Si, Pb, As, Sb, (Bi), S, (Se), (Te), (Mo), W, (F), Mn, Fe, (Co), and (Ni). Elements in parentheses have a low to very low frequency of occurrence. The Au/Ag ratio in the ores is generally greater than 1.

Deposits in essentially sedimentary terranes are widespread throughout the world. In Canada, examples occur in the Archean Yellowknife supergroup in the Yellowknife district, Northwest Territories (Ptarmigan, Thompson-Lundmark and Camlaren mines); in the Paleozoic Cariboo group at Wells, British Columbia (Cariboo Gold Quartz mine); and widespread in the Ordovician Meguma group of Nova Scotia. Elsewhere in the world deposits of this type occur in the auriferous Appalachian "Slate Belt" of the United States (Paleozoic); Salsigne mine, Montagne Noire, France (Paleozoic); Sovetskoe deposit, Yenisey region, USSR (Proterozoic); Muruntau deposit, Uzbec SSR (Paleozoic); Bendigo deposits, Victoria, Australia (Paleozoic); and the Pilgrims Rest and Sabie goldfields in the Transvaal System, South Africa (Proterozoic ?).

6. Gold-silver and silver-gold veins, lodes, stockworks, and silicified zones in a complex geological environment, comprising sedimentary, volcanic, and various igneous intrusive and granitized rocks

Deposits in this category combine nearly all the epigenetic features described in categories 4 and 5. Quartz is a predominant gangue, and some deposits are marked by moderate developments of carbonates. The orebodies constitute principally quartz veins, lodes, and silicified and carbonated zones. The gold is commonly free but may be present as tellurides and disseminated in pyrite and arsenopyrite. The Au/Ag ratio of the ores is variable depending upon the district and often upon the deposit.

The deposits have a widespread distribution throughout the world in rocks ranging in age from Precambrian (Archean) to Tertiary. Examples in Canada include the Precambrian (Archean) deposits in the Kirkland Lake and Little Long Lac-Sturgeon River districts of Ontario and in the Jurassic volcanics of the Rossland gold-copper camp in the West Kootenay district of British Columbia. Elsewhere, gold is won from deposits of this type in Alaska at the Alaska Juneau mine (Mesozoic); Grass Valley and Nevada City auriferous districts, California (Mesozoic); and the Central City district, Colorado (Tertiary).

7. Disseminated and stockwork gold-silver deposits in igneous, volcanic, and sedimentary rocks

Three general categories can be recognized within this type:

1. Disseminated and stockwork gold-silver deposits in igneous bodies
2. Disseminated gold-silver and silver-gold occurrences in volcanic flows and associated volcaniclastic rocks
3. Disseminated gold-silver deposits in volcaniclastic and sedimentary beds: (1) deposits in tuffaceous rocks and iron formations (2) deposits in chemically favorable sedimentary beds.

The principal economic element in these deposits is gold, with small amounts of silver. A few deposits yield the base metals, but they are generally not known as base-metal deposits. The grade of the deposits is highly variable. Most are relatively low grade (generally less than 15 g Au/ton), but are characterized by large tonnages. The elements commonly concentrated in these deposits (omitting those in the common gangue minerals such as quartz, silicate, and carbonate minerals) are: Cu, Ag, Au, (Ba), (Sr), Zn, Cd, Hg, B, (Sn), Pb, As, Sb, Bi, V, S, Se, Te, Mo, W, (F), Fe, Co, and Ni. Elements in parentheses are infrequent or occur only in certain deposits. The Au/Ag ratio of most deposits is greater than 1.

The deposits in the first category occur in igneous plugs, stocks, dykes, and sills that have been intensively fractured or shattered and infiltrated by quartz, pyrite, arsenopyrite, gold, and other minerals. Most of the deposits are stockworks or diffuse irregular impregnations. The alteration processes vary with the types of host rock. In granitic (felsic) rocks, sericitization, silicification, feldspathization (development of albite, adularia, etc.), and pyritization are predominant; in intermediate and mafic rocks, carbonatization, sericitization, serpentinization, and pyritization prevail. Alunitization may occur in both felsic and mafic rocks in places. The Au/Ag ratio in most deposits is greater than 1.

Deposits of this type are common in Canada, particularly in the Canadian Shield and Cordillera. Examples are the Howey and Hasaga mines at Red Lake, Ontario, in an Archean quartz porphyry dyke; the Matchewan Consolidated and Young Davidson mines in Archean syenite plugs and dykes in the Matachewan district of Ontario; and the Camflo mine in an Archean porphyritic monzonite stock near Malartic, Quebec. Elsewhere typical examples occur in the Beresovsk auriferous district, Urals, USSR (Paleozoic); Twangitza mine, Kivu Province, Zaire (Precambrian); and the Morning Star mine, Woods Point, Victoria, Australia (Paleozoic).

The disseminated gold-silver occurrences in volcanic flows and associated volcaniclastic rocks in the second category are relatively common, but commercial deposits of this type have not been worked. Most occurrences are very low grade, commonly less than 0.01 oz Au/ton (0.3 ppm). Silver contents are higher in places, averaging in some cases 3.5 oz Ag/ton (120 ppm).

The disseminated occurrences in the second category are in reality large irregular and diffuse zones of alteration manifest mainly in rhyolites, andesites, basalts, and their associated volcaniclastic rocks. These zones of alteration constitute silicification, sericitization, epidotization, argillization, or alunitization, commonly associated with pyritization and carbonatization. In the mafic and intermediate rocks, the effects are commonly collectively called propylitization. Large volumes of the volcanic country rocks are affected, giving them a bleached and altered aspect. Locally diffuse silicified zones, quartz veins, alunite veins, and pyrite veins and segregations ramify through the altered rocks.

Occurrences of this type are generally enriched in S, Ba, B, Hg, Sb, As, Pb, Zn, W, Mo, Se, Te, and Ag. The Au/Ag ratio is variable, but most occurrences exhibit values less than 1.

The disseminated gold-silver deposits in volcaniclastic and sedimentary beds in the third category are usually conformable with the sedimentary and volcaniclastic beds, although in some cases their limits may infringe irregularly on overlying or underlying rocks. Some are large and of relatively high grade; others are commonly too low grade or not of sufficient tonnage to merit attention.

Two general subcategories of these deposits can be recognized: (1) those developed in tuffaceous rocks and iron-formations within volcanic and sedimentary terranes and (2) those resulting from extensive infiltration or replacement of chemically favorable beds, particularly carbonate rocks or calcareous pelites. The first is especially common in Precambrian terranes, although there is no reason why they should not occur in rocks of younger age; however, examples of the latter are rare to date. The second can apparently occur in rocks of any age.

Gold deposits in tuffaceous and other volcaniclastic rocks and in iron-formations in volcanic and sedimentary terranes are particularly common in the Archean greenstone and associated sedimentary belts of the Canadian Shield and in other similar rocks throughout the world. Orebodies in tuffaceous rocks are generally large-tonnage, irregular disseminated bodies containing essentially pyrite, pyrrhotite and arsenopyrite, with much secondary fine-grained quartz and various silicates. Elements exhibiting an enrichment in these types of deposits include Cu, Ag, Zn, Cd, B, Pb, As, Sb, Bi, Te, and S. Less common are Ba, Sr, Hg, Sn, V, Mo, W, Co, and Ni. The gold is usually free in the matrix of the rock or present in a finely divided state in the sulfide and arsenide minerals. A typical example of this type of deposit is the Madsen mine in Archean tuff at Red Lake, Ontario. The deposits now being developed at Hemlo, Ontario, may also belong in this category.

Auriferous deposits in iron-formations are of two types—disseminated bodies similar to those just described, and zones of quartz veins or stockworks traversing the constituent rock members of the iron-formations. These bodies and zones contain essentially quartz with pyrite, pyrrhotite, and arsenopyrite; the gold is generally present in the native state, and the enriched elements are similar to those mentioned for the deposits in tuffaceous and other volcaniclastic rocks. Typical examples of deposits in iron-formations are the Central Patricia mine and the Pickle Crow mine in the Archean Crow River greenstone belt of northern Ontario, the Detour Lake mine in northeastern Ontario, and the Cullaton Lake mine in eastern Northwest Territories. Elsewhere deposits in iron formations include probably the Homestake mine, Lead, South Dakota (Precambrian?); the Morro Velho mine, Minas Gerais, Brazil (Precambrian); and a number of deposits in the Precambrian (Archean) iron-formations of Zimbabwe.

Gold deposits resulting from extensive infiltration or replacement of chemically favorable beds (the second subcategory) are developed principally in calcareous and dolomitic pelites and psammites and in thin-bedded carbonate rocks invaded by granitic stocks and porphyry dykes and sills; a few occur in porous sandstones. Most deposits are characterized by one or more of silicification, argillization, pyritization, and arsenopyritization, and introductions of elements such as Au, Ag, Hg, Tl, B, Sb, As, Se, Te, and the base metals. The gold is usually disseminated through the altered rocks in a very finely divided form (< 5 μ) and is generally, although not always, rich in silver.

Deposits of this type have a widespread distribution throughout the world and are commonly referred to as "Carlin type" because of their occurrence in Silurian silty limestone and dolomitic siltstone in the Carlin-Gold Acres district of Nevada. Similar deposits have been recognized in British Columbia (Specogna), and in USSR (Kuranakh).

8. Gold deposits in quartz-pebble conglomerates and quartzites

These constitute the largest and most productive of the known auriferous deposits, producing some 50% of the annual gold production of the world; some deposits are also economic sources of uranium, thorium, and rare earths. Typical examples are the Witwatersrand deposits in South Africa; other deposits include those in the Tarkwaian System of Ghana and at Jacobina, Bahia, Brazil. The orebodies in the quartz-pebble conglomerate deposits are marked by the presence of abundant pyrite (or hematite) with variable and usually minor to trace amounts of a host of other sulfides, arsenides, sulfosalts, and minerals such as uraninite, thucholite, and brannerite, principally in the matrix of the conglomerates or quartzites. The gold is mainly present as the native metal in a very finely divided form (< 80 μ) essentially in the matrix of the conglomerates or quartzites; minor amounts of the element also occur in the pyrite and in the various other sulfides, arsenides, sulfosalts, and so forth. Elements concentrated in the quartz-pebble conglomerate type of deposit are variable. Most orebodies report enrichments of Fe, S, As, Au, and Ag; some are marked by above average amounts of U, Th, rare earths, Cu, Zn, Pb, Ni, Co, and platinoids. The average Au/Ag ratio in the ores is 10.

9. Eluvial and alluvial placers

These modern placers produce both gold and silver, the latter metal being present usually as a small percentage content of the gold dust and nuggets. Accompanying heavy minerals commonly include variable quantities of monazite, magnetite, ilmenite, cassiterite, wolframite, scheelite, cinnabar, and platinoid minerals. The Au/Ag ratio in placers is generally greater than 1.

Fossil (lithified) equivalents of both eluvial and alluvial placers are known, but few are economic. Here we exclude the quartz-pebble conglomerates of the Witwatersrand and other similar deposits already mentioned, which appear to be modified paleo-placers, although other origins have been suggested (discussed in chap. 14). Placers have been worked for centuries in most countries of the world. The placers of the Pactolus, a tributary of the Gediz (Sarabat) in Anatolia, Turkey, and of the Maritsa (Hebrus) in Thrace were famous in ancient times; in modern times the placers of Colombia, California, Victoria (Australia), Alaska, Yukon, British Columbia, and the far eastern USSR have produced large amounts of gold.

10. Miscellaneous sources

Gold is won from a number of miscellaneous sources, mainly as a byproduct from nickel, copper, and other base metal ores as follows:

1. Nickel-copper ores associated with basic intrusives—Sudbury type
2. Massive sulfide deposits containing essentially Fe, Cu, Pb, and Zn sulfides in volcanic and sedimentary terranes
3. Polymetallic vein and lode deposits containing essentially Fe, Cu, Pb, Zn, Ag sulfides in volcanic and sedimentary terranes
4. Kurokô (black ore) sulfide deposits, occurring mainly in Japan, of which some are greatly enriched in both gold and silver
5. Disseminated deposits—porphyry Cu-Mo type (relatively large sources of gold (and silver) especially in the United States, New Guinea, and the USSR)
6. Certain types of uranium (pitchblende) deposits (e.g., Jabiluka, Northern Territory, Australia)

In these varied deposits, gold usually occurs as the native metal in a very finely divided state, or as tellurides but can also occur in a finely divided form or be present as a lattice constituent in pyrite, arsenopyrite, chalcopyrite, and other base metal sulfides, arsenides, sulfosalts, and selenides.

REFERENCES AND SELECTED BIBLIOGRAPHY

Bache, J.-J., 1982. Les gisements d'or dans le monde, *BRGM Mem. 118,* 101p.
Bateman, A. M., 1950. *Economic Mineral Deposits,* 2nd ed., John Wiley & Sons, New York, 916p.

Boyle, R. W., 1979. The geochemistry of gold and its deposits, *Canada Geol. Survey Bull. 280,* 584p.
Canadian Institute of Mining and Metallurgy, 1948, 1957. *Structural Geology of Canadian Ore Deposits,* 2 vols., Canadian Inst. Min. Metall., Montreal.
Cumenage, E., and F. Robellaz, 1898. *L'or dans la nature,* P. Vicq-Dunod et Cie, Editeurs, Paris, 106p.
Curle, J. H., 1905. *The Gold Mines of the World,* 3rd ed., George Routledge & Sons, London, 308p.
Dunn, E. J., 1929. *Geology of Gold,* Charles Griffin & Co., London, 303p.
Emmons, W. H., 1937. *Gold Deposits of the World,* McGraw-Hill, New York, 562p.
Foster, R. P., ed., 1984. *Gold '82: The Geology, Geochemistry and Genesis of Gold Deposits,* A. A. Balkema, Rotterdam, 753p.
Friedensburg, F., 1953. *Die Metallischen Rohstoffe,* heft 3, *Gold,* F. Enke Verlag, Stuttgart, 234p.
Gmelin, 1950-1954. *Gmelins Handbuch der anorganischen Chemie,* System-nummer 62, *Gold,* pts. 1 and 2, Weinheim/Bergstrasse, Verlag Chemie, GMBH, 406p.
Hodder, R. W., and W. Petruk, eds., 1982. Geology of Canadian gold deposits, *Canadian Inst. Min. Metall. Spec. Vol. 24,* 286p.
Jensen, M. L., and A. M. Bateman, 1979. *Economic Mineral Deposits,* 3rd ed., John Wiley & Sons, New York, 593p.
Lindgren, W., 1933. *Mineral Deposits,* 4th ed., McGraw-Hill, New York, 903p.
Maclaren, J. M., 1908. *Gold, Its Geological Occurrence and Geographical Distribution,* Mining Jour. London, 687p.
Mellor, J. W., 1923. *A Comprehensive Treatise on Inorganic and Theoretical Chemistry,* vol. 3, pp. 491-618, Longmans, Green & Co., London.
Park, C. F., and R. A. MacDiarmid, 1970. *Ore Deposits,* W. H. Freeman & Co., San Francisco, 522p.
Petrovskaya, N. V., 1973. *Native Gold* (in Russian), Izd. "Nauka," Moscow, 347p.
Proust, G. P., 1920. *L'or—prospection, gisement, extraction,* Gauthier-Villars et Cie, Editeurs, Paris, 319p.
Raguin, E., 1961. *Géologie des gîtes minéraux,* Masson & Cie, Editeurs, Paris, 686p.
Ridge, J. D., ed., 1968. *Ore Deposits of the United States, 1933-1967,* 2 vols., Am. Inst. Min. Metall. Petrol. Eng., New York.
Routhier, P., 1963. *Les gisements Métallifères,* t. I and II, Masson et Cie, Paris.
Schneiderhöhn, H., 1955. *Erzlagerstätten,* Gustav Fischer Verlag, Stuttgart, 375p.
Shilo, N. A., 1981. *Fundamentals of the Study of Placers,* "Nauka," Moscow, 383p.
Smirnov, V. I., 1976. *Geology of Mineral Deposits,* Mir Publishers, Moscow, 520p.

CHAPTER

2

Gold During the Pre-Classical (Primitive) Period (5000 B.C.-600 B.C.)

And the gold of that land is good.
Genesis 2:12

Gold was probably the first metal known to the early hominids who, on finding it as nuggets and spangles in the soils and stream sands, were undoubtedly attracted by its intrinsic beauty, great maleability, and virtual indestructibility. As tribal development progressed through the Paleolithic, Mesolithic, and Neolithic ages, and as people congregated into civilized centers, the metal appears to have taken on a sacred quality because of its enduring character (immortality), being worn initially probably as amulets and later fashioned into religious objects (idols). By the time of the early Indus (Harappa, Mohenjo-Daro, etc.), Sumerian, and Egyptian civilizations (3000-2000 B.C.) gold had not only retained its sacred quality but had become the symbol of wealth and social rank (the royal metal). Homer (c. 1000 B.C.), in the *Iliad* and *Odyssey,* the epic poems of ancient Greece, mentions gold repeatedly both as a sign of wealth among mortals and as a symbol of splendor among the immortals.

Early references to the first discovery of gold are essentially legendary or mythical. Thus Cadmus, the Phoenician, is said by some early writers to have discovered gold; others say that Thoas, a Taurian king, first found the precious metal in the Pangaeus Mountains in Thrace. The *Chronicum Alexandrinum* (c. A.D. 628) ascribes its discovery to Mercury (Roman god of merchandise and merchants), the son of Jupiter, or to Pisus, king of Italy, who, quitting his own country, went into Egypt. Similar legends and myths concerning the initial discovery of gold are extant in the ancient literature of the Hindus (the *Vedas*) as well as in that of the ancient Chinese and other peoples. In actual fact the discovery of the element we call gold is lost in antiquity.

The principal source of gold in primitive times was undoubtedly stream placers, although there is considerable evidence in certain gold belts (e.g., Egypt and India (Kolar)) that eluvial deposits, auriferous gossans, and the near-surface parts of friable (oxidized) veins were mined. The eluvial and alluvial placers were worked in the crudest manner by panning or the simplest form of sluicing. The auriferous gossans and exposed parts of friable veins were simply grubbed out, gophered, trenched, or pitted along their strike length with the crudest of tools—stone hammers, antler picks, and bone and wooden shovels. Only rarely were adits, simple shafts, and drifts attempted and then only in the soft rocks of the zone of oxidation. Firesetting was probably employed by the ancient Egyptians, Semites, Indians, and others to break up the hard quartz veins, although there is only limited evidence to support this contention. Size or grade of deposit made little difference; both small and large deposits that showed free gold visibly or in the pan were worked, a circumstance permitted by the low cost of maintenance of slaves, convicts, and prisoners of war who were assigned by those in authority to the gold placers and mines.

GOLD DEPOSITS IN PRIMITIVE TIMES

Early references to the geology, mining, and metallurgy of gold appear in ancient Egyptian codes, on stelae, and in pictograms and inscriptions in the tombs of the Pharaohs. In the code of Menes (c. 3100 B.C.), founder of the first Egyptian dynasty, it was decreed that "one part of gold is equal to two and one half parts of silver in value," an indication of the marked abundance of gold and the relative scarcity of silver at the time. The inscription in the temple at Edfu, Egypt, depicting an epistle to Seti I (nineteenth dynasty, c. 1320 B.C.) from the Sun God reads, "I have given thee the gold countries: given thee what is in them of electrum, lapis lazuli, and malachite," a citation recording the extensive prospecting and mining for gold carried out by Seti I in Egypt, Nubia, and Sinai.

The most ancient geological map known, the famous *Carte des mines d'or* in the Turin Museum, is a Rameside papyrus and fragments depicting a gold mining region active at or about the time of Seti I (c. 1320 B.C.). On it are located roads, miners' houses, gold mines, quarries, auriferous mountains, and so on (Frontispiece). The exact site shown on the map is problematical. Some authors have suggested the mines represented are those of the Wadi Kareim or the Wadi Hammamat, on the Qena-Qoseir road (Gardiner, 1914). Ball (1942) and Derry (1951) say that the area represented on the map is the Wadi Fawakhir in which the El Sid gold mine is situated.

The ancient Sumerian, Akkadian, Assyrian, and Babylonian civilizations utilized gold extensively, but their sources of the precious metal are relatively uncertain. Placers in the upper reaches of the Tigris and Euphrates rivers were probably the principal source, although acquisition through trade with the early civilizations of Arabia, Iran (Elam), the Oxus, Altai Mountains, and India cannot be ruled out. The ancient civilizations of Harappa, Mohenjo-Daro, and others of the Indus Valley also knew and used gold, its source being probably placers in the upper reaches of the Indus River and its various tributaries or through trade with the ancient peoples of Afghanistan, Baluchistan, and northern, eastern, and southern India.

References to gold and gold mining are numerous in the Old Testament of the Hebrews. In fact, gold is the first metal mentioned in the Hexateuch, which includes Genesis, the narrative of which was probably first cast into written form in the tenth century B.C. Six sources of gold are mentioned in the Old Testament (Havilah, Ophir, Sheba, Midian, Uphaz, and Parvaim); the exact locations of all six are problematical and have given rise to much speculation.

Some authorities claim that all six sources are Arabian; others have suggested locations much farther afield.

In Genesis 2:10-12* it is written:

> And a river went out of Eden to water the garden; and from thence it was parted, and became into four heads.
>
> The name of the first is Pison: that is it which compasseth the whole land of Havilah, where there is gold;
>
> And the gold of that land is good: there is bdellium and the onyx stone.

There has been much speculation as to the location of the land of Havilah, the most probable from a geological viewpoint being that the river Pison is the modern Coruh, which drains into the Black Sea near Batumi, and that Havilah is the Pontic goldfield near Trabzon, Turkey. This field is also probably one of the sites where Jason and the Argonauts sought the Golden Fleece, because within historical times placer miners used sheep's fleeces in this and other fields to catch the gold in their crude sluices. Another location often suggested is a site north of the ancient site of Babylon between the Euphrates and Tigris rivers. The statement in Genesis that the gold was "good," probably meaning relatively pure, suggests a placer source for the metal. No placers, so far as can be ascertained, ever existed near Babylon, although it should be remarked that placers may have been worked near the headwaters of both the Euphrates and Tigris in the mountainous areas of Armenia (eastern Turkey). Of these various locations the Pontic goldfield would seem to be the more probable location of Havilah.

There has also been much speculation as to the location of Ophir, the fabulously rich land of gold from which King Solomon's Phoenician (Tharshish) navy brought large amounts of the metal (some 34 metric tons) to his kingdom. In Genesis 10 it is associated with Havilah and Seba; the former, as just noted, was probably the Pontic goldfield on the south shore of the Black Sea, a location that may account for the long period of time, some three years, to make the voyage from Ezion-geber at the head of the Gulf of Aqaba to Ophir and back (I Kings 10:22). The cargos mentioned—almug (sandalwood) trees, precious stones, ivory, apes, and peacocks—suggest circumnavigation of Africa. Tharshish or Tarshish (a region centered on Cadiz) suggests that the gold may have come from Spain, and specifically from the oxidized deposits of the Huelva region where the modern mining town of Tharsis, often equated with Tarshish, is located. Supportive evidence is found in the first book of the Maccabees (I Macc. 8:1-3) that mentions the gold and silver mines of Spain. Other possibilities are East Africa, principally Zimbabwe and specifically the ruins of Great Zimbabwe, where some think King Solomon's mines and metallurgical plants were located. Still other possibilities suggested for the site of King Solomon's mines are southern Turkey (Taurus Mountains), northwest Saudi Arabia (the land of the ancient Midians and possibly the Eldorado of the Hebrews), Sudan (ancient Nubia; *Nub* means *gold* in ancient Egyptian), Altai (Purington, 1903), Ethiopia (along the coast between ancient Adulis and Bab el

---

*All biblical quotations are from the authorized King James version of the Holy Bible, edition of 1611, or from the Vulgate edition in the case of the Books of the Maccabees.

Mandeb, whose natives called themselves *Aphar*), India (possibly the Kolar region), Cuba, Peru, the Far East (particularly Japan), Arctic Canada, and a hundred other places. The story of Ophir is well told by Rickard (1932) and Sutherland (1969). The history of the ancient Zimbabwian (Rhodesian) gold mines and of the ruins of Great Zimbabwe are related in fascinating detail by Summers (1969).

Sheba, whose queen brought King Solomon great store (6 metric tons) of gold (I Kings 10:10), corresponds to modern Yemen where ancient eluvial placers may have occurred in association with oxidized zones of copper and lead sulfide deposits. It seems more probable, however, that the Queen of Sheba obtained much of her gold from Punt, which can be equated with the auriferous areas in the countries bordering the Red Sea and the Gulf of Aden (Sudan, Ethiopia, Djibouti, Somalia) and possibly also with Zimbabwe. It was from these areas in Punt that the Egyptian navy in Queen Hatshepsut's time (1503-1482 B.C.) and later brought great stores of gold and stibium (stibnite) to Egypt.

Midian, often considered the Eldorado of the Hebrews, occupies the northernmost coastal district of Hejaz, Saudi Arabia, on the Red Sea and its Gulf of Aqaba. This area apparently abounded in gold in Biblical times, as witnessed by the quotations in Numbers 31:50-54 relating to the spoil of gold taken by the Israelites after the first Midianite war and by the statements in Judges 8:24-27 describing the golden tribute accepted by Gideon after the conquest of Midian. Much of the gold of Midian appears to have come from oxidized gold quartz deposits that were worked to considerable depths in ancient times (Burton, 1979).

The two other sources of gold mentioned in the Old Testament, Uphaz (Jeremiah 10:9; Daniel 10:5) and Parvaim (II Chronicles 3:6) cannot be identified from any of the references. It seems probable that they were located in the auriferous regions of western Arabia.

The numerous references to gold and silver in the Old Testament attest to the importance of the metals in Biblical times. In the majority of cases when the two precious metals are mentioned together silver comes first, reflecting perhaps a very early period when gold was less valued than silver, a situation perhaps confirmed by the fact that most of the gold in very ancient times came from placers whose dust and nuggets contained only minor amounts of silver. Later, as argentiferous galena deposits were worked (probably in Punt, Arabia, Attica, the Aegean, Asia Minor, Thrace, Macedonia, and elsewhere) silver apparently became plentiful so that by Solomon's time the metal was "nothing accounted of" (I Kings 10:21) and the king "made silver to be in Jerusalem as stones" (I Kings 10:27).

Geological references to gold and silver are relatively rare in the Old Testament. In the book of Job it is stated, "Surely there is a vein for the silver, and a place for gold where they fine it" (Job 28:1), and "As for the earth . . . it hath dust of gold" (Job 28:5-6)—quotations that contain, albeit naively, two of the loftiest truths concerning the occurrence of gold and silver, namely in veins and in placers.

There is some evidence from ancient writings and workings that gold placers and residual (oxidized) deposits were exploited sporadically in antiquity in the many islands of the Aegean (Thasos, Samos, Siphnos), in Anatolia (Lydia) and the Troad (Troy), in Thrace, Macedonia, and Arcadia, in the area bordering the southern shore of the Black Sea (Pontus Euxinus), in Cappadocia (central

Turkey), in Bactria (upper reaches of the Oxus River), in middle Asia (Tien Shan and Altai Mountains), and perhaps in Dacia (Transylvania). Much gold also appears to have been won in Spain, probably from a variety of deposits and areas (Huelva, Almeria).

India, particularly southern India, has long been known for its aurificity, where in ancient times much gold was won from eluvial and alluvial placers and from the oxidized outcrops of veins. Diodorus Siculus, in his *Bibliotheca historica* written in the first century B.C., says that in India the earth "contains rich underground veins of many kinds, including many of silver and gold . . . ," Likewise in China gold was sought and utilized during the early Shang civilization (1800-1027 B.C.) of the Huang-Ho (Yellow) River, the precious metal being obtained principally from placers in the hinterlands of this great river system and possibly also from placers in Mongolia. Mills (1916) suggests that gold mining (placering) was probably introduced into Korea in 1122 B.C. by the followers of Ki-ja, who migrated from China. From Korea the methods of eluvial and alluvial placering for gold were taken to Japan, probably as early as 660 B.C. (Bromehead, 1942).

Gold was known to the early Amerindians, but the metal was not held in high regard in the period covered in this chapter. Later, during the first centuries of the Christian Era, gold assumed much greater importance in the Olmec, Zapotec, Mayan, Aztec, and other civilizations of Mexico and Mesoamerica and in the Inca civilization of South America. Gold was not prized by the Amerindians of Canada and the United States, and the aborigines of Australia seem not to have paid any attention to the precious metal.

To summarize: Gold was probably the first metal known to humankind, and references to it have appeared almost from the birth of writing. All of the first civilizations prized and utilized gold and sought the precious metal in their lands and suzerainties or through trade. References to the geological setting of gold deposits are sparse in the pre-Classical literature, but the most ancient of geological maps known portrays an auriferous region in Egypt.

REFERENCES AND SELECTED BIBLIOGRAPHY

Aitchison, L., 1960. *A History of Metals*, 2 vols., Macdonald & Evans, London, 647p.

Andrée, J., 1922. *Bergbau in der Vorzeit,* C. Kabisch, Leipzig, 72p.

Ball, J., 1942. *Egypt in the Classical Geographers,* Survey of Egypt, Cairo, Government Press, Bulaq, 203p. (See Appendix 1).

Bromehead, C. N., 1942. Ancient mining processes as illustrated by a Japanese scroll, *Antiquity* **16:**193-207.

Browne, C. A., 1935. The chemical industries of the American aborigines, *Isis* **23** (2):406-424.

Burton, R. F., 1979. *The Gold-mines of Midian and the Ruined Midianite Cities (1878),* Falcon-Oleander, London, 238p.

Cambridge Ancient History, 1982-, 12 vols., Cambridge Univ. Press, Cambridge.

Cheyne, T. K., and J. S. Black, Eds., 1899-1903. Encyclopaedia Biblica, 4 vols., Macmillan, New York.

Davies, O., 1935. *Roman Mines in Europe,* Clarendon Press, Oxford, 291p.

Derry, D. R., 1951. A 3000 year old gold mine, *Canadian Mining Jour.,* **72** (pt. 2, no. 8):68-72.

Dominian, L., 1912. History and geology of ancient gold-fields in Turkey, *Am. Inst. Min. Eng. Trans.* **42:**569-589.

Forbes, R. J., 1971. *Studies in Ancient Technology,* 2nd ed., vol. 8, E. J. Brill, Leiden, Holland, 295p.

Gardiner, A. H., 1914. The map of the gold mines in a Ramesside papyrus at Turin, *Cairo Sci. Jour.* **8:**42-46.
Gardiner, A., 1961. *Egypt of the Pharaohs,* Oxford Clarendon Press, 461p.
Garland, H., and C. O. Bannister, 1927. *Ancient-Egyptian Metallurgy,* Charles Griffin & Co., London, 214p.
Hawkes, J., ed., 1974. *Atlas of Ancient Archaeology,* McGraw-Hill, New York, 272p.
Healy, J. F., 1978. *Mining and Metallurgy in the Greek and Roman World,* Thames & Hudson, London, 316p.
Healy, J. F., 1985. The processing of gold ores in the ancient world, *Canadian Inst. Min. Metall. Bull.* **78** (874):84-88.
James, T. G. H., 1972. Gold technology in Ancient Egypt, *Gold Bulletin* **5** (2):38-42.
Jenness, D., 1960. The Indians of Canada, 5th ed., *Canada Natl. Mus. Bull. 65,* 452p.
Kidder, A., 1964. South American high cultures, in *Prehistoric Man in the New World,* Univ. Chicago Press, pp. 451-486.
Levey, M., 1959. *Chemistry and Chemical Technology in Ancient Mesopotamia,* Elsevier, Amsterdam, 242p.
Lucas, A., 1948. *Ancient Egyptian Materials and Industries,* 3rd. ed., Longmans, London, 580p.
Martin, P. S., G. I. Quimby, and D. Collier, 1947. *Indians before Columbus,* Univ. Chicago Press, Chicago, 582p.
Mills, E. W., 1916. Gold mining in Korea, *Royal Asiatic Soc. Korea Branch Trans.* **7:**5-39.
Needham, J., 1954-1983. *Science and Civilization in China,* 6 vols., Cambridge Univ. Press, Cambridge.
Partington, J. R., 1936. *Origins and Development of Applied Chemistry,* Longmans, Green & Co., London 597p. Reprint Arno Press, New York, 1975.
Patterson, C. C., 1971. Native copper, silver, and gold accessible to early metallurgists, *Am. Antiquity,* **36** (3):286-321.
Purington, C. W., 1903. Ancient gold mining, *Eng. and Mining Jour.* **75:**437.
Ray, P. C., 1905. *A History of Hindu Chemistry from the Earliest Times to the Middle of the Sixteenth Century A.D.,* vol. 1, Bengal Chemical & Pharmaceutical works, Calcutta, 312p.
Rickard, T. A., 1932. *Man and Metals,* 2 vols., McGraw-Hill, New York, 1068p.
Rosenfeld, A., 1965. *The Inorganic Raw Materials of Antiquity,* Frederick A. Praeger, New York, 245p.
Sagui, C. L., 1930. Economic geology and allied sciences in ancient times, *Econ. Geology* **25:**65-86.
Shepherd, R., 1980. *Prehistoric Mining and Allied Industries,* Academic Press, London, 272p.
Summers, R., 1969. Ancient mining in Rhodesia, *Natl. Mus. Rhodesia Mem. 3,* Salisbury, 236p.
Sutherland, C. H. V., 1969. *Gold: Its Beauty, Power, and Allure,* 2nd rev. ed., McGraw-Hill, New York, 196p.
Tylecote, R. F., 1962. *Metallurgy in Archaeology,* Edward Arnold, London, 368p.
Tylecote, R. F., 1976. *A History of Metallurgy,* The Metals Society, London, 182p.
Weeks, M. E., and H. M. Leicester, 1968. *Discovery of the Elements,* 7th ed., Jour. Chem. Educ., Easton, Pa., 896p.
Wheeler, M., 1968. *The Indus Civilization,* Cambridge Univ. Press, Cambridge, 144p.
Whitehouse, D., and R. Whitehouse, 1975. *Archaeological Atlas of the World,* W. H. Freeman & Co., San Francisco, 272p.

CHAPTER
3

Gold During the Classical Period

The metals obtained by mining, such as silver, gold, and so on come from water.
Theophrastus, Peripatetic philosopher, 3rd century B.C.

Science in the Classical period was confined mostly to speculation. The pre-Aristotelians explained the origin of the universe and natural phenomena in the light of four basic elements—earth, water, air, and fire. This theory probably originated in the ancient Indus Valley civilization (Harappa?), was taken up by the early Babylonian natural scientists, and passed on to the early Greek philosophers, among whom Empedocles (c. 490–430 B.C.) is usually credited with refining the concepts of the theory. Aristotle embraced the four element theory and added another concept, that of the ether.

During pre-Hellenistic and Hellenistic times gold and silver were mined extensively throughout the Mediterranean, in Asia Minor, and elsewhere in western and northern Europe and in Africa and Asia. A large literature exists on ancient mining in these areas, all admirably summarized and synthesized by Rickard (1932), Davies (1935), and Healy (1978).

Initially it would seem from the evidence presented by many of the deposits and dumps in Egypt, the Aegean, Turkey (Anatolia), Iran, India, and contiguous regions that most of the gold came from eluvial and alluvial placers; only later as these placers approached exhaustion were the oxidized zones of gold-quartz and sulfide deposits exploited, first by open-cut methods and then by underground workings. Firesetting seems to have been widely employed by the Greeks and the Romans in underground mining. Pliny, in his *Historia naturalis,* claimed that vinegar was more efficacious than water in quenching and disintegrating the hot rock. Because of water and ventilation problems the depth of exploitation of bedrock deposits was limited, probably to 200 m or less in most auriferous regions. All mining was done by slaves and convicts. Prospecting knowledge was crude and was based principally on visual signs of the presence of gold in quartz float, in exposures in quartz outcrops, in gossans, nearby soils and weathered residuum, and in the sediments of streams and rivers. Much "gophering" was employed in prospecting, evidenced by an abundance of pits and shallow shafts in most of the ancient mining areas of the Aegean, Turkey, Egypt, India (Kolar), and elsewhere. The gold pan appears to have been employed for testing gossans, soils, and alluvium and for winning the metal since the earliest times. Likewise, the rocker and sluice with riffles, animal fleeces, or ulex for trapping the gold seem to have been in use almost from the beginning of placer gold mining. The animal (sheep or goat) fleeces were dried and the gold was then shaken out of them into pans for further concentration. When ulex (a prickly plant of the furze or gorse family) was used it was burned and the gold washed out of the ashes.

Some placer operations employed the boulder-riffle method of concentrating the gold; in this method boulders were arranged in such a way that as the water rushed along carrying the gold, it swirled around the boulders, depositing the nuggets and dust in the slack-water zones around and between the boulders. After an appropriate interval the sluicing water was turned off or diverted to allow the cleanup. Stretches of streams with natural riffles, such as slate and schist beds and folia oriented at right angles to the stream direction, also appear to have been employed in some placer areas. In large placer operations, especially where high-level (terrace) gravels were exploited by the Romans, as along the Sil in Northwestern Spain and the Vrbas in Yugoslavia, hushing (booming) was employed. This method frequently required aquaducts and canals several kilometers in length for transport of the water to the crude monitors. Where bedrock quartz deposits were exploited, the separation of the native gold from the dross, mainly quartz, was accomplished by hand picking, followed by crushing and grinding in stone mortars, querns, and crude "hour glass" and other types of mills, and finally by washing on sloping boards or flat rocks. These washeries can be seen in pictorial form on walls and tablets in Egyptian tombs and in a dilapidated condition in a number of ancient mining regions in Greece, Egypt, and the Middle East.

GOLD DEPOSITS AND THEORIES OF THEIR ORIGIN

In early Classical times ancient gold placers and mines were known on many of the Aegean Islands, particularly Siphnos, in mainland Greece, along the southern shores of Pontus Euxinus (Black Sea), and near the western coast of Asia Minor; most were small and soon worked out by the fifth century B.C. Prospecting farther afield, particularly in the region of Mt. Tmolus (the modern Boz Dag), revealed the rich electrum placers of the rivers Pactolus and Hermus (the modern Gediz). Legend has it that the Pactolus is the river in which Midas, the mythical founder of the Phrygian kingdom, on the advice of Bacchus bathed in its waters to rid himself of the fatal faculty of turning everything he touched into gold. From the Pactolus came large stores of placer gold won mainly by the Lydian kings, of whom Ardys (c. 650 B.C.) minted at Sardis the earliest gold coins extant. In the course of time the Lydian monarchs became the richest princes of their age, especially Croesus who followed Alyattes (605-560 B.C.) on the throne and whose name we associate today with enormous wealth. Another source of gold described in early Classical and later Classical times was located in Thrace and Macedonia, where the mines at Dysoron and on Mt. Pangaeus provided great wealth for Thracians, Athenians, and Philip II of Macedonia, founder of Philippi near Mt. Pangaeus. It appears that much gold also reached the Aegean area in early Classical times from Egypt, Armenia (Turkey), Dacia (Romania, Hungary) and from as far afield as Spain, Siberia, and India.

During Hellenistic times many of the gold mines of Macedonia and Thrace were exploited intensively as were also some of those in various auriferous regions of Asia Minor. The Ptolemies, the dynasty of Macedonian Kings that ruled Egypt (323-30 B.C.), prospected extensively in Egypt, Nubia (Sudan), and probably also in Arabia, winning considerable gold from these ancient mining regions.

Many of the early Greek authors mention gold (and silver) but without any particular geological reference. The *Iliad* and *Odyssey* of Homer (c. 1000 B.C.), the traditional epic poet of Greece, refer to gold (and silver) in numerous contexts and locations, the latter probably authentic in many cases. We have also the Greek myth of Jason and the Golden Fleece (Fig. 3-1). According to

Figure 3-1. The Argonauts examining the Golden Fleece at Colchis on the Euxine (Black Sea). (From the woodcut in Agricola, De Re Metallica, 1556).

this myth, the Golden Fleece was taken from the ram on which Phrixus and Helle escaped from being sacrificed. It was hung up in the grove of Ares in Colchis and recovered from King Aeetes by the Argonautic expedition under Jason, with the help of the sorceress Medea, the king's daughter. In actual fact the Argonauts were early prospectors who sought the source of the ancient placers on the Black Sea. At that time (1200 B.C.) the workers of auriferous placers recovered the gold by trapping the metallic particles on sheep's fleeces placed in crude sluices. The fleeces were then hung up to dry in nearby trees and were later shaken to collect the gold.

Gold is a subject touched upon by many of the Classical Greek authors, but again the geological references to the precious metal are mostly vague and nonspecific. Among these Greek writers we may mention Herodotus (c. 482-425 B.C.), author of *History of the Persian Wars*; Thucydides (c. 460-400 B.C.), author of *History of the Peloponnesian War*; Xenophon (c. 430-355 B.C.), historian, man of letters and author of many books and tracts, including *De vectigalibus,* which is of some interest in the matter of mining; Plato (428-347 B.C.), pupil of Socrates, renowned for his great philosophical works, of which *Timaeus* is of most interest to natural scientists; Aristotle (384-322 B.C.), disciple of Plato, teacher to Alexander the Great, and author of many famous philosophical treatises, of which the *Meteorologica* is of particular interest in the present context; and Theophrastus (371-288 B.C.) of whom more details will be discussed later.

Little of interest concerning gold deposits can be found in the writings of Thucydides, Xenophon, Plato, and Aristotle, although in the works of the last two there are some speculations on the origin of mineral (gold) veins.

Plato's view of the origin of mineral veins is complex and far less intelligible than other aspects of his philosophy. In the dialogue of the *Timaeus* a pythagorean view of the constitution of matter is given in which the minute particles of the four fundamental elements are considered in geometrical terms (e.g., earth a cube, air an octahedron, water an icosahedron, etc.) that undergo changes (transmutations) into one another in definite ratios by resolution into triangles and reassociation of these triangles. The metals were evidently considered to be composed of various associations of "fusible" water (liquid), a theory that originated with Thales. Aristotle, likewise, embraced the four element theory of material things in the *Meteorologica* and considered that these elements underwent changes (or transmutations) within the earth that were actuated by the deep penetration of the sun's rays. As a result exhalations emanated from the earth, some fiery and dry that produced stones (rocks) and others moist (watery) that gave rise to metals. One can see in these views three fundamental concepts: (1) transmutation of the elements (especially the hope of transmuting the base metals into gold), a possibility that intrigued philosophers and particularly alchemists for centuries and that we now know to be a fact, exemplified by the atomic transmutation of uranium into lead in the earth; (2) the dry and fiery exhalations, which constitute the basis of the magmatic theory of the formation of igneous rocks and certain mineral deposits; and (3) the moist exhalations, or hydrothermal solutions that are considered by many to give rise to most mineral veins.

Herodotus, the "father of history," was born in Asia Minor at Halicarnassus, a Greek city then under Persian rule. He seems to have traveled widely throughout the Persian Empire as well as to Babylon, Egypt, Greece, Libya, Thrace, Macedonia, Scythia, and the Black Sea region. Most of his observations appear to be based on first hand knowledge, although some of his statements are obviously derived from hearsay. None of his observations are detailed, and few give any information of geological interest. In his *History of the Persian Wars* he describes the great gold deposits near Mt. Pangaeus in Thrace in only one sentence: "Mt. Pangaeus is a great and high mountain abounding in mines of gold and silver . . . ."

Theophrastus, the "father of botany," has left us *De lapidibus* (On stones), an important tract that is a landmark in geological science. Another tract, *On Mines,* has unfortunately been lost. His works *On the History of Plants* and *On the Causes of Plants* are classics in botany.

Theophrastus was Aristotle's favorite pupil and his successor as head of the Lyceum (Peripatos) at Athens. His observations on things mineral are acute, considering the state of science in the last half of the fourth century B.C. He mentions gold in a number of contexts in *De lapidibus,* an existing fragment of what was evidently a much larger work. Many translations of this fragment exist, the latest in English with an extensive commentary by Caley and Richards (1956).

Theophrastus mentions the gangue stones of gold and silver (pyrite and galena) and writes at some length on the touchstone (probably chert or cherty sediment) and its ability to test what we now refer to as the fineness or carat of gold. Of particular interest, however, is the opening statement in *De lapidibus* as given in the translation by Caley and Richards (1956, p. 45).

> Of the substances formed in the ground, some are made of water and some of earth. The metals obtained by mining, such as silver, gold, and so on, come from water; from earth come stones, including the more precious kinds, and also the types of earth that are unusual because of their color, smoothness, density, or any other quality.

The idea advanced by Theophrastus that gold is made of water probably stems directly from the Platonic view that gold was some kind of dense congealed water, an ancient postulate originating with the "father of Greek philosophy," Thales (624-546 B.C.) of Miletus, who argued that the fundamental matrix (Arché) of all matter was water. Thales was probably led to this conclusion by his observation, and that of others, that mineral matter was commonly precipitated at the orifices of springs. Interpreting broadly we have, therefore, in the statement of Theophrastus the first exposition of the hydrothermal origin of gold deposits.

The Romans mined gold extensively in the metalliferous regions of their empire but, while advancing the art and science of metallurgy, did relatively little beyond simple description with the science of mineral deposits. Diodorus Siculus (first century B.C.), the later Hellenistic historian writing in his *Bibliotheca historica* at Rome, mentions gold placers and mines in many regions of the Roman Empire and gives lurid accounts of the working of gold and silver mines by slave labor at the time of Julius and Augustus Caesar. The gold placers of Gaul are described, as are those of the Rhine; brief accounts are given of the gold deposits in Arabia and as far afield as India. Describing the gold deposits of Egypt and neighboring countries he states (Booth, 1700):

> In the confines of Egypt and in the neighbouring countries of Arabia and Ethiopia there is a place full of rich gold mines, out of which, with much cost and pains of many laborers, gold is dug. The soil here is naturally black, but in the body of the earth run many white veins, shining with white marble, and glistering with all sorts of other bright metals; out of which laborious mines those appointed overseers cause the gold to be dug up by the labor of a vast multitude of people.

Diodorus in this excerpt and in that quoted in Healy (1978, p. 84) is speaking of gold quartz stockworks in greenstones (amphibolites), or perhaps also in black slates or schists. The reference to white marble (calcite) is probably only partly correct. It would seem that perhaps quartz, the most common gangue in gold veins, was meant.

Strabo (63 B.C.-A.D. 20), another late Hellenistic historian, traveled extensively and made many personal observations on gold deposits throughout the Roman Empire. In his *Geographia* he mentions briefly the gold placers and mines of Spain and Portugal (Lusitania, Baetica), Macedonia (Paeonia), Italy, Arabia, Egypt, the Caucasus, and India. Writing of the gold of the Caucasus Strabo remarked: "The mountain torrents are said to bring down gold, and these barbarians (Soanes) catch it in troughs (sluices) perforated with holes and in fleecy skins." Strabo's reference is probably to the river Phasis (the present Rioni in Georgian SSR, draining from the Caucasus), which drained the Colchis region, supposedly one of the goals of the Argonauts in the legend of Jason. Describing the gold deposits of Egypt and Nubia Strabo quotes Agatharchides of Cnidus (c. 132 B.C.) freely and gives us an insight into the geological setting of some of the deposits. In an interpretative translation we learn: "There (in Egypt and Nubia) the rock is black and full of streaks and veins of a mineral with a remarkable whiteness the lustre of which surpasses the most brilliant natural materials." Obviously Strabo, like Diodorus Siculus, is speaking of gold-quartz stockworks in greenstones (amphibolites) or in black slates or schists. The white mineral is probably milky quartz. These interpretations are supported by recent work in Sudan (Fletcher, 1985; Gaskell, 1985).

A number of Roman writers have left works of interest to geologists and mineralogists. Lucretius (99-55 B.C.) author of *De rerum natura,* a famous philosophical work, describes many natural phenomena but mentions gold and mineral deposits in only a cursory manner. Vitruvius (c. 27 B.C.), the renowned Roman military engineer and architect, likewise treats the subject of mineral deposits and gold only briefly in his *De architectura.* Pliny the Elder (A.D. 23-79) was the only Roman to engage in extensive natural history studies; of all the classical writers he alone has given us a documentation of the works of earlier writers and a summary of the knowledge of minerals and mineral deposits extant at the beginning of our era. His extensive travels, as a onetime cavalry officer and later prefect (admiral) of the Roman Fleet under Vespasian, are recorded in thirty-seven books in his encyclopedic *Historia naturalis.* This famous treatise was suddenly terminated by Pliny's death as a martyr to science when he was suffocated by volcanic gases as he sought to study the cause of the violent eruption of Vesuvius in A.D. 79.

The *Historia naturalis* is a great assemblage of fact, fancy, and fiction written both from observation and hearsay. Where observation prevails the narrative is precise and logical; where hearsay predominates much of the text is fanciful and sometimes bizarre. Book 33 treats of the natural history of metals, and in it the first section is reserved for gold.

After a long discourse on the avarice of man for gold, the wearing of gold rings and other golden adornments, coinage, other uses of gold, and the special qualities of gold, Pliny turns to methods of discovering and mining gold ores. In the translation of Rackman (1968, p. 53) with my explanations in parentheses we read:

> Gold in our part of the world—not to speak of the Indian gold obtained from ants or the gold dug up by griffins in Scythia—is obtained in three ways: in the detritus of rivers, for instance in the Tagus in Spain, the Po in Italy, the Maritza (Hebrus) in Thrace, the Sarabat (Pactolus) in Asia Minor and the Ganges in India; and there is no gold that is in a more perfect state, as it is thoroughly polished by the mere friction of the current. Another method is by sinking shafts; or it is sought for in the fallen debris of mountains. Each of these methods must be described.
>
> People seeking for gold begin by getting up (searching for) *segullum*—that is the name for earth that indicates the presence of gold. (Beneath) this is a pocket of sand, which is washed, and from the sediment left an estimate of the vein is made. Sometimes by a rare piece of luck a pocket is found immediately, on the surface of the earth, as occurred recently in Dalmatia when Nero was emperor, one yielding fifty pounds weight of gold a day. Gold found in this way in the surface crust is called *talutium* if there is also auriferous earth underneath. The otherwise dry, barren mountains of the Spanish provinces which produce nothing else whatever are forced into fertility in regard to this commodity.
>
> Gold dug up from shafts is called 'channelled' or 'trenched' gold; it is found sticking to the grit (gangue) of marble, not in the way in which it gleams in the lapis lazuli of the East and the stone of Thebes and in other precious stones, but sparkling in the folds of the marble. These channels of veins wander to and fro along the sides of the shafts. . . . The third method will have outdone the achievements of the Giants. By means of galleries driven for long distances the mountains are mined by the light of lamps.

In a following passage Pliny deals with the constitution of gold. In the translation by Rackman (1968, p. 53) we read:

> All gold contains silver in various proportions, a tenth part in some cases, an eighth in others. In one mine only, that of Callaecia called the Albucrara mine, the proportion of silver found is one thirty-sixth, and consequently this one is more valuable than all the others.

> Wherever the proportion of silver is one-fifth, the ore is called electrum; grains of this are found in 'channelled' gold.

The reference to the Indian gold ants and the Scythian griffins comes from Herodotus and is a tale spun through the ages down to the present day, being repeated and analyzed by Strabo and Pliny and by T. A. Rickard as late as 1930. Adams (1938) discusses the probable origin of the tale in detail. My personal opinion is that in ancient times someone found eluvial gold particles in ant (termite) mounds somewhere in Asia, from whence sprung the fable of the golddigging ants. In this respect it was well known by the old prospectors that ants will collect splendent particles of metals and minerals such as gold and galena. Some birds have a similar penchant. The griffin tale from Scythia may be similarly based on the circumstance that the spoil near certain animal burrows (e.g., marmot) in auriferous regions may contain particles of gold.

The material called *segullum* (or *segutilum* in some Latin versions) should probably be translated as *gossan. Segullo* is still used by Castilian (Spanish) prospectors to define the gossanous material capping certain auriferous deposits. I can find no translation for the Latin word *talutium* (or *talutatium* in some versions); I suspect that nuggety or eluvial was meant. The use of *marble* for the gangue of gold veins is evidently an error; quartz was obviously meant. *Channelled* gold in the second excerpt should be translated as *vein* gold.

In a later passage Pliny deals with the oxidation processes of copper minerals in gold veins. In the translation given by Bailey (1929, pt. 1, p. 105) it is stated:

> Chrysocolla is an exudation found in the shafts which we have already mentioned. It oozes down the vein of gold, and its muddy substance is congealed under the influence of the winter cold till it is as hard as pumice. It is well established that the best quality forms in copper mines and the next best in silver mines. It occurs in lead mines also, but this kind is inferior to that found in gold mines. In all these mines, however, it is also prepared artificially (though its worth is far below that of the natural product) by leading a gentle stream of water into the vein all through the winter till the month of June, and evaporating the extract in June and July. It is therefore perfectly clear that chrysocolla is none other than a decomposition product of a vein of metal.

There has been considerable discussion by classic scholars about the identity of chrysocolla. The modern chrysocolla is a copper silicate, and Pliny may have meant this mineral in some cases; more generally, however, the mineral in question was probably malachite or azurite, the green and blue copper carbonates. The reference to the precipitation mechanism as due to winter cold follows from the ancient idea that many minerals were congealed (frozen) from water analogous to the formation of ice. This passage is one of the first references in the geological literature to oxidation phenomena in mineral (gold) deposits (see also chap. 16).

The last of the classical Roman writers to briefly refer to gold deposits was Seneca (A.D. 3-65) contemporary of Pliny and tutor to Nero. In his *Quaestiones naturales* he says in (an interpretative) translation: "In the earth there are some kinds of moisture that harden when fully formed. From these arises all metalliferous deposits from which our avarice seeks gold and silver." This passage suggests that Seneca followed Theophrastus in his view of the origin of gold (deposits) from water.

Early Indian (Hindu) manuscripts mention gold in many contexts but deal only in the simplest of generalities with the origin of the various types of deposits from which the precious metal was obtained. The *Rig-Veda,* first of the books of the sacred Sanskrit scriptures of the Hindus, composed about 1500 B.C., contains several references to gold and silver but gives only the barest details about the geology of the metals (Ray, 1904, 1925; Bhagvat, 1933). In the *Arthashastra* of Chanakya or Kautilya (fourth century B.C.) there are further details about gold and silver, but these are concerned principally with the mining, metallurgy, and testing of the two metals. Mention is made, however, of the occurrence of gold ores in mountains and of the different types of auriferous ores—unmixed yellow (probably pyritic ore?), red and reddish yellow (limonitic ore?), and some with a bluish color on fractures (oxidized copper sulfide ore). Nodules (nuggets) and needles of gold are also described from placers. Several color types of gold are mentioned—that with a blue shade, silvery (electrum), thorn-applelike (red), and pure (the color of the pollen of the lotus). The early Indian peoples mined gold extensively from bedrock deposits and also obtained the metal from placers in many of the rivers of the Indian Peninsula. Mining of the bedrock deposits was carried out principally in the oxidized zones of auriferous veins and disseminations in sheared zones and iron-formations, as is evident from very ancient workings in the famous Kolar field in south India. Some of these workings probably provided gold to the early Indus Valley civilization of 4000-3000 B.C. (Ray, 1948).

Allchin (1962) has reviewed the evidence of early gold mining in India and the methods employed in winning the precious metal. He has concluded from archaeological evidence and C-14 dating that placer mining in India was widespread long before the Christian Era and that large-scale bedrock mining (principally in the oxidized zones but also in some cases in the hard rock) in southern India began with the Mauryan colonization of the Deccan about the end of the fourth century B.C. The Hutti and Kolar fields appear to have been first prospected and mined about the beginning of the Christian Era.

It seems strange that the Greeks, Romans, and Indians never developed any precise scientific theories on the origin of gold and other types of deposits despite the fact that gold was mined from its principal deposits (excluding the quartz-pebble conglomerates) throughout their empires for many centuries. It has been said that the Roman in general, and the Roman (mining) engineer in particular, was a very practical man and not given to speculation and theories; the same can be said of the Greek but only in part. Natural phenomena interested the Greeks but their philosophers were concerned more with speculations on the great manifestations of nature rather than with mundane things such as gold veins. Another factor, it is said, revolves about the use of slaves in Greek, Roman, and Indian mining ventures. To engage in the earthy tasks of mining (and geology) was deemed to be below the elite and generally considered to be disparaging of one's station in life. As a consequence Greek, Roman, and Indian writers provide little if any observational detail on geological and geochemical processes. Their knowledge of minerals was limited to only a few species and their familiarity with rocks was essentially negligible. Therefore they could not have logically evolved any systematic theories for the origin of (gold) veins and other types of deposits.

REFERENCES AND SELECTED BIBLIOGRAPHY

Adams, F. D., 1938. *The Birth and Development of the Geological Sciences,* Williams & Wilkins, Co., Baltimore, 506p.

Allchin, F. R., 1962. Upon the antiquity and methods of gold mining in ancient India, *Jour. Econ. Social History of the Orient,* **5** (pt. 2, 197):195-211.

Bailey, K. C., 1929. *The Elder Pliny's Chapters on Chemical Subjects,* part 1, Edward Arnold & Co., London, 249p.

Bhagvat, R. N., 1933. Knowledge of the metals in ancient India, *Jour. Chem. Educ.* **10:**659-666.

Booth, G., 1700. The historical library of Diodorus the Sicilian in 15 books, etc., London.

Caley, E. R., and J. F. C. Richards, 1956. *Theophrastus on Stones,* Ohio State Univ., Columbus, Ohio, 238p.

Davies, O., 1935. *Roman Mines in Europe,* Clarendon Press, Oxford, 291p.

Fletcher, R. J., 1985. Geochemical exploration for gold in the Red Sea Hills, Sudan, in *Prospecting in Areas of Desert Terrain,* Inst. Min. Metall., London, pp. 79-94.

Gaskell, J. L., 1985. Reappraisal of Gebeit gold mine, Northeast Sudan: a case history, in *Prospecting in Areas of Desert Terrain,* Inst. Min. Metall., London, pp. 49-58.

Healy, J. F., 1978. *Mining and Metallurgy in the Greek and Roman world,* Thames & Hudson, London, 316p.

Rackham, H., transl., 1968. *Natural History,* by Pliny, vol. 9, Libri 33-35, Harvard Univ. Press, 421p.

Ray, P. C., 1904, 1925. *History of Hindu Chemistry. . . ,* 2 vols. Chuckervarty and Chatterjee, Calcutta. Rev. ed. in one vol., as *History of Chemistry in Ancient and Medieval India,* Indian Chemical Society, Calcutta, 1956.

Ray, P. R., 1948. Chemistry in ancient India, *Jour. Chem. Educ.* **25:**327-335.

Rickard, T. A., 1932. *Man and Metals,* 2 vols., McGraw-Hill, New York, 1068p.

CHAPTER

4

Gold During the Middle Ages

Gold is of the metals the most precious.
Geber, ninth century

The fall of the Roman Empire in the west during the latter part of the fifth century was followed by widespread political and economic chaos that existed in Europe for more than four centuries (the Dark Ages) and was only slowly terminated by the institution of the feudal system in the ninth century. This social order passed into decline near the end of the thirteenth century and was gradually replaced during the next three centuries by nation-states with monarchial and nobility systems, coteries of salaried civil servants, and rudimentary parliaments.

The social chaos, incessant warfare, plagues, and general economic instability during the Early Middle Ages (fifth to the eleventh centuries) resulted in a marked reduction in mining and placering operations for gold. With increasing stability in the High and Late Middle Ages (eleventh to the sixteenth centuries) mining activity increased, and gold was widely sought and won from many of the auriferous regions of western and central Europe, the Middle East, and Middle Asia. Gold mining and placering in China are very old occupations that were pursued extensively in medieval times, judging from the records noticed by Needham (1959). Gold mining in Korea began in the year 1079, but placers had been worked as early as 1122 B.C. (Mills, 1916); in Japan gold mining and placering probably began on a small scale about the beginning of our era, increasing only slowly during the Middle Ages. By 1601 gold mining on Sado Island (west coast of Japan) was well advanced, as witnessed by a unique scroll some 6 meters long and $^1/_3$ of a meter wide illustrating the technique of mining and refining of gold (Bromehead, 1942). During the last centuries of the Early Middle Ages the Moslem Arabs opened or reopened many of the gold placers and mines under their suzerainty in Spain, Africa, and the Middle East. Much gold also reached the Arab caliphates of North Africa over the trans-Sahara caravan routes through Timbuktu to Fez, Tunis, and Tripoli (Barbary) from the golden land of Wangara (Bovill, 1958). Wangara can probably be equated with the placer belts in the ancient kingdoms of Ghana, Mali, and Songhai, specifically the Bambuk-Buri, Lobi, and Ashanti goldfields. The auriferous Hausa states (in northern Nigeria) may also have been part of Wangara. The secret of the source of the gold in Wangara was guarded for centuries; the Arabs (Moors) and those who followed in the Late Middle Ages and in modern times, the Portuguese, French, British, Dutch, and Spanish, all sent expeditions southward from their lands to find the fabled golden land of Wangara.

In central Europe the pagan Avars, Czechs, and Saxons mined gold in Bohemia, Transylvania, and the Carpathians. This particular mining revival, led mainly by the Saxon and other Germanic peoples, flourished during the High and Late Middle Ages, particularly in central Germany (Harz Mountains and Bohemia), France, Italy, and Britain. Events of this period included the emancipation of the miner from slavery and serfdom as the Saxon miner became a free agent whose services were in demand from Britain to Transylvania. Also during this period there were major advances in mining technology, mining geology, and metallurgy, subjects recorded by Calbus (physician and burgomaster of Freiberg), Biringuccio (master founder of Siena), Ercker (superintendent of mines at Annaberg), and Agricola (physician of Joachimsthal and Chemnitz) during the Renaissance (see chap. 5).

During the Middle Ages gold placers were worked in much the same way as in Roman times, but considerable improvements in the methods of booming, hydraulicking and sluicing were introduced, especially the use of "long toms" and rockers. Similarly in bedrock gold mining numerous innovations and improvements of methods and machines utilized by the Romans for centuries were introduced, particularly in underground drainage by employing better archimedean screws, waterwheels, and force pumps and in ore crushing and grinding by the introduction of waterwheels and windmills. Improvements were also made in the miner's tools and in the techniques of open-cut mining, shaft sinking, drifting, stoping, timbering, ventilation, lighting, mine surveying and so on. Hoisting up shafts and inclines was made less onerous by improved versions of the windlass, often employing horses rather than men. Rock and ore were, however, still mainly broken by hand by chipping, wedging, or grubbing, generally after firesetting. The technique of making black powder reached Europe, probably from China, in the Late Middle Ages, but it is doubtful if the explosive was extensively used for blasting rock until much later. All these various improvements in undergound mining permitted exploitation of many bedrock gold deposits below the oxidized zones and in some mines, where drainage adits could be driven or where improved pumps could be installed, well below the water table.

THEORIES OF THE ORIGIN OF GOLD DEPOSITS IN MEDIEVAL TIMES

Intellectual activity in much of Europe during the Early Middle Ages (fifth to the eleventh centuries) and High Middle Ages (eleventh to the fourteenth centuries) was confined principally to the cloister and consisted mainly of theological speculations. The only glimmer of scientific progress during this long dismal period came from the Arabic schools founded in the ninth and tenth centuries at Baghdad, Damascus, Alexandria, Cordova, and Seville after the Moslem conquests of the Middle East, North Africa, and Spain. The greatest and most influential philosophers, alchemists, and physicians at these schools were Jabir ibn Hayyan (Geber) (c. 721-815), Abu-Bakr Muhammad ibn Zakariya Ar-Razi (Rhazes) (c. 865-932), and Abu-Ali al-Husayn ibn Abdullah ibn Sina (Avicenna) (980-1037). All held views on certain geological subjects and on the origin of metals and veins in the earth; those of Geber and Avicenna merit brief mention.

Geber's writings appeared only in the thirteenth century and have been variously ascribed to Geber himself and to a number of classical Latin writers who assembled various Arabic alchemical works. Geber's writings show a keen interest in chemical methodology and in natural chemistry. In the latter context his musings on gold are of interest. The translation given here is by R. Russell from his *Works of Geber*, re-edited by E. Holmyard in 1928 and reprinted in Schwartz and Bishop (1958, p. 190).

Of Sol, or Gold.

> We have already given you, in a General Chapter, the Sum of the Intention of Metals; and here we now intend to make a special Declaration of each one. And first of Gold. We say, Gold is a Metallick Body, Citrine, ponderous, mute, fulgid, equally digested in the Bowels of the Earth, and very long washed with Mineral Water; under the Hammer extensible, fusible, and sustaining the Tryal of the Cupel, and Cement. According to this Definition, you may conclude, that nothing is true Gold, unless it hath all the Causes and Differencies of the Definition of Gold. Yet, whatsoever Metal is radically Citrine, and brings to Equality, and cleanseth, it makes Gold of every kind of Metals. Therefore, we consider by the Work of Nature, and discern, that Copper may be changed into Gold by Artifice. For we see in Copper Mines, a certain Water which flows out, and carries with it thin Scales of Copper, which (by a continual and long continued Course) it washeth and cleanseth. But after such Water ceaseth to flow, we find these thin Scales with the dry Sand, in three years time to be digested with the Heat of the Sun; and among these Scales the purest Gold is found. Therefore, We judge, those Scales were cleansed by the benefit of the Water, but were equally digested by heat of the Sun, in the Dryness of the Sand, and so brought to Equality. Wherefore, imitating Nature, as far as can, we likewise alter; yet in this we cannot follow Nature.
>
> Also Gold is of Metals the most precious, and it is the Tincture of Redness; because it tingeth and transforms every Body. It is calcined and dissolved without profit, and is a Medicine rejoycing, and conserving the Body in Youth. It is most easily broken with Mercury, and by the Odour of Lead. There is not any Body that in act more agrees with it in Substance than Jupiter and Luna; but in Weight, Deafeness, and Putrescibility, Saturn, in Colour Venus; in Potency indeed Venus is more next Luna than Jupiter, and then Saturn: but lastly Mars. And this is one of the Secrets of Nature. Likewise Spirits are commixed with it, and by it fixed, but not without very great Ingenuity, which comes not to an Artificer of a stiff neck.

Geber's idea of the formation of gold scales and nuggets in situ in alluvial sands seems to follow from the Greek (Thales, Theophrastus) postulate that gold originated from water (in Geber's case from cupriferous waters). Onto that postulate has been grafted the alchemical idea (also borrowed from the transmutation concepts of the Greeks) that the sun was capable of transmuting the base metals (e.g., copper) into gold, as will be discussed in this text. Geber's theory that gold nuggets are formed in situ in eluvial and alluvial sands, when shorn of its alchemical fantasies, is not unlike present-day in situ accretion theories of the origin of gold dust and nuggets in placers (see also chap. 15).

Geber's description of the properties of gold accord reasonably well with the known facts. He finds that gold is always closely associated with *Luna* (silver); but the association with *Jupiter* (tin) is not entirely correct in natural situations. It should be remarked, however, that gold alloys readily with tin, but natural alloys with this base metal have not yet been recorded.

Avicenna, the great Persian physician and translator of Aristotle, in his treatise *De congelatione et conglutatione lapidum (de mineralibus)* grouped minerals in a relatively modern way as stones (rocks), sulfur minerals, metals, and salts. About veins he had little to say in detail, but he disagreed with Aristotle and the alchemists of his time about the role of transmutation of metals in the earth, holding to the idea that each metal was a specific type of earth (element).

It is interesting at this point to digress briefly and consider the various alchemical theories on the origin of mineral (gold) deposits. Alchemy is thought by some historians of science to have developed in China in the early centuries of our era and to have diffused through India to the Middle East and Alexandria; others consider a contemporaneous development in China and Alexandria more probable. The word *al-chemi* is evidently of Arabic origin and is said by some to mean *black land* (Egypt) in reference to the dark silty soil of the Nile delta. Another version contends that *al-chemi* is derived from the Coptic and means *black art*, as practised by the early chemists, who dealt essentially with the reduction of ores, the making of glazes and glasses, and the concoction of medicinal potions, all mysterious operations quite beyond the ken of most people of the time.

The early Western alchemists were strongly influenced by Aristotelian views as well as by those of the astrologers, believing that the center of the earth was a holocaust of fire produced by the focus of the rays of all the seven known planets upon the earth, which was considered at the time to lie at the center of the universe. Thus, Apollo or Sol (the sun) gave rise to gold, Diana or Luna (the moon) to silver, Mercury to quicksilver, Venus the metal copper, Jupiter, tin, Saturn, lead, and Mars, iron. These views were held by many of the great alchemists of the time, Geber, Rhazes, Paracelsus, and Norton, and by many of the great scholastic philosophers of the period among whom may be mentioned Roger Bacon (1214-1292), Vincentius Bellovacensis (1190-1264), Albertus Magnus (1200-1280), and his illustrious pupil, Thomas Aquinas (1225-1274).

Albertus Magnus (Albert of Cologne), Dominican, saint, and patron of the natural sciences, was one of the foremost philosophers of the Middle Ages and wrote extensively on most aspects of philosophy and on a great variety of scientific subjects. During his long life Albert traveled widely, visiting many of the gold placers and mines in central Europe, as is readily apparent from his many writings. His knowledge of alchemy (chemistry) was, however, limited. One of Albert's works, *De mineralibus* (On minerals), written about 1260, is of particular interest in the present context. In this work Albert embraced the

Aristotelian view of nature, basing his science of mineralogy on the "four eternal causes": material, efficient, formal, and final. As regards the material cause, the matter of which the minerals are made, his basis is the four elements (earth, air, water, and fire). After ranging over the field of rocks he turns to the metals such as copper, lead, silver, and gold and discusses the places where these metals are produced and how they originated in deposits. His theory is essentially Aristotelian, onto which have been grafted many of the ideas of the medieval alchemists, as can be seen from the following passage taken from Wyckoff's (1967, pp. 182-183) translation.

> The natural scientist seeks to understand the cause of all these things; and, as we have said in the science of stones, the place produces things located in that place because of the properties of heaven poured into them by the rays of the stars. For as Ptolemy says, in no place does any of the elements receive so much of the rays of all the stars as in Earth, because [Earth] is the invisible centre of the whole heavenly sphere; and the power of the rays is strongest where they all converge; and therefore Earth is productive of many wonderful things.
>
> In order to know the cause of all the things that are produced, we must understand that real metal is not formed except by the natural sublimation of moisture and Earth, such as has been described above. For in such a place, where earthy and watery materials are first mixed together, much that is impure is mixed with the pure, but the impure is of no use in the formation of metal. And from the hollow places containing such a mixture the force of the rising fume opens out pores, large or small, many or few, according to the nature of the [surrounding] stone or earth; and in these [pores] the rising fume or vapour spreads out for a long time and is concentrated and reflected; and since it contains the more subtle part of the mixed material it hardens in those channels, and is mixed together as vapour in the pores, and is converted into metal of the same kind as a vapour.

In a later passage Albert discussed his ideas on the precipitation of gold. Further, in Wyckoff's translation (p. 233) we read:

> For almost everywhere gold is found, as we have said, in the form of dust or grains. And the reason for this is that the material is subtle, and it is driven out and sublimed. Evidence of this is that [gold] is found [that looks] like hardened droplets. For in the pores of the natural vessels the concentrated vapour is repeatedly doubled back upon itself and converted into fluid which takes [the form of] rounded drops. And if sometimes they are hollow, elongated, and [look] as if they were made up of smaller ones, this is because in the neck of the natural vessel the vapour is not converted or hardened all at once, but a bit at a time; and thus a second [drop] is added to the first, and sometimes a third to the other two, just as happens in the formation of hail.

Albert's concept of the origin of placer gold is of considerable interest because he was evidently the first to clearly outline an in situ chemical accretion for gold dust and nuggets in alluvial deposits. Again from Wyckoff's translation p. 184 we read:

> But gold which is formed in sands, as a kind of grains, larger or smaller, is formed from a hot and very subtle vapour, concentrated and digested in the midst of the sandy material, and afterwards hardened into gold. For a sandy place is very hot and dry; but water getting in closes the pores so that [the vapour] can not escape; and thus it is concentrated upon itself and converted into gold. And therefore this kind of gold is better. And there are two reasons for this: one is that the best way of purifying Sulphur is by repeated washing, and the Sulphur in watery places is repeatedly washed and purified; and for the same reason the earthy Quicksilver is often washed and purified and rendered more subtle. Another reason is the closing of the pores underneath the water along the banks; and thus the dispersed vapour is well-compressed and condensed, and is digested nobly into the substance of gold, and hardens into gold.

Albert's view of the origin of placer gold was challenged by Biringuccio in his *Pirotechnia,* written in 1540, and by Agricola in his *De re metallica,* published in 1546 (see chap. 5).

Vincentius Bellovacensis (Vincent de Beauvais), French Dominican, compiled an immense encyclopedia, the *Speculum Majus,* of which one part of three, the *Speculum naturale,* in 32 books and 3,718 chapters, is a summary of the natural history known to the western Europeans at the middle of the thirteenth century. In this great work Vincent discusses geology and mineralogy, mostly following the thought of Avicenna and Albertus Magnus (his contemporary). Vincent believed in the alchemical sulfur-mercury theory of the constitution of metals, in the transmutation of metals, and in other alchemical doctrines about the natural upgrading of base metals with time, to yield silver and ultimately gold, as the most noble of all metals, shown by the following statements (see Needham, 1959, p. 639).

> Gold is produced in the earth with the aid of strong solar heat, by a brilliant mercury mixed with a clear and red sulphur, digested and ripened for more than a hundred years. . . . White mercury, fixed by the virtue of incombustible white sulphur, engenders in mines a matter which fusion changes to silver. . . . Tin is generated by a clear mercury and a white and clear sulphur, digested and ripened for a short time subterraneously. If the digestion and ripening process is very prolonged, it becomes silver.

The general theory as enunciated by the later alchemists is best stated in the words of Aurelio Augurelli (1454-1537), alchemist of Venice, as freely translated from *Vellum aureum et chrysopoeia,* published in Venice in 1515, as follows:

> The origin of the metals is the center of the earth . . . penetrated by the sun's rays and by other celestial rays which ripen and mature the

> assembled vapours that then pass upward and fill fractures in the crustal rocks. Where the vapours are condensed and cannot move farther they solidify into those unripe metals that fill veins in the earth's crust. . . . Finally with time Nature transforms these metals into gold, silver, copper, and so on. . . .

Some of the later alchemists thought that mineral veins were offshoots of a giant treelike body rooted deep within the earth. The mineral veins were considered to be the branches of this great tree, and the metals were supposed to have risen like sap. The treelike body was represented as growing under the stimulus or influence of some celestial body (in the case of gold, the sun), and base metals such as lead and copper were constantly being transmutated into the noble metals. This belief in the transmutation and growth of metals within the earth's crust was widespread in Europe in medieval times and had many fanciful modifications. One of these implied that the minerals and metals were male and female and produced seeds (the petrific and metalline seeds) by which they reproduced themselves.

Many interesting accounts of the spontaneous generation and growth of metals, particularly gold, in the earth's crust were extant in the late Middle Ages and were carried into the literature of the sixteenth to eighteenth centuries. Several alchemists describe how sprigs and dendrites of gold grow among the vines of Hungary and Romania. Others relate how the "golden tree" shed spangles and nuggets of gold like leaves, and seeds, and how these particles of gold gathered in the soils and alluvium—an interesting origin indeed for eluvial and alluvial gold placers. Adams (1938) mentions a thesis in the University of Halle entitled *De auro vegetabili pannoniae,* written by a certain Huber in 1733, which presents a comprehensive study of the literature on the vegetable growth of gold. The translation of a part of this thesis as given by Adams (p. 295) is as follows:

> And just as in these places there are growing plants, members of the Vegetable Kingdom, so not infrequently, by a natural spectacle which is altogether wonderful and delightful, it comes to pass that gold, as if joined with these vegetable growths by a bond of consanguinity, laying aside, as it were, its own metallic character, grows after the fashion of plants out of the same lap of Mother Earth. Between the gold and the vine, indeed, these observers relate that there exists so close an intercourse: that the gold not only embraces the vine externally under the form of threads after the fashion of a climbing plant: but that even the vine sometimes puts forth little shoots and tendrils of pure gold, sometimes little berries of the same metal between its leaves. Gold is found intimately associated not only with the vine but with other vegetable growths: occurring either twisted up in various manners with their roots, or else growing near them in the form of little strings or threads. And this species of gold springing after the manner of vegetable growths, or in the midst of them, we designate by the name of Vegetable Gold.

Huber apparently espoused the opinion then current that the gold grew like a plant, in some cases in dendritic or treelike forms, about the roots of vines. Gold associated with the roots of plants, especially those that are decaying, is not uncommon and is obviously the result of the reduction of auriferous soil solutions by decaying vegetation. In some places it is apparent that distorted nuggets of gold grew in this manner. Furthermore, it is interesting to note that the observation by the medievalists of enrichments of gold below trees has been amply confirmed by modern analysis of humus and mull developed in the A horizon of soils beneath heavy forest cover.

A number of alchemists and early mineralogists believed that each metal began as a soft plastic material that they called *gur* or *bur*. This material oozed out of fissures and was probably ordinary limonite, fault gouge, and other unctuous substances like kaolinite and precipitated carbonates. There were still others who connected the veins and metals with the breathing of the earth, which was thought to function and exhale like a giant animal. This theory was current in the seventeenth century and was adopted by Kepler (1571-1630), the great German astronomer. The idea of the growth and transmutation of metals, however, held the general stage and greatly influenced the views of the old mining geologists. On finding bismuth, cobalt, and zinc in the veins instead of gold and silver, they said "We have come too soon," implying that the baser metals had not yet had sufficient time to be transmuted (ripened) into the noble metals. These and a host of other fantastic ideas were held from early medieval times until the birth of modern chemistry in the latter part of the seventeenth century. Even Robert Boyle, the "father of modern chemistry," in his *Sceptical Chymist* (published in 1661) could not quite bring himself to discard the views of the ancients about the magical growth and spontaneous generation of minerals in the earth.

These Western ideas about the origin of mineral (gold) deposits had a parallel development in the Indian and Chinese civilizations that merits brief mention.

The early Indian (Hindu) and Chinese philosophers perceived matter in terms of four material elements, earth, water, air, and fire (light), in much the same way as those in Asia Minor and the Mediterranean. Alchemy was also practiced at an early stage in both the Indian and Chinese civilizations, and as mentioned previously may actually have originated in China before our era. Like Western alchemists, the Chinese also believed in the transmutation of the elements in the earth.

According to Allchin (1962) bedrock gold mining in India declined in southern India in the early part of the third century A.D., evidently due to the breakup of slavery as a social institution. Another factor, particularly in the Hutti and Kolar fields, appears to have been that the rich auriferous oxidized zones were worked out and water problems were encountered with which the miners could not cope. Placer mining, however, continued over much of India during medieval times but only on a small and local scale. The decline in gold mining is reflected in the literature of the Gupta period (A.D. 320-500), the period of some seven centuries marked by an endless succession of internal wars and foreign invasions, and by the Delhi Muslim Sultanate, founded in A.D. 1206 and ended in 1526 with the establishment of the Mogul Dynasty. Nowhere, so far as I can find, does the literature of these periods deal with gold and its deposits in more than a cursory manner.

Professor J. Needham (1959, p. 650) has given us an admirable account of the geological sciences in ancient China in his *Science and Civilization in China*. In this work there is a passage dealing with the formation of mineral deposits written by Chêng Ssu-Hsiao, who died in A.D. +1332.

> In the subterranean regions there are alternate layers of earth and rock and flowing spring waters. These strata rest upon thousands of vapours *(chhi)* which are (distributed in) tens of thousands of branches, veins and thread (-like openings). (There are substances there) both soft and firm, ever flowing back and forth, and undergoing transformations. (The veins are) slanting and delicate, like axles interlocking and communicating. (It is like a) machine *(chi)* rotating in the depths, (and the circulation takes place as if the veins had) intimate mutual connections (and as if) there were piston bellows *(tho yo)* (at work). The mysterious network *(hsüan kang)* spreads out and joins together every part of the roots of the earth. The (innermost parts of the earth are) neither metal nor stone nor earth nor water (as we know them). Thousands and ten thousands of horizontal and vertical veins like warp and weft weave together in mutual embrace. Millions of miles of earth are as if hanging and floating on a sea boundlessly vast. Taking all (including land and sea) as earth, the secret and mystery is that the roots communicate with each other. The natures, veins, colours, tastes and sounds, both of the earth, the waters, and the stones, differ from place to place. So also the animals, birds, herbs, trees and all natural products, have different shapes and natures in different places.
>
> Now if the *chhi* of the earth *(ti chhi)* can get through (the veins), then the water and the earth (above) will be fragrant and flourishing . . . and all men and things will be pure and wise. . . . But if the *chhi* of the earth is stopped up *(sai),* then the water and earth and natural products (above) will be bitter, cold and withered . . . and all men and things will be evil and foolish. . . .
>
> The body of the earth is like that of a human being. In men there is much heat in and under the watery abdominal organs *(shui tsang);* if this were not so, they could not digest their food nor do their work. So also the earth below the aqueous region is extremely hot; if this were not so, it could not 'shrink' all the waters *(so chu shui)* (i.e. evaporate them and leave mineral deposits), and it could not drive off all the (aqueous) Yin *chhi (hsiao chu Yin chhi).* Ordinary people, not being able to see the veins and vessels which are disposed in order within the body of man, think that it is no more than a lump of solid flesh. Likewise, not being able to see the veins and vessels which are disposed in order under the ground, they think that the earth is just a (homogeneous) mass. They do not realise that heaven, earth, human beings, and natural things, all have their dispositions and organizations *(wên li).* Even a thread of smoke, a broken bit of ice, a tumbledown wall or an old tile, all have their dispositions and organizations. How can anyone say that the earth does not have its dispositions and organizations?

In this passage we have an exposition, albeit rather convoluted, of the precipitation of mineral matter from aqueous solutions and an intimation of the theory of lateral and metamorphic secretion. The comparison of the manifestations of the earth with the human body is interesting in that similar analogies were made by a number of medieval European alchemists.

Gold is mentioned in many of the works of the early Chinese philosophers.

Needham (1959, p. 674) quotes two that are of interest. The first is from the *Kuan Tzu* book.

> Huang Ti said, 'I should like to know about these things.' Po Kao answered, 'Where there is cinnabar above, yellow gold will be found below. Where there is magnetite above, copper and gold will be found below. Where there is *ling shih* above, lead, tin, and red copper will be found below. Where there is haematite *(chê)* above, iron will be found below. Thus it can be seen that the mountains are full of riches.'

and the second from the *Pên Tshao Shih I* of A.D. +725.

> Generally one sees those who search for gold dig down into the earth for several feet until they come to a stone called *fên tzu shih* ('tangle-stone') (which accompanies the gold). This is always in black lumps, as if charred, and underneath it is the gold-bearing ore, also in lumps, some as large as one's finger, others as small as beans, and coloured a mulberry yellow. When first dug out it is friable.

Geologists will recognize that the statements in both quotations represent probably the first attempts to describe zoning in gold deposits. The "tangle stone" is evidently limonite and/or wad from the description. Quite frequently gold is greatly enriched where black manganese or wad and limonite are developed in the oxidized zones of gold deposits (see Paper 16-1). The gold referred to as a mulberry yellow color is evidently secondary (or mustard) gold, which is often friable or pulverulent.

Needham (1959, p. 675) also quotes some surprising information about early Chinese knowledge concerning the association of plants and mineral deposits and about the use of geobotanical and biogeochemical prospecting in China as far back as A.D. 800. The statement about the plant indicators of silver and gold in the *Yu-Yang Tsa Tsu* is of interest.

> When in the mountains there is the *tshung* plant (the ciboule onion), then below silver will be found. When in the mountains there is the *hsiai* plant (a kind of shallot), then below gold will be found. When in the mountains there is the *chiang* plant (ginger), then below copper and tin will be found. If the mountain has precious jade, the branches of the trees all around will be drooping.

Professor Needham goes on to discuss how the early Chinese noted the deleterious (chlorotic) effects produced in plants by excesses of elements such as copper and lead; finally he remarks on how the early Chinese discovered that certain plants accumulate metals, for example, gold by the rape-turnip (*Brassica rapa-depressa* R477), and so on. I myself (Boyle, 1979, p. 83) have attempted to explain these phenomena on the basis of the sulfur content (derived from auriferous sulfides) of the various plant species that indicate or accumulate gold, but later research has shown that this answer is not complete.

The historians tell us that the Late Middle Ages (fourteenth and fifteenth centuries) in Europe were marked by great confusion and chaos fomented by economic depression, wars (the Hundred Years War), schisms in the Roman Church, rebellions and revolutions, and widespread plagues (the Black Death). Science progressed little, and few writings of this desperate period are of much interest to geologists. Nevertheless, people had to eat, and agriculture, mining, and industry continued, often in a desultory manner, as armies, disgruntled barons, and rebellious mobs fought out their quarrels across many European lands. In art, literature, and science there was much threshing of old grain from which few kernels of new approaches and knowledge emerged. In science, challenges to the Aristotelian (deductive, speculative) concept of nature as promulgated by Thomas Aquinas and others continued, especially initiated by Oxfordians Robert Grosseteste (1175-1253) and his pupil Roger Bacon (1214-1292), who sought to lay bare the secrets of nature by the methods of experimental study, thereby introducing the modus operandi of modern science.

In technology, advances in metallurgy, especially in alloys, led directly to the invention of movable type and printing in the middle of the fifteenth century. This invention was to have far-reaching effects in all human endeavors, especially science and technology. The Greek and Latin classics became readily available, and numerous technical manuals made their appearance in the latter part of the fifteenth century and during the sixteenth century, especially in the fields of mineral deposits, mining, and metallurgy. The invention of cannon, which reduced the castles (fiefdoms) of the nobles and strengthened the hands of the monarchs, led to greater emphasis on metallurgy and a great demand for metals such as copper, tin, and zinc. To meet these requirements the miners of central Germany (Bohemia), the principal source of metals in Europe at the time, invented piston pumps for dewatering their mines, thus allowing deeper and more sophisticated mining methods. The old records also show that considerable prospecting proceeded during late medieval times particularly for gold and silver, metals increasingly required for specie as payment for the spices, silks, and other goods avidly sought from India and the Orient. We read that bedrock mines were developed in many parts of Bohemia, at Kremnitz (Kremnica) and Schemnitz (Banska Stiavnica), in the Vosges, Auvergne, and Pyrenees in France, and elsewhere in Europe. Many of these mines produced principally silver, but much gold was also won from their ores. Gold placers were worked along the Rhine, Rhone, Garonne, Po, and other European rivers. Rickard (1932) for instance tells us that Aeneas Sylvius, writing in 1458, boasted of the mineral wealth of Germany, saying: "Gold dust sparkles in the waters of the Rhine; there are rivers in Bohemia in which the Taborites find lumps of gold the size of peas." Finally, improvements in navigation, in ocean-going ships, and in the knowledge of geography were to lead to the tracing of routes around Africa to India and the Far East, the rediscovery of the Americas, and the circumnavigation of the globe in the late fifteenth century and early sixteenth century, events that were to transform the civilizations of the world beyond all recognition.

REFERENCES AND SELECTED BIBLIOGRAPHY

Adams, F. D., 1938. *The Birth and Development of the Geological Sciences,* Williams & Wilkins Co., Baltimore, 506p.

Allchin, F. R., 1962. Upon the antiquity and methods of gold mining in ancient India, *Jour. Econ. Social History of the Orient* **5** (pt. 2, 197): 196-211.

Bovill, E. W., 1958. *The Golden Trade of the Moors* Oxford Univ. Press, London, 281p.
Boyle, R. W., 1979. The geochemistry of gold and its deposits, *Canada Geol. Survey Bull. 280,* 584p.
Bromehead, C. N., 1942. Ancient mining processes as illustrated by a Japanese scroll, *Antiquity* **16:**193-207.
Emmons, W. H., 1917. The enrichment of ore deposits, *U.S. Geol. Survey Bull. 625,* 530p.
Gimpel, J., 1976. *The Medieval Machine.* Penguin Books, New York, 274p.
Mills, E. W., 1916. Gold mining in Korea, *Korea Branch Trans. Royal Asiatic Soc.* **7:**5-39.
Needham, J., 1959. *Science and Civilization in China, vol. 3, Mathematics and Sciences of the Heavens and the Earth.* Cambridge Univ. Press, London, 877p.
Rickard, T. A., 1932. *Man and Metals.* 2 vols., McGraw-Hill, New York, 1068p.
Schwartz, G., and P. W. Bishop, 1958. *Moments of Discovery.* 2 vols., Basic Books, New York.
Wyckoff, D., 1967. *Book of Minerals.* by Albertus Magnus, Oxford Univ. Press, New York, 309p.

CHAPTER

5

Gold During the Renaissance

Metallic gold is used by the alchemists
to prepare a liquid that they affirm
will restore youth when drunk.
Agricola, De natura fossilium, 1546

The Renaissance, rebirth, or more strictly speaking the intellectual revival of Europe in the fifteenth and sixteenth centuries, was marked by the advent of Humanism, a revolution in art, sculpture, and letters but with relatively little progress in natural science, and during much of the sixteenth century by the Reformation led by the son of a miner, Martin Luther (1483-1546). The Humanists devoted themselves to the study of the language, literature, and antiquities of ancient Greece and Rome, hoping to find in the past a novel form of thought about nature for the future. They considered themselves in rebellion against the scholasticism of medieval times and were preoccupied with man in relation to human society rather than to God. The foundation of the Vatican Library at Rome by Pope Nicholas V was a landmark of the Renaissance, as were also the writings of Alberti, Castiglione, Machiavelli, Erasmus, More, Shakespeare, and Luther in the fields of social and political thought, literature, and religious doctrine. In music, architecture, sculpture, and art it was the period of Orlando di Lasso, Palestrina, Brunelleschi, Raphael, Donatello, Botticelli, Titian, Cellini, Leonardo da Vinci, Durer, Holbein, van Eyck, Breughel, and Michelangelo. As one gazes up at Michelangelo's Creation of Man on the ceiling of the vault of the Sistine Chapel of St. Peter's in Rome, one sees in Adam a veritable symbol of awakening Renaissance man marveling at all about him.

Capitalism, the monetary system whereby talent and ability, not origin and estate, are the qualifying factors for its aristocracy, appeared in its first manifestations during the early Crusades (eleventh and twelfth centuries), grew slowly during later medieval times, and expanded rapidly during the Renaissance with the establishment of banks in Genoa, Florence, Augsberg, Lyon, and Antwerp, all controlled by powerful families. Among these were the Fuggers, initially weavers in Augsburg, of whom one, Jakob Fugger the Rich (1459-1525), banker to the Hapsburgs and the popes, created a financial empire through extensive investment in mining in Austria, Hungary, and Spain, thereby monopolizing the silver, lead, copper, and quicksilver production of Europe. This expansion of capitalism and investment in mining in central Europe, the Tyrol, and Spain stimulated mining and metallurgical technology and the publication of various tracts dealing with these subjects, some of which are described next.

The Humanists contributed little to the progress of natural science because they were on the whole more interested in the relationship of man to man than in that of man to nature and more absorbed in literature than in chemistry, physics, and mathematics. Nevertheless, as scientists we owe them a debt of gratitude for accurate translations of many Greek and Roman scientific treatises that were to form the bases for the advance of science in modern times. These treatises were to have a great influence on Leonardo da Vinci, Copernicus, and Galileo. We also find this influence reaching out to Calbus, Biringuccio, Agricola, and Ercker, the most celebrated of the earth scientists and metallurgists of the Renaissance.

The advanced technology utilized in gold mining and placering, and described by the Renaissance writers, was all developed during the High and Late Middle Ages and are briefly mentioned in the previous chapter. Blasting techniques for breaking rock and ore using black powder seem to have been experimented with in the late years of the Renaissance, but explosives did not find widespread use in mining until modern times.

THEORIES OF ORIGIN OF GOLD DEPOSITS DURING THE RENAISSANCE

The invention of printing in the midfifteenth century was followed during the Renaissance by many small treatises on various technical arts, among which those dealing with mining and metallurgy are of great interest to us in the context of gold. Here belong *Bergwerk-und-Probierbüchlein* by Calbus of Freiberg, *De la pirotechnia* by Biringuccio of Siena, *De re metallica* by Agricola of Joachimsthal and Chemnitz, and *Beschriebung Allerfürnemisten Mineralischen Ertzt und Berckwercksarten* by Ercker of Annaberg. All of these works became standard references immediately on their publication and remained so for more than a century.

In the dialogue of *Bergwerk-und Probierbüchlein* (written about 1497 by Calbus (Ulrich Rülein von Kalbe), learned doctor and onetime burgomaster of Freiberg in Saxony) the master miner Daniel (probably the first mining geologist) explains to his apprentice Knappius the nature and origin of mineral deposits. In the translation by Sisco and Smith (1949, p. 19) we read:

> It should be realized that for ores to grow or to be born requires an agent to exert an influence, and a passive thing or matter that is qualified to be influenced. In the words of the naturalists, the common maker of ore and all other things that are born is Heaven with its movement, radiance, and influence. The influence of Heaven is diversified by the movement of the firmament and the countermovement of the seven planets. In this way each metallic ore receives an influence from its own particular planet, specifically assigned to it because of the characteristics of the planet and the ore, and also because of their conformity in warmth or frigidity, moisture or

dryness. Thus, gold is made by the Sun or his influence, silver by the Moon, tin by Jupiter, copper by Venus, iron by Mars, lead by Saturn, and quicksilver by Mercury. That is why Hermes and other learned men often call the metals by these names, that is, they call gold sun, in Latin *sol,* and silver moon, in Latin *luna*. . . .

But the passive thing, or the common matter of all metals, is, according to the opinion of the philosophers, sulfur and quicksilver, which, through the movement and influence of Heaven, must be joined and hardened into a metallic body or an ore. Some think that through the movement and influence of Heaven vapors or fumes (called *exhalationes minerales*) of sulfur and quicksilver are pulled up from the depth of the earth, which, when ascending through fissures and fractures [which become the veins and stringers] are united by the influence of the planets and are made into ores. But there are others who do not believe that metals are made from quicksilver because metallic ores occur in many locations where no quicksilver is found. They assume, instead of quicksilver, a moist, cold, muddy matter, without any sulfur, that exudes from the earth as if it were its sweat, and think that all metals are made by its commingling with sulfur.

But never mind; if you understand and interpret them correctly, both theories are right; that is, ore or metal is made of the moisture of the earth, called matter of the first order, and of vapors and fumes, called matter of the second order, both of which shall here be called quicksilver. Thus, in the mingling or union of quicksilver and sulfur in ore, sulfur acts as the male seed and quicksilver as the female seed in the birth or conception of a child. That is the story of sulfur as a special, qualified maker of ores or metals.

We see in this account of the origin of ores a somewhat muddled version of the views of the Aristotelians, the alchemists, and the astrologers. Reading further in the *Bergbüchlein* we learn about methods of prospecting for veins, and in the fifth chapter we have an exposition on gold ore. From the translation of Sisco and Smith (1949, p. 39) we read:

On Gold Ore

Gold, however, according to the opinion of the phiolosophers, is made from the very finest sulfur—so thoroughly purified and refined in the earth through the influence of Heaven, especially the Sun, that no fattiness is retained in it that might be consumed or burnt by fire, nor any volatile, watery moisture that might be vaporized by fire—and from the most persistent quicksilver, so perfectly refined that the pure sulfur is not impeded in its influence on it and can thus penetrate and color it from the outside to its very core with its persistent shade of citrine. And thus the two, sulfur and quicksilver, being the mineral matter, are joined into a metallic body in the most powerful and enduring union through the influence of Heaven, delegated to the Sun, and through the fitness of the location, through which the mineral exhalations of sulfur and quicksilver wind and drive and break their way. And such union cannot be dissolved even by the most violent and most powerful effort of fire.

Gold occurs in different ways. Some, in ordinary river sand, some under the overburden near swamps, some in pyritic deposits, some, as the native metal, in stringers and veins, and some in various ores and alteration products contained in veins and stringers, whether these are schists, or black, brown, grey, blue, or yellow alteration products, or clayey ores. The gold generated in river sand is the purest and most exalted kind because its matter is most thoroughly refined by the flow and counterflow of the water and also because of the characteristics of the location where such gold is found, that is, the orientation of the river in which such placer gold is made.

The most suitable location for a river is one between mountains in the north and a plain in the south or west. And the most suitable direction of the current is from east to west. The next best is from west to east, with mountains located as described before. The third best is from north to south, with mountains in the east. But the worst, as far as the generation of gold is concerned, is from south to north if high mountains rise in the west. The possible directions of the flow of water are as manifold according to the quarters of the earth, as those of the strike of veins, which was described earlier in the chapter on silver ores. And each direction is judged better or worse in the measure as it approaches or deviates from what has been said above.

The better to recognize such locations and streams that carry gold, it should be remembered that in general gold is likely to be born in streams in which precious stones are found, such as amethysts, rubies, rock crystals, and other highly refined pebbles, which are an indication of the fitness of the place. According to the opinion of Albertus Magnus, hot and dry fumes or exhalations are seldom extracted from the earth without being accompanied by warm, moist vapors. The gem stones are wrought and born of dry fumes; and the clearer, finer, and nobler the fumes are, the more beautiful and the better and harder will be the gems. Metals are wrought and made from moist vapors, and how strong and good the metal is will depend on how clear, pure, and well-digested the matter is from which the vapors or mists are extracted. Since moist and dry exhalations rise together, but each is hardened according to its own nature, it is a very reliable indication of the occurrence of gold, as said before, if precious stones are found in a river. Also, where you find in a river or nearby little crystals of tourmaline of a dense, fine structure a gold occurrence is not far off. It is, however, essential that the crystals be very fine because where the coarse kind is found, there is little hope for an occurrence of the best and finest of the metals, the gold. The value and actual gold content of the gold that is generated under the overburden near swamps depends on how much of the grey or black [magnetite] sand that together with the little leaves or grains of gold constitutes the schlich is mixed up with it in smelting. In many places this schlich contains more silver than gold, and sometimes even copper, so that the gold is less valuable through admixture with silver. This sand may also contain an impurity which darkens the noble and exalted color of the gold so that it gives the impression of being low-grade gold. In reality this subtracts only a little from its value since by some minor deft manipulation such impurity can be removed from the gold to restore its exalted color.

For more information on a likely place for the generation of this kind of gold you should know that gold schlich can probably be coaxed out of spots where many little weathered furrows (i.e. riffles) are found under the soil where the placer gold occurs. These resemble the little veins or cracks that are sometimes found to run through the loam in loam pits. And where the little furrows bunch or multiply, they multiply and increase the mineral Power from the earth so that more gold is generated.

Gold that is generated in a pyritic deposit is mixed with many and varied worthless impurities because pyrites is made from contaminated sulfur and an impure earthy matter. But through the influence of the Sun and Heaven, and given enough time, the finest part of the pyrites is gradually cleansed and boiled into a persistent gold ore, which must be separated from the impure pyrites by the industrious application of strong fire.

Such gold-bearing pyrites is found in some places as bedded deposits that extend through the rock as a complete stratum; according to regional custom, these are sometimes called horizontal veins. Others occur in the form of [fissure] veins, that is, as upright veins that have hanging and footwalls.

The flat-lying pyritic deposits are very low in gold content because the influence of Heaven, owing to the lack of fitness of the position, can exert itself but little.

The gold-bearing pyrites that occur in veins are supposed to increase in richness and gold content in the measure as the country rock of the hanging and footwalls of a vein becomes finer and richer. And depending also on whether or not the strike and outcrop of a vein are in the right direction and whether a vein encounters other stringers that enrich it, as was explained in the chapter on silver ores, the occurrence will vary in quality and gold content. Of the gold that is generated in other than pyritic veins some is found as native gold attached to the rock, some in a yellow clay, some in a brown, fine alteration product, and some finally mixed and worked in with quartz.

Where this brown alteration product occurs as a vein, the prospects are very good; because, with added ore from hanging-wall stringers, it will become very rich at depth.

Similarly, where the yellow clay occurs as a vein, it is promising to mine, provided the vein has a fine country rock in its hanging and footwalls. Furthermore, where native gold is found in stringers that run near a vein, it should be observed where the stringers join the vein; and there you may confidently start to mine and sink a shaft. If, however, such stringers swerve away from the vein, you are likely to be disappointed unless they join another vein. Wherever hanging-

> wall stringers that contain native gold leave a vein either sideward or downward, it is advisable to explore for other veins; by such foresight the stringers and the veins may be worked together.

Disregarding some rather fanciful ideas such as the influence of Heaven, we have in this description a relatively modern account of the occurrence of gold. Calbus credits Albertus Magnus with some of the ideas in his narrative and follows him in considering that alluvial gold accretes in situ in streams. The reference to the association of tourmaline, especially the very fine crystalline type as opposed to the coarse crystalline variety, is remarkable because I have found this fact to be true in many auriferous deposits personally investigated. The mention of the association of gold and pyrite and also of contaminated sulfur in the pyrite is of interest. The contaminated sulfur, or "bastard" sulfur as many alchemists referred to the substance, is none other than arsenic, and the type of pyrite meant is probably arsenopyrite. The remark that the stratiform (massive) pyrite deposits are low in gold is a general truism; why, we do not yet know. Mention of the brown alteration product (limonite) and the yellow clay (scorodite) suggests that Calbus was aware that some oxidized gold veins are enriched at depth, often immediately below the gossan. This mention is one of the first detailed references in the modern literature on gold to the secondary enrichment processes in auriferous deposits (see also chap. 16).

Vannoccio Biringuccio (1480–1539), metallurgist, master founder, and munitions advisor to the Petruccis of Siena and to Pope Paul III in Rome wrote the *Pirotechnia,* probably in 1538, and it was published in 1540, the year following his death. The work consists of ten books, most of which are concerned with metallurgy, pottery, and munitions. In the preface to the first book a guide to prospecting, developing, and mining mineral deposits is outlined, and in the first chapter, entitled, "Concerning the Ore of Gold and its Qualities in Detail", we read from the abbreviated translation of Smith and Gnudi (1959, p. 28) the following:

> Because I cannot say that I have seen with my own eyes mountains which contain gold ores or places where the practice of such work is carried on, I shall tell you only what I have been told by trustworthy persons as I carefully tried to understand, or else what I have learned by reading various authors. From these I have gathered that it is true that most of this metal is found in Scythia and in those regions called oriental, perhaps because the sun seems to shine forth with greatest vigor in those places. Among these it is said that the Indies hold first place, particularly those islands which as we hear are called Peru, recently discovered by the naval armada of the sacred King of Portugal and of His Majesty the Emperor, and still other places. Also, gold is found in many localities in Europe such as Silesia, many parts of Bohemia, Hungary, in the Rhine, and in the Apsa. Pliny says that it is also found in Asturia and in Lusitania and that the Romans extracted twenty-three pounds every year.
>
> And speaking thus of this precious metal I believe that it is certain that it is and can be generated in all those places where the heavens influence the elemental dispositions and causes. And here wishing particularly to tell you what I have heard concerning this, I say that

> gold is generated in various kinds of rocks in the most rugged mountains that are completely barren of soil, trees, and grasses. And of all the rocks for such metal the best is a blue stone called *lapis lazuli,* which has a blue color similar to the sapphire, but is neither so transparent nor so hard. It is also found in orpiment and even more it is found associated with the ores of other metals. Much is also found in the river sands of many regions. That which is found in mountains is in the form of veins between one stratum and another, united with the blue rock, and indeed is much mixed in with this. They say that such ore is better the heavier and the more full of color it is, and the more flecks of gold appear in it. They also say that it is generated in another rock similar to saline marble but of a duller color, and in still another rock whose color is yellow with many red specks in it. They also say that it is found in certain black rocks, scattered loosely about like small stones in a river. And furthermore they say that it is likewise found in a certain bituminous earth of color similar to clay and that such earth is very heavy and has a strong sulfurous odor. The gold extracted therefrom is very beautiful and almost completely pure, but it is very difficult to get out because it is of the finest grain, almost like atoms, so that the eye distinguishes it with the greatest difficulty. Nor can one proceed as with lapis lazuli or other rocks or as one treats river sands, for when it is found there, and even more when it is washed, it falls only with difficulty to the bottom, and, growing vitreous on melting, it becomes pasty with the matrix and its earthy matter. Nevertheless, in the end it is possible to recover it using the greatest patience with one method or another and finally with mercury.
>
> As I told you before gold is also found in the sands of various rivers, as in Spain in the Tagus, in Thrace in the Hebrus, in Asia in the Pactolus and the Ganges, in various rivers in Hungary, Bohemia, and Silesia and in Italy in the Ticino, the Adda, and the Po. It is not, however, found throughout their beds but only in particular places where in certain bends there is some bare gravel, or where the water in times of flood leaves a certain sandy sediment in which gold is mixed in tiny particles like scales or even smaller than a grain of flour. In the winter when the floods pass they take and carry them almost beyond the bed of the river so that when the waters return to their normal state they cannot easily take them away again, and thus they form mounds.

Continuing with the origin of placer gold we read further in the translation by Smith and Gnudi (1959, p. 30)

> But now let us cease speaking of these things because here perhaps you or someone else might like to know why such gold is carried by the water into these river sands and woods and whether indeed it is produced therein. I have often thought about this, greatly marveling, particularly in regard to the Ticino, the Adda, and the Po, but the reason is not clear to me, since although I told you before that great floods of water carry it to where it can be extracted, there is no

gold mine near those places or even one of any other metal that I know of. I am also confused because I have seen several authors who believe that it is generated in the very place where it is found; and if this were true it would not be true that the waters had brought it. But that it is generated there seems to me a very difficult thing to comprehend, since I do not understand whether it is produced by the innate properties of the waters or of the earth or indeed of the heavens, for it appears reasonable that if the cause were any of these it would be found everywhere in the bed of a given river, and, seeking, one would find it everywhere at all times. If the influence of the heavens is the powerful cause that produces gold, it seems to me that it would necessarily have to operate instantaneously because it is not possible otherwise to perceive the order that Nature uses in generating metals. It would have to produce it first in the open in a place where there is a continuous flow of water, and then it would have to have the power to remove the earthy materials from place to place and also to mix with it the greatly different qualities of cold and humidity. And even if this composition and order begun by the waters of the river should not change, it seems to me that the rains or floods which pass over it would completely soften, break, and entirely spoil anything that might be conceived therein. Furthermore, if this material is generated there, I wish to be told why it is generated only in these and not in other places, and why silver, copper, lead, or one of the other metals similar to gold is not likewise generated in a similar manner, for these substances are perhaps even easier for Nature to form than gold because of the many concordances and ultimate perfections that gold requires. Moreover, in many places in the countryside near Rome particles of iron are found in the sands of several small rivers and I would like to know why this also is conceded only to certain particular parts of the river and not to all parts.

For these reasons and visible phenomena it seems more probable that gold is carried by the water than that it is generated there. Nor does our dilemma help us to ascertain the truth. For, speaking just between ourselves—not with firm conviction, but only to tell you what I think—I say that I incline toward either of two theories. Of these one is that this phenomenon occurs only in very large rivers which receive much water from springs, streams, and other rivers and so it often happens that, with the melting of snows or the coming of heavy rains, they wash the banks and the slopes of near-by mountains, in which it may be that there is earth that, by its own particular nature, contains the substance of gold; or else the ores are located in some peak or surface where men have not yet taken the trouble to go or where access is difficult, and may then be exposed to insemination by the sun or by the coldness of the snows or by the waters, and broken up because in heavy rains anything is easily worn away and carried off to the rivers. Alternatively it might be that such earth is inside the near-by mountains or indeed in the same principal stream that has its bed hidden from our eyes. Since it is never dried up or free from continuously running water, it is not strange that in so many centuries the true origin and knowledge of such a thing should not be understood by those who live near by.

But however it may be, in the end the truth is that gold is found in the sands of many rivers, particularly, according to my information, in those mentioned above. Therefore, if I have marveled at this thing, I deserve to be excused, because where it is impossible to understand the certain cause of things either by reason or by direct observation, doubt always exists and new reasons for wonder are born. But I marvel even more greatly at what I have heard told many times as the truth by various persons: that in several places in Hungary at certain times the purest gold springs from the earth like grass and wraps itself like the stems of convolvulus around the young dry shoots. It is about as thick as a piece of string and about four *dita* long or even a *palmo*. Apparently Pliny in the thirty-third book of his *Natural history* refers to this or a similar thing when he speaks of ores, incidentally referring to the fact that in his time this same thing occurred in Dalmatia. If what is said be true, then indeed would the farmers in the fields reap the fruits of celestial instead of terrestrial sowing, and they would be considered blessed, since such previous and pleasing fruits would be produced by God, by the heavens, or by Nature, without any labor or skill on their part. This would indeed be a unique grace, since among all the vast amount of earth and number of possessions that are cultivated by living creatures, none but these regions are worthy of such a harvest.

What shall I say of what Albertus Magnus writes in his work *De Mineralibus,* where he says that he has seen gold generated in the head of a dead man? He says that when this was dug up by chance and found to be extraordinarily heavy, it was seen to be full of very fine sand. Because of its weight those who saw it thought that it was metal and by experimenting finally found it to be of the purest gold. It seems to me that his words have no other significance than that the ready disposition of the thing and the great influence of the heavens had generated this precious metal. Since I heard it thus, I wanted to pass it on to you. To tell the truth it is not easy to believe this, and certainly to me it seems incredible, yet considering who tells of it and thinking how great are the forces of superior causes and of Nature, we can receive it, having faith and respect for the learning to those who relate it, since by ourselves we lack full understanding of the causes of things.

We see from these passages that Biringuccio, while clinging to the astral (astrological) theory of the origin of gold, nevertheless gives us a fairly accurate account of the natural occurrences of the precious metal. In later passages (p. 41) he questions the veracity of the alchemists and advises Messer Bernardino di Moncelesi of Salo, to whom the first book is addressed, as follows:

For this reason I tell and advise you that I believe the best thing to do is to turn to the natural gold and silver that is extracted from ores

> rather than that of alchemy (i.e. the transmutation of base metals into gold), which I believe not only does not exist but also, in truth has never been seen by anyone, although many claim to have seen it.

The astral theory that gold is found in greatest abundance in those lands (between the tropics) where the sun shines with greatest vigor had many ramifications in medieval and Renaissance history and indeed in more modern history. For examples: geographical exploration during the fifteenth and sixteenth centuries was stimulated in part by the search for gold, and the tropics were considered the most favorable zones in which to find the precious metal, as witnessed by the statement of Columbus recorded in his log book as he approached Watling Island in the Caribbean in 1492: "From the great heat which I suffer, the country must be rich in gold." Further, it will be recalled by those interested in the history of North America that a dispute (Nootka Sound controversy) arose between Spain and Great Britain over the sovereignty of the lands bordering the northwest coast of America. Spain contended she possessed sovereignty by authority of the Papal Bull of Alexander VI in 1493, but Britain took the view that rights of sovereignty could be obtained only through trade and the establishment of colonies. Spain meanwhile had established a settlement on Nootka Sound and in 1789 seized four British ships in the sound. This act nearly precipitated a war but was finally resolved in favor of the British viewpoint in a convention signed on October 28, 1790. It appears probable that among the factors that influenced the Spanish decision was the advice given the Spanish king, Charles IV, that gold was unlikely to occur in the northern regions of America because it was thought that the element was generated only in those regions most influenced by the sun.

The references to the very fine-grained high-purity gold associated with a heavy sulfurous bituminous earth in Biringuccio's text is unclear as to just what type of deposit is meant. I suspect that the fine gold found in certain secondarily enriched sulfide zones below the zone of oxidization is intended. If so, this reference is the first in the literature to the very finely divided, nearly pure gold often found with fine-grained supergene (often black) sulfides (pyrite, marcasite, chalcocite) in the secondarily enriched zones of gold deposits. Finally, Biringuccio ponders the origin, as many have since his time, of gold in alluvial sands. After some discourse he concludes that the gold is transported by streams from oxidized and disintegrated bedrock deposits, an opinion opposed to that of Calbus and Albertus Magnus, who considered a chemical accretion theory for the origin of alluvial gold more probable.

Georgius Agricola (Georg Bauer or George the Farmer, 1494–1555), a name familiar to all miners and earth scientists, was a native of Glauchau in Saxony and later physician in Joachimsthal (Jachymov) and burgomaster of Chemnitz (Karl-Marx-Stadt), two of the foremost mining towns in Renaissance Europe. Agricola, as a classical scholar and humanist, made many contributions to medicine, chemistry, mathematics, theology, and history, but his most important works were in mineralogy, geology, and mining. His views in earth sciences mark a transition from those of medieval to those of modern times. Among his writings on mineral deposits those of greatest interest in the context of gold include *Bermannus, Sive de re metallica dialogus* (1530), a short work (dialogue) on the mines and ores of the Erzgebirge; *De ortu et causis subterraneorum* (1546), a work that deals among other subjects with the origin of mineral deposits; *De natura fossilium* (1546), generally credited as the first systematic textbook on mineralogy, the system employed being based on the physical properties (e.g., color, luster, taste, etc.) of minerals; and *De re metallica* (1556) dealing with the prospecting, developing, mining, and metallurgy of mineral bodies. Many of the observations made in these works are Agricola's own although he sometimes borrows from the *Bergbüchlein* and other now nonexistent treatises. The classic translation of *De re metallica,* with appended notes on most of Agricola's other writings, is that by Herbert Clark Hoover, geologist and mining engineer and onetime president of the United States, and his wife, Lou Henry Hoover, scholar and classicist (Hoover and Hoover, 1912).

Agricola clearly recognized the difference between rocks and minerals, and in his work on ore genesis, *De ortu et causis subterraneorum,* he recognized that many types of mineral deposits (veins) are concentrated in openings (fissures) which he called *canales*. These he thought were late events (i.e., later than the strata), caused by the solution effects of circulating underground waters. Agricola recognized three types of veins: normal steep-dipping veins, (*vena profunda*), composite vein systems, impregnations, and stockworks, (*vena cumulata*), and bedded veins (*vena dilatata*).

Agricola, while retaining his belief in the Aristotelian concept of the four elements, discarded the alchemical idea of the transmutation of elements in the earth and inveighed against the astral theory of the generation of metals and the fundamentalist views of the origin of all things instantaneously at Creation, according to Genesis. On the contrary he concluded from his observations of ground waters and springs that mineral veins were deposited by circulating undergound waters. Stripped of extraneous verbiage. Agricola's theory is that surface waters percolate downward, become heated, dissolve mineral matter, rise again, and deposit their mineral matter in the "canales." He considered the source of the heat to be deep-seated layers of burning bitumen (essentially coal), a fantastic idea, but one believed by many during the Renaissance and even later. Agricola was, therefore, a meteoric water secretionist, and it is a tribute to his genius to note that modern isotopic research has shown that certain types of epithermal gold deposits probably originated in the manner he suggested (see chap. 13).

Agricola speaks of gold in many contexts, about human avarice for the precious metal, about its ores, and about its metallurgy. In *De re metallica* he classifies vein gold ores as follows (Hoover and Hoover, 1912, p. 107):

> Now we may classify gold ores. Next after native gold, we come to the *rudis,* of yellowish green, yellow, purple, black, or outside red and inside gold colour. These must be reckoned as the richest ores, because the gold exceeds the stone or earth in weight. Next come all gold ores of which each one hundred *librae* contains more than three *unciae* of gold; for although but a small proportion of gold is found in the earth or stone, yet it equals in value other metals of greater weight. All other gold ores are considered poor, because the earth or stone too far outweights the gold. A vein which contains a larger proportion of silver than of gold is rarely found to be a rich one. Earth, whether it be dry or wet, rarely abounds in gold; but in dry earth there is more often found a greater quantity of gold, especially if it has the appearance of having been melted in a

> furnace, and if it is not lacking in scales resembling mica. The solidified juices, azure, chrysocolla, orpiment, and realgar, also frequently contain gold. Likewise native or *rudis* gold is found sometimes in large, and sometimes in small quantities in quartz, schist, marble, and also in stone which easily melts in fire of the second degree, and which is sometimes so porous that it seems completely decomposed. Lastly, gold is found in pyrites, though rarely in large quantities.

In footnotes, *rudis* is translated by the Hoovers as *crude,* and they state that what is really meant is perhaps ores very rich in gold. In a further commentary they suggest that Agricola apparently believed that there were various gold minerals manifest by different colors, such as green, yellow, purple, and black. One wonders if Agricola was not here referring to the various tints that native gold may have in certain deposits. For instance I have seen yellowish gold, greenish gold, reddish gold, and black gold in veins and particularly in the oxidized zones of gold deposits. The *librae* is a measure of weight equal to 12 *uncia,* and the *uncia* is equal to 412.2 troy grains. The Hoovers give the value of the gold ore mentioned by Agricola as 72 oz 18 pennyweights per short ton.

The earth having an appearance of being melted in a furnace is obviously scoriaceous limonite (geothite), and the mineral in scales resembling mica is probably jarosite. Both minerals are common in the gossans of surface enriched auriferous deposits, especially those with abundant pyrite and other sulfides.

In *De ortu et causis subterraneorum* Agricola disagrees with the chemical accretion theory of Albertus Magnus for placer gold, maintaining that the gold is torn away from its parent veins and stringers and collects mechanically in the streams and rivers.

Lazarus Ercker (1530-1594), onetime assayer at Dresden, and later resident of Annaberg where he was chief superintendent of mines and comptroller of the Holy Roman Empire and Kingdom of Bohemia, published the treatise *Beschriebung Allerfürnemisten Mineralischen Ertzt und Berckwercksarten* (Description of ore processing and mining methods) in 1574. This treatise is a systematic review of the analytical and assay methods then in use, some of which are still employed today in gold assaying; in addition the work contains long sections on the occurrence of the ores of gold, silver, copper, and other metals. Many translations of this famous treatise have appeared, the first in English in 1683 by Sir John Pettus while incarcerated in the Fleet prison in London and entitled *Fleta minor* or *The Laws of Art and Nature.*

In the section on gold ores Ercker discusses the occurrence of the precious metal in some detail. From the translation by Sisco and Smith (1951 p. 93) we learn:

> Beautiful native gold occurs most frequently in a white quartz; somewhat less often in a blue or yellow hornstone and also in blue, ferruginous, and yellow schists, but only very fine and in flakes. At the gold mine at Knin, located two leagues from Eule (Jilové) in Bohemia in the direction toward the setting sun, there occurs a grayish, argentiferous pyrites in a hard quartz, which, after crushing and washing, yields a beautiful high-grade native gold which is first

not visible in the pyrites. At present I know of no place where more valuable gold is extracted or recovered directly from the ore.

In addition, there is good native gold in all the auriferous placer ores, which are usually sandy but which are otherwise not all alike: in some, the gold occurs massive and in grains; in others, as flakes and light particles. The washing of almost all this placer ore also yields a heavy schorl or wolfram and in some cases grains of tin and ironstone. These have traveled much and far; together with the gold, they were torn from veins by the Flood, swept away, and collected together in such a marvelous and characteristic way that the color and distinctive appearance of placer deposits can be clearly and easily recognized. This is how rivers and streams flowing over such deposits became inseminated [with gold] so that at many localities, not only in far-away kingdoms and countries but also here in Germany, native gold is now washed from them and extracted. However, most of these occurrences are poor and will not repay the expense of washing.

I cannot agree with those among the old writers who claim that it was the River Nile, which flows into the sea in Egypt, that inseminated and flooded the streams and rivers with native gold at the time of the Deluge, when all the sands became mixed up. Because, even if the aforesaid river is very large and does flow through vast Ethiopia (also called India), where much gold is reputedly found, and is supposed to be the mightiest of all the rivers, flowing the farthest, I still think that it is much too small to have been so rich in alluvial gold that it could have scattered gold into the sands and streams of so many places throughout the world.

Then you hear a lot of talk here in Germany about various kinds of pebbles that are found in many parts of the country, in mountains and streams, and are carried away by foreigners and wayfarers. Many resemble gravel; some are brown, yellowish, or black and look like glass on the inside; usually they are round or squarish. It is said that gold is made from them. Personally I do not believe it because I have assayed these pebbles in various ways, in the fire and otherwise, but have never been able to find any gold in them. I learned this much, however, from trustworthy people, who heard the whole story from these wayfarers, that the pebbles do not contain gold, nor is gold made from them; but they are carried by the wayfarers for pay to Italy and other places where they are used as an addition in making beautiful pigments and enamels. Such pigments and enamels are there esteemed as highly and sold as dear as if they were gold. All of which is reasonable and credible, especially since there are other minerals here in Germany that yield enamels and pigments.

Furthermore, besides native gold, there sometimes occurs in the quartz of the gold mines at Eule in the kingdom of Bohemia a fine, gray, scaly ore, which on account of its color is called ironman (hematite). This is rich in gold, which, however, contains silver, so that it cannot be compared with the other native gold occurring in quartz. There are many gold pyrites that contain not only gold but

> also silver, and usually more silver than gold; and pyrites that are very rich in copper and also contain silver, which silver is rich in gold; and white pyrites that contains no copper and very little silver and is yet auriferous. The copper-bearing pyrites whose silver contains gold are usually interspersed with fine quartz.
>
> Concerning the marcasite, of which many make fables and have written that it is a pyrites so rich in gold that it loses less than one-fourth in the fire and becomes more beautiful the longer it is roasted and kept red-hot, I have searched for it often and persistently but have never obtained it; neither have I ever encountered anybody who has seen such pyrites. As far as I can figure it out, this marcasite can and must be nothing but a very good, rich gold ore; whether it is given this name or another makes no difference.

Ercker's mention of the Flood (the Noachian Deluge) as producing all of the gold placers in Europe and elsewhere is of interest because it is the first reference to this particular origin for placers that I can find. Of course the Flood was later to play an important role in the arguments about the origin of many types of mineral deposits, as we shall see in later chapters. The various kinds of pebbles mentioned by Ercker were probably tektites (moldavites), according to my observations in the old placer areas of Bohemia. The auriferous white pyrites is arsenopyrite.

To summarize, we can say that the Renaissance was the period when more modern theories on the origin of auriferous veins were considered, and when the origin of placer (alluvial) gold was debated. The Renaissance writers could not quite rid themselves of the Aristotelian and alchemical concepts of matter, but with Agricola a definite trend developed toward acute observation and the formulation of theories more in agreement with the facts presented by auriferous vein deposits.

REFERENCES AND SELECTED BIBLIOGRAPHY

Bandy, M. C., and J. A. Bandy, trans., 1955. De natura fossilium, by Georgius Agricola, (1546), *Geol. Soc. America Spec. Paper 63,* 240p.

Hoover, H. C., and L. H. Hoover, trans., 1912. De re metallica, by G. Agricola, (1556), *Mining Mag. London,* 637p.

Sisco, A. G., and C. S. Smith, trans., 1949. *Bergwerk-und-Probierbüchlein,* Am. Inst. Min. Metall. Eng., New York, 196p.

Sisco, A. G., and C. S. Smith, trans., 1951. *Beschriebung allerfürnemisten mineralischen Ertzt und Berckwercksarten, u.s.w.* (Lazarus Ercker's treatise on ores and assaying), Univ. Chicago Press, Chicago, 360p.

Smith, C. S., and M. T. Gnudi, trans., 1959. *De la Pirotechnia,* by Vannoccio Biringuccio, M. I. T. Press, Cambridge, Mass., 477p.

CHAPTER

6

Gold During the Transition to Modern Scientific Views

For I know by good and long experience and by many accurate trialls that Quick-silver the most friendly mineral to the royall mettalls, can by no means or Artifice whatsoever be fixed and coagulated into either of the Royal Mettalls.
Gabriel Plattes, 1639

The interval between the publication of Agricola's *De re metallica* (1556) and the great controversy in earth sciences, instigated by Werner and Hutton and lasting into the early years of the nineteenth century, marks a time span of nearly three centuries, during which many changes took place in politics, social structure, economics, and science in Europe and the newly discovered (i.e., by Europeans) lands of the New World (Americas), Africa, the Far East, Siberia, and Australasia. The period witnessed in turn the great events of the aftermath of the Reformation (1517-1560), the exploration and exploitation of the newly found lands (Americas, Siberia, Africa, Australia, etc.), the continued expansion of capitalism and mercantilism, the rise and later decline of monarchial absolutism in Europe, the emergence of parliamentary monarchy in England, the Seven Years War (which removed French dominion from America 1759), the American Revolution (1775-1783), the French Revolution (1789-1795), and the rise and fall of Napoleon Bonaparte (1815).

Amid all the vicissitudes of the period 1556–1820 there was steady progress in science. By the closing years of the period a large body of verifiable knowledge had accumulated in the fields of mathematics, astronomy, physics, chemistry, and medicine. In physics and astronomy the revolution begun by Copernicus (1473-1543) in his *De revolutionibus orbium coelestium* culminated in Newton's (1642-1727) monumental *Principia,* in which the geocentric Ptolemaic views of the solar system were abolished forever. Similarly in chemistry the pioneer work of Boyle (1627-1691) in demolishing the Aristotelian four element theory and establishing the clear definition of an element and compound and in challenging the current fanciful alchemical theories in his *Sceptical Chymist* led ultimately to the modern view of oxidation-reduction reactions, enunciated first by Lavoisier (1743-1794); these efforts finally overthrew the phlogiston theory that had sidetracked chemistry for almost a century. Technology also advanced greatly during the seventeenth and eighteenth centuries, particularly in mining methods, ventilation, and drainage as the demand for metals, especially the precious metals, grew. Most of this progress in mining technology took place in central Europe (Erzgebirge) and in England, and from these regions many of the ideas of the period about ore-forming processes would soon originate.

Prospecting and discovery of gold deposits, remained largely a matter of chance during the transition period, although references in the literature of the period point to a greater appreciation of the role of indicators of auriferous deposits, such as quartz, gossans, mineralized springs, red earths, and depression lineaments along mineralized faults and fractures. Mining methods were improved over those utilized in the Late Middle Ages and Renaissance, including particularly the employment of horsepower on a wider scale during the seventeenth and early part of the eighteenth centuries and the introduction of the steam engine for pumping water from deep mines and other tasks during the last half of the eighteenth century and the early part of the nineteenth. Gunpowder (black powder) was used more extensively for blasting toward the end of the period, and there was much improvement in drilling, using handsteel methods for driving drifts and extracting the ore. Improvements in mine ventilation, drainage, timbering and lighting, and in many of the metallurgical techniques led into those of the modern era.

THEORIES OF ORIGIN OF GOLD AND OTHER MINERAL DEPOSITS

During the seventeenth century the study of ore deposits suffered a period of abeyance. In England Francis Bacon (1561-1626), onetime lord chancellor under James I, was the foremost philosopher of the time and undoubtedly one of the greatest prophets of science. He sought to reform all knowledge and thus create a new learning, and he was a firm advocate of experimental method and inductive reasoning. Yet in all of his many works there is little of interest for the geologist, except perhaps his speculations on continental drift, and nothing specifically about mineral deposits. Rather one has to turn to an agriculturist, Gabriel Plattes, and to a South American curate, Alvaro Alonzo Barba, to discover the state of the theories of the origin of ore deposits.

In Plattes's *A Discovery of Subterraneall Treasure* (1639, p. 34)), we read in the first part of chapter 8 dealing with gold the following interesting discourse.

> And first, whereas it (gold) is oftentimes found in the sand in Rivers, let no man thinke that it could be generated there, but that the swift motion of the water from the high Mountaines, brought it thither, with earth and altogether, till such time as the motion of the Water grew more slow: and so according to its property, being not able to carry forward still both the substances, did still carry the earth with it, and let the heavier body sinke.
>
> Therefore I would have those that have occasion to deale in the hot Countries where gold is usually generated, to make triall in all such Rivers which runne from great Mountains with a swift course in such places, where the motion of the water beginneth to grow slow.

And later in the chapter (p. 36) we learn Plattes's views on the epigenetic origin of gold from subterranean vapors.

> Now whereas I have formerly affirmed that all mettals in general are generated of the clammy and gluttenous part of the subterraneall vapours, arising from *Bituminous* and *Sulphurous* substances, kindled in the bowells of the earth: it behooveth me to shew how gold, such a fixed substance can be found pure of it selfe, and not mixed with other base mettals.
>
> And the reason of this can be no other, but because that all other mettalls whatsoever will putrifie in the earth in length of time, and turne to earth againe; but *gold* wil never putrifie by reason of his *excellent composition,* being made of a *Balsamick Sulphur,* or fatnes, which is incombustible, and differeth from the *Sulphure* or *fatnes* contained in the other *mettals,* even as naturall *Balsome* differeth from all other *oyles,* and fat substances: so that though it be an *oyle* in shew, yet it will sink in water, whereas all other *oyles* wil swimme upon the top of the water.
>
> . . . .
>
> Now whereas the substance of *gold* is not subject to putrifie in the earth by any length of time, it is probable enough that other metalls might be *generated* with it at the first, and afterward *putrified and consumed* from it in length of time, leaving the gold pure.
>
> . . . .
>
> And the reason why the hotter the Country is, the richer the Minerals are, can be no other but the same, that roasted meates are sweeter than boyled meates, or raw meates: the reason whereof is plaine, for that the rawish and unsavory part is exhaled by the heate of the fire, leaving the sweeter part behind.
>
> Even so in hot Countries, all that part of the subterraneall vapours, which here is condensed into Lead, and other base mettalls, can there have no leave to congeale, by reason of the heate: but is all or most part therof exhaled out of the Mines, leaving behind the royall metalls, whose property is to coagulate with heat: whereas the property of the base metalls is to evapourate with heate and to congeale with cold.
>
> The contrary opinion to this; namely that the substance of the best metals are convertible into royal mettals by heate and digestion, hath filled the world with false Books and receipts in *Alchimy,* and hath caused many men to spend much money, labour, study, and charges to no purpose.

We learn from this chapter that Plattes still adhered to the astral theory that postulated that the sun had considerable influence on the localization of gold deposits, for he repeatedly advises search for the precious metal in the hot (tropical) countries. His explanation of the nature of gold is partly alchemical although he denies the alchemical theories of transmutation. His subterraneous vapor (exhalative) theory of the origin of gold veins seems to have been derived from the speculations of the Greeks. For the origin of placer gold Plattes comes down explicitly on the side of the mechanical weathering origin of placer gold.

A contemporary of Plattes, Edward Jorden (1569-1632), studied hot springs extensively, and in his great work *Discourse of Naturall Bathes and Minerall Waters* (1631) came to the conclusion that the internal heat of the earth was due to natural causes and not to a vast internal coal-fired conflagration or to the penetration of the sun's rays. For the origin of vein minerals he advocated a fermentation process based on the "metallic seed" or "seminary spirit" of the minerals of the earth.

> There is a Seminarie Spirit of all minerals in the bowels of the earth, which meeting with convenient matter, and adiuuant causes, is not idle, but doth proceed to produce minerals, according to the nature of it, and the matter which it meets withall: which matter it workes upon like a ferment, and by his motion procures an actuall heate, as an instrument to further his work; which actuall heate is increased by the fermentation of the matter. (p. 51)

There is in this statement, shorn of its alchemical connotations, the seeds of the metamorphic secretion theory of endogenic mineral deposits.

Views similar to those of Jorden were published in 1671 by John Webster in his *Metallographa* (A History of Metals). He inclined to the view that metals, including gold, grew (or were generated) in the earth; he called it the *vegetability* of metals. Webster (p. 70) gave a number of reasons for his views, among which the following may be mentioned:

> A third reason I take to be this, To prove that Metals are generated: That whosoever hath diligently considered the manner how most metals do lie in their wombs, or beds, which for the most part are hard Rocks, Cliffs, and Stones, or things equivalently as hard as they, as lank and spare, must necessarily conclude, that they could never have penetrated the Clefts, Chinks, and porous places of such hard bodies, but that before their entrance into those cavities, they were in *principis solutis,* either in form of water, or vapours, and steams. And then were those steams, or that water produced before their induration into a Metalline form, and after concocted and maturated into several forms of Metals; which is an analogous, if not an univocal generation; otherwise they could never be found in such streight passages, and narrow cavities, as all experience doth testifie they are.

One sees from this excerpt that Webster, despite his other quaint ideas, had a grasp of the rudiments of the hydrothermal theory. Later in his treatise he asked a number of pertinent questions about the growth or generation of metals; some of these questions have been elucidated in modern times by a knowledge of oxidation and reduction, secondary enrichment, and other processes in mineral deposits.

The first treatise on the geology and metallurgy of ores originating in the Americas is the *El arte de los metales* written by Alvaro Alonzo Barba in 1637 at Potosi, Bolivia, and printed in Madrid, Spain, in 1640 (Fig. 6-1). Padre Barba was curate of the parish of San Bernardo, Potosi, for many years; he read widely on natural science and traveled extensively in the silver and gold camps in and near his parish. His great work was reprinted twice, in 1675 and in 1729, and was translated into English, French, and German; in English the last translation is by Douglass and Mathewson (1923). Barba's work fell under the harsh scrutiny of the Inquisition in the latter part of the seventeenth century, was banned, and was burned, evidently because of his (alchemical) ideas on transmutation of the metals and because in certain passages he used the collective word *Nature* rather than *God* as the creator of ores. One of these particular passages in chapter 18 on the creation of metals follows from the translation by Douglass and Mathewson (1923, p. 43).

> Many of the generality of People, in order to avoid profound Discussion, say that in the beginning of the World God created Ores in the

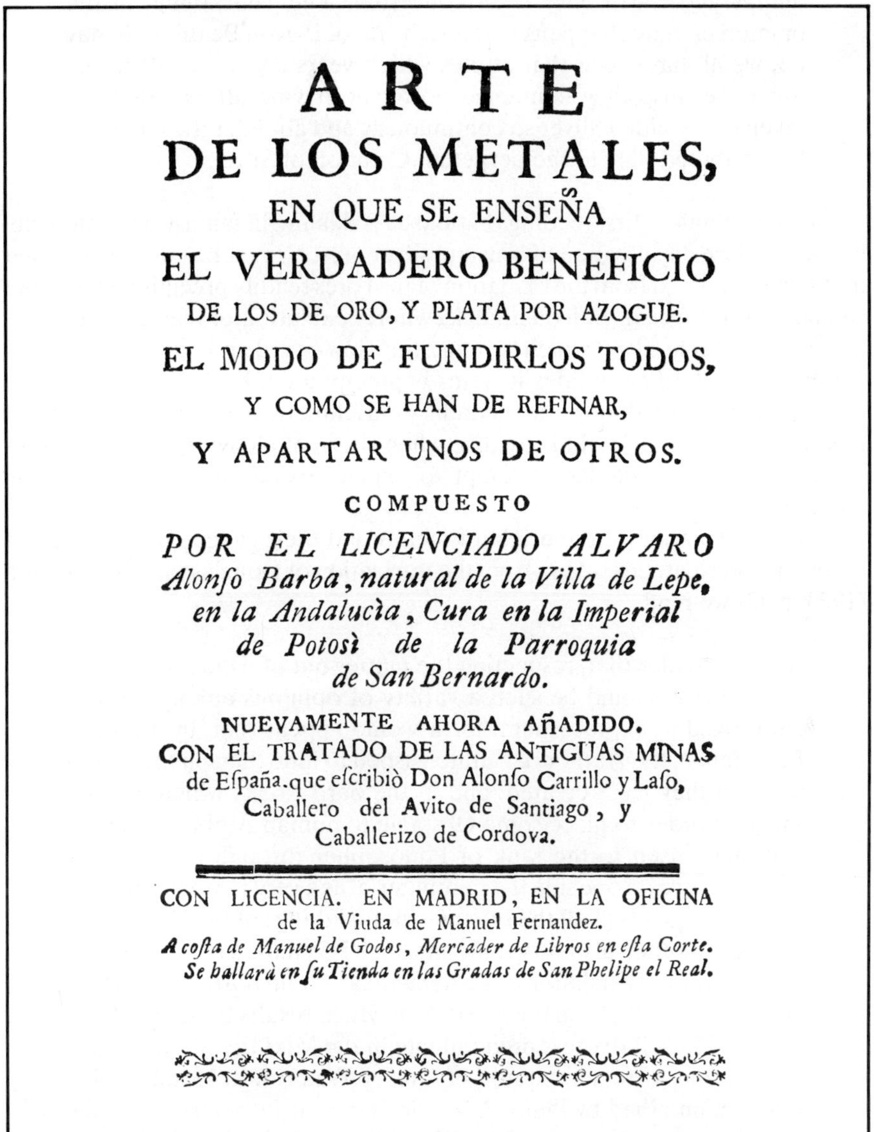

ARTE
DE LOS METALES,
EN QUE SE ENSEÑA
EL VERDADERO BENEFICIO
DE LOS DE ORO, Y PLATA POR AZOGUE.
EL MODO DE FUNDIRLOS TODOS,
Y COMO SE HAN DE REFINAR,
Y APARTAR UNOS DE OTROS.
COMPUESTO
POR EL LICENCIADO ALVARO Alonſo Barba, natural de la Villa de Lepe, en la Andalucìa, Cura en la Imperial de Potosì de la Parroquia de San Bernardo.
NUEVAMENTE AHORA AñADIDO.
CON EL TRATADO DE LAS ANTIGUAS MINAS de Eſpaña.que eſcribiò Don Alonſo Carrillo y Laſo, Caballero del Avito de Santiago, y Caballerizo de Cordova.

CON LICENCIA. EN MADRID, EN LA OFICINA de la Viuda de Manuel Fernandez.
A coſta de Manuel de Godos, Mercader de Libros en eſta Corte. Se hallarà en ſu Tienda en las Gradas de San Phelipe el Real.

Figure 6-1. Title page of Alvaro Alonzo Barba's treatise on the art of the metals written at Potosi, Bolivia, and published in Madrid, Spain, in 1640. This work is the first treatise on ore deposits written in the Americas.

> form in which they to-day exist and are found in their Veins. This is an offence to Nature, denying to her, without any Reason, the productive Virtue she possesses in all other sublunar things. Furthermore, experience in many parts of the World has proved the contrary. As an example and proof of this, it is sufficient to note what is brought to pass before the eyes of all, in Ilua (Elba), an island near Tuscany, where Iron abounds. After the Men have worked the Veins to the greatest possible Depth, they return the Earth and dumps to the workings; within a period of not more than ten or fifteen Years great quantities of Ore are taken out, into which the dumps and Earth have been converted. The same thing, in the opinion of many, happens in this rich Hill of Potosi. Be this as it may be, we all have seen that Stones which years ago were left in the mines because they contained no Silver, having afterwards been taken out, yielded Silver so continuously and abundantly that it can be attributed only to the perpetual Creation of Silver.

One may think at first reading that Barba is dealing in fantasy when describing the modern day creation of iron and silver ores. Not so. It is well known that iron springs derived from the oxidation of iron ores readily precipitate limonite; similarly with silver in Bolivia and elsewhere, oxidation of lean argentiferous ores yields soluble silver, which on migrating downward and coming in contact with pyrite and other sulfides in veins is precipitated. Barba mentions many such examples of oxidation and secondary enrichment in his famous treatise. In fact the geological part of his treatise when shorn of alchemical connotations represents the first modern attempt to explain oxidation and reduction in metallic veins.

In his explanation of the origin of metals and their gangue minerals Barba follows the Peripatetics. Again from the translation of Douglass and Mathewson (1923, p.43) we read:

> It is no wonder that, respecting the matter out of which Metals are made, there should be such a variety of opinions among Persons authorized to express them; for it would appear that, by a special Providence, the Author of Nature wished to hide them in the Depth wherein they were created, and in the hard Rocks which enclose them, in order to place some Obstacle to human Ambition. Those who have risen to the rank of Philosopher through the study of Causes, leave on one side the raw material as a very remote Principle of Metals, as it is of all the other corporeal things of the World, and indicate another, also very remote, which is in part, a certain Humid and Unctuous Exhalation, together with a certain portion of viscous and greasy Earth, from the Mixture of which results both Metal and Stones. Thus, if dryness predominate in the Mixture, Stone results; and if it has a greater amount of Greasy Water in it, Metal results: this is maintained by Plato, Aristotle and their followers. From the abundance of this pure, shiny Humidity is derived the Lustre of Metals, in which, among the other elements, Water, as is well known, predominates; and thus they flow and melt under the action of Heat. From the varied Temperament and Force of the matter

> referred to is derived the Diversity of the metals, of which the purest of all, and Nature's chief effort, is Gold.

Later in his treatise Barba describes the geological and mineralogical occurrence of gold in many parts of the Cordillera of South America. Of the deposits of the Charcas district of Bolivia he says (p. 65):

> There is no one who has not heard of Carabaya, which is a Region famous for the abundance and purity of its Gold, its Metal being as fine as the celebrated Gold of Arabia. It is Gold of twenty-three carats and three grains. The quantity which has been, and is still being extracted, is incredible, although that which has been taken out so far has been found by People who have looked for it only in the Rivers; only now are they commencing to work the many rich Veins. Carabaya is situated alongside of Larecaja, where auriferous Ores abound. In some of the streams in those Parts, Gold is found in the Form and Colour of small lead shot, which, when they have been melted, take the red Colour of Gold, with but little loss through the disappearance of the Crust which covered them when found. The man who discovered the Gold in this form did not know what it was until he was enlightened by a Friend whom I had advised of its true Nature.
>
> From Larecaja to Tipuani is a region occupied by savage Indians, the same having been invaded over twenty years ago by an Expedition sent to La Paz, I forming part thereof. There so much Gold is found that the accounts of it would be incredible, were it not for the many Witnesses who have seen it, and vouch for the fact.
>
> The real name of the city of La Paz, *Chaquiyapu,* which we have corrupted into *Chuquiabo,* means, in the native language of this Country "Farm" or "Estate" of Gold. Many mine Workings exist there, dating from the time of the Incas. It is a land well known to be fertile in Gold; and, during the rainy Season boys find Nuggets in the Streets, especially in that one which descends by the Monastery of the Dominicans towards the river. In Coroyco, and other parts of the Andes of Chuquiabo, there is Gold also in many ravines, grey on the outside like Lead.

Finally, Barba gives many practical hints on how and where to prospect for gold veins; for instance (p. 59):

> There is no sure or infallible rule which, without further experiment or examination, will enable the particular kind of metal contained in a hill to be determined simply by the Colour of the earth on the surface. And thus, although the material in which Gold is found is most frequently Red or dark Yellow, like to brick which has been very much burnt, veins thereof are frequently found in white *Calichal* (White Earth, Kaolin) as in Oruro and Chayanta.

Anyone who has seen the primary, and in places supergene, white and yellow (clay mineral, alunitic, and propylitic) alteration and the supergene red (limonitic) and yellow (jarositic) oxidation products associated with (Tertiary) gold deposits will recognize instantly the value of these guides.

Late in the seventeenth century John Woodward (1665-1728), the English physician, antiquarian, and collector of fossils, sought to explain the features of the globe in terms of a flood (the universal Deluge) released through fractures (perpendicular intervals) from inside the earth in *An Essay Toward a Natural History of the Earth and Terrestrial Bodies, Especially Minerals* (1695) and later in *The Natural History of the Earth, Illustrated, Inlarged, and Defended* (1726). In that second work, concerning the origin of mineral veins he states: "Water takes up the particles of metall, which lay before loose, and separate, in the interstices, and pores of the strata of stone, and thence carries them into the perpendicular fissures of the strata."

Later Pryce (1778, p. 6), after examining various theories of origin, explained the tin veins of Cornwall in the following way:

> From this we may reasonably infer, that water, in its passage through the earth to the principal fissures, imbibes, together with the natural acids and salts, the mineral and metallick particles, with which the different strata are impregnated; and meeting, in those fissures, matters which have nearer affinities with the acid, of course disengages it, in whole or in part, from the metallick and mineral particles, which it had held dissolved; and which, on being so disengaged, by the natural attraction between its parts, forms different ores, more or less homogeneous, and more or less rich, according to the different mixtures, which the acid had held dissolved, and the nidus in which it is deposited. The acid, now impregnated with a new matter, passes on; till meeting with some other convenient nidus, it lodges in that, and thereby acquires a fresh impregnation, perhaps at last totally unmetallick; or, for want of meeting with a proper nidus, appears at the surface, weakly or strongly tinctured with those principles it had last imbibed.

These are clearly statements of the lateral secretion theory; they differ only in detail from the theory postulated by Agricola. Nils Steensen (Nicholas Steno) (1638-1686), the Danish physician and theologian, likewise espoused a lateral secretion theory. In his *De solido intra solidum naturaliter contento dissertationis prodromus* (1669) he disputed the Vulgar opinion and the Aristotelian and alchemical notions about the origin of veins at Creation and hypothesized that uplifting earth movements created clefts, fissures, and interstices within the rocks in which mineral matter from the rocks was precipitated. To quote Steno, "It is more than probable that all those minerals which fill either the clefts or expanded spaces of rocks had as their matter the vapour forced from the rocks themselves. . . ." (Winter, Hobbs, and White, 1968, p. 236).

René Descartes (Renatus Cartesius) (1596-1650), the greatest of the French philosophers of the seventeenth century, sought to geometrize all nature. Writing in his *Principia philosophiae* (1644), he considered the earth to be a cooled star, the crust of which had been chilled, leaving a hot enterior. Below the crust he envisaged a metallic shell from which were exhaled the mineral matters, including gold, that were chilled and solidified in the faults and fissures in the disrupted outer crust. It is interesting to observe that these ideas still have some adherents. The widespread concept that the earth is composed of various shells (crust, mantle, iron core, etc.) has a similar basis. Likewise, the idea that some of these shells are sulfidic and metallic and give rise to mineral deposits has been postulated in recent years, principally by J. S. Brown in his *Ore Genesis, a Metallurgical Interpretation* (1948) and by Quiring (1954), who advocates a gaseous and liquid ascent of sulfides from a deep sulfide shell (chalcosphere).

Germany, particularly the region of the Erzgebirge, was the site of intensive mining during the eighteenth century. Most of the deposits were veins, and the origin of the fractures in which they occur and the derivation of their mineral content was the subject of much speculation among the mining geologists of the time. Much of this speculation was centered at Freiberg (in Saxony), which later was the site of the founding of the world famous Bergakademie (1765), among whose celebrated scholars and students may be mentioned Abraham Gottlob Werner, Alexander von Humboldt, and Mikhail V. Lomonosov. Only a brief summary of the various speculations on the origin of veins by a few of the German mining geologists are considered here, more or less in chronological order.

Henckel (1725, 1727) wrote that mineral veins were derived from emanations (vapors) derived from three sources: (1) saline and sulfurous waters arising from deep sources within the earth, the vapors from these waters having been partly derived from oceanic water that had gained access to the central fires, was vaporized, and expelled toward the surface; (2) from various bodies disseminated in the earth's crust, which upon comingling reacted and yielded vapors that produced the metallic minerals; and (3) from the *witterung* or exhalations of certain minerals already in existence within the earth, such as pyrite, which when exposed to oxidation becomes hot and gives off fumes.

One can see in the three sources postulated by this theory the germs of certain modern concepts concerning the origin of certain types of gold deposits: (1) the idea of recycled oceanic and meteoric water as the source of hydrothermal solutions, particularly with respect to certain Tertiary (epithermal) gold deposits, (2) metamorphic secretion, and (3) oxidation and secondary enrichment.

C. F. Zimmerman (1746), a student of Henckel, anticipated the concepts of metasomatism and replacement and the nineteenth century view of lateral secretion when he postulated that mineral veins were formed by the transformation of the rocks enclosing them. His theory was based on accurate observations of the alteration of the country rocks along fissures and the complete change of the mineral constitutents of the host rock and their partial replacement by ore and gangue minerals. A little later Von Oppel (1749) clearly recognized the fundamental difference between veins and bedded ore deposits. He also emphasized that veins are localized by structures such as fissures and faults formed by the dessication of soft sedimentary rocks or by movements in the earth's crust, and that these structures existed prior to the formation of the mineral veins and have definite structural patterns. Lehmann (1753, 1756) at the middle of the century hypothesized that vein minerals were deposited in fissures by vapors and exhalations from the earth's interior, a view similar to that advocated a century earlier by Descartes. Lehmann further considered veins to be comparable to the branches of a tree, the roots of which were deeply embedded in the core of the earth (the "Golden Tree"). Finally, he held that certain locales and conditions in the earth were more susceptible to metallization than others; these locales were referred to as the matrices or wombs of the metals by his predecessors and contemporaries. He called them the "mothers of metals" or *Metalmüttern.* This concept is the first intimation in the literature on mineral deposits about what we now call metallogenic provinces.

Later in the eighteenth century, C. T. Delius (1770, 1773) denounced the alchemists for their fantastic theories, inveighed against Lehmann's metalmüttern theory, and advocated a lateral secretion mechanism for the concentration of metals in veins, stating that meteoric water heated by the sun's rays within the rocks was the collector of sparsely disseminated mineral matter from the country rocks. Carl Gerhard (1781) held similar ideas but omitted the effect of the sun. Delius held some unusual views about the continuation of gold veins at depth. Noting that those of the Siebenburger are rarely auriferous below about 100 m he concluded that the sun's heat could not penetrate below this depth, and thus a mechanism for mobilizing the gold from the enclosing rocks was absent. Delius was, however, an astute observer and recognized the surface alteration of vein deposits, the development of gossans, and the formation of a zone of secondary enrichment under favorable circumstances. He thought these phenomena were due to the "burning action" of the sun, an incorrect interpretation, and it was not until early in the present century that the processes of secondary enrichment were understood and worked out satisfactorily (see chap. 16). In Delius's zonation scheme gossanous material marked by the copper minerals azurite and malachite, was followed downward by zones respectively rich in gold, silver, and finally lead, each metal requiring in turn with depth less heat from the sun to develop.

During the closing years of the eighteenth century, the effects of metamorphism seem to have impressed certain writers. Among these were Charpentier, a professor of mining at Freiberg, and Von Treba, a mining official of Brunswick, who explained vein deposits as due to *Gährung* (fermentation) and *Faulniss* (decomposition) or metamorphism of the country rocks. Charpentier (1778, 1799), basing his theory on observations made on the veins of Saxony, conjectured that solutions or vapors migrating through the rock along minute fissures and cracks had effected changes in them, developing new minerals and changing strips of the country rock into veins. Von Treba (1785) held similar views and thought that veins were simply one of the manifestations of the processes that changed sediments into schists and gneisses, shales to slate, and gneisses to granite. These early writers made the first attempts to explain the relationship of vein deposits to metamorphic processes. Daubrée (1879) and T. Sterry Hunt (1873*a,* 1873*b,* 1897) added significantly to this concept at a later date, but most investigators that followed, especially those writing on the magmatic hydrothermal theory, neglected the important part metamorphism plays in the concentration of the elements in veins and replacement deposits.

In the latter part of the eighteenth century, two novel and rival theories made their appearance. Their advocates were two famous geologists, Abraham Gottlob Werner, a Saxon, and James Hutton, a Scotsman.

Werner was professor of mineralogy and geology at the world-famous mining academy of Freiberg, in Saxony, Germany, established in 1765 to stimulate the study of minerals and mineral deposits. He attributed the origin of all rocks, sediments, basalts, and granites to chemical precipitation and mechanical sedimentation processes from the primeval ocean. Thus was born the classical

Neptunist theory. Late in his career Werner wrote a small book, *Neue Theorie von der Entstehung der Gänge* (1791), in which he discredited the lateral secretion theory and the suggestion that the minerals of veins were precipitated from vapors rising from the hot interior regions of the earth. Instead, he vehemently advocated the theory that the minerals were deposited from descending, percolating water from the primeval ocean.

Werner's theory was widely accepted at the time, especially by his students who disseminated his ideas throughout many countries. It was, however, challenged by Hutton, who published his *Theory of the Earth* in 1785 as a paper in the *Transactions of the Royal Society of Edinburgh* and later in 1795 as a book of two volumes with the same title. In this great work Hutton ascribed the origin of basalts, granites, and similar rocks to igneous agencies, and gneisses and schists to the effects of metamorphism. His Plutonist or magmatic theory led him to the conclusion that veins resulted from the forceful injection of a hot molten mineral mass that solidified in faults and fractures. He argued that silicates and sulfides were insoluble in water and could be transported only in a hot fused state, a postulate that Kirwan (1793, 1799) criticized severely for not being in accord with chemical facts. Hutton allowed water no scope in his theory, and he was thus an extreme advocate of what later became known as the ore-magma theory of mineral deposits (see chap. 9).

These two theories, the Neptunist and Plutonist, were debated for many years; finally it was recognized during the first part of the nineteenth century that both were extreme. Work on volcanic phenomena and the presence of sublimated mineral products at the orifices of fumaroles and other volcanic vents convinced a number of investigators of mineral deposits, especially Ami Boué (1820, 1822), that the transport of ore and gangue elements took place mainly by highly heated vapors. On cooling, these vapors deposited their mineral content by sublimation. Thus arose the sublimation theory, of which there are still some adherents. Dolomieu (1783), Breislak (1811), and particularly Scrope (1825), however, after extensive studies on volcanic phenomena, concluded that water and other volatiles were largely responsible for the fluidity of lavas and might play a part in mineral transport. These three geologists, however, held only vague ideas on the genesis of vein deposits, although it should be mentioned that Breislak gave an early exposition of magmatic segregation. He thought that the metallic substances would separate from a fused mass because of their mutual affinities and high specific gravities, and that when the rock solidified, the metallic substances (sulfides and oxides) would appear as veins that would cut one another because each substance would separate at a different time.

During the transition to modern views we see that observations on the nature of veins and their host rocks forced certain constraints on the theories of origin of these mineral bodies. Secretion theories prevailed in some quarters and emanation theories in others. Metamorphism and replacement processes were noted, and the effects of the surface oxidation of veins were recognized. Toward the end of the period the extreme views of the Neptunists and Plutonists were supported by many geologists, but later studies on volcanism suggested modifications of these extreme theories, ultimately leading to the magmatic hydrothermal theory, elucidated in greater detail in chapter 10.

REFERENCES AND SELECTED BIBLIOGRAPHY

Boué, A., 1820. *Essai géologique sur l'Ecosse,* Paris.

Boué, A., 1822. Mémoire géologique sur l'Allemagne, *Jour. Phys.* **94:**297-312, 345-379; **95:**31-48, 88-112.

Breislak, S., 1811. *Introduzione alla geologia,* Milan.

Brown, J. S., 1948. *Ore Genesis; A Metallurgical Interpretation; An Alternative to the Hydrothermal Theory,* Hopewell Press, Hopewell, N.J., 204p.

Charpentier, J. F. W., 1778. *Mineralogische Geographie der Chursächischen Lande,* Leipzig.

Charpentier, J. F. W., 1799. *Beobachtung über die Lagerstätte der Erze, hauptsächlich aus den sächsischen Gebirgen,* Leipzig.

Daubrée, G. A., 1879. *Etudes synthétiques de Géologie expérimentale,* Paris.

Descartes, R., 1644. *Principia philosophae,* Amsterdam.

Delius, C. T., 1770. *Abhandlung von dem Ursprunge der Gebürge und der darinne befindlichen Erzadern,* Leipzig.

Delius, C. T., 1773. *Anleitungen sur der Bergbaukunst nach ihrer Theorie und Ausübung,* Vienna.

Dolomieu, D., 1783. *Voyage au Îles de Lipari fait en 1781, ou notices sur les Îles Aeoliennes,* Paris.

Douglass, R. E., and E. P. Mathewson, trans., 1923. *The Art of the Metals,* by Alvaro Alonzo Barba (1640), John Wiley & Sons, New York, 288p.

Gerhard, C. A., 1781. *Versuch einer Geschichte des Mineral-Reichs,* Berlin.

Henckel, J. F., 1725. *Pyritologia oder Kieshistorie,* Leipzig.

Henckel, J. F., 1727. *Mediorum Chymicorum non ultimum conjunctionis primum appropriatio, etc.,* Dresden.

Hunt, T. Sterry, 1873*a*. The geognostical history of the metals, *Am. Inst. Min. Eng. Trans.* **1:**331-346.

Hunt, T. Sterry, 1873*b*. The origin of metalliferous deposits, *Am. Inst. Min. Eng. Trans.* **1:**413-426.

Hunt, T. Sterry, 1897. *Chemical and Geological Essays,* 5th ed., Scientific Pub. Co., New York, 489p.

Hutton, J., 1785. Abstract of a dissertation concerning the system of the earth, its duration, and stability; Theory of the Earth, reprinted in *Contributions to the History of Geology,* G. W. White, ed., vol. 5, Hafner Pub. Co., Darien, Conn., 1970.

Hutton, J., 1795. *Theory of the Earth with Proofs and Illustrations,* 2 vols., Edinburgh. Reprint Hafner Pub. Co., New York, 1959.

Jorden, E., 1631. *Discourse of Naturall Bathes, and Minerall Waters,* Thomas Harper, London.

Kirwan, R., 1793. Examination of the supposed igneous origin of stony substances, *Royal Irish. Acad. Trans.* **5:**51-81.

Kirwan, R., 1799. *Geological Essays,* London.

Lehmann, J. G., 1753. *Abhandlung von den Metal-Müttern und der Erzeugung der Metalle,* Berlin.

Lehmann, J. G., 1756. *Versuch einer Geschichte des Flötzgerbirgen,* Berlin.

Plattes, G., 1639. *A Discovery of Subterraneall Treasure,* London. Reprint Inst. Min. Metall., London, 1980.

Pryce, W., 1778. *Mineralogia Cornubiensis.* Reprint D. Bradford Barton, Truro, U.K., 1972, 331p.

Quiring, H., 1954. Schalenbau der Erde und sphärogene Erze, *Internat. Congrès Géol. 19é, C. R.* Sec. 13, pt. 3, Fas. 15, pp. 431-438.

Scrope, G. P., 1825. *Considerations on volcanoes . . .,* London.

Von Oppel, F. W., 1749-1752. *Anleitungen zur Markscheidekunst nach ihren Anfangsgründen und Ausübungen kurtzlich entworfen,* Dresden.

Von Oppel, F. W., 1772. *Bericht von Bergbau,* Leipzig.

Von Treba, F. W. H., 1785. *Erfahrung über das Innere der Gebirge,* Leipzig.

Webster, J., 1671. *Metallographia: or An History of Metals,* London.

Werner, A. G., 1791. *Neue Theorie von der Entstehung der Gänge,* Frieberg. (Trans. by C. Anderson as *New Theory of the Formation of Veins,* Edinburgh, 1809.)
Winter, J. G., W. H. Hobbs, and G. W. White, 1968. *The Prodromus of Nicolaus Steno's Dissertation concerning a Solid Body . . .,* Hafner Pub. Co., New York, 283p.
Woodward, J., 1695. *An Essay Towards a Natural History of the Earth and Terrestrial Bodies, Especially Minerals, etc.,* R. Wilkin Printer, London.
Woodward, J., 1726. *The Natural History of the Earth, Illustrated, Inlarged, and Defended,* London.
Zimmerman, C. F., 1746. *Untererdischen Beschreibung der Meissnischen Erzgebirges,* Obersächische Bergakademie, Dresden.

CHAPTER

7

Gold in the Modern Era

L'or est de tous les âges.
Proust, 1920

The period 1820 to the present has been molded by the consequences of great scientific discoveries and technological developments. The Industrial Revolution, begun in Britain in the latter part of the eighteenth century, spread during the nineteenth century to mainland Europe and to the Americas and other parts of the world. Initially based on the steam engine, that is, on iron and coal, the Industrial Revolution was further stimulated by the development of electrical power, the internal combustion engine, the telegraph, telephone, radio and television, and at the midpoint of the twentieth century by the jet engine and atomic power, all combining to create the vast industrial enterprises and communication and transportation systems familiar to everyone today. In this development geology, especially coal, ore, and petroleum geology, has played an indispensable role.

The Industrial Revolution and its sequel modern industrialization have had far-reaching effects on the patterns of the politics and economics of the nineteenth and twentieth centuries. The first effect in all countries was a population shift and increase creating a whole new class, an urban proletariat espousing a socialist philosophy; this change was accompanied by the growth of a large liberalist middle class and a general decline of the conservative monarchies, the landed aristocracy, and the Church (the Old Order). To these social movements was added the growth of intense nationalism.

The stresses and strains between the various political factions in the nineteenth century led to at least three waves of revolutions and reforms in Europe, to civil wars in the Americas, and finally, in the twentieth century, to the Russian and Chinese revolutions. Nationalism promoted a number of unifications (e.g., Germany, Italy) in the nineteenth century, but unfortunately fomented a number of wars in the late nineteenth century and two devastating wars in the twentieth. As a result of these social upheavals and adjustments the world now experiences a measure of uneasy democracy.

The nineteenth and twentieth centuries have witnessed major advances in science and technology stimulated by a progressive philosophy initiated in the eighteenth century by the Enlightenment and continued through the following two centuries by the philosophy of rationalism, a conviction that all things can be accomplished by reason through the exact determination and exploration of facts (positivism) and their translation into practical application (pragmatism).

Major landmarks in chemistry and physics include the statement of the modern atomic theory and the invention of chemical symbols by John Dalton (1766-1844); the formulation of the laws of electrolysis by H. Davy (1778-1829) and M. Faraday (1791-1867); the great improvements in precise analytical techniques by J. J. Berzelius (1779-1848); the invention of spectroscopy by R. W. Bunsen (1811-1899) and G. R. Kirchhoff (1824-1887); the founding and development of physical chemistry, thermochemistry, and organic chemistry; the discovery of X-rays by W. C. Roentgen (1845-1923); the elucidation of the periodic system of the elements in its modern form by D. I. Mendeleyev (1834-1907) and J. L. Meyer (1830-1895); the discovery of the electron by J. J. Thomson (1856-1940); and the discovery of radioactivity by A. H. Becquerel (1852-1908), the most significant scientific discovery ever made. That significant discovery permitted P. and M. Curie (1859-1906 and 1867-1934), E. Rutherford (1871-1937), F. Soddy (1877-1956), H. G. J. Moseley (1887-1915), N. Bohr (1885-1962), A. Einstein (1879-1955), M. Planck (1858-1947), and many others to elucidate the structure and bonding of matter, the existence of isotopes, and the equivalence of mass and energy, and ultimately to transmute the elements, the alchemists' dream. In the life sciences the most significant events were the elucidation of some of the aspects of biological evolution by J. B. P. A. Lamarck (1744-1829) and later by C. R. Darwin (1809-1882) and A. R. Wallace (1823-1913), and the formulation of the laws of heredity by G. J. Mendel (1822-1884) and of mutation by H. De Vries (1848-1935). All these advances in physics, chemistry, and life sciences had a profound influence on geology and geochemistry and also in many respects on the theories of origin of mineral deposits.

Only a brief account need be given here about the great advances in prospecting for gold deposits and in the mining and metallurgical techniques applied to the extraction and winning of gold during modern times. Geological knowledge of the occurrence of gold in its varied deposits has accumulated at an ever increasing rate since the seventeenth century, and this knowledge has been applied by prospectors and miners in their search for and development of auriferous deposits throughout the world. Geophysical methods, particularly those based on magnetic, electrical, and electromagnetic principles and developed mainly during the twentieth century, have rendered assistance in the discovery and tracing of many blind auriferous deposits. Similarly, geochemical methods developed after the Second World War have given great impetus to prospecting for gold because of their definitive nature for localizing auriferous concentrations.

In placering and in open-pit and underground mining the introduction of the steam engine, followed by the internal combustion engine and the electric motor, has removed most of the manual drudgery of shoveling, drilling, drifting, crosscutting, stoping, benching, mucking, loading, transporting, and hoisting. The introduction of compressed air machines was a major advance in drilling techniques for hard rock mining in the nineteenth century, and these techniques have been improved in the twentieth with the development of multiple drilling machines (jumbos). Developments in drilling were accompanied by improvements in loading muck, from handloading to mechanical shovels.

Haulage, likewise, evolved from human and animal portage to electric and diesel locomotives, large capacity rubber-tired vehicles, and conveyor systems. Explosives, in the form of (nitroglycerine) dynamite, gelatin dynamite, and ammonium nitrate introduced by Alfred Nobel in the 1860s to replace black powder revolutionized hard rock mining in the nineteenth century, and more specialized explosives such as Sprengel mixtures (e.g., nitrobenzene mixed with potassium chlorate) have done likewise in the twentieth. Drainage technology and the ventilation (and cooling) of mines first saw modern improvements in the eighteenth and nineteenth centuries and have undergone such sophistication today that some gold mines in the Witwatersrand of South Africa now mine at depths of 10,000 ft (3,050 m) and more. Gold placer operations are now highly mechanized operations using where practicable great mechanical chain-bucket dredges, first introduced in the early years of the twentieth century; smaller operations likewise are generally highly mechanized, using high-pressure water monitors for hydraulicking, bulldozers, draglines, highly efficient sluices, gravity-concentrating tables, and so on.

The modern techniques of the treatment of auriferous gravels still use many of the ancient methods of winning the gold, such as gravity separation and amalgamation, although the tables, spirals, blankets, and plates are now highly sophisticated. Some auriferous ores containing essentially free gold likewise receive many of the treatments devised by the Greeks and Romans but now mechanized, including crushing, grinding, washing, and separation of the gold by gravity separation and amalgamation. More generally, however, the winning of gold in a modern goldmining operation involves giant powered crushers, grinders, and cyanidation plants; such plants, invented by J. S. MacArthur and R. W. and W. Forrest in 1887, utilize a weak alkali cyanide solution to extract the gold. Other plants treating complex ores have flotation circuits, roasting furnaces, heap-leaching facilities, carbon-in-pulp circuits, and so on. Finally, the gold bars produced at the mining plants are refined principally by the chlorine gas process or by electrolysis at mints and other refineries.

The mining and production of gold in the modern era has shown a general increase with time, such that today it annually exceeds 1,300 metric tons (Table 18-2). Five peaks appear in the production record of modern times: 1492-1600, 1600-1800, 1820-1880, 1890-1920, and 1933-1939. The first marks the great influx of gold into Europe after the discovery of America; the second registers the increased production following the discovery of the great Colombian and other South American placers and lode deposits by the Spaniards; the third followed the discovery of the great placer and lode gold areas in Siberia, California, Australia, and New Zealand; the fourth reflects the discoveries of placers and bedrock deposits in Alaska, Yukon, Central Canada, and the fabulous deposits of the South African Witwatersrand, the most productive of all auriferous deposits; and the last registers an increase in gold mining during the Great Depression of the 1930s. At the time of writing (1985) a further peak appears imminent as a result of an increase in the price of gold.

OUTLINE OF THEORIES OF ORIGIN OF MINERAL DEPOSITS IN THE MODERN ERA

The first few years of the nineteenth century, as noted in chapter 6, were dominated by the theory of the Neptunists, who argued that descending waters from the universal primeval ocean had been mineral carriers, and by the theory of the Plutonists, who argued that veins had been deposited from an igneous (fused) mass of mineral matter. As more observations on volcanism, igneous phenomena, and mineral deposits became available these extreme views were greatly modified, particularly by Necker (1832), Scheerer (1847), and Élie de Beaumont (1847), who pointed out the apparent association of veins and lodes to igneous rocks, especially granites and porphyries. Élie de Beaumont published a famous paper in 1847 entitled *Note sur les émanations volcaniques et métallifères,* which ranks as one of the noblest classics in the science of mineral deposits. This paper contains a vast amount of data about volcanic products and mineral deposits and deserves special attention because its author attributed the source of thermal waters to magmatic agencies and attached much importance to mineralizers that carried off metalliferous matter from crystallizing igneous intrusions and deposited it in nearby fissures and faults. Thus was born the modern magmatic hydrothermal theory of mineral deposits.

An alternate theory based on the principles of secretion of ore and gangue elements from the host rocks of mineral deposits was developed at the midpoint of the nineteenth century. This theory, originally postulated by Agricola (see chapter 5) received impetus as a result of the analytical work of Forchhammer (1855), who demonstrated that ordinary rocks contain trace amounts of base metals and the other elements often found in veins. K. Gustav Bischof (1847-1855), the renowned geochemist, also supported the lateral secretion theory, and similar ideas were held by Daubrée (1860, 1879, 1887*a*, 1887*b*), T. Sterry Hunt (1873*a*, 1873*b*, 1897), Phillips (1884), Sandberger (1882-1885), S. F. Emmons (1886, 1887), G. F. Becker (1888), and C. R. Van Hise (1901, 1904) during the nineteenth century and the early part of the present century. All these geologists ascribed the source of the metals to the host rocks of the vein deposits but differed in their opinions as to whether the metals were deposited by descending or ascending meteoric water. Some investigators, including Daubrée, thought that meteoric water penetrated deeply into the crust where it effected widespread metamorphism, dissolved the metals and gangue elements, and rose again to deposit them as vein minerals in fissures. Other investigators including S. F. Emmons, claimed that the veins were deposited from descending meteoric water that had leached metalliferous and gangue material from the neighboring rocks. W. Wallace (1861) held yet another opinion, explaining the origin of the lead veins in the Carboniferous limestones of North England by deposition from descending meteoric solutions that had picked up mineral matter from exposed land surfaces.

In summary, during the latter part of the nineteenth century and during the present century three principal theories to explain the origin of epigenetic mineral (gold) deposits have gained widespread scientific recognition—the igneous (plutonist) theory, the magmatic hydrothermal theory, and secretion theories.

REFERENCES AND SELECTED BIBLIOGRAPHY

Adams, F. D., 1938. *The Birth and Development of the Geological Sciences,* Williams & Wilkins, Co., Baltimore, 506p.

Becker, G. F., 1888. Geology of the quicksilver deposits of the Pacific Slope, *U.S. Geol. Survey Monograph 13,* 486p.

Bischof, K. G., 1847-1855. *Lehrbuch der chemischen und physikalischen Geologie,* vols. 1-3, suppl. by F. Zirkel, 1871, A. Marcus, Bonn. (Trans. by B. H. Paul and J.

Drummond, 1854-55, as *Elements of Chemical and Physical Geology,* 2 vols., Cavendish Society, London).
Crook, T., 1933. *History of the Theory of Ore Deposits,* Thos. Murby & Co., London, 163p.
Daubrée, G. A. For a listing of Daubrée's voluminous works see De Lapparent, 1897, *Soc. Géol. France Bull.,* 3rd ser., **25:**245-284.
Daubrée, G. A., 1860. Etudes et expériences synthétiques dur le métamorphisme et sur la formation des roches crystallines, *Acad. Sci. Paris Mém.* **17:**pp. 1-127.
Daubrée, G. A., 1879. *Etudes synthétiques de géologie expérimentale,* Paris.
Daubrée, G. A., 1887*a*. *Les eaux souterraines, à l'époque actuelle,* Paris.
Daubrée, G.A., 1887*b*. *Les eaux souterraines aux époques anciennes,* Paris.
Élie de Beaumont, J. B., 1847. Note sur les émanations volcaniques et métallifères, *Soc. Géol. France Bull.,* 2nd ser., **4** (pt. 2):1249-1334.
Emmons, S. F., 1886. Geology and mining industry of Leadville, Colorado, *U.S. Geol. Survey Monographs,* vol. 12, 770p.
Emmons, S. F., 1887. The genesis of certain ore deposits, *Am. Inst. Min. Eng. Trans.* **15:**125-147.
Forchhammer, J. G., 1855. Ueber den Einfluss des Kochsalzes auf die Bildung der Mineralien, *Poggendorf's Ann. Phys. Chem.* **95:**60-96.
Hunt, T. Sterry, 1873*a*. The geognostical history of the metals, *Am. Inst. Min. Eng. Trans.* **1:**331-346.
Hunt, T. Sterry, 1873*b*. The origin of metalliferous deposits, *Am. Inst. Min. Eng. Trans.* **1:**413-426.
Hunt, T. Sterry, 1897. *Chemical and Geological Essays,* 5th ed., Scientific Pub. Co., New York, 489p.
Necker, A. L., 1832. An attempt to bring under general geological laws the relative position of metalliferous deposits. . . , *Geol. Soc. London Proc.* **1:**392-394.
Phillips, J. A., 1884. *A Treatise on Ore Deposits,* Macmillan, London, 651p.
Sandberger, F., 1882-1885. *Untersuchungen über Erzgänge* 2 vols., C. W. Kreidel, Wiesbaden.
Scheerer, T., 1847. Discussion sur la nature plutonique du granite et des silicates cristallins qui s'y rallient, *Soc. Géol. France Bull.,* 2nd ser., **4** (pt. 1):468-498.
Van Hise, C. R., 1901. Some principles controlling the deposition of ores, *Am. Inst. Min. Eng. Trans.* **30:**27-177.
Van Hise, C. R., 1904. Treatise on metamorphism, *U.S. Geol. Survey Monographs,* vol. 47, 1286p.
Wallace, W., 1861. *The Laws Which Regulate the Deposition of Lead Ore in Veins, Illustrated by the Mining Districts of Alston Moor,* London, 258p.

CHAPTER
8

Geochemistry of Gold

Gold. Found in nature as the free metal and in tellurides. Very widely distributed and under a great variety of conditions, but almost invariably associated with quartz or pyrite. Gold has been observed in process of deposition, probably from solution in alkaline sulphides, at Steamboat Springs, Nevada. It is also present, in very small traces, in sea water.
F. W. Clarke, 1908

The early hominids had no knowledge of chemistry and did not differentiate between the pure and impure (electrum) gold nuggets they recovered from stream sands. With the onset of civilization and the congregation of early man into urban communities some unknown worker in gold (the first metallurgist) discovered that gold nuggets were variable in composition and could be purified by fire employing a cementation process using common salt. After firing his crude charcoal smelting furnace the pure molten gold was run off leaving a dross, mainly silver chloride, which when mixed with charcoal and again fired was reduced to metallic silver. From the weights of the separated gold and silver a crude fineness value was calculated. This procedure represented the first attempt to analyze native gold and provided the first data on the geochemistry of the element. Exactly when this experiment occurred we do not know, but can presume that it took place about 3500 B.C. at the Sumerian sites of Ur, Kish, and Lagash in Mesopotamia, at the early sites of Mohenjo-daro and Harappa in the Indus Valley, in the valley of the Nile in Egypt, or possibly at the early Sinian sites in the valley of the Hwang-Ho (Yellow) River in China, the four most ancient centers of early civilization. Certainly by Biblical times the analysis of native gold was well advanced as indicated by the first mention of the metal in Genesis 2:10-12, where it is said that "the gold of that land (Havilah) is good," meaning that the gold was of high purity.

That the assaying of gold was well advanced in Pharaonic times is attested by the complaint of Burraburiash, the king of Babylon, to Egyptian Pharaoh Amenophis IV (1377-1358 B.C.), about receipt of poor (probably adulterated) gold (Irving, 1974, p. 3): "Your Majesty did not examine the samples of gold which were sent to me last time, for after putting them into the furnace the gold was less than its weight." Apparently only 5 minas of gold were recovered from 20 minas of alloy after trial in the fire, according to the hieroglyphics of the tablet.

The later civilizations of India, China, Asia Minor, Egypt, Greece, and Rome practiced metallurgy extensively and brought the winning of gold and silver from both oxidized (supergene) and hypogene ores to a fine art. The method utilized for complex sulfide ores, liquation, was known by 2500 B.C. and underwent much improvement by the Hittites, Persians, Egyptians, Greeks, and Romans in later centuries. Other methods of winning gold and silver from complex ores were cupellation with lead and amalgamation with mercury, both extensively practiced by the Greeks and Romans. Parting of gold and silver by acids was apparently unknown to the Romans; the first mention of that method appeared in the Moslem alchemical treatises of the thirteenth century.

The origin of the methods of fabricating gold and silver alloys is lost in antiquity, but can probably be related to the period around 3500 B.C. when working of gold began in the early civilizations just mentioned. Certainly the alloying methods were well known to the early Egyptians, Greeks, and Romans, as it was described in their pictographs and hieroglyphs and later by many of their classical writers. Many Biblical references allude to the refining of the precious metals, such as Proverbs 17:3, which states that "the fining pot is for silver, and the furnace for gold." Techniques of assaying the fineness of both natural and artificial gold alloys by fire and by the touchstone and touch needles (Au-Ag alloys of known composition) were also known early in the Classical world and are described in some detail by Theophrastus in his *De lapidibus* (On stones). From the translation by Caley and Richards (1956, p. 54) we read:

> The nature of the stone which tests gold is remarkable, for it seems to have the same power as fire, which can test gold too. On that account some people are puzzled about this, but without good reason, for the stone does not test in the same way. Fire works by changing and altering the colors, and the stone works by friction, for it seems to have the power of picking out the essential nature of each metal.
>
> They say that a much better stone has now been found than the one used before; for this not only detects purified gold, but also gold and silver that are alloyed with copper, and it shows how much is mixed in each stater. And indications are obtained from the smallest possible weight. The smallest is the *krithē,* and after that there is the *kollybos,* and then the quarter-obol, or the half-obol; and from these weights the precise proportion is determined.
>
> All such stones are found in the river Tmolos. They are smooth in nature and like pebbles, flat and not round, and in size they are twice as big as the largest pebble. The top part, which has faced the sun, differs from the lower surface in its testing power and tests better than the other. This is because the upper surface is drier, for moisture prevents it from picking out the metal. Even in hot weather the stone does not test so well, for then it gives out moisture which causes slipping. This happens also to other stones, including those from which statues are made, and this is supposed to be a peculiarity of the statue.

The weights referred to as krithe, kollybos, and so on are said by Caley and Richards (1956) to be equivalent to 0.06, 0.09, 0.18, and 0.36 g respectively.

These passages by Theophrastus represent the first account of a method of determining the quantitative composition of an alloy, or for that matter of fact, of any material. They are, therefore, of great interest in the history of assaying and analytical chemistry (see Caley and Richards, 1956, pp. 150-156).

Later, Pliny (A.D. 79) remarked on the touchstone in his *Historia naturalis* (Bailey, 1929, Pt. 1, p. 127):

> An account of gold and silver is incomplete without a mention of what is called the whetstone (touchstone), found formerly, according to Theophrastus, only in the river Tmolus, but now known to be widely distributed. It is called by some the Heraclian, by others the Lydian stone. These stones are of moderate size, not exceeding four inches by two, and the side that was exposed to the sun is better than the side next the ground. In testing, workers skilled in the use of the whetstone (touchstone) take with it a scraping from an ore, as one might with a file, and can then tell straight off, to the nearest scruple, how much gold, silver, or copper it contains, by this wonderful method which never fails them.

The touchstone in ancient times was usually a black chert or black slate although other rocks such as black trap (basalt) may have been used in the early Indus civilizations. Reference to the river Tmolus in Theophrastus's account is of interest because it is near the auriferous river Pactolus and the ancient Lydian city of Sardis, the site of the mint where King Alyattes (610-561 B.C.) minted the first staters. The accuracy with which the early assayers applied the touchstone to ores, as related by Pliny, explains how the Greek and Roman metallurgists were able to extract gold and silver from their ores so efficiently. But there were some in ancient times who questioned the accuracy of the touchstone in the hands of experts (see Smith and Gnudi, 1959, p. 203, footnote). We cannot assume, therefore, the unqualified veracity of the ancient rhyme from the *Times Whistle:*

> All is not golde that hath a glistering hiew,
> But what the touchstone tries and findeth true.

Early quantitative determinations of gold in ores and rocks were based on fire assay methods, the principles of which appeared first in the works by Calbus (*Bergbüchlein,* 1497) (Sisco and Smith, 1949), Biringuccio (*Pirotechnia,* 1540) (Smith and Gnudi, 1959), and especially Lazarus Ercker (*Beschriebung aller furnemisten Mineralogischen Ertzt und Berkwerksarten,* 1574) (Sisco and Smith, 1951). This work is the first detailed analytical chemical treatise; it describes the techniques for the analysis of gold and silver ores as well as of many other elements. Some of the assay methods for gold and silver described by Ercker are still used in modern assay and analytical laboratories. Modern methods for the estimation of gold in all types of earth materials are based on combinations of assay and spectrochemical techniques and on neutron activation analysis. Depending on the size of the sample, assay and spectrochemical techniques are capable of detecting down to 0.001 ppm Au in most types of geological materials; neutron activation analysis can detect gold values as low as 0.0005 ppm in rocks and other types of earth materials.

GEOCHEMISTRY OF GOLD

Early geochemical knowledge of the distribution of gold in the various materials of the geospheres (lithoshere, pedosphere, hydrosphere, biosphere) was rudimentary, principally because of the imprecision of the assay methods employed. On the other hand, the distribution of gold (and silver) was relatively well known in the various types of mineral deposits (principally veins) in which enriched quantities of the precious metals occurred.

The first geologist-geochemist to marshall the known facts about the geochemistry of gold and its distribution was K. Gustav Bischof, professor of chemistry and technology in the University of Bonn. In his *Lehrbuch der chemischen und physikalischen Geologie,* comprising three volumes (1847-1855), we find a good general summary of the distribution and nature of gold in its principal types of deposits in various parts of the world as known at the midpoint of the nineteenth century (Paper 8-1). No data are given on the distribution of the element in the principal rock types; the first efforts to obtain these data came only at the turn of the nineteenth century with improved methods of analysis (assaying) of the precious metals.

The last half of the nineteenth century was marked by steady progress in defining the mineralogy and geochemistry of both gold and silver as well as that of the platinoids. Several treatises appeared that dealt in part with the mineralogy and chemistry of the gold minerals and the general mineralogy and chemistry of gold deposits. Among these writings are *Chemical and Geological Essays* by T. Sterry Hunt (1897), *Allgemeine und chemische Geologie* by J. Roth (1879-1893), and *Chemische Mineralogie* by R. Brauns (1896).

The last quarter of the nineteenth century and the first half of the twentieth century can truly be called the "golden age of geochemistry," both in a figurative and literal sense. Modern methods of trace analysis, especially those based on the optical spectrograph, stimulated a number of investigators at the turn of the nineteenth century to investigate the distribution of the less-abundant elements in all types of earth materials. Among the most active of these investigators we may mention A. de Gramont, Sir William Crookes, W. N. Hartley and H. Ramage, and H. L. Lundegardh. Those who investigated the distribution of the precious metals, particularly gold and silver, include A. Liversidge, F. Lauer, L. Wagoner, and J. R. Don. Their results are summarized by F. W. Clarke (1924).

Frank W. Clarke (1847-1931) was chief chemist to the United States Geological Survey for some forty years and published over 300 papers dealing with a wide range of subjects in chemistry, mineralogy, and geochemistry. His greatest work, *The Data of Geochemistry,* was first published in 1908 with four later editions, the last appearing in 1924. This great compendium ranks as a landmark in modern geochemistry and is the first systematic record of the chemical constitution of the earth in a modern sense. In the opening chapter of the 1924 edition the general geochemistry of gold is succinctly described in the classical quotation cited at the beginning of this chapter.

Later in the chapter on metallic ores (chapter 15, 1924 ed., pp. 656-663) Clarke gives a masterful summary of the geochemistry of gold as then known (Paper 8-2).

(Text continues on page 104.)

8-1: GEDIEGENES GOLD

Dr. Gustav Bischof

Reprinted from *Lehrbuch der chemischen und physikalischen Geologie,* 3 vols., Adolph Marcus, Bonn, 1855, pp. 2050-2057.

Gold, auf Quarz-, Brauneisenstein- und Eisenkies-Gängen, im Thon-, Grauwacken-, Glimmer- und Hornblendeschiefer, im Granit, Gneifs, Syenit, Quarzporphyr, Gabbro, Diorit, Diabas, Aphanit, Serpentin und Dolomit, in Geschieben und im Sande vieler Flüsse in Körnern, Blättchen und Staub, meist begleitet von Quarz, Brauneisenstein, der zum Theil durch Zersetzung von Eisenkies entstanden ist, nicht selten auch begleitet von Zirkon, Magneteisen, Iserin, Spinell u. s. w. In *Rio Atrato* in *Peru* fand man Geschiebe bis zu 26 Pf. Gewicht, und die gröfste Goldmasse von 86 Pf. im Goldsand-Lager von *Alexandrowsk* bei *Miask*, 3 Meter tief auf Diorit liegend. In *Australien* fand Kerr eine Goldmasse von 106 Pf. Gewicht im anstehenden Quarz *). In *Chile* stammt alles Gold, welches in grofsen Körnern und etwas ansehnlicheren rundlichen Stücken gefunden wird, aus dem goldführenden Schuttlande, und diefs rührt ohne Zweifel von der Zerstörung der obersten Theile der Gänge her; nur äufserst selten trifft man in jetziger Zeit in einiger Teufe auf dieser Lagerstätte noch Goldkörner von einiger Gröfse an (Domeyko) **).

Sein Vorkommen in kleinen Krystallen und zu Drusen verbunden, so wie in draht-, haar-, moos- und baumartigen Formen führt auf Reductionsprocesse von Goldverbindungen, sein so häufiges Auftreten in Quarzgängen führt auf Absätze solcher Goldverbindungen aus Gewässern, welche Kieselsäure abgesetzt haben. Welche Goldverbindungen aber im Mineralreiche exi-

---

*) Berg- und hüttenmännische Zeitg. 1853. No. 23.

**) N. Journ. f. Mineral. u. s. w. 1847. S. 238.

stiren mögen, davon kann man sich keine genügende Vorstellung machen. Ein lösliches Schwefelgold wird durch Schmelzen von Fünffach-Schwefelkalium mit überschüssigem Golde erhalten; ein solcher Procefs kann aber im Mineralreiche nicht vorausgesetzt werden. Schwefelgold geht mit mehreren elektronegativen Schwefelmetallen lösliche Verbindungen ein. Wäre daher die Existenz des Schwefelgoldes im Mineralreiche nachgewiesen: so würde damit die Möglichkeit solcher Verbindungen und ihr Absatz aus wässerigen Flüssigkeiten gegeben sein. Die so leichte Reduction der Goldverbindungen überhaupt würde es dann begreiflich machen, wie sich daraus gediegenes Gold abscheiden könnte. Auf diese Weise würde dann auch die Bildung des Tellurgoldes (S. 1956) aus einer Verbindung mit Schwefeltellur zu erklären sein.

Schwefelgold aus einer Goldauflösung durch Schwefelwasserstoff gefällt, und sorgfältigst ausgewaschen, wurde mit einer grofsen Menge reinen Wassers in einem verschlossenen Gefässe behandelt, und die filtrirte Flüssigkeit zur Trockne abgedampft. Ein Rückstand in der Porzellanschale war kaum wahrzunehmen; als aber nur einige Tropfen Salzsäure zugesetzt wurden: so entstand eine goldgelbe Auflösung. Um sicher zu sein, dafs nicht etwa eine Spur zurückgebliebenen Königswassers einen Theil des Schwefelgoldes während des Abdampfens wieder aufgelöst habe, wurde das ausgewaschene Gold mit 3 Pfd. Wasser behandelt und diefs drei Mal wiederholt; es zeigte sich indefs jedes Mal jene Reaction mit Salzsäure. Die noch deutlich nachzuweisende Löslichkeit des Schwefelgoldes in reinem Wasser ist daher unzweifelhaft; ob es sich als solches gelöst, oder ob es sich auf Kosten des vom Wasser absorbirten Sauerstoffs oxydirt habe, wurde nicht ermittelt. Diese Löslichkeit würde daher alle Schwierigkeiten beseitigen, welche der Fortführung des Goldes in wässriger Lösung entgegenstehen, wenn nur irgend eine Erscheinung vorläge, welche auf die Bildung des Schwefelgoldes im Mineralreiche schliefsen liefse.

Alles gediegene Gold, wovon sehr zahlreiche Analysen vorliegen, ist silberhaltig. Im sogenannten Elektrum erreicht das Silber das Maximum von 36 Proc. (Klaproth). Nachstehende Zusammenstellung giebt eine Uebersicht von der Zusammensetzung des gediegenen Goldes in seinen verschiedenen Fundorten.

Silber	Kupfer nach Proc.	Eisen	
2—35,8	—	—	vorzugsweise aus *Südamerika* (Boussingault nach 17 Analysen).
0,16—38,7	fast kleine	alle Mengen	vom *Ural* (G. Rose nach 23 Analysen).
3,53—28,5	kaum	bis 1	Goldwäschen von *Katharinenburg* (Awdejew nach 16 Anal.).
3 —15	etwas	etwas	Waschgold von *Chile* (Domeyko nach 4 Anal).
6,7 —12,9	bis	0,86	von *Californien* (C. Brunner, Henry, Oswald, Rivat, Teschemacher, Hofmann, Levol nach 10 Anal.).
5,9 —15,3	bis 0,9	—	vom *Senegal* (Levol).
3,6	—	—	*Australien* (Kerl).
6,2	—	0,8	Grafschaft *Wicklow* in *Irland* (Mallet).
4,7 —6,9	—	—	im Eisenkies von *Piemont* (Michelotti nach 5 Anal.).
	Palladium		
4,2	9,9	—	von *Porpez* in *Südamerika* (Berzelius).

Bestimmte Mischungsverhältnisse zwischen Gold und Silber finden, wie man sieht, nicht statt; aber wichtig ist die Thatsache, dafs selbst im Elektrum das Silber nur ungefähr $\frac{1}{3}$ beträgt.

Im Vorhergehenden ist ein geringer Goldgehalt in verschiedenen Erzen angegeben worden. Namentlich sind Bleiglanz, Kupferkies und Eisenkies nicht selten goldhaltig. Von letzterem giebt es, nach Gahn, keinen, der nicht bei genauer Prüfung Spuren von Gold zu erkennen gäbe, und dieser, wenn auch immer nur geringe Goldgehalt ist manchmal so bedeutend, dafs seine Gewinnung lohnend wird. So kommen bei *Trinidad* und in der Umgegend von *Santa Rosa* im *Valle de Osos* goldhaltige Eisenkies- und Quarzgänge vor (Degenhardt) *). In *Chile* sind die Kiese, welche Gänge von 6 bis 9 Fufs Mächtigkeit bilden, bei weitem reicher an Gold, als die Quarzgänge (Domeyko). Es ist nicht zu ermitteln, ob das Gold im Eisenkies mit Schwefel verbunden, oder metallisch vorhanden sei; die Annahme ist indefs nicht unwahrscheinlich, dafs es darin als Schwefelsalz existire. Würden solche Schwefelgold-haltige Eisenkiese durch Oxydation zersetzt: so würde sich das Gold, welches keinen Antheil an diesen Zersetzungsprocessen nähme, metallisch ausscheiden, und einen um so feineren Staub bilden, je geringer seine Menge wäre. Durch Gewässer fortgeführt, würde sich dieser Goldstaub im Alluvium absetzen. Was vom Eisenkies gilt, hat auch Bezug auf die goldhaltigen Kupferkiese und Bleiglanze.

In *Chile* sind Quarz, Brauneisenstein und Eisenglanz die einzigen Mineralien, welche am Ausgehenden der Goldführenden Gänge vorkommen, und eben diese Mineralien finden sich auch als Begleiter des Goldes im Alluvium (Domeyko). So wie im Eisenkies, so ist auch hier das Gold an Eisen geknüpft, und defshalb würde zu begreifen sein, wie bei Bildung des Eisenkieses aus goldhaltigem oxydirten Eisen das edle Metall in seine Mischung treten könnte. Sollte es von Eisenkies auf nassem Wege in ein Schwefelsalz ebenso umgewandelt werden, wie es auf trocknem Wege vom fünffach-Schwefekalium in ein solches umgewandelt wird: so würde sogar seine Concentration im Eisenkies denkbar sein. Die eigentlichen Golderze kommen in *Chile* nur in oberen Teufen oder ganz am Ausgehenden der Gänge in zerfressenem, mit Eisenoxydhydrat und ochrigem Thone gemengten Quarz, also auch hier wieder in Begleitung mit oxydirtem

*) Karsten's und v. Dechen's Archiv Bd. XII. S. 14.

Eisen vor. Die goldhaltigen Eisenkiese gehen dagegen in den Gängen bis zu bedeutenden Tiefen nieder. Wahrscheinlich ist es daher, dafs die äufserst dünnen Goldblättchen in oberen Teufen zugleich mit den Materialien, woraus sich Eisenkies bildete, durch Gewässer in die Tiefe geführt wurden.

Das Gold aus anstehendem Gestein von *Californien* findet sich im Quarz, das von *Australien* im Brauneiseinsteine (Breithaupt) *).

Der Chemiker findet im Silber der Münzen stets mehr oder weniger Gold. In den Silbererzen, aus denen dieses Silber gewonnen wurde, würde er aber den geringen Goldgehalt nicht mehr nachweisen können; denn wenn selbst das Silber in manchen Erzen, wie z. B. im Bleiglanz, ein sehr kleiner Bruchtheil ist: so würde die geringe Menge Gold in diesem Silber, selbst nicht mehr durch die empfindlichsten Reagentien in solchen Erzen zu erkennen sein. Während daher das Silber im gediegenen Golde immer noch eine namhafte Gröfse ist, dürfte das Gold in vielen Erzen eine kaum mehr zu bestimmende Gröfse sein.

Es wurde wiederholt nachgewiesen (S. 2018), dafs, bei der Zersetzung silberhaltiger Erze, das Silber an den Oxydationsprocessen Antheil nimmt und in löslichen Verbindungen fortgeführt wird. Sind solche Erze goldhaltig: so bleibt das Gold ungelöst zurück. Werden solche Erzgänge, nach erfolgter chemischer Zersetzung ihrer Gangmassen, mechanisch zerstört: so wird das ausgeschiedene Gold mit den übrigen Zersetzungsproducten durch die Gewässer mechanisch fortgeführt. Der Silbergehalt im gediegenen Golde beweiset indefs, dafs die Scheidung beider Metalle von einander nicht vollständig erfolgt, und dafs sich daher ihre Verwandtschaft zu einander im Mineralreiche, wie bei unseren künstlichen Scheidungsprocessen, geltend macht.

Setzt man zu einer Lösung von Goldchlorid eine Lösung von kieselsaurem Kali: so verschwindet die gelbe Farbe der ersteren fast ganz. Nach einer halben Stunde wird die Flüssigkeit blau und nach einiger Zeit entsteht ein gallertartiger tief dunkelblauer Niederschlag, der sich fest an das

*) Berg- und hüttenmännische Zeitung 1853. No. 35.

Glas anlegt. Auf ihm bilden sich nach mehreren Tagen moosartige Gestalten, die als Efflorescenzen aus der gallertartigen Masse erscheinen. Als dieser Niederschlag unter Wasser mehrere Stunden lang den Sonnenstrahlen ausgesetzt wurde, zeigte sich auch nicht eine Spur einer Reduction des Goldes. Nachdem er aber fünf Monate unter Wasser stehen geblieben war, und ausgewaschen wurde, zeigten sich im kieselsauren Goldoxyd viele, zum Theil mikroskopisch kleine Pünctchen reducirten Goldes. Die von diesem Silicate abfiltrirte Flüssigkeit war farblos und reagirte alkalisch; kieselsaures Kali war daher im Ueberschusse angewandt worden. Als Salzsäure bis zur sauren Reaction zugesetzt wurde, färbte sich die Flüssigkeit gelb und eine Lösung von schwefelsaurem Eisenoxydul bewirkte eine stark blaue Färbung im durchgehenden und eine braune im reflectirten Lichte. Kieselsaures Goldoxyd war daher aufgelöst gewesen. Nachdem das Auswaschen fortgesetzt worden, bis die alkalische Reaction verschwunden war, zeigte das Abwaschewasser keine Spur von Gold mehr. Da zu vermuthen war, dafs das überschüssige kieselsaure Kali auflösend auf das kieselsaure Gold gewirkt habe: so wurde davon wieder etwas zugesetzt; die Flüssigkeit zeigte aber keine Spur von Gold. Es gelang mir nicht, die Ursache dieses Verhaltens aufzufinden; ich behalte mir jedoch weitere Untersuchungen vor.

Diese, wenn auch noch unvollkommenen, Versuche zeigen indefs, dafs kieselsaures Goldoxyd unter gewissen Umständen in deutlich erkennbarer Menge aufgelöst und reducirt werden kann.

Das bei weitem am häufigsten in Quarzgängen und im aufgeschwemmten Lande, in Begleitung mit Quarzsand vorkommende Gold zeigt eine unverkennbare Beziehung zur Kieselsäure. Der Ursprung der Kieselsäure erscheint daher auch als der des Goldes. Jenen kennen wir: es sind die Silicate im Gebirgsgesteine, durch deren Zersetzung der Quarz in die Gänge geführt wird. In diesen Silicaten haben wir daher auch das Gold zu suchen, und es liegt nahe, zu vermuthen, dafs es darin gleichfalls als Silicat vorhanden sei. Diefs durch das Experiment zu constatiren, liegt ausser den Grenzen der Möglichkeit; denn wenn es auch gelingen sollte, in Silicaten ebenso Gold nachzuweisen, wie bereits unedle Metalle darin aufgefunden worden sind: so wird man doch nie ermitteln können, ob es darin regulinisch oder mit irgend einer Substanz verbunden war. Da wir im Stande sind, die ausserordentlich geringen Minima von Gold im Eisenkies noch nachzuweisen: so steht der Hoffnung, dafs es auch gelingen werde, es in Silicaten aufzufinden, wenig entgegen, und um so weniger, da uns Reagenzien zur Entdeckung des Goldes zu Gebote stehen, deren Empfindlichkeit kaum von irgend einem Reagenz auf andere Metalle übertroffen wird. Es ist gleichgültig, ob das Gold im Eisenkies mit Schwefel und in Silicaten mit Kieselsäure verbunden oder als Metall vorhanden sei, die Reagenzien, welche es nachweisen, sind jeden Falls bei weitem empfindlicher, als diejenigen, welche die Existenz seiner Begleiter oder die Nichtexistenz derselben erkennen lassen.

Dafs Processe, welche wir zur künstlichen Darstellung des kieselsauren Goldoxyd anwenden, im Mineralreiche nicht statt finden, ist klar. Es liegt nur die Alternative vor, ob Gold als gediegenes Metall, oder in irgend einer Verbindung geschaffen worden ist. Die Analogie mit anderen Metallen spricht für das letztere, seine geringe Verwandtschaft zu anderen Substanzen für das erstere. Berücksichtigen wir auf der andern Seite, dafs die Verwandtschaftsverhältnisse mancher Stoffe in feinster Zertheilung ganz andere sind, als in Massen: so erscheint es annehmbar, dafs auch Gold, welches, wenn es als solches in Mineralien existiren sollte, gewifs feiner zertheilt ist, als vielleicht irgend eine andere Substanz, in diesem Zustande der Zertheilung Affinitäten äussern könne, welche dieses Metall in Massen nicht besitzt. Das durch Reduction von Eisenoxyd mittelst Wasserstoff dargestellte Eisen zeigt schon in gewöhnlicher Temperatur eine so starke Verwandtschaft zum Sauerstoff, dafs es, durch die Luft fallend, verbrennt, während es als feinste Feilspäne in trockner Luft seinen metallischen Zustand nicht verändert. Es ist daher denkbar, dafs Gold, in seiner feinsten Zertheilung in Gesteinen, durch Gegenwart von Kieselsäure zur Verbindung mit Sauerstoff und hierauf mit dieser disponirt werden könne.

Man braucht nur einzuräumen, dafs kieselsaures Goldoxyd im Mineralreiche existire, sei es als eine ursprüngliche, oder als eine spätere Bildung: so erklärt sich das Vorkommen desselben in Quarzgängen ganz einfach. Gewässer führten dieses Silicat mit Kieselsäure in Gangspalten, und nach dem Absatze beider Substanzen reducirte es sich ebenso, wie es sich in dem beschriebenen Versuche reducirt hatte. Da selbst in den reichsten Quarzgängen das Gold nur ein überaus kleiner Bruchtheil des Quarzes ist: so begreift man, dafs der Einführung des kieselsauren Goldes nichts entgegen stehen würde, selbst wenn sich die Löslichkeit desselben zu der der Kieselsäure eben so verhalten sollte, wie die Menge des Goldes zu der des Quarzes in den Gängen. Sollten indefs weitere Versuche ergeben, dafs die Löslichkeit des kieselsauren Goldoxyd in Wasser durch Gegenwart von kieselsaurem Kali befördert wird: so treten von dieser Seite her unserer Annahme die geringsten Schwierigkeiten entgegen.

Goldamalgan, in kleinen, weifsen, leicht zerdrückbaren Kugeln im columbischen Platinerz, besteht aus Quecksilber 57,4, Gold 38,39, Silber 5,0 (Schneider).

8-2: GOLD

Frank Wigglesworth Clarke

Reprinted from The data of geochemistry, *U.S. Geological Survey Bull. 770,* 1924, pp. 656-663.

Although gold is one of the scarcer elements, it is widely diffused in nature. It is found in igneous rocks, sometimes in visible particles; it accumulates in certain detrital or placer deposits; it also occurs in sedimentary and metamorphic formations, in quartz veins, and in sea water.[4] A notable amount of gold is now recovered from copper ores, during the electrolytic refining of the copper. A. Liversidge[5] found traces of gold in rock salt from several localities, in quantities of about 1 to 2 grains per ton. F. Laur,[6] in Triassic rocks taken from deep borings in the department of Meurthe-et-Moselle, France, found both gold and silver. The maximum amount in a sandy limestone, was 39 grams of gold and 245 of silver per metric ton, but most of the assays ran much lower.

Gold has been repeatedly observed as a primary mineral in igneous or plutonic rocks. G. P. Merrill[7] reports it in a Mexican granite, embedded in quartz and feldspar. W. Möricke[8] found visible gold in a pitchstone from Chile; and O. Schiebe[9] discovered it in an olivine rock from Damara Land, South Africa.

In a series of assays of rocks collected at points remote from known deposits of heavy metals, L. Wagoner[10] found the following quantities, in milligrams per metric ton, of gold and silver. The samples are Californian, except when otherwise stated.

---

[3] For a list of the Survey publications on gold and silver see Bull. 470, 1911.

[4] See ante, p. 124.

[5] Jour. Chem. Soc., vol. 71, 1897, p. 298.

[6] Compt. Rend., vol. 142, 1906, p. 1409. Also in Compt. rend. Soc. ind. minérale, Sept.-Oct., 1906.

[7] Am. Jour. Sci., 4th ser., vol. 1, 1896, p. 309. Compare W. P. Blake, Trans. Am. Inst. Min. Eng., vol. 26, 1896, p. 290.

[8] Min. pet. Mitt., vol. 12, 1891, p. 195.

[9] Zeitschr. Deutsch. geol. Gesell., vol. 40, 1888, p. 611. For other examples see Stelzner-Bergeat, Die Erzlagerstätten, pp. 69-70. See also J. Catharinet, Eng. and Min. Jour., vol. 79, 1905, p. 127, on gold in the pegmatite of Copper Mountain, British Columbia. R. W. Brock (idem, vol. 77, 1904, p. 511) reports gold in British Columbia porphyries. On primary gold in a Colorado granite, see J. B. Hastings, Trans. Am. Inst. Min. Eng., vol. 39, 1909, p. 97. An association of gold with sillimanite is reported by T. L. Watson. Am. Jour. Sci., 3d ser., vol. 33, 1912, p. 241. A. Lacroix found disseminated gold in a biotite gneiss from Madagascar; see Compt. Rend., vol. 132, 1901, p. 180.

[10] Trans. Am. Inst. Min. Eng., vol. 31, 1901, p. 808.

Gold and silver in rocks from California, Nevada, etc.

[Milligrams per metric ton.]

	Au.	Ag.
Granite	104	7,660
Do	137	1,220
Do	115	940
Syenite, Nevada	720	15,430
Granite, Nevada	1,130	5,590
Sandstone	39	540
Do	24	450
Do	21	320
Basalt	26	547
Diabase	76	7,440
Marble	5	212
Marble, Carrara	8.63	201

In a later investigation Wagoner [1] determined gold and silver in deep sea (Atlantic Ocean) dredgings. In six samples assayed the gold ranged from 15 to 267 milligrams per metric ton, and the silver from 304 to 1,963 milligrams.

These figures suggest a very general distribution of gold in rocks of all kinds. J. R. Don,[2] however, in an extended investigation of the Australian gold fields, found that the deep-seated rocks contained gold only in association with pyrite. When pyrite was absent, gold was absent also. The country rocks of the vadose region, on the other hand, were generally impregnated with gold, even at a distance from the auriferous reefs, and Don supposes that the metal was probably transported in solution. This point will be discussed later.

Gold occurs principally in the free state or alloyed with other metals, such as silver, copper, mercury, palladium, rhodium, bismuth, and tellurium. Leaving detrital or placer gold out of account, its chief mineral associates are quartz and pyrite. Its connection with pyrite is so intimate that some writers have argued in favor of its existence as gold sulphide,[3] but the evidence in favor of that belief is very inadequate. No unmistakable gold sulphide has yet been found as a definite mineral species, nor is it likely to form except in an environment entirely free from reducing agents. The compounds of gold are exceedingly unstable, and the metal separates from them with the greatest ease.

[1] Trans. Am. Inst. Min Eng., vol. 38, 1907, p. 704

[2] Idem, vol. 27, 1897, p. 564. A. R. Andrew (Trans. Inst. Min. Met., vol. 19, 1910, p. 276) questions the trustworthiness of many such assays of country rock. He thinks that gold as an impurity in litharge accounts for most of the reported findings.

[3] See, for example, T. W. T. Atherton, Eng. and Min. Jour., vol. 52, 1891, p. 698, and A. Williams, idem, vol. 53, 1892, p. 451. Williams cites an auriferous pyrite from Colorado which yielded no gold on amalgamation, but from which gold was extracted by solution in ammonium sulphide. Gold sulphide is soluble in that reagent. Hence the inference that it may have been present in the ore. See also a paper by W. Skey, Trans. New Zealand Inst., vol. 3, 1870, p. 216.

On the petrologic side gold is most commonly associated with rocks of the persilicic type, such as granite and its metamorphic derivatives. I refer now to its primary occurrences. It is not rare in association with dioritic rocks, but in rocks of subsilicic character it is exceedingly uncommon. Its very general presence in quartz veins is testimony in the same direction and suggests the probability that gold is more soluble in silicic magmas than in those richer in bases. The auriferous quartz veins were probably formed in most instances from solutions; but J. E. Spurr [1] has argued that in some cases they are true magmatic segregations. This view was developed by Spurr in his studies of gold-bearing quartz from Alaska and Nevada, but it has been questioned by C. R. Van Hise [2] and others.

The composition of native gold is variable. The purest yet found, from Mount Morgan, Queensland, according to A. Leibius,[3] assayed as high as 99.8 per cent, the remainder being mainly copper, with a trace of iron. Gold commonly ranges from 88 to 95 per cent, with more or less alloy of the metals already mentioned. The following analyses well represent the character of the variations:

Analyses of native gold.

A. Gold from Persia. Analyzed by C. Catlett in the laboratory of the United States Geological Survey.

B. Electrum, Montgomery County, Virginia. Analysis by S. Porcher, Chem. News, vol. 44, 1881, p. 189.

C, D, E. Gold associated with native platinum, Colombia. Analysis by W. H. Seamon, Chem. News, vol. 46, 1882, p. 216.

F. Amalgam, Mariposa County, California. Analysis by F. L. Sonnenschein, Zeitschr. Deutsch. geol. Gesell., vol. 6, 1854, p. 243. Specific gravity, 15.47. Near $AuHg_3$.

G. Palladium gold. Taguaril, Brazil. Analysis by Seamon, Chem. News, vol. 46, 1882, p. 216. See Wilm, Zeitschr. anorg. Chemie, vol. 4, 1893, p. 300, on palladium gold from the Caucasus. Also E. Hussak, Zeitschr. prakt. Geologie, 1906, p. 284, on palladium gold in Brazil.

H. Maldonite, or "black gold," Maldon, Victoria. Analysis by R. W. E. MacIvor, Chem. News, vol. 55, 1887, p. 191. An alloy near Au_2Bi.

	A	B	C	D	E	F	G	H
Au	93.24	65.31	84.38	80.12	84.01	39.02	91.06	65.12
Ag	6.65	34.01	13.26	2.27	7.66		Trace.	
Cu	None.	.14	1.85	15.84				
Hg					7.06	60.98		
Pd							8.21	
Bi								34.88
Fe	.11	.20		Trace.			Trace.	
Quartz		.34						
	100.00	100.00	99.49	98.23	98.73	100.00	99.27	100.00

[1] See papers in Trans. Am. Inst. Min. Eng., vol. 33, 1903, p. 288; vol. 36, 1906, p. 372. Also Econ. Geology, vol. 1, 1906, p. 369.

[2] A treatise on metamorphism: Mon. U. S. Geol. Survey, vol. 47, 1904, pp. 1048–1049. See also J. B. Hastings, Trans. Am. Inst. Min. Eng., vol. 36, 1906, p. 647. Hastings regards the Silver Peak ores as deposited by ascending waters along lines of fracturing.

[3] Proc. Roy. Soc. New South Wales, vol. 18, 1884, p. 37.

The tellurides [1] containing gold are also variable in composition, partly because most of them contain silver, and often other metals, which may be only impurities. Kalgoorlite and coolgardite, for example, which are tellurides of gold, silver, and mercury, are mixtures of the mercury compound, coloradoite, with other species.[2] Calaverite and krennerite approximate to gold telluride alone. Sylvanite, petzite, muthmannite, and goldschmidtite are tellurides of gold and silver. The following analyses are sufficient to indicate the composition of the more important of these minerals:

Analyses of tellurides containing gold.

A. Calaverite, $(AuAg)Te_2$, Cripple Creek, Colorado. Analysis by W. F. Hillebrand.

B. Krennerite, $(AuAg)Te_2$, Nagyag, Hungary. Analysis by L. Sipöcz, Zeitschr. Kryst. Min., vol. 11, 1886, p. 210.

C. Sylvanite, $(AuAg)Te_2$, Grand View mine, Boulder County, Colorado. Analysis by F. W. Clarke, Am. Jour. Sci., 3d ser., vol. 14, 1877, p. 286.

D. Potzite, $(AuAg)Te_2$, Norwegian mine, Calaveras County, California. Analysis by Hillebrand.

	A	B	C	D
Au	38.95	34.77	29.35	25.16
Ag	3.21	5.87	11.74	41.87
Cu	....	.34	....	....
Fe	....	.59	....	....
Te	57.27	58.60	58.91	33.21
Se	....	....	....	Trace.
Mo	....	....	....	:08
Sb	....	.65	....	....
Fe_2O_3	.12	....	....	....
Insoluble	.33	....	....	....
	99.88	100.82	100.00	100.32

There has been much discussion over the tellurides of gold. B. Brauner [3] asserts that crystalline "polytellurides" can be formed, which dissociate upon heating, leaving the compound Au_2Te as an end product. Theoretically, the telluride Au_2Te_3 should also be capable of existence. According to V. Lenher,[4] the tellurides of gold are probably not definite compounds, but more in the nature of alloys. Attempts at the synthesis of a distinct compound failed. T. K. Rose,[5] however, who studied the alloys of gold and tellurium, obtained a definite compound, $AuTe_2$, identical with the natural calaverite. The same result was also obtained by G. Pellini and E. Quercigh.[6] W. J. Sharwood [7] has pointed out the very general association of bismuth with tellurium gold ores.

[1] For a general review of the tellurides, with references to literature, see J. F. Kemp, Min. Industry, vol. 6, 1898, p. 295.

[2] L. J. Spencer, Mineralog. Mag., vol. 13, p. 268, 1903. Spencer gives a good bibliography of the Australian tellurides.

[3] Jour. Chem. Soc., vol. 55, 1889, p. 391.

[4] Jour. Am. Chem. Soc., vol. 24, 1902, p. 358. See also R. D. Hall and V. Lenher, idem, p. 919.

[5] Trans. Inst. Min. Met., vol. 17, 1908, p. 285.

[6] Rend. R. accad. Lincei, 5th ser., vol. 19, 1910, p. 445.

[7] Econ. Geology, vol. 6, 1911, p. 22.

Although gold is primarily a magmatic mineral, it is also transported in and deposited from solutions. Many occurrences of gold indicate this fact very plainly. O. Dieffenbach,[1] for instance, mentions gold incrusting siderite at Eisenberg, near Corbach, in Germany. O. A. Derby[2] reports films of gold on limonite, from Brazil. A. Liversidge [3] found it in recent pyrite, which formed on twigs in a hot spring near Lake Taupo, New Zealand. J. C. Newbery [4] mentions gold in a manganiferous iron ore coating quartz pebbles, the quartz itself being free from gold. In the sinter of Steamboat Springs, Nevada, G. F. Becker [5] found both gold and silver; 3,403 grams of sinter gave 0.0034 of gold and 0.0012 of silver. Gold is also reported by J. M. Maclaren [6] in the siliceous sinter of the hot springs at Whakarewarewa, New Zealand. R. Brauns [7] has described gold as a cement joining fragments of quartz. The specimen of cinnabar from a fissure in Colusa County, California, mentioned by J. A. Phillips,[8] which was covered by a later deposit of gold, is also suggestive. According to R. W. Stone,[9] the coal of Cambria, Wyoming, contains appreciable quantities of gold. All of these occurrences are best interpreted on the assumption that the gold was precipitated from solution; and, indeed, they can hardly be explained otherwise. The fact that gold actually exists in natural solutions has already been shown in regard to sea water and certain mine waters, and J. B. Harrison [10] has detected it in the water of Omai Creek, British Guiana. He has also found gold in the ash of ironwood; in confirmation of an earlier observation by E. E. Lungwitz. The occurrence of gold as a volcanic emanation has also been reported by W. H. Goodchild.[11] The crater of the volcano La Sufral, in Colombia, is periodically filled by jets of steam from which sulphur is deposited; and that sulphur contains an easily determinable proportion of gold. This mode of occurrence seems so far to be unique.

The natural solvents of gold appear to be numerous—that is, if the recorded experiments are all trustworthy. G. Bischof [12] found that gold was held in solution by potassium silicate, and Liversidge [13] was able to dissolve the metal by digesting it with either potassium or sodium silicate under a pressure of 90 pounds to the square inch.

---

1 Neues Jahrb., 1854, p. 324.
2 Am. Jour. Sci., 3d ser., vol. 28, 1884, p. 440.
3 Jour. Roy. Soc. New South Wales, vol. 11, 1877, p. 262.
4 Trans. Roy. Soc. Victoria, vol. 9, 1868, p. 52.
5 Mon. U. S. Geol. Survey, vol. 13, 1888, p. 344.
6 Geol. Mag., 1906, p. 511.
7 Chemische Mineralogie, 1896, p. 406.
8 Quart. Jour. Geol. Soc., vol. 35, 1879, p. 390. On the natural associations of gold, see F. C. Lincoln, Econ. Geology, vol. 6, 1911, p. 247.
9 Bull. U. S. Survey No. 499, 1912, p. 63.
10 Geology of the gold fields of British Guiana, 1908, p. 209.
11 Mining Mag., vol. 19, 1918, p. 191.
12 Lehrbuch der chemischen und physikalischen Geologie, 2d ed., vol. 3, p. 843.
13 Proc. Roy. Soc. New South Wales, vol. 27, 1893, p. 303.

C. Doelter [1] claims that gold is perceptibly soluble in a 10 per cent sodium-carbonate solution, and also in a mixture of sodium silicate and bicarbonate. Solutions of alkaline sulphides have been found by several authorities, notably by W. Skey,[2] T. Egleston,[3] G. F. Becker,[4] and A. Liversidge,[5] to be effective solvents of gold; and Skey reports that even hydrogen sulphide attacks the metal perceptibly. All of these solvents occur in natural waters.

Solutions of ferric salts are also capable, under proper conditions, of dissolving gold. According to H. Wurtz,[6] ferric sulphate and ferric chloride are both effective. P. C. McIlhiney [7] found that the chloride acted on the metal only in presence of oxygen, which serves to render the ferric salt an efficient carrier of chlorine. Some experiments by H. N. Stokes [8] in the laboratory of the United States Geological Survey, showed that ferric chloride and also cupric chloride dissolve gold easily at 200°. The reactions are reversible, and gold is redeposited on cooling. Ferric sulphate, according to Stokes, does not dissolve gold unless chlorides are also present. Perhaps the pseudomorph of gold after botryogen, a basic sulphate of iron, described by W. D. Campbell,[9] may have originated from some solution in ferric salts.

F. P. Dewey [10] has found that finely divided gold is perceptibly soluble in nitric acid, but that observation has little bearing upon its natural solution. W. J. McCaughey [11] has reported its solubility in hydrochloric acid solutions of iron alum and cupric chloride. With rising temperature the solubility increases rapidly. N. Awerkiew [12] finds that gold is also dissolved by hydrochloric acid in presence of organic matter.

The usual laboratory solvent for gold, aqua regia, owes its efficiency to the liberation of free chlorine. T. Egleston [13] asserts that traces of nitrates with chlorides in natural waters can slowly dissolve the metal. J. R. Don[14] found that weak hydrochloric acid, 1 part in 1,250 of water, in presence of manganese dioxide, would take gold into solution. R. Pearce [15] heated gold and a solution containing 40

[1] Min. pet. Mitt., vol. 11, 1890, p. 328.
[2] Trans. New Zealand Inst., vol. 3, 1870, p. 216; vol. 5, 1872, p. 382.
[3] Trans. Am. Inst. Min. Eng., vol. 9, 1880–81, p. 639.
[4] Am. Jour. Sci., 3d ser., vol. 33, 1887, p. 207.
[5] Proc. Roy. Soc. New South Wales, vol. 27, 1893, p. 303.
[6] Am. Jour. Sci., 2d ser., vol. 26, 1858, p. 51.
[7] Idem, 4th ser., vol. 2, 1896, p. 293.
[8] Econ. Geology, vol. 1, 1906, p. 650.
[9] Trans. New Zealand Inst., vol. 14, 1881, p. 457. Campbell's observations need to be verified. The specimen was found in the Thames gold field, New Zealand.
[10] Jour. Am. Chem. Soc., vol. 32, 1910, p. 318.
[11] Idem, vol. 31, 1909, p. 1261.
[12] Zeitschr. anorg. Chemie, vol. 61, 1909, p. 1.
[13] Trans. Am. Inst. Min. Eng., vol. 8, 1879–80, p. 454.
[14] Idem, vol. 27, 1897, p. 564. According to Don, ferric salts are not effective solvents for gold.
[15] Idem, vol. 22, 1893, p. 739.

grains of common salt to the gallon, with a few drops of sulphuric acid and some manganese dioxide, and obtained partial solution. T. A. Rickard [1] treated a rich Cripple Creek ore, which contained manganic oxides, with a solution of ferric sulphate, sodium chloride, and a little sulphuric acid, and practically all of the gold dissolved. On immersing in this solution a fragment of black, carbonaceous shale, the gold was reprecipitated. How far solutions of this kind can be produced in nature is uncertain; but the extreme dilution of the solvents may be offset by their prolonged action. The laboratory processes all tend to accelerate the reactions. V. Lenher's observation,[2] that strong sulphuric acid, in presence of oxidizing agents, such as the dioxides of manganese and lead, dissolves gold, is probably not applicable to the discussion of natural phenomena. W. H. Emmons,[3] however, from a study of the experiments already cited, and also of the association of manganese oxides with gold in nature, has shown that the manganese plays an important part in the formation of auriferous deposits. Its effect is due to its interaction with acid solutions of chlorides, with which it generates chlorine; chlorine being the actual solvent of gold. In the presence of alkaline solutions, or of calcite, free chlorine can not appear, and the manganese oxides become inoperative.[4] Calcite, however, and also magnesite have been shown by V. Lenher [5] to be effective precipitants of gold.

The experiment by Rickard just cited, is especially suggestive as illustrating the ease with which gold is redeposited from its solutions. So far is gold is concerned, the reducing agents are numberless, and many of them occur in nature. Organic matter of almost any kind will precipitate gold, and such matter is rarely, if ever, absent from the soil. Gold, therefore, although it may enter into solution, is not likely to be carried very far. On mere contact with ordinary soils it would be at once precipitated.[6]

Gold is also thrown out of solution by ferrous salts, by other metals, and by many sulphides, especially by pyrite and galena.[7] According to Skey one part of pyrite will precipitate over eight parts of gold. The sulphides of copper, zinc, tin, molybdenum, mercury, silver, bismuth, antimony, and arsenic, and several arsenides, all act in the same way. So, too, does tellurium, according to V. Lenher,[1] and also the so-called tellurides of gold. If the latter were definite compounds, they could hardly behave as precipitants for one of their constituent elements.

---

[1] Trans. Am. Inst. Min. Eng., vol. 26, 1896, p. 978.

[2] Jour. Am. Chem. Soc., vol. 26, 1904, p. 550.

[3] Bull. Am. Inst. Min. Eng., 1910, p. 767, and Jour. Geology, vol. 19, 1911, p. 15. See also A. D. Brokaw, Jour. Geology, vol. 18, 1910, 10p. 32, 1.

[4] See F. T. Eddingfield, Philippine Jour. Sci., vol. 8A, 1913, p. 125; Econ. Geology, vol. 8, 1913, p. 498.

[5] Econ. Geology, vol. 13, 1908, p. 161. See also H. H. Morris, Jour. Am. Chem. Soc., vol. 40, 1918, p. 917.

[6] On the relations of vegetation to the deposition of gold see E. E. Lungwitz, Zeitschr. prakt. Geologie, 1900, pp. 71, 213.

[7] See C. Wilkinson, Trans. Roy. Soc. Victoria, vol. 8, 1866, p. 11; W. Skey, Trans. New Zealand Inst., vol. 3, 1870, p. 225; vol. 5, 1872, pp. 370, 382; A. Liversidge, Proc. Roy. Soc. New South Wales, vol. 27, 1893, p. 303; C. Palmer and E. S. Bastin, Econ. Geology, vol. 8, 1913, p. 140; F. F. Grout, idem, p. 407; A. D. Brokaw, Econ. Geology, vol. 21, 1913, p. 251.

[1] Jour. Am. Chem. Soc., vol. 24, 1907, p. 355. See also R. D. Hall and V. Lenher, idem, p. 919. Later papers by Lenher are in Econ. Geology, vol. 7, 1912, p. 744; vol. 9, 1914, p. 523. On the relations of colloidal gold to ore deposition, see Bastin, Jour. Washington Acad. Sci., vol. 5, 1915, p. 64.

Clarke was primarily a chemist, but he also had a profound knowledge of geological processes. It was said that he had the natural ability to consider fully and appreciatively the views of others, at times not in accord with his own views (he was a magmatist), and to present what seemed to him to be the truest interpretation of geochemical phenomena. Anyone who has read his great work on geochemistry will recognize this ability almost immediately; it is readily apparent in the previous excerpt.

Contemporaries of Clarke in the USSR include particularly the geochemist V. I. Vernadsky (1863-1945) and his disciple, A. E. Fersman (1883-1945). Vernadsky, the "father of biogeochemistry and radiogeology," published hundreds of papers on mineralogy, radiogeology, biogeochemistry, and geochemistry, many reprinted in his *Selected Works,* comprising five volumes. Gold is mentioned at numerous places in these volumes, usually in context with other elements or in discussions of geochemical processes, all summarized later by A. E. Fersman.

A. E. Fersman was a pupil of Vernadsky at Moscow University and collaborated with him for more than forty years. Fersman wrote extensively on all aspects of geochemistry, his great work being *Geochemistry,* published in four volumes in 1934, 1937, and 1939. Fersman stressed the paramount importance of field studies in all of his works. All of the concepts and ideas he developed came from a long study of the field observations carefully integrated with the known physico-chemical facts. In the fourth volume of his great work dealing with the geochemistry of individual elements, he summarized the known facts about the geochemistry of gold (Paper 8-3).

8-3: GOLD (Au—At. Wt. 79)

A. E. Fersman

This article was translated expressly for this volume by the Translation Bureau, Department of the Secretary of State, Government of Canada, from *Geochemistry,* vol. 4, Leningrad, 1939, pp. 262-271.

The geochemical properties of gold are poorly known. We have only a series of empirical facts that are largely uncoordinated with the common chemical properties of the metal. Only a comprehensive study of the chemistry and geochemistry of the element can provide us with the correct basis for understanding the migration characteristics of gold and the formation of its deposits.

Some fundamental features of gold follow as an introduction.

Gold has two valencies, 1 and 3. Trivalent gold is the more stable of the two, but the element in both valence states is capable of forming complex ions, particularly the trivalent state. Gold salts are characterized by their easy decomposition, resulting in the release of metallic gold. Because of this fact free ions of gold can exist in aqueous solutions only in minute, often indeterminable amounts. On the other hand, the complex compounds of gold, e.g., cyanides, are exceptionally stable in aqueous solutions.

Since the entire industry of recovering gold from its deposits is based on stable complex compounds of the element, it would seem that the geochemical migration of gold in nature depends similarily on the formation of complex salts, most probably those of the alkalies, sulphur, and perhaps also silicic and sulphuric acids, etc. On the whole gold is markedly siderophile and has also a chalcophile character (V. M. Goldschmidt).

The monovalent radius of gold is not well known, and the value of 1.27 Å is questionable. The marked polarization of this ion is responsible, for its EK being almost twice as much as that of potassium (0.65 as against 0.36), despite the fact that the radius of the monovalent gold ion is close to that of potassium. However, these values, obtained indirectly by calculating the lattice energies of only two compounds, are probably not valid, and a further study of the energy of other gold compounds is imperative. (EK is an empirical constant calculated for each element representing the contribution of a particular element to the lattice energy of its compounds (See Fersman, A. E., 1935. The EK system; Compt. Rend. Acad. Sci. U.S.S.R., v. 2, p. 558-566)).

The atomic radius of gold is 1.44 Å, practically the same as that of silver. Because of this both metals form alloys at all ratios, a fact well known in nature.

According to N. Bohr the structure of the gold atom is typically odd number, i.e. it has one electron in the P shell; hence the valency of 1 is completely comprehensible, and this produces a very stable type of ion. Less comprehensible is its tri-valency.

In addition to the indicated capacity of forming stable alloys with silver (and partly copper), another property having great practical significance should be stressed. This is the specific gravity (19.3), which depending on the content of silver varies between 15.6 and 19.3 (the pure metal). Only (platinum), osmiridium, iridosmium, and pure iridium have higher specific gravities.

The specific gravity, stability of the metallic state, and chemical 'nobility' of gold are the causes of its characteristic accumulation in placers and its great significance in the supergene zone.

It would be, however, an error to think that the high specific gravity determines the mechanical stability of this metal. On the contrary, the combined properties of its metallic lattice, the strong polarization of its ions in compounds, its exceptional softness, and its low valency are the reasons why gold is mechanically dispersed, and why the metal readily yields colloidal solutions.

The mineralogy of gold has been poorly studied, and likewise there is little basic data on the details of the distribution of gold in the earth's crust, despite the fact that Buffon at the end of 18th century called gold the *ubiquitous* element despite its rarity and value.

The principal gold minerals are:

1. Native gold—pure and in solid solutions with Ag, Cu or Bi (electrum, cupreous gold, bismuthaurite).
2. Gold in solid solution and double-compounds with the platinum group metals (aurosmiridium, porpezite, rhodite, platinum-gold).
3. Tellurides—calaverite, krennerite, sylvanite, hessite, petzite, nagyagite.
4. Selenides—rather rare and poorly known.
5. Gold—concentrated in sulphides, particularly in pyrite and arsenopyrite. The nature of gold in the lattices of these minerals is unknown.

The clarkes of gold are not precisely known; commonly the figures given are: weight clarke, 5×10^{-6}; atomic clarke, 4×10^{-7}; weight clarke in meteorites, 2.5×10^{-6} (all in %).

V.M. Goldschmidt found that Co-arsenides and especially Ni-arsenides of the hydrothermal phases I-K contain from 10^{-5} to 10^{-2}(%) gold.

The gold content of placers is 1×10^{-4}%. (Gold is analytically easy to detect in quantities as low as 1×10^{-5}% and less.)

The geochemistry of gold: At present a complete lack of data prevails in the literature on the problem of the geochemical migration of gold. While the two schools of American scientists led by Spurr and Lindgren are sparring with each other in vain, no advance is made on the problem because a true geochemical analysis of gold deposits is still lacking. Nevertheless, some single precise observations, especially those of F. Buschendorf and N. Gornostaev, provide some valuable data and allow us to summarize to a first approximation the geochemistry of gold, and to plot its fundamental geochemical diagram (Fig. 1).

I stress the significance of this diagram, not only because it offers a scientifically justified theoretical basis for understanding the manner of the migration of gold, but also because the conclusions drawn from it have great significance for the prospecting and development of our primary gold deposits.

I pass now to an analysis of the fundamental conclusions that can be drawn from the geochemical diagram. The geochemical analysis follows the usual scheme, taking as its basis the study of the various derivatives of granodioritic magma. These are indicated by the letters A, B, C, etc., which mark the definite geochemical geophases during the cooling of a granitic magma; the boundaries between C and D and G and H are marked respectively by the upper critical point, corresponding to about 600°C (for a granodioritic magma) and by the critical point of water, at about 400°C. The granitic magma upon cooling yields a pegmatitic residue practically free of gold, and a series of high-temperature gaseous emanations (I, II, III), which become spatially separated from the granite (or from its effusive facies). Upon cooling these give rise to distillates, four of which are characteristically important in the formation of gold deposits.

I Differentiates from effusive rocks containing Te, Se and Ag.

II High temperature quartz veins with sulphides whose crystallization begins near 400-450°C and forms typical gold deposits.

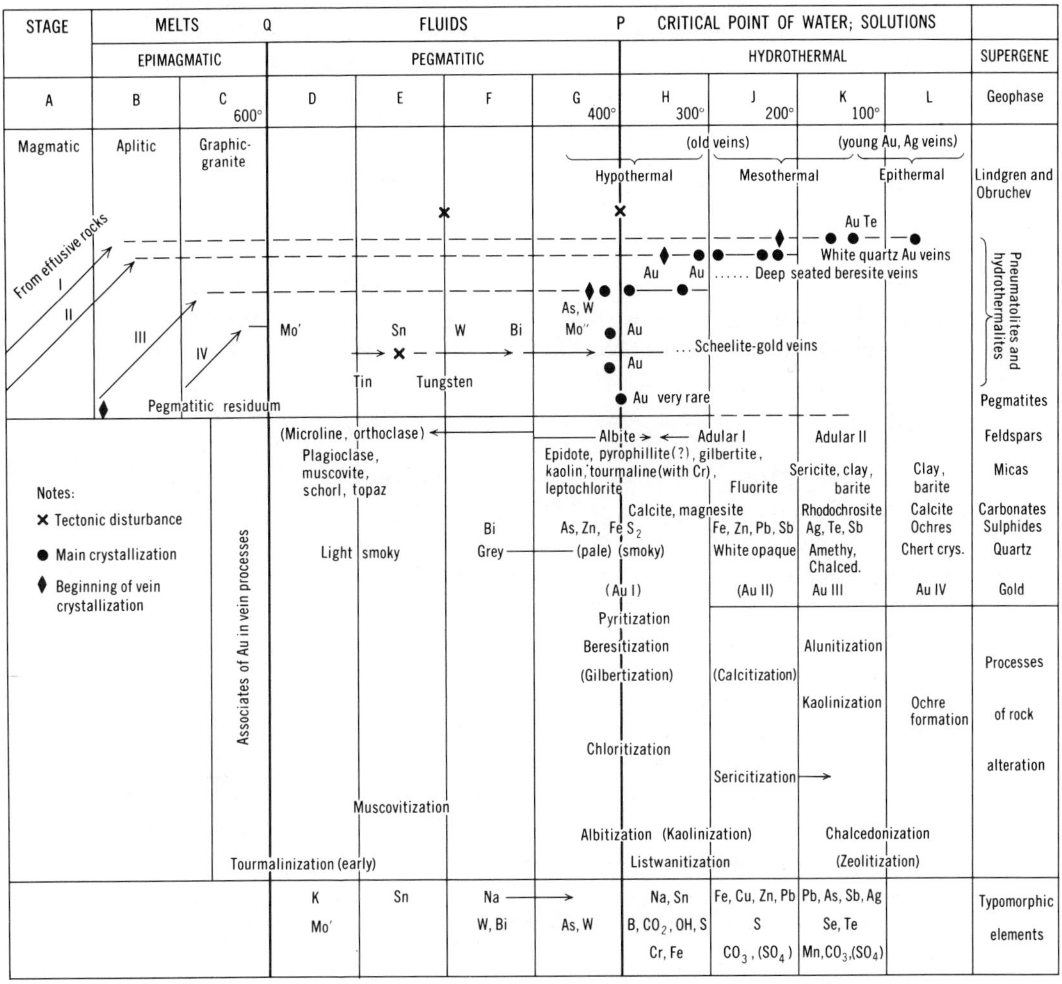

Figure 1. Geochemical diagram of gold

III Apatite-beresite veins; these can be considered as the spatial fraction of geophase B; upon further cooling quartz veins are formed in the beresites and at the same time the entire complex (of rock) is gilbertized. This is the type that Spurr and Vogt are inclined to consider as very high temperature deposits. In fact, this is a high temperature type but only at the moment of its separation, since according to the processes of formation and the time of precipitation (crystallization) it actually belongs to another type.

IV In the sequence of tin and tungsten distillates (normally following each other) we recognize a specific scheelite-gold vein type.

Bramall and Dowie (1936) gives the following weight clarkes for gold:

in igneous rocks	1×10^{-5}-0.1 g/t
in pegmatites	$\sim 5 \times 10^{-5}$-0.5 g/t
in acid migmatites	1.5×10^{-4}-1.5 g/t

When we examine the initial time of gold deposition, we see that on the whole it corresponds rather definitely to the boundary of geophases G-H, i.e. exactly to the lower critical point of water (approx. 400°C). However, maximum gold deposition belongs to the later periods of geophase I and even K. There ensues from this position of gold in the diagram a series of individual aspects in its paragenesis and in its vein types. The first generation of gold is located at the point of very mobile equilibrium, located at the boundary of fluidal (gaseous) and hydrothermal systems; therefore, gold shares the features of both systems, or very often is of intermediate nature. The rather definite characteristics of veins of this generation indicate that the "closed" nature of this system was more or less preserved, and that it was little altered by the later effects of other solutions and wall rocks. The position at the boundary of two physico-chemical stages is also characteristic of the concentration of many highly dispersed minerals like albite, muscovite, tourmaline needles often containing some chromium, etc. Tourmaline-bearing copper veins are partly formed at this stage.

The paragenesis of gold, sulphides, green (partly Cr-bearing) micas, and adularia (feldspar) is very characteristic; the field to the left of gold in the diagram (i.e. left of the boundary between geophases G and H) is occupied by minerals of bismuth, tungsten, and arsenic (e.g., aikinite at Beresovsk and its replacement of native gold); contemporaneous with gold are gilbertite, leptochlorite, tourmaline, iron sulphides, feldspar, (rare) carbonates, and quartz; post gold minerals are quartz, zinc and lead sulphides, and many others of geophases K and L. Quartz is particularly characteristic of gold paragenesis; the specific properties of gold-quartz are well known to gold miners and prospectors and have been scientifically treated for the first time by Buschendorf and his school for gold prospecting in the Soviet Union.

In my scheme, quartz definitely preserves its properties in the most diverse types of ore deposits; in those of gold it is mainly quartz (III), characterized at the initial stages either by a greasy, grey hue, or by small crystals, frequently displaying a specific smoky hue (e.g., Berezovsk in the Urals, Dauphine in the French Alps).

The constant and characteristic association of gold with quartz and chalcedony is one of the typical features of the geochemistry of gold, and this fact induced Lindgren in 1937 to conceive the intriguing idea concerning the joint migration of colloidal solutions of silica and gold salts. This point of view has been confirmed in many recent publications (e.g., F. Ahlfeld). The coagulation of SiO_2 (and simultaneously of gold) is stimulated by various factors, in particular by lowering the temperature, a sharp change in pH toward neutrality, and the presence of electrolytes acting as coagulators. Frondel's (1938) experiments have shown that at the beginning of the process of emanation gold is carried in acid emanations as an electrolyte; with decrease in temperature gold is enveloped by SiO_2-sols that act as a protective colloid.

Furthermore, it is interesting to note that gold also dissolves in solutions containing Cl_2 or CN, in solutions of NaHS (but not Na_2S), and forms the sulphoaurate NaAuS. O. E. Zvyagintsev emphasizes the solubility of Au in selenic and probably in telluric acids, and perhaps also in humic substances (Freize, Kozhevnikov). From a geochemical viewpoint there are the interesting observations of O. E. Zvyagintsev, indicating that it is possible to establish a rather sharp difference between the parageneses of the Baleisky and Kluchevsky deposits; all elements in the first deposit have a VEK not exceeding 1-1.1; in the second, however, elements are known with VEKs ranging from 0.25 to 2.28 (e.g., ruthenium). (VEK—*See* note above describing the EK system.)

I shall not dwell upon the entire geochemical nature of gold, but mention only the most characteristic and most definitive features. On the whole the geochemistry of gold must be reassessed after the vast amount of material on gold veins has been studied anew, especially that available on the gold deposits of the Soviet Union (Berezovsk, Kochkar, Yenisey Taiga, Aldan River, Kolyma River, and Transbaikalia). Another observation is also interesting, namely, the more to the right lies the gold complex in the diagram (Fig. 1), the lower is the temperature of the first precipitation of gold, the less its purity, and the more its silver content. This observation is in full agreement with those made on other minerals of sulphide veins (e.g., fahlores, etc.). However, in every given complex of veins with several generations of gold there is an inverse regularly, namely, that the gold becomes increasingly purer with falling temperature. This phenomenon of self-purification is well known in the course of the formation of pegmatite veins; it is also characteristic of gold especially when the analysis of gold nuggets from placers is considered.

Thus, we can conclude that:

1. Prospecting for and development of primary gold deposits can be carried out only effectively on the basis of the geochemical study of the processes forming their veins.
2. For this purpose a planned and systematic study of gold veins in the Soviet Union from the point of view of modern geochemical and mineralogical concepts is required.
3. In most vein-forming processes gold is associated with various volatile emanations from granitic magmas, but the element crystallizes comparatively late in three definite generations.
4. The initial point at which gold is precipitated during the formation of quartz veins lies near the critical point of water, i.e. about 400°C, whereas the maximum precipitation of gold occurs apparently at lower temperatures (of the order of 200°C, or lower).

Analysis of The Geochemical Diagram: I have given the corrected and my supplemented geochemical diagram (Fig. 1) for the year 1931. Many new schemes have been suggested in recent years, which on the whole confirm this diagram.

Petrulian (1931) has given for the first time a precise, and at the same time, a true geochemical description of gold in Transylvania, particularly from Verespatak. The marked telescoping habit of gold lies on the diagram precisely in geophase I; sphalerite accompanies gold during the initial stage and galena at the end of the telescoping; sulphosalts appear much later, after MnS and $MnCO_3$. The typical gold II variety fits my scheme. Arsenopyrite appears before gold.

Buschendorf (1926) has drawn several schemes for the Fichtelgebirge in Germany and for the gold veins of S. America; I have summarized them in the diagram in Fig. 2. The interpretation of Buschendorf's interesting and precise descriptions leads to the conclusion that early gold is associated with geophase H, but begins to be precipitated at geophase G and continues to do so during geophase I. This paragenesis is in full agreement with my scheme.

Gukovksy records two types of gold in the deposits of the Kuznetski Ala Tau: a low temperature type corresponding to geophases H-I, and a high temperature type (I) appearing together with bismuth and pyrrhotite, but before chalcopyrite and especially galena. According to Gukovsky the last type of gold is formed under high pressure.

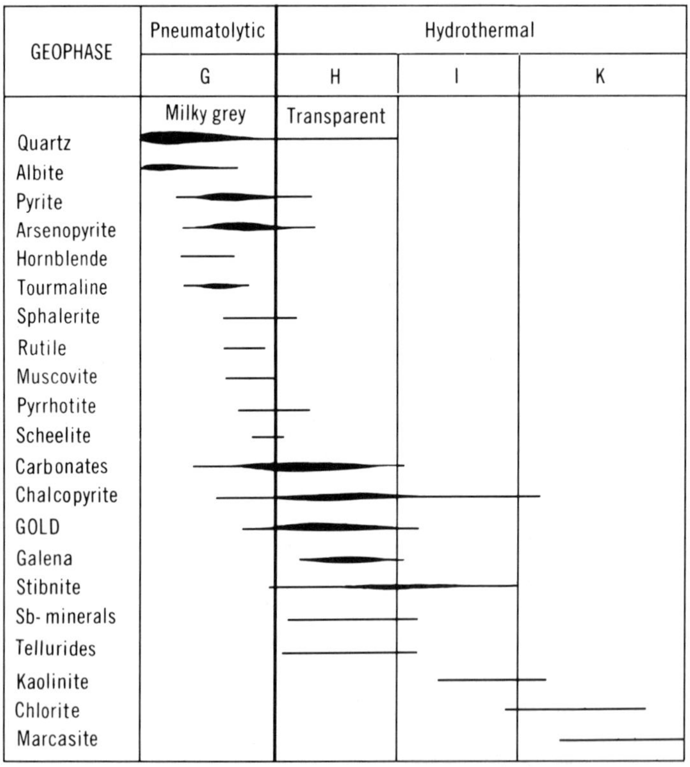

Figure 2. Geochemical diagram (After F. Buschendorf, 1926).

Having analysed the gold deposits of western Siberia, Gukovsky draws the conclusion in his recent work in 1938 that there are two fundamental types of precipitated gold, one corresponding to geophases G-H and the other I-K.

In the first case the gold deposits are associated with small intrusions; the crystallization is quite regular and consecutive with milky quartz and bismuth. The gold is of high fineness.

In the second case the gold deposits are associated with large intrusions with distinct borders, variable composition of solutions, superposition of zones, and low fineness gold; the crystallization of gold is later than galena and is generally closely associated with sulphides. The quartz is fine-grained, grey, and accompanied by calcite. This scheme is in full agreement with my geochemical diagram.

Gukovsky's work merits attention as regards the association of precipitation processes with tectonics, and likewise for stressing the complexity of the problem (viz. the association of gold in geophases G-H, with scheelite, wolframite, and cassiterite).

S. A. Yushko (1936), having analysed the association of gold in sulphide deposits of the Urals, finds only late gold of geophase K, appearing after chalcopyrite and galena.

Newhouse (1936) gives a detailed description of the gold veins of Nova Scotia and comes to the conclusion that around the granitic aureoles there are several zones carrying gold, the maximum concentration of the element being observed at a distance of 3-5 km from the granite.

The author (Newhouse) also notes the presence of gold veins of the high temperature geophases, having partly the nature of pegmatites and containing much molybdenite, andalusite, etc. Accompanying the veins of the high-temperature geophases (D-G) there are also veins of geophase H and finally low-temperature veins of geophase K (with stibnite and native antimony). Veins of geophases I-K yielded more than 55 percent of the total gold produced in Nova Scotia.

F. Ahlfeld (1937), having analyzed various types of Bolivian gold deposits, distinguishes six vein types of geophases F-K. Each type is characterized by its own paragenesis in each of the above indicated geophases. An interesting fact is the gradual increase in the gold content as compared with that of silver upon passing from the hotter to the cooler geophases (with Se and Te). According to the author gold migrates in association with colloidal silica.

Natarov (1937) has presented an interesting study of the Ayakhtinsky gold deposit, integrating the geochemical processes with tectonics and the geological history, and has compiled a detailed geochemical diagram. Gold in this deposit is the late type (III).

Valuable data have been accumulated in the works of NIGRI on gold and gold prospecting.

The chief value of these data is the coordination of the geochemical succession with the tectonic and geological history of an auriferous district. The following conclusions drawn from the studies, e.g., with respect to the Sovietski deposit in the N. Yenisey Taiga, can be applied, *mutatis mutandi* to the majority of gold deposits (especially of the Pacific type):

1. The formation not only of a deposit but also of each individual generation of minerals lasted a long time.

2. The majority of mineral generations in the veins, including the ore minerals, were formed by metasomatism (replacement) of preceding generations by subsequent ones. The mechanism for this kind of process is particularly interesting when both generations—the replaced and the replacing—belong to the same mineral species (e.g., quartz).
3. The growth of all crystalline substances, with the exception, perhaps, of the last generation of chalcedonic quartz is accomplished by the precipitation of substances from molecular rather than colloidal solutions.
4. The formation of a mineral deposit takes place over a considerable temperature range, the most important events being related closely to the extreme temperatures. Thus, the main mass of quartz filling the veins crystallized during the pneumatolytic phase and belongs to the pneumatolytes; the gold, on the other hand, was deposited from the coolest hydrothermal solutions.
5. The analysis of the literature leads one to the conclusion that gold, in most of its deposits, is generally the last of the ore minerals to be deposited appearing after galena, and before mercury and antimony sulphides; on the whole gold is a low-temperature hydrothermal mineral. Probably the zone of high temperature in the geochemical scheme of Emmons (1937) should be abolished, and gold should be assigned its proper place in the fourth zone, after lead and silver and before mercury and antimony. (Note that the typical Sovietski ore deposit falls according to the usual scheme into the 11th zone of Emmons.)
6. The first steps in applying the methods under development for recognizing the various generations of quartz in gold deposits of diverse types have already revealed, firstly, the great fruitfulness of this method, and secondly, the exceptional diversity of vein-quartz, which forms both the long and the short series generations while at the same time differing somewhat in its overall characteristics. At the same time this whole diversity reveals a clear regularity, which is closely associated with the general geological conditions. The preservation of a certain parallelism with the generations of pegmatite quartz, as recognized by A. E. Fersman, is interesting.

To this we should add:

The characteristics of ore deposits change depending largely on the depth of a granite body; when the latter occurs close to the surface (hypabyssal type) the phenomenon of strong telescoping and early deposition of gold is observed.

The early gold mentioned above contains more silver than later generations, which is in full agreement with the idea of self purification (i.e., late generations of gold in the same deposit contain progressively less silver).

Of the above statements only the fifth requires modification, since there doubtlessly exists a higher temperature gold phase; moreover, the graduation of the diagram is debatable.

Finally, we give below a summary of the genetic conditions for the precipitation of gold as outlined by Emmons (1937); he distinguishes 8 types.

	Geophases
1. Magmatic segregations	
2. Pegmatites	
3. Contact-metamorphic, (dispersed)	
4. Hypothermal veins	H
5. Mesothermal veins	I
6. Epithermal veins	K
7. Precipitates from cold water	L
8. Placers (mechanical)	L

From this it can be seen that the last five types completely fit into my scheme (on supergene gold—see below); however, the position of the first three types remains questionable. The analysis of examples, given by Emmons, reveals that the first three types are really deposits associated with or in the indicated formations, but the time of precipitation of the gold and the ratio of gold to other minerals in these deposits remains a problem.

Thus, my scheme (Fig. 1) would seem to apply to the majority of gold deposits.

The relation of gold with magmas: On the whole, gold is bound to acid granitic magmas, and in particular the metal is associated with SiO_2. The most typical gold carriers are the granodiorite and granite facies (less often the diorite facies), and concentrations of gold are associated with the apical parts, cupolas, and tops of intrusions. Emmons (1937) states, that as a rule the following zonation can be outlined around a batholith:

1. No large deposits occur farther than 2 km from a granite batholith;
2. Individual deposits occur beyond 1 km distance from the batholith;
3. Near a granite batholith many deposits may occur; and
4. The richest deposits are associated with individual cupolas and small intrusions.

Gold migrates in the volatile components of a magma, and the apical and lateral parts of a batholith's summit are the richest in gold, since gas emanations migrate into these sites from the batholith. Marked enrichment of gold is observed (similar to that of tin) in chimneys through which the volatile components drain from a batholith. From this it follows that gold is confined mainly to the upper erosional profile of magmatic cupolas (according to Emmons' terminology, the acrobatholithic part).

The association of gold with sodic granites, granodiorites or albitites is petrographically and geochemically important (contrary to that of tin which is most often associated with potassic granites). This phenomenon has not been explained, but the association with sodium is unequivocal.

Supergene migration: The solubility of gold in geochemical environments results from its general properties, but these are poorly known and conjectural; the only indisputable processes occur in gossans, where gold is either dissolved by mercury (thus forming amalgams), or NaCl reacts with MnO_2, releasing Cl_2, which then dissolves the gold. Hence the specific mobility of gold in NaCl-rich deserts, and in deposits where $MnCO_3$ is oxidizing. I offer the hypothesis that gold may be transferred under subtropical conditions in solutions of salts containing natural CN and CNS; the formation of these salts has been proven by Schoep in the Congo and is probably widespread. In general, the dissolution of gold demands the presence of chlorides and peroxides (MnO_2, Co_2O_3 or Ni_2O_3). The solubility in sulphate solutions has been refuted.

Neither the pH nor the oxidation potential of gold precipitation are known. The common geochemical association of gold with SiO_2 has still not been explained; the character of the colloidal dispersion of gold is remarkable (up to 1-3 microns).

Gold migrates in placers and in the supergene zone, but on the whole less than silver, which for example is removed from the surface parts of gold nuggets, rendering them higher in fineness. The opinion that larger gold nuggets were formed in situ in placers is not very credible; nevertheless, the partial recrystallization of such nuggets is indisputable. The phenomenon of rejuvenation of gold observed by prospectors in old dumps or placers is explained by the transformation of the elusive colloidal gold into microcrystalline gold which can be recovered by repanning the dumps, placers, etc. When gold tellurides are oxidized the tellurium forms iron tellurite or the mineral tellurite, and gold is split out as a dark-brown, highly comminuted powder.

Practical Suggestions: I do not have any personal experience in investigating gold deposits, and hence I can only outline some general practical suggestions.

On the whole, prospecting for gold and the exploration of primary gold deposits should be conducted on the basis of the proposed geochemical diagram (Fig. 1), whose parts converge and overlap each other as a result of telescoping when the parental granitic batholith solidified at a shallow depth, and when the temperature decreased rapidly and in paroxysms, simultaneously with violent tectonic faulting, fracturing, etc. The timing and course of the processes of gold deposition were changeable depending on the external pressure.

Emmons (1937) offers in his great monograph on gold much practical advice which has been deduced from the analysis of several hundreds of deposits throughout the world. Repeated below are some of his fundamental conclusions:

Gold fields extend over large regions of the continents, including those in the great Pacific ring (of Tertiary and Post-Mesozoic age), and a number of great fields occur within the old Precambrian shields (Canadian, South African, Siberian, and Australian). F. Blondel (1936) remarks that 80% of the gold deposits of the world are associated with old shields, including 14% situated within the shields, 14% in their metamorphic series, and 52% in deposits of secondary derivation (e.g., quartz-pebble conglomerates of the Witwatersrand, S. Africa). Rocks of Cimmerian and Alpine ages account for 14% and those of Hercynian and Alpine for 6% of gold production as calculated from the yearly production records, but not including the reserves.

Emmons (1937) indicates further that the most valuable gold deposits are associated with cupolas and intrusions of granites and granodiorites and partly with explosion pipes and chimneys. He remarks on the associations of gold with yellow quartz rich in sulphides, this fact being one of the surest prospecting indicators.

Various accessory minerals occurring with gold in placers serve as guides in the search for auriferous deposits; this fact merits special attention. Thus, two types of gold deposits (primary) can be discovered by these guides:

1. The gold-scheelite type with W, Sn, Mo, Bi, Co (?), B (schorl),
2. Gold-telluride type with Te (Se), Bi (stannite).

The great importance of gold deposits in the Soviet Union, and the favourable prospect of discovering new gold fields, provide an impetus for Soviet geologists and geochemists to implement an extensive study of these deposits.

Geologists and petrographers have the task of studying the type and tectonics of gold-bearing granites, and the geochemists should concentrate their studies on their mineralogy and geochemical diagrams. The processes of supergene migration of gold should be particularly studied.

Valuable results can be obtained by combining tectonic, petrographic, and geochemical data, not only in prospecting for new ore deposits but also during exploration for deep deposits.

S. S. Smirnov (1936) presents a series of interesting practical remarks on supergene gold.

MAJOR REFERENCES

[As given in the original text]

Bulynnikov, A. Ya. Auriferous Formations of the Kuznetsk Ala-Tau. A Symposium, dedicated to Acad. Usov, Tomsk, 1933.
Vernadsky, V. I. Descriptive Mineralogy, 1914, I, pp. 264-402.
Zvyagintsev, O. Ye. The Geochemistry of Gold. Symposium, Univ. of Phys. and Chem. Moscow, 1939, pp. 48-64.
Zvyagintsev, O. Ye, V. A. Volkova, and E. L. Pisarzevskaya. Chem. Res. on the Gold of the Balei Mine. Proceedings of the Acad. of Sci., 1938. Chem. Ser., No. 3, pp. 509-518.
Zhemchuzhny, S. F. Native Gold: Phys. and Chem. Invest. Transactions, Inst. II, 1922, Issue 1.
Kropachev, G. P. The Distribution of Gold in Nature. Sov. Gold Ind., 1935, No. 8, p. 46.
Monich, V. K. Geol. Outline of the Berikul'skiy Gold-Mining Area. Transactions of Gold Prospecting Enterprise, 1937, VII, pp. 82-83, 96, 107.
Natarov, V. N. The Geochem. of the Ayakhta Ore Deposit. Transactions, Gold Prospecting Enterprise, Moscow, 1937, IV, pp. 3-47.
Pedashenko, A. I. Ores of Ore Deposits: Genesis and Paragenesis. Materials of a Study of the Okhotsk-Kolyma Territory. 1936, Series I, IX, pp. 75-102.
Smirnov, S. S. The Oxidation Zone of Sulfide Deposits. Dept. of Sci. and Techn. Info., 1936, pp. 170-178.
Smirnov, A. A. The Minerals of the Native Group. Minerals of the USSR, 1938, I.
Fersman, A. Ye. The Geology of Gold. Proceed. Acad. Sci., USSR, 1931, pp. 199-204.
Yushko, S. A. The Mineralogic Form of Gold and its Associations in Pyritic Ores of the Urals. Proceed. Acad. Sci., USSR, 1936, No. 2-3, pp. 435-440 (contains an interesting geochem. diagram).
Brammall, A. and D. L. Dowie. Distribut. of Gold and Silver in Crist. Rocks. Miner. Magaz. 1936, XXIV, p. 260.
Gibson, C. S. Recent Investigat. in the Chemistry of Au. Nature. 1938, 142, p. 339.
Niggli, P. Ore Deposits of Magmatic Origin. 1929.

Gold Resources of the World (XV Int. Geol. Congr. 1928). Johannesburg. 1931, p. 1-457.

Buschendorf, F. Primärzone alter Goldquarzgänge. Zeit. f. pr. Geol. 1926, XXXIV, p. 1-16.

Buschendorf, F. Die primären Gold-Erze Ficht. Neues Jahrb. f. Min. 1930. BB.LXII.

Petrulian, N. Gisement aurifère de Rosia Montana (Verespatak), Ann. Inst. Geol. al Romaniei. XVI, 1931, p.1-43.

Dunn, E. Geology of Gold. London, 1929.

Idriese, I. L. Prospecting for Gold. Sydney, 1936.

Blondel, F. La géologie et les mines des vieilles platteformes. Paris. 1936, p. 194, 214.

Goldschmidt, V. M. und Cl. Peters. Zur Geochemie d. Edelmetalle. Nachr. Geselsch. d. Wiss. Gottingen. 1932, p. 377-401.

René van Aubel. Géochimie de l'or. Annuaire Soc. Géol. Belgique 1934, 57, 131-150.

Krusch. Die metallischen Rohstoffe. V. Gold. 1938.

Lindgren, W. (on colloidal solutions of gold). Trans. Amer. Inst. Min. Eng. 1937, p. 126, p. 356-376.

Emmons, W. H. Gold Deposits of the World. 1937, p. 1-562.

Berg, G. Vorkommen und Geochemie d. miner. Rohstoffe. 1929, p. 274-291.

Savage, E. M. Prospecting for Gold and Silver. 1936, p. 1-300.

Bibliography on gold, see Mineral Resources. 1937, XLV, p. 300-306.

Anderson, I. Essential Criteria in Examining Gold-Quartz Mines. Eng. Min. Journal. 1935, CXXXVI, p. 445-449.

Ahlfeld, F. Typen boliv. Goldlagerstätten. Centralbl. f. Min. 1937 (A), p. 240-255.

Lindgren, W. Success. of Min. and Temp. in Ore-deposits of Magmat. Affiliat. Amer. Inst. Min. Met. Engin., Mai 1936.

Hewett, D. F. Zonal Relation of Gold. Amer. Inst. Min. Met. Eng. Trans. 1931, 96, p. 305-346.

Ogryzlo, S. P. Econ. Geology. 1935, 30, p. 400.

Nemec. Gold in Zea-Mays. Ber. d. deut. Botan. Ges. 1935, 53, p. 560.

Newhouse, W. H. A Zonal Gold Mineralization in Nova Scotia. Econ. Geol. 1936, XXXI, p. 505-832.

Frondel, C. Stability of Colloid Gold under Hydrotherm. Conditions. Econ. Geol., 1938, XXXIII, p. 1220.

Fersman emphasized that the chief purpose of geochemistry is not simply the chemical analysis of earth materials, but also the investigation and study of the distribution of the elements in the various earth spheres. This approach is evident in his *Geochemistry* and particularly, as shown above in the translated passages, in his attempt to relate the occurrence of gold deposits to magmatic and hydrothermal systems.

Victor Moritz Goldschmidt (1888-1947) was one of the foremost geochemical investigators during the first part of the present century. Goldschmidt's contributions to geochemistry are many. He gave us the form of the phase rule that bears his name, and his work on crystal chemistry and on the sizes of the atoms and ions gave a better appreciation of why certain elements are in certain specific crystallographic combinations, and how one element can replace another in minerals. His detailed studies on the geochemistry of individual elements and his estimates of their abundance will long remain as a monument to him. All these contributions, and many others too numerous to mention here, appeared in a classic series of papers entitled *Geochemische Verteilungsgesetze der Element (1-9, 1923-38),* and in the seventh Hugo Muller lecture to the Chemical Society of London, entitled *The Principles of Distribution of Chemical Elements in Minerals and Rocks,* published in 1937. His magnum opus, *Geochemistry,* appeared after his death largely through the editorial and compilative labors of Dr. A. Muir, Dr. A. Kvalheim, and others who knew Goldschmidt well during his lifetime.

Two works on gold by Goldschmidt are notable; one, *Zur Geochemie der Edelmatalle,* written with Cl. Peters, appeared in 1932; the other in his *Geochemistry* is reproduced as Paper 8-4.

During the period 1920-1955 a number of compendia and summary papers on the geochemistry of gold appeared, including works by Mellor (1923), Aubel (1934), Zvyagintsev (1941), Rankama and Sahama (1950), Friedensburg (1953), and Gmelin (1950-1954). Others are mentioned in Boyle (1979). The compendia by Mellor, Gmelin, and Friedensburg give extensive data on both the chemistry and geochemistry of gold. The work by van Aubel is of interest and covers the distribution of gold in the earth spheres, the fate of the metal in magmatic processes, the localization of hypogene gold deposits, and the distribution of the metal in these deposits. The paper is largely geological in nature and provides an up-to-date geochemical-geological summary of gold deposits at the midthirties of this century.

(Text continues on page 127.)

8-4: GOLD

V. M. Goldschmidt

THE geochemistry of gold is only moderately well known. Adequate analytical methods are now available and some of the main problems may be posed, but a great number of data have still to be collected. The most suitable method for geochemical investigations on small amounts of gold, silver, and the platinum metals in rocks in minerals, as developed in the author's laboratory, consists in a combination of an assay process with a subsequent spectrographic investigation of the resulting metal alloy.

The average content of gold in the earth's crust is best estimated in relation to that of silver. All mineralogists agree that the proportion, by weight, of silver to gold, based on the average contents of these metals in important ores, cannot be less than 20:1. The estimate of 10:1 given by I. and W. Noddack (1934) for magmatic rocks is decidedly too much in favour of gold. Accepting the proportion 20:1, we may estimate the average amount of gold from the 0·02 p.p.m. Ag in magmatic rocks (see the chapter on silver), obtaining 0·001 p.p.m. or 1 mg. per ton of rock.

Gold has only a very slight tendency to enter into oxygen compounds, as clearly indicated in the thermo-chemical chapter in the first part of this book. The occurrence of gold in hydrothermal deposits in association with sulphides and related minerals might have suggested that this metal is rather chalcophil. The investigations by the author and Cl. Peters (1932 *b*) on gold in the nickel-iron and in the troilite (iron monosulphide) of meteorites clearly demonstrated that it is strongly siderophil, which could also have been deduced from chemical and thermochemical data. Gold collects very markedly in fused iron or iron alloys. If we take an average of the data by different authors we get the following figures:

	Au *p.p.m.*	
	Nickel-iron	*Troilite*
Goldschmidt and Peters (1932 *b*) .	5	1
I. and W. Noddack (1931–4) .	3·4	0·3
Average	4	0·7

For our average meteorite, i.e. 10 parts of silicate, 2 parts of nickel-iron, and 1 part of troilite, we find 0·7 p.p.m., or 0·7 g. gold per ton, which is about 700 times more than in the average magmatic rock.

[E. Goldberg, A. Uchiyama, and H. Brown (1951), using the method of neutron activation for the determination of gold in a wide range of iron meteorites, report an average value of 1·44 p.p.m. Au, with a range of 0·094 to 8,744 p.p.m. The gold concentration in general increases with increasing nickel concentration.

Using the value 1·4 p.p.m. together with the above average for troilite gives 0·27 p.p.m. Au for the average meteorite, i.e. only 300 times greater than for the average of magmatic rocks. The ratios to silicon and iron given by Goldschmidt have been left unchanged pending revised values for the other constituents of meteorites.]

The following table illustrates the proportion of gold to silicon and to iron; no data for the solar atmosphere or for stellar atmospheres are known.

Proportion of Gold to Silicon and Iron

	Au:100 Si	
	By weight	*By atoms*
Magmatic rocks	0·00000037	0·00000005
Meteorites, average	0·0004	0·000057
	Au:100 Fe	
Magmatic rocks	0·000002	0·00000057
Meteorites, average	0·00022	0·000063

The scarcity of gold in the upper lithosphere is thus due to its siderophil character, the bulk of it concentrating in an iron phase, presumably in the innermost core of the globe. Gold also occurs in inclusions of native iron in basalt in amounts from about 0·5 to 3 p.p.m.

Gold in magmatic and hydrothermal rocks

The geochemical distribution of gold in the magmatic rocks of the lithosphere indicates that it is not exclusively confined to one single type of magmatic rock. We know of its occurrence in association with olivine rocks, although the deposits do not seem to be frequent or to attain extensive dimensions. In chromite from various peridotites V. M. Goldschmidt and Cl. Peters (1932*b*) found 0·2 p.p.m. Au. In a contact deposit of gabbro with limestone gold and auriferous arsenopyrite occur at Hedley, Alaska. We know gold to be present in the pyrrhotite-pentlandite ore magmas and in the high-temperature hydrothermal iron pyrite and chalcopyrite, both associated with gabbroid magmas. The amounts of gold in this kind of deposit are very small, usually much less than 1 p.p.m. In the first, gold is much subordinate to the platinum metals, in the second the reverse is the case. Despite the small absolute amounts of gold in such deposits, what is present is recovered from the anode slimes during the electrolytic refining of nickel and copper, but it is only an unimportant by-product of the mining and metallurgical operations. Of much greater economic importance are hydrothermal gold deposits associated with andesites and dacites. Gold is here found in part as the metal, in part in combination with tellurium. More rarely it occurs with selenium, but is very often associated with tellurides of bismuth, sometimes also with ore minerals of antimony and sometimes with silver minerals. Important gold deposits of this type are known in western North America and in Transylvania. A few hydrothermal deposits of gold consisting of rich gold tellurides are connected not with volcanics of the andesite-dacite families, but with volcanics of a more alkaline type, phonolites and trachytes, such as at Cripple Creek, Colorado, U.S.A.

In contrast to those hydrothermal deposits of gold connected with volcanic rocks are those equally important hydrothermal deposits connected with plutonic rocks. Among these is a group of deposits characterized again by the association of gold with tellurium, bismuth, and antimony that is found with diorites, quartz-diorites, granodiorites, and sometimes granites. That gold is normally present in granitic residual magmas has long been known. V. M. Goldschmidt and Cl. Peters (1932*b*) found small amounts, 0·2–0·5 p.p.m. Au, in nearly all cassiterites. They also regularly observed small amounts of gold in molybdenite, MoS_2, another mineral associated with pneumatolytic mineralization from granites, the amount varying between 0·1 and 5 p.p.m. Au.

From the point of view of a mining geologist the scarcity of gold in the lithosphere is further enhanced by the distribution of the lithosphere's small stock among a great variety of magmatic rocks and their ore deposits. In this way gold is in marked contrast to nickel, a metal which has been almost entirely concentrated in only one or two types of ore deposits. But in the case of nickel also the bulk of the metal has been lost irretrievably to the iron core of the earth.

Other hydrothermal gold deposits which are connected with the residual solutions from granite magmas and also granodiorites are characterized by the association of gold with iron pyrites in quartz veins and very often in the same manner with arsenopyrites, FeAsS or $FeAs_2$.

It follows from general geological considerations that hydrothermal deposits are primarily located at higher levels of the lithosphere when connected with volcanics than when they are connected with plutonic intrusions. In consequence of the long interval of geological time which is needed for deep rock erosion, the veins connected with plutonics may become visible at the surface only very long after their formation. The hydrothermal deposits connected with volcanics become uncovered very much sooner and will be removed much more quickly by further erosion. Thus the hydrothermal gold deposits of volcanics that we know are mostly from such a young geological formation as the Tertiary. Deposits connected with plutonic intrusions are known mainly from older formations such as Precambrian and Palaeozoic, but in some cases as young as Mesozoic. In this way the mining geologist distinguishes between the 'young' and 'old' types of hydrothermal gold deposits.

A most interesting problem is why gold is so often associated with tellurium, bismuth, and antimony (or less frequently selenium), in contrast, for instance, to silver, which is most frequently associated with compounds of sulphur and especially with lead sulphide. Another problem is the association of gold with hydrothermal quartz veins from residual solutions of granite magmas, and why the gold there is so often closely associated with iron pyrite or with arsenopyrite.

Crystal chemistry of gold in relation to ore deposits

Too little is known about the crystal chemistry of simple gold compounds at various temperatures to venture upon a definite explanation of all the various element, and mineral, associations in which gold takes part. Nevertheless, some data are worth considering even in a tentative discussion of the main problems involved.

The atomic radii of gold and silver are very similar—one of the effects of the 'lanthanide contraction'—both elements in the metallic state with the coordination number 12 having the radius 1·44 Å. The two metals therefore form a continuous series of solid solutions, and in nature solid solutions in the proportions Au:Ag = 1:1 and 3:1 are well known. At present we know very little about the size of gold ions, because gold tends very strongly to form covalent bonds. Even in such simple compounds as the tellurides of formula $AuTe_2$ (krennerite and calverite) molecular lattices are observed, probably with covalent bonds between the atoms in the molecule $AuTe_2$.† The author considers it possible that at higher temperatures there may prevail packings of tellurium anions into which gold may enter as a univalent or trivalent cation. For univalent gold Pauling has calculated an ionic radius (1·37 Å) which is definitely larger than the radius of univalent silver (1·26 Å).

Quite a number of useful data relating to host minerals for gold have recently been published by G. M. Schwartz (1944).

† Confirmed by Tunell and Pauling (1952).

One tentative explanation for the association of rich gold ores with tellurium, bismuth, and antimony might be that Te^{2-} and Bi^{3-} are anions of large ionic radius, which amounts to about 2·2 Å for Te^{2-} and Sb^{2-} whereas Bi^{3-} is still larger. One might consider the possibility that mineral packings containing tellurium, antimony, or bismuth might capture ions or perhaps atoms of gold by some mechanism similar to that which captures silver in packings of sulphur ions, especially in galena.

The mineral nagyagite, a telluride and sulphide of gold and lead, seems to furnish an excellent example of the collection of gold in a tellurium-sulphur packing. According to the rather incomplete investigation by B. Gossner nagyagite may be essentially a tetragonal variant of the rock-salt structure of a solid solution of lead sulphide and lead telluride. In this solid solution gold has been collected in the interstices of the sulphur-tellurium packing perhaps with some rearrangement of the metal-metalloid ratio. The atomic proportion between lead and gold is about 6:1 (Gossner, 1935).

If the captures were made in a state of metallic atoms, one might suspect that gold and silver should behave exactly alike, because of the similarity of their atomic radii, but experience clearly indicates a difference between the two elements with regard to their capture in tellurides and sulphides. We observe that gold enters only into tellurides to any extent, not into sulphides, but silver is collected by sulphides and also in some cases, without doubt, by tellurides. This indicates that the particles of gold are larger than those of silver.

Of course the present minerals of the hydrothermal gold deposits may represent later low-temperature conditions.

It will be well worth while to look for β-forms of the gold tellurides and tellur-bismuthides at somewhat elevated temperatures, comparable to high and medium hydrothermal conditions. Such high-temperature forms may represent the tellurium or tellurium-bismuth packings in which gold has been originally 'collected'. At low temperatures the gold tellurides have molecular lattices of low symmetry containing molecules of $AuTe_2$. Perhaps similar features may be discovered in the low temperature modifications of Ag_2S, Ag_2Te, or Cu_2S.

There seems to be only one indication in the literature that a gold compound of this group has an α–β transition which may possibly be related to the α–β transitions of copper and silver sulphides. This refers to the compound $AuSb_2$ which, however, is reported to have an ordered fluorite structure.

The power of bismuth to collect precious metals is indicated also by the observation of V. M. Goldschmidt and Peters that all metallic or semi-metallic bismuth minerals have collected small amounts of elements of the palladium and platinum groups. It is quite likely that the common denominator in the association of tellurium, bismuth, and more rarely selenium and antimony in the collection of gold is to be sought in the ability of these elements to form packings of such large ions or atoms that gold can be collected in their interstices.

Another independent problem in the geochemistry of gold is connected with the very frequent association of small amounts of gold with minerals of the iron pyrite-arsenopyrite families. Usually only quite subordinate amounts of gold, up to 100 p.p.m., are found in these minerals. The gold is usually visible under the ore-microscope in polished sections, but very probably represents exsolution from original solid solutions which have been formed at elevated temperatures. The author considers it possible that reversible transformations may be expected in the lattices of these minerals due to thermal rotation and even thermal dissociation of such complex particles as S_2, SAs, or SbAs. This would lead to exceptionally strong thermal expansion in the temperature interval when thermal rotation of the complex particles starts. Thereby the AX_2 and AXY lattices would become most suitable host crystals not only for gold but also for the metals of the palladium and platinum groups. V. M. Goldschmidt and Cl. Peters (1932*b*) showed that gold, platinum, and palladium, besides silver, had collected in such hydrothermal minerals as CoAsS, $CoAsS_2$, NiAsS, NiSbS. The amounts of gold were generally from 0·3 to 5 p.p.m. in the cobalt minerals, and 0·5 to 20 p.p.m. in the nickel minerals, while very rarely in some samples of NiAs nearly 100 p.p.m. Au were found. It is probable that the siderophil properties of gold (and platinum) are involved in the collection of these metals in the host crystals of the pyrite and arsenopyrite families.

Gold in processes of weathering and soil formation

In processes of weathering gold ores are known to react very differently from the ores of other metals such as zinc, copper, or lead. In most cases the bulk of gold remains in the zone of oxidation. This is true both for compact primary metallic gold (reef-gold) and for gold liberated by the oxidation of tellurides and the like, when the metal often forms a very fine brown powder. Only part of the gold from the primary ores finds its way in solution down to the zone of cementation where it is again precipitated. Transfer of gold into this zone, resulting in a 'secondary enrichment', occurs especially when the element is associated with iron pyrite, because both ferric sulphate and chloride are active in dissolving metallic gold. In some cases the gold from oxidized pyrite remains as metal in the oxidation zone. The dissolved gold may again be precipitated by practically any sulphide, arsenide, or sulpho-salt, or simply by ferrous ions, in the zone of cementation.

These experiences are reflected also in the geochemistry of soils, except that here the action of organic constituents has also to be considered.

In many cases metallic gold, as solid particles of varying size, may be transferred from products of weathering directly into soils, and may remain there more or less unchanged. Since the specific gravity of gold or of gold alloys is very much higher than that of most ordinary sand and soil minerals (by a factor of about 6–8), the gold will tend to collect in the lowest soil horizons, often immediately above bed-rock or above some solidly compacted soil horizon. Much more interesting from a geochemical viewpoint are the phenomena connected with the solution and re-precipitation of gold in soils. Some types of gold particles in soils, sands, gravels, and consolidated sediments clearly reveal their origin as primary 'reef gold' by their jagged outlines, by their often plate-like form, or by their intergrowth with vein quartz or other hydrothermal minerals. Other particles, either fine dust, larger particles, or even nuggets, have evidently grown in soil. The gold may have been brought into solution by the action of strongly oxidizing ions, for instance those of trivalent iron or quadrivalent manganese, or by oxidizing anions such as nitrate or nitrite, and afterwards have been precipitated again at places where conditions were favourable for reduction to the metallic state.

J. B. Harrison (1908) refers to evidence produced by E. E. Lungwitz that ashes of ironwood trees which have grown on auriferous laterite may contain 2–10 grains of gold per ton. In the ash of the upper part of the trunks near the branches he found as much as 28 grains Au per ton of ash.

Du Bois working on material from Surinam could not confirm these results. J. B. Harrison (loc. cit., p. 210) found 1 grain of Au per ton of ash from the bark (4·78 per cent. ash) and 7–10 grains per ton in the ash from the wood (0·67 per cent. ash). The chemical evidence shows that such 'secondary' gold usually contains much less silver than the reef gold from the same region, since the natural electrolytic refining process eliminates the silver which is less easily reduced again to the metallic state after dissolution.

There is ample evidence that dissolved gold from soils may enter the circulation of living plants under very different conditions of climate or temperature. In particular it has been reported that species of horsetails (*Equisetum arvense* and *E. palustre*) may concentrate gold together with silica among their mineral constituents (Nemec *et al.*, 1936). The occurrence of gold in the corn plant (*Zea mays*) has also been reported (Nemec, 1935). Convincing evidence for the circulation of dissolved gold in living plants is the concentration of gold in the uppermost humus layer of forest soils in Central Germany, observed by Goldschmidt and Peters (1933 *c*). It is possible that this observation may be used as an indication of the occurrence of gold in lower levels of the soil.

No data on gold in agricultural soils have been published, nor have any physiological effects of deficiency or excess of gold been reported for agricultural crops or domestic animals.

In the ashes of coal V. M. Goldschmidt and Cl. Peters repeatedly observed the concentration of gold in amounts up to 1 p.p.m. of the ash, in association with small amounts of metals of the platinum and palladium groups.

Gold in the sequence of sediments

Metallic gold occurs in residual sediments such as sands, sandstones, gravels, and conglomerates, concentrated by reason of its chemical inertness and its very high density. Gold is not very resistant to mechanical attrition by hard minerals and therefore is often comminuted to very fine dust during transportation and concentration. Gold in residual sediments tends to collect either upon bed-rock or upon other bottom horizons, as it does in soils. This very significant feature of all 'placer' deposits of gold and of platinum metals is due to the high specific gravity of the metal compared with the other minerals of sediments. Gold is often associated with other heavy minerals, for instance in the 'black sands'.

Gold deposits in residual sediments are known from many countries and from many geological periods from recent times back to the Precambrian. Most famous are the deposits in the Precambrian quartz-conglomerates of the Witwatersrand in South Africa, where amounts of 3 to 15 p.p.m. gold are found in the pyritiferous matrix between the old quartz pebbles. The deposits are generally considered to have originated as placer gold in residual gravels, though an epigenetic hydrothermal origin of the gold has also been advocated.

We do not know much about the occurrence of gold in normal marine hydrolysate sediments, nor are reliable data available about the average amounts of gold in oxidate sediments.

In sulphide-bearing marine sediments, such as the black muds formed under anaerobic conditions in stagnant sea water, gold is found to be concentrated, together with silver, as in the mud from Walfish Bay on the coast of south-west Africa. In fossil reduzate sediments we find gold concentrated in amounts of 1 p.p.m. as in the bituminous copper shale of the Mansfeld district.† Small amounts of gold, about 0·05–0·1 p.p.m., are found in the lower Ordovician black shales of the *Dictyograptus* horizon in Norway and Sweden.

Very elaborate investigations have been made by F. Haber (1928) on the occurrence of gold (and silver) in sea water. Earlier investigations by various analysts had suggested quite considerable amounts of gold in sea water. Haber developed an excellent method for micro-assay and found that these older data were evidently erroneous, either because of unsuitable methods or through contamination from impure analytical chemicals. The amounts of gold in sea water are exceedingly small, averaging 0·000006 p.p.m. It is not certain whether this amount of gold is all present as a solute, presumably as chloro-aurate anion, or whether at least part of it is in suspension, as is indicated by irregular variations between different samples.

I. and W. Noddack (1939) report 0·000008 p.p.m. Au from sea water, and also from seven out of nine marine animals. No gold was found in an ascidian or in a shark, which means less than 0·007 p.p.m. in the dry matter. The highest amounts, 0·024 and 0·03 p.p.m., were found in a holothurian and in a fish.

REFERENCES

[As given in the original text]

Goldberg, L., Uchiyama, A. and Brown, H., 1951. Geochim. et Cosmochim. Acta 2, 1.
Goldschmidt, V. M. and Peters, C., 1932b. Nachr. Ges. Wiss. Göttingen, math.-phys. K1.377.
Goldschmidt, V. M. and Peters, C., 1933c. Nachr. Ges. Wiss. Göttingen, math.-phys. K1. 371.
Gossner, B. 1935. Cbl. Min. Geol. A, 321.
Haber, F., 1928. Z.d. Ges. f. Erdk. Erganz. Hft. 3, 3.
Harrison, J. B., 1908. Book: Geology of the British Guiana Goldfields (London).
Nemec, B., 1935. Ber. deutsch. bot. Ges 53, 560.
Nemec, B., Babicka, J., and Osorsky, A., 1936. Bull. internat. Acad. Sci. Bohème, 7.
Noddack, I., and W., 1931a. Z. phys. Chem. 154.
Noddack, I., and W. 1931b. Ibid. Bodenstein Festbd. 890.
Noddack, I., and W., 1934. Svensk. Kem. Tidskr. 46, 173.
Noddack, I., and W., 1939. Arkiv. Zool. 32A, No. 1, Art. 4.
Schwartz, G. M., 1944. Econ. Geol. 39, 371.
Tunell, G. and Pauling, L., 1952. Acta Cryst. 5, 375.

† I. and W. Noddack (1931 *a*) in their publication on the geochemistry of rhenium reported 10 p.p.m. Au in the Mansfeld shale, an amount which is very much too high.

The book *Geochemistry of Gold* by O. E. Zvyagintsev is an extensive account of the geochemistry of the element as known in 1941. The author discusses the distribution and migration of gold in the various geospheres, and gives data on the content of gold in meteorites, in natural waters, in plants, in coal and in the anthroposphere (sphere of man). The chemical processes whereby gold is concentrated in hydrothermal veins and in placers are discussed in considerable detail.

The summary account of the geochemistry of gold (and silver) given by K. Rankama and Th. G. Sahama (1950) and reproduced as Paper 8-5, from their *Geochemistry*, gives the status of the geochemistry of the element as known at the midpoint of the present century.

For some twenty years after 1950 interest in gold and gold mining declined throughout the world, and consequently few papers and monographs appeared on the geochemistry of gold. During the 1970s, however, a price increase for the metal stimulated further publication, and three monographs appeared, one in the USSR by Petrovskaya (1973), one in Canada by Boyle (1979), and another in France by Bache (1982).

The work by Petrovskaya, *Native Gold,* reviews the geochemistry of gold, discusses the metallogeny of the metal, and describes the geological and mineralogical features of endogenic and exogenic (supergenic) gold deposits in part 1. Parts 2 and 3 deal with the details of native gold (e.g., morphology, alteration) in all of its occurrences in endogenic and exogenic deposits. Chapter I of part 1, translated and presented as Paper 8-6, is a review of the geochemistry of gold as understood in the Soviet Union in 1973.

(Text continues on page 133.)

8-5: SILVER, GOLD

Kalervo Rankama and Th. G. Sahama

ABUNDANCE AND GEOCHEMICAL CHARACTER

AS IN the case of the platinum metals and copper, very small amounts of silver and gold, if any, are present in silicate minerals constituting the normal igneous rocks. Their abundance in igneous rocks and in the various meteorite phases is presented in Table 37.1, which is based on the values given by Goldschmidt (1937*b*), partly according to Noddack and Noddack and partly according to previous researches of Goldschmidt and Peters. However, some of the

TABLE 37.1

ABUNDANCE OF SILVER AND GOLD

MATERIAL	Ag	Au
	g/ton	
Nickel-iron	4	4
Troilite	18	0.5
Silicate meteorites	(0)	(0)
Igneous rocks	0.10	0.005

values recorded appear unreliable. As stated by Goldschmidt (1937*b*), the values for igneous rocks might be too high.

Geochemically, silver and gold differ sharply from each other. Silver is strongly enriched in the sulfide phase of the meteorites. This fact and the manner of occurrence of silver in the upper lithosphere afford proof of its typically chalcophile character. Gold, on the other hand, is very pronouncedly concentrated in the nickel-iron. It is true that gold readily accompanies chalcophile elements in the upper lithosphere, but this circumstance is due to the difference in the redox potential in the superficial parts of the Earth and in the meteorites (see chap. 1). With reference to its terrestrial manner of occurrence, gold, contrary to previous belief, is siderophile, being enriched in the nickel-iron core. It therefore very much resembles platinum in this respect.

The presence of silver has been established in the solar atmosphere, and gold is probably also present therein.

MANNER OF OCCURRENCE IN THE UPPER LITHOSPHERE

Silver, gold, and copper all belong to the same subgroup of the Periodic System. Therefore, the manner of occurrence of silver and gold in the upper lithosphere notably resembles that of copper. Copper and silver are sulfophile and occur mostly combined with sulfur and, partly, with selenium in the form of various sulfides and selenides. Silver forms, in addition, a number of tellurides and sulfosalts, the latter chiefly with copper, germanium, tin, lead, and manganese. In spite of its siderophile character, gold follows copper and silver in the upper lithosphere: it is sulfophile. However, it does not form any independent sulfide minerals. Gold accompanies selenium and, particularly, tellurium and becomes enriched together with these elements. It is found in Nature in the native state, alloyed with silver and the platinum metals, and as tellurides in many sulfide deposits. The tendency of copper and silver to combine with sulfur and the affinity of gold for tellurium is probably partly due to the ionic properties of these elements. Their ionic radii are the following:

Cu^+,	0.96 kX	S^{2-},	1.74 kX
Ag^+,	1.13 kX	Se^{2-},	1.91 kX
Au^+,	1.37 kX	Te^{2-},	2.11 kX

Consequently, one would expect the cation with the greatest radius, Au^+, preferably to combine with the biggest of the negatively charged ions, viz., Te^{2-}.

Silver and especially gold are enriched in telluric iron separated from basalt: Goldschmidt and Peters (1932*b*) report up to nearly 10 g/ton Ag and 5 g/ton Au in such irons. Like copper, silver and gold also readily become enriched in sulfides separated during the early stages of magmatic differentiation. In the case of silver, which is geochemically chalcophile, this behavior is rather natural. Gold, on the contrary, affords another example of the general rule, valid for many siderophile elements like the platinum metals and rhenium, that, in the absence of a metal phase, such elements are concentrated in the sulfide phase. However, there are other elements, more or less pronouncedly siderophile, which are carried over to the silicate phase, e.g., phosphorus and tungsten.

Among the early-separated oxides, only chromite contains gold and silver: Goldschmidt and Peters (1932*b*) report contents of 1 g/ton Ag and 0.2 g/ton Au.

During the main stage of crystallization the content of silver and gold actually present in the separated minerals is probably still smaller than that of copper. Like copper, the two metals are enriched in the late magmatic products. The separation of gold, in particular, often begins prior to that of copper during the pegmatitic stage of crystallization. The pegmatitic and hydrothermal formations are the most characteristic abodes of silver and gold. Minerals genetically connected with pneumatolytic tin deposits may also carry notable amounts of silver and gold, such as cassiterite (up to 0.5 g/ton Au and nearly 100 g/ton Ag, according to Goldschmidt and Peters, 1932*b*), molybdenite (in excess of 100 g/ton Ag, nearly 10 g/ton Au), and triplite. The two metals are often enriched also in pegmatite minerals, such as ilmenorutile, tantalite, and samarskite (Lunde, 1927; Lunde and Johnson, 1928).

With the exception of native silver, which, like native copper, is often of secondary origin, the most important minerals of silver include the following sulfides and sulfosalts:

Argentite and acanthite,	Ag_2S
Stephanite,	$5Ag_2S \cdot Sb_2S_3$
Polybasite,	$8(Ag,Cu)_2S \cdot Sb_2S_3$
Pyrargyrite,	Ag_3SbS_3
Proustite,	Ag_3AsS_3

All the minerals listed are important ore minerals of silver, argentite being probably the most important primary silver mineral. The argentian tetrahedrite (freibergite), $(Cu,Ag)_3(Sb,As)S_{3-4}$ (?) may also be an important ore of silver. Still another argentiferous mineral of technical importance is galena, which may contain up to 2 per cent silver. The galenas formed at high and intermediate temperatures are richer in silver than those crystallized at low temperatures (Oftedal, 1940). The silver content of galena is usually due to the presence of admixed silver minerals separated from the galena structure. Sometimes there occur small amounts of argentite or matildite (schapbachite), $AgBiS_2$, which form isomorphic mixtures with galena. Pure sphalerite is always devoid of silver (Oftedal, 1940).

Native silver is never pure; it usually contains gold, copper, and other metals as impurities. Silver also occurs alloyed with mercury as kongsbergite, α-(Ag,Hg), and moschellandsbergite, γ-(Ag,Hg).

The most important gold tellurides are calaverite, $AuTe_2$, and sylvanite, $AuAgTe_4$. Like native silver, native gold also is always impure. Silver, copper, iron, and the platinum metals are the foremost impurities. The gold amalgam, (Au,Hg), is rare.

In the simultaneous presence of a sulfide and an arsenide (antimonide) phase, gold, because of its siderophile behavior, is preferentially concentrated in arsenides and antimonides, which possess a pronouncedly metal-like character. Likewise, gold, as a siderophile element, prefers the metal-like minerals of the pyrite and marcasite-loellingite groups. It is generally associated with pyrite and arsenopyrite.

According to Goldschmidt and Peters (1932*b*), silver predominates over gold in most sulfides and sulfosalts found in hydrothermal veins, as in chalcopyrite, bornite, and tetrahedrite, in which the Ag : Au ratio is 500–2,000, whereas gold is preferentially concentrated in the hydrothermal cobalt arsenides and especially nickel arsenides. If both chalcopyrite and pyrrhotite (pentlandite) are formed in a sulfide deposit, silver prefers the former. Lunde (1927) reports 11.5 g/ton Ag in the chalcopyrite from Knaben, Norway, and only 4.0 g/ton in the associated pyrrhotite. In late hydrothermal veins gold accompanies quartz, whereas silver minerals are usually found in carbonate veins (van Aubel, 1934). Gold seems to prefer the sodic albite pegmatites connected with acidic and intermediary magmas, whereas tin is found in potassic pegmatites. In the magmatic solutions gold, being chemically rather inert, is at least partly carried as dissolved or colloidal native metal.

No reliable values are available to show the average content of silver and gold in igneous rocks. Lunde (1927) and Lunde and Johnson (1928) found up to 2.1 g/ton Ag in dunite, up to 9.4 g/ton in peridotites, and 0.3 g/ton in eclogite. In eclogite, silver is contained in the garnet, whereas the pyroxene is devoid thereof. Goldschmidt and Peters found 0.5 g/ton Ag in a norite; Wagoner (according to Clarke, 1924) reports 0.5 g/ton Ag in a basalt, 7 g/ton in a diabase, and 0.9–7.7 g/ton in granites. The average found by Goldschmidt (1926) for the alkalic igneous rocks of the Oslo province is 0.12 g/ton Ag. The gold analyses are still scarcer. Leutwein (1939) found 0.1 g/ton Au in pyroxenite, and 0.1–0.2 g/ton in amphibolites of magmatic origin. The value for amphibolites of sedimentary origin was 0–0.1 g/ton. Wagoner found 0.026 g/ton Au in a basalt, 0.076 g/ton in a diabase, and 0.1–1.1 g/ton in granites.

CYCLE OF SILVER AND GOLD

Gold and silver also differ notably from each other in their behavior during weathering. Like copper, silver is brought into solution as sulfate. However, the solubility of silver sulfate is lower than that of cupric sulfate. In the cementation zone of ore deposits, silver is reprecipitated, usually as the chloride chlorargyrite, AgCl, which is an ore mineral of silver, or as sulfide. Bromides and iodides of silver are similarly found in the upper parts of silver deposits as products of secondary reactions caused by descending waters. The silver-bearing solutions formed during weathering deposit their silver as either sulfide, chloride, or sulfate in the hydrolyzate sediments. Sometimes silver is deposited in sandstones as chloride closely associated with plant fossils and clay galls (Utah, United States). Another example of the action of biological processes in the deposition of silver is afforded by the *Dictyonema* shales of Scandinavia, which may contain several g/ton Ag (Goldschmidt, 1931*a*). The marine hydrolyzates are sometimes remarkably rich in silver. Goldschmidt (1926) reports the averages given in the accompanying table. These values afford proof

Material	Ag (g/ton)
Norwegian Quaternary clays	0.05
Deep-sea sediments	0.11
Terrigenous sediments, western coast of South Africa	0.66

of considerable regional variations in the silver content of various parts of the Earth.

According to Wagoner (quoted by Clarke, 1924), sandstone contains 0.44 g/ton Ag and limestone about 0.2 g/ton. Silver becomes concentrated in marine oxidates also: Goldschmidt and Peters (1932*b*) found 1 g/ton Ag in a manganese nodule.

Because gold is chemically more inert than silver and because its compounds are readily reduced to metal, it remains largely in the native state and becomes concentrated in the resistates. Owing to its high specific gravity, gold often forms placer deposits with a relatively high gold content. These deposits represent one of the most important types of gold ores. Gold is readily dissolved and transported, probably largely in colloidal solutions, which may deposit it even at low temperatures. Any gold dissolved from gold ores in the zone of oxidation is redeposited in the cementation zone, in which a secondary enrichment may sometimes take place. Only negligible amounts of gold are carried into the seas; gold is finally removed by adsorption on hydrolyzate sediments. Like silver, gold may become enriched in the oxidates. Wagoner (according to Clarke, 1924) found an average of 0.028 g/ton Au in some sandstones and 0.005–0.009 g/ton in limestones. Goldschmidt and Peters (1932*b*) report 0.2 g/ton Au in a manganese nodule.

MANNER OF OCCURRENCE IN THE BIOSPHERE

Silver is known as a widely, but very sporadically, distributed microconstituent in various terrestrial and marine organisms. Up to 1.5 g/ton Ag are reported in the ashes of marine algae. Sea water rich in plankton organisms is reported to be particularly high in silver and gold, and it has been suggested that these metals are not present in true solution in sea water but are adsorbed on the surface of marine organisms. Silver is also reported to occur in fungi, and algae are believed to precipitate silver from thermal waters. A maximum of 10 g/ton Ag is reported in coal ashes (see chap. 8), in which silver is consequently considerably enriched. According to Mitchell (1944), the silver content of soil may be as high as 2 g/ton.

Gold is known to become concentrated by certain plants particularly in seeds, but its biological role in plants is unknown. Only few plants seem to be able to concentrate gold. The ashes of *Equisetum* growing in gold-rich areas may contain up to 610 g/ton Au, corresponding to a content of approximately 60 g/ton in the plant. Humus soil contains 0.1–0.5 g/ton Au (v. Thyssen-Bornemisza, 1942). Up to 0.5 g/ton Au may be present in coal ashes. Gold is also found in some marine animals. Fine gold particles derived from sand are sometimes found in the stomachs of birds.

The action of man as a concentrating agent of silver and gold is of geochemical interest. Zviaginzev (1941) has calculated that the concentration of gold in the anthroposphere is already of the order of 0.001 per cent of the total amount of gold in the Earth.

REFERENCES

[As given in the original text]

Aubel, René van (1934) Géochimie de l'or. Ann. Soc. geol. Belg. 57, p. B131.

Clarke, Frank Wigglesworth (1924) The data of geochemistry. 5th ed. U.S. Geol. Survey Bull. 770.

Clarke, Frank Wigglesworth and Washington, Henry Stephens (1924) The composition of the earth's crust U.S. Geol. Survey, Profess. Paper 127.

Goldschmidt, V. M. (1926) Probleme und Methoden der Geochemie. Gerlands Beitr. Geophys. 15, p. 38.

Goldschmidt, V. M. (1931a) Der Kreislauf der Metalle in der Natur. Metallwirtschaft 10, p. 265.

Goldschmidt, V. M. (1937b) Geochemische Verteilungsgesetze der Elemente. IX. Die Mengenverhaltnisse der Elemente und der Atom-Arten. Skrifter Norske Videnskaps-Akad. Oslo. I. Mat.-naturv. Klass, No. 4.

Goldschmidt, V. M. and Peters, Cl. (1932b) Zur Geochemie der Edelmetalle. Nachr. Ges. Wiss. Göttingen, Math.-physik. Klass III, :24, IV:26, p. 377.

Leutwein, Friedrich (1939) Über das Vorkommen einiger seltener Elemente in metamorphen Gesteinien des Schwarzwaldes. Vorläufige Mitteilung. Zentr. Mineral. Geol. A, p. 123.

Lunde, Gulbrand (1927) Über das Vorkommen des Platins in norwegischen Gesteinen und Mineralen. Z. anorg. allgem. Chem. 161, p. 1.

Lunde, Gulbrand and Johnson, Mimi (1928) Vorkommen und Nachweis der Platinmetalle in norwegischen Gesteinen. II. Z. anorg. allgem. Chem. 173, p. 167.

Mitchell, R. L. (1944) The distribution of trace elements in soils and grasses. Proc. Nutrition Soc. Engl. and Scot. 1, p. 183.

Oftedal, Ivar (1940) Untersuchungen über die Nebenbestandteile von Erzmineralien norwegischer zinkblendeführender Vorkommen. Skrifter Norske Videnskaps-Akad. Oslo. I. Mat.-naturv. Klasse, No. 8.

Thyssen-Bornemisza, Stephan V. (1942) Geochemische und pflanzenbiologische Zusammenhange im Lichte der angewandten Geophysik. Beitr. angew. Geophys. 10, p. 35.

Zviaginzev, E. (1941) The geochemistry of gold (in Russian). Moscow. Referred to inTomkeieff (1944). (Nature 154, p. 814).

8-6: AN OUTLINE OF THE GEOCHEMISTRY OF GOLD

N. V. Petrovskaya

This chapter was translated expressly for this volume by the Translation Bureau, Department of the Secretary of State, Government of Canada, from *Native Gold,* Idz. "Nauka", Moscow, 1973, pp. 8-20.

Gold, to quote V. I. Vernadskii (1927), is an element that is "ubiquitous, and present in all soils and rocks . . . and occurs in perpetually dynamic equilibria in aqueous solutions." The reasons for its extremely wide dispersion in nature and its concentration in local zones of the Earth's crust remain far from being understood. A. E. Fersman (1939) stressed that "the geochemical properties of gold have been inadequately studied; as regards the geochemical migration of gold, confusion has hitherto reigned supreme." O. E. Zvyagintsev (1941) concluded that "the available precise scientific observations and experiments are by no means sufficient to deal fully with the geochemistry of gold."

The stages in the development of the geochemistry of gold are closely related to the evolution of methods of determining the low content of this element in rocks and minerals. Investigators achieved initial success by using combined chemical and X-ray spectral methods with a sensitivity extending to $10^{-5}\%$, i.e. 0.1 g/t. Improved methods of assaying have lowered this value to 0.02 g/t (Plaksin, 1958).

Scientists have established a new landmark by employing neutron activation methods, which permit determination of $10^{-9}\%$ gold in a weighed portion of as little as 0.5 g, i.e. 0.02 mg/t of gold (Vincent and Crocket, 1960; Shcherbakov and Perezhogin, 1964; and others). Specialists have in recent years developed combined chemical-spectral methods that involve either preliminary leaching of the precious metals from samples weighing 5-10 g, or obtaining a bullion bead by assay methods (studies by P. Barnet, N. T. Voskresenskaya, A. Spachkova, O. B. Fal'kova, and others). The sensitivity of gold determination by these methods extends to $2.10^{-8}\%$. The improvement of quantitative and semiquantitative spectral analysis methods has permitted the detection of as little as $10^{-6}\%$ gold. Researchers are developing chromatographic and ion-exchange methods of separating metals. Together with the well-known manuals on the analytical chemistry of gold and other precious metals is the recent monograph by F. Beamish (1966) covering his long experience with the analysis of the precious metals.

The number of determinations of small amounts of gold in rocks has markedly increased from year to year. I. and W. Noddack, in their 1930 summary, give a couple of dozen such determinations; the study by V. Goldschmidt and Cl. Peters in 1932 contains more than 100. Today the number of analyses for gold is practically incalculable.

The data derived have enabled investigators to clarify the behaviour of gold in natural processes, as shown in studies on the general problems of geochemistry (Vinogradov, 1956; Mason, 1966; and others) and in a number of works on certain aspects of the geochemistry of gold (Vincent and Crocket, 1960; Shcherbakov and Perezhogin, 1964; Phan, 1965; Razin and Rozhkov, 1966; Shcherbakov, 1967, and other scientists).

The chemical properties and structure of the gold atom. Gold (atomic number 79) is in the Ib subgroup of the sixth period of Mendeleev's periodic table of the elements. The presence of an 18 electron outer shell determines its close association with the other elements of this subgroup.

The valences of gold (1 and 3) are determined by its capacity to donate one outer electron from its P orbit and two others from its O orbit, giving the stable electron configuration of 2-6-8. Univalent ions form less stable compounds than trivalent ions; the latter more readily form soluble complexes. Investigators suppose that under certain conditions, as in seawater, that the univalent gold complex $AuCl_2^-$ is stable (Peshchevitskii and others, 1965).

Affiliation with the Ib subgroup of elements, in which chalcophile properties are characteristic, is plainly evident in the geochemical association between gold and its nearest neighbours in the periodic table, silver and copper. Likewise, the siderophile nature of gold is clearly apparent as Goldschmidt was the first to note. Many other authors have enlarged on these features (Mason, 1966; Shcherbakov, 1967; and others).

The radius of the gold atom, determined from interatomic distances, is 1.44Å; the radius of Au^+, calculated from indirect data, is 1.37Å; this value is approximate owing to the strong polarization of the ion (Fersman, 1939). The radius of Au^{3+} is 0.85 Å (Ahrens, 1964). The capacity of trivalent gold to form complex anions is distinctly evident.

We know of 14 isotopes of gold with mass numbers from 192 to 206 (Kay and Libby, 1962), but only one is stable: ^{197}Au; the others are radioactive, with half-lives ranging from a few seconds to several days (transmutations involving nuclear capture mainly from the K-shell); ^{198}Au is used in the radioactivation methods of analyzing gold-bearing rocks and minerals.

In terms of its main properties, gold occupies an extreme position compared with its neighbouring elements in the subgroups of Mendeleev's table. Its atomic weight is 196.967, the greatest among the elements of the Ib subgroups of periods 4 to 6, and the least among the elements of subgroups II to VIIb of the 6th period. It is characterized by an extremely low shielding of the atomic nucleus by intermediate non-valent electrons. As a result of this feature the strength of the bond between the nucleus and the outer electron (located on the O orbit with the orbital quantum number 6s) is considerable, while the ionization potential is extremely high—9.223 electron volts (Ahrens, 1964). The electronegativity of gold is higher than that of other metals.

The properties of gold determine its chemical inertness, the instability of almost all of its compounds, and its clear tendency to occur in the metallic state. F. A. Letnikov (1963), in describing different types of compounds in connection with his concept of "series of native [e.g., metal] status," notes that gold lies at the end of each such series, the reason being the minimal isobaric potentials for the formation of its compounds, except for chlorides and tellurides. Such, for example, is the native status series with respect to sulfur (at 300°): Sn-Pb-Fe-Pt-Cu-Os-Bi-Sb-Hg-As-Ag-Au. The electrode potential of gold in aqueous solutions is extremely high; according to Latimer, $E_0 = 1.68$ for the reaction $Au = Au^+ + e$; for the reaction $Au = Au^{3+} + 3e$ the value is 1.50 millivolts. Under natural conditions electrons cannot be transferred from gold to an oxidizing agent.

The melting point of gold is 1,063°C. Its volatility is negligible, but is increased in a reducing medium (by a factor of two in a carbon monoxide atmosphere as against air), and at high temperatures. At 1,250°, 0.2% of gold evaporates as minute droplets—a "metallic mist." Additions of Cu, Pb and Bi increase the volatility of gold by a factor of 3-4, and As, Hg, Sb, Zn and Fe, by a factor of 8-10 (Boitsov and others, 1946). The heat of sublimation of gold at 298° is 87.7 kilocalories/mol.

When we assess the mode of migration of gold, we must remember the experimentally investigated capacity of its atoms to diffuse in different solid media, including quartz crystals. Gold has a strong tendency to form colloidal solutions whose micelles bear a negative charge; its sols are unstable and readily coagulate when exposed to sulfides.

The principal geochemical associations and compounds of gold. Having regard to the composition of gold-bearing mineral parageneses (ores), natural gold compounds, and the various constituents of the native metal, we must include many of the elements in the Mendeleev (periodic) table as geochemical associates of gold. Among these, however, are some that exhibit a constant association with gold; these include the metals with which gold forms solid solutions and intermetallic compounds, and the various components of the anionic complexes of gold. The former determine the main features of the mineralogy of gold; the latter determine the solubility and hydrothermal transfer of gold under natural conditions.

Among the metal associates silver followed by copper are the most important; both are close neighbours of gold in the Ib subgroup of the 4th to 6th periods of Mendeleev's table of elements. Prominent in the latter is a kind of "copper-silver-gold axis" (Fig. 1); to its left are the metals (Ni, Pd, Pt) whose

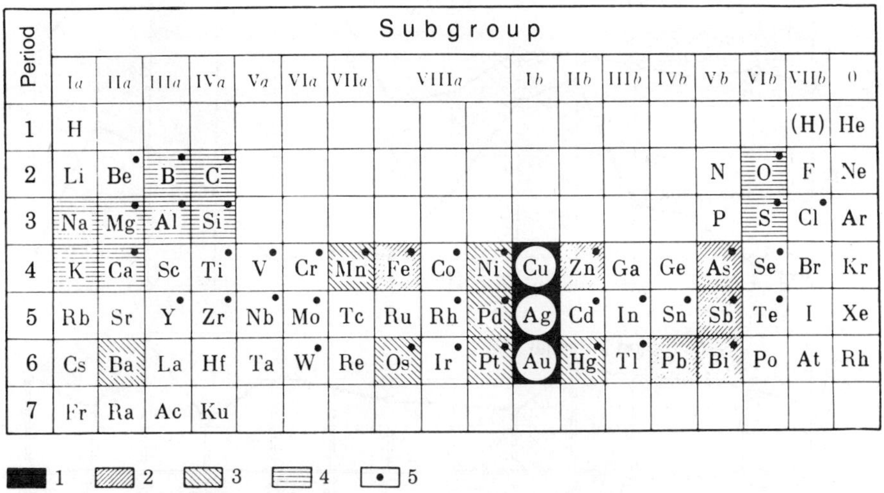
Subgroup
Period
Ia IIa IIIa IVa Va VIa VIIa VIIIa Ib IIb IIIb IVb Vb VIb VIIb 0
1 H (H) He
2 Li Be B C N O F Ne
3 Na Mg Al Si P S Cl Ar
4 K Ca Sc Ti V Cr Mn Fe Co Ni Cu Zn Ga Ge As Se Br Kr
5 Rb Sr Y Zr Nb Mo Tc Ru Rh Pd Ag Cd In Sn Sb Te I Xe
6 Cs Ba La Hf Ta W Re Os Ir Pt Au Hg Tl Pb Bi Po At Rh
7 Fr Ra Ac Ku
1 2 3 4 5

Figure 1. Geochemical table of elements associated with gold. 1. Elements universally associated with gold; 2. Elements typical of minerals commonly associated with gold; 3. Elements concentrated in gold-bearing mineral associations of individual orebodies; 4. Elements characteristic of gold ores only; 5. Trace elements commonly found in gold and its compounds (including artificial impurities).

concentrations are associated with magmas of basic composition; following Fersman we are inclined to assign the metals on the "right flank" (Zn, Cd, Hg) to geochemical associations that frequently accompany moderately acid and "mixed" magmatic formations.

As the degree of constancy of association with gold decreases, the following series of elements emerges to a first approximation: Ag-Cu-Fe-Sb-Te-Bi-Pb-Hg-Zn. Figure 2 compares the principal chemical properties of these elements with those of gold, platinum, and the other platinoids. The similarity of their ionic radii, which determines the geochemical relationship of the elements, is most clearly evident within the range of the Au-Ag-Cu field of the graph. The boundaries of this field are fixed by the fact that with the elements to the right the association with gold weakens with an increase in the radii of the elements (from Sb to Bi), while to the left the same situation prevails with a decrease in these values. Curves plotted to compare other properties of the elements are likewise symmetrical. The differences noted appear in the isolated nature of the geochemical associations of gold.

Silver and copper, the elements most commonly associated with gold, only partly resemble it in a number of properties, including ionization potential (much less than that of Au); this results in a certain independence in their behaviour in some natural processes, especially in the zone of hypergenesis (oxidation and weathering). The association of the other elements that determine the mineralogical features of auriferous ore deposits and the composition of gold minerals, are examined below in the pertinent chapters.

Among the elements forming soluble compounds with gold, and thus determining its mobility under natural conditions, the most prominent are the halogens, sulfur, oxygen and carbon.

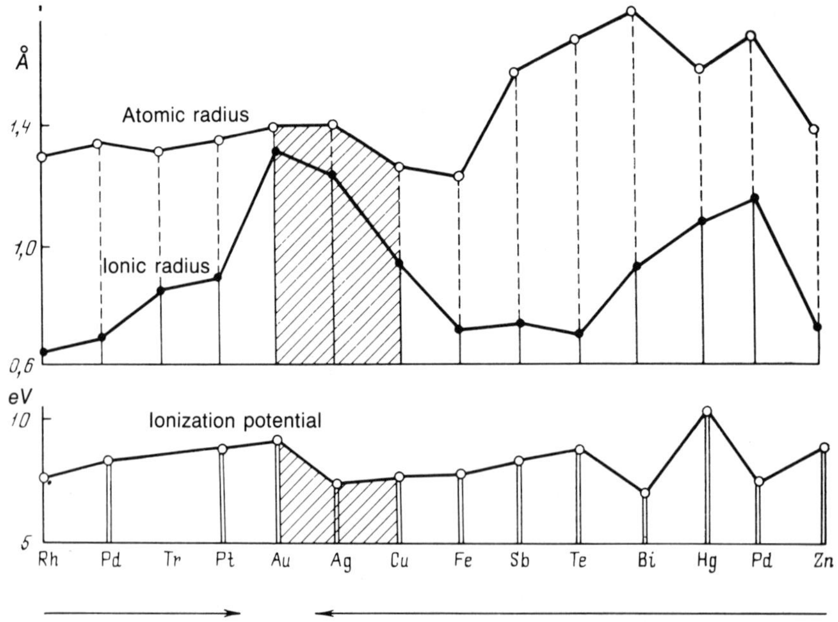

Figure 2. Comparison of the properties of elements associated with gold. The shaded areas denote the properties of the constant associates of gold. The arrows indicate the directions of intensification of elemental bond formation with gold.

Halogens. Chlorine, whose compounds have a wide distribution in nature, is of prime interest to researchers dealing with gold.

Aurous chloride is the least stable; when it breaks down it forms chloroaurate with the simultaneous release of a metallic precipitate ($3AuCl \rightarrow AuCl_3 + 2Au$). In the presence of oxidizing agents $AuCl_3$ may also form when gold is dissolved in iron and copper chlorides. When dissolved in water, chloroaurate forms the complex compound $H_2[AuCl_3]$, which can be classified as a type of aquoacid (Maslenitskii and Chugaev, 1972). Investigators have obtained complexes including those with ammonia ($AuCl \cdot NH_3$, etc.), metals ($Me[AuCl_4]$), and organic compounds. Among the halides of gold, AuF_3, $AuBr_3$, and AuI_3 are known; the last two may develop under natural conditions (Chukhrov, 1947).

Sulphur. Gold is closely associated with this element in much of its geochemical history. It is axiomatic to state that there are no gold ores completely devoid of sulfide minerals. Scientists link the natural migration of gold in many ways to the development of soluble gold-sulphur compounds. Such compounds are known to include those both with and without oxygen.

Sulfoaurates (MeAuS) appear when gold is reacted with hydrosulfides of alkaline metals, promoted by a rise in the temperature (Ogryzlo, 1935; Zvyagintsev, 1941). The literature mentions the gold sulfides Au_2S and Au_2S_3 as products of the reaction of hydrogen sulfide with chloroaurate or gold cyanide (Plaksin, 1958); these sulphides have, however, been little studied. Thiosulfate compounds of gold are known: these break down readily when exposed to acids. The literature mentions binary thiosulfates of gold and U, K, Ca, Ba and Mg of the type $Au_2S_2O_3 \cdot 3Me_2S_2O_3$. Many investigators have linked the supergene migration of gold to its solubility in iron sulfates.

Gold-sulfur compounds are mainly metastable and break down with the release of their metallic (gold) phase. Only certain complexes are relatively stable within the specific limits of the acidity and alkalinity of the natural environment; we shall examine their composition in the sections devoted to the characteristic genetic features of gold deposits.

Oxygen. One of the commonest oxidizing agents, oxygen, is involved in the formation of gold complexes with sulfur and other elements. Experimenters have obtained aurous oxide Au_2O and aurous hydroxide Au(OH) by the decomposition of aurous chloride. Salts of aurous oxide—the aurites—are soluble, but break down readily, forming a hydrous gold oxide.

Auric oxide Au_2O_3 and the corresponding hydroxide $Au(OH)_3$ are precipitated by the action of potassium carbonate on a hot chloraurate solution:

$$2AuCl_3 + 3K_2CO_3 + 6H_2O = 2Au(OH)_3 + 3H_2O + 3CO_2 + 6KCl.$$

The heat of formation of Au_2O_3 equals 12.3 kilocalories per gram/mol, i.e. its heat of formation is greater than that of free gold; this indicates the instability of the compound (Zvyagintsev, 1941). When auric hydroxide and solutions of caustic alkali or alkali carbonate react, aurates are formed; the gold enters anionic complexes: AuO_2^-, $Au(OH)_5^{2-}$, and $Au(OH)_6^{3-}$. The composition of these complexes depends on the alkalinity of the medium.

Oxygen compounds of gold are unstable (they readily form explosive mixtures); Garrels and Christ (1965) determined the conditions under which they develop. Such compounds are not found in nature. Researchers have speculated that such oxygen compounds occur as oxide films on gold grains

(Plaksin, 1958), but the data in the following chapters show that such films usually consist of iron hydroxides. Oxygen is present on the surface of gold in the form of extremely thin absorptive layers.

Carbon. Organic carbon, humic acids, and carbon dioxide are of great interest for understanding the geochemistry of gold. Researchers have noted a link between gold and organic matter as far back as the last century. In recent decades, investigators have studied the solubility of gold in humic acids and have concluded on sound evidence that native organic gold compounds of the chelate type exist (Hausen and Kerr, 1968; Radtke and Scheiner, 1970).

Carbon dioxide bound as carbonates accompanies nearly all natural gold occurrences. The data below suggest that it plays an essential role in the transportation and precipitation of gold.

The solubility of gold in cyanide solutions, with the formation of $HAu(CN)_2$ and the subsequent precipitation of gold cyanide—AuCN—is widely used in the practical hydrometallurgical treatment of ores. Fersman (1931) postulated the occurrence of natural cyanide solutions capable of dissolving gold, but this hypothesis has not been borne out.

Silicon. Certain concentrations of gold are associated with silica, as indicated by the essential role of quartz in the vast majority of gold deposits. The nature of this association has not yet been fully clarified. Researchers are of the opinion that when hydrothermal activity is in progress, Au and SiO_2 migrate and concentrate under similar conditions.

Since the last century, scientists have postulated that natural gold silicates exist, but adequate proof is lacking. The literature mentions synthetic compounds arising from the reaction of auric oxide with $NaSiO_3$; these are stable in the presence of free alkali and break down readily when exposed to acids and steam at high temperatures (Boitsov et al., 1946; Plaksin, 1958). The resultant substances have not been described and apparently have not been studied in any detail. Artificial compounds of gold and silicon are extremely unstable, as are also solid solutions of these elements. The sorption of gold by silica gels has been established experimentally; gold atoms can also be captured when silicate solutions are partly polymerized.

The above data show that only a few of the compounds of gold are stable, a limiting factor determining the number of possible agents for its transport.

It should be noted that the ionic forms of gold in solutions have been insufficiently studied; even in the laboratory their nature cannot be safely predicted owing to their variability, resulting from a multitude of factors including the pH and temperature of solutions, the duration of action with reagents, and so on (Beamish, 1966). They probably vary even during the course of natural processes.

The distribution of gold and its clarke. Gold is one of the elements in which the Earth's crust is markedly deficient. Its average content in the crust is 20 times less than that of silver, 40-100 times less than that of mercury, antimony and bismuth, and several thousand times less than that of copper, zinc, and lead (Mason, 1966).

The value of the clarke of gold (first determined by I. Vogt in 1898), according to the data of most investigators, is approximately $4.10^{-7}\%$, i.e. 4 mg/t (4 ppb) (according to A. E. Fersman it is 5, to A. P. Vinogradov 4.3, and to G. Taylor 4 mg/t). Citing these values, writers sometimes refer to gold as "an extremely rare element" (Zvyagintsev, 1941; Phan, 1965). If, however, we remember that the element is widely dispersed this characterization is obviously imprecise. It would be more correct to refer to gold as a "low-clarke element."

Many investigations at the end of the nineteenth century and in the first half of the twentieth century were devoted to determining the distribution of gold in different natural materials, beginning with the studies of I. D. Nuerbern, A. Karnozhitskii, F. Lincoln, V. I. Vernadskii, V. Goldschmidt, and others. Mason, Vinogradov, Vincent and Crocket, and Yu. G. Shcherbakov have published new data.

Researchers have established that the Au content of meteorites is much higher than that of any of the rocks in the Earth's crust and ranges from 50 to 1440 mg/t (Phan, 1965). Because the composition of meteorites has been compared with that of the core of the Earth, first Goldschmidt, and then Shcherbakov, viewed these facts as proof of a considerable accumulation of gold together with certain other heavy metals in the core.

Gold occurs universally in the Earth's crust and hydrosphere, but only in trace amounts; its content in most rocks is lower than the clarke value. The following series of natural materials exhibit an increasing gold concentration: seawater—sedimentary rocks—acid igneous rocks—basic and ultrabasic igneous rocks—chromitites of basaltic rocks—hydrothermal ores (Table 1).

The gold content of seawater is on the average two, and according to some data, three, orders of magnitude lower [by factors of $10^2/10^3$] than the clarke value (Zvyagintsev, 1941; Mason, 1966), but the total amount of gold in the oceans is enormous. On the basis of old estimates by S. Arrhenius, Kharitonov (1936) determined the total quantity to be 8 billion tons; on the basis of more recent data, the quantity is 10-20 million tons (Babicka, 1943) or 5-6 million tons (Shcherbakov, 1967).

Table 1. Average Gold Content of Natural Materials

Geologic Materials	*Au (1×10^{-7}%) (ppb)*	*Literature Source*
Meteorites:		
iron	50.0	Goldschmidt and Peters, 1932
chondrites (iron)	100.0	Goldschmidt and Peters, 1932
chondrites	170.0	Vinogradov, 1962
The Earth's crust as a whole	4.3	Vinogradov, 1962
The Earth's hydrosphere	0.01	Zvyagintsev, 1941
Seawater	0.8 to 5.0	Zvyagintsev, 1941
Sedimentary rocks (clays and shales)	1.0	Vinogradov, 1962
Igneous rocks:		
acidic	4.5	Vinogradov, 1962
acidic	2.0	Vinogradov, 1962
basic	4.0	Vinogradov, 1962
basic	11.0	Shcherbakov, 1969
basic	1.0 to 9.0	Moiseenko et al., 1971
ultrabasic	5.0	Vinogradov, 1962
ultrabasic	8.2	Shcherbakov, 1969
Metamorphic rocks (slates)	0.7 to 4.2	Moiseenko et al., 1971

Numerous attempts to develop methods for the industrial extraction of gold from seawater, begun by eight years of work by F. Haber, have not yet been successful.

The gold content of seawater is variable: in the polar seas it is as little as 0.05 mg/t; off the coast of Europe 1-3 mg/t, and in the offshore zones of Australia, California, and East Asia it is as much as 50 mg/t (Kropachev, 1935; Zvyagintsev, 1941). The enrichment of seawater in gold results from its removal from the continents by rivers. The forms in which gold occurs in seawater have been little studied. The difficulty of extracting gold from seawater, such as methods involving carbon filters, indicates the presence of certain highly stable gold complexes, probably chlorides. On the basis of thermodynamic calculations, Krauskopf (1951) concluded that the complexes $[AuCl_4]^-$ and $[AuCl_2]^-$ were present in seawater; some investigators (Peshchevitskii and others, 1965) consider that the latter complex predominates.

Colloidal and coarser gold suspensions are present together with soluble compounds in places, especially in the coastal zones of marine basins. It is also recognized that the metal may be concentrated in certain organisms (Phan, 1965). According to Babicka (1943), seawater contains 10 times more silver than gold.

The waters of many rivers carry gold in variable amounts, especially in the case of rivers draining gold-bearing regions. Thus, according to the data of G. P. Kropachev (1935), 1 tonne of the water in the Unda River in Southern Transbaikalia has been found to hold 1 gram of gold or more. In the Kuranakh region, Central Aldan, the waters contain 2 mg/t of dissolved and 1-9 mg/t of "suspended" gold (Razin and Rozhkov, 1966). Such instances are numerous.

Numerous hot springs in regions of active volcanism have a higher gold content. S. I. Naboko and S. F. Glavatskikh (1969) found 0.015 mg/litre of dissolved gold in the mineralized waters of Kamchatka (along with compounds of Cu, As, Sb, Hg, Pb and Zn). Investigators have detected minute particles of native gold in the precipitates of these waters, along with quartz, realgar, orpiment, arsenopyrite, sphalerite, galena, chalcopyrite, and mercury minerals. Others have observed gold accompanied by adularia in a similar association in the sediments of certain hot springs in Japan, Nevada, and New Zealand (White, 1955). Spring waters usually contain less than 1 mg/litre, but in some cases as much as 56 mg/litre Au.

The gold content of soils is low, but in gold-bearing regions it rises to 5.7 10^{-5}% and in places to $5.7 \cdot 10^{-4}$% (Vinogradov, 1957; Razin and Rozhkov, 1966). Investigators have noted that plants are enriched in gold in such regions. According to Kropachev (1935), the cortex of trees that have grown on the tailings of old placer mines has an established content of 0.1-2.00 mg Au per 2-7 kg of wood (0.1-1.5 g/t). Studies by Babicka (1943) proved the constant presence of gold in certain plants: the marsh horsetail *(Equisetum palustre),* beech *(Fagus),* and corn (*Zea mays* (L.)) growing in gold-bearing regions. This investigator, followed by Vinogradov (1957), observed small specks of lamelliform and dendritic gold in the ashes of corn kernels (regions of Slovakia, Svanetia, and others).

O. E. Zvyagintsev (1941), on the basis of published analyses of coals from England and the Ruhr, held that coal does not concentrate gold. This conclusion is probably true only for certain regions. Some areas (in the states of Wyoming and Utah, the island of Borneo, and others) are known where investigators have established a high gold content in carbonaceous sediments. Sampling coal for gold is likely to be a sound practice, especially within the confines of gold-bearing provinces.

Gold has been found in animal bones, blood, and hair (as much as 4 g per tonne of ash, according to Babicka), and in certain microorganisms. Researchers have identified microorganisms specifically of the fungus mold type that absorb gold from solutions. Similar experiments are being continued in various countries.

The Au content of sedimentary rocks, not including placers, is on the whole comparatively low and is higher only in certain sedimentary strata; both clastogenic and chemogenic gold are present. Authors have repeatedly discussed the value of the latter in the literature in connection with the behaviour of gold in the processes of its precipitation from ocean water. P. A. Kharitonov (1936), who actively supported theories of the possible concentration of chemogenic gold in sediments, cites the facts of the discovery in salt evaporated from seawater of 457 mg/t of Au and 54.4 g/t of Ag, and the presence of gold in silt samples dredged from the sea bottom. Subsequent studies have yielded conflicting results. Researchers have established that the gold content of Atlantic Ocean sediments is extremely low (Anoshin et al., 1969), but that certain deepwater red clays are higher in the element (Phan, 1965). Investigators have found unusually high Au concentrations (as much as 470 mg/t) and Ag (52.9 g/t) in the sediments associated with thermal (56-36°C) brines in depressions of the Red Sea (Tooms, 1970).

The highest concentrations of gold in sediments occur in thin, muddy fractions whose metal content depends on the presence of organic matter, carbonate, and pyrite (Polikarpochkin et al., 1968; Anoshin et al., 1969). Phosphorites and glauconites are gold-bearing (0.03-0.6 g/t), particularly in deposits of the Russian platform. Researchers postulate the sorption of gold by the phosphate minerals and glauconite (Rozhkov et al., 1967).

Clastogenic gold (usually finely divided) is present here and there in sands, sandstones, and conglomerates. We have information on a higher gold content in certain glacial deposits, including some near Moscow (data of A. P. Karpinskii, B. M. Dan'shin, and others). A vast literature deals with the gold-bearing conglomerates of South Africa, Siberia, the Caucasus, and other regions.

The dependence of the gold content of sediments on their composition and facies varies in different regions. Thus, in the Lena and Chukchi-Anadyr areas chemogenic gold is enriched in the clays and shales; in siltstones and sandstones chemogenic gold is concentrated only in places where organic matter, carbonates, pyrite, and iron hydroxides have accumulated (Nifontov, 1957; Polikarpochkin et al., 1968). Proterozoic phyllites and the dark grey shales of the Uderei formation of the Yenisey Ridge contain up to 0.2-0.3 g/t, and in some places as much as 1-3 g/t of gold (Petrov, 1969). Here, as in other regions, a correlation between the gold content and that of organic carbon is evident. On the other hand, the Altai clay shales and phyllites are poor in gold (0.0019-0.0029 g/t), whereas appreciable amounts of the metal are present in the sandstones (Shcherbakov, 1967). The sandy and shaly rocks of the Muruntau region in Western Uzbekistan contain gold in amounts ranging from thousandths to hundredths of a gram per tonne (Chebotarev, 1969).

Similar examples can be multiplied. Researchers have not yet analyzed these in detail, but a correlation of average values for gold content where the data are highly scattered, is inadequate. It seems an inescapable fact that some of the gold in sedimentary rocks is to a certain extent related to hydrothermal dispersion haloes.

Placers, a major genetic class of gold deposit, are a special type of sediment in which clastogenic gold is concentrated. They are associated with auriferous drainage sediments, which have no commercial value, but serve as indicators of the generally higher gold content of the pertinent areas.

The effusive and effusive-sedimentary rocks of a number of regions are richer in gold than normal sedimentary rocks. Thus, investigators have detected appreciable amounts of gold in the reworked tuffaceous sandstones and shales of some regions in Kazakhstan. Volcanic ash near Santiago (Chile), containing as much as 1 g/t of gold, has been mined on a small scale. One researcher (Koeberlin, 1934) has advanced the theory that volcanoes eject gold and ash at the same time, but it seems more probable that the loose tuffaceous materials served merely as a collector for the gold precipitated from aqueous solutions.

Yu. G. Shcherbakov (1967, 1969) noted a relative enrichment of gold in the tuffogenic and sedimentary-tuffogenic rocks of the Altai ore provinces.

Scientists have everywhere noted the presence of gold in igneous rocks, ranging from traces to 10 mg/t and more; (see Table 1). Most investigators acknowledge that the average content of gold tends to increase from acidic to basic rocks, a feature first noted by Goldschmidt, confirmed by subsequent studies (Shcherbakov and Perezhogin, 1964; Phan, 1965), and thoroughly substantiated by Shcherbakov (1967). The gold content of intrusives resulting from the differentiation of basic and ultrabasic magma is 10-15 times greater than the clarke value (Schneiderhohn, 1955; Godlevskii et al., 1970).

In a number of regions a higher gold content is characteristic of acidic rocks, which contain 2-5 times more gold than basic rocks. Examples include the auriferous regions of Kazakhstan (Voskresenskaya and Zvereva, 1968), Central Asia (Palei et al., 1967), etc. Similar features have been described in publications outside the USSR. Apparently the variations in the gold content of intrusives reflect some hitherto unexplained regional influences.

Nor are we certain of the reason for the discrepancy between the tendency towards the primary accumulation of gold in basic rocks and the regular association of most gold deposits with granitoid intrusive complexes (Bilibin, 1947; Bulynnikov, 1968, and others). It may be, as Shcherbakov (1967) has noted, that magmas of different basicity have to varying degrees donated to solutions the gold that they contained.

An analysis of the published data leads to the conclusion that the gold content of magmatic formations depends in a complex way on many factors, including the completeness with which the gold has been removed beyond the limits of the intrusive, or, on the other hand, the addition of the metal to such formations by emanations from more plutonic areas of the Earth's crust. Other factors include the extraction of gold from the enclosing rocks; the hybridization of magmas; the concentration of gold in the upper parts of intrusives; the development of superimposed hydrothermal mineralization, and so on.

We have little information on variations in the gold content of intrusive bodies in different plutonic facies. We know only that the gold content of basalts in the oceanic crust is almost triple that of basalts in the continental crust (Phan, 1965). Geologists in Kazakhstan have noticed that xenoliths of plutonic origin in rock massifs of Upper Ordovician age are greatly enriched in gold (Voskresenskaya and Zvereva, 1968). This information agrees with the above mentioned theories of the relative enrichment of plutonic areas of the Earth (possibly the upper mantle) with gold, probably unevenly distributed.

Gold is not usually concentrated in pegmatites (Fersman, 1931). Only isolated pegmatite deposits with relatively high gold concentrations are known, examples being the Passagen deposit in Brazil, which by 1931 had yielded about 50 tonnes of the metal, and the Natas deposit in Southwest Africa, producing scheelite and gold (Schneiderhöhn, 1955); the latter is associated with hydrothermal sulfides, including pyrrhotite, chalcopyrite, and arsenopyrite.

In some deposits the auriferous ore bodies are intersected by later pegmatite veins, near which the ores are noticeably depleted. An example is the Antelope deposit in Southern Rhodesia (Phan, 1965); specialists assume that the gold was mobilized by the fluids of the pegmatite melt.

Metamorphic rocks contain gold in amounts on the whole comparable with their concentrations in sediments. In a number of cases an increase in the degree of metamorphism of rocks has been accompanied by an increase in their gold content; investigators have noted this specifically in certain regions of Canada (Boyle, 1961) and in the Selemdzhin area of the Amur basin (Moiseenko, 1965). On the other hand, regions are known where metamorphosed rocks contain less gold than the original rocks. Thus, the clay shales of the Vitim-Patom Highland contain 0.9 mg/t Au, whereas their derivatives in the sericite-chlorite and biotite-chlorite metamorphic facies contain 0.3–0.5 mg/t. Variable gold contents, in places as much as 1.4 mg/t, are typical of the rocks of the epidote-chlorite facies (Petrov et al., 1970). Such examples are not unique.

Gold distribution is uneven in zones of contact metamorphism. In the ore fields of the Selemdzhin region the gold content near intrusives is almost the same as in the original unmetamorphosed rocks; mainly silver has been removed (Moiseenko, 1965). Cases are known where gold-bearing rocks during metamorphism to amphibolites have lost their gold (Boyle, 1961). Apparently the mobility of gold depends on the degree to which the rocks have been heated, and the extend of the penetration of emanations into them.

Concentrations of gold in hydrothermal deposits attain maximum values exceeding the clarke by many orders of magnitude. A characteristic feature is the sharply pronounced unevenness in the distribution of gold in the different products of hydrothermal activity; along with areas containing numerous occurrences of slightly gold-bearing and "barren" hydrothermal veins and veinlets there exist deposits in which gold is concentrated in amounts ranging up to tens, hundreds, and even thousands of grams per tonne.

Gold is an accessory element in the ores of many other metals, principally copper, lead, zinc, nickel and platinum; it is also found in deposits of tin, antimony, mercury, molybdenum, uranium, etc.

A characteristic feature of gold is its dispersion in the haloes of hydrothermally altered rocks associated with gold deposits. The gold content of these altered rocks is in most cases one to two orders of magnitude higher than the background values; investigators have noted higher gold contents both near ore bodies and at distances of tens and hundreds of metres outward from them in places where jointing and hydrothermal action on the rocks are marked. Narrow zones with low gold contents are present within the limits of such haloes in a number of gold fields (Berikul'skii, Tsentral'nyi, and Darasun). These border gold-quartz veins and are considered to be the result of the removal of gold from the rocks into the vein sites (Roslyakova et al., 1970). This pattern is not common, since frequently the zones around ores are enriched with gold and provide ore.

The above information indicates that gold is very widely dispersed in nature, and that large quantities of the metal occur outside the sphere of influence of the processes whereby gold ores are concentrated.

If we assume, according to Zvyagintsev's calculations (1941), that the total gold content of the Earth's crust is approximately 40 million tonnes, and that the rocks of the floors of the seas and oceans, according to Babicka (1943), contain 10–20 million tonnes, it follows as a rough estimate that gold deposits and ores from which the metal can be recovered as a by-product, contain only 0.003–0.005% of the total amount of gold available in the upper parts of our planet. Somewhat less than half of this amount has already been won by mankind, has entered the gold stock of the anthroposphere, and has thus been withdrawn from the geochemical cycle. This stock may increase sharply when it becomes technically feasible to extract gold from seawater and rocks containing very large amounts of the dispersed metal.

The main cycles of gold migration. As our brief review of existing concepts has shown, many areas in the geochemistry of gold are still unknown. Scientists have hitherto shown little agreement concerning the capacity of gold for migration under natural conditions. V. I. Vernadskii (1922) stressed that "it (gold) goes readily into solution and is just as readily precipitated . . . ." Another comparatively widespread viewpoint is that gold does not belong to the mobile elements (Kharitonov, 1936; S. S. Smirnov, 1936). Both views are to a certain extent true: gold sparsely dispersed in rocks and ores is readily and in large amounts drawn into a broad geochemical cycle, whereas major concentrations (deposits) of the metal usually undergo only partial, mainly mechanical, dispersal during weathering.

The nature of the endogenic cycle of the migration of gold remains controversial. Scientists consider that the cycle includes the gold of the subcrustal parts of the planet, of the deep ultrametamorphic regions, and of the ordinary crustal magmatic chambers and their enclosing rocks.

The entry of gold from subcrustal depths is borne out by its constant presence in higher amounts in rocks and ores that are differentiates of basaltic magmas whose sources are apparently located within the limits of the upper mantle (Godlevskii et al., 1970). We should also keep in mind the vast dimensions of the auriferous belts that can be related to the position of the ultradeep planetary zones of faulting in the Earth's crust. The idea of the active role of ultrametamorphism in the generation of gold-bearing plutonic solutions is based on investigations of granitization processes. The gold depletion of highly metamorphosed rocks mentioned above is considered as proof that the metal was removed from the rocks during high grade metamorphism.

A number of investigators have defended the hypothesis of the metamorphogenic migration and concentration of gold originally dispersed in the rocks of auriferous regions (Schneiderhöhn, 1955; Shcherbakov and Perezhogin, 1964; Badalov, 1965; Buryak, 1968, and others). The proof of this hypothesis can be summarized as follows: Researchers have established that many gold deposits are confined to sedimentary-volcanic rocks containing a higher gold clarke, and they have recognized that the gold content of rocks depends on the degree of their metamorphism. Furthermore, the increased mobility of gold at high temperatures has been proven experimentally. These data support the conclusion that the capacity of gold to migrate is increased under the conditions of high-temperature metamorphism. However, these data do not solve the problems of (a) the scale of migration of gold or (b) the reasons for major concentrations of the metal. Certain data oppose such a possibility. In many gold-bearing regions geologists have not observed any relationship between the concentration of gold in deposits and the degree of metamorphism of the host rocks. For example, the auriferous belt of the Yenisey Ridge intersects zones of phyllites, hornfels and highly metamorphosed rocks of biotite grade schists containing garnet, andalusite, and so on. Gold concentrations are also known in extremely little-altered sandstone-conglomerate deposits (Transbaikalia). The auriferous enrichment of rocks in gold-bearing regions indicates rather that the element is introduced. The rearrangement of gold (the depletion phenomenon near post-ore dikes and intrusions) has been established only in very local situations. This whole problem needs further research.

Ideas on the behaviour of gold in magmatic processes have remained largely hypothetical. According to present-day concepts, magmas may have accumulated gold in the course of their differentiation, or they may have absorbed the metal from surrounding rocks, as well as from currents of subcrustal emanations along whose path of movement intrusives were located. When the magmatic chambers cooled, the gold distribution changed regularly. Originally it was dispersed throughout the mass of the melt together with petrogenic elements. An analysis of the variations in the gold content of magmatic rock complexes, and thermodynamic calculations (Vincent and Crocket, 1960), agree with this conclusion, one which was formulated in the early stages of research on the geochemistry of gold (Vernadskii, 1922).

When the magma cooled, some of the gold in the form of atomic groupings was trapped in an unstable state in the defects of the crystal lattices of silicate minerals. We consider the assumptions of some investigators that the presence of stable silicates of gold and isomorphic replacements in rock-forming silicates (Shcherbakov, 1967) to have little basis in fact. The diffusive mobility of gold, stimulated by high temperatures, should have caused the breakdown of these compounds and the separation of the native metal. I.K. Davletov (1970) came to a similar conclusion.

The concentration and removal of gold from magmatic chambers took place only as the melt cooled, concomitant with an increasing concentration of fluid components. Further migration and precipitation of gold resulted from changes in the composition and temperature of thermal springs during their ascent and are closely related to the mobility and accumulation of silica, sulfur and alkalis, whose chemistry varied at different depth levels in the earth. We shall return to a consideration of the forms of gold transfer in connection with questions of the origin of gold deposits. We would only note here that in the course of hydrothermal processes gold has a protracted mobility, and that the gold once deposited underwent repeated dissolution and regrouping.

The exogene cycle of gold in its deposits begins with the formation of a zone of oxidation, the leaching out of gold, its transfer in the form of new complex compounds, and, under certain conditions, its secondary concentration. The oxidation and mechanical disintegration of gold ores creates conditions for the release of particles of the native metal, with their subsequent transportation in drainage systems and their accumulation in placers. The erosion of auriferous rocks is accompanied by the removal of large amounts of finely dispersed gold, which is transported by rivers to sea and ocean basins, where the metal takes part in sedimentation processes.

The associates of gold, originating in the early cycles of its migration, are dissimilar. During the endogenic cycle, gold compounds are diverse and generally not very stable, while gold associations are the most complex in terms of composition. At high temperatures the volatility of gold, and its capacity for forming complexes, are plainly evident, and at low temperatures gold often forms gels (colloids). The principal features of the exogenic cycle are governed by the processes of mechanical redistribution of gold and its accumulation by gravity. The soluble compounds in the exogenic cycle include the stable chloride, or possibly oxygen compounds; sometimes gold-organic complexes predominate, whereas the sulfur compounds (including those bound as alkalis) are unstable.

The succession of cycles leads to the redistribution of gold, and their recurrence in the course of time yields deposits of complex origin.

REFERENCES

Ahrens, L. H. The significance of the chemical bond for controlling the geochemical distribution of the elements, part 1. Physics and Chemistry of the Earth, 1964, v. 5.

Anoshin, G. N., Emel'yanov, E. M., and Perezhogin, G. A. Gold in present-day sediments of the northern Atlantic Ocean Basin. Geokhimiya, 1969, No. 9.

Babicka, J. Gold in Lebewessen. Microchemie, 1943, Bd. 31, N4.

Badalov, S. T. Some considerations on the origin of quartz-gold ore veins. Problems of Postmagmatic Ore Formation, Vol. II, Prague, 1965.

Beamish, F. E. The analytical chemistry of the noble metals. Oxford, Pergamon Press, 1966.

Bilibin, Yu. A. Some features of gold metallogeny. Journal of the All-Union Mineral Society, 1947, No. 1.

Boitsov, A. V., Boitsova, G. F., and Avdonin, N. A. Noble metals. Metallurgy Press, 1946.

Boyle, R. W. The geology, geochemistry and origin of the gold deposits of the Yellowknife district, Canada. Geol. Survey Canada, Mem. 310, 1961.

Bulynnikov, A. Ya. The genetic relationship between gold mineralization and magmatic rock groups and complexes. Bulletin of the Tomsk Polytechnical Institute, 1968, Issue 134.

Buryak, A. A. The effect of regional metamorphism processes and host rock composition on the development of different types of gold mineralization (using the Baikal region as an example). In the collected papers entitled Transbaikal Mineral Deposits: Geology and Exploration. Chita, 1968.

Buryak, V. A., and Popov, N. P. Types of genetic zoning of mineralization in Precambrian gold-bearing provinces (Patoma highland). Izv. Tomsk. politekhn. in-ta, 1968, Vol. 134.

Chebotarev, G. M. Gold distribution in the sandy-shale (slate) rocks of Muruntau. Reports of the Academy of Sciences, Uzbek Soviet Socialist Republic, 1969, No. 6.

Chukhrov, F. V. Gold migration in an oxidation zone. Journal of the Academy of Sciences USSR, Geological Series, 1947, No. 4.

Davletov, I. K. The behaviour of gold during the crystallization of intrusive rock, based on the example of the Chatkal zone in the Tien Shan. Bulletin of the Moscow Naturalists Society, Geochemistry Division, 1970, No. 5.

Fersman, A. E. A contribution to the geochemistry of gold. Dokl. AN SSSR, Series A, 1931, No. 8.

Fersman, A. E. Geochemistry, Vol. IV. Amalgamated State Scientific and Technical Publishing House for Chemical Literature, Leningrad, 1939.

Garrels, R. M. and Christ, C. L. Solution, minerals and equilibria. N.Y., 1965.

Godlevskii, M. N., Razin, L. V., and Konkina, O. M. The gold content of Noril'sk-type differentiated intrusions. Proceedings of the Central Scientific Research Institute for the Geological Exploration of Nonferrous and Noble Metals, 1970, Issue 87.

Goldschmidt, V. and Peters, C. Zur Geochemie der Edelmetalle. Nach. Gess. Wiss. zu Göttingen, Math. Phys.k. 1, 1932.

Hausen, D. M. and Kerr, P. F. Fine gold occurrence at Carlin, Nevada. In: Ore Deposits of the United States, 1933-1967. Amer. Inst. Mining Metallurg. and Petroleum Engrs., Inc. N.Y. 1968.

Karnozhitskii, A. N. Gold currency before the court of natural history. Scientific Review, 1898, Nos. 2 and 3.

Kay, J., and Libby, T. L. Tables of physical and chemical constants. State Publishing House for Physics and Mathematics Literature, 1962.

Kharitonov, P. A. Sedimentary gold deposits as a geochemical problem. Problems of Present-Day (Soviet) Geology, 1936, No. 6.

Koeberlin, F. K. An hypothesis as to origin of gold in volcanic ash. Engng. Mining J. 1934, v. 135, N9.

Krauskopf, K. B. The solubility of gold. Econ. Geol., 1951, v. 48.

Kropachev, G. P. The occurrence of gold in nature. Soviet Gold Mining, 1935, No. 8.

Letnikov, F. A. Principles governing the appearance of native elements. Journal of the Academy of Sciences, Kazakh Soviet Socialist Republic, Geological Series, 1963, Issue 6/57.

Maslenitskii, I. N., and Chugaev, L. V. The metallurgy of noble metals. Metallurgy Press, 1972.

Mason, B. Principles of geochemistry, N.Y., 1966.

Moiseenko, V. G. Metamorphism in Amur Valley gold deposits. Khabarovsk, 1965.

Moiseenko, V. G., Shcheka, S. V., Fat'yanov, I. I., and Ivanov, V. S. Geochemical features of gold distribution in the rocks of the Pacific Ocean Belt. Izd-vo Nauka, (Science Press), 1971.

Naboko, S. I., and Glavatskikh, S. F. Mineral-forming thermal springs in regions of active volcanism. Reports of the Academy of Sciences USSR, Geological Series, 1969, No. 1.

Nifontov, R. V. The chemogenic gold content of sedimentary rocks in some mining districts. In the collected papers entitled, Methods of Investigating Raw Minerals. State Scientific and Technical Publishing House of Literature on Geology, Geodesy, and the Conservation of Mineral Resources, 1957.

Ogryzlo, S. P. Hydrothermal experiments with gold. Econ. Geol., 1935, v. 30, N4.

Palei, L. Z., Murovtsev, A. V., and Borozenets, N. I. A contribution to the geochemistry of gold in the Sultanuizdag. Uzbek Geological Journal, 1967, No. 6.

Petrov, V. G. Endogenic deposits in the northern Enisei Ridge. Author's Abstract of Thesis for Candidate's Degree, Novosibirsk, 1969.

Petrov, V. G., Krendelev, F. V., Bobrov, V. A., and Tsimbalist, V. G. The behaviour of radioactive elements and gold during the metamorphism of sedimentary rocks in the Patoma highland. Annual report of the V.I. Vernadskii Institute of Geochemistry and Analytical Chemistry, Order of Lenin, Siberian Branch, Academy of Sciences USSR for Papers in 1969, Irkutsk, 1970.

Peshchevitskii, B. I., Anoshin, G. N. and Erenburg, A. M. Chemical forms of gold in seawater. Reports of the Academy of Sciences USSR, 1965, Vol. 162, No. 4.
Phan Kiêu Duong. Enquite sur l'or dans les roches. Origine de l'or des gisements, Chronique mines et rech. minière, 1965, v. 33, N343.
Plaksin, I. N. The metallurgy of noble metals. Metallurgizdat (Metallurgy Press), 1958.
Polikarpochkin, V. V., Korotaev, I. Ya., and Gapon, A. E. The concentration of chemogenic gold in sedimentation. Ezhegodnik GEOKhI SO AN SSSR, Irkutsk, 1968.
Radtke, A. S. and Scheiner, B. J. Studies of hydrothermal gold deposition; Carlin gold deposit Nevada: the role of carbonaceous materials in gold deposition. Econ. Geol., 1970, v. 65, N2.
Razin, L. V., and Rozhkov, I. S. The geochemistry of gold in the weathering crust and biosphere of Kuranakh-type gold ore deposits. Moscow, Izd-vo Nauka, 1966.
Roslyakova, N. V., Roslyakov, N. A., and Zvyagin, V. G. Gold behaviour in the primary haloes of some veined gold ore deposits. Izv. Tomsk. politekhn. in-ta, 1970, v. 239.
Rozhkov, I. S., Nikitin, N. M., and Yasyrev, A. P. New data on the gold content of sedimentary strata in the central Russian platform. Dokl. AN SSSR, Vol. 173, 1967, NO. 5.
Schneiderhöhn, H. Erzlagerstätten, Kurzvorlesungen zur Einfuhrung und zur Widerholung. Jena, 1955.
Shcherbakov, Yu. G. Gold distribution and concentration conditions in ore provinces. Izd-vo Nauka, 1967.
Shcherbakov, Yu. G. The distribution of elements in the earth, and the ore content of magmas. Proceedings of the Siberian Scientific Research Institute of Geology, Geophysics and Raw Minerals, 1969, Issue 90.
Shcherbakov, Yu. G., and Perezhogin, G. A. A contribution to the geochemistry of gold. Geokhimiya, 1964, No. 6.
Smirnov, S. S. The oxidation zone of sulfide deposits. Izd-vo AN SSSR. Moscow and Leningrad, 1936.
Tooms, J. S. Metal deposits in the Red Sea. Their nature, origin and economic worth. Underwater Sci. and Technol. J., 1970, v. 2, N1.
Vernadskii, V. I. Descriptive mineralogy: Background, Vol. 1. St. Petersburg, 1922. Izbr. soch. (Selected Works), Vol. II, Izd-vo An SSSR (Academy of Sciences USSR Press), 1953.
Vernadskii, V. I. History of minerals in the earth's crust, Vol. I. Leningad, 1927. Izbr. soch. (Selected Works), Vol. IV, Izd-vo AN SSSR, 1959.
Vincent, E. A. and Crocket, J. H. Studies in the geochemistry of gold. Geochim. et cosmochim. acta, 1960, v. 18, N 1-2.
Vinogradov, A. P. Patterns in the distribution of chemical elements in the earth's crust. Geokhimiya, 1956, No. 1.
Vinogradov, A. P. The geochemistry of rare and trace elements in soils. Academy of Sciences USSR Press, 1957.
Vinogradov, A. P. The average chemical element content of the main types of igneous rock in the earth's crust. Geokhimiya, 1962, No. 7.
Voskresenskaya, N. T. and Zvereva, N. F. Some aspects of the geochemistry of gold in connection with the mineralization of magmatic complexes in northern Kazakhstan. Geokhimiya, 1968, No. 4.
White, D. E. Thermal springs and epithermal ore deposits. Econ. Geol. Fiftieth Anniversary Volume, 1955, v. 2.
Zvyagintsev, O. E. The geochemistry of gold. Izd-vo AN SSSR, Moscow and Leningrad, 1941.

The monograph by Boyle (1979), *The Geochemistry of Gold and Its Deposits,* contains a comprehensive compilation of the geochemistry of the element, describes the principal types of auriferous deposits, and discusses their origin. In addition a chapter deals with geochemical prospecting for all types of gold deposits. An abstract of the geochemical part of the monograph is given in chapter 1, including the abundance of gold in the upper lithosphere and the average content of the precious metal in the common rock types, normal soils, and natural waters.

The monograph by Bache (1982) contains a short section on the geochemistry, mineralogy, and associates of gold, and is followed by chapters on the classification and description of gold deposits and their distribution throughout the world.

Summarizing this part of the chapter it can be said that the assaying of native gold and artificial gold alloys by pyrometallurgical methods was practiced with relatively high precision as early as 2000 B.C. and probably earlier. The touchstone, likewise, is an ancient method of analyzing native gold nuggets and artificial alloys, being employed as early as 600 B.C. Major improvements in assaying were introduced during late medieval times and were described in great detail during the Renaissance by Biringuccio and Ercker. Many of the methods described by these two authors, with some modifications, continued to be used until the middle of the twentieth century for auriferous ores. During the last decade of the nineteenth century and first half of the present century determination of the content of gold in meteorites and the materials of the five spheres of the earth were carried out principally by a combination of assay and spectrochemical methods. The sensitivity of these methods was relatively high, permitting determinations of the order of a few parts per billion. Thus the figure for the terrestrial abundance of gold (1-5 ppb) given by a number of geochemists at the turn of the century, and for several decades thereafter, differs little from that given by geochemists using neutron activation and other sophisticated methods in later decades of the present century.

SPECIALIZED PAPERS ON THE GEOCHEMISTRY OF GOLD

A number of specialized papers dealing with certain critical aspects of the geochemistry of gold have appeared since the midpoint of the nineteenth century. Some deal with the distribution of gold in the oceans; others present data and discuss the various modes of migration of gold in hypogene (endogene) and supergene environments. Only a few of the latter are considered here; references to other papers of no less importance on the subject are given in Gmelin (1950-1954), Petrovskaya (1973), and Boyle (1979).

The French chemist, J. L. Proust (1754-1826) was the first to report the presence of gold in sea water, and the English chemist S. Sonstadt carried out the first quantitative analysis finding less than a grain of gold per ton (about 65 ppb). His research, reported in the *Chemical News* in 1872, is of historical and landmark importance. In the introduction to his paper *On the Presence of Gold in Sea-water* (p. 159) he indicates the content of the metal present and the difficulties of determining the amount of gold in sea water.

> I have used three entirely different methods for the detection of gold in sea-water, but all the methods were applied to the water itself, not to the residue left on evaporation. The experiments have been made upon specimens collected at different times from different parts of Ramsey Bay, Isle of Man, and the results obtained from the different specimens have been in entire accordance. The proportion of gold contained in sea-water (certainly less than one grain in the ton) is much too small to admit of separation, or even detection, by the usual tests applied in the usual manner. Besides the difficulty of detection arising from the small proportion of gold present, there is another difficulty of a graver kind, due to a continuous re-solution of the gold after it has been separated in the metallic state. This re-solution is owing to the separation of iodine under the influence of reducing agents upon the iodate of calcium, which, in a paper published in the *Chemical News,* I have shown to exist in sea-water. Even if the reducing agent is added in very large excess, oxidation takes place so rapidly under the continuous re-forming power of iodate of calcium, that, sooner or later, according to the excess of reducing agent used, the stage arrives at which iodine is set free, and the suspended gold is re-dissolved.

Many investigators of the gold content of the oceans were to follow the lead of Sonstadt. Fritz Haber (1927, 1928) carried out investigations of the gold content of ocean water for a period of some eight years principally for the purpose of winning gold from the oceans to pay the enormous reparations imposed on Germany after the First World War. He found much less gold (0.003-4.8 ppb) than formerly estimated, and noticed wide variations in the oceans both on a local and areal basis and attributed them to a variety of causes, mainly to the presence of gold in various forms, such as the chloroauric ion, in colloidal form and absorbed by both organic and inorganic matter. In recent years a number of investigations of the gold content of ocean water, based on selective extraction and neutron activation methods, have shown that the average gold content of the oceans is about 0.0112 ppb and that of silver about 0.3 ppb, giving an Au/Ag ratio of 0.04 (Boyle, 1979).

The nature of gold in endogene (hydrothermal) solutions and the mode of its solution and precipitation have been the subjects of considerable study since 1850. Two general concepts have prevailed: one that the gold migrates in true solution and the other that gold is transported as a colloid. Both concepts are reviewed in detail in Boyle (1979).

Among the most significant early papers dealing with the transport of gold in endogene (hydrothermal) solutions are those by Bischof, Doelter, Skey, Egleston, Becker, Liversidge, Stokes, and Don. Their results are summarized by F. W. Clarke in Paper 8-2. All of these investigators concluded that gold migrated under endogene conditions in true solution in solvents (waters) that included such components as potassium or sodium silicate, sodium carbonate, alkaline sulfides, or chlorides.

Two papers by V. Lenher (1912, 1918) summarize his research and that of others on the transportation and deposition of gold in nature and provide landmarks for the early years of the present century. Paper 8-7 is a summary of the facts as known in 1912.

8-7: THE TRANSPORTATION AND DEPOSITION OF GOLD IN NATURE

Victor Lenher

Reprinted from *Econ. Geology* 7:744-750 (1912).

Of all the factors which play an important part in the genesis of ore deposits, the agency of solution is perhaps the most fundamental for it is by solution that the chemist is able to study and to attempt to imitate the chemistry of ore deposition as carried out in nature. Any information that will throw light on the character of possible ore-bearing solutions may be expected to aid in studying the many problems incident to the solution, transportation, and deposition of the metallic ores.

In connection with some recent chemical studies which have been made with gold, certain solutions have been worked with, the deportment of which toward various reagents as well as with certain minerals, indicates a degree of stability which appears to be of geological significance. Indeed certain gold solutions possess a stability from the purely chemical standpoint which one would not be likely to expect from our general knowledge of the ease with which gold is deposited out of most of its solutions by even the mildest reducing agents.

How gold is dissolved and transported in underground waters has not been clearly shown. The suggestive work of Stokes (Economic Geology, I., p. 650, 1906) on the solubility of gold in cupric chloride or in ferric chloride solutions at 200° with the redeposition of metallic gold on cooling, appears to afford a possible means of transportation of gold solutions at elevated temperatures with the subseqiest deposition of the gold by lowering of the temperature. The work of Emmons on "The Agency of Manganese in the Superficial Alteration and Secondary Enrichment of Gold-Deposits in the United States" (*Trans. Amer. Inst. Min. Eng.*, 1910, 767) together with the work of McCaughey (*Jr. Amer. Chem. Soc.*, 31, 1,263, 1909), and Brokaw (*Jr. of*

Geol., Vol. 18, 321, 1910), as well as the earlier work of Pearce (*Trans. Amer. Inst. Min. Eng.*, 22, 739, 1893), Rickard (*Trans. Amer. Inst. Min. Eng.*, 26, 978, 1896), McIlhiney (*Amer. Jr. Sci.*, 1896, 293), and Don (*Trans. Amer. Inst. Min. Eng.*, 27, 599, 1897), apparently require that free chlorine is the solvent for the gold in the first instance, or if not free chlorine, a solvent whose powers are practically equivalent to that of free chlorine. When the chloride solution of gold is the transporting solution, it seems obvious that the gold is subsequently deposited as metal by a reducing agent. The agency of manganese in the solution of gold and its transportation is consistent in many gold deposits with the accompanying manganese deposits and throws considerable light on the superficial transportation of gold.

The solubility of gold in such media as the alkaline cyanides can hardly be deemed of material importance from the viewpoint of a natural transporting solution. The action of concentrated sulphuric acid or strong phosphoric acid in the presence of oxidizing agents on gold will cause solution (Lenher, *Jr. Amer. Chem. Soc.*, 26, 550, 1904) but this solvent action requires a higher concentration of acid than can be expected in nature. Similarly though hydrochloric acid under pressure (Lenher, ECON. GEOL., 4, 562, 1909) and nitric acid (Dewey, *Jr. Amer. Chem. Soc.*, 32, 318, 1910) at atmospheric pressure dissolve gold, the solution occurs only in concentrated acids, and the facts are of no importance in seeking for natural solvents for gold.

From all of these acid solutions the precipitation of the gold is usually assumed to take place by ferrous sulphate, metallic sulphides such as pyrites, or, in the case of Stokes' experiments, by lowering of the temperature.

The alkaline solutions which can dissolve and carry gold have not as a rule received as serious consideration as transporting media as the better known chloride solutions. Indeed the literature on alkaline gold solutions is very meager. This is particularly true in regard to the alkaline sulphide solutions.

That gold can be brought into solution by means of the alkaline sulphides has long been known, but it is doubtful if the geological significance of the resulting solutions has been fully appreciated.

Probably at least as early as the time of Glauber it was known that gold can be rendered soluble by fusion with liver of sulphur. Stahl in the seventeenth century ("Observations Chymico-Physica Medicæ") is the first to bring out the fact clearly, and in doing so suggests that Moses burned the golden calf with sulphur and alkali and gave the solution to the children of Israel to drink.

In more recent times, Skey (*Trans. New Zealand Institute,* 5, 382, 1872) in studying the formation of gold nuggets in drift suggests the solution of gold in the alkaline sulphides as the medium by which gold can be carried. Eggleston (*Trans. Amer. Inst. Min. Eng.,* 9, 640, 1880–1881) in studying the formation of gold nuggets and placer deposits found spongy gold to be soluble in the alkaline sulphides. Becker (*Amer. Jr. Sci.* (3), 33, 207, 1887), in his studies on the mercury deposits of the Pacific coast, has shown that gold dust dissolves in sodium sulphide. He believed that some of the gold veins bear so considerable a resemblance to the quicksilver deposits that like the latter they were formed by precipitation from solutions of the soluble double sulphides. Liversidge (*Journal of the Royal Society of New South Wales,* 27, 303, 1893), in studying the question of the origin of gold nuggets, reviews the earlier work on the solubility of gold and finds it to be soluble in sodium sulphide.

These solubility experiments have been repeated in our laboratory and the fact corroborated that metallic gold is soluble in solutions of the alkaline sulphides. More significant, however, appears the fact that from these solutions of gold in the alkaline sulphides, iron pyrites will not throw out the gold.

As is well known, the metals, the metallic sulphides, and even many kinds of organic matter will precipitate gold from the solution of gold chloride. In the case of the alkaline sulphide solutions containing gold, neither pyrites nor metallic iron will precipitate the gold, but on the other hand, gold deposits out of these solutions by exposure of the solution to the air, under which condition the sulphide oxidizes.

The sulphide solutions of gold are permanent stable solutions to the ordinary reducing agents, that is, to such reducing agents as precipitate gold from the chloride solution. These alkaline sulphide solutions deposit their gold content by contact with acid or by exposure to oxidation. Not only are the alkaline sulphide solutions of gold stable to the metallic sulphides, but experiments made in sealed tubes have demonstrated that sodium, potassium, ammonium, or calcium sulphide solutions will dissolve gold leaf in the presence of pyrites without any deposition whatever of gold on the pyrites.

It is therefore obvious that through the agency of the alkaline sulphides it is possible for gold to be transported in alkaline sulphide solution through a bed of pyrites without deposition of metallic gold, and indeed it is possible to think of such a water passing through a bed of gold-bearing pyrites actually enriching itself by solution of the gold from the pyrites. To follow such a solution farther, it can be conceived that gold can be carried through a reduced zone and later the gold can be deposited by meeting acid in the reduced zone, or in absence of acid can be carried indefinitely until it reaches a zone of oxidation when the metal would be deposited.

When sodium thiosulphate is allowed to act on gold in the presence of oxygen, the double thiosulphate of gold and sodium is formed. This double thiosulphate is also formed when auric chloride and sodium thiosulphate are brought together in solution. This salt possesses remarkable stability in that the dilute acids, hydrochloric or sulphuric, do not at once decompose it, nor does ferrous sulphate or oxalic acid, two of the most common precipitating agents for gold, reduce it to metallic gold at once. All of these reagents do in time or in stronger solutions precipitate the gold from this thiosulphate compound. The extraction of metallic gold from silver ores in the thiosulphate extraction process depends on the formation of the double thiosulphate (Stetefeldt, "The Lixiviation of Silver Ores with Hyposulphite Solutions," pp. 15, 38).

These thiosulphate solutions are reasonably stable to iron pyrites, but on standing, metallic gold slowly deposits on the pyrites.

The sulphite solutions of gold are another example of a means in which gold can be held in alkaline solution. Curiously enough the double sulphite of gold and ammonium is quite stable, while the double gold sulphites of the alkalis are not nearly so stable.

Von Haase (*Chemiker Zeitung,* 535, 1869) has studied the double sulphite of gold and ammonium. Haase worked in ammoniacal solution and was able to crystallize the double salt out of solution. The ammoniacal solution as prepared by Haase has been made and has been found to be quite stable to iron pyrites and to metallic iron. Ammoniacal sulphite solutions have been preserved for months in stoppered flasks and when tested from time to time by withdrawing small portions and acidifying, metallic gold is instantly precipitated. These ammoniacal sulphite solutions when sealed in tubes with iron pyrites or with metallic iron deposited no gold in months.

Both the sodium and potassium gold sulphites have been described by Haase. In solution, these salts are more unstable than the corresponding ammonium compound. The sodium as well as the potassium gold sulphite solutions yield gold to pyrites by a few minutes' contact. In reality a small quantity of the potassium or sodium gold sulphite added to the ammonium gold sulphite solution increases very much the instability of the latter toward reducing agents; indeed it is only necessary to add a small quantity of the sodium or potassium compound to the ammonium salt to cause the latter to lose its gold to pyrites practically as readily as though no ammonium salt had been present. This tendency on the part of the fixed alkalis to increase the instability of the ammonium gold sulphite solutions would seem to indicate that the sulphite solutions are not so plausible a means of transportation of gold in underground waters, inasmuch as in most natural waters sodium and potassium salts are present, while ammonium salts are found only in traces. Indeed natural ammoniacal solutions free from the alkalis are quite unknown.

The alkaline solutions of gold in the lower form of valence of gold, namely the aurous state, present some interesting phenomena when considered from the viewpoint of transportation of gold.

The aurous state of gold is produced when the ordinary or auric salts are reduced in a certain definite manner. The agencies by which this lower state of oxidation of gold can be produced are limited.

Aurous chloride and bromide are produced by the action of a moderate heat on the ordinary auric chloride or bromide. Experiments made recently in this laboratory show that by proper control sulphur dioxide, one of the best laboratory reducing agents and one of the common reagents used to precipitate gold from solution, can be used to effect the reduction of gold from the auric to the aurous state.

When sulphurous acid is added to a neutral or acid solution of gold without having present some other salt, it is very difficult to stop the reduction of the auric form of gold at the aurous state, the tendency being to produce complete reduction with the precipitation of metallic gold. If, however, a large excess of any of the alkaline chlorides, calcium, magnesium or zinc chlorides, be present, the reduction of the auric form of gold to the aurous state by means of sulphurous acid can be readily controlled. When an ordinary auric chloride solution is treated with a large excess of one of the above mentioned chlorides, and the solution then treated with a solution of sulphurous acid, the amber yellow color of the auric chloride gradually fades until the solution is rendered completely colorless. This colorless stage represents the existence of the aurous form of the gold, and the gold exists in this solution in all probability as a double aurous chloride.

Solutions of gold prepared in the manner indicated are far more stable under certain conditions than the ordinary auric chloride solutions. These conditions in which marked stability has been observed are somewhat curious. When such a solution is kept out of air contact and when no free acid or at most very little free acid is present, the solutions are fairly stable; if, however, the solution is exposed to the air, gold begins slowly to deposit. As far as the experiments have gone on this line, it seems as though the precipitation of metallic gold from this partially reduced solution when exposed to the air is due to autoreduction to metallic gold by the oxygen of the air.

The partially reduced or aurous solution out of air contact is far more stable to pyrites than the ordinary or auric compounds. Here again is a suggestive solution so far as transportation of gold is concerned.

Of the various means of solution and transportation of gold which have been observed, it would appear that the alkaline sulphide solution may with study solve some at least of the problems of gold deposits. At all events, a solution is known which will not lose its gold to pyrites and yet will transport gold. The alkaline sulphides can easily be conceived as important natural solvents, and while other alkaline solutions such as the aurous solutions possess considerable stability, yet their formation in nature would not appear to be so likely as the sulphide solution.

In the paper entitled *Further Studies on the Deposition of Gold in Nature,* Lenher (1918, pp. 161-163, 183-184) gives data on the solubility of gold hydroxide at elevated temperatures 100°C-200+°C and discusses the behavior of gold chlorides toward calcite and magnesite under the action of heat and pressure. The introduction and observations of this paper are of considerable interest.

Introduction

The various explanations which have been offered for the deposition of gold in nature have accounted in a satisfactory manner for the formation of certain deposits. Unquestionably, the alkaline sulphides play a very important part in the transportation of gold. The alkaline sulphides can, however, transport gold only in a zone free from oxidizing agents, since the oxidizing agents as well as the acids precipitate metallic gold from its sulphide solution. From the sulphide solutions gold is not deposited by the common reducing agents, but on the contrary, its precipitation takes place either by oxidation or by acidification.

Presumably a large proportion of the gold which has been transported and deposited in the zone of oxidation has been carried in solution as the chloride or double chloride. From such a solution the gold is in a large measure deposited by reducing agents as pyrites or some other sulphide, by a ferrous mineral or by an organic reducing agent. Reduction from the chloride solution or from alkaline solutions is also accomplished in the zone of oxidation by the oxidized ores of manganese or by such oxidizing agents as will deposit gold by the phenomenon of autoreduction.

Some of the more important gold deposits, however, cannot be explained by the theories proposed. For example, the low-grade gold ores in quartz commonly consist of a vein or bed of quartz containing very finely divided gold disseminated through it. This gold may appear in particles which are sufficiently large that they can be observed with the naked eye, but for the most part the gold particles are usually so small that they are indistinguishable. Moreover, the low-grade gold ores are usually remarkably uniform so far as their gold content is concerned; that is to say, their precious metal content varies comparatively little, and, expressed in actual percentage of gold present, the variation is insignificant.

Again, the various theories offered for the transportation and deposition of gold fail to account for the general presence of silver with the gold. For example, the alkaline sulphide solutions, which are so important in the transportation of gold, do not appear to deport themselves similarly with silver. Silver sulphide is according to our experiments quite insoluble in the alkaline sulphide solutions and in the sulphaurate solutions. Hence the transportation of gold by the alkaline sulphides cannot explain the transportation of silver, nor can this method of transportation of gold in the light of our present knowledge be reconciled by the presence of silver in gold deposits. Up to the present time the presence of silver in gold is usually accounted for by the general principle that metallic silver will precipitate metallic gold from a gold solution as the two metals stand in that order in the electrochemical series, a fact which is so commonly evidenced in photography in the "toning" of a silver print by means of a gold solution.

Experiments recently conducted in our laboratory have shown quite another side to the deposition of gold in nature. The oxide of gold is rather remarkable for its stability. The oxide of silver, as is well known, like many of the silver compounds, breaks down rather easily into metallic silver and oxygen, under the action of only a moderate degree of heat. Gold oxide, on the other hand, is remarkably stable under the action of heat up to moderately high temperatures, when it gradually dissociates into metallic gold and oxygen. The fact that the oxide of silver and the oxide of gold are decomposed by sufficient elevation of the temperature and the fact that the oxide of gold can be formed so readily in nature, probably are of considerable importance in the formation of certain gold deposits.

Observations

From our experiments it is evident that the compounds of gold are more resistant to high temperatures when the pressure is that exerted by the expansion of steam, than they are at atmospheric pressures.

The action of calcium carbonate or magnesium carbonate, which are plentiful in nature both in the form of carbonate rocks and dissolved in the natural waters, is to first precipitate auric hydroxide when they come in contact with a water containing gold chloride; then if the temperature becomes higher than 310° under the pressure which steam would exert at those temperatures, which is approximately 100 atmospheres, crystals of metallic gold are produced.

It should be borne in mind that while the precipitation of auric hydroxide takes place readily when a gold chloride solution comes in contact with the alkaline earth carbonates, auric oxide or auric hydroxide do not occur in nature. No oxidized gold compound occurs in nature, and although we have repeatedly called attention to the stability of gold oxide, it must be distinctly borne in mind that its stability is only relative, and that in general the compounds of gold are the most easily broken down into metal of the compounds of any of the metals.

The action of various salts, such as magnesium chloride and sodium chloride, is to somewhat lower the decomposition point while the action of calcium chloride appears to have a slight tendency to prevent this decomposition.

Inasmuch as the alkaline chlorides are solvents to a slight degree for silver chloride, it is possible that the existence of silver in the native gold may be accounted for in this way. While experimental data are lacking on the coprecipitation of gold and silver by this procedure, yet we have ample evidence that calcium carbonate or magnesium carbonate in all of their natural forms precipitate silver carbonate, which at the boiling point of water or in the autoclave at 150° is completely reduced to metal.

The above experiments, in which an attempt has been made to

> imitate as closely as possible the conditions existing in nature, appear to be sufficiently suggestive to be offered as an explanation of at least how it is possible that certain gold deposits have been produced.

Transport of gold in hypogene (hydrothermal) solutions as chloride complexes has appealed to a number of investigators, particularly Ogryzlo (1935), Krauskopf (1951), and Helgeson and Garrels (1968). Ogryzlo (1935) examined the solubility of gold in water containing a number of components including chlorides, sodium carbonate, sodium sulfide and sodium bisulfide. The summary and conclusions from his paper *Hydrothermal Experiments with Gold,* (pp. 423-424) follow:

Summary and Conclusions

> 1. 20 per cent. hydrochloric acid vapor in a dynamic system has no action on gold between 250° and 600°C.
> 2. When chlorine and steam are passed over gold, $AuCl_3$ is formed. The volatilization of $AuCl_3$ in the presence of chlorine and steam begins below 125°C, increases rapidly to a maximum at 200°C, and then decreases rapidly to almost nothing at 400°C. The decrease in the amount of volatilization above 200°C is due to the fact that around this temperature the dissociation pressure of the $AuCl_3$ increases much more rapidly than the vapor pressure. It is possible that the presence of water vapor may have an effect on the temperature at which maximum volatilization takes place. Pressure would tend to prevent dissociation and would, therefore, affect the temperature of maximum volatilization.
> 3. Gold dissolves in weak hydrochloric acid at high temperatures and pressures. The amount dissolved increases with the concentration of the acid and the temperature and pressure. Pressure is an important factor in causing solution. Larger quantites of gold are dissolved in the presence of air than in the presence of carbon dioxide or nitrogen.
> 4. The writer did not find that aqueous solutions of alkali chlorides have an appreciable solvent action on gold in the presence of air or nitrogen and under high temperature and pressure conditions, although Lenher reports that they have.
> 5. Sodium carbonate was not found to have solvent action on gold at high temperatures and pressures. This is not in agreement with Doelter's work.
> 6. Stokes found that gold dissolves in hydrochloric acid solutions of ferric chloride at high temperatures and pressures and in the absence of oxygen. This was confirmed by the writer.
> 7. The writer found that Na_2S solutions dissolved only traces of gold at high temperatures and pressures, and none at room temperature. These results do not agree with those of Becker, Egleston, and Lenher.
> 8. On the other hand, considerable quantities of gold are dissolved by solutions of NaHS at high temperatures and pressures. Some gold is also dissolved by NaHS at room temperature.

In this investigation an attempt was made to imitate as closely as possible the conditions existing in nature. The experimental evidence shows that gold may be transported either in acid or slightly alkaline aqueous solutions and also in the vapor phase as $AuCl_3$. In acid solutions, gold is probably carried as a chloride. Precipitation from chloride solutions would result when the solutions come in contact with one of the many reducing agents which occur in nature. In hot alkali hydrosulphide solutions, gold is probably carried as a double sulphide of the alkali metal and gold. Oxidation or acidification of the solution would cause precipitation.

In a closed system, the partial pressures of $AuCl_3$ and its dissociation products would tend to prevent both volatilization and dissociation of the chloride after an equilibrium has been reached. Therefore, under high temperature and pressure conditions $AuCl_3$ is probably not carried in the vapor state. However, under somewhat reduced pressure conditions when the solutions and gases are in communication with the earth's surface, gold may be transported in the vapor phase as $AuCl_3$. It is difficult to place limits on the temperature at which volatilization would take place in Nature, as the pressure factor is uncertain.

Krauskopf (1951) calculated the solubilities of gold from electrode potential and free energy data and found results that agreed well with the experimental results reported in the literature. The summary from his paper *The Solubility of Gold,* (pp. 869-870) follows.

Summary

1. Solubilities of gold calculated from thermodynamic data agree reasonably well with experimental results.
2. In acid solutions gold may be transported as the ion $AuCl_4^-$, provided that reducing agents are absent. The metal is dissolved by an acid solution provided that the solution contains Cl^- and that a fairly strong oxidizing agent is present. At high temperatures and pressures the hydrogen ion of the acid is a sufficiently strong oxidizing agent; at low temperatures a substance like MnO_2, O_2, Fe^{+++}, or Cu^{++} must be present in addition. The requirement that an oxidizing agent be present, or at least that a reducing agent be absent, probably means that gold would not be transported in acid vein solutions at low temperatures, since such solutions would contain reducing agents like H_2S and Fe^{++}. On the other hand, solution and transportation of gold in acid solution is probably the mechanism of supergene movement of gold.
3. The solubility of gold in naturally-occurring alkaline solutions which do not contain sulfide is negligible.
4. Gold may be transported in alkaline sulfide solutions, even in dilute solutions near the neutral point. The gold is probably present as the very stable ion AuS^-.
5. Gold may be precipitated from solution by any one of a number of mechanisms, but there is no need to call on one more complicated than a fall in temperature or pressure or both.

6. If a solution of gold at high temperature and pressure is cooled, part of the metal may be forced out of solution but may appear as a sol rather than as a precipitate. In this form it may remain in suspension down to low temperatures.
7. Because gold may be transported in solutions of various compositions, the presence of gold in a vein deposit gives little information as to the character of the vein fluids.
8. In sea water gold is probably present as both AuO_2^- and $AuCl_4^-$. A limit to the amount dissolved may be set by the formation of these ions from metallic gold with the aid of dissolved oxygen.

Helgeson and Garrels (1968) also applied thermodynamic calculations to the solubility, hydrothermal transport, and deposition of gold. They concluded that sufficient gold can be carried in solution as aurous chloride complexes to account for hydrothermal gold ore deposits precipated above 175°C. The concluding section of their paper *Hydrothermal Transport and Deposition of Gold* follows.

Concluding Remarks

The solubility of gold has been the subject of repeated alchemical, experimental, theoretical, economic, industrial, and geologic investigation for thousands of years. Despite all of this attention, the origin of hydrothermal gold deposits is still poorly understood. Most theories of hydrothermal gold deposition fail to account for the geologic characteristics of gold ore deposits. Much of the pertinent experimental work has been semi-quantitative at best, and the fugacity of oxygen and pH are rarely controlled and/or monitored in such experiments. As a result, high-temperature gold solubilities predicted from thermodynamic data cannot be checked adequately against experimental results reported in the literature. As least for the present, the credibility of thermodynamic predictions of gold solubilities at high temperatures rests with the extent to which such predictions agree with the geologic occurrence of gold. In this respect, and from a thermodynamic and chemical standpoint, the geochemical model of gold deposition presented above appears to be realistic.

It has been demonstrated that gold can be dispersed in significant concentrations in aqueous solutions as a stable colloid, [Frondel, C., 1938. Stability of colloidal gold under hydrothermal conditions, *Econ. Geology* **33:**1-20.] as alkali thioaurate or gold sulfide complexes in alkali sulfide solutions [Krauskopf, K. B., 1951. The solubility of gold, *Econ. Geology* **46:**858-870; Smith, F. G., 1943. The alkali sulphide theory of gold deposition, *Econ. Geology* **38:**561-590.] and as $AuCl_4^-$ in highly oxidizing acid solutions [Cloke, P. L., and W. C. Kelly, 1964. Solubility of gold under inorganic supergene conditions, *Econ. Geology* **59:**259-270; Kelly, W. C., and P. L. Cloke, 1961. The solubility of gold in near-surface environments, *Mich. Acad. Sci., Arts, and Letters Papers* **46:**19-30; Krauskopf, K. B., 1951. The solubility of gold, *Econ. Geology* **46:**858-870.]. The results of this study suggest that none of these is a requirement for the hydrothermal transport and deposition of gold. Thermodynamic considerations indicate that gold is carried in hydrothermal solutions primarily in the aurous state and geologic observations suggest that less than 0.02 ppm gold in solution is sufficient to account for major gold ore deposits.

The order of magnitude of the mass transfer involved in the formation of major gold deposits can be determined by making a few simple calculations. For example, to precipitate a quartz vein 1,000 feet high, 1,000 feet long, and one foot wide over a temperature profile from 300° to 200°C requires 2.3×10^8 short tons or $7.6 \times 10^9 ft^3$ of solution. Assuming the solution is saturated with respect to gold along curve *a* (based on geologic consideration) or curve *g* (predicted from thermodynamic calculations) in Figure 9 [and Figure 10], 4.25 short tons of gold and 13,000 tons of pyrite would be precipitated from solution along with the quartz. At $35.00 per troy ounce of gold, the quartz vein would constitute a $4,400,000 ore body. Even assuming a conservative flow rate of 10,000 $ft^3\ day^{-1}$, it would take no more than 2,000 years to form this ore body, and the flow velocity in the vein would only be 0.2 cm $minute^{-1}$.

Hydrothermal ore deposition is an irreversible process of mass transfer that can be evaluated quantitatively for most types of ore deposits at high temperatures from a thermodynamic, chemical and mathematical standpoint. Geologic application of such calculations to a given ore deposit is hindered primarily by the lack of key geochemical data. These data are, specifically, fluid inclusion compositions and temperatures of filling, compositions and mass ratios of gangue, ore, and alteration minerals, and mineral associations in ore deposits and altered wall rocks adjoining veins (p. 634).

Figure 9. (Top). Computed solubility of gold at elevated temperatures. Curve *a* corresponds to curve *a* in Figure 2, which is based on geologic considerations. Curves *g* and *h* represent solubilities computed from thermodynamic data and mass transfer calculations for cooling hydrothermal solutions with different initial compositions at 300°C (see text). All three curves are for a solution in equilibrium with quartz, pyrite, and gold.

Figure 10. (Bottom). Schematic illustration of the gold-quartz vein precipitated along curve *g* in Figure 9. The relative proportions of quartz and pyrite are indicated by the hatched and clear areas, respectively. The gold values given above are based on a gold price of $35.00 per troy ounce and the vein width corresponds to the stoping width of ore.

(See over for figures.)

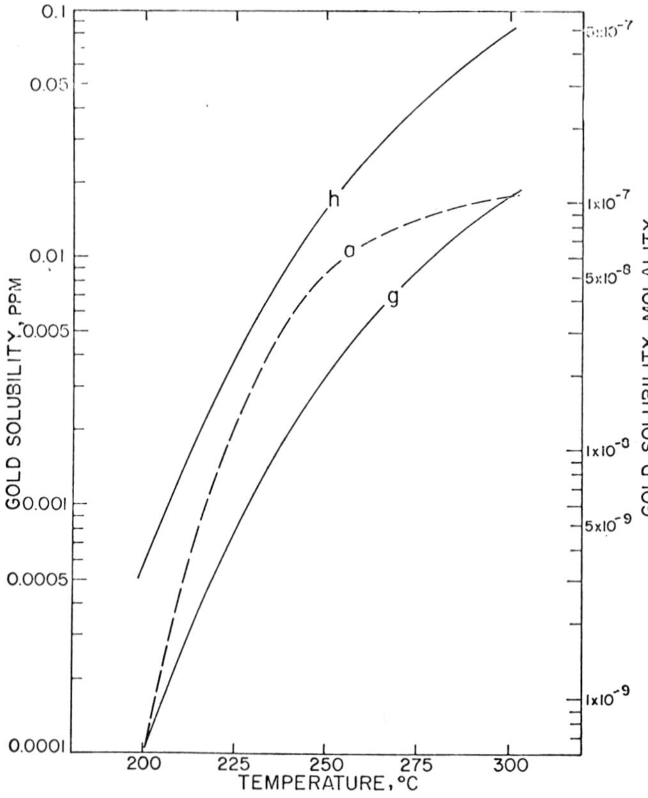
GOLD SOLUBILITY, PPM
0.1
0.05
0.01
0.005
0.001
0.0005
0.0001
GOLD SOLUBILITY, MOLALITY
1x10^{-7}
5x10^{-8}
1x10^{-8}
5x10^{-9}
1x10^{-9}
h
a
g
200
225
250
275
300
TEMPERATURE, °C

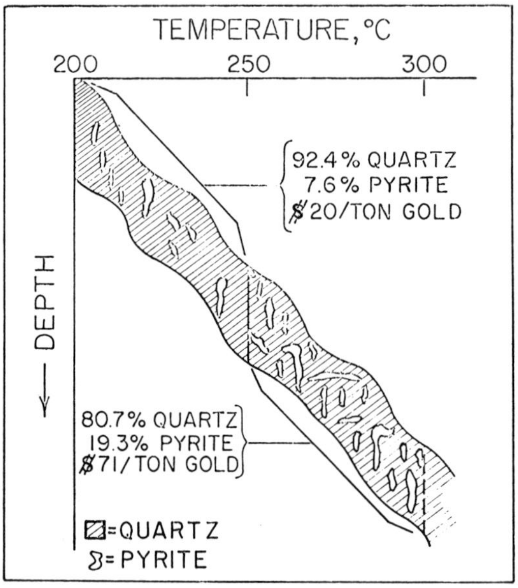
TEMPERATURE, °C
200
250
300
DEPTH
92.4% QUARTZ
7.6% PYRITE
$20/TON GOLD
80.7% QUARTZ
19.3% PYRITE
$71/TON GOLD
=QUARTZ
=PYRITE

Boyle (1969) has criticized this proposed acid-chloride transport mechanism from a number of angles, pointing to the general low content of chloride now present in gold veins and their wall rock alteration zones, the alkaline character of the wall rock alteration associated with gold deposition, the low transfer percentages of alumina in alteration zones associated with gold deposits, the general presence in gold deposits of carbonates that are difficult to precipitate from highly acid solutions, and the general alkaline nature of hot spring waters presently precipitating gold. Others have also criticized the chloride transport of gold under hydrothermal conditions, among whom may be mentioned Rytuba and Dickson (1977) and Weissberg (1970), the latter having shown experimentally and by examination of thermal waters presently precipitating gold in the Broadlands, Ohaki and Waiotapu areas, New Zealand, that the $[AuS]^-$ complex is more than adequate to account for the transport of gold in these near-neutral low-salinity hydrothermal solutions.

One cannot doubt that gold is solubilized as a chloride complex, as proven by both past and more recent data (Vilor and Shkarupa, 1971; Henley, 1973). Excepting conditions where minerals such as alunite are precipitated in hydrothermal gold deposits, an indication of relatively high acidity, it appears more probable, however, that gold is transported under hydrothermal conditions as a sulfide or sulfide-arsenide-antimonide complex.

Alkali sulfide, bisulfide, sulfide-arsenide-antimonide, and telluride complexing in hydrothermal solutions seems most probable considering the mineralogy and wall rock alteration effects in gold deposits. It has long been known that gold is soluble in alkali hydrogen sulfide (e.g., NaHS) solutions at low temperatures, and that gold sulfide is soluble in solutions of alkali sulfide and polysulfide in excess at low temperatures. At high temperatures gold is also soluble in concentrated alkali sulfide solutions. There is some agreement that the principal complex ion in such solutions is $[AuS]^-$, although some $[AuS_2]^-$ may also be present in the systems. In a recent paper Seward (1973), determined the solubility of gold in aqueous sulfide solutions from pH4 to 9.5 in the presence of a pyrite-pyrrhotite redox buffer at temperatures from 160°C to 300°C and 1 kbar. Maximum solubilities were obtained in the neutral region of pH. It was concluded that three gold complexes contributed to the solubility: $[Au_2(HS)_2S]^{2-}$ predominated in alkaline solution, $[Au(HS)_2]^-$ in the near neutral pH region, and $[Au(HS)]^0$ with less certainty in the acid pH region. The stabilities of the first two complexes are much greater than chloroaurate (I) species, a feature suggesting that the chloride complexes play only a subsidiary part in the transport of gold in neutral and alkaline hydrothermal solutions.

The sulfide and polysulfide complexes of gold are stable in aqueous solutions between a pH of 6 and at least 10. The $[AuS]^-$ complex is also relatively stable to metallic sulfides and a number of other reductants. Because many hot spring waters are alkaline, and other effects such as certain types of wall rock alteration suggest alkaline solutions, transport of gold as soluble alkaline sulfide complexes has appealed to a number of investigators as a transport mechanism for gold during the formation of hypogene deposits.

Smith (1943, p. 582-589) reviewed the literature and carried out an extensive series of experiments involving the solution and transport of gold, silver, and tellurium in alkali sulfide solutions. He was able to show experimentally that gold is soluble in, and can be crystallized from, such solutions, and that two other common gold minerals, electrum (Au, Ag) and calaverite ($AuTe_2$), can be similarly synthesized. The discussion and conclusion on the transport and deposition of gold, silver, and tellurium in his classic paper *The Alkali Sulphide Theory of Gold Deposition* deserve special attention because of their probable importance in the formation of hypogene gold ores (Paper 8-8).

(Text continues on page 173.)

8-8: THE ALKALI SULPHIDE THEORY OF GOLD DEPOSITION

F. G. Smith

Discussion

The solubility of metallic sulphides in aqueous solutions of sodium sulphide has been shown to be dependent on the sodium sulphide concentration. Thus anhydrous fusions of sodium sulphide and polysulphide are considerably better solvents for metallic sulphides than aqueous solutions of sodium sulphide and polysulphide. It was also shown that the simple metallic sulphides are precipitated if the sulphide concentration of solutions of sodium thiometallic salts is lowered sufficiently, and if it is very low, metals rather than their sulphides, can be made to separate. Since it has been postulated that natural ore-bearing solutions contain alkali sulphide, the geological conditions which will bring about a lowering of the sulphide concentration should be considered.

The factors in nature which lead to a lowering of the sulphide concentration of ore-bearing solutions may be highly complex, but a few of the simpler ones may be discussed. Loss of hydrogen sulphide by separation of the ore-bearing solution into two phases, one liquid and one gaseous, is considered to be of relatively minor importance, especially in the deeper types of ore deposits. It may be of importance, however, in the low-pressure regions of near-surface vulcanism. In such regions, escape of hydrogen sulphide has been repeatedly noticed, the sulfurous smell of fumaroles and hot springs in such regions generally being due to hydrogen sulphide. Of probably greater geological significance is the loss of sulphide ion due to the action of the ore- bearing solutions on the minerals of the rocks traversed. There are three general reactions which are of importance in this connection. One, not so common as the others, is due to oxidation of sulphide ion to sulphate ion by oxidizing minerals in the wallrock. The effect of this has been noticed at Keweenaw Point by Butler, Burbank, *et al.* (1929) [Butler, B. S., W. S. Burbank, et al., 1929. The copper deposits of Michigan, *U.S. Geol. Survey Prof. Paper 144,* 238p.]. Hematite in the lava flows through which the solutions circulated was converted into ferrous minerals and native copper was simultaneously precipitated. The apparent rarity of this type of reaction is probably due to the infrequent occurrence of minerals in a high state of oxidation making up a large percentage of crustal rocks. The second process, and doubtless the most important, is the conversion of iron minerals in the wallrock to sulphides, predominantly pyrite. In most ore deposits pyritization of the wallrock has occurred, sometimes in very large amounts, though the pyrite is usually disseminated in rather small crystals through the body of the altered rock. The third process, not as important, is the conversion of aluminum minerals in the wallrock to alunite $(Na,K)_2Al_6(OH)_{12}(SO_4)_4$, or minerals of that general type. This reaction does not appear to be very common in the ore deposits formed at relatively high pressures, but is very extensive in some regions, notably in the Goldfield district, Nevada (Ransome, 1907). Apparently in cases of alunitization, the sulphate concentration of the solutions responsible for the alteration is high, and this has been shown to take place in complex alkali sulphide solutions when the hydrogen sulphide is allowed to escape. This, of course, would be much more common under low-pressure than under high-pressure conditions. In reality, the first and third processes are two expressions of the same reaction, with concurrent loss of hydrogen sulphide and increase in sulphate ion. Another factor in reducing the sulphide concentration of ore-bearing solutions would be dilution by meteoric water. A condition might exist in which hydrothermal solutions of magmatic origin would enter the zone of abundant groundwater. The mixing of the two solutions would result in cooling, partial oxidation of sulphide ion to sulphate ion, and displacement of hydrolysis equilibria in such a way that the sulphide ion concentration would be greatly reduced. All of these changes would tend to break down the soluble alkali thiometallic compounds and precipitate metallic sulphides, or free metals. This is more than a hypothetical possibility, since it has been observed in the mercury deposits in the Terlingua District, Texas, by C. P. Ross (1941) [Ross, C. P., 1941. The quicksilver deposits of the Terlingua region, Texas, *Econ. Geology* **36:**115-142.], but it is not likely of great importance in most ore deposits.

Although gold is very soluble in alkali sulphide solutions, forming complex thioaurites, it separates from such solutions as the metal, and not as a sulphide, when the sulphide ion concentration is reduced to the point where the complex is broken down, even though the concentration of free sulfur (polysulphide ion) is high compared to the concentration of the sulphide ion. This is apparently due to the great instability of gold sulphide.

When tellurium is present in homogeneous systems including gold and sodium sulphide, the concentration of free sulfur (as polysulphide ion) influences the course of events as the sulphide ion concentration is reduced. If the amount of free sulfur is small, then some thiotellurite ion is decomposed, forming polytelluride ion, and this appears to react with thioaurite ion, precipicating gold polytelluride ($AuTe_2$). If the free sulfur concentration is large, all the thioaurite ion is decomposed before the thiotellurite ion, and thus metallic gold is precipitated before metallic tellurium, and gold telluride is not formed. It would therefore appear that the conditions of deposition of calaverite are somewhat restricted, and in nature these conditions are apparently not often satisfied, since metallic gold occurs much more frequently than the gold telluride. If both occur in a gold ore, metallic gold is generally found to have been precipitated after gold telluride.

There may be other methods of gold precipitation operative in nature. For instance, adsorption by colloids may be the reason why ubiquitous minerals such as pyrite and quartz often are auriferous. Some experimental support for this was obtained, and it is hoped that further data will permit publication of this supplementary information.

Conclusion: A Theory of Gold Ore Deposition

The types of gold ores which have been considered and to which the following theory of formation can be applied, are those which are generally referred to as "primary". They are generally in, or related to, faults in the rocks in which they occur, and contain many minerals which are not stable under atmospheric conditions, the most important being the sulphides of the heavy metals.

Following the generally accepted theory, it is believed that metalliferous ores including gold ores are derived from igneous intrusives. The process of separation of ore-bearing fluids from the magma is undoubtedly highly complex, but the general sequence of events is believed to be somewhat as follows. A large igneous intrusive is forcibly emplaced into rocks cooler than itself. The magma at the time of emplacement is at least partly in the liquid condition, the liquid part containing the elements necessary for the formation of the common rock-forming silicates, and in addition, the components of the magma which are present in relatively small amounts and which do not readily separate along with crystalline silicates. Of these, water is the most important, but compounds containing sulfur, chlorine, and the heavy metals are also present along with minor amounts of practically all of the common and rare elements. As the magma cools, silicates separate and the residual liquid becomes richer in the rare components we have mentioned. At a late stage in the crystallization of the magma, the residual liquid may be considered to be an aqueous solution containing some of the elements of the latest silicate minerals that have crystallized. The latest silicate minerals being generally alkali alumino-silicates, the aqueous solution in contact with them is believed to be strongly alkaline in the ordinary use of the term. This aqueous residual solution is expelled from the magma by forces which may include the continuation of those which intruded the parent magma, and also the force of increasing vapor pressure of the residual solution as it becomes richer in the volatile compounds, especially water. Channels of escape into the surrounding rocks may be provided by local crustal disturbance associated with the now nearly solid intrusive mass. The solutions derived from the magma in this way are believed to be the normal ore-depositing agents. This process of formation of ore-bearing solutions rests on the presumption that the pressure prevailing at all times is so high that only condensed systems are possible, so that no distinction need be made between solutions which, if the pressure were released, would expand indefinitely with no separation of a second fluid phase and those which would form a second fluid phase.

The composition of the ore-bearing solutions when first separated from the magma is believed to be alkaline, and the most characteristic compounds which they deposit on cooling are alkali silicates, quartz, and metallic sulphides. The oxycompounds are believed to be carried in the form of double oxides with alkali metal, such as sodium silicate, and breakdown of the complex double oxides precipitates simple oxides such as silica. Since the solutions also transport and deposit metallic sulphides, a consideration of the reactions necessary for this condition in alkaline solutions leads to the conclusion that alkaline sulphide must be present in abundant excess. As is well known, alkali sulphide readily forms double sulphides with metallic sulphides. It is believed that the metals are carried in the double sulphide form and are precipitated as the simple sulphides when the double complex is broken down. Thus the development of a theory of deposition of any one of the metals will depend largely upon the chemical properties of the double sulphides of that metal and alkali metals. Since sodium appears to be the most common alkali metal present in remnants of the ore-depositing fluids, such as liquid inclusions in minerals, and hot springs, the chemical data upon which the following theory of gold deposition is based will be largely the properties of the double sulphides of gold and sodium.

Processes in nature which affect the sulphide concentration of ore-bearing solutions will control the deposition of the metals. Action of the solutions on the minerals of the rocks through which they pass, escape of hydrogen sulphide, and dilution by meteoric water are probably the most important reactions which reduce the sulphide concentration. The wallrock may lower the sulphide and polysulphide concentration by changing iron-bearing minerals to pyrite. Less frequently the sulphate concentration may be lowered by the alteration of aluminum-bearing minerals to alunite. Such lowering of the sulphate concentration displaces the homogeneous equilibrium in complex alkali sulphide solutions in such a way that there is a concurrent lowering of the sulphide and polysulphide concentration. Loss of hydrogen sulphide by boiling at relatively low pressures is probably not common in nature. Escape of hydrogen sulphide, however, would displace the equilibrium mentioned above in such a way that the sulphide and polysulphide concentration would decrease and the sulphate concentration would increase. Dilution by meteoric water would lower the sulphide concentration and in addition would in some cases oxidize some sulphide ion to sulphate ion. Thus the deposition of metallic sulphides carried in ore-bearing solutions as complex double sulphides would result from pyritization and alunitization of the wallrock, loss of hydrogen sulphide, and dilution by ground water.

Gold is much more soluble than most of the common metals in alkali sulphide solutions. It is included in the group of metals whose sulphides are very soluble in dilute aqueous solutions of alkali sulphide at ordinary temperature. This group also contains mercury, bismuth, antimony, arsenic and tellurium. In the group of metals whose sulphides are soluble in aqueous solutions of alkali sulphide only at elevated temperature, the most soluble metal appears to be silver. Thus the natural association of minerals containing gold which are deposited late in the ore sequence with minerals containing silver, bismuth, and tellurium is explained. When the soluble double sulphides of gold and alkali are broken down by a lowering of the sulphide concentration, gold deposits as the metal, since its sulphides are very unstable above 40°C. In solutions containing both silver and gold, if the sulphide concentration is reduced to the point where metallic gold separates, it does so along with considerable silver in solid solution, forming argentiferous gold or electrum.

If tellurium is present along with gold in natural alkali sulphide solutions, gold tellurides may or may not precipitate as the sulphide concentration is reduced, depending on the state of the tellurium in the solutions. If the free sulfur concentration in the polysulphide form is high, then the tellurium is present as alkali thiotellurites, and when in this form metallic gold precipitates first and much later the thiotellurite complex is broken down and metallic tellurium is precipitated. If the free sulfur concentration is low, then the tellurium is present in part as polytelluride ion, and gold tellurides are first precipitated, followed later by the excess of metallic gold. These reactions have been duplicated by experiment, the order of solubility in alkali sulphide solutions being found to be, starting with the most insoluble—gold telluride, gold, tellurium. This explains the natural paragenesis of gold tellurides, gold, and tellurium, which apparently deposit in the same order.

Weissberg (1970) has reinvestigated the solubility of gold in alkaline sulphide solutions and drawn upon natural situations in gold-precipitating hot springs to substantiate his claim that the most effective gold transport mechanism at near-neutral pH is by aqueous solutions containing HS^- ions, as indicated next in one of the sections taken from his paper *Solubility of Gold in Hydrothermal Alkaline Sulphide Solutions* (pp. 554-555).

The Transport of Gold in Natural Hydrothermal Solutions

The experimental evidence indicates that the most effective gold transport in aqueous sulphide solutions occurs in near-neutral pH solutions containing a high proportion of sulphide as HS^- ions. In more strongly acid or alkaline solutions, where the sulphide species present are dominantly H_2S or S^{--} respectively, the solubility of gold is much lower and these solutions also seem less likely to play an important role in the hydrothermal transport of gold from the geological point of view. Some geological evidence for the chemistry of gold transport may be derived from a few natural hydrothermal systems available for study where ore grade concentrations of gold are currently being deposited as precipitates from hot springs and also from geothermal drillhole discharges. These hot springs occur in areas of recent volcanism and high heat flow where the geology is similar in many respects to that typical of many epithermal gold-silver deposits.

At Steamboat Springs, Nevada, some of the siliceous muds recently deposited by the hot springs contain as much as 10 ppm Au as well as 400 ppm Ag, 45 ppm Hg, and 4% of Sb [White, D. E., 1967. Mercury and base-metal deposits with associated thermal and mineral waters, in *Geochemistry of Hydrothermal Ore Deposits,* H. L. Barnes, ed., Holt, Rinehart & Winston, New York, pp. 575-631]. The metal-rich mud is transported to the surface by a near-neutral sodium-chloride-bicarbonate water that contains only 1-8 ppm total sulphide sulfur. At depth, the water temperature is about 175°C and the metals in the water are probably carried in true solution. Rapidly decreasing temperature near the surface probably accounts for the deposition of metastibnite (amorphous Sb_2S_3) and the gold and silver coprecipitated with it. Analyses of the water at Steamboat Springs for gold and silver are not available yet, but the water does contain appreciable amounts of arsenic (about 1.7 ppm) and antimony (0.3 ppm) and perhaps as much as 0.1 ppm of mercury [White, 1967 ibid.].

In New Zealand, precipitates rich in gold and silver are currently being deposited from hot springs and drill-hole discharges in three separate thermal areas along the eastern margin of the Taupo Volcanic Zone [Weissberg, B. G., 1969. Gold-silver ore-grade precipitates from New Zealand thermal waters, *Econ. Geology* **64:**95-108]. These precipitates occur as amorphous red-orange sulphides with opaline silica and contain up to 85 ppm Au, 500 ppm Ag, 2,000 ppm Hg, 1,000 ppm Tl, 2% As, and 10% Sb, although the transporting waters carry only 4×10^{-5} ppm Au, 6×10^{-4} ppm Ag, and 7×10^{-3} ppm Tl, 8 ppm As, 0.3 ppm Sb, and 120 ppm total sulphide sulfur.

The pH of this water, at the surface after cooling, ranges from 7.0 to 8.8 pH units. At depths near 1 km, the hydrothermal solutions are sodium-potassium-chloride-bicarbonate waters at 200-290°C (near the boiling point versus depth curve) with a calculated pH of 6-6.5, slightly alkaline with respect to pure water. The metals transported by the water at depth are probably in true solution.

The solubility of gold at depth under the natural conditions can be calculated from the known chemistry and mineralogical equilibria involved. An approximate value for the equilibrium constant of the reaction, $Au + HS^- = AuS^- + \frac{1}{2}H_2$, was calculated at 250°C, from the gold solubilities experimentally determined in near neutral bisulphide solutions at 250°C giving $K_{250°C} = 10^{-3.3}$ (based on $AuS^- = 10^{-3}$ m, $HS^- = 0.3$ m, and assuming that hydrogen, 5×10^{-4} m or $P_{H2} = 0.02$ atm, was formed only by the solution reaction with no diffusion of hydrogen through the walls of the gold container, and the HS^- activity at 250°C in the experimental system was equal to the sulphide molality at room temperature). Because the proposed solution reaction of gold with sulphide solutions involves the oxidation of gold to the aurous state and consequently the reduction of hydrogen ion to hydrogen, the solubility of gold in natural systems depends on the hydrogen fugacity, which is controlled principally by equilibria between the minerals pyrrhotite, pyrite, magnetite, and hematite. At 250°C, the hydrogen fugacity is about 10^{-4} atm in equilibrium with the assemblage pyrite-magnetite-hematite and may increase up to about 0.1 atm in equilibrium with the assemblage pyrite-magnetite-pyrrhotite [Holland, H. D., 1965. Some applications of thermochemical data to problems of ore deposits II. Mineral assemblages and the composition of ore-forming fluids, *Econ. Geology* **60:**1101-1166; Raymahashay, B. C. and H. D. Holland, 1969. Redox reactions accompanying hydrothermal wall rock alteration, *Econ. Geology* **64:**291-305]. Browne and Ellis [Browne, P. R. L., and A. J. Ellis, 1970. The Ohaki Broadlands hydrothermal area, New Zealand: Mineralogy and related geochemistry, *Am. Jour. Sci.* **269:**97-131.] showed that at Broadlands, New Zealand, f_{H2} was about 0.1 atm. At pH 6, the concentration of HS^- will be about 1/10 that of H_2S (assuming that pK_1 of H_2S is 7 at 250°C, and (S^{--}) is negligible). Since the total sulphide concentration at Broadlands, New Zealand, is 4×10^{-3} molal, the HS^- concentration is approximately 4×10^{-4} molal.
Hence, at 250°C,

$$(AuS^-) = \frac{K_{250°C} \times (HS^-)}{(f_{H2})^{1/2}} = \frac{10^{-3.3} \times 4 \times 10^{-4}}{(f_{H2})^{1/2}}$$
$$= 2 \times 10^{-5} \text{ m or 4 ppm Au for } f_{H2} = 10^{-4} \text{ atm}$$
$$= 1 \times 10^{-6} \text{ m or 0.2 ppm Au for } f_{H2} = 0.1 \text{ atm.}$$

These calculated solubility values are in excess of the measured concentrations of gold in the hydrothermal waters from the drill-holes at Broadlands, which are 4×10^{-5} to 10^{-3} ppm Au [Weissberg,

1969, op. cit.; Browne, P. R. L., 1969, Sulfide mineralization in a Broadlands geothermal drill hole, Taupo Volcanic Zone, New Zealand, *Econ. Geology* **64:**156-159, respectively] and indicate that, at depth, where the temperatures are about 250°C, the waters are undersaturated with respect to gold. These calculated values are also a few orders of magnitude greater than the solubilities of gold in acid, chloride-rich solutions at 250°C calculated by Helgeson and Garrels [Helgeson, H. C., and R. M. Garrels, 1968. Hydrothermal transport and deposition of gold, *Econ. Geology* **63:**622-635.] and suggest that the formation of gold sulphide complexes in near-neutral solutions may be the dominant transport mechanism, especially in the formation of the lower temperature epithermal deposits, where gold is associated with adularia in some deposits, and in others, with the sulphides of arsenic, antimony and mercury—minerals apparently requiring near-neutral to alkaline conditions for their transport or formation.

The nearly constant association of sulfur, arsenic, antimony, and gold in deposits, and the presence of aurostibite, $AuSb_2$, in some auriferous deposits, has led the writer (Boyle, 1969, 1979) and others (Seward, 1973; Sorokin, 1973) to postulate that arsenic and antimony complexes are important agents in the hypogene transport of gold. The complexes are probably of the type $[Au(AsS_2)]^0$, $[Au(AsS_3)]^{2-}$, $[Au(Sb_2S_4)]^{1-}$, and $[AuSb]^{2-}$. Similarly the common occurrence of gold tellurides in auriferous deposits also suggests that telluride complexes probably play a part in the transport of gold in hydrothermal solutions. These complexes are probably of the type $[Au(Te)_2]^-$.

From time to time a number of other complexes have been advocated for the hydrothermal transport of gold, including the hydroxide, thiosulphate and other oxidized thio species, cyanide and thiocyanate complexes, organometallic complexes, and colloids. With the exception of colloids space precludes consideration of these modes of transport here. They are discussed in some detail in Boyle (1979).

Because of the relative ease with which gold colloids are formed they have repeatedly been suggested as a mode of transport of the precious metal in hydrothermal solutions. Hatschek and Simon (1912) experimented with silica gels throughout which soluble gold salts were uniformly dispersed; they concluded from their experiments that when a fracture was filled with gelatinous silica the dispersed gold salts would be reduced and precipitated, forming some of the textures seen in gold-quartz veins.

In a classic work on the role of colloidal solutions in the formation of mineral deposits Boydell (1924) referred to the occurrence of much fine-grained quartz, chalcedony, and opal in gold deposits and to the presence of sulphides and gold in many deposits in an extremely fine state of division. He thought that these phenomena indicated colloidal transport of gold and silica, flocculation of the colloids to form gels, and finally recrystallization of the constituents of the gels to form the quartz and gold as they are now found. Lindgren (1933) advocated a somewhat similar origin for certain types of gold-quartz deposits. More recently Zhirnov (1972) has described spherical brownish colloidal-sized particles of gold (0.1 μ and smaller) in colloform bands in the Kaul'da gold deposit of the Almalyk ore field, USSR. The fineness of these particles was 794. Zhirnov concluded that these particles initially represented colloidal gold protected by silica sols, and that they were deposited from solutions by a rapid drop of temperature at the margins of the veins where pyrite assisted in the coagulation of the gold colloids. Wright (1969) figures numerous textures of ore and gangue minerals in the auriferous Tennant Creek (Australia) deposits, which he believes proves that the minerals were at one time in the colloidal state. Subsequent to deposition they have crystallized, and some have been partly mobilized. He suggests that the mobilization process took place by thixotropic reliquefaction of the gelatinous parent materials of the minerals now present.

Frondel (1938) experimented with colloidal gold under hydrothermal conditions and found that gold sols were protected by silica colloids against electrolytes (NaCl) and spontaneous coagulation by increase of temperature. He also noticed an increase in the stability of protected gold sols with increase in temperature, and observed that the protected sols were stable up to 350°C. From the experimental data in his classic paper *Stability of Colloidal Gold under Hydrothermal Conditions* Frondel drew the following conclusions.

Summary

1. The stability of unprotected gold sols toward electrolyte (NaCl) increases with increasing temperature.
2. The stability of unprotected gold sols toward electrolyte increases with increasing dilution of the sol. The increase is not linear, but drops off at high dilutions. The temperature rate of change of stability remains about the same with dilution.
3. Unprotected sols containing no added electrolyte coagulate spontaneously between 150° and 250° with increasing temperature.
4. Colloidal silica protects colloidal gold both against electrolyte and against spontaneous coagulation by increase of temperature. The protection is accompanied by a reversal of charge of the sol in the presence of increasing concentrations of electrolyte.
5. The protecting action of silica against electrolyte for sols in the negative zone of stability increases with increasing temperature, and increases more rapidly than the increase of stability with temperature of the unprotected sols.
6. The protecting action of silica against electrolyte increases with increasing dilution of the gold sol.
7. Protected sols containing no added electrolyte are stable at 350°C, but coagulate spontaneously at 410°.
8. Protected sols in the negative zone of stability are tolerant to NaOH and sensitive to HCl, and the opposite is true of sols in the positive zone of stability. Unprotected sols behave similarly to negative protected sols, but the tolerance to NaOH is less, and the sensitivity to HCl is greater. The stability of negative protected sols toward electrolyte is increased by adding NaOH to the protected sol, within a certain range of NaOH concentrations, but is decreased by HCl.
9. Silica sols are colloidal at 25° and 100°, and very probably up to 350°, in dilutions considerably under the solubilities recently reported by Hitchen and by Gruner. It is suggested that these investigators

have measured a peptization equilibrium that simulates a true solubility, of the sort found for other hydrophilic colloids.

10. The origin and subsequent history of hydrothermal solutions is discussed with reference to the state in which gold is carried. In acid solutions the gold is very probably transported in true solution. Part or all of the gold salt, however, may be carried as adsorbed electrolyte on colloidal silica present in the solution. With approaching neutrality or alkalinity of the originally acid solution, the gold is precipitated out as a sol, protected by colloidal silica. The factors which may effect coagulation of the sols are briefly discussed. (pp. 18-20).

The supergene migration of gold in meteoric waters has been repeatedly proven, but the form of the gold in such waters remains largely unknown. Suggested dissolved states for gold include: hydroxide, chloride complexes, thiosulphate and other oxidized thio species, sulphate, arsenate, antimonate, polysulphides, cyanide and cyanate complexes, organometallic complexes, and colloids. Most of these modes of solution, excluding the last three, have already been discussed.

Natural cyanide, thiocyanate, and thiosulphate complexes exist in nature and have been suggested as effective solubilizers of gold under supergene conditions. The first investigator to suggest such a mode of solution was Lungwitz 1900*a*, 1900*b*). Much later Lakin et al. (1974) concluded that gold cyanide offers the most feasible form of soluble gold in soils. Paper 8-9, the summary from their bulletin *Geochemistry of Gold in the Weathering Cycle,* is an excellent exposition of the behavior of gold under supergene conditions.

Organometallic complexing as a supergene transport mechanism for gold was first suggested by Lungwitz (1900*a*, 1900*b*) and Harrison (1908). Freise (1931) was the first to show experimentally that humic acids "dissolve" gold, as discussed in Paper 8-10.

(Text continues on page 192.)

8-9: SUMMARY OF CONCLUSIONS

H. W. Lakin, G. C. Curtain, A. E. Hubert, H. T. Shacklette, and K. G. Doxtader

Reprinted from Geochemistry of gold in the weathering cycle, *U.S. Geol. Survey Bull. 1330,* 1974, pp. 71-75.

The chemistry of gold is the chemistry of its complex compounds. The stability of complex gold anions is reflected in the ease with which gold metal is oxidized in the presence of the complexing ion. The order of increasing ease of oxidation of gold to form the corresponding gold complex anion studied in this report is Cl^{-1}, Br^{-1}, CNS^{-1}, I^{-1}, and CN^{-1}, with $S_2O_3^{-2}$ probably between CNS^{-1} and I^{-1}.

The removal of gold from solution by various minerals either by sorption or more probably by reduction of the gold complex to yield gold metal showed a suspectibility to removal of $AuCl_4^{-1} > AuBr_4^{-2} > AuI_2^{-1} > Au(CN)_2^{-1} > Au(CNS)_4^{-1} > Au(S_2O_3)_2^{-3}$. A marked difference in stability of the halide complexes to the cyanide, thiocyanate, and thiosulfate complexes was particularly evident; the halide complexes were much less stable and, therefore, much less mobile in a natural environment.

A study of the absorption of gold by plants showed a strong tendency for the reduction of the halides by plant roots. Colloidal gold was not absorbed by the plants used in these experiments, although colloidal gold has been suggested as the probable form of gold taken up by plants. The most readily absorbed complex was the cyanide.

The solubility of gold in acid halide solutions also shows an increasing ease of solution with increasing stability of the complex. Thus, gold chloride is readily formed in strongly acid solutions in the presence of manganese dioxide; gold bromide is formed in acid solution in the absence of manganese dioxide but is more readily dissolved in the presence of manganese dioxide; however, with gold iodide, the solution of gold is inversely proportional to the amount of manganese dioxide present, being highest in the absence of manganese dioxide.

A comparison of the solubility of gold in the input solutions used in leaching copper from porphyry copper ore to which a given amount of chloride, bromide, or iodide has been added shows increasing solubility of gold from chloride to iodide; however, for a given concentration of the halide the solubility of gold decreases with increasing iron, aluminum, and copper. Some gold dissolved in the input solution would be lost from solution in the output solution. Input chloride solutions capable of dissolving gold failed to maintain gold in solution in the presence of an oxidized Carlin gold ore. The reduction of gold chloride by many minerals, plus the complexing of chloride by iron and copper, hampers the recovery of gold from copper ores by present leaching processes. The lack of recovery of gold from copper ores by leaching processes will greatly decrease the production of gold in the United States if present leaching processes gain wide use.

To dissolve about the same amount of gold in a given time (1.5 mg/l in 3 weeks) in input leach solutions, the solutions must also contain 1,800 ppm chloride ion, or 400 ppm bromide ion, or 13 ppm iodide ion, giving an equivalent effectiveness of 100 Cl^{-1} to 22 Br^{-1} to 0.7I^{-1}. The relative crustal abundance of these elements is 100 ppm Cl to 1.3 ppm Br to 0.24 ppm I. Thus, considerable enrichment of bromine and iodine would be required for these elements to be as effective as chloride in solubilization of gold in acid solutions. Bromine and especially iodine are enriched in certain plants and the leaching of plant residues gives rise to unusual concentration of these halides. Iodine leached from kelp has produced waters in Java that contain 150 ppm iodine; they also contain 68 ppm bromine. Organic-rich soils of Japan contain as much as 850 ppm bromine. Gold chloride, gold bromide, and gold iodide, therefore, may be formed locally under unusual but possible circumstances. When they are formed, they are readily decomposed by slight changes in their surroundings; consequently, they are transient forms.

The thiosulfate ion is a metastable species that can be formed by the action of water on native sulfur in basic solutions, by the reaction of hydrogen sulfide with oxygen, and by the catalyzed reaction of the sulfide ion with oxygen in soils by the enzyme oxidase. The thiosulfate anion as a complex former with gold, therefore, may be considered a transition anion from the simple single-element halides to the complex anions largely of biological origin.

Gold is soluble in thiosulfate solution over a wide range of hydrogen ion concentration. It is particularly of interest that gold is soluble as the gold thiosulfate in the normal pH range of soils and that once gold thiosulfate is formed it is remarkably stable. In the weathering of pyritic gold ores in limestones, the gold may be mobilized as the thiosulfate.

The formation of gold thiocyanate takes place readily over a wide range of pH. Free gold is dissolved in the presence of oxygen and thiosulfate, but gold in massive pyrite and in gold telluride is not attacked. The thiocyanate ion is of biochemical origin and is relatively rare. In microsystems within the soil it may reach concentrations adequate for complexing gold, but it does not seem a particularly important facet in the solution of gold.

Gold cyanide offers the most feasible form of soluble gold in soils. Gold reacts with oxygen in the presence of the cyanide ion to form a stable complex. The presence of the cyanide ion in soils is widespread and often of considerable concentration. Over 1,000 species of plants are known to contain cyanogenetic substances, as do many arthropods and moths; certain fungi release gaseous hydrogen cyanide in remarkably large volumes. A few varieties of flax excrete significant quantities of hydrogen cyanide into the soil from their root system.

Nutrient solution in which these varieties were grown contained as much as 37 mg of hydrogen cyanide per plant. In order to determine the abundance of plant species that are cyanogenic, plants were sampled from the Front Range of the Rocky Mountains in Colorado to north-central Nevada and less completely in some other areas. Of the 150 species tested, 116 species (78 percent) gave no test for cyanide, 15 species (10 percent) gave questionably positive tests for cyanide, and 19 species (12 percent) gave positive tests for cyanide. Leaf gold was dissolved by an aqueous suspension of macerated plant species which gave positive tests for cyanide.

The fact that cyanide, the byproduct of many life processes, forms such a strong complex with gold would be unique if cyanide were, in fact, the only organic ion formed by natural processes that produced stable gold compounds. The work reported in the present paper by Kenneth G. Doxtader on solubilization of gold by microorganisms and organic substances demonstrates the need for more careful research on the solution of gold by the complexing action of naturally occurring organic substances. It is in this area that knowledge of the behavior of gold in the weathering cycle is notably lacking. An organic substance might be found which has the complexing property of cyanide but which lacks the poisonous character of cyanide; it would be a boon to the recovery of gold from low-grade deposits.

The analytical determination of gold in geologic materials is made difficult because of the strong probability that a representative sample is not available to the analyst. A relative standard deviation of less than 50 percent cannot be obtained on 10-g sub-samples if the gold particles are greater than 0.01 mm in diameter. Gold wires of a few millimeters in length and consisting of less than 1 ppm of the sample are so few that the chances of finding one piece of gold in a 10-g sample are remote. With few exceptions the materials analyzed in this work contained fine gold, and, therefore, the results are reasonably reproducible.

Gold in some soluble form, perhaps as gold cyanide, is absorbed by plants but is not used as a nutrient by plants. It is therefore found accumulating as a reject, like barium, in the woody parts of the plant. The decomposition of plant debris results in the reduction of the gold in the plant material, and gold accumulates in the humus horizon of the soil. This horizon, commonly mixed with colluvial inorganic material, constitutes what we have called mull, and mull has been proved to be a very satisfactory sample for gold exploration. A vein only a few inches thick was located by mull sampling at 200-foot intervals on a ridgetop; the closest reconnaissance sample was more than 50 feet from the gold-bearing vein.

The distribution of gold within soil profiles may be grouped in three categories: (1) high gold content in the surface humus horizon; (2) high gold content in the surface and bottom horizons; (3) high gold content in the bottom horizon.

Gold may become enriched in the surface soil because of sheet erosion removing lighter material. It may also become enriched by biogeochemical cycling of gold, a process in which gold is leached from the soil and bedrock, is absorbed by the vegetation, and is eventually concentrated in the mull and upper layers of the soil as the vegetation decays. This process is presumed to be the dominant one in those profiles in which the gold content is high in the ash of mull.

The high density of gold particles that permits the residual enrichment of the particles in sheet erosion also permits their accumulation at bedrock surfaces by gravity separation in the soil profile by much the same process as in stream sediments. When gravity separation is combined with biochemical or erosional surface enrichment one finds soil profiles with high gold content at both the surface and the rock bottom.

When no biogeochemical or erosional enrichment of the surface soil is occurring, the gravity effect is dominant and the soil shows high gold content only near bedrock or at the interface with a clay pan.

The distribution of gold with particle size is a function of the completeness of the weathering of the ore. Inasmuch as many gold ores are emplaced with silicification, the ore may be more resistant to weathering than are the wallrocks. In such places, fine gold may be enclosed in coarse particles in the soil. This fact must be considered in soil or stream sampling in gold exploration.

8-10: THE TRANSPORTATION OF GOLD BY ORGANIC UNDERGROUND SOLUTIONS

Fred W. Freise

ALL mining companies or individual miners who have worked alluvial gold deposits in Brazil realize that gold placers thoroughly exhausted may after a period of years once more be panned and yield a profitable amount of newly accumulated gold. The native gold digger maintains that every gold placer within ten years is again valuable enough to be worked over once more and that the *pinta que paga, i.e.* the "paying spark" reappears the sooner if the exhausted gold field has been hidden from the sun by vegetation or other means.

The author worked gold placers at the eastern boundary of the State of Minas Geraes in the districts of Palma and Muriahé in 1908 and 1909 and for the second time in 1926; the first time monazitic sands were the principal object of the mining work, but the gold contents of the gravel (8.5 grams per ton) were recovered, since they almost defrayed the pay roll. But when, in 1926, the same places were opened again, an average of 4.85 grams of gold per ton was realized near the bottom rock, the metal being quite different from the original gold, both in color, purity, coarseness, and affinity to mercury. The nature of the territory and the sequence of the strata precludes the hypothesis of mechanical transfer from a higher point; the occurrence suggested that this gold had been brought to its place by chemical transportation.[1]

In the eastern part of the State of Rio, the author opened up some gold placers in 1912; the gold content of the gravel was 11.6 grams per ton, and the average yield was 10.85 grams. When, in 1926, the old diggings, quite overgrown with "caapœira"

[1] Details concerning the above-mentioned gold and monazitic deposits were published by the writer in *Zeit. f. d. Berg-, Huetten- u. Salinenwesen i. Preuss. Staate*, vol. LVII., pp. 47–64, 1910.

(second growth), were reopened, the lowest layers immediately above the granitic bed rock yielded 4.66 grams per ton, of a greenish variety of gold which chemically behaved quite differently from the commonly known allotropic modifications of gold.

Finally, an opportunity to see gold transported chemically by underground solutions was given to the writer in 1927, when he had occasion to examine the large tailing heaps accumulated by an important gold mining company in the center of Minas Geraes. These tailings generally showed 0.48 grams of gold per ton, but in certain parts overgrown by shrubberies there was found as much as 3.69 grams per ton. Since for the last twenty years or more the average ore treated assayed 9.5 to 10.2 grams and the recovery was 9.0 to 9.5, this high content in the tailings can hardly be ascribed to losses in milling and treatment, but must be attributed to accumulation after the tailings had been dumped.

These observations suggested an investigation into the nature of the agents that might have caused the transport of gold from its original point to lower levels. The physical and chemical properties of the original gold and the "new gold," as it may be called, were investigated. The specific gravity was determined by means of the specific-gravity bottle; the color and the surface were observed under the microscope; differences in hardness or in toughness could not be investigated since the particles of the new gold were too small. To examine the affinity to mercury, *i.e.* to see whether the new gold is a "free milling" or a "refractory" one, the gold samples were passed over copper plates of 4 to 6 inches, coated with a thin film of mercury. To determine the action of cyanide solutions on the two varieties of metal, carefully weighed portions of gold of the same degree of fineness were exposed to solutions of cyanide of potassium of various strengths for many days. The gold in solution was determined at equal intervals for both kinds of gold under treatment. The results of these tests are shown in Table I.

TABLE I.

PHYSICAL AND CHEMICAL PROPERTIES OF THE DIFFERENT KINDS OF GOLD OBSERVED.

Kind of Gold.	Spec. Grav-ity.	Color.	Surface.	% Metal Caught on Mercury.	% Gold Dissolved in KCN-solution of . . . % .5 When in Contact during . . . Hours		1.0		2.5	
					24	144	24	144	24	144
Common placer gold, 98.3% Au, 1.7% Cu	19,562	Yellow	Scarred	66.45	16.2	33.5	22.4	55.7	43.7	88.8
"New" gold, 100% Au.	19,222	Greenish	Plain	18.38	3.5	11.8	10.1	48.2	21.4	44.5
"Black" gold, after crust was taken off. 100% Au.	19,217	Greenish	Plain	16.66	2.9	11.5	9.9	40.4	19.7	43.9

It will be noted that the new gold is very refractory against mercury but is relatively more soluble in cyanide solutions.

It is remarkable that the physical and chemical properties of the new gold are, even in minute details, identical with those of the so-called "black gold" (*ouro preto*) found in Brazil at many places near the ancient State capital of Minas Geraes, which drew its name from the numerous deposits of "black gold" in its vicinity. These were worked out by the first discoverers in the districts of Itabirito, Diamantina, Sabará, Santa Barbara, São Gonçalo do Sapucahy, Carangola, all in Minas. Black gold is also commonly found in the districts of Rio Verde, Formosa, Santa Luzia in the State of Goyaz, and in the district of Santa Rita do Araguaya in Matto Grosso. The native prospector generally inadvertently throws it away with the heavy residuals in the pan, such as titanite, black garnet, magnetite, rutile, wolframite, and tourmaline; only by chance is it detected in the pan when the surface of the gold is given thorough attention. Black gold, as it is found at these places, is gold covered with a dark brown to dull black coating of 2 to 25μ in thickness; where larger grains of metal are found, they prove to be composed of several smaller individuals each of which preserves its own coating. A washing with a 5 per cent. solution of K_2CO_3 at a temperature of 35°–45° is sufficient in most cases to make the coating disappear;

where the coating is thicker, a cautious heating to 300°–330° is required, which is followed by a washing with dilute sulphuric acid. The sulphate of iron which is then formed has to be removed by washing before cyanide or other agents can be applied to the metal, which then shows its customary yellow color. The dark varieties of "black gold" show 0.22 to 0.35 per cent. of Fe_2O_3 and the browner ones, 2.85 to 3.5 per cent., in the coating. In so far as the writer's investigations go, it cannot be said whether there is only one kind of coating or several kinds; the brown one which was investigated can be considered as $C_2O_7H_{12}Fe$ (humate of iron).

The above considerations as to the character of the coating of the "black gold" flakes suggest that waters charged with organic acids formed by the decomposition of vegetable matter were responsible for the transportation of gold from higher to lower levels. To check this idea, several series of tests were made, first with an artificial product prepared from bituminous brown coal found in the vicinity of Carangola, Minas Geraes, and afterward with "black water" generated in virgin forests from centers of humification of decayed vegetation.

The brown coal contained 21.33 per cent. water, 15.62 per cent. ash, 36.28 per cent. fixed carbon, and 26.77 per cent. of volatile matter, and held in its combustible matter about 68 per cent. of substances soluble in alkali. According to the process indicated by Simek[2] this raw product was used to isolate a dark brown substance which, after a long period of drying, first in air, then in vacuum at 55° C., showed upon analysis 67.21 per cent. carbon, 4.98 per cent. hydrogen, 1.23 per cent. nitrogen, and 7 per cent. ash, of which the most important constituents were Fe_2O_3 and P_2O_5. Separation by solubility of the different components of this raw product was not tried; for the experiments described, solutions in water were used.

In order to determine the minimum concentration of such suspensions that would act upon free gold, carefully weighed quantities of gold in the form of plates of uniform granulation were exposed to solutions of known concentration contained in glass

[2] See *Brennstoff Chemie*, vol. 9, 1, 12, p. 381, 1928.

TABLE II. Reactions of Humic Acids on Gold Under Different Conditions.

Strength of Solution.	Kind of Water.	Gold Experimented on.		Conditions.	Duration of Experiment. Hours.	Dissolved.	
		Shape.	Fineness.			Mg.	%
% .10	Rain water, airless	Plate, .01″	850/1000	No Motion	24	8,354	.05
					48	15,228	.91
		Dust, .01″	850/1000	"	24	11,182	1.14
					48	23,075	2.35
					72	25,553	2,59
.50	Rain water, airless	Plate, .01″	1000/1000	Constant Motion	24	2,134	.005
					48	2,664	.0062
	Rain water, aerated	Dust, .005″	1000/1000	"	24	Traces	0
					48	Traces	0
2.50	Dist. water, airless	Dust, .005–.1 mm.	900/1000	No Motion	24	32.551	3.68
					48	65.001	7.21
					72	97.161	11.08
4.00	Rain water, airless	Same material as before		Slow Motion	24	27.101	3.11
					96	111.402	12.50
					192	168.249	19.28
					300	175.455	20.21
10.00	Rain water	Dust, .002–.08 mm.	600/1000–975/1000	Slow Motion + air	24	23.450	1.56
					48	42.761	2.84
					72	70.111	4.66
					240	166.369	11.00
5.00	Dist. water	Gold precipitated by $FeSO_4$		No Motion	48	Traces	
					96	Little more	
10.00	Dist. water with 1 % of K_2CO_3	Plate, .01″	900/1000	"	24	Traces	
					30	Reaction stopped	
10.00	Dist. water with 1 % of K_2CO_3, 1 % of $NaHCO_3$	Same material and same conditions of experiment			Suspension without effect		
10.00	Aq. dist. with .5 % KNO_2, .5 % KNO_3, .5 % NaCl	Same material and same conditions of test			After 12 h. without any effect		
10.00	Water with .5 % Na_2SO_4, .5 % $CaHCO_3$	Identical material, under same conditions			Merely traces dissolved; after a few hours no effect whatever		

bottles that were kept in slow, continuous motion. In some instances distilled water free from air was used; in others, free access to the air was allowed. Another series of experiments was conducted under free admission of carbonic acid, and a third was carried on with water mineralized with one or more salts in the same concentration as in subsoil water. The results are shown in Table II.

The data of Table II. allow of the following conclusions:

1. Gold is attacked by humic acids, even those of a very dilute character, provided sufficient time is allowed and oxygen is excluded. Oxygen rapidly destroys the organic combination, as may readily be seen by the clarifying of the water from dark brown to chestnut and reddish yellow to gold yellow; clear brown suspensions are absolutely innocuous to gold metal, even in 20 per cent. solutions.

2. Distilled water and water freed from air maintain the organic acid solutions in their full activity for a considerable time; rain water is only slightly inferior. With natural waters, the different minerals contained in them act differently on the organic acids; e.g. the carbonates and bicarbonates are the first salts to destroy the humic acids in the water, then follow sulphates, bisulphates, and nitrates; chlorides seem to have but little effect on the stability of the organic acids. The writer has not as yet determined whether the susceptibility of the metal to solution varies with different temperatures; all experiments have been conducted at field temperatures.

3. The fineness of the original metal does not appear to influence the solubility in the humic acids, at least so far as silver, copper, and palladium, as constituents of the gold ore, are concerned. Differences can be noted, however, between the reactions of the acids on natural gold and on gold produced by precipitation. In the latter case, solubilities vary according to the precipitant used; metal precipitated by ferrous sulphate, for example, is less rapidly attacked than gold precipitated by oxalic acid. The causes of these differences have not yet been investigated; probably the different precipitates are different allotropic varieties of gold.

Following these investigations more elaborate experiments were made, utilizing moor water or black water from the Serra dos Aymorés, the divide between the States of Minas Geraes and Espirito Santo and one of the thickest virgin forest districts of the whole country.[3]

Observation shows that these waters, which are dark yellow to brown or black in color, absolutely clear and transparent, low in oxygen or free from it, and of a strong acid reaction, quickly attack all the easily soluble components of the rocks such as the combinations of K_2O, Na_2O, CaO, and MgO. A relative concentration of the iron and manganese hydroxides and oxides is followed by the destruction of these constituents; the alumina and phosphoric anhydride combinations, if not destroyed and transported, are at least affected. Finally there remains merely silicic acid free from any metallic accessories except oxides of tin, tungsten, titanium, and zirconium. Gold, silver, and palladium disappear among the first constituents destroyed.

There can be no doubt that the black waters transport metallic combinations in the form of definite chemical compounds and not by adsorption; formulas can be established for the majority of metals, *i.e.* iron, copper, and manganese, for their combination with humic acids, and this is true of gold also.

After the black water has dissolved the metals, contact with the open air is sufficient to oxidize the metals and to cause them to form a scum on the water surface, in the case of iron, copper, and manganese (and probably of some other common metals also). In the case of gold, however, other conditions are necessary to bring about its reappearance. Simple contact with the atmosphere is not sufficient to separate it, since oxygen does not act

[3] Under normal conditions, an acre of ground produces in this region about 3,500 cubic feet of timber, 8 to 11 per cent. of which decays annually (about 7.5 to 10.8 tons); 80 per cent. of this raw material disappears by quick fermentation which leaves only 5 per cent. or even less, equal to one eighth inch, as humus on the ground, while 20 per cent. suffers transformation under water, giving origin to the so-called black waters (*rios negros* or *rios pretos*) commonly noted on geographical maps of the tropics. Although these black waters are not so extensive in this part of Brazil as in the Amazon districts, or in Sumatra or Borneo, they carry annually thousands and thousands of tons of rock material in suspended colloidal form to the ocean or to deeper levels alongside the Rio Doce.

intensely enough on the gold humate; it is indispensable for the precipitation of gold that the gold humate in its underground circulation meet strongly mineralized waters such as carbonate or sulphate waters.[4]

While investigating a small placer gold deposit in one of the side valleys of the Rio Pancas, one of the affluents of the Rio Doce on the left which comes down from the center of the Serra dos Aymorés, the writer verified in an extensive series of experiments the processes of gold solution, transport, and redeposition that go on in nature in the manner described above. For this purpose a wooden flume six feet wide, four feet deep, and 50 feet long, was built with an inclination of one inch in three feet; this flume was filled for nine-tenths of its length with thoroughly washed sand of determined fineness, or with mixtures of sand and clay, or with pure kaolin, according to the details shown in Table III. The filling was evenly packed each time by means of water jets. The upper tenth of the flume was finally filled with sand, sand and clay, or pure kaolin, as in the longer part of the flume, but in this section there was admixed a carefully weighed amount of gold dust of known fineness and grain that could be recognized at once if panned out. Into this end of the flume was delivered a constant stream of black water derived from a nearby rivulet that carried from 6 to 10 per cent. of raw humus. Before the black water was introduced into the head of the flume, the whole flume was covered with water-tight boards. The amount of water allowed to sink into and pass through the flume filling was measured at the lower end by means of a tank fed by a pipe that passed through the bottom boards of the flume. This tank also permitted samples to be taken of the water that sank through the sand or other filling.

With each filling, one experiment lasted sixty days; at ten-day

[4] The conglomerates known by the name of *tapanhoancanga* or *canga* (nigger head) in Minas Geraes, famous for their gold and diamonds, are partly cemented by gold that originally circulated in solutions formed in the manner here described. It is highly probable that the deposits of lignite known and partly explored in the State of Minas are the remainders of those virgin forests that gave origin to the black waters which transported the gold solutions mentioned here.

TABLE III. Transportation of Gold.

Flume Filling and Gold Admixed in mg./kg. of Filling	% of pores.	Gold contents in mg./kg. verified after the following days of water run:																	
		10			20			30			40			50			60		
		at the following distances in feet from water inlet:																	
		15	30	45	15	30	45	15	30	45	15	30	45	15	30	45	15	30	45
Sand; 50 mg./kg.; dust of .1–.5 mm.	22	31	11	3	26	11	7	16	15	12	13	18	13	11	22	13	10	25	13
Sand; 20 mg./kg.; dust of .05–.1 mm.	24	10	7	1	6	7	2	5	8	3	3	?	5	2	?	9	tr.	7	15
Sand and clay 50/50; 25 mg./kg.; .05–.15 mm.	16	17	3	0	14	5	2	11	?	3	9	8	5	Not investigated					
Sand and clay 25/75; 30 mg./kg.; .08–.2 mm.	12	21	7	tr.	16	9	1	13	12	2	10	14	4	7	17	4	Not investigated		
Pure kaolin 45 mg./kg.; .1–.5 mm.	10	30	10	2	22	14	4	No observations made						17	11	7	?	?	?

intervals samples were taken at definite points of the filling and at the water outlet; the results of these investigations are set down in Table III. The conclusions that may be drawn from these results are:

1. The rate of transportation of gold (and other metals also) from higher to deeper levels, depends upon the permeability of the soil; the more densely the soil is packed, the more difficultly the organic agents permeate; the finest slimes as they float away from concentrating tables, after they have settled in a tank are hardly penetrable by the organic waters.

2. Strata free from limestone and other carbonates give better results; *i.e.* they are more rapidly freed of their gold contents than limestone and other carbonate strata. The reason for this is probably the greater affinity of humic acids to these carbonates than to gold; the same fact is observed with iron-containing constituents of the stratum.

3. There can be no doubt that such organic suspensions as here described and applied are effective agents for transporting metals from one part of a mineral deposit to another, and that by them even the most minute traces of metallic matter are dissolved and transported.

4. For iron and manganese humates the process of re-deposition of the metals is rather evident but in the case of gold the re-precipitation depends upon the character of the mineral solutions that are circulating in the subsoil, and these cannot always be readily determined.

In soil science, the fact that iron components of the soil are transferred to deeper levels has become evident by the " hard pan " formation found in many places. In geology, the circumstances under which the later iron and manganese deposits have been formed must be linked with the activity of such organic waters as have been considered here. That, even to-day, similar agents may move metals, is a matter that can be investigated only under very favorable climatic conditions. It is likewise true in the case of gold deposits that intense field and laboratory research is bound to accumulate further evidence concerning the possibility that black water on, and in, the ground, may form new workable deposits even to-day, where old deposits have been given up as exhausted.

It is not impossible that the observations and experiments here recorded may eventually have an economic bearing in providing a scientific basis for the finding of low-grade mineral deposits derived from placer or other sources. That question, however, will not be considered here.

Investigations by Garces (1942) and Fetzer (1934, 1946), however, showed that gold is not dissolved by humic acids.

In recent years Ong and Swanson (1969) have carried out extensive investigations of the effect of organic acids on gold. They summarize their findings in the abstract from their paper *Natural Organic Acids in the Transportation, Deposition, and Concentration of Gold* (pp. 395-396).

> Interactions of various forms of gold, as plates, coarse particles, colloids, and ions ($AuCl_4^-$), with different types of natural organic acids show that gold is not oxidized and complexed by the organic molecules as reported by some investigators. To the contrary, organic concentrations in the range of 3-30 ppm have the capacity to reduce gold chloride solutions to negatively charged colloids of metallic gold. For the 30 ppm organic acid solutions, the reduction process is accomplished by the formation of a protective coating of hydrophilic organic molecules around the hydrophobic gold sol making the gold very stable for at least 8 months and not easily coagulated by cations. The gold sols so formed are less than 10 mμ in size. This protective layer is also formed when colloidal gold is mixed with organic acids. For the 3-ppm organic acid, the organic matter concentration is too low to form the protective coating and the colloidal gold precipitates.
>
> Two of the organic acids used in this study were extracted from peat and humate-cemented sand, and one was collected from a brown lake water. All three are similar in composition and in physical properties to organic acids isolated from soils and surface waters. These natural organic acids can be considered as hydrophilic colloids having a particle size in the 2.4-10 m μ range and having negative charges due to the dissociations of the carboxylic and phenolic groups.
>
> The significances of this study in geochemical processes involving the mobility of gold are: (1) In a very acidic solution, for example near an ore deposit, where gold is soluble as its chloride and organic acids are not soluble, the presence of solid organic matter causes the reduction of gold chloride, and metallic gold of colloidal size will be precipitated and intimately associated with the organic matter; (2) in slightly acidic to basic waters, for example in the general range of composition of natural waters, gold can only be transported as stable organic protected colloids; and (3) these organic-protected gold colloids are precipitated when they enter a different chemical environment, such as sea water or brackish water with their abundant ions or clay colloids, or when they enter an acid environment ($pH < 3$).

The transport of gold as a colloid in surface waters and supergene oxidizing groundwaters has long been advocated by a number of geologists, including Bastin (1915), Boydell (1924), and Lindgren (1924, 1933). More recently Goni Guillemin, and Sarcia (1967) have carried out an investigation of gold colloids and their stability. The abstract and conclusions from their paper *Géochimie de l'or exogène* (pp. 259, 267) follows.

Abstract

The stability of colloidal suspensions of gold, their transport and flocculation have been investigated; in addition, the experimental formation of gold nuggets under natural conditions was studied. The measurements of pH, Eh and the concentration of ionic gold, proved that the possibility of transportation of ionic gold in nature was overestimated. Ionic or even metallic gold may give stable colloidal suspensions. It may be easily flocculated and give by compaction gold nuts similar to natural gold nuts. By diffusion of gold solutions in silica gel we obtained direct gold films and also some textures frequently present in gold deposits.

Conclusions

Ces travaux, qui ne représentent qu'une phase préliminaire d'un programme plus vaste, nous permettent de dégager certaines conclusions essentielles:

1. La possibilité de transport de l'or dans la nature sous forme ionique a été surestimée. En effet, les solutions auriques sont très instables et se décomposent rapidement dans les conditions d'acidité des eaux naturelles ou règnant dans la zone d'oxydation des gisements métalliques surtout sulfurés.
2. Sous forme ionique ou même métallique, l'or peut donner facilement des suspensions colloidales dont la stabilité permet la migration à grandes distances. Il peut être floculé par un changement des conditions physico-chimiques, surtout du pH et du Eh, ces paramètres étant une conséquence des variations géologiques: zones marécageuses chargées de matière organique, contact des eaux de différentes salinités, corrosion des roches encaissantes, suivie d'oxydation, de kaolinisation, etc. . . . Cet or floculé, encore très fin, possède des propriétés de réactivité chimique particulière mises en évidence par les essais d'amalgamation.
3. Par compaction due à des mouvement tourbillonnaires, si fréquents dans les zones des placers aurifères, nous avons reproduit des pépites d'or (pepites synthetiques) morphologiquement similaires aux pépites naturelles. Ceci peut expliquer la naissance des pépites secondaires, leur croissance par nourrissage à partir des nouvelles venues d'or colloidal, et finalement leur polissage dynamique sous eau; tous ces processus s'accomplissent dans des conditions géologiques de surface. Le phénomène de la régénération de placers aurifères, décrit par *Freise,* trouve ainsi une explication.
4. Par l'application du processus de réduction des solutions auriques dans un milieu composé de gel de silice, nous avons obtenu la formation directe de films d'or, reproduit des structures en cocarde, rythmiques, et même bréchiques grâce à l'interposition d'obstacles à la diffusion normale. Signalons qu'un fragment de quartz hyalin, utilisé comme obstacle à la diffusion, a été moulé par une très fine pellicule d'or; ses fissures renfermant de plus à la fin des expériences, des lamelles d'or bien visibles à faible grossissement (60×).

Nous poursuivons actuellement l'étude du dépôt par diffusion dans le gel de silice, et d'autres métaux tels que ceux du groupe du platine, de l'argent, du cuivre etc. . . .

Des mesures de la vitesse de diffusion à différents isothermes, ainsi que de la surface réelle du gel par la méthode BET, nous permettront de préciser l'influence des facteurs tels que la granulométrie, la température, la viscosité, etc.

GENERAL SUMMARY

In nature gold may be solubilized as halogen, sulfide, polysulfide, antimony-arsenic-sulfur, telluride, telluride-sulfide, cyanide, thiosulfate, thiocyanate, organometallic and various other complexes. Under restricted conditions gold may also be rendered mobile as a colloid.

Under hypogene hydrothermal conditions the precipitation of gold is effected by changes in temperature, pressure, and concentration of mobile species (complexes). For instance, precipitation may take place as the result of decomposition of gold complexes, as for example when alkali sulfide solutions bearing gold are neutralized or diluted, when the alkali component (K,Na) is isolated from the system as a result of reaction with wall rocks (e.g., the formation of sericite or albite), or when S^{2-}, As and Sb of complexes react with the ferrous iron of wall rocks to form minerals such as pyrite, pyrrhotite, arsenopyrite and gudmundite. When gold is carried as a colloid, coagulation and/or coprecipitation by interaction with a great many natural substances may take place.

Under supergene conditions the precipitation of gold can take place by a great variety of reduction mechanisms, by adsorption, absorption, and coprecipitation with both inorganic and organic (humic) colloids, and by destabilization of the many complexes (chloride, sulfide, organic, etc.) that gold may form under supergene conditions.

Many of the mechanisms for the transportation (migration) and precipitation of gold under both hypogene and supergene conditions have been outlined in this chapter; others are discussed by Boyle (1979).

REFERENCES AND SELECTED BIBLIOGRAPHY

Aitchison, L., 1960. *A History of Metals,* 2 vols., Macdonald & Evans, London, 647p.

Aubel, R. van, 1934. Géochimie de l'or, *Soc. Géol. Belg. Annal.* **57:**B131-B150.

Bache, J.-J., 1982. Les gisements d'or dans le monde, *BRGM Mem. 118,* 101p.

Bailey, K. C., 1929. *The Elder Pliny's Chapters on Chemical Subjects,* pts. 1 and 2, Edward Arnold & Co., London, 548p.

Bastin, E. S., 1915. Experiments with colloidal gold and silver, *Washington Acad. Sci. Jour.* **5:**64-71.

Bischof, K. G., 1847-1855. *Lehrbuch der chemischen und physikalischen Geologie,* vols. 1-3, suppl. by F. Zirkel, 1871. (Trans. by B. H. Paul and J. Drummond 1854-55, as *Elements of Chemical and Physical Geology,* 2 vols., Cavendish Society, London).

Boydell, H. C., 1924. The role of colloidal solutions in the formation of mineral deposits, *Inst. Min. Metall. Trans.* **34** (pt. 1):145-337.

Boyle, R. W., 1969. Hydrothermal transport and deposition of gold, *Econ. Geology* **64:**112-115.

Boyle, R. W., 1979. The geochemistry of gold and its deposits *Canada Geol. Survey Bull. 280,* 584p.
Brauns, R., 1896. *Chemische Mineralogie,* Chr. Herm.Tauchnitz, Leipzig, 460p.
Caley, E. R., and J. F. C. Richards, 1956. *Theophrastus on Stones,* Ohio State Univ., Columbus, Ohio, 238p.
Clarke, F. W., 1924. The data of geochemistry, *U.S. Geol. Survey Bull. 770,* 841p.
Dunn, E. J., 1929. *Geology of Gold,* Charles Griffin & Co., London, 303p.
Emmons, W. H., 1937. *Gold Deposits of the World,* McGraw-Hill, New York, 562p.
Ercker, L., 1574. See Sisco and Smith, 1951.
Fersman, A. E., 1931. *The Gelogy of Gold,* Doklady Acad. Sci. USSR, Moscow, pp. 199-204.
Fersman, A. E., 1934-1939. *Geochemistry,* 4 vols., Leningrad.
Fetzer, W. G., 1934. Transportation of gold by organic solutions, *Econ. Geology* **29:**599-604.
Fetzer, W. G., 1946. Humic acids and true organic acids as solvents of minerals, *Econ. Geology* **41:**47-56.
Freise, F. W., 1931. The transportation of gold by organic underground solutions, *Econ. Geology* **26:**421-431.
Friedensburg, F., 1953. *Die Metallischen Rohstoffe,* heft 3, *Gold,* F. Enke Verlag, Stuttgart, 234p.
Frondel, C., 1938. Stability of colloidal gold under hydrothermal conditions, *Econ. Geology* **33:**1-20.
Garces, H., 1942. Solubilidad del oro en acides humicos, *Prim. Cong. Panamericano Ing. Min. Geol. Anal.* **3:**1135-1138.
Gibson, C. S., 1938. Recent investigations in the chemistry of gold, *Nature* **142**(3590):339.
Gmelin, 1950-1954. *Gmelins Handbuch der anorganischen Chemie,* System-nummer 62, *Gold,* pts. 1 and 2, Weinheim/Bergstrasse, Verlag Chemie, GMBH, 406p.
Goldschmidt, V. M., 1937. The principles of distribution of chemical elements in minerals and rocks, *Jour. Chem. Soc.* p. 655-673.
Goldschmidt, V. M., 1954. *Geochemistry,* Clarendon Press, Oxford, 730p.
Goldschmidt, V. M., and Cl. Peters, 1932. *Zur Geochemie er Edelmetalle,* Nachr. Ges. Wiss. Göttingen, Math.-physik. Klasse, III:24, IV:26, p. 377-401.
Goni, J., C. Guillemin, and C. Sarcia, 1967. Géochimie de l'or exogène, *Mineralium Deposita* (Berlin) **1:**259-268.
Haber, F., 1927. Das Gold im Meerwasser, *Angew. Chemie Zeit.* **40:**303-314.
Haber, F., 1928. Das Gold im Meere, *Ges. Erdkunde Berlin Zeit.* Erg. **3:**3-12.
Harrison, J. B., 1908. *The Geology of the Goldfields of British Guiana,* Dulau & Co., London, 320p.
Hatschek, E., and A. L. Simon, 1912. Gels in relation to ore deposition, *Inst. Min. Metall. Trans.* **21:**451-480.
Helgeson, H. C., and R. M. Garrels, 1968. Hydrothermal transport and deposition of gold, *Econ. Geology* **63:**622-635.
Henley, R. W., 1973. Solubility of gold in hydrothermal chloride solutions, *Chem. Geology* **11**(2):73-87.
Hunt, T. Sterry, 1897. *Chemical and Geological Essays,* 5th ed., Scientific Publ. Co., New York, 489p.
Irving, H. M. N. H., 1974. *The Techniques of Analytical Chemistry,* H.M. Stationery office, London, 36p.
Krauskopf, K. B., 1951. The solubility of gold, *Econ. Geology* **46:**858-870.
Krusch, J. P., 1938. *Die Metallischen Rohstoffe: Gold,* F. Enke Verlag, Stuttgart.
Lakin, H. W., G. C. Curtin, A. E. Hubert, H. T. Shacklette, and K. G. Doxtader, 1974. Geochemistry of gold in the weathering cycle, *U.S. Geol. Survey Bull. 1330,* 80p.
Lenher, V., 1912. The transportation and deposition of gold in nature, *Econ. Geology* **7:**744-750.
Lenher, V., 1918. Further studies on the deposition of gold in nature, *Econ. Geology* **13:**161-184.
Lindgren, W., 1924. The colloidal chemistry of minerals and ore deposits, in *The Theory and Application of Colloidal Behavior,* R. H. Bogue, ed. McGraw-Hill, pp. 445-465.
Lindgren, W., 1933. *Mineral Deposits,* 4th ed., McGraw-Hill, New York, 930p.
Lungwitz, E. E., 1900*a*. The lixiviation of gold deposits by vegetation, *Eng. and Mining Jour.* **69:**500-502.

Lungwitz, E. E., 1900*b*. Der geologische Zusammenhang von Vegetation und Goldlagerstätten, *Prakt. Geol. Zeit.* **8**:71-74.
Mellor, J. W., 1923. *A Comprehensive Treatise on Inorganic and Theoretical Chemistry,* vol. 3., Longmans, Green & Co., London.
Niggli, P., 1929. *Ore deposits of Magmatic Origin,* H. C. Boydell trans., Thos. Murby & Co., London, 93p.
Ogryzlo, S. P., 1935. Hydrothermal experiments with gold, *Econ. Geology* **30**:400-424.
Ong, H. L., and V. E. Swanson, 1969. Natural organic acids in the transportation, deposition and concentration of gold, *Colorado School Mines Quart.* **64**(1):395-425.
Petrovskaya, N. V., 1973. *Native Gold* (in Russian) Izd. "Nauka," Moscow, 347p.
Pliny the Elder (Gaius Plinius Secundus), 79 A.D. Historia naturalis, Libra XXXIII, Aurum, Roma.
Rackham, H., trans., 1968. *Natural History,* by Pliny, vol. 9, Libri 33-35, Harvard Univ. Press, Cambridge, Mass., 421p.
Rankama, K., and Th. G. Sahama, 1950. *Geochemistry,* Univ. Chicago Press, Chicago, 912p.
Ransome, F. L., 1907. The association of alunite with gold in the Goldfield District, Nevada, *Econ. Geology* **2**:667-692.
Rickard, T . A., 1932. *Man and Metals,* 2 vols., McGraw-Hill, New York, 1068p.
Roth, J., 1879-1893. *Allgemeine und chemische Geologie,* 3 vols., Berlin.
Rytuba, J. J., and F. W. Dickson, 1977. Reaction of pyrite + pyrrhotite + quartz+ gold with $NaCl-H_2O$ solutions, 300-500°C, 500-1,500 bars, and genetic implications, in *Problems of Ore Deposition,* vol. 2, B. Bogdanov, ed., Bulgarian Acad. Sci., Sofia, pp. 320-326.
Sarton, G., 1945-1948. *Introduction to the History of Science,* 3 vols., Carnegie Inst. Washington, Williams & Wilkins Co., Baltimore, 2155p.
Savage, E. M., 1934. *Prospecting for Gold and Silver,* McGraw-Hill, New York, 300p.
Seward, T. M., 1973. Thio complexes of gold and the transport of gold in hydrothermal ore solutions, *Geochim. et Cosmochim. Acta* **37** (3):379-399.
Sisco, A. G., and C. S. Smith, transl., 1949. *Bergwerk-und-Probierbüchlein,* Am. Inst. Min. Metall. Eng., New York, 196p.
Sisco, A. G., and C. S. Smith, trans., 1951. *Lazarus Ercker's Treatise on Ores and Assaying* (in German), Univ. Chicago Press, Chicago, 360p.
Smith, C. S., and M. T. Gnudi, trans., 1959. *Pirotechnia,* by Vannoccio Biringuccio (1540), MIT Press, Cambridge, Mass., 477p.
Smith, F. G., 1943. The alkali sulphide theory of gold deposition, *Econ. Geol.* **38**:561-590.
Sonstadt, E., 1872. On the presence of gold in sea water, *Chem. News* **26** (671):159-161.
Sorokin, V. N., 1973. On the probable form of gold transportation in hydrothermal solutions, *Geokhimiya,* **12**:1891-1894.
Vernadsky, V. I., 1914. *Descriptive Mineralogy,* I, Moscow, pp. 264-402.
Vernadsky, V. I., 1954-1960. *Selected Works,* 5 vols., Izd. Akad. Sci. USSR, Moscow.
Vilor, N. V., and T. A. Shkarupa, 1971. Fine-disperse gold dissolution in hydrothermal solutions at high temperatures and pressures, *Internat. Geochem. Congr. Abstracts of Reports I,* Moscow, pp. 280-281.
Wedepohl, K. H., 1969-. *Handbook of Geochemistry,* Springer-Verlag, Berlin.
Weissberg, B. G., 1970. Solubility of gold in hydrothermal alkaline sulfide solutions, *Econ. Geology* **65**:551-556.
Wright, K., 1969. Textures from some epigenetic mineral deposits of Tennant Creek, central Australia, in *Remobilization of Ores and Minerals,* P. Zuffardi, ed., Cagliari, Italy, pp. 219-251.
Zhirnov, A. M., 1972. Hypogene colloidal gold in the Kaul'da gold ore deposit (central Asia), *Uzb. Geol. Zk.* **1**:93-95.
Zvyagintsev, O. E., 1941. *Geochemistry of Gold* (in Russian), Izd. Akad. Nauk USSR, Moscow, 114p.

CHAPTER

9

The Origin of Epigenetic Gold Deposits—the Ore-Magma Theory

The igneous theory is sufficient to account for the primary forms and the arrangements of minerals in lodes.
Thomas Belt, 1861

Before proceeding to the main part of this chapter it should be recalled that four general theories for the origin of vein-type gold deposits had been postulated by the midnineteenth century: the ore-magma theory involving siliceous melts, the magmatic-hydrothermal theory, secretion theories, and abyssal theories. The ultimate source of the gold, other metals, and gangue elements (minerals) in gold deposits is mentioned only briefly in these early theories and usually with little elaboration. It was generally assumed that the magmas, the basic materials of the first two theories, were developed by some process involving the melting (granitization) of preexisting crustal rocks, generally sediments, although some early investigators allude to melting of deeper rocks (the mantle?). How the magmas became enriched in gold (and other metals) was usually not mentioned, and indeed today the details are still bypassed by most writers on the subject. The secretionists placed the source of the gold (and other metals) in the host rocks of the deposits, some in the rocks immediately enveloping the deposits (lateral secretion), and others in the general piles of sediments and volcanics hosting the deposits (diagenetic and metamorphic secretion). Advocates of abyssal theories relegated the source of the gold and other metallic vapors, solutions, or melts to the great depths of the earth, often to deep-seated metallic, sulfidic, or other mineral spheres far below the plane of observation.

THE ORE-MAGMA THEORY OF THE ORIGIN OF AURIFEROUS VEINS

James Hutton (1726–1797) seems to have been the first to suggest that mineral veins were formed by igneous melts, basing his theory on the (erroneous) assumption that siliceous matter (quartz) and sulfides were insoluble in water and could, therefore, only be emplaced by an injected fused mass. This extreme

Plutonist view is elaborated in his *Theory of the Earth* (1795) and later by John Playfair (1802) in his *Illustrations of the Huttonian Theory.* In this exposition Playfair wrote (p. 244):

> The slate also in which gold and silver are often found pervading masses of quartz, and shooting across them in every direction, furnishes a strong argument for the igneous origin, both of the metal and the stone. From such specimens, it is evident, that the quartz and the metal crystallized, or passed from a fluid to a solid state, at the same time; and it is hardly less clear, that this fluidity did not proceed from solution in any menstruum: For the menstruum, whether water or the *chaotic fluid,* to enable it to dissolve the quartz, must have had an alkaline impregnation; and, to enable it to dissolve the metal, it must have had, at the same time, an acid impregnation. But these two opposite qualities could not reside in the same subject; the acid and alkali would unite together, and, if equally powerful, form a neutral salt, (like sea-salt), incapable of acting either on the metallic or the siliceous body. If the acid was most powerful, the compound salt might act on the metal, but not at all upon the quartz; and if the alkali was most powerful, the compound might act on the quartz, but not at all on the metal. In no case, therefore, could it act on both at the same time. Fire or heat, is sufficiently intense, is not subject to this difficulty, as it could exercise its force with equal effect on both bodies.
>
> The simultaneous consolidation of the quartz and the metal is indeed so highly improbable, that the Neptunists rather suppose, that the ramifications in such specimens as are here alluded to, have been produced by the metal diffusing itself through *rifts* already formed in the stone [Kirwan, R., 1799. *Geological Essays,* p. 401]. But it may be answered, that between the channels in which the metal pervades the quartz, and the ordinary cracks or fissures in stones, there is no resemblance whatever: That a system of hollow tubes, winding through a stone, (as the tubes in question, must have been, according to this hypothesis, before they were filled by the metal), is itself far more inconceivable than the thing which it is intended to explain; and lastly, that if the stone was perforated by such tubes, it would still be infinite to one that they did not all exactly join, or inosculate with one another.
>
> The compenetration, as it may be called, of two heterogeneous substances, has here furnished a proof of their having been melted by fire. The inclusion of one heterogeneous substance within another, as happens among the spars and drusens, found so commonly in mineral veins, often leads to a similar conclusion. Thus, from a specimen of chalcedony, including in it a piece of calcareous spar, Dr. Hutton has derived a very ingenious and satisfactory proof, that these two substances were perfectly soft at the same time, and mutually affected each other at the moment of their concretion [Hutton, J. 1795. *Theory of the Earth,* vol. 1, p. 93].
>
> Each of these substances has its peculiar form, which, when left to itself, it naturally assumes; the spar taking the form of rhombic

> crystals, and the chalcedony affecting a mammalated structure, or a superficies composed of spherical segments, contiguous to one another. Now, in the specimen under consideration, the spar is included in the chalcedony, and the peculiar figure of each is impressed on the other; the angles and planes of the spar are indented into the chalcedony, and the spherical segments of the chalcedony are imprinted on the planes of the spar. These appearances are consistent with no notion of consolidation that does not involve in it the simultaneous concretion of the whole mass; and such concretion cannot arise from precipitation from a solvent, but only from the congelation of a melted body. This argument, it must be remarked, is not grounded on a solitary specimen, (though if it were it might still be perfectly conclusive), but on a phenomenon of which there are innumerable instances.
>
> According to this theory, veins were filled by the injection of fluid matter from below; and this account of them, which agrees so well with the phenomena already described, is confirmed by this, that nothing of the substances which fill the veins is to be found any where at the surface. It is not with the veins as with the strata, where, in the loose sand on the shore, and in the shells and corals accumulated at the bottom of the sea, we perceive the same materials of which these strata are composed. The same does not equally hold of metallic veins: "Look, says Dr. Hutton, into the sources of our mineral treasures? Ask the miner from whence has come the metal in his veins? Not from the earth or air above, nor from the strata which the vein traverses: these do not contain an atom of the minerals now considered. There is but one place from whence these minerals may have come; this is the bowels of the earth; the place of power and expansion; the place from whence has proceeded that intense heat, by which loose materials have been consolidated into rocks, as well as that enormous force, by which the regular strata have been broken and displaced." [Hutton, J. 1795. *Theory of the Earth,* vol. 1, p. 130]

The Plutonist theory was challenged by Kirwan (1793, 1799), a Neptunist, who dwelt at length on the solubility of silica, sulfur, and other mineral components in water, drawing his evidence from hot spring waters, siliceous sinters, and the known chemical facts of the solubility of silicon and sulfur compounds. Kirwan (1799, p. 401) wrote the following about gold in his *Geological Essays:*

> **Native Gold**
>
> Of all metals gold is most frequently found native. According to Bergman, it is more universally diffused than any other metal, except iron; this may be a consequence of its great divisibility and want of affinity to other substances, as oxygen, sulphur, etc. hence at the emersion of primeval mountains, it remained entangled or dispersed through the stony masses of many of them wherever these were permeable to water; the golden particles were, however, in a course of ages, washed and carried down in minute rills into the

> neighbouring plains, until arrested by some obstacle long enough to suffer the gold to deposit; these minute particles being thus brought into contact in the minutest state of division, united with each other by virtue of their integrant affinity, sometimes involving sandy particles, and thus formed those shapeless masses, of various sizes, which are sometimes met with in various countries, and lately in the county of Wicklow; that these lumps were never in fusion is evident from their low specific gravity, and the grains of sand found in the midst of them. I found the specific gravity of a lump found in the county of Wicklow, of the size of a nutmeg, to be only 12,800, whereas after fusion it was 18,700, and minute grains of sand appeared on its surface. Hence many rivers were anciently auriferous, which now cease to be so; as the Tagus, Po, Pactolus, Heber. Pliny lib. xxxiii, cap. 4, and though in France some are still auriferous, yet it appears, by the testimony of Diodorus, that they were much more abundantly so in former ages (Lib. v. cap. 19.). Hence also native gold is seldom alloyed with any metal, except silver or copper, to which it has the greatest affinity, and which are also least liable to a combination with sulphur or acids.
>
> It is oftener found in iron ores than in any other, because these are far more universal and abundant than any other, particularly in a more or less indurated and brittle, brown or reddish brown iron stone; though originally it was deposited in primeval mountains, yet from them by subsequent operations of nature it has frequently been deposited in secondary masses, yet still it is most frequently found in quartz, feldspar, etc. sometimes in gypsum, baroselenite, etc.

J. F. Fournet (1801-1869) professor of geology at the University of Lyon, France, was the next to advocate an ore-magma theory for mineral veins. From careful field observations and petrological correlations in Aveyron and elsewhere in France he concluded that the metalliferous veins were the magmatic end results of a process that initially gave rise to granite followed by porphyry dikes and veins. In a summary of his last papers on the subject, *Suite des Aperçus rélatifs à la theorie des filons, métallifères,* (Fournet, 1856 pp. 899-900) he considered that the processes of ore deposition were due to two causes—magmatic injection and weathering or oxidation—and concluded:

> En résumé, l'organisation complète des filons me conduit à la conclusion qu'ils sont le résultat de deux causes, savoir: les action plutoniques qui ont opéré par la voie de la fusion, et les action atmosphériques qui ont remanié les produits antérieurs. La fusion, aidée de la pression, de la surfusion, de la cristallisation et de quelques effets mécaniques, peut expliquer tous les phénomènes de gîtes. La théorie que j'admets a d'ailleurs l'avantage de raccorder la formatin des filons avec celle des roches éruptives. Elle explique parfaitement les transitions insensibles qui unissent ensemble les filons à silicates à ceux dont les gangues sont purement salines ou quartzeuses. J'attends, d'ailleurs, des objections autres que celles qui m'ont déjà été posées pour y répondre d'un seul coup. Les action atmosphériques ou superficielles sont trop naturelles, trop

> bien en rapport avec les principes chimiques, pour ne pas êtres acceptées. Elles généralisent les effets de la kaolinisation en les faisant passer du domaine des roches siliceuses à celui des matières filoniennes. Par leur caractère, ces deux théories satisfont plus que toutes les autres au grand principe de Newton: *Natura simplex est, et superfluis non luxuriat causis.* C'est donc encore cette simplicité qui m'enhardit à soutenir ma manière de voir et qui me porte spécialement à rejeter les complications tubulaires de M. Durocher, sans compter qu'elles ne sont, en aucune façon, démontrées par la constitution des filons.

Similar views were held by A. Petzholdt (1810-1889) in his *Geologie* (1845). He expressed the opinion that most metalliferous veins originated from magmatic injections that derived their metals from the earth's metallic interior.

Thomas Belt (1832-1878), mining engineer and renowned naturalist, was the first to make extensive and detailed studies of gold-quartz veins in many parts of the world. He worked in Australia while the great deposits of Victoria were under development and in production, and later examined many of the gold deposits in Nicaragua, Mexico, Brazil, the United States, North Wales, Nova Scotia, Canada, and Russia. Belt wrote two books, *The Naturalist in Nicaragua* (1874) and his classic work *Mineral Veins: An Enquiry into their Origin, Founded on a Study of the Auriferous Quartz Veins of Australia* (1861). Belt was greatly impressed by the massive nature of many quartz veins, their close relationship to granites and porphyries, their frozen contacts, their suspended wall rock fragments, and other features, all of which led him to postulate an injected molten siliceous magma with entangled metallic vapors as the mother material of the veins. The conclusion of his 1861 study of the Australian (Victoria) veins merits reprinting in full.

> **Conclusion**
>
> I have thus shown that the production of fissures, and their injection with fused matter, are the natural results of plutonic action, and that the igneous theory is sufficient to account for the primary forms and the arrangements of minerals in lodes. The following is a brief summary of the conclusions arrived at:
>
> 1st. That the auriferous quartz veins of Australia are filled with minerals, which are not liable to be decomposed by the action of water, and which apparently now exist in the same state as they were originally deposited.
>
> 2nd. That in these veins the distribution of the gold, and the structure and arrangement of the quartz, are explained by the theory that they are fissures that have been filled with molten silica, containing entangled metallic vapours.
>
> 3rd. That mineral veins are constantly found in connection with igneous rocks, and that in some cases, as in Cornwall and Wicklow, a regular sequence of events have followed the intrusion of molten granite, by which granitic, porphyritic, and mineral veins have been successively formed.
>
> 4th. That the fusion of rocks in the bowels of the earth, and their subsequent consolidation, supply the requisite conditions for the

> rending open of the superincumbent rocks, and the filling of the rents so formed with fluid matter, varying in composition according to the comparative depth from which it has been projected.
> 5th. That the objections raised against the igneous theory of quartz veins and of granite are not tenable, being based either on a misapprehension of the theory, on a misinterpretation of observed facts, on experiments where the natural conditions were not fulfilled, or on the obscurity in which certain delicate chemical questions are still involved.
> 6th. That the investigation of the origin of the lodes of the baser metals in Europe has been impeded by the confusion arising from the mixing up of the results due to secondary agencies, with those referrible to original deposition.
> 7th. That mineral veins and trappean dykes have many features in common, and that the points in which they vary may be explained by a reference to the different conditions under which the igneous matter has been developed. (pp. 51-52)

Belt's later observations in Nicaragua, Nova Scotia, and elsewhere caused him to modify his views only slightly while maintaining his original thesis of an ore magma as revealed in his book *The Naturalist in Nicaragua,* where in chapter 6 (pp. 79-80) on quartz lodes he summarized their origin as follows:

> I shall only now give a brief *résumé* of the conclusions I have arrived at respecting the origin of mineral veins.
> 1. Sedimentary strata have been carried down, by movements of the earth's crust, far below the surface, covered by other deposits, and subjected to great heat, which, aided by the water contained in the rocks and various chemical reactions, has effected a re-arrangement of the mineral contents of the strata, so that by molecular movements, the metamorphic crystalline rocks, including interstratified granites and greenstones, have been formed.
> 2. Carried to greater depths and subjected to more intense heat, the strata have been completely fused, and the liquid or pasty mass, invading the contorted strata above it, has formed perfectly crystalline intrusive granites and greenstones.
> 3. As the heated rocks cooled from their highest parts downwards, cracks or fissures have been formed in them by contraction, and these have been filled from the still-fluid mass below. At the beginning these injections have been the same as the first massive intrusive rocks, either granite or greenstone; but as the rocks gradually cooled, the fissures reached greater and greater depths; and the lighter constituents having been drawn off and exhausted, only the heavier molten silica, mingled with metallic and aqueous vapours, has been left, and with these the last-formed and deepest fissures have been filled. These injections never reached to the surface—probably never beyond the area of heated rocks; so that there have been no overflows from them, and they have only been exposed by subsequent great upheaval and denudation.

4. Probably the molten matter was injected into the fissures of rocks already greatly heated, and the cooling of these rocks has been prolonged over thousands of years, during which the lodes have been exposed to every degree of heat, from that of fusion to their present normal temperature. During the slow upheaval and denudation of the lodes, they have been subjected to various chemical, hydro-thermal, and aqueous agencies, by which many of their contents have been re-arranged and re-formed, new minerals have been brought in by percolation of water from the surrounding rocks, and possibly some of the original contents have been carried out by mineral springs rising through the lines of fissures which are not completely sealed by the igneous injection, as the contraction of the molten matter in cooling has left cracks and crevices through which water readily passes.
5. Some of the fissures may have been re-opened since they were raised beyond the reach of molten matter, and the new rent may have been filled by hydro-thermal or aqueous agencies, and may contain, along with veinstones of calcite derived from neighbouring beds of limestone, some minerals due to a previous igneous injection. Crevices and cavities, called *vughs* by the miners, have been filled more or less completely with crystals of fluor spar, quartz, and various ores of metals from true aqueous solutions, or by the action of super-heated steam.
6. By these means the signs of the original filling of many mineral lodes, especially those of the baser metals, have been obscured or obliterated; but in auriferous quartz lodes both the metal and the veinstone have generally resisted all these secondary agencies, and are presented to us much the same as they were first deposited, excepting that the associated minerals have been altered, and in some cases new ones introduced, by the passage of hot springs from below or percolation of water from the surface.

The reader will recognize from this excerpt that Thomas Belt, held quite modern views on metamorphism, the origin of granitic rocks, and the source of the metals via anatexis (granitization).

Josiah Edward Spurr (1870-1950), one of the great American economic geologists of the first part of the twentieth century, was a strong advocate of ore magmas. Spurr based his opinions on a lifelong experience with ore deposits in North America, having studied gold and other metal deposits of many types and ages in Canada, the United States, and Mexico. His great work *The Ore Magmas* (1923) is seldom read today by economic geologists, a regrettable situation, for it contains a multitude of factual observations, the interpretation of which may be questioned but the authenticity of which there can be no doubt. Spurr observed what since has repeatedly been noticed in gold-bearing districts, a sequence of igneous rocks and veins beginning with granitic rocks and progressing through pegmatites, aplites, and quartz-feldspar porphyries to gold-quartz veins. Many of the last Spurr called "veindikes," and he reasoned that they were originally differentiated magmas intruded into the country rocks. The composition of the veindike magmas was siliceous, fluxed with variable quantities of water, carbon dioxide, and other volatiles that effected the wall rock alteration observed adjacent to many types of gold-quartz veins. Fragments of wall rock suspended in gold-quartz veins are considered as the main proof of the viscous (magmatic) character of the vein-forming media. In the synopsis of chapter 1 of his *Ore Magmas* and in the preliminary part of chapter 16, Spurr outlines his ideas in a succinct manner (Papers 9-1*A* and 9-1*B*).

(Text continues on page 209.)

9-1A: THE ORIGIN OF ORE MAGMAS OR SOLUTIONS: VEINDIKES

J. E. Spurr

Reprinted from *The Ore Magmas,* McGraw-Hill, New York, 1923, pp. 1-3.

Whatever will explain the close association, in veins, of quartz and metals will illuminate ore deposition. In 1908, I observed in Alaska that quartz veins carrying gold could be traced into pegmatites and aplites consisting of quartz and feldspar. These passed into various types of hornblende-feldspar rocks, and all these varieties I accounted for (and still do) by what is called *magmatic differentiation.* Assuming this magmatic differentiation origin for veins and rocks, I understood for the first time the association of quartz and gold, both being results of extreme differentiation. Therefore, there is no hard and fast line between the magma solutions which have deposited alaskites and those which deposited the related auriferous quartz veins—both are magma solutions.

Water must be present in all magmas, which are fluid more on account of solution than heat, important though the latter is. On rock solidification, water is expelled. The gradation of granites and alaskites into pegmatites shows that the pegmatite magma is a variation or residuum of the granite magma; and the freedom of segregation of like minerals in the pegmatites proves that the pegmatite magma contains more water and other mobile elements than does the granite magma. Rare minerals are thus frequently concentrated in pegmatites—such as apatite, topaz, and beryl.

Igneous rocks contain metals—like gold and copper—universally distributed, perhaps as silicates; the metals also occur in rocks as oxides and sulfides, such as magnetite, ilmenite, pyrite, probably chalcopyrite, and perhaps molybdenite. In pegmatites metallic minerals are also present—pyrite, chalcopyrite, molybdenite, and gold are frequently found.
Such pegmatites may show all gradations from a quartz-feldspar to a pure quartz rock.

It is difficult to choose between the words "dike" and "vein" for these types. Therefore, I propose to call these borderland types "vein-dikes"; they are intrusive, but the intrusive magma differed from that typical of the usual igneous rock.

Scheerer pointed out, as early as 1846, the dependence of granitic magma solutions on contained water. In 1884, Lehmann confirmed this view and argued that pegmatites were injections, and that the quartz veins or dikes transitional into the pegmatites must have had a similar origin. Howitt, in 1887, came to the same conclusion in Australia. In the United States, Lane in 1894, and Crosby and Fuller in 1897, maintained that quartz "veins" with an obvious relation to pegmatites were the end product of magmatic differentiation.

More difficultly came the acknowledgment that those quartz veins which contained metallic minerals had this origin by magmatic differentiation. Howitt did not acknowledge it for the Australian types. I published this conclusion in 1898, for the Yukon gold-quartz veins (that they were the end products of granitic magma differentiation), and a little later in the same year Hussak described an auriferous quartz vein in Brazil as an intrusive ultra-siliceous granitic dike.

Let us remember that all magmas—even rock magmas—are solutions.

The theory of the formation of mineral veins by emanations from volcanic and other igneous rocks dates back at least to Élie de Beaumont in 1847; but this is not the magmatic-differentiation theory which I am propounding. The magmatic-differentiation theory for auriferous quartz veins was, indeed, propounded before me, by Thomas Belt, in 1867 and 1871. He noted the transition from granite to quartz veins, and believed this to be the origin of gold-quartz veins in general.

In the Silver Peak district, Nevada, I found further proof of the transition from alaskite to gold-quartz veins.

This differentiation of granitic or alaskitic magmas into gold-quartz veins as final end products must take place at a depth of several miles—my impression is that it is many miles in some cases. As to temperature, artificial formation of granitic minerals, and other criteria, indicate that granites and pegmatitic granites crystallized between 575 and 800°C; while the pegmatites and related (pegmatitic) quartz veins crystallized from 575° to a somewhat lower temperature.

Metallic sulfides in pegmatites are usually not important commercially; but in quartz veins of close pegmatitic affiliation the metallic sulfides are more frequently commercially valuable. The commonest ores which I have seen of this kind are those of tungsten; tin also commonly occurs in veins or veindikes of this type.

Closely allied to these, but not so closely allied to pegmatites, is a certain type of quartz vein carrying free gold, auriferous pyrite, and other sulphides, such as those I described in Alaska and at Silver Peak, and those of California, Canada, the Appalachians, and Australia. These are "free-milling" (i.e., easily amalgamated) medium to low-grade ores; and they occur only in regions which have undergone deep erosion.

These ores are often true veindikes, and have intruded their country rock, pressing the fissure walls apart by their inherent or "telluric" pressure. They assume lenticular forms when formed at great depths; formed at somewhat lesser depths, they have the more tabular vein form, and contain typically rather more gold.

9-1B: THE ORIGIN OF FISSURE VEINS

J. E. Spurr

Reprinted from *The Ore Magmas*, McGraw-Hill, New York, 1923, pp. 707-709.

This chapter sets forth that although a considerable proportion of mineral veins are veindikes—that is, that they are intruded under their own gaseous tension—while many have been formed by replacement or impregnation by thinner and more aqueous ore magmas, yet the locus and shape of all these veins is determined by pre-existing fissures in the rocks. Certain fissure veins are confined to the areas of restricted intrusive bodies, making it certain that the fissures were developed as the result of adjustment of the igneous rock after intrusion. Such fissures may cut across intrusive necks or bodies, from side to side, or form, roughly, parallel to the contacts. In some instances two sets of fissures due to adjustment of the intrusive rock have been noted within the rock, the major set parallel to the long axis of the intrusion, the minor set transverse to it. Fissures form along dikes; indeed, a fissure once opened is apt to be reopened; and this may give rise to compound veins or, indeed, to compound dikes. Usually the fissures which are filled by veins are of very slight relative movement or displacement. Where a vein occupies a fault of considerable displacement, it belongs in many cases to a later intrusion episode than the fault. But faulting may result from the upward and outward shove of intrusion, or, on the other hand, from contraction and sagging after intrusion; and in the former case considerable faulting may elapse before the ore deposition, and still the two may belong to the same intrusion episode. Horizontal movement is as common in faulting as the vertical, and often predominates. But the inception of the average fissure is probably from a vertically exerted force, as indicated by the very common dip of 60-70°. The common east-west trend of veins in the Western United States is perhaps due to the east-west trend of subcrustal magma invasions. Conjugated vein fissures are believed to owe their origin to force exerted by an intrusion, either upward or downward.

This chapter is, in a way, partly a restatement of some of the conclusions arrived at in Chapter VIII, but from a somewhat different point of view. In general, I have assumed that magmas are, more or less, in a state of gaseous tension, by virtue of which they differentiate and by virtue of which they have intrusive power; and this applies to all magmas, whether silicate magmas, which are the common rock magmas, or to those highly specialized magmas which are the products of differentiation, and which I have called ore magmas. The rock magmas are present in enormous quantity in many cases; they intrude as immense masses, and also as dikes of varying width. The ore magmas, like all extreme magmas, are produced only in small quantity; and they are characteristically intruded in dike-like form, which we may call veindikes or veins. It is likely that a considerable proportion of the mineral veins are not far different from the pegmatites in mode of origin, in that they are intrusive under their own gaseous tension, which has been strong enough to push the intruded rocks aside; and this is probably true at all zones of ore deposition. Not only the pegmatites, and the gold-quartz veins, but the tungsten-quartz, and the tin-quartz veins, show the characteristic features of this origin—frequent wide regular veins with parallel walls (which could never have represented the filling of pre-existing open fissures) and included angular unsupported fragments of the wall rocks. And the same features persist in the wide quartz veins of all stages, up to those in Tertiary lavas, such as I have described at Tonopah and Aurora. Such veins are often fifty or a hundred feet wide, and very long and regular. Empty fissures of this size could never have existed. An origin of the veins by replacement is in very many cases quite out of the question, for the contact with the wall rocks is sharp, and the included fragments have not only sharp contacts with the vein quartz, but the original angular corners are not rounded by corrosion. The forcing apart of the walls of these veins by the pressure of crystallization has been argued repeatedly, but the application of such a supposition is misty. In the majority of these cases the vein-quartz filling will be found to be homogeneous and granular. How did the filling get into the vein fissure before it crystallized, for obviously it all crystallized in place as we find it? No distinct analysis or picture of the workings of this crystallization hypothesis seems possible. The walls of the veins have certainly been thrust apart by pressure, but the thrusting aside was accomplished before crystallization; and the thrusting power, therefore, resided in the fluid quartz magma, which, accordingly, was under pressure, quite like the pressure of a rock magma which solidifies, after penetration, as a rock dike. Such pressure, I have assumed, depends on the gaseous tension which resides in the magmas in each case, on account of their being in more or less a gaseous condition, or partly made up of gaseous elements.

These characteristics, once grasped, and their significance as well, will be found to be true of many sulfide veins, as well as quartz veins, at all stages; and, moreover, of many veins of barite, fluorite, celestite, dolomite-siderite-magnesite-rhodochrosite, and even of calcite.

The above, while I believe it is most typical of veins, is not true of all veins. Many veins have formed, of course, by replacement along fracture zones, and by impregnation of porous zones or rocks of all descriptions. A difference in the aqueous content and in the gaseous nature and gaseous content of different vein magmas, even different quartz-vein magmas, must be granted, as deduced from these varying characteristics. Some ore magmas are more aqueous than others—consist more and more of water, and come accordingly to prefer to soak through broken rocks rather than to shove them asunder.

Suspended wall rock fragments in gold-quartz veins have been one of the principal arguments for an ore-forming viscous (magmatic) medium. Rollin Farmin used this feature to substantiate his magmatic thesis for the origin of the auriferous veins at Grass Valley, California. The gold-quartz veins of Grass Valley are hosted mainly by grandiorite and diabase. The veins contain essentially coarse-grained, milky quartz, ankerite, calcite, and minor amounts of metallic sulfides, sericite, chlorite, and gold. Many of the veins contain wall rock inclusions that are altered to masses of carbonate, chlorite, talc, sericite, mariposite, and so forth. The wall rocks are similarly altered; the carbonated, sericitized, and chloritized wall rocks form envelopes about the quartz veins. In two classic papers Farmin (1938, 1941) described and figured the wall rock fragments in the veins and concluded that their presence was due to an injected vein-forming magmatic solution. His abstract and conclusions from his paper follow.

Abstract

The mesothermal gold-quartz veins at Grass Valley, California, contain many inclusions of country rock in quartz, some of which are broken and the segments separated. When the quartz is removed from the inclusions they can be reassembled into larger, original units of country rock. Adjacent walls of the veins have not shared in the movements that dislocated the inclusions and no evidence of faulting is found in the massive quartz matrix, which exhibits features of original texture. Dislocation of the fragments is attributed to thrust from an injected, vein-forming solution, which forcibly entered rock fractures and spread apart their walls. [p. 579]

Conclusions

Vein matrix exhibiting original texture includes fragments of wall rock that have been dislocated without corresponding dislocation of the enclosing quartz and wall rock. Most of the current explanations of inclusions do not readily account for the dislocations; the writer attributes them to pressure from an injected, vein-forming, magmatic solution that was relatively concentrated. [p. 599]

In his second paper Farmin (1941) extended his observations to other deposits of the Sierra Nevada, California. His interpretative discussion and mechanism of gold-quartz vein emplacement (Paper 9-2) merit close study by economic geologists.

The last investigator to espouse the ore-magma theory for the origin of gold-quartz veins was W. J. Bichan. In one of the discussions of L. C. Graton's (1940) paper on the nature of the ore-forming fluid, Bichan (1941) disagreed with the origin of gold-quartz veins from hydrothermal solutions, asked a number of searching questions, and postulated a *silicothermal solution* as the medium from which the auriferous veins were deposited. He asked fourteen questions and answered them as presented in Paper 9-3.

(Text continues on page 227.)

9-2: HOST-ROCK INFLATION BY VEINS AND DIKES AT GRASS VALLEY, CALIFORNIA

Rollin Farmin

ABSTRACT.

Structural deformation of the rocks of the western Sierra Nevada has developed zones of faulting and of minor partings that are only slightly dilatant, appropriate in deep-seated environment. Gold-quartz veins and dikes of igneous rocks contain inclusion breccias and occupy partings of highly dilatant character—features that may result from the inflation of deep-seated rocks by an injected fluid but would only be associated with near-surface fault fissures. The folding of inter-vein and inter-dike wall rock during inflation of the partings is discordant with an open-fissure environment. Most of the structures found in the gold-quartz veins are likewise found in igneous dikes of comparable size—parallel mechanisms of emplacement are indicated.

[*Editor's Note:* Material has been omitted at this point.]

INTERPRETIVE DISCUSSION.

The Mechanism of Dike Emplacement.

Volcanology supplies a reliable background of knowledge about the upper ends of dikes and about the near surface activity [12] of magma. Volcanic material of various types rises to the surface and is extruded, along with entrained xenoliths, from partings in the country rock. Whatever may be its nature, the volcanic motivating force is sufficient to elevate the magma to mountain tops, to lift spines as at Pelèe, to blast away a cubic mile of cover over a crater or to split new partings through the base of a volcanic cone if the former vent is blocked.

By geologic backtracking, extinct volcanoes and their underlying, dike-filled vents are traced downward toward the zone where deep-seated dikes and ores are emplaced. For example, 30 miles north from Grass Valley, feeder dikes are intermittently exposed in mines from a point level with the floor of extrusion on down to a depth of 2400 feet below that floor.[13] These dikes occupy dilatant partings similar to those recently described at Cornucopia, Oregon.[14] Because changes in form and texture of dikes are gradual in exposures ranging from near-surface to deep-seated environments, it seems reasonable that the mechanism of emplacement is not greatly different at depth—that the magmatic fluid is

[12] Zies, E. G.: The surface manifestations of volcanic activity. Trans. Amer. Geophys. Union, 19th Ann. Meeting, pp. 10–23, 1938.

[13] Farmin, R.: *op. cit.*, p. 594. Details of the occurrence are described.

[14] Goodspeed, G. E.: *op. cit.*, p. 176, 1940.

propelled through the partings of the host rock by a fully adequate telluric force. Where this is the case, the structures of the dilatant dikes are readily explained as resulting from a progressive magmatic inflation of the host rock along zones of sheeting, along folia and along irregular partings. The process involves uplift of overlying blocks and involves thrust faulting, minor folding and metamorphism of the wall rocks. The dike-forming minerals crystallize from the fluid without allowing much deflation of the distended fissures, judging from the lack of collapse breccia and relaxational faults.

The concept that the injected magmatic fluid is a wet mush of partly formed crystals [15] seems to fit many features of the deposits, but the comb textures and mineral banding are not readily explained by it unless they have resulted from recrystallization.

Reasons for concluding that the mechanism of dike emplacement is a forcible intrusion, not limited to loci of pre-existing openings, are (1) the general inflation of the host rock by dikes is not limited to partings in the "tension" position, (2) intra-magmatic fault offsets are either reverse or simply dilatant, (3) the slicing and separation of the country rock into slabs in lit-par-lit and book structure and the folding of the inter-dike "straps" of country rock could not develop along the unloaded surfaces of an open fissure, (4) the inclusion by dikes of angular fragments of the wall rocks indicates intrusion because these rocks nowhere are found to be brecciated by the structural deformation along barren faults and folds.

The Mechanism of Vein Emplacement.

Several mechanisms have been suggested to explain emplacement of quartz veins in the Sierra Nevada gold belt. The first three discussed here are variations of the "hydrothermal hypothesis," the last is the hypothesis of a magmatic injection.

1. *Replacement.*—Although the replacement mechanism has been advanced for the veins of Grass Valley,[16] its general effect-

[15] Sosman, R. B.: Evidence on the intrusion-temperature of peridotites. Amer. Jour. Sci., V, XXXV–A: 359, 1938.

[16] Howe, E.: The gold ores of Grass Valley, Calif. Econ. Geol., 19: 595–622, 1924.

tiveness must be very limited in view of the almost universally dilatant character of the partings occupied by the quartz. The metamorphosis of the wall rocks adjacent to the quartz veins to an ankerite-chlorite-sericite-talc rock involves abundant addition of carbon-dioxide (by diffusion ?) but perhaps does not involve the withdrawal of an equivalent amount of material because the system shows increase in volume, shows increase in specific gravity of the altered rock and because no elements seem to have been eliminated by the process.

2. *Lateral Secretion.*—The derivation of some of the materials for vein filling from wall rocks (not necessarily adjacent to the point of deposition) has been suggested by Knopf [17] as an auxiliary mechanism in vein formation. Even if it is coupled with the linear force of growing crystals, lateral secretion does not offer a mechanism for the transportation of xenolithic inclusions nor for the folding of "straps" and therefore cannot be more than an incidental feature of an hydrothermal deposition.

3. *Accretion from Hydrothermal Deposition in Reopened Partings.*—The accretion mechanism (progressive deposition from dilute hydrothermal solution in partings opened by the bridging of fault walls) has been the backlog for most geologic interpretations of the quartz veins. It seems to explain some segments of veins and the alteration of wall rocks admirably; and it offers much freedom for explaining the complex microscopic textures to which many geologists not actively engaged in mining devote considerable attention.

Along the larger veins the specific information about pre-, intra-, and post-mineral faulting movements seldom is sufficiently well known to permit a rigorous test of the accretion hypothesis. Fortunately, however, satisfactory measurements of these movements are available along the subordinate veins, spurs and stringers that have contributed a respectable proportion of the gold mined in California. The partings that encase these parts of the veins commonly are mere slits in the rock, showing only microscopic lateral offsets but containing quartz an inch or two wide

[17] Knopf, A.: The mother lode system of California. U. S. Geol. Surv. Prof. Pap. 157: 32, 1929.

that tapers to the very tips of the partings; beyond the tips the counry rock is unbroken, as has been verified by thin sections and by acid leaching of the carbonatized rock. Obviously, diagonal thrusting does not explain the development of these openings, which are one hundred times wider than the total lateral movement involved. Therefore, the structural preparation of openings that is fundamental in the accretion mechanism is found to be lacking in most of these stringers.

Further structural difficulties are encountered when the accretion mechanism is offered to explain the " straps " of country rock lying between veins or to explain the flexed slabs of wall rock in the " centipede " veins. The dense rocks that are host to the veins cannot be flexed appreciably except under great restraining load—which is absent in an open fissure. The partings occupied by the quartz are smooth, shear-type joints that are discordant with open fissures and, although similar smooth partings are abundant in the barren country rocks, tensile openings are not found in them, nor are breccias found consequent to the collapse of such openings.

At first glance, both book and lit-par-lit structures seem to be well explained by the accretion mechanism. According to the statement by Hulin:[18]

> Portions of the Mother Lode contain innumerable thin oriented parallel fragments of the foliated wall rocks. At times these are spaced only a small fraction of an inch apart through vein thicknesses of several feet. This "book structure" could conceivably have resulted only through countless reopenings of the vein fissure contemporaneous with the mineralization, each reopening tearing loose a thin film of the foliated wall rock which had frozen to the vein.

A criticism of this mechanism is that the foliated country rocks always lie at an angle to the veins and only locally are dragged into approximate parallelism; therefore, the rock wafers are not folia pulled apart by vein opening because they are cross-grained and would open only as ragged masses. Instead, the wall rocks have been self-sliced in a shear-type deformation requiring great restraining load and subsequently have been leafed apart to receive the quartz. Many of the slabs are flat-lying, cross-grained masses

[18] Hulin, C. D.: Structural control of ore deposition. Econ. Geol., 24: p. 30, 1929.

of large area (Fig. 13) that would not be self-supporting for an instant if a tensile opening were to rob them of support and most assuredly would not withstand the slicing process except under load. Whether the slicing and opening by the accretion mechanism is thought of as a progressive or as an alternating process, it seems to be geologically discordant and mechanically unsound.

It may be suggested that ground water penetrates to the zone of sliced ground and supports the rock slabs during an accordion-like opening and until the arrival of the hydrothermal fluid. Although some support would be delivered to the walls in this case, the flat-lying slabs nevertheless would break and sink in a water free to displace upward in the fissures. Furthermore, the interchange of support from one fluid to another without damaging the slabs would be a remarkable feat of geologic juggling.

A certain combination of diverse structures occasionally is found in ribboned veins: angular inclusions of the several country rocks, together with the " late " sulphides, occur in a central panel of the quartz vein, sandwiched between marginal panels that exhibit delicate book structure but lack the sulphides. By the usual statement of reopening, the septa of country rock are pulled successively from the walls and therefore the marginal panels should be younger than the central ones, instead of older as is indicated by the mineralogy. Moreover, the angular inclusions will not refit into the smooth walls nor into the adjoining smooth, included septa; and in flat-lying veins the fragments could not tumble far in the narrow spaces available between ribbons. The angular inclusions cannot be remnants from incomplete replacement of a large central slab of included country rock because they are of heterogenous rock types. Therefore, the accretion mechanism does not seem to offer a good explanation for this complex structure.

A further test for the accretion mechanism is in the occurrence of quartz in continuous shoots that maintain fairly uniform dimensions and lenticular cross-section for impressive lengths along the rake. Because the veins are limited to fault openings by the hypothesis, it is necessary that the faults—while traversing hard and soft rocks of various sorts—maintain a succession of arched openings that are aligned, side to side, without interruption by a compression belt, through the mile or two known to us from mining. Moreover, the unbroken tension bridge must have extended on down to the igneous hearth and on up to the surface; and it must have endured through the very long period of accretion of the quartz and until the " late " metallic sulphides and gold were added. Bearing on this last requirement is the fact that extensive mining of many quartz veins at Grass Valley shows a persistent 1 to 3 per cent tenor of these seemingly-late minerals in the ores; perhaps half of all of the quartz that has been exposed contains these " sulphurets " to the amount of 1 per cent. Therefore, the shoots must have been open or reopened throughout most of the time in order to receive so widely the " late " minerals in overlapping sequence. Thus, the entire accretion mechanism depends on the favorable alignment of coincidences for each second it operates—that chance would permit the accretion of more than one such shoot in a district seems a very remote possibility.

A further stumbling block is the fact that sizeable deposits of the " late " minerals without much quartz are so rare along the intra-mineral faults. Surely if separate solutions [19] brought these materials, at intervals they should have escaped into newly formed fault openings, especially if reopening of faults is made a basis for the mechanism.

Even the " plumbing " for a hydrothermal " circulation " is obscure. How are the vein trunks filled—from top to bottom or bottom to top, from side walls inward or from a central panel outward? These questions bear on a possible blockage of the hydrothermal flow and the consequent incomplete filling of the veins (to be expected far in excess of the small, ellipsoidal, disconnected vugs that are found in them). It seems difficult to avoid a blockage of flow of the solution from the trunk channels into the broad, leaf-shaped stringers, which commonly are attached to the sides of the vein trunk only by narrow stems. Stringers of this shape present a real problem in the " plumbing " for a con-

[19] Hulin, C. D.: Structural control of ore deposition: the effects of mineral sequence. Abstract, ECON. GEOL., 34: 471, 1939.

tinued "circulation" of the hydrothermal fluid throughout the long period of vein accretion and enrichment. Lindgren[20] discounts the importance of filling by diffusion through stagnant solution and Graton[21] stresses the necessity for a long-continued flow of the solution past all points of deposition. Many veins and stringers are "frozen" fast to their walls and seem as impervious as the dense country rocks; a system of "plumbing" for packing them so tightly from end to end by hydrothermal flow should not be taken for granted.

Evidence for "openings" from bridging fault surfaces should be found if they have been formed on a scale to accommodate the rotation and transportation of six-foot slabs of country rock as vein inclusions; however, such openings, and the breccias that would develop by their collapse, are lacking. Instead, the zones of dilation along the thrust faults apparently were filled at all times with gouge, fluccan and similar products; and the fault blocks glided along the undulating surfaces somewhat after the manner of moving glaciers rather than as inflexible masses. In this geologic environment actual openings are discordant; and volumes such as those occupied by the veins are not passively yielded by the country rocks. Gouge is here pictured not only as the product of abrasion but also as a fluff of rock splinters[22] spalled from the decompressed fault surfaces where diagonal movement tends to separate them. The irregular masses of gouge and fluccan measure the dilation involved in a fault movement and fill the only "openings" made available for the accretion of vein matter.

4. *Host-Rock Inflation by a Vein-Forming Magmatic Fluid.*—A few geologists have advocated vein formation by the mechanism of an intrusive, magma-like fluid, or slush, from which the gold-quartz veins crystallize. In an excellent paper[23] on the nearby veins of Alleghany, Ferguson partially endorses such a

20 Lindgren, W.: Mineral deposits. 3rd ed., p. 201, 1928.

21 Graton, L. C.: The nature of the ore-forming fluid. Econ. Geol., 35: 320, 1940.

22 Willis, B. & R.: Geologic structures. 2nd ed,, pp. 446–8, 1929. A discussion of the work of Bridgeman and others on the failure of rocks under pressure.

23 Ferguson, H. G. and Gannett, R. W.: Gold quartz veins of the Alleghany district, California. U. S. Geol. Surv. Prof. Pap. 172: 85, 1932.

mechanism, conditional upon the non-recrystallized character of the vein textures. He puts his finger on the basis for most of the current objections to the intrusive hypothesis (p. 83) as follows:

> Another difficulty . . . seems to lie not in its failure to fit the observed conditions but in the question whether the existence of a magma composed essentially of silica . . . is possible.

These objections are based on laboratory experimentation under conditions that do not even approximate the magmatic environment; patently, no finality attaches to them.

The geologic parallel between the aplitic-pegmatitic dikes and the gold-quartz veins is strikingly close in the Grass Valley district. Some of the features exhibited by both are:

(1) Mineral banding, sheeting, schistose, gneissoid, book and lit-par-lit structures.

(2) Mineral constituents in common: quartz, orthoclase, albite, epidote, muscovite, pyrophyllite, chlorite, scheelite, pyrite, gold, molybdenite.

(3) Deposits of the same size and shape, surrounded by envelopes of altered wall rocks. Along the Sierra Nevada belt the tonnage of quartz veins probably equals or exceeds the tonnage of the aplitic-pegmatitic dikes.

(4) Continuity of deposits from top to bottom, along persistent, raking panels, with constituent minerals rather well distributed throughout.

(5) Drag-folded wall rock along the partings occupied; localized zones of folding of walls during the deposition, as around blunt spur stringers and as in the "straps" between *en echelon* stringers.

(6) The inflation of sets of shear-type joints to accommodate lit-par-lit deposits; and the inflation of all partings occupied without restriction to those in the "tension" position in the ellipsoid of strain.

(7) A parallel structural response to variations in the character of the country rock traversed—simple, tabular forms in granodiorite, opposed to the book structure, ramified and lit-par-lit forms in the jointed, metamorphic rocks.

(8) Contact or inclusion breccias of angular fragments of country rock, in part xenolithic, are developed during the emplacement of the deposits in country rocks that elsewhere are not brecciated by faulting.

(9) The thin wafers of country rock included as book structure in veins and dikes commonly are sliced at an angle to the foliation of the rock, implying the operation of a mechanism other than reopening.

The foregoing, many-sided geologic parallel between dikes and the gold-quartz veins constitutes a strong argument for parallel origins and parallel mechanisms of emplacement for both. This deduction by analogy may be weak, as reasoning goes, but the "hydrothermal hypothesis" is equally vulnerable, being based on an analogy between our water chemistry of the laboratory and the complex physical chemistry of magmatic processes plus the questionable assumption that sizeable openings of specialized shape remain open for vein deposition, from the magmatic hearth to the earth's surface, for a long span of time.

For each of the structures that are exhibited both by dikes and veins, the inflation mechanism is a convincing explanation, proved workable by the magmas that come within our range of observation. Aside from possible differences in the phase and composition of the injected fluids, the structures and textures of the resulting deposits are parallel and continuous. Geologically, dikes and the quartz veins form a series of deposits that should be graded by composition only. No break is apparent in the series, such as would be expected where a dilute solution separates from the concentrated parent magma and where the process ceases to be intrusive and becomes permissive.

The complex, microscopical detail of vein textures may not be matched throughout, in counterpart, by dike textures although the parallel seems to be nearly complete. If the microscopic space relations between the mineral particles in a few cubic millimeters of ore is expanded into a geologic history that seems to be capped by the late, almost-accidental addition of the gold and metallic sulphides, then some error surely has crept into the interpretation.

Such an interpretation ignores the widespread cohabitation of these seemingly-late minerals with the gangue, in tremendously ramifying deposits, and the consequent implication of a joint deposition. Furthermore, the paragenetic " sequence " of minerals is much the same in most of the gold-quartz deposits of the world, many of which are of pre-Cambrian age and now are of metamorphic character. Since the " sequence " survives metamorphism, it probably is not the order of importation of these materials to the site of deposition but instead expresses some property of the minerals that governs paragenesis during the primary crystallization under hydrostatic stress and also during recrystallization under differential stress.

The paragenetic " sequence " is also a list of the minerals present in order of decreasing hardness [24] and it may be expanded to include many of the non-metallic gangue minerals, some of which are " later " than the gold—and softer. Polished sections of rich ore show gold and petzite not in veinlets but rather in spidery and holly-leaf masses, nested between quartz and ankerite grains. These masses are discontinuous and are neither the fillings of dilatant fissures nor the expectable forms of replacements or vug fillings. Instead, they suggest the forms that would develop in soft minerals squeezed between hard minerals during a leisurely metamorphism. The observed paragenetic relation thus could result from the schistose or gneissoid deformation of minerals that had first crystallized nearly simultaneously. Certain fractions of the deposits show little evidence of a general deformation, however, so a paragenetic control geared to mineral hardness needs to operate likewise at the time of deposition, under nearly hydrostatic pressure.

The suggestion of a primary control of paragenesis by some property geared to mineral hardness is offered principally to illustrate the possibility of finding a simple explanation for the world-wide " sequence " of ore and gangue minerals in veins. Some such new interpretation is needed to explain the seemingly-late

[24] Bandy, M. C.: A theory of mineral sequence in hypogene ore deposits. Econ. Geol., 35: 375. 1940. " The physical properties of the sulphides were examined but, aside from hardness, no other property was generally consistent."

paragenesis of gold in the Sierra Nevada, at the Witwatersrand, and other districts, where the late addition of an independent gold seems inconsistent with the general geology of the deposits.[25]

The writer can offer no firm conclusion concerning the composition of the fluid from which deep-seated quartz veins are deposited. A rank guess could be based on the ratio of "filling" of quartz to the volume of material added to the altered wall rocks. On this basis, the fluid may contain 10 to 40 per cent SiO_2 before the supplementary, volatile fraction presses on into the pores of the wall rock, perhaps as a CO_2-rich gas phase. The SiO_2 did not join in this penetration and therefore may have been restricted to the liquid and solid phases.

SUMMARY OF CONCLUSIONS.

1. Geologic deformation of the country rocks in the Grass Valley region has not developed extensive open partings of the size and character necessary to accommodate the abundant dikes and quartz veins, nor has faulting produced brecciated rock fragments of the sort that now are angular inclusions in the veins and dikes.

2. Portions of the röck encasing the veins and dikes have been folded concurrently with the inflation of the partings occupied—a type of deformation discordant with open fissures.

3. Deposition of the veins by the mechanism of a quiet, "hydrothermal accretion" in reopened faults is considered improbable because in many of the veins the necessary structural preparation is lacking.

4. Vein deposition through inflation of partings in the country rock by an injected, quaisi-magmatic fluid is the mechanism which best explains the occurrence. The many-sided, geologic parallel between the closely associated veins and dikes of the region embraces most of their structures, textures, mineral species and habits of environment; this parallel constitutes a strong argument for

25 Ödman, O. H.: Late gold and some of its implications. Econ. Geol., 33: 772–5, 1938. duToit, A. L.: Developments on and around the Witwatersrand. Econ. Geol., 35: 106, 1940.

parallel origins and mechanisms of emplacement. For each of the parallel features, the dikes offer a concrete proof of the workability of the inflation mechanism that is suggested for the emplacement of the veins.

5. The paragenetic "sequence" of ore and gangue minerals does not necessarily imply an independent, late importation of the gold and other seemingly-late materials into the vein. Instead, the "sequence" may express some fundamental property of the minerals, geared to hardness, which survives metamorphism. The hypothesis of a late, independent introduction of the gold, metallic sulphides and the softer gangue minerals is inconsistent with their wide distribution through tremendously ramifying vein systems.

Idaho Maryland Mines Corp.,
Grass Valley, Cal.,
Sept. 15, 1940.

9-3: NATURE OF THE ORE-FORMING FLUID

W. J. Bichan

[*Editor's Note:* In the original, material precedes this excerpt.]

1. What differences exist between the quartz of igneous rocks and of pegmatites, and that of the hypothermal ore veins, which would prove that the former crystallized from a magma and that the latter was deposited from a solute consisting predominantly of water?

2. A similar query is in order regarding those other minerals common to the magmatic stage in pegmatites and to the hypothermal veins, *e.g.* tourmaline and albite; also those of magmatic segregations such as the common sulphides.

3. In view of the important work of R. E. Gibson [3] on the elevation of the inversion temperature of quartz with increased pressure, and the discovery by V. B. Meen [4] of high-quartz in certain of the Canadian hypothermal veins, are we still to accept Lindgren's estimate of the temperature ranges of deposition of the various classes of deposits commonly called hydrothermal, without making necessary adjustments dependent upon the depth at which mineralization took place?

4. How can high-quartz be deposited at a temperature above 832° C. (at a depth of 30 km.) from a solution that ceases to be liquid several hundred degrees below that temperature?

5. Relying upon the solubility figure for silica in water at 335° C. determined by Hitchen [5] to be 0.21 parts of silica in 100 parts of water at a pressure of approximately 136 atmospheres, it may be asked whether the quantity of water required to hold in solu-

[3] Gibson, R. E.: The influence of pressure on the high-low inversion of quartz. Geophys. Lab. Pap. No. 663. (Reprinted from Jour. Phys. Chem., 32: 1197–1205 and 1206–1210, 1928.)

[4] Meen, V. B.: The temperature of formation of quartz and some associated minerals. Univ. of Toronto Studies, Geol. Ser. No. 38, pp. 61–68, 1935.

[5] Hitchen, C. Stansfield: A method for the experimental investigation of hydrothermal solutions, with notes on its application to the solubility of silica. Trans. Inst. Min. Met., 44: 277, 1935.

tion the amount of quartz introduced in the vicinity of the major hypothermal ore-bearing regions of the world is not equally unreasonable as that cited by Graton for the amount of gas involved in the chlorine transfer of iron at Cerro de Pasco?

6. According to a calculation by H. C. Boydell,[6] mineralizing solutions in a given case could lose less than 2° C. of their temperature in 5,000 feet of ascent along a vein zone. From Hitchen's experiments it can be found that the difference in silica solubility over a range of 2° C. at 335° C. is less than 0.003 parts of silica in 100 parts of water. Thus, each ton of quartz deposited *per foot of depth* would require the circulation through the vein zone of almost *170 million tons* of hydrothermal solution. In the case of a typical mine milling 1,000 tons daily, the amount of solution thus presumed to have been engaged in the deposition process of its normal ore reserve can be multiplied appropriately to give a figure of *170 billion tons,* or an even more untenable figure than that of Graton in the Cerro de Pasco case.

7. How can a magma yield "aqueous solutions rich in silica"[7] in view of Hitchen's determination of the low solubility of that substance in water?

8. Hitchen's determinations show that the solubility of silica in alkaline solutions decreases with rising temperatures; would not such solutions tend to rob the wall-rocks of quartz as they attained cooler horizons in the earth's crust?

9. What happens to Graton's "special ore fluid" if it finds no avenue of release from the magma chamber when once it has separated, by whatever process, from the main body of the magma?

10. What are the solid products of solutions derived from the magma when the latter is tapped slightly in advance of or during the separation of the supposed "special ore-forming fluid"?

11. What product appears when the supposed immiscible liquids within the magma are tapped simultaneously, and what is

[6] Boydell, H. C.: Temperature of formation of an epi-thermal ore deposit. Trans. Inst. Min. Met., 41: 499, 1932.

[7] Lindgren, Waldemar: *op. cit.*, p. 125.

to prevent both such liquids becoming channelized in the same vein system?

12. What happens to the vast quantities of water used in mineral transport under a hydrothermal system when the solutions are unable to reach the surface after depositing the bulk of their mineral content?

13. Would not hydrothermal solutions containing the amount of carbon dioxide essential to the formation of the carbonates be competent to decompose pyrite as soon as it had been formed, even in greater degree than this mineral undergoes alteration by surface waters?

14. Would not aqueous carbonated mineralizing solutions effect kaolinization of the wall-rock feldspars in the same manner that similar solutions of surface origin give rise to china-clay, yielding a totally different wall-rock alteration from that usually described as hydrothermal.

The entire absence of facts suggesting any condition of immiscibility within the magma from which the "hydrothermal" deposits are derived, and the lack of an abrupt change in the differentiation processes of the magma sufficient to account for the sudden production of hydrothermal solutions in adequate quantity at the end of the pegmatitic stage, render it vitally necessary to consider the products obtainable from the magma, should the latter be tapped at any given moment throughout the time-range of its existence.

It is suggested that the magmatic stage, in the formation of the quantitatively minor products of the magma, persists farther than the stage of production of "normal" pegmatite; in fact, it may persist to the leptothermal stage of ore-deposition, or close to the limits at which depositional banding of the quartz, and residual open-space within the veins (both taken to be indicative of crystallization from aqueous solution), become apparent.

The work of Gibson, previously mentioned, in conjunction with the known facts regarding the upper limit of quartz crystallization, bring the limiting temperatures for the formation of high-quartz, and permit some low-quartz to be formed, well within the magmatic range of temperature and at reasonable depth, and tend to show that the concept of ore-magmas is as tenable a theory as any yet evolved in explanation of the processes of emplacement of such magmatic derivatives.

To avoid confusion, it is proposed to limit the term magma to the fluid within the magma chamber and its major embayments or apophyses, and to substitute the term *silicothermal solution* for such portions as are drawn off along dike-fissures and vein-zones to form the pegmatites and certain ore-deposits.

When, during the course of progress towards the surface, the bulk of the silica composing a silicothermal solution has been deposited, and conditions have been attained at which the water present assumes a liquid state, then only do the hydrothermal processes become active in the production of some of the leptothermal deposits, as well as those classified as epithermal and telethermal.

Such a theory as that proposed would:

1. Reconcile the common characteristics of the quartz of the igneous rocks, pegmatites, and ore-deposits of hypothermal and mesothermal character, by contrast with the banded variety of the lower temperature stages.

2. Account for the contrast between the compact vein-filling of the higher-temperature "hydrothermal" deposits with the drusy development of the lower-temperature types.

3. Yield a single deposition process for the sulphides found in both the magmatic segregations and many veins of the higher temperature category, and for the tourmaline common to igneous rocks, pegmatites, and many of the hypothermal veins.

4. Support a broader temperature range of ore-deposition consistent with the depth-range apparent for many ore-deposits whose general characters persist with little change for many thousands of feet vertically.

5. Clarify the position with regard to the occurrence of high-quartz in some hypothermal veins and in numerous pegmatites.

6. Obviate the practical difficulties involved in the transport of large quantities of silica in an aqueous solution.

7. Explain the usual graduated suite of magmatic derivatives dependent upon the tapping of the magma chamber at successively later stages throughout the differentiation process.

8. Conform with the known field-facts of geology in all respects connected with ore-deposits in general and their associated phenomena such as wall-rock alteration.

Bichan elaborated his theory in two papers (1944, 1947) following W. H. White's (1943) classic paper detailing the late generation of gold in most auriferous veins (see Paper 17-1). Two stages in the deposition of gold-quartz veins were postulated by Bichan (1944, p. 238).

> This writer considers quartz deposition to have taken place generally in two stages: first, a dynamic stage during which quartz and other minerals were being slowly deposited from silicothermal solutions moving through the channelways, and under the influence of falling temperature. Quartz and the minerals of early formation (the "magmatic stage" in currently popular theories of pegmatite formation) grew outward from the walls of the vein, confining the fluid residuum to smaller spaces within the vein. The relative proportions of residuum still in the fluid state depends upon the length of time during which the mineralizing fluids were traversing the channelways and assisting in the accretion of fresh crystalline material. Gold and its associates had often begun to be separated in the fluid state, mainly on account of local pressure changes. The second stage followed the eventual blocking of the channelways in the upper reaches; the silicothermal solutions came to rest, and a period of comparatively rapid crystallization ensued under static conditions, wherein the fluid was at rest. The quartz of this stage exhibits a felsitic, or pseudo-cataclastic texture, and the still molten gold association remained as the last group of constituents to crystallize, within the pseudo-cataclastic sections noted by White. Thus the "lateness" of gold is apparent, whatever may have been the stage at which it separated from the transporting fluid.

The ore-magma theory of the origin of gold-quartz veins has had a history of nearly two centuries and has undergone modification from Hutton's idea of a dry magma to a silicothermal solution fluxed with water and other volatiles. The theory finds few advocates today, as the features of most gold-quartz veins are now explained by hydrothermal solutions, diffusion, replacement, remobilization of gold, and other mechanisms of deposition and reworking of the quartz and gold. However, the early advocates of the magmatic theory noted in most gold districts a general petrological sequence that is true: granite—aplite—pegmatite—quartz-porphyry—gold-quartz veins. The first four types of rocks are generally considered to be of magmatic origin; could not, therefore, some gold-quartz veins be similarly derived?

REFERENCES

Belt, T., 1861. *Mineral veins: An Enquiry into their Origin, Founded on a Study of the Auriferous Quartz Veins of Australia,* John Weale, London, 52p.

Belt, T., 1874. *The Naturalist in Nicaragua,* J. M. Dent & Son, London, 403p.

Bichan, W. J., 1941. Nature of the ore-forming fluid, *Econ. Geology* **36:**212-217.

Bichan, W. J., 1944. Gold deposition, *Econ. Geology* **39:**234-241.

Bichan, W. J., 1947. The ubiquity of gold values and their relation to zones of tensional dilation, *Econ. Geology* **42:**396-403.

Farmin, R., 1938. Dislocated inclusions in gold-quartz veins at Grass Valley, California, *Econ. Geology* **33:**579-599.

Farmin, R., 1941. Host-rock inflation by veins and dikes at Grass Valley, California, *Econ. Geology* **36:**143-174.

Fournet, J. F., 1856. Aperçus rélatifs à la théorie des gîtes métallifères et Aperçus rélatifs à la théorie des filons, *Acad. Sci. Paris Comptes Rendus* **42:**1097, **43:**345-352, 842-849, 894-900.

Graton, L. C., 1940. Nature of the ore-forming fluid, *Econ. Geology* **35**(suppl. to no. 2):197-358.

Hutton, J., 1795. *Theory of the Earth with Proofs and Illustrations,* 2 vols., Edinburgh. Reprint Hafner Pub. Co., New York, 1959.

Kirwan, R., 1793. Examination of the supposed igneous origin of stony substances, *Royal Irish Acad. Trans.* **5:**51-81.

Kirwan, R., 1799. *Geological Essays,* London.

Petzholdt, A., 1845. *Geologie,* C. B. Lorck, Leipzig, 645p.

Playfair, J., 1802. *Illustrations of the Huttonian Theory of the Earth,* William Creech, Edinburgh. Reprint with introduction by G. W. White, Univ. Illinois Press, Urbana, Ill., 1956.

Spurr, J. E., 1923. *The Ore Magmas,* 2 vols., McGraw-Hill, New York, 915p.

White, W. H., 1943. The mechanism and environment of gold deposition in veins, *Econ. Geology* **38:**512-532.

CHAPTER

10

The Origin of Epigenetic Gold Deposits—the Magmatic Hydrothermal Theory

Il est probable que les sources thermales les plus chaudes, les sources thermales principales, émanent directement des roches éruptives.
J. B. Élie de Beaumont, 1847

Gold deposits are rarely found in regions where there are no exposed igneous bodies.
W. H. Emmons, 1937

The hydrothermal theory of the origin of epigenetic mineral deposits has had a long gestation and has many variants. Hydrothermal origin without qualifiers means simply transport and deposition of the mineral matter of veins and similar deposits by and from hot waters. With qualifiers such as meteoric, metamorphic, or magmatic the theory is more specific, indicating respectively that the hot waters are derived as a result of near-surface, metamorphic, or magmatic processes. The last is the subject of most of this chapter; the first two processes are discussed in greater detail in subsequent chapters.

The gist of the magmatic hydrothermal theory is that as a magma crystallizes certain fractions separate and crystallize in accordance with the thermodynamics of the prevailing system. Some of the constituents may separate early (e.g., Cr, Pt, Fe) and give rise to magmatic segregations. Still later other fractions may separate and form pegmatites with their suite of rare elements and volatiles (e.g., Li, Ta, U, F). The last fractions to separate contain a high proportion of volatiles (e.g., H_2O, CO_2, S), most gangue elements (e.g., Si, Fe, Al, Ca), and the metals (e.g., Cu, Ag, Au, Zn); these move upward and outward as acid, neutral, or alkaline gases or liquids (hydrothermal solutions) from the crystallizing magmatic (principally granitic or granodioritic) masses into the surrounding host rocks where they may replace certain rocks (contact metamorphic and skarn deposits) or concentrate and deposit in fissures, faults, and so on as veins, stockworks, and similar bodies.

To trace this theory as regards gold deposits is a difficult problem because of the many variants suggested during the course of the history of the theory. One must, therefore, generalize; there is no possibility of including all of the many variants in the space available here.

When geologists first recognized that gold veins might be a phase of igneous hydrothermal activity is uncertain. Certainly the ancient naturalists as far back as the early Greek civilization observed the exhalations, hot waters, and precipitates associated with hot springs and connected them with volcanic (igneous) processes. I can find no mention, however, of the association of gold with hot springs in the ancient and more modern writings, until the Plutonist-Neptunist controversy during the first half of the nineteenth century. The Huttonians held no particular views about the origin of gold veins, although James Hutton in his writings connected the magmatic origin of quartz veins with igneous processes as noted in chapter 9. The Wernerians, on the other hand, relegated the formation of mineral veins solely to the agency of water. In his treatise *Neue Theorie von der Entstehung der Gänge* Werner (1791) stated, "All true veins were originally, and of necessity, rents open in their upper part, which have been afterwards filled up from above." He then went on to expound on his Neptunist idea of the formation of stratified rocks by precipitation from oceanic waters, and that when the rocks (mountains) shrank as a result of desiccation, or were affected in some cases by earthquakes, rents were formed. He then concluded that "the same precipitation, which in the humid way formed the strata and beds of rocks, (also the minerals contained in these rocks), furnished and produced the substance of veins; this took place during the time, when the solution from which the precipitate was formed, covered the already existing rents, and which were as yet wholly or in part empty, and open in their upper part." Werner was, therefore, one of the first of the descensionists.

When both the Neptunist and Plutonist theories were evaluated in the light of the extensive work of a number of volcanologists, particularly Dolomieu (1783), Breislak (1811), and Scrope (1825), it was generally realized that hydrothermal activity often accompanied volcanic (igneous) events.

One of the first geologists to recognize the joint relationship of igneous and aqueous processes in the formation of mineral veins was Ami Boué (1794-1881), German petrologist and geologist. In his *Essai géologique sur l'Ecosse* (1820) and *Mémoire géologique sur l'Allemagne* (1822) he discussed gases and vapors in metamorphism and the relationship of igneous dykes and metalliferous veins, concluding that both igneous and aqueous action played a role in vein formation. He was followed by A. L. Necker (1786-1861), Swiss mineralogist and petrologist, who worked in Switzerland and Scotland. In a communication to the Geological Society of London (Necker, 1832) he mentioned the work of Boué and also that of Alexander von Humboldt, who noted the intimate relationship of certain auriferous veins to granitic, syenitic, and porphyritic bodies in the Ural and Altai ranges of USSR. Necker augmented these observations with others in many countries of Europe, North America, and South America where he considered it obvious that metalliferous (auriferous) veins are closely associated with granitic rocks. Murchison, de Verneuil, and von Keyserling (1842, p. 748) reiterate this view in their great memoir on the Urals.

> The Ural-tau or crest is to a very great extent a wall of schist and quartz rock diversified by points of igneous rocks, and though of no great altitude, it is very remarkable that throughout 17 degrees of latitude this water-shed is not broken through by any great transverse valley. The Ural-tau marks, in fact, one long line or fissure of eruption. With the exception of the gold mines near Bissersk, on its west flank, all the gold alluvia of the chain occur on its eastern flank; and when it is stated that this circumstance is connected with the

> fact, that all the great masses of igneous rocks have been evolved on the eastern flank, it will at once be seen (as insisted upon so well by Humboldt) that there is an intimate connexion between the eruption of plutonic rocks and the formation of the gold mines whence the local alluvia have been derived. That this connexion exists in regard to other mineral veins, is also equally apparent in the Ural mountains; for with very rare exceptions, it is only on their eastern or eruptive side that copper veins, malachite, platinum and magnetic iron prevail.

Finally, among the older investigators we may note the significant observations of Charles Darwin (1809-1882) during the voyage of the *Beagle*, concerning the relationship of gold and other metalliferous veins, intrusive rocks, and metamorphism in Chile and elsewhere. In his *Geological Observations in South America*, Darwin (1846, p. 237) concludes:

> Finally, I may observe, that the presence of metallic veins seems obviously connected with the presence of intrusive rocks, and with the degree of metamorphic action which the different districts of Chile have undergone. Such metamorphosed areas are generally accompanied by numerous dikes and injected masses of andesite and various porphyries: I have in several places traced the metalliferous veins from the intrusive masses into the encasing strata. Knowing that the porphyritic conglomerate formation consists of alternate streams of submarine lavas and of the debris of anciently erupted rocks, and that the strata of the upper gypseous formation sometimes include submarine lavas, and are composed of tuffs, mudstones, and mineral substances, probably due to volcanic exhalations,—the richness of these strata is highly remarkable when compared with the erupted beds, often of submarine origin, but *not metamorphosed*, which compose the numerous islands in the Pacific, Indian, and Atlantic Oceans; for in these islands metals are entirely absent, and their nature even unknown to the aborigines.

At the midpoint of the nineteenth century two fundamental papers appeared that provided great impetus to the hydrothermal theory. The first was a memoir by T. Scheerer (1847), German geologist. Water was given an important role in the formation and crystallization of granitic magmas in this memoir; aqueous solutions, exuded from the cooling and crystallizing granitic magma, were deemed to play a vital role in the formation of certain mineral veins. The second paper, referred to as a note in the *Bulletin de la Société Géologique de France*, is in actuality a long paper of some 85 pages by the eminent French structural geologist and petrologist, Élie de Beaumont (1798- 1874), entitled *Note sur les émanations volcaniques et métallifères.* This paper, long regarded as one of the milestones in economic geology, details the geochemical processes associated with volcanism and igneous (granitic) intrusion. Élie de Beaumont (1847) noted the intimate relationship of thermal (hot spring) waters and volcanism and remarked, "Il est probable que les sources thermales les plus chaudes, les sources thermales principales, émanent directement des roches éruptives." [*Trans.* It is probable that the principal and the hottest thermal springs, emanate directly from the (intrusive) rocks.] Of equal importance in the context of the present chapter is the explicit statements by Élie de Beaumont regarding the effect of mineralizers (e.g., H_2O, CO_2, Cl, F, S, etc.) in the crystallization history of granitic magmas and their role in the subsequent history of the transport of the metals, including gold into vein systems. The idea of mineralizers as effective mobilizers (complexation agents) in hydrothermal systems was taken up by the French geologist G. A. Daubrée, one of the fathers of experimental geochemistry. In his *Etudes synthétiques de géologie expérimentale* (1879) and later in his great treatises on underground waters, *Les eaux souterraines à l'époque actuelle* (1887*a*) and *Les eaux souterraines aux époques anciennes* (1887*b*), he detailed the function of mineralizers in mineral vein formation and placed great emphasis on the role of water in regional metamorphism and vein formation. It is of interest to note that Daubrée considered that the waters (hydrothermal solutions) involved in vein formation were essentially of atmospheric (superficial) origin. These waters, he opined, had penetrated deeply into the earth by capillary action, had become heated, produced the metamorphism of the mineral belts, and on rising through fissures deposited their mineral lode. As we shall see subsequently, this idea based on the evidence of hydrogen and oxygen isotopes, has recently been evoked for the origin of many of the (propylitic) Tertiary and younger gold deposits.

By 1860 three general theories of the origin of gold veins were extant: secretion mechanisms including lateral secretion *sensu strictu* and metamorphic secretion *sensu lato*; igneous (melt) injection; and hydrothermal with its many variants including descending or ascending mineral-laden waters, the latter usually having a source in crystallizing igneous (granitic) magmas, and ascending mineral-laden steam and other gases of similar derivation. These three theories and their many variants were discussed at length by B. Von Cotta, professor of geology in the Royal School of Mines at Freiberg, Saxony, in his *Treatise on Ore Deposits* (1870), the first edition of which appeared in 1859. This famous treatise was the first of its kind on ore deposits, and it remained a popular textbook, influencing the thinking of economic geologists in Europe and America for more than two decades.

Von Cotta presented concise information on all aspects of the gold deposits then known in Europe, discussing the auriferous lodes and placers of various mineral belts in Germany (Fichtelgebirge, Eisenberg, Rhine Valley), the northern Carpathians (Borsa), Transylvania (Offenbanya, Nagyag), what was then part of Hungary (Schemnitz, Kremnitz), the Alps (Salzburg Tauren; La Gardette, France), Urals (Beresov, USSR), and elsewhere. He noted the vertical zonations often present in the veins of some auriferous belts, namely, a gold zone near the surface grading into a deeper lead-silver zone, which in turn passed into a very deep pyritic copper zone, and he attributed this distinctive zonation to different deposition temperatures and pressures for each metal. As regards the origin of gold veins he leaned toward the magmatic hydrothermal theory, although he was not dogmatic in this respect.

Secretion theories, as discussed in chapter 13, held the stage for a brief span during the latter part of the nineteenth century and early part of the twentieth, being supported by a number of geologists among whom may be mentioned Sandberger, T. Sterry Hunt, S. F. Emmons, and Van Hise. These theories were in due course strongly attacked by A. W. Stelzner, F. Posepny, and L. De Launay.

A. W. Stelzner (1879, 1881), German geologist, opposed the lateral secretion theory of Sandberger on the grounds that the metals in the rocks near veins were probably added and not subtracted (or leached) as the secretion theory demanded. The same argument was advanced by F. Posepny (1894), Czechoslovak economic geologist, in his historic paper *The Genesis of Ore-Deposits.* In this paper he discussed the Tertiary gold deposits of Hungary (Schemnitz), Transylvania (Dacia, Verespatak, Vulkoj), and the Comstock Lode, Nevada, concluding that the gold and other elements in these classic veins and lodes were brought into place by ascending juvenile thermal waters issuing from the deep regions (barysphere) of the earth. The intimate relationship of the mineral-laden solutions to igneous intrusions was not stressed. In the long discussion that followed Posepny's paper (*Am. Inst. Min. Eng. Trans.* **24:**942-1006) there was some agreement with his theory and some skepticism. T. A. Rickard noted the lack of intrusives in some gold belts (e.g., Otago, New Zealand), and he and others delivered a number of polemics on the mechanics of water circulation in the earth. Still others wondered about the presence of open fissures deep in the earth, and considered that some great gold-silver deposits (e.g., Comstock Lode) were essentially the result of metasomatic processes. Finally, Le Conte doubted the existence of a metalliferous barysphere and placed the source of the metals in the thermosphere (the heated country rocks undergoing metamorphism).

The exposition of the magmatic hydrothermal theory as expressed by L. de Launay in his great treatise *Traité de métallogénie: Gîtes minéraux et métallifères* (1913) was representative of the opinions held by a number of economic geologists at the turn of the nineteenth century and early part of the twentieth. De Launay considered vein deposits to have been deposited by ascending thermal waters and attached considerable importance to the intimate association of vein deposits and igneous rocks that had brought the metals from deep (baryspheric) sources. All of the metals, including gold, receive extensive attention in De Launay's treatise. The known auriferous deposits in the many mineral belts of the world are described in considerable detail, including their mineral associations, morphology, geological setting, and so on. The text is remarkably modern and deserves reading by those presently interested in gold.

During the present century the magmatic hydrothermal theory has gained prominence. The original theory of Élie de Beaumont and earlier investigators has undergone numerous modifications and refinements as a result of the work of a host of economic geologists and geochemists. The number of papers available on the subject would fill many volumes and certainly cannot be referred to here. Some recent textbooks and papers that summarize most aspects of the theory can, however, be mentioned, including those by Beyschlag, Vogt, and Krusch (1914-1919), Lindgren et al. (1927), Lindgren (1933), Niggli (1929), Emmons (1937, 1940), Bateman (1954), Park and MacDiarmid (1970), Smirnov (1976), and Jensen and Bateman (1979).

The magmatic hydrothermal theory as it now stands postulates that as a magma crystallizes certain fractions separate and crystallize out in accordance with the prevailing thermodynamics of the system. Some of the constituents may separate early and give rise to magmatic segregations. Still later other fractions may separate and form pegmatites. The last fractions will contain a large proportion of the volatiles, gangue elements, and metals, and these will move upward and outward as acid, neutral, or alkaline gases or liquids from the crystallizating (granitic) magma into fissures, faults, and other structures where they are deposited as late stage (replacement) pegmatites, contact metamorphic deposits, veins, disseminated (porphyry copper-type) deposits, and replacement deposits of all types. A veritable multitude of ideas on the composition and nature of the ore-forming fluid exuded from magmas are extant. Some investigators assume acid properties, others neutral properties, and still others alkaline properties. Many postulate continuous change as the liquids or gases react with the wall rocks of the mineral deposits. Space precludes any discussion of these views here; interested readers should refer especially to the papers and books by Graton (1940), Ingerson (1954), Korzhinskiy (1964), Roedder (1965), Barnes (1967, 1979), and Smirnov (1976), the recent contributions to the subject by numerous authors in the symposia published by the International Association on the Genesis of Ore Deposition, and in the reviews edited by Henley and co-workers (1984) and Berger and Bethke (1985). Succinct outlines of the geochemistry of magmatic hydrothermal processes respecting gold are given by Smith (1943) and Petrovskaya (1973) (see papers 8-6 and 8-8). The role of colloids is discussed by Lindgren (1924) and Boydell (1924). Ridge (1968) has given a masterful summary of the changes and developments in concepts of ore genesis from 1933 to 1967. Included are various experimental data and derived postulates concerning the vicissitudes of the magmatic hydrothermal (and other) genetic theories of mineral (gold) deposits from Bowen (1933) to White (1967).

THE MAGMATIC HYDROTHERMAL THEORY OF THE ORIGIN OF AURIFEROUS DEPOSITS

The number of papers advocating a magmatic hydrothermal origin for epignetic gold deposits is so overwhelming that a choice of landmark papers is extremely difficult and really quite arbitrary. Most of the papers selected were published in the hey-day of the magmatic hydrothermal theory. The reader will note, however, that ideas about the source(s) of hydrothermal solutions have undergone considerable changes; initially, crystallizing magmas were considered to be the source; later, sources such as magmatic-meteoric, metamorphic, vadose and meteoric (including sea water), or combinations of these have been ascribed. More recent authors simply term the solutions hydrothermal (usually ascending) without qualifiers. As will be seen, the source of the gangue minerals (especially quartz) in epigenetic auriferous deposits has also presented problems, as has the source of the gold and its accompanying metallic minerals.

The classic treatise that advocates a magmatic hydrothermal origin for epigenetic gold deposits is that by W. H. Emmons (1937). In the introduction to this work the part "Gold and Igneous Rocks" (Paper 10-1) is a landmark and deserves careful study by all geologists interested in gold deposits.

(Text continues on page 260.)

10-1: GOLD AND IGNEOUS ROCKS

William Harvey Emmons

Reprinted from *Gold Deposits of the World,* McGraw-Hill, New York, 1937, pp. 12-37.

Gold Deposits Independent of Igneous Rocks.—Gold deposits are rarely found in regions where there are no exposed igneous bodies. Small amounts of gold are found in northern Illinois[3] far from any outcrop of igneous rocks, but the deposit is not workable, and such an occurrence is uncommon. The auriferous deposits of Otago, New Zealand, have been referred to as examples of gold lodes with no igneous affiliations.

Association of Gold Deposits with Igneous Rocks.—The association of auriferous lodes with igneous rocks is practically universal. In areas of late metallization the lodes are associated with lavas or with intrusions formed near the surface. In areas that have older intrusions which have been subject to greater erosion, the deeper seated and generally more coarsely grained intrusives are likely to crop out. In such areas the gold lodes are almost invariably associated with granitic rocks or with the porphyritic phases of such rocks. Not only are gold lodes associated chiefly with acidic intrusions, but they are found chiefly in certain positions. The great shield areas of the earth particularly are instructive. These areas are deeply eroded and in general show relations of lodes and intrusives more clearly than the areas of young lodes where in many regions the intrusives are buried.

In the shield areas the granitic rocks have brought in the ore, yet large areas of the granite are barren. The ores are found within the "islands" or other areas of invaded rocks and in granite within a mile or less from the contacts of invading and invaded rocks. In these areas which supply a large part of the world's lode gold, the relations are as follows: Gold pro-

[3] Hershey, O. H., *Am. Geologist,* vol. 24, pp. 240–244, 1899.

duction from invaded rocks and from intruding rocks within a mile of contact, 99.9+ per cent; gold production from intruding rocks more than a mile from contact, 0.1− per cent (see figures on page 20). Where the invaded rocks are intruded by small stocks of granitic rocks or by their porphyries, gold lodes are likely to be found in or near the small intrusives. These are the most favorable places for gold deposits, although many deposits are found where no stock is exposed.

In the shield areas the gold-bearing regions may be classified in accordance with increasing productiveness as follows:

1. Granitic intrusion more than a mile from invaded rocks—important deposits essentially wanting.

2. Granitic intrusions within a mile of invaded rocks—a few valuable deposits.

3. Invaded rocks—many deposits.

4. Small stocks of intruding rocks and invaded rocks near them—deposits greatly concentrated.

These relations and also relations that are observed in areas less deeply eroded than the shield areas lead to the conclusion that in such areas most gold lodes are associated with granitic batholiths and their satellitic stocks. The metalliferous lodes of essentially all developed gold-bearing areas of the earth have been plotted on geologic maps, and from these tracings have been made showing (1) the intrusive igneous rocks, (2) rocks older than the intrusives, (3) rocks younger than the intrusives, (4) the metalliferous lodes. The lodes are subdivided into (*a*) those deposits that have supported important mining operations, (*b*) deposits that have been worked but without much profit, (*c*) deposits of little or no proved value. This work has shown that valuable deposits are found in the larger granitic intrusives as far as 3 miles from the contact at places high on the roofs of batholiths, but in deeply eroded areas such as the shields the valuable lodes, as stated, are essentially wanting in the intrusives at places more than 1 mile from the contact. This line, or rather this warped plane below which valuable deposits are absent, is called the "dead line." The part of the batholith above this line is its "hood," and the part below it is the "barren core." The roof of the batholith is the invaded rock which lies above it (Figs. 6, 7, 8).

A batholith is a great, irregular, deep-seated igneous mass that broadens downward. The term is used in this sense by Suess, Daly, Barrell, Butler, Grout,[1] and others. Unlike laccoliths batholiths have no visible floors but extend downward beyond the range of observation. Their roofs, moreover, are much less regular than the roofs of laccoliths. They are highly undulatory. In composition the batholiths are acidic,

[1] GROUT, F. F., Petrography and Petrology, McGraw-Hill Book Company, Inc., 1932. Full references to literature, p. 486.

ranging from quartz-bearing diorite through quartz monzonite and granodiorite to granite.

The molten rock or magma which, cooling, forms the batholith rises by thrusting the invaded rocks aside, by melting or assimilating them, and, according to Daly, by stoping or prying off pieces of the invaded

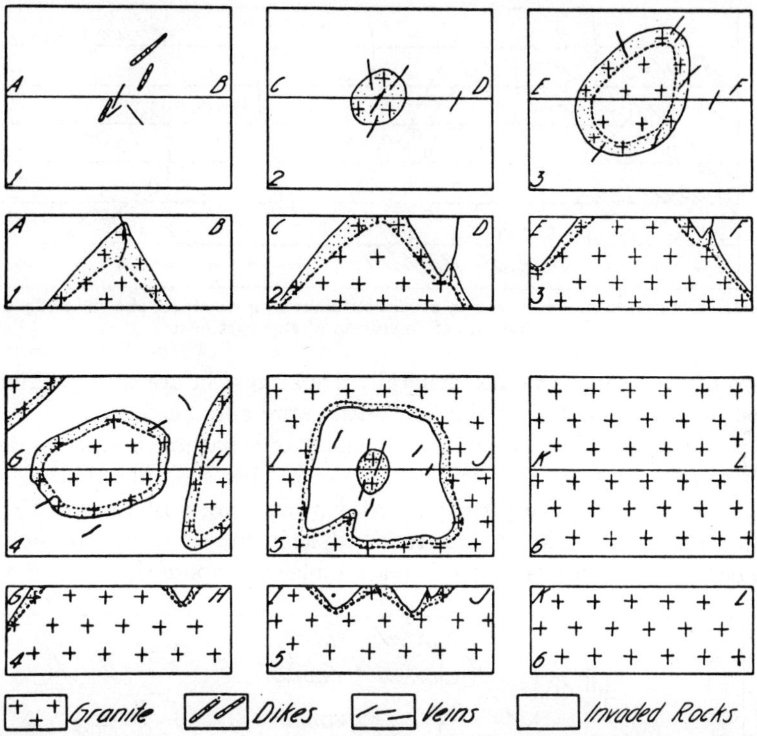

FIG. 6.—Six stages of erosion of a batholith. 1. Cryptobatholithic stage: Batholith concealed; only roof rocks with dikes, sills, and veins crop out. 2. Acrobatholithic stage: Summits of cupolas and stocks exposed, but the "dead line" lies below the present surface. 3. Epibatholithic stage: central disk of the barren core exposed. 4. Embatholithic stage: Roof rocks constitute the frame of the picture, as in preceding stages, but invading rocks nearly equal invaded rocks. 5. Endobatholithic stage: Intruding granitic mass forms the frame of the picture. Roof rocks are islands in invading rocks. 6. Hypobatholithic stage: erosion has gone so deep that roof rocks are practically eliminated. No large deposits of precious metals found.

rocks which sink to depths where probably some are melted. When cooling begins, the magma is probably more basic than the final product.[1] The dark heavy minerals form early and sink. Lighter minerals rise, and as a result the upper part of the batholith—the part that comes under observation—is acidic. This process, called magmatic differentiation, seems to account for the acidic or granitic composition of batholiths.

[1] EMMONS, W. H., Primary downward changes in ore deposits: Am. Inst. Min. Eng. *Trans.*, vol. 70. pp. 964–994, 1924.

There is much evidence that batholiths, in general, slope outward, or that they increase in size with depth. At many places the contacts are observed to dip away from the exposed mass, and few are seen to dip inward toward the central portions of the batholith. The mapping of areas containing granitic intrusives the world over shows that the out-

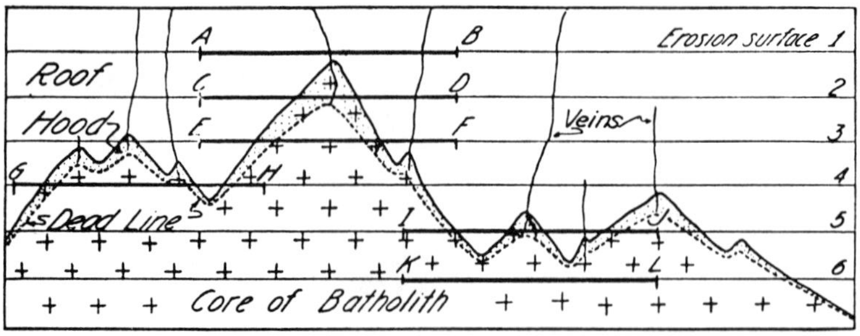

FIG. 7.—Diagrammatic cross section of batholith showing six stages shown in Fig. 6 in one batholith, before erosion of any part of it.

crops of the granitic bodies generally are largest in areas that have been most deeply eroded since the batholiths were emplaced.

No one may observe the emplacement of a batholith and the deposition of ores that follows its emplacement. We see only the batholiths and their associated ore deposits at various stages of erosion (Fig. 6). At places only dikes and veins are found, as in stage 1 (cryptobatholithic). No batholith is evident, but it is reasonable to suppose that the dikes rose

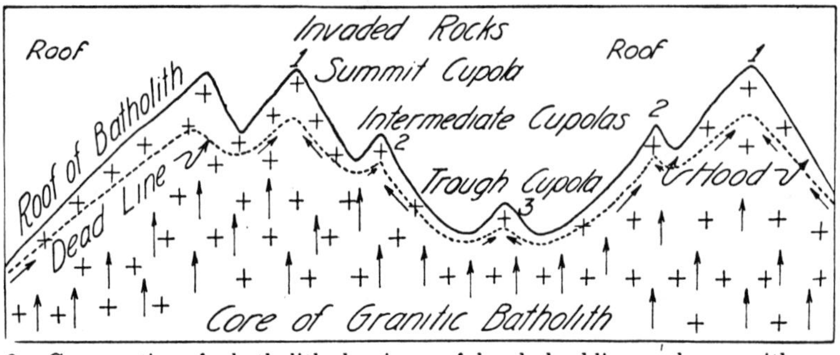

FIG. 8.—Cross section of a batholith showing roof, hood, dead line, and core, with summit, intermediate, and trough cupolas. Arrows show probable movement of volatile matter.

from an underlying mass. At stage 2 (acrobatholithic) the top of the batholith is eroded, and veins are in and around the truncated top. At stage 3 (epibatholithic) the batholith is eroded deeply, and part of the barren core is exposed. At stage 4 (embatholithic) the invaded rocks frame the picture, but invading rocks make up nearly half of the area. At stage 5 (endobatholithic) islands of invaded rocks are surrounded by intruding rock. The deposits essentially are confined to the islands and to the granitic rock within a mile of the contact. At stage 6 (hypobatholithic) islands are removed by erosion, and ore deposits are essentially wanting. A cross section of a batholith before erosion is represented by Fig. 7. The cupolas at the tops of batholiths are their summit cupolas; the lower ones are intermediate cupolas; and the lowest ones are trough cupolas (Fig. 8).

Petrology of Hood of Batholith.—The contact zone or hood of the batholith at places offers no sharp contrast, petrographically, with the core. At other places it differs considerably.

1. It may be more basic than the core.
2. It may be more finely crystalline than the core.
3. It may be more schistose than the core.

1. If a batholith becomes more acidic during cooling on account of differentiation, the part that solidifies first will be more basic than the part that crystallized last. The hood or marginal portion is formed at the earlier stage and may be more basic than the core, since it represents the part that solidified before differentiation had gone far. There are scores of examples of intrusives with marginal phases that are more basic than the central portions.

2. At places there are chilled, fine-grained marginal zones on the borders of batholiths. On emplacement the intrusive is likely to cool quickly around the margin. At many places, however, the border phase is a uniform coarse-grained granite which differs little from the core. It is not unlikely that the more basic fine-grained phases formed at early stages of cooling and that this phase later was dissolved or stoped out by a subsequent rise of the magma.

3. In general, the earlier phases of intrusion are marked by the rise of stocks and dikes into the roof of the batholith which lies above the hood and, subordinately, into the hood. Some of these stocks are cupolas, and others are long, irregular masses which are not known to broaden downward. In general, they are more basic than the core of the batholith, and also they are either porphyritic or of finer grain than the core of the batholith. Some of them show foliation and gneissic or schistose texture. That is true also of border phases of certain batholiths.

Intrusion, Folding, and Dynamic Metamorphism.—Batholithic intrusions, mountain folding, and dynamic metamorphism often are closely associated in time and place, although there are some batholiths that intrude rocks that are not highly folded, and certain highly folded areas are not known to contain batholiths. Neither is dynamic metamorphism confined to areas of intrusion, and certain areas of intrusion show relatively little dynamic metamorphism of rocks associated with the intrusives. In many areas, however, the roof rocks of batholiths at places are strongly dynamically metamorphosed.[1] The batholiths doubtless underlie great areas at depths of a few miles. The roof of the batholith and the hood, if one has formed, are in critical positions if the rocks below them are still liquid, for they are not supported below by solid foundations. A small amount of lateral pressure will accomplish movement more readily, and more dynamic metamorphism will result, for the prism of solid rock that is deformed extends to no great depth and rests on liquid. The aqueous solutions that rise from the batholith, moreover, may aid the processes of dynamic metamorphism by forming the platy minerals which may give the roof rocks cleavage.

Certain small stocks not more than 2 miles wide have well-defined schistose borders with the schistosity parallel to its contacts, whereas 100 feet within its borders the rock is without schistosity. An example is the granitic stock 4 miles southwest of Goudreau, Ontario. It is probable that the upward thrust of the partly liquid body developed the schistosity of the rim (page 63).

Gases in Magmas.—In the San Juan Mountains, Colorado, and at many other places there are deposits of volcanic tuffs thousands of feet thick. Much of this material is finely comminuted, and evidently it was blown to small fragments by volcanic gases. Such deposits and the violent eruptions of certain volcanoes are evidence of large amounts of gases in certain magmas.[2]

The belief that certain cooling magmas release large quantities of gas under high pressure is based also on observations. Recently at frequent intervals vast quantities of "autoexplosive lavas," which are lavas highly charged with gas under high pressure, have issued from Mount Pelee.[3] These lavas rise from the crater and expand into fiery clouds. The lava chills in part to form solids, but it rolls down hillsides like water at a rate as high as 100 feet per second. There is so much gas that it lubricates the entire stream. At intervals lavas, nearly gas free, issue; and generally when the gas-free lavas issue, there are no issues of autoexplosive lavas. It is not improbable that partial crystallization is one of the causes that result in the accumulation of the material highly charged with gas at a high place in an underlying magmatic chamber. After this escaped to the surface to form the flaming cloud, an issue of material only slightly charged with gas would follow.

Cooling and Vapor Pressure.—Morey made a series of experiments in which he observed the cooling of a molten system containing H_2O, 9.1 per

[1] Barrell, J., Relation of subjacent igneous invasion to regional metamorphism: *Am. Jour. Sci.*, 5th ser., vol. 1, pp. 178–186, 1921.

[2] Fenner, C. N., Ore Deposits of the Western States: Am. Inst. Min. Eng., Lindgren vol., pp. 61–73, 1933.

[3] Perret, F. A., The Eruption of Mount Pelee, 1929–1932, Carnegie Inst. of Washington, pp. 1–125, 1935.

cent; K_2O, 17.3 per cent; and SiO_2, 73.6 per cent. This system, confined in a bomb, was fluid at 500°C., and exerted no vapor pressure. Cooling to 420°C. a large part of it crystallized, and at that temperature much water had separated from the crystals and remained as steam, which had a pressure of 4,998 pounds per square inch, a force sufficient to lift nearly a mile of granite.[1] If granite (Fig. 9) cooling at 600° expels vapor in the

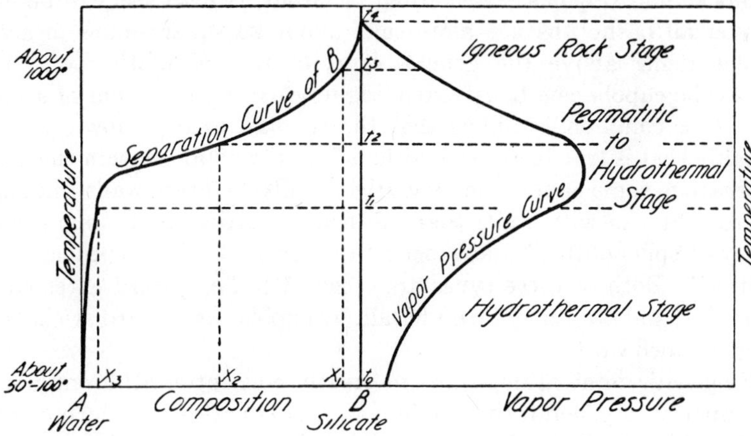

FIG. 9.—Diagram illustrating the effect of cooling on a system containing a volatile substance (water) and a refractory substance (a silicate), based chiefly on a diagram by *Niggli.* With cooling, the refractory substance crystallizes out of the system and, simultaneously, the liquid residue builds up a strong vapor pressure. The line AB = 100 per cent of the mixture. A is volatile and B is nonvolatile substance. The curve shows the percentage of A and B present in the molten mixture when B crystallizes out from the system. Vapor pressure curve shows pressure at temperature stated in a gastight container.

process of crystallization, it is reasonable to suppose that greater vapor pressure would result, and it is thought that the vapor expelled from crystallizing granitic batholiths probably has a pressure sufficient to lift two or three miles of granite. This vapor, collecting near the pinnacles of the cores of batholiths, is believed to exert sufficient pressure to fracture the hoods and roofs of batholiths near cupolas. If that is true, one would expect to find that fracture systems above cupolas are arranged in patterns which depend in part on the shapes of cupolas.

The mechanics of the situation require that conical cupolas should have radial fracture patterns and that elongated stocks or cupolas should have fracture systems in which the fractures lie nearly parallel to the directions

[1] MOREY, G. H., Development of pressures in magmas as a result of crystallization: *Jour.* Wash. Acad. Sci., vol. 12, pp. 219–310.

NIGGLI, P., Ore Deposits of Magmatic Origin, trans. by H. C. Boydell, pp. 1–93, London, 1929.

EMMONS, W. H., On the mechanism of the deposition of certain metalliferous lode systems associated with granitic batholiths, in Ore Deposits of the Western States (Lindgren Volume), pp. 327–349, Am. Inst. Min. Eng., 1933.

of the elongations of the cupolas. Study of essentially all the cupolas of the earth that have associated important vein-filled fractures shows that this is generally true. This very common relation seems to support the theory that such fracture systems may owe their origin to the vapor pressure that is generated by the volatile fluids expelled from cooling and crystallizing granitic batholiths.

Stocks and Cupolas.—According to Grout,[1] stocks seem to be essentially small batholiths not more than about 20 square miles in area or cupolas rising above the general level of the batholith roof. Daly[2] defines the cupola as a body extending upward from the roof of a batholith. It is commonly implied that the cupolas broaden downward, and probably that is true in general, although at the depths within the zone of observation some stocks have vertical walls or even walls that slope inward. Stocks with walls that are nearly vertical or that are wedge-like are "sphenoliths," and stocks that contract with depth are "ethmoliths." Both of these types are believed to be upward extensions of batholiths, and both may be underlain by cupolas or upward swells which broaden downward.

Epigenetic mineral deposits are greatly concentrated[3] around stocks and cupolas; in general, the sulphide ores of contact-metamorphic origin are found around small stocks. Valuable deposits of this group are rarely, if ever, found around stocks that are more than 5 miles in diameter.

Illustrations of various types of stocks and of other upward swellings of the hoods of batholiths are shown in Figs. 10 and 11. Five of these show elongated outcrops of intrusives; one is circular. In Fig. 10 at *A* is shown an ethmolith rising from a normal cupola. This figure shows a structure something like one that would be outlined by the frame of a greatly elongated sawbuck. In the examples that are available the parts that lie below the necks are not developed; but since the roofs of batholiths generally slope outward from the stocks, it is inferred that the funnel-like body in depth will generally broaden. *B* shows a sphenolith rising from a cupola, *C* illustrates the normal cupola with definite elonga-

[1] GROUT, F. F., Petrography and Petrology, McGraw-Hill Book Company, Inc., 1932.

[2] DALY, R. A., Igneous Rocks and Their Origin, McGraw-Hill Book Company, Inc., 1914.

[3] STEINMANN, G., Ueber gebundene Erzgänge in der Kordillere Südamerikas, Int. Kongr. Düsseldorf, Abt. IV, Vortrag 20, pp. 172–181, 1910.

SINGEWALD, J. T. JR., The Erzgebirge tin deposits: *Econ. Geol.*, vol. 5, pp. 166–177, 265–272, 1910.

FERGUSON, H. G., and A. M. BATEMAN, Geologic features of tin deposits: *Econ. Geol.*, vol. 10, pp. 209–262, 1915.

BUTLER, B. S., Relations of ore deposits to different types of intrusive bodies in Utah: *Econ. Geol.*, vol. 10, pp. 101–122, 1915.

tion, and *D* shows the batholith at a deeper stage of erosion. In Fig. 11, *A* is the normal cupola like *C* in Fig. 10; *B* is a larger body of which the interior is barren, but on the right side there is a marginal roof pendant. Lodes are found in the granite between the roof pendants and the contact

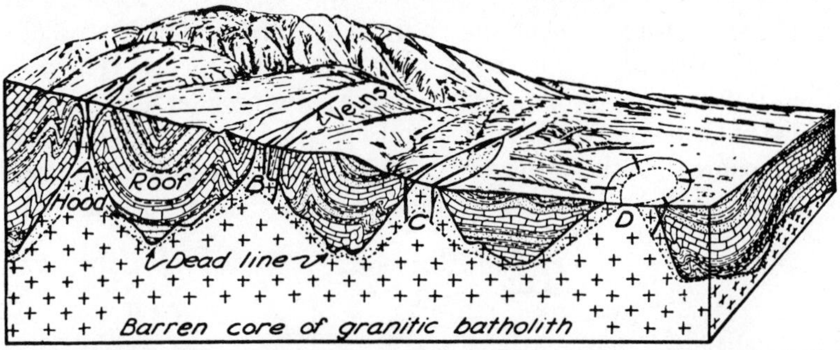

FIG. 10.—Diagram of a region truncated along inclined plane to illustrate batholith exposed at different elevations. *A*. Funnel-like intrusion (ethmolith), with elongated outcrop, rising from cupola; veins lie nearly parallel to long axis of its outcrop. *B*. Wedge-like intrusion (sphenolith) rising from cupola; veins in and near the wedge strike approximately in direction of long axis of outcrop. *C*. Cupola with veins in and near it striking parallel to its long axis. *D*. Conical stock with circular plan and veins in hood and roof which radiate from core.

of the granite with the main body of the invaded rock. The lode system is the roof-pendant, border-zone type. This body has also a vein in a "finger." At *C* (Fig. 11) a large apophysis of the granite cuts across roof

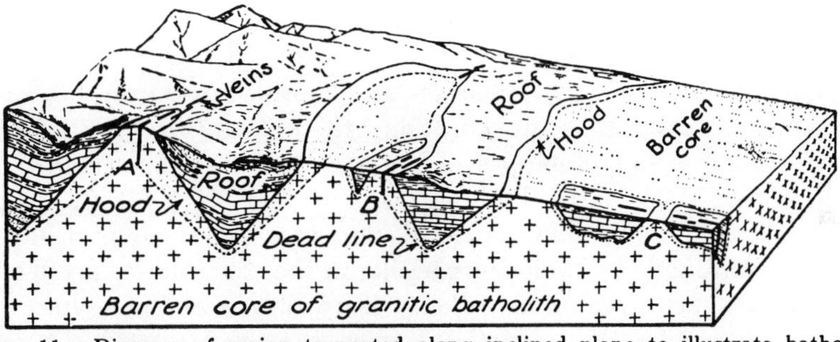

FIG. 11.—Diagram of region truncated along inclined plane to illustrate batholith exposed at different elevations. Metalliferous deposits are in hood and in roof; core is barren. Concentrations are in and near: *A*, cupola, *B*, border pendant, *C*, hourglass structure. *A*. High cupola; veins lie in and near it and are parallel to long axis of outcrop. *B*. Border roof pendant; veins lie in granite between it and border. *C*. Veins developed on hourglass structure with neck of granite that lies between roof pendants. Veins lie in and near neck of hourglass, but many strike across it. All these deposits lie in or near upward bulges of hood.

pendants, making what is called the "hourglass" structure. Lodes are deposited in and near the neck of the hourglass.

Each of the six types described is an expression of a similar feature, namely, an upward swell of the batholith. Stocks as at *A* and *B* (Fig. 10) with deeper erosion will probably be shown to join cupolas, and features like *B* and *C* (Fig. 11) would have shown either cupolas or narrow fingers before erosion had gone so deep. At *D* (Fig. 10) the lodes are radially disposed about the outcrop of the intrusion. In all other figures they are shown to lie rudely parallel to the long axis of the stock except at *C* in Fig. 11, where they lie near the hourglass neck but strike parallel to the contact of the main granite mass and the invaded rocks. The fingers of the batholiths (Fig. 12) like cupolas carry lodes parallel to their long axes.

Rise of Gases.—The relations established by Goranson's[1] experiments seem to show that the water of the earth is very greatly concentrated in the earth's shell.[2] The observations of Fenner[3] seem to justify the con-

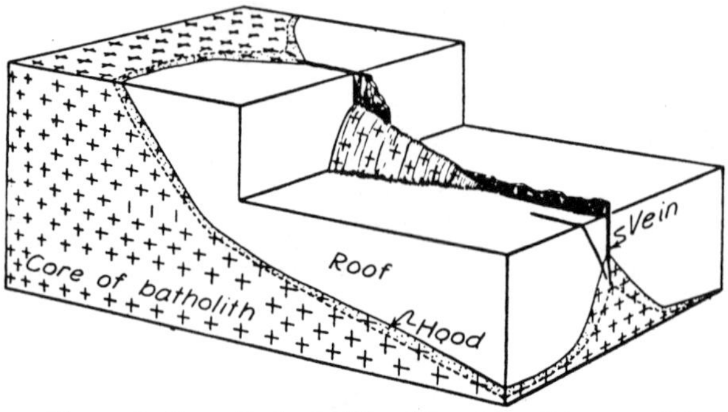

FIG. 12.—Finger-like projection of a batholith with vein striking parallel to the long axis of the projection.

clusion that large amounts of water and other volatile substances are present in certain magmas.

Large volumes of water, chlorine, and fluorine are known to pass from volcanic centers. In some eruptions these quantities are enormous. Zies estimated that fumaroles of the Valley of Ten Thousand Smokes evolved 1,250,000 tons of hydrochloric acid and 200,000 tons of hydrofluoric acid annually, and Calkins calculated from the amount of scapolite developed around the Philipsburg batholith that the chlorine given off was measurable in cubic miles at ordinary temperatures and pressures.

It is probable that certain magmas that rise toward the earth's surface are saturated with water and other volatile substances in solution. Pressure would decrease, owing to ascent; and decrease in pressure lowers the capacity of the magma to carry water in solution. Escape

[1] GORANSON, R. W., The solubility of water in granite magma: *Am. Jour. Sci.*, vol. 22, pp. 481–502, 1932.

[2] EMMONS, W. H., The basal regions of granitic batholiths: *Jour. Geol.*, vol. 41, pp. 1–11, 1933.

[3] FENNER, C. N., Pneumatolytic Processes in the Formation of Minerals and Ores, Ore Deposits of the Western States: Lindgren vol., Am. Inst. Min. Eng., pp. 58–106, 1933.

of gases could take place at this stage with contact metamorphism of the invaded rocks. Probably certain ores of contact-metamorphic origin are formed at this early stage, but the chief period of ore deposition is later.

When crystallization of the magma begins, femic crystals, which in general have higher melting points, will form first. These sink, by gravity, and the lighter minerals, forming later, rise. There is not much concentration of water in these products, for basic rocks are about as high in water as acidic ones. In a magma with 2 or 3 per cent of water, the water will be concentrated in the liquid remainder of the magma because most crystalline rocks—acidic and basic—carry less than 2 per cent water, the average being 1.15 per cent.

In the early stage of cooling of a batholith probably most of the water is held in solution. It is doubtful whether any steam bubbles would then be present, for as the temperature of the magma is lowered its capacity to hold water is increased.

It is probable that bubbles of steam rising through a magma serve as vehicles by which metals are carried upward, for into these bubbles are distilled from the surrounding magma gases with lower vapor pressure. As pointed out by Fenner, they act as vacua in relation to such gases and could serve as collection chambers for other gases.

If a magma moderately high in water had become 90 per cent crystalline, the residue would contain a relatively high concentration of water. Even though its capacity to hold water is increased by lowering of temperature and rising of pressure, by crystallization of silicates, a point is reached where steam bubbles could form. These, gathering gases of lower vapor pressure, could rise by gravity. This probably is the chief stage of the final separation of metalliferous solutions from the magma of the batholith. The bubbles would rise to the base of the hood, pass to the pinnacles of the core, and there form gas cones or cones highly charged with gas below cupolas of the batholith.

When, as a result of crystallization, the vapor pressure is sufficient to fracture the cupolas and the parts of the roof near them, the fractures will become the channels that lead the gases to the hood and roof of the batholith. The fluid system—chiefly gas as it enters the fractures of the hood—becomes partly liquid as it passes upward and then chiefly liquid as it rises into the fractures in the upper parts of the hood and in the roof where fissures, permeable beds, easily replaceable beds, anticlinal structures, older faults, fractured pipes, or subsidence pipes supply the channels that direct the solutions.

If gas under high pressure accumulates at the high points of the undulating top of the core of the batholith, this pressure would be exerted in all directions, but it would tend to open the areas above the cupolas of the batholith because in these the roof of the batholith is thinnest. If the cupola is a cone, and pressures were exerted against its walls in all directions, radial cracks should form, provided load and strength of overlying rocks were equal. Or possibly a "perforation pipe" or pencil-like chimney may form.

Orientation of Fractures.—There are few conical upward projections from batholiths. The Lands End massive of Cornwall is nearly circular, and veins in it are rudely radial. The north end of the Arbus massive of Sardinia is nearly circular, and it has radial lodes. If pressure were exerted from within a conical mass, radial fissures would be formed (Fig.

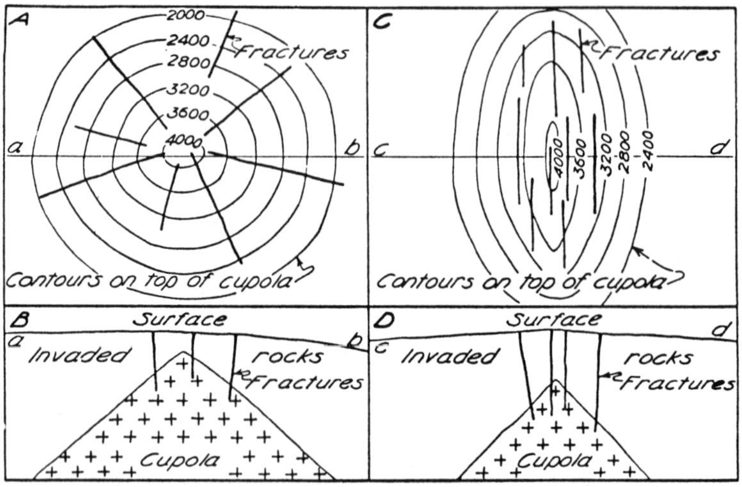

FIG. 13.—*A*. Contours showing the top of a conical cupola with radial fractures. *B*. Cross section of same. *C*. Contours showing a cupola having the shape of an elongated dome, with fractures nearly parallel to long axis of cupola. *D*. Cross section of *C*. *A* is represented by only two or three important mining districts; *C* is represented by scores of important districts.

13*A*, *B*). If pressure were exerted outward from an elongated dome or cupola, fissures would form parallel to the long axis of the cupola (Fig. 13*C*, *D*). This type of intrusive is represented by scores of districts. If the elongated cupola is symmetrical (Fig. 14*A*), with dips from its major axis nearly equal, a relatively simple system of openings should form extending from the cupola to the surface. If it is asymmetrical (Fig. 14*B*), the vertical force on one side would be greater than on the other, and a shearing stress would be developed. Since rocks are weakened by the presence of joints, the pressure may merely move the blocks rather than break them, although fracturing and comminution on a considerable scale might follow readjustment.

The systems of rudely parallel veins are more common than all other vein systems combined. Some such systems are not connected with exposed igneous intrusives, but a very large number of them are. Most stock-like intrusives are distinctly elongated, and many of the larger intrusives have finger-like projections extending outward from them. Veins associated with the small elongated stocks and cupolas generally lie nearly parallel to the long axes of the stocks, and veins in and near the finger-like projections of the larger bodies generally lie parallel to the long axes of the fingers (Fig. 12).

The finger of acidic rock commonly points to the deposit, or else the deposit lies in it. Many lodes are nearly parallel to the long axis of the finger. Some of these fingers are known to have sloping roofs, and

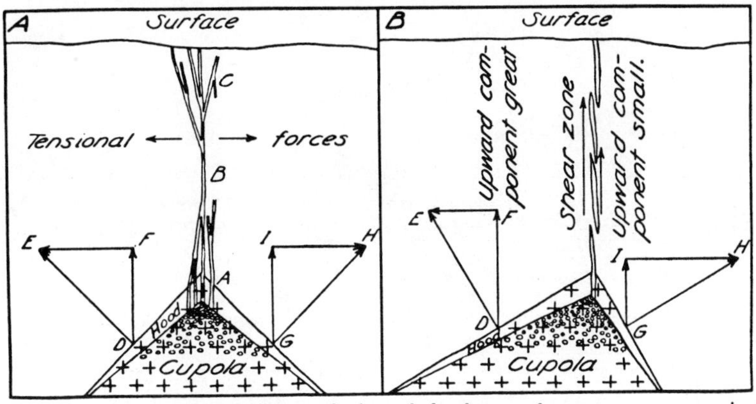

FIG. 14.—*A*. Diagram showing resolution of the forces of vapor pressure against the walls of an elongated cupola. The compressive stress *DE* is resolved into a vertical force in the direction of *DF* and a horizontal one in the direction of *FE*. *GH* is resolved into *GI* and *IH*. The forces *FE* and *IH*, although due to compressive stress, push in opposite directions and away from *A*, where the effect is tensional. *B*. Diagram showing how a shear zone may form as a result of vapor pressure acting on the unsymmetrical walls of an elongated cupola. The vertical component of *DE* is *DF*, which is greater than the vertical component of *GH*, which is *GI*. Unequal upward pressure results in shearing.

the axis of a finger commonly pitches outward from the main body of the batholith. The finger in a certain sense is a cupola which extends outward and upward from the side rather than from the top of the batholith. As stated, the associated lodes as a rule strike rudely parallel to the axis of the finger. The relation is such as would be expected if fracturing were caused by pressure originating in the magma of the batholith, this pressure being exerted below the two walls of the finger that slope away from its axis. Likewise a cupola that resembles in shape the top of a hayrick would be fractured on lines nearly parallel to the elongation of the cupola by pressure exerted from below on the sloping roof.

Elongated Stocks and Lode Systems.—Examples of elongated stocks with veins striking approximately parallel to the long axes of stocks include those of Philipsburg, Montana (Fig. 15); Grass Valley (Fig. 16) and Four Hills (Fig. 17), California (page 26); Bridge River (page 100) and Rossland (page 104), British Columbia; and many others.

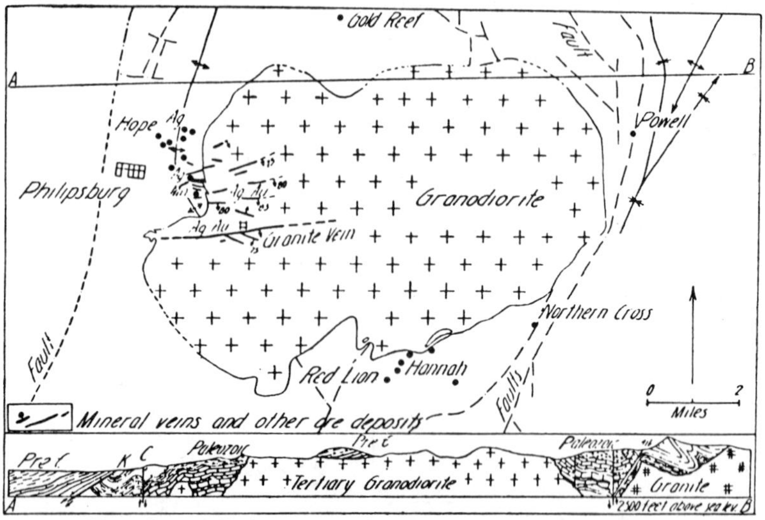

FIG. 15.—Plan of granodiorite mass and geologic cross section of Philipsburg district, Montana. The lodes are around the batholith and on its west side, and extend well toward the center. The mass is 5 miles wide. (*After W. H. Emmons and F. C. Calkins, U. S. Geol. Survey.*)

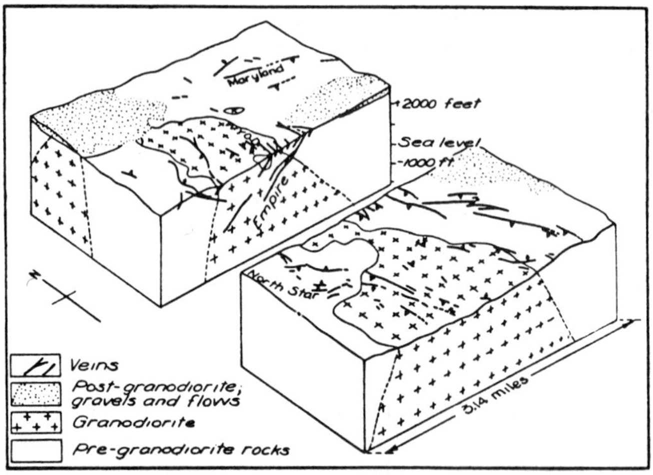

FIG. 16.—Block diagram of Grass Valley, California. Cut on shaft of Empire mine to show relations of veins to cupola. (*After Johnston and Cloos, Econ. Geol., vol. 29, 1934, with surface from Lindgren.*)

The pre-Cambrian shield areas contain many examples of small elongated cupolas and fingers extending from the batholiths with veins striking nearly parallel to their long axes. That is because erosion has been so deep that parent igneous rocks are generally exposed in the ore-bearing areas, and patterns may be interpreted more easily. At Porcupine, Ontario, the lodes lie at the end of and near the Pearl Lake intrusive, but the age relations of the latter and the lodes are in controversy. Kirkland Lake (Fig. 18, page 27), Swayze (page 46), Moss (Huronian), Sultana, central Manitoba, and many other districts show small stocks with veins approximately parallel to the long axes of the stocks.

The Bourlamaque batholith (Fig. 19) and its satellites in western Quebec, show a striking relation between the vein systems and the intrusives. Of eight vein systems developed, seven groups lie nearly parallel to the elongations of outlying intrusives or to the elongation of projecting fingers of the central Bourlamaque batholith.

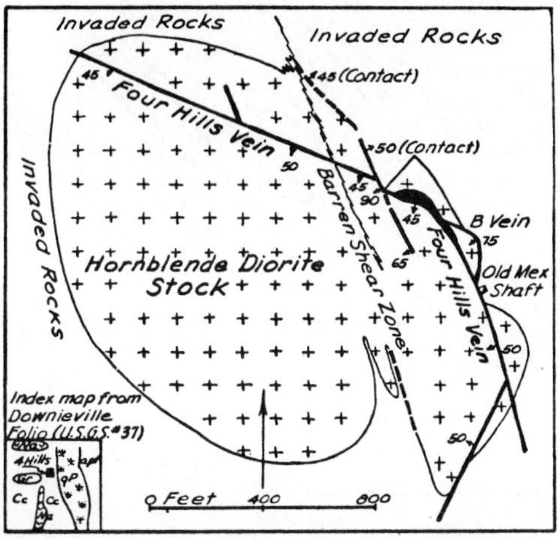

Fig. 17.—Surface map of Four Hills area, Sierra County, California. The fracture system bends to conform with the finger of the intrusive. (*Courtesy of W. I. Gardner.*)

Southern Rhodesia, like Canada, has many deposits that lie in or near small elongated intrusives and strike nearly parallel to the long axes of the intrusives. Examples are Shamva, Sherwood Star, Felixburg, Salisbury Belt, and many others. The belt of lodes that are worked in the great Globe, Phoenix, and Gaika mines, near the contact of schist and intruding granite, are partly in the granite and partly in invaded schists. Maufe's map shows a small marginal roof pendant of schists, and a short distance east of the Gaika mine and at Globe there are small roof pendants. These and other small bodies of schist in the granite suggest that just east of the mines there was a long body of schist, part of which is eroded. The lodes are evidently of the marginal roof-pendant main-contact type (Fig. 11*B*). They recall the relations at Ophir, California (Fig. 20). In the Battlefields district, also, several veins have been worked in a narrow granite neck between the main contact to the west and a roof pendant to the east.

In Western Australia the majority of the most valuable lodes lie off the ends of small stocks and of fingers of larger masses. The lodes of

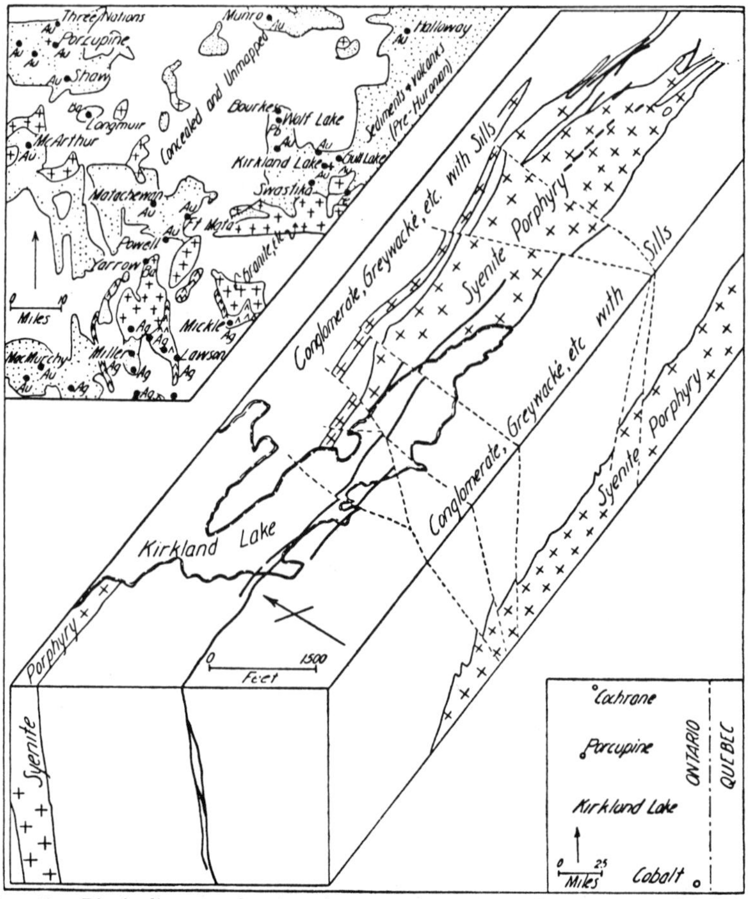

FIG. 18.—Block diagram showing the general relations of the auriferous deposits of Kirkland Lake, Ontario, to invading and to invaded rocks. Many of the smaller sills are omitted. (*Based on a diagram by E. W. Todd, Ontario Department of Mines.*) Upper inset shows general relations of mining districts of part of Ontario to intruding and to intruded rocks. (*Inset after Miller, Knight, Collins, Quirke, and others.*)

the Golden Mile, Kalgoorlie, of Paddy's Flat, Meekathara, and the Sons of Gwalia lode of the Leonora district are similarly situated with respect to small stocks and projecting fingers of larger granitic masses. The deposits of Sir Samuel, Golden Ridge, and other small mining districts also are located on or near the long axes of small intrusives. So are the great copper-gold lodes of Moonta in South Australia. The deposits of Wallaroo near by lie close to the end of the Moonta stock but strike across its axis.

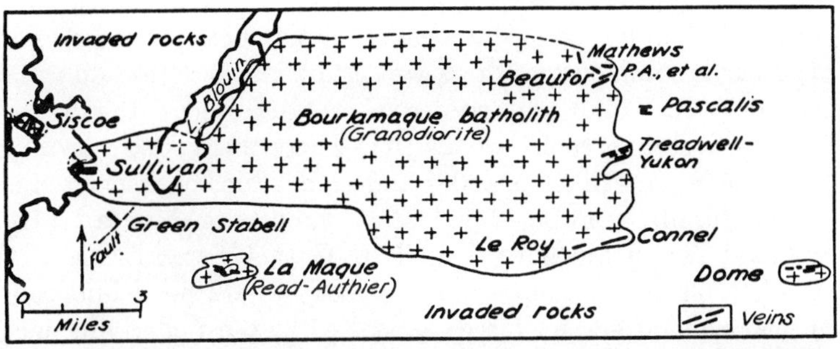

Fig. 19.—Map of Bourlamaque, Quebec, granodiorite batholith and satellitic stocks showing gold-bearing veins in fingers of the batholith striking nearly parallel to the long axes of the fingers, and in satellitic stocks, nearly parallel to the long axes of the stocks. (*Based chiefly on maps by Hawley.*)

In the Pilbara district in the northwestern part of Western Australia there is a striking relation of the mining districts to cupolas and fingers (page 490).

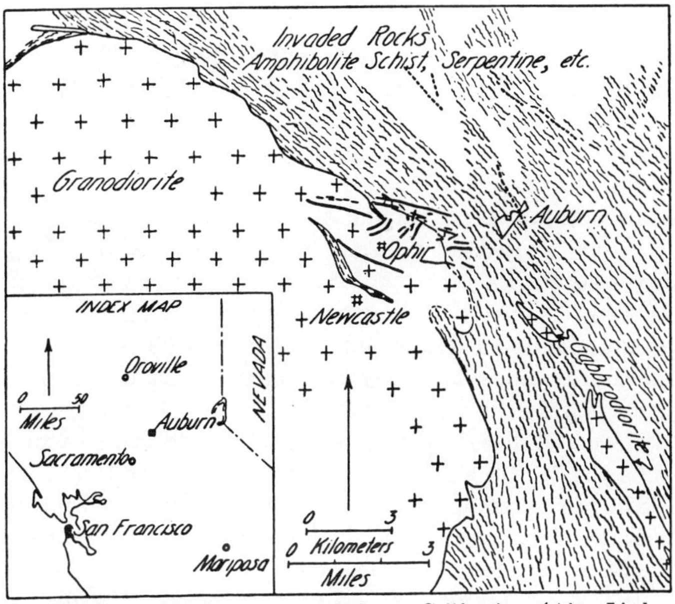

Fig. 20.—Map of Ophir district, near Auburn, California. (*After Lindgren.*) The veins near Ophir are concentrated at the edge of a batholith, near a marginal roof pendant. Before erosion, the mineralized area was probably below the summit of a cupola.

Lodes Crossing Cupola Strike Lines.—There are certain areas in which the lodes lie in and near stocks and satellitic cupolas of batholiths where there is no marked parallel relation of strike of lodes and the long axes of the intrusives. Such relations are found where the lodes were formed in late geologic time and presumably near the surface at the time of deposition. In other areas where lodes were formed in connection with batholiths that had flat roofs the parallel relation between lodes and long axes of stocks is common, yet there are certain districts that do not show this relation. In Idaho and western Montana, where the batholiths had comparatively flat roofs, are districts where there is no very marked parallelism between strikes of lodes and elongations of associated intrusives. The parallel relation is evident at Philipsburg, Montana; Hailey, Idaho; and elsewhere but not strikingly at Garnet, Scratch Gravel Hills, and at other places. In homogeneous rocks where hood and roof are broken near the surface by forces generated in the underlying magma, the surface configuration of the area might control, for in valleys the rock column overlying the top of the core of the batholith would be thinner than elsewhere, and many of these valleys would cross the long axes of the underlying stocks. In other districts ancient lines of weakness such as faults and shear zones are, doubtless, controlling factors, for these may be reopened to form new channels for rising fluids. The parallel relation of lodes and elongated cupolas is most consistent in the pre-Cambrian shields and in other areas where the batholiths had thick roofs, because in such areas the topography at the line of fracturing would have had less effect in localizing fractures, whereas with thin roofs this topography might have been a controlling factor. In areas eroded so deeply that the true coarse-grained cupolas are exposed it is found that lodes generally lie parallel to long axes of stocks, and this relation should be useful in prospecting areas around them. In areas where the granite cupolas are not exposed the relation is not so helpful, yet in some such areas there is an obvious relation between shallow-seated intrusives and the lodes. In Tiawan lodes lie parallel to the elongation of a belt of intrusive andesites. In Transylvania the lodes lie parallel to elliptic areas of rock alteration which may be connected with underlying cupolas (see pages 320, 321).

It may be urged that the long axes of stocks and the major mineralized fractures lie parallel because both follow lines that are loci of weakness. That may be true of some districts, but it is not universally true, for at places both the long axes of the intrusive stocks and the veins strike across the lines of tectonic deformation. At Philipsburg, Montana, for example, the major faults and the axes of folds strike nearly north, whereas the long axis of the Philipsburg batholith strikes east nearly at right angles to the lines of major deformation of the area. In this district the major veins strike approximately parallel to the long axis of the Philipsburg batholith and nearly at right angles to the lines of major deformation as represented by axes of folds and of major faults (Fig. 15).

Mineralization of Major Fractures.—It is well known that few faults of great throw are mineralized, even in areas of strong mineralization. That may be due to impermeable gouge developed in them by movements, or it may be that they do not extend to the sources of mineral-bearing solutions.

There are, however, a considerable number of faults with moderate throw that are mineralized, and some of these carry gouge, as do the veins of the Blue vein system of Butte, Montana. Certain vein systems that have no obvious relation to the axes of cupolas or to fingers of stocks near by may nevertheless lie in fractures that have been reopened by pressure of the vapor released by cooling magmas. If strong fissures suitably located were available, these, regardless of their orientation, might be reopened with less pressure than would be required to form new ones, and they would be utilized as channels for escaping aqueous fluids which deposit lode ores. In other vein-filled fractures vapor pressure probably played no part.

One of the regions, where no clear-cut relation between the lodes and the shapes of batholiths exists, is in central Germany, particularly near the Harz Mountains and in the Saxon Erzgebirge. In this region the lodes generally strike northwest parallel to the direction of the Hercynian fracturing or northeast parallel to the direction of Variscian folding. Apparently earlier lines of weakness determined the direction of the ore-bearing fractures.

Stocks Contracting Downward.—There are certain stocks that become narrower in depth. These may broaden into typical cupolas at greater depth, but that is not proved. Stocks intruded near the surface are likely to flare out into funnel-shaped bodies because with pressure decreasing near the surface there is less resistance to lateral expansion. The intrusive stock of Nagyag, in the Transylvanian Erzgebirge (page 322), and the Potosi stock of Bolivia are well-known examples of stocks that narrow downward like inverted cones. These intrusives are of very late age and were formed near the surface. That may be true of Ely, Nevada, where deep exploration has shown that the stocks contract with depth and where, according to Locke, only a few hundred feet have been eroded since the deposition of the ore. The Fierro-Hanover intrusive in New Mexico, as shown by Schmitt,[1] narrows with depth.

In the few examples of metallized upward-flaring stocks, the veins or mineralized zones generally lie rudely parallel to long axes of the intrusives. That is true in the Transylvanian Erzgebirge, at Ely, and of some of the stocks in Bolivia, although maps of only a few of the vein systems in Bolivia are available. In general, the parallel alignment of

[1] Schmitt, H., The Central mining district, New Mexico: Am. Inst. Min. Eng. *Contribution* 39, pp. 1–22, 1933.

veins and of the long axes of cupolas is more marked in and around deep-seated cupolas than in stocks formed near the surface.

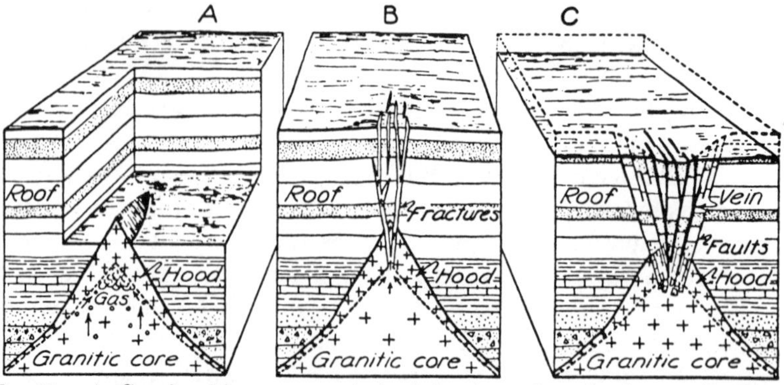

FIG. 21.—*A*. Cupola with gas accumulation below it at pinnacle of core. *B*. Fractures formed by failure of hood and roof above core pinnacle. *C*. Graben or area of subsidence caused by escape of gas from pinnacle of core.

Small Subsidence Areas.—In certain regions of strong mineralization there are small areas of subsidence where rocks are broken and at places faulted downward. Locke[1] has explained such areas by "mineralization stoping." He believes that the solutions first entering ore channels were strongly dissolving solutions and that the ore-depositing and rock-cementing solutions came later. That explanation may be a true one for certain deposits; but if gas cones form at pinnacles of the cores that lie below cupolas, it is possible that small areas of subsidence may form by overlying material's slumping downward to fill space formerly occupied by gas (Fig.21).

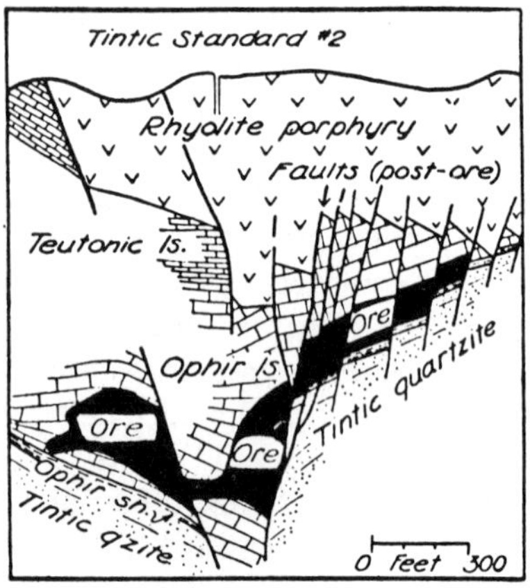

FIG. 22.—Cross section of Tintic Standard mine, East Tintic, Utah. (*After G. W. Crane.*) The ore bodies form a V-shaped series of fault blocks or a graben.

Examples[2] of small areas of subsidence marked by strong mineralization include Tintic Standard mine, Utah (Fig. 22); Ophir, Utah (Fig. 23); Las Pilares, Nacozari, and Cusi-Mexicana, Mexico; Waihi, New Zealand (Fig. 24); and many others. Spurr[3] believes that many of the ore pipes are explosion vents.

[1] LOCKE, A., The formation of certain orebodies by mineralization stoping: *Econ. Geol.*, vol. 21, 431–453, 1926.

[2] EMMONS, W. H., On the origin of certain systems of ore-bearing fractures: Am. Inst. Min. Eng. *Trans.*, vol. 115, pp. 9–35, 1935.

[3] SPURR, J. E., The Ore Magmas, pp. 858–891, McGraw-Hill Book Company, Inc., 1923.

Zoned Lode Systems.—As metal-laden fluids rise, the metals are deposited along channels, at many places in standard order,[1] the least soluble ones being deposited nearest the source. This order is shown in

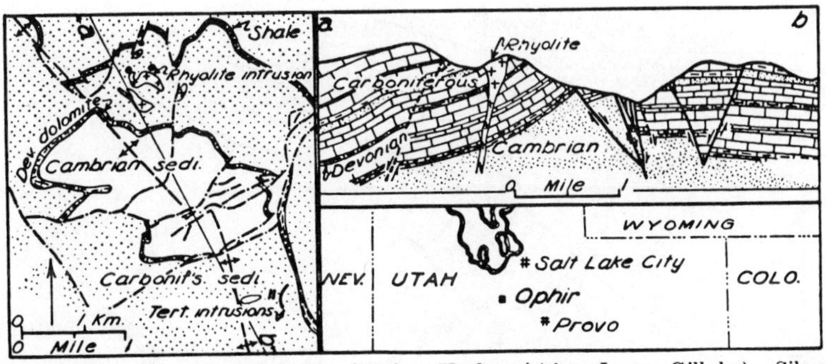

Fig. 23.—Map and section of Ophir district, Utah. (*After James Gilluly.*) Silver-lead deposits occur in the fault fissures and replace limestone beds in the faulted area.

the table (page 34) and the same succession is expressed in veins from depths to the surface, as is observed in horizontal sections from a center of

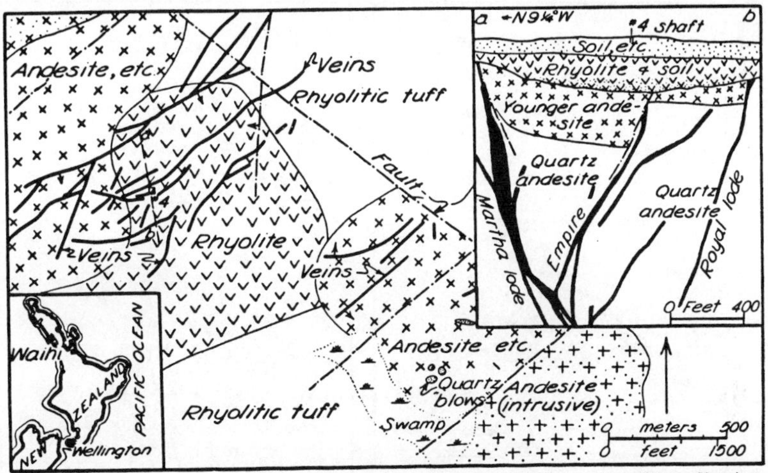

Fig. 24.—Map and cross section on line *AB* of part of Waihi gold-mining district, New Zealand. The lodes in andesite crop out as shown, but they are covered over by the rhyolite where their positions near the base of the rhyolite are shown on map. The cross section shows a graben or down-faulted block about 1500 feet across. (*Data from Morgan.*)

mineralization outward (Figs. 25, 26). The entire succession is never seen in a single vein or in a single district. The table is a composite made up

[1] Rastall, R. H., Metallogenic zones: *Econ. Geol.*, vol. 18 p. 115, 1923.

Spurr, J. E., A theory of ore deposition: *Econ. Geol.*, vol. 2, pp. 781–795, 1907; The Ore Magmas, pp. 604–624, McGraw-Hill Book Company, Inc., 1923.

Emmons, W. H., Primary downward changes in ore deposits: Am. Inst. Min. Eng. *Trans.*, vol. 70, pp. 964–992, 1924.

of observations in many different lodes. At places zones are lacking, and there is much overlapping. There are few reversals, however, and most

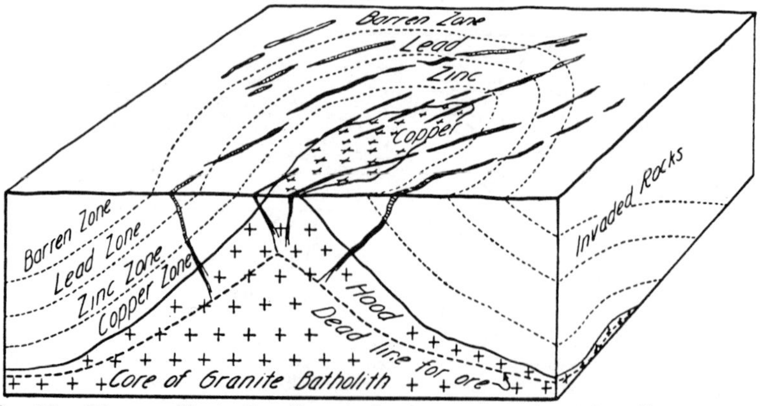

FIG. 25.—Diagram illustrating the zonal arrangement of metals. Copper ores were deposited in lodes in and near the cupola, zinc ores farther out and above copper ores, and lead ores above zinc and still farther out.

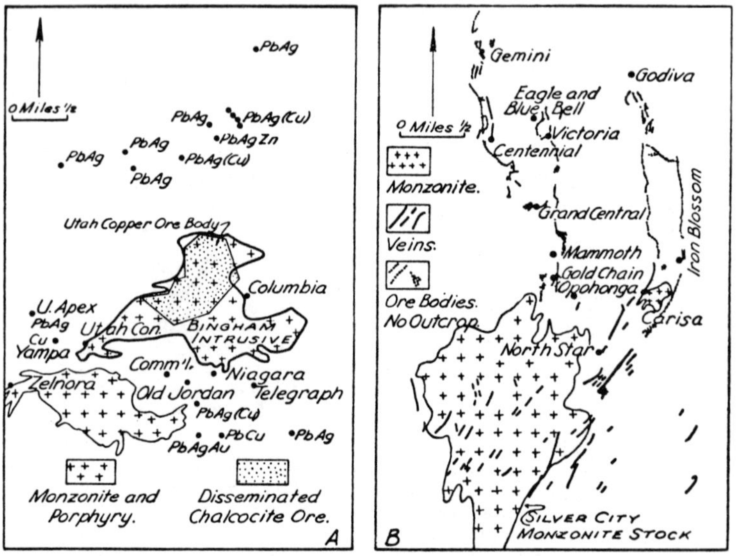

FIG. 26.—*A*. The Bingham stock, Utah. (*After Boutwell.*) The zones are (1) copper-gold in and near the stock; (2) lead-silver farther away. *B*. The Tintic stock, Utah. (*After Lindgren and Loughlin.*) The zones are (1) copper-arsenic near the stock; (2) copper-lead-silver farther away; (3) lead-silver still farther away.

of these may be explained by supergene concentration or by two periods of deposition.[1]

[1] EMMONS, W. H., Hypogene zoning in metalliferous lodes: Sixteenth Int. Geol. Cong. *Rept.*, vol. 1, pp. 417–432, Washington, 1936.

Gold appears in large amounts at two places in the series, and that renders the interpretation of its zoning difficult. The gold lodes of the shields do not exhibit clear-cut examples of zoning of the metals. That is probably because few metals other than gold, copper, and zinc are present in important amounts. It is the writer's belief that fluids that

A Reconstructed Vein System, from the Surface Downward

1. Barren zone:	Chalcedony, quartz, barite, fluorite, etc. Some of the veins carry a little mercury, antimony, or arsenic.
2. Mercury:	Cinnabar deposits, commonly with chalcedony, marcasite, etc. Barite-fluorite veins.
3. Antimony:	Stibnite deposits, often passing downward into galena with antimonates. Some carry gold.
4. Gold-silver:	Bonanza gold deposits and gold-silver deposits. Argentite with arsenic and antimony minerals common. Tellurides and selenides at places. Relatively small amounts of galena, sphalerite, and chalcopyrite are present; gangue includes quartz, adularia, alunite, with calcite, rhodochrosite, and other carbonates.
5. Barren:	Nearly consistent barren zone; represents the bottoms of many Tertiary precious-metal veins. Quartz, carbonates, etc , with small amounts of pyrite, chalcopyrite, sphalerite, and galena.
6. Silver:	Argentite veins, complex silver minerals with antimony and arsenic, stibnite, some arsenopyrite, etc.; quartz gangue, at places with siderite.
7. Lead:	Galena veins, generally with silver; sphalerite generally present, increasing with depth; some chalcopyrite. Gangue of quartz with carbonates.
8. Zinc:	Sphalerite deposits; galena and some chalcopyrite generally present. Gangue is quartz and in some deposits carbonates of calcium, iron, and manganese.
9. Copper:	Tetrahedrite, commonly argentiferous; chalcopyrite present. Some pass downward into chalcopyrite. Enargite veins, generally with tetrahedrite.
10. Copper:	Chalcopyrite veins, most with pyrite, many with pyrrhotite. The gangue is quartz and at some places carbonates and feldspar. Orthoclase and sodic plagioclase not rare, but high-calcium plagioclase very rare; generally carry precious metals. Uranium, probably main horizon of uraninite.
11. Gold:	Deposits with pyrite, commonly arsenopyrite. Quartz, carbonates, and some with feldspar gangue. Some with tourmaline. Tellurides not uncommon and at places abundant. At places zones 10 and 11 are reversed.
12. Arsenic:	Arsenopyrite with chalcopyrite, etc.
13. Bismuth:	Bismuthinite deposits. Native bismuth, quartz, pyrite, etc.
14. Tungsten:	Veins with tungsten minerals, arsenopyrite, pyrrhotite, pyrite, chalcopyrite. Tungsten occurs in higher zones in fairly large amounts, but this is the main horizon.
15. Tin:	Cassiterite veins with quartz, tourmaline, topaz, feldspar, etc.
16. Barren:	Quartz, feldspar, pyrite, carbonates, and small amounts of other minerals.

deposit lodes generally leave the batholith from cupolas and that at the high pressures prevailing conditions for zoning are not favorable. It is around the summit cupolas (acrobatholithic) (Fig. 8) that zoning is most clearly expressed. In the intermediate horizons (embatholithic) there is a regional zoning of lodes with gold deposits closely spaced near the border zones of great batholiths and in and around marginal cupolas, and farther away from the batholiths copper deposits with less gold prevail. This is true of the Alaskan Arc, the South Alaskan arm, the Sierra Nevada batholith in California and of the gold and copper deposits of the Southern Appalachian Mountains.

Mineral Character of Gold Lodes.—Hydrothermal metamorphism or wall-rock alteration is brought about by solutions that move through fissures or other channels and soak into the wall rock and change it. The walls of essentially all lodes that are formed by hot waters show these changes. In some the wall rock is altered only a fraction of an inch from the fissure; in others the alteration extends many feet away. In many the altered wall rock is ore. It is not always possible to determine whether a mineral or an ore has been formed by filling an opening or by replacing the wall rock. Although some valuable deposits show only narrow zones of alteration, in general a strong and thoroughgoing wall-rock alteration is regarded as a favorable indication. In forming an opinion as to the value of a vein, the character of its mineralization and the wall-rock alteration should be taken into consideration. Mineral character and wall-rock alteration depend upon the character of solutions, the character of the country rock, depth, temperature, and pressure. It is a complicated problem, yet by the study of deposits formed in various rocks and at various depths certain types may be set up, and these may serve crudely as standards with which an undeveloped lode may be compared. The table presented on page 36 shows the great variety of minerals present in valuable gold lodes. No mineral or group of minerals may be regarded as significant of persistence, value, or lack of value. Yet in certain restricted provinces experience has shown that certain types of lodes in that province are likely to prove valueless. Thus in some areas carbonate veins, or great "bull-quartz" veins, or quartz feldspar veins have proved uniformly valueless. In a partly developed area containing valuable lodes one is justified in giving much weight to the mineral character and wall-rock alteration of other lodes under investigation.

In many lodes the richer gold ore is associated with "graphite." Hydrocarbons which are commonly present in sedimentary rocks are reducing agents and probably have reduced solutions that carried gold, the reduction resulting in its precipitation. Graphite is an inert substance, but in many lodes probably it should be regarded as the resulting product of the actively reducing hydrocarbons. Thus graphite may be associated with gold because it is a residual of a hydrocarbon that reduced gold-bearing solutions.

DISTRICTS WITH EPITHERMAL GOLD DEPOSITS

DISTRICT	Quartz	Adularia	Sericite	Chlorite	Ca-Mg-Fe Carbonates	Rhodochrosite	Barite	Fluorite	Zeolites	Kaolinite	Alunite	Pyrite	Marcasite	Chalcopyrite	Sphalerite	Galena	Sb-Compounds	Tellurides	Selenides	Visible Gold
Cripple Creek, Colo.	●	●	•	•	•	•	•	●		●	•	•	•		•	•	•	●		•
Goldfield, Nev.	●		•	•	•		•			●	●	●	•		•	•	●	•		●
Republic, Wash.	●	●	•	•	●			•	•			●		•			•	•	●	•
Waihi, New Zealand	●	●	●	•	●	●			•	●		●	•	•	•	•	•	?	•	●
Lebongstreek, Sumatra.	●	•	•	•	●	●			●	●		●		•	•	•	•		●	•
Nagyag, Roumania.	●		●	•	●	●	•			●		●	•	•	•	•	●	●		•
Brad, Roumania	●		●	•	●	●	•		•	●		●	•	•	•	•				•
Telluride, Colo.	●	●	●	•	●	●	●	●				●		•	•	•	●			●
Comstock Lode, Nev.	●		●	•	•	•			•	●		●		•	•	•	●			●
Tonopah, Nev.	●	●	●	•	●	•	•		•			•		•	•	•	●		•	•

DISTRICTS WITH MEZOTHERMAL GOLD DEPOSITS

DISTRICT	Quartz	Albite	Sericite	Chlorite	Carbonates	Mariposite	Pyrite	Pyrrhotite	Chalcopyrite	Arsenopyrite	Sphalerite	Galena	Tetrahedrite	Stibnite	Molybdenite	Tellurides	Selenides	Magnetite	Scheelite	Visible Gold
Mother Lode, Cal.	●	•	●	●	•	•	●	•	•	•	•	•	•	•	•	•			?	●
Grass Valley, Cal.	●	•	●	•	•	•	●	•	•	•	•	•	•		•	•			?	●
Bridge River, Brit. Col.	●	•	●		•	•	●	•	•	●	•	•	•	•		•				•
Kirkland Lake, Ont.	●		●	●	●		•	?	•		•	•			•	●				•
Nova Scotia	●	•	●	•	●		●	•	•	●	•	•		•	•				•	●
Haile Mine, N Car	●	•	●	●	•		●	•		•	•				•					•
Bendigo, Victoria.	●	•	•	•	•		•	•	•	•	•	•	•	•	•			•		●
Charters Towers, Queensland	●		●	•	●		●		•	•	●	●				•				•
Mount Morgan Queensland.	●		•	●	•		●	●	●											•
Boliden, Sweden	•		●	•	•		●	•	●	●	•	●					●	●		•

DISTRICTS WITH HYPOTHERMAL GOLD DEPOSITS

DISTRICT	Quartz	Feldspar	Sericite	Biotite	Chlorite	Amphibole	Garnet	Tourmaline	Carbonates	Fluorite	Pyrite	Pyrrhotite	Arsenopyrite	Chalcopyrite	Sphalerite	Galena	Molybdenite	Tellurides	Selenides	Magnetite	Scheelite	Visible Gold
Homestake Mine, S.D.	●	•		•	●	●	•		•	?	●	●	●	•	•	•		•		•	?	•
Juneau, Alaska.	●	●	•	●	•			•	•		●	●	•	•	•	•				•		•
Porcupine, Ontario.	●	●	●		●		•	●	●		●	•	•	•	•	•	•	•			•	●
Noranda, Quebec.	●	•	●		●				●		●	●		●	•			•		•		•
Red Lake, Ontario.	●	•	●		•			•	●		●			•	•	•		•			•	•
Dahlonega, Georgia.	●		●		●	•	●		•		●	•		●	●					•		•
Rossland, Brit. Col.	●		●	●	●	•	•	•	●		•	●	•	●	•	•	•			•		•
Morro Velho, Brazil.	●	●	•		•				●		•	●	●	•						•	•	●
Kalgoorlie, W. Australia.	●	●	●		●			•	●	•	●			•	•	•		●		•		•
Meekathara, W. Aust.	●		●		•			•	●		●		●							•		●
Kolar, Brit. India.	●	•	•	●	•			•	•		•	•	•	•	•	•				•	•	•

IMPORTANT CONSTITUENTS THUS ● ACCESSORY CONSTITUENTS THUS •

Table showing minerals present in auriferous deposits.

Classic papers dealing with hydrothermal deposits of specific auriferous regions are numerous. The following choices are landmarks in the magmatic hydrothermal theory.

E. S. Moore (1879-1966) was professor of geology in the University of Toronto from 1922 to 1949. His researches in the Precambrian rocks of Canada dealt with many aspects, one of which was the origin of the Archean batholiths and associated rocks and their relationship to deposits of the gold-quartz type, such as those at Timmins and Kirkland Lake, Ontario. His presidential address to the Society of Economic Geologists at Minneapolis in 1939 represents many years of study of lode gold deposits of the Canadian Shield (Paper 10-2).

(Text continues on page 274.)

10-2: GENETIC RELATIONS OF GOLD DEPOSITS AND IGNEOUS ROCKS IN THE CANADIAN SHIELD[1]

E. S. Moore

INTRODUCTION.

Most geologists are still sufficiently conservative to believe in the genetic relations of ore deposits and igneous rocks that are closely associated in the field. The opinion of Holmes,[2] recently expressed so forcefully, that lead deposits originate independently of the magmas which supplied any of the associated igneous rocks has shaken the confidence of but few geologists in the value of such an association as a criterion in prospecting. The consensus is overwhelming that the Precambrian gold deposits may properly be tied to magmas of certain ages and to certain igneous rocks derived from them, which show many similarities in petrographic characters and, in a broad way, in magmatic sequences. In dealing with an area of over 1,800,000 square miles, the remarkable similarity in many of the features, both major and minor, in the areas where gold occurs, is very impressive. Such a situation is partly the result of long periods of erosion reducing the shield to a level where the granitic rocks, which almost everywhere seem to underlie the continental masses at great depths, prevail, and conditions have progressed far toward uniformity. The hypothermal portions of the vein systems alone remain and in many respects the gold deposits throughout the shield appear to be more

1 Presidential address delivered before the Society of Economic Geologists at a joint session of the Society and the Geological Society of America, Minneapolis Meeting, December 29, 1939.

2 Holmes, A.: The origin of primary lead ores. Econ. Geol., 32: 763–782, 1937 and 33: 829–867, 1938.

uniform in many of their characters than those formed outside the shields where a greater diversity of geological conditions exist.

MAJOR DIVISIONS OF THE PRECAMBRIAN.

The difficulties encountered by the Precambrian geologist in dealing with chronology are too familiar to require enumeration. He is often dependent for chronological data on igneous sequences, and widespread batholithic injections of granite with assumed concomitant mountain-building, which would lead to erosion and deposition now represented by mere remnants of sediments.

It is not my intention to enter here on a detailed discussion of correlation problems of the Precambrian, because this would lead into many blind alleys and raise questions bristling with controversy. I would like to have it distinctly understood that in making the following statements it is not considered that the ages of all the granites and gold deposits mentioned have been settled beyond all doubt. Some more accurate means of determining the ages of Precambrian rocks than we now possess is necessary before we can feel sure of them, and the conclusions drawn in this paper are only the results of the acceptance of opinions regarding ages expressed by a large number of men who have worked in the areas concerned. It must be recognized that there are fashions in geology, as in other matters, and at this stage in our geological studies it may be fashionable to assign gold deposits to the Algoman system, just as it was at an earlier period fashionable to call practically all of the Precambrian granites Laurentian. An abbreviated classification that presents the divisions most concerned in this discussion is as follows:

Keweenawan:	Sediments; volcanics; diabase and gabbro on a large scale; granite batholiths, and minor gold deposits.
Animikie:	Sediments and minor volcanics.
Huronian:	Cobalt sediments. Bruce sediments.
Matachewan:	Diabase and gabbro.
Algoman:	Igneous rocks dominantly acid but with large quantities of intermediate and relatively small quantities of basic rocks. Granite batholiths and gold deposits of major importance.
Haileyburian and other post-Timiskaming:	Dominantly basic and ultra-basic rocks, with some diorite.

Timiskaming:	Dominantly sedimentary with volcanics in some areas. Sediments of wide derivation.
Laurentian:	Almost entirely granitic rocks in batholitic injections, with minor gold deposits.
Keewatin:	Chiefly volcanic rocks with minor quantities of sediments mostly of local derivation.

In this classification certain recognized series have been omitted to avoid difficulties in correlating them, but it shows an important feature that is generally recognized, and that is the existence of three systems with granite batholiths and gold deposits. Of these, the Laurentian and Algoman granites had a much wider distribution than those of the Keweenawan, although the Laurentian granite is not now found in place in many areas. Its distribution is attested by the widespread occurrence of boulders of this rock in the Timiskaming conglomerate. The latter rock is intruded by the Algoman granites and questions that naturally arise are: what became of the great masses of earlier granite, and does the Algoman represent the remelted Laurentian, or how otherwise could it have been displaced? If such remelting occurred, did it play any part in furnishing the more abundant gold to the Algoman magmas?

PERIODS OF GOLD DEPOSITION.

In perusing a large number of geological reports dealing with areas and gold camps in the Canadian shield, seventy-five have been considered in which the rock formations were chronologically arranged in accordance with the classification given above. In others the data were insufficient to make assignments of the rocks and gold deposits on a basis of relative age. Every gold deposit mentioned in the seventy-five reports was regarded as belonging to one of the three systems; Laurentian, Algoman and Keweenawan, with their granite batholiths.

The close genetic and structural relationships of gold deposits and batholiths, have been so clearly demonstrated by W. H. Emmons in his many papers on this subject, not only in the Canadian shield but in other shields as well, that they are generally accepted without hesitation by those who have studied the subject in the field. There are, however, many problems concerned with the occurrence of the gold that have not been settled. For example, it is estimated that over 95 per cent of all the gold so far found in the Canadian shield has come from the Algoman. In this estimate, Noranda, in Quebec, has not been included because there is still a decided difference of opinion among geologists who have studied the ore deposits as to whether most of the gold is younger or older than a diabase dike that cuts some of the ore. This dike is probably Keweenawan and if the ore is younger than the dike it might be Keweenawan and the estimate mentioned would be changed materially.

Why this dominance of gold in the Algoman? Is it due to wider distribution of rocks of this age; more extensive and elaborate differentiation in magmas producing greater concentration of the metal; or better structural control?

As stated, the distribution of Algoman granites in place is much greater than that of the Laurentian or Keweenawan, and the number of gold deposits that have attracted some attention is greater per given area for the Algoman than for the Laurentian. In the case of the Keweenawan the difference in number for equal areas is, however, about the same, and the main factor is the much greater size of a considerable number of the Algoman deposits. The situation is well illustrated in southeastern Ontario where a large number of gold deposits are grouped around an area of granite regarded as Algoman, whereas relatively few occur with the older granites.

In the extent of differentiation, with the production of a diversity of igneous rocks, the Algoman, as will be shown when magmatic sequences are discussed, far surpasses the other two systems, which reveal comparatively little diversity in rock species. And yet, the number of gold deposits which occur with the Killarney granite of the Keweenawan system, in a relatively small area, leads one to the conclusion that a great diversity of differentiation products does not appear to be the only factor in the production of gold. There must be something else that exerts a controlling influence and that is apparently structure. The question might be asked as to why the Laurentian granites were not intruded under as favorable structural conditions as the Algoman.

The difference is probably due to the fact that when the latter rocks were injected there were many more favorable structures in existence than when the first great batholithic intrusions with accompanying mountain-building occurred. Both old and new structures were available for the formation of deposits. Then again the Keweenawan granites were injected mostly into the Huronian sediments some of which are shaly in nature and not very favorable ground.

MAGMATIC SEQUENCES.

Since an early date, petrographers have generally recognized that as differentiation of magmas progresses, the products tend to become, on the whole, more and more acid in character. The progress toward more acid products does not always proceed uniformly but in certain cases by steps, and in places the process seems to show a descent from the basic to the acid along two converging lines that are followed by the more femic and more salic constituents. This situation appears to produce reversals in the order of formation of the salic and femic rocks and alternations in order of intrusion result. This tendency to alternate is probably best explained by applying the *reaction principle* so ably expounded by Bowen.[3] In the cases where plutonic rocks are concerned, the greatest diversity in the character of the rocks is found in the satellitic bodies that tend to develop toward the closing stages in the crystallization of the magma. It is here that are found the complementary dikes and diaschistic dikes, and such rocks as lamprophyres, aplites, alaskites, pegmatites, albitites and quartz veins.

It is of interest to examine individually the three systems discussed above to see whether they conform to the accepted plan of differentiation in plutonic bodies.

Laurentian.—In few areas is there any definite evidence of plutonic basic rocks preceding the great injections of granite. The immense extrusions of basic to intermediate lavas of the Keewatin might possibly be regarded as a phase of the granite magmas and

[3] Bowen, N. L.: The Evolution of the Igneous Rocks. Princeton Univ. Press, 1928.

there are also coarse-grained basic intrusives in the Keewatin, but these can not be definitely tied to those magmas. The numerous occurrences of acid lavas and dikes of porphyry that are found in the Keewatin in some areas might also be considered as the shallow manifestations of the same magmas. As for diverse end products of a sequence, there are scarcely any except pegmatites and quartz veins, the latter of which are mostly barren of metals. In a great many areas only granite is mapped as Laurentian, in others quartz porphyry and granite, and in some syenite is mentioned with the granite.

Keweenawan.—In dealing with this system, because of lack of time, discussion of the more complex situation in the Lake Superior basin, where sediments and volcanics, and intrusives of several types occur together, will be omitted. Attention will be confined chiefly to the remarkable series of intrusions of diabases and granite. The former extend over a large section of the shield and the latter, so far as known, is confined mainly to the Lake Huron and Lake Superior regions where the orogeny that developed the Lacloche mountains occurred. The sequence begins with intrusions as dikes and sills of quartz to normal diabase and gabbro, with norite phases represented by the Nipissing diabase and the Sudbury nickel intrusive. This diabase is separated from the Matachewan quartz diabase by the Cobalt sediments and on the whole is fresher in appearance. Important ore deposits, such as those of nickel-copper, cobalt-silver, and copper, and some smaller lead-zinc deposits are connected with this epoch of igneous activity. The silver-cobalt ores are believed to have been derived from a phase of the diabase magma more alkalic than the diabase itself.[4] Three small gold deposits in the Lake Huron region—the Haviland, Payton and Crystal, have been assigned by Collins [5] to this epoch, since they occur in the diabase and differ from those associated with the Killarney granite nearer the lake. They contain considerable silver, and barite gangue, whereas those with the

[4] Moore, E. S.: Genetic relations of silver deposits and Keweenawan diabases in Ontario. Econ. Geol., 29: 725–756, 1934, and Bastin, E. S.: The nickel-cobalt-native silver ore type. Econ. Geol., 34: 39–40, 1939.

[5] Collins, W. H.: North shore of Lake Huron. Can. Geol. Surv. Mem. 143: 116–120, 1925.

granite are highly arsenical quartz deposits. They, and the Bruce Mines copper deposits, which carry some gold and occur in the diabase, may be regarded as being somewhere intermediate in origin between the cobalt-silver and the arsenical gold deposits, the former being derived from a magma relatively richer in alkali and poorer in silica than the latter.

Following the diabase, the Killarney granite was intruded along the North Shore and its injection was accompanied by strong orogeny. It is much more restricted in distribution than the diabase. Many small arsenical gold deposits are genetically related to this granite, which on the whole is a coarse-grained and relatively low-silica type. It produced a goodly number of pegmatites and pegmatitic quartz injections, and some porphyry dikes, but practically no end-phase differentiation products except pegmatites, quartz veins and ore. The granite was followed over a much wider area which is more or less coterminous with that of the quartz diabase, though somewhat smaller, by olivine diabase intrusions, mostly in the form of dikes. These show relatively little metallization and they are very uniform in petrographic character over a large area. It seems difficult to conceive of this olivine diabase representing a differentiate of the granite magma. Its uniformity in character, its extent, and the lack of any regard shown by the intrusions for local structure suggest rather a source in some deep-seated layer such as the postulated basaltic layer.

Algoman.—This system shows a marked contrast to the Keweenawan and Laurentian in the great diversity in rock types and their numerous occurrences in so many different areas. One gets the impression that a very long period of time was involved in the production of these rocks and that they were not all produced in the different areas of the shield at the same time. They should be regarded rather as the result of a series of similar processes operating continuously from area to area through a long period. The nearest approach on this continent to the conditions in the Algoman system seems to be those found in the Sierra Nevada and Coast Range batholiths. In addition to the occurrence of a great variety of rock types, porphyritic texture is a characteristic feature of many of the rocks of this system. These features, and the number of intrusions in some of the sequences seem to indicate much movement of magma and numerous disturbances in the magmatic chambers. This no doubt was a factor in providing opportunities for escape of the ore-bearing solutions.

In scanning the geological reports previously mentioned, it was found that 65 per cent of the areas described show the presence of Haileyburian or other post-Timiskaming and pre-Algoman igneous rocks with no intervening sediments. In some areas these rocks seem to be so closely related to the Algoman granites that they may readily be regarded as the first phase in a sequence. They are dominantly basic and range from peridotites to diorites. Where such rocks precede the granites there is a strong tendency for basic rocks in relatively small volume to reappear in the later phases of the sequence, whereas the later phases are generally acid, where the sequence begins with acid rocks as it does in about 35 per cent of the areas.

In a previous paper [6] it has been pointed out that the great majority of the batholiths consist of two types of granite, an earlier, gray, somewhat more basic type forming an outer zone with a reddish and more acid type in the interior. In some places the latter type intrudes the former, but generally there is a gradation, and in all cases they were undoubtedly derived from the same magma. The cause of this difference in sections of the batholith has been explained by different geologists as due to differentiation, and to assimilation of intruded rock. The evidence in most cases seems to favor the former explanation.

In his descriptions of the space relations of these batholiths to the gold deposits, Emmons has demonstrated that no deposits may be expected in the main batholiths where exposed in areas of more than 12 to 15 miles in diameter, because they have been removed by erosion. The veins are assumed to have formed in and around cupolas, domes and other projections on top of the batholith by rising solutions. In the Precambrian areas the gold commonly has been formed from solutions rising upward along limbs of synclines, in roof pendants and other related structures in the

[6] Moore, E. S. and Charlewood, G. H.: Two-granite batholiths in the Precambrian. Trans. Roy. Soc. Can., 27: Sec. IV, 1933.

troughs between the main, exposed portions of batholiths, but in the vicinity of apparent apophyses of the batholith. Some of these troughs in which the ore deposits originate extend thousands of feet below the original peaks on the batholiths. At what depth then, within the magma, does the rest-magma in which the metals are concentrated develop? It must be very deep, and is it not probable that it escapes more frequently through the thinner shell of crystallized granite in the trough than through the higher reaches of the batholith where the shell would be thicker? It would appear from field relations that the main function of the cupolas and domes is a structural one in producing channelways by intrusion and through squeezing of other rocks around these resistant masses during movements that occur during the last phase of the igneous activity and which accompany ore deposition in so many of the fields.

In addition to the granites there are found in the Algoman diorites, granodiorites, monzonites, syenites, quartz porphyries, feldspar porphyries, aplites, alaskites, albitites, pegmatites, lamprophyres, and in a few areas diabase dikes. In at least three important gold fields albitite dikes occur, and they are associated with either diorite or monzonite intrusions in each case, indicating that they arise from magmas comparatively rich in soda. In several fields the gold-quartz veins are closely related to pegmatites. The syenites are found mainly in dikes and small bosses and not in direct gradational contact with the granites. Aplites and lamprophyres occur together but more comonly not, and the aplites seem to be more common in areas where the granites were not preceded in the sequence by basic rocks. These conditions indicate that the idea of the complementary dikes forming by a simple splitting of magmas into acid and basic counterparts is scarcely valid.

The relation of certain quartz-diabase dikes in some of the Algoman sequences is not so easily explained as that of the other rocks mentioned. There is a large area in Ontario and Quebec, centering roughly near Kirkland Lake, where such dikes are numerous. They were at first thought to represent an independent epoch of igneous activity and designated Matachewan diabase, although they have never been found separated from the Algoman rocks by intervening sediments. Todd,[7] however, mentions three of these dikes in the Kirkland Lake area, and one of them on the Teck-Hughes property is displaced at least 1300 feet by the fault on which the ore bodies occur. This indicates that the dikes preceded the ore which for various reasons is regarded as Algoman in age. In an unpublished thesis for the University of Toronto, A. W. Derby has presented the results of a field and laboratory study of the igneous rocks of the Kirkland Lake area, and an area extending to the southwest of it. He concludes that there are gradational phases between the Algoman and Matachewan rocks as shown by accessory minerals and field relations. In other areas they give the impression of being quite independent of each other.

These diabases are practically devoid of any evidences of having played a part in metallization. They are generally porphyritic and uniform in character. Like the olivine diabases in the Keweenawan, previously described, they can scarcely be considered end products of the granite magmas as are the lamprophyres. They occur over a large area and seem to have been drawn from a large magma of fairly uniform character. Where they grade into the Algoman rocks there must have been a common center of eruption.

THE RÔLE OF THE LAMPROPHYRES AND OTHER BASIC ROCKS IN ORE DEPOSITION.

The close field relations of lamprophyres and other basic rocks to ore deposits has frequently been observed. In 1925 Spurr[8] pointed out the common occurrence of such rocks in magmatic sequences and postulated a temperature of formation of basic rocks associated with copper-zinc-lead ores as 400–500° C. This is in keeping with recent figures given by Sosman[9] for a peridotite dike cutting coal seams in Pennsylvania. A study of the heating effects of this dike on the coal indicated that the coal could not

[7] Todd, E. W.: Kirkland Lake gold area. Ont. Dept. Mines, 37: Part 2, 1928.

[8] Spurr, J. E.: Basic dike injections in magmatic sequences. Bull. Geol. Soc. Amer., 36: 545–582, 1925.

[9] Sosman, R. B.: Evidence on the intrusion-temperature of peridotites. Amer. Jour. Sci., V, 35–A: 353–359, 1939.

have been heated to more than 550° or to less than 480° C. Since petrographers have considered that dry melts of highly basic rocks require temperatures of about 1900° C. to liquify them, and ore deposits mostly form at temperatures below the maximum mentioned by Sosman, these figures are of special interest to students of ore deposits. Sosman suggests that the peridotite dike may have been intruded as a hot, plastic, fluid mud in which crystals were lubricated by films of liquid.

Hulin [10] has been a strong advocate of the derivation of ores from basic phases of magmas. He concludes, after an investigation of a large number of mining camps, that the minor acid intrusives are generally followed by minor basic intrusives and these by the ores. He has also expressed the opinion that where the basic intrusions are the last in a sequence, gold is late in the paragenesis of the ore deposits and where the acid intrusions are the later formed rocks, the gold is earlier.

In a recent paper, Spurr [11] observes that Hulin's deduction regarding the genetic relations of the ores and basic magmas is difficult to apply and that in many cases no such definite relations are found. It is generally accepted by geologists that gold is almost always late in gold-quartz deposits and Mawdsley [12] has recently discussed several gold camps in the Canadian shield in which the evidence of late gold is quite definite. An examination of the rock sequences in these camps does not appear to show that the presence or absence of basic rocks has affected the situation.

The common occurrence of lamprophyres in the Algoman sequences has been a matter of considerable interest to those who have studied these rocks. In my study of the Algoman sequences, lamprophyres were found in 50 per cent of the areas investigated, and allowance should no doubt be made for the possible failure of the geologist to find such dikes in all cases since they are mostly small and in many places much weathered. They occur more than one and one-half times as frequently as aplites. In three areas

[10] Hulin, C. D.: Metalization from basic magmas. Univ. Cal. Pub. Geol. Sci., 18, No. 9: 233–284, 1929.

[11] Spurr, J. E.: Diaschistic dikes and ore deposits. Econ. Geol., 34: 41–48, 1939.

[12] Mawdsley, J. B.: Late gold and some of its implications. Econ. Geol., 33: 194–210, 1938.

they cut the quartz veins and in 35 per cent of the areas they are earlier than the veins but closely related in sequence. In another 28 per cent they are earlier than some of the acid rocks and in an equal number of areas they are among the earliest rocks of the sequences.

The lamprophyres of the Algoman, like those of later geological systems, constitute a group of rocks of quite varied composition. The minettes, or mica-lamprophyres, are the most common, with hornblende types next in order, and augite and olivine varieties much rarer. Vogesite, kersantite, camptonite, and syenitic and diabasic lamprophyres have all been mentioned. Gradations from basic syenite to lamprophyre is found at Kirkland Lake and Cooke [13] has stated that in the Larder Lake area these rocks occur in series, the older dikes being more basic than the younger.

The origin of lamprophyres has received much attention from petrographers. Bowen [14] states that Niggli and Berger, after a special study, concluded that they resulted from remelting of local accumulations of early crystals, thus producing liquids of lamprophyric composition. He indicates that lamprophyres are such a broad and ill-defined group that it is difficult to make a general statement concerning them, to which exception might not be taken. He emphasizes their femic and porphyritic character and alkalic groundmass, and proceeds to show how olivine types might be developed in accordance with the *reaction principle.* According to this principle no liquid of the bulk composition of the rock could be developed but the end product of the process would still give an olivine lamprophyre from early-formed olivine crystals supplemented by alkalic liquid. The suggestion of Sosman for the origin of the peridotite dike in Pennsylvania seems to satisfy the requirements for the origin of the olivine lamprophyres when they occur very late in the sequence, and they may be regarded as having been expelled just before the door of the magma was finally closed.

A consideration of the other types of lamprophyres found so commonly in the Algoman indicates that they fall satisfactorily

13 Cooke, H. C.: Kenogami, Round and Larder Lake areas, Timiskaming district, Ontario. Can. Geol. Surv. Mem. 131 : 43, 1922.

14 *Op. cit.,* p. 258.

into Bowen's [15] *reaction series,* since biotite should form most abundantly with increasing alkalic character of the solutions, and the hornblende next in order of abundance.

As to the rôle of the lamprophyres and related basic rocks of the Algoman system in the formation of the gold deposits, the evidence points to a situation comparable to that so well described by Buddington [16] in the Ore Deposits of the Western States, where he says "but the hypothesis of any but a subordinate genetic connection between lamprophyres and metallization, except in so far as both are connected with the same magmatic cycle and related structural dynamics, is unnecessary, and is without adequate support at the present time." In Porcupine, the largest gold camp in Canada, there are no basic rocks that one could consider closely related to the gold deposits, and the same is true of other important camps. The fact that the materials forming the lamprophyres escaped from the magma indicates that channelways were also available for the ore-bearing solutions. Further, these rocks commonly exert a marked structural influence in the formation of the gold deposits because of their toughness and lack of fractures, a feature well illustrated at Kirkland Lake and in other gold camps. Dikes of these rocks have served in a number of areas to impound the ore-bearing solutions and increase concentration of gold adjacent to them. The conclusion that one may draw from their occurrence in the Canadian shield is that they, and also the minor acid intrusives, are simply indicators of the progressive steps in the progress of differentiation toward a more alkalic and siliceous magma with simultaneous concentration of the metals. The metals remain in solution to the end of the sequence, to be finally expelled from the magma under great pressure along with the remnants of any other constituents, femic or salic. Whether these are acid or basic does not seem to affect profoundly the results in so far as the gold deposits are concerned, so long as the end products are rich in alkalies and free silica.

15 *Op. cit.*, p. 59.

16 Buddington, A. F.: Ore Deposits of the Western States, Lindgren Volume, A. I. M. E., p. 379, 1933.

The great gold-quartz lodes of the Homestake mine at Lead, South Dakota, were discovered in 1876 and have produced a large proportion of the lode gold of the United States (more than 30 million oz). The Homestake deposit comprises a complex of inclined mineralized pipes, pods, veins, saddles, and lenses developed along the crests of plunging folds within the Precambrian Homestake formation. The auriferous orebodies comprise abundant veins and masses of quartz in chloritized portions of cummingtonite or sideroplesite schist of the Homestake formation, all containing disseminations of pyrrhotite, pyrite, arsenopyrite, and gold. D. H. McLaughlin, onetime consulting geolgist for the Homestake Mining Company and later its president, described the geology of the Homestake deposit and concluded the following as regards its origin (McLaughlin, 1931 pp. 328-329):

> *Origin*—The Homestake orebodies have been considered by the majority of geologists who have studied them to be of deep-seated origin and of pre-Cambrian age. The mineral assemblage and textures, the remarkable uniformity in character of ore from the surface to the present lowest workings, the interruption and displacement of the orebodies by the (Tertiary) rhyolite dikes, the presence of fragments of quartz and arsenopyrite in rhyolite breccias, the occurrence of detrital gold (as well as later introduced gold) in the Cambrian conglomerate and sandstone, and the contrast between the Homestake ore and the Tertiary siliceous gold ores in the northern Black Hills are among the arguments that force acceptance of pre-Cambrian origin.
>
> The hypothesis that the ores were of Tertiary age and related to the rhyolite intrusives has been warmly supported by a few geologists. The distribution of rich ore with respect to the rhyolite dikes, the richness of orebodies in blocks within the dike zone, the occurrence of gold ores of unquestioned Tertiary age in the overlying Cambrian dolomites, the presence of gold in some pyritic concentrates from the rhyolite dikes (about 0.1 oz. per ton of pyrite), and the occurrence of arsenic in both Homestake and Tertiary ores are the major arguments that have been advanced. The points made are indeed suggestive, but on careful analysis none of them is convincing or can stand against the opposing evidence.
>
> The ultimate source of the solutions from which the gold was derived is unknown. The evidence indicates, however, that it was probably deep-seated and was tapped only by a few outlets along specific structures. Channelization of rising solutions along relatively permeable rock in the accessible folds is believed to be responsible for the localization of the ore. The arches of dense Ellison slate over the plunging anticlines of Homestake formation, the intense crumpling of the somewhat brittle layers of sandy quartzite in the Homestake formation, and the incipient fracturing related to the folding are all significant factors in providing the channelway and confining the gold to it. Furthermore, the rocks of the Homestake formation afford conditions favorable for alteration by the hydrothermal agents and for the precipitation of gold. The remarkably selective replacement in the Homestake beds is a fact of observa-

> tion, though the full explanation is perhaps still to be given. The presence of cummingtonite, which appears to be necessary for the formation of chlorite, and the presence of chlorite, which certainly was an important factor in the precipitation of gold, are probably necessary mineralogical conditions for ore.

The reader will note from this discussion that a controversy exists as regards the age of the mineralization of the Homestake lodes. Slaughter (1968, p. 1459) comments as follows on this aspect of the origin of the gold:

> Gold, with its accompanying mineralization, is localized in the Homestake Formation in zones of cross folding. The cross folding increased permeability of the schists and thus provided access to various parts of the folded band of sideroplesite and/or cummingtonite schist by the mineralizing solutions.
>
> The differences in chemical composition between the Homestake Formation, on the one hand, and the other formations in the district, on the other, probably were sufficiently important essentially to have confined the deposition of gold and most of the other hydrothermally introduced minerals to the Homestake Formation. The emplacement of gold and the minerals associated with it in much the same abundance in both sideroplesite and cummingtonite schists suggests that the iron-magnesium ratio in the Homestake Formation was a more important control than were the actual rock-forming minerals present. Had the Homestake Formation, however, not been opened through dilation in the zones of cross folding, the favorable character of the beds of that formation would have been largely, if not entirely, wasted.
>
> Gold is associated with all four stages of mineralization. It is later than any mineral with which it is associated. It can be interpreted as having been deposited in the fourth stage only, replacing minerals of the earlier stages. Whereas some gold accompanies the fourth stage of mineralization which is later than the Tertiary dikes, it is most abundant in its association with quartz, chlorite and arsenopyrite of the first stage.
>
> Much of the vast amount of vein quartz present in the ore bodies was derived from the Homestake Formation and moved only short distances.
>
> The question of age of the gold and related minerals cannot be positively answered at this time. It is either all Tertiary or part Tertiary and part Precambrian. That the gold is all Precambrian in age seems quite improbable.

Recently, on the basis of lead isotopes in galena from quartz veins, Rye, Doe, and Delevaux (1974) concluded that the ore in the Homestake mine was emplaced by metamorphic secretion from a 2.5 billion-year-old source (i.e., the Homestake formation) during regional metamorphism and intrusion that occurred 1.6 billion years ago. The lead isotopes in Tertiary galenas from Paleozoic and Precambrian host rocks were found to have had different growth histories. The

Tertiary galenas in Paleozoic rocks did not derive their lead entirely from Tertiary intrusive rocks in the area, but rather the lead may have been obtained mainly or in part from the Paleozoic rocks. Tertiary galenas in Precambrian rocks, on the other hand, seem to have obtained their lead largely from the transected metasedimentary host rocks or from the underlying basement, with little or no contribution from other sources (see also chapter 13).

Gold-quartz deposits, mainly veins, stockworks, and saddle reefs, essentially in graywacke-slate assemblages, are common in auriferous Precambrian, Paleozoic, and Mesozoic belts throughout the world. Classic examples occur in Paleozoic rocks in the Bendigo and Ballarat areas in Victoria, Australia, and in the Paleozoic Meguma Group in Nova Scotia, Canada. They have commonly been referred to as "Bendigo type" (see chap. 1).

These deposits have been variously interpreted as due to consolidation from igneous melts (Belt, 1861; see chap. 9), lateral secretion (Gilpin, 1888), and precipitation from magmatic hydrothermal solutions. E. R. Faribault (1860-1953) of the Geological Survey of Canada mapped and studied the gold-quartz deposits of Nova Scotia for more than fifty years, and considered them to have originated from ascending hydrothermal solutions, the source of which was uncertain. His detailed views on the origin of the deposits have been compiled by Malcolm (1976) and are presented as Paper 10-3.

(Text continues on page 279.)

10-3: GOLD FIELDS OF NOVA SCOTIA

W. Malcolm

Reprinted from *Canada Geol. Survey Mem. 385,* 1976, pp. 52-54.

CHAPTER IV

Genesis

The veins were formed in the openings produced by the movements of the strata. During the folding of the interstratified beds of slate and quartzite, or shale and sandstone, there was a certain amount of slipping of one bed over another. This slipping produced openings along the bedding planes, which were in general widest at the apex of the fold and decreased in width down the limb until at a depth of a few hundred feet they pinched out. During or subsequent to the formation of these openings, which took place within the less resistant beds, the vein filling was introduced by solutions. Thus is explained the dependence of vein distribution on rock structure. The arching of the rocks on closely folded symmetrical domes produced fissures passing over the apex down each limb; on broad domes the arches were not strong enough to sustain themselves and the fissures were formed only on the limbs; on unsymmetrical domes the slipping of the strata was such as to produce fissuring along the bedding planes of the limb with the higher angle of dip; and subordinate flexures in which the strata were given a curve of less radius than ordinary were especially favourable to the production of fissures.

The process of folding was long continued, and the deposition of vein matter probably took place during the process. Small fissures were formed along the bedding planes and filled with quartz, only to be followed by other parallel openings between the quartz sheet and the slate and further precipitation of quartz in the new openings. Films of slate adhering to the quartz forming the wall of the new fissure thus became embedded in the vein. A succession of such events produced the laminated character of the interstratified veins. Another explanation that has been given of the laminations is that quartz was deposited in the slate along a number of parallel planes lying close together in an area of minimum pressure and that the quartz films increased in thickness through a widening of the spaces either by the folding of the strata or by metasomatic replacement.

The origin of the corrugations is dependent on the rock folding and the following explanation has been suggested. Many veins were formed long before the folding processes were completed and during the subsequent stages they were subjected to the same forces as the rocks. The main forces that brought about the folding were horizontal, and produced a tendency towards a thinning of the beds on the limbs and a proportionate thickening at and near the apex. This expressed itself in a motion of the more plastic beds from the limbs towards the apex. Any quartz veins already formed in the slate partook of the same lateral motion and became corrugated. These corrugations are of the nature of drag-folds, the higher bed of quartzite on an anticline having moved upward with regard to a lower bed and effected a dragging motion in the intervening slate and quartz.

Origin of Mineral-Laden Solutions. Three different opinions have been held as to the origin of the minerals by which the fissures were filled: (1) that they were deposited from descending solutions; (2) that they were dissolved out of the country rock; (3) that they were deposited from ascending solutions. Little evidence has been adduced in favour of the first, and the two most generally held are the second and third.

The lateral secretion theory found an exponent in Gilpin, (Proc. and Trans. Royal Soc. Canada, vol. VI, sec. IV, p. 63) who points to the fact that of the two kinds of country rock the slates alone carry gold of any appreciable amount. He also points out that most of the interbedded veins occur in slate and the richer parts of the fissure veins commonly follow the intersection of the vein with slate beds. He expresses the opinion that "so far as the subject has received attention the slates appear to be the source of the gold. The metal, in common with various metallic compounds, may have been carried and deposited in the layers as they were forming. That which fell in the sand would, presumably, for the most part, accumulate in the underlying bed of denser material, forming the first stage in the concentration now presented." He suggests "that the gradual deposition of gold from currents in the beds of clay or mud and sand might, through special currents, be accelerated or specially increased at certain points, and that from this enriched material the veins derived their 'pay-streaks'."

Woodman (Proc. Boston Soc. Nat. Hist., vol. 28, No. 15, pp. 391 and 395) expresses the opinion that most of the quartz of the interstratified veins was introduced rapidly by ascending hot solutions, and that possibly a small proportion of the gold had a similar origin. Regarding the origin of the sulphides in the country rock and in the veins, and the genetic relations existing between the two, the evidence adduced so far has been insufficient to lead to any final conclusion. But the "method of occurrence of gold in the veins of this series, its distribution in the country rock, and its relations to sulphides point strongly to the conclusion that at least a large part was deposited in the sediments and has been long in process of concentration in veins by water which comes downward from the surface. It is possible that not all the gold in a region of so complicated a history has the same source; but while some may have been brought up with the quartz, the facts so far observed do not show that more than a small share of it had that origin."

These remarks do not apply to the productive cross veins or fissure veins, for in these "the structure of the veins and the character and positions of the accompanying minerals point strongly to a deep-seated origin for the metal."

Faribault and others are of the opinion that the veins were filled by ascending solutions. These found a passage upward through the fractured portions of the domes. A fracturing across the bedding, as well as fissuring along the bedding planes, seems to have been necessary for the formation of veins and ore deposits; veins are not commonly found along straight, nonplunging anticlines, although there was, no doubt, a great deal of fissuring along the bedding planes; on the other hand, where the anticlines plunge and the rocks were fractured across the bedding, veins are abundant. The cross fractures are themselves filled with quartz forming the angulars entering and leaving the interbedded veins. The cross fractures seem, therefore, to have provided channels for the passage of solutions across the beds of quartzite and slate to the interbedded fissures along which deposition took place. That the solutions entered by way of the angulars is borne out by the fact that the rich parts of interbedded veins are those parts lying between the line of entrance of an angular and the line along which it leaves the main lead.

The source of the ascending solutions is not known. Near the outlet of Moose lake in the western end of Mooseland gold district a few auriferous interbedded veins have been traced to the granite, but without increase in size or other irregularity. At other places, as Country Harbour and Forest Hill, interbedded veins are cut by dykes of granite, and the proximity of the intrusion appears to have had little or no effect on the size or richness of the veins.

Dawson (Acadian Geology, Third Edition, Supplement, p. 85) expresses the opinion that the granite intrusion and the formation of the gold veins may have been "roughly contemporaneous." It has been suggested, also, that, as the cooling and solidification of the granite was long continued, the auriferous veins may have been formed by solutions given off from one portion of the granite mass and afterwards cut by dykes given off from other portions of the mass that were a little later in solidifying. Or it may be that veins were formed from solutions given off by one granitic intrusion and cut by dykes from a somewhat later intrusion.

Further light may be thrown on the problem by a study of those deposits where it is said feldspar forms a part of the gangue, as at Lower Seal Harbour, or where there are numerous veins carrying mica as at Cochrane Hill, Crows Nest, and Forest Hill. A study of the genetic relations of the interbedded scheelite-bearing veins of Moose River may also be of service in this connexion.

In conclusion it must be said that, although certain field relations indicate that the veins were formed prior to the granite intrusion, the question of the source of the solutions is still open.

Precipitation. Little study has been given to the cause of the precipitation of the metallic contents of the veins. Certain slate beds apparently exercised a greater precipitating effect than others. These are generally black and are frequently impregnated with arsenopyrite, pyrite, or pyrrhotite, and the interstratified veins that can be worked with profit are usually found in beds of this character. In some cases the cross veins also are found to be enriched where they lie in contact with strata of this class.

With only slight modifications this summary succinctly describes the origin of Bendigo-type deposits throughout the world. The precise relationship of the gold-quartz deposits to the granitic bodies has long been the subject of much discussion. On this problem, following studies in Nova Scotia and elsewhere, I have remarked (Boyle, 1979, p. 280) as follows:

In detail the precise relationships of the gold deposits in this category to granitic bodies invading the sedimentary terranes are often not entirely clear in many auriferous belts. In many sedimentary terranes the metamorphic, granitization and vein-forming processes appear to be coeval or nearly so. In other regions, however, particularly in greywacke-slate terranes, some of the quartz deposits appear to be truncated by the granitic bodies, and xenoliths of the quartz veins containing arsenopyrite and pyrite occur in the granitic rocks. In a few places (e.g., Yellowknife) some of the auriferous quartz veins are cut by granitic dikes or pegmatites. Such relationships are often observed where there are two ages of granitic rocks in the terrane as in the Precambrian Yellowknife Supergroup and in the Ordovician Meguma Group of Nova Scotia. Commonly the first phase is a biotite granite whereas the second is a muscovite granite usually with a plethora of associated pegmatites. Where these geological features are present the sequence of events seems to have been:

1. Initial stage of folding of greywacke and slate with the concomitant effects noted in 2.
2. Rise of the geotherms through the sedimentary pile due to metamorphism and granitization at depth. This is attended by the migration of silica, sulphur, arsenic, gold and other constituents into available dilatant zones to form the saddle reefs, leg reefs and interbedded veins with their contained auriferous sulphides and native gold.
3. Continued folding resulting in the corrugation of the quartz veins, formation of quartz boudins, ptygmatic veins and a general crushing and recrystallization of the quartz. During this period remobilization and reconcentration of the initial gold in dilatant sites (shoots), commonly as the native metal, may take place in the quartz bodies. In addition further increments of gold, sulphur, arsenic, silica and other constituents may be provided by the prevailing diffusion currents.
4. Continued granitization at depth followed by the injection of high level granitic stocks and small batholiths. The development of pegmatites commonly takes place during this stage.
5. Late stage minor folding and faulting.

The Mother Lode system of California comprises two-score mines or more in a belt some 120 mi long and about 1 mi wide on the western foothills of the Sierra Nevada (Knopf, 1929; Clark, 1970). The deposits are quartz veins and bodies of pyritized and mineralized country rock in a system of linked and anastomosing faults, fractures, and shear zones cutting a complex of Jurassic and Paleozoic (Carboniferous) greenstones, serpentinites, slates, schists and granodiorite, the last probably of Cretaceous age. The principal minerals are coarse-grained milky quartz, ankerite, calcite, mariposite, pyrite, arsenopyrite, galena, sphalerite, chalcopyrite, tetrahedrite, scheelite, molybdenite, native gold, and various tellurides. Alteration is intense in the greenstones and serpentinites, being mainly carbonatization (ankeritization), pyritization, and sericitization.

In his great memoir on the gold deposits of the Mother Lode system Knopf (1929) discusses their origin as given in Paper 10-4.

(Text continues on page 286.)

10-4: THE MOTHER LODE SYSTEM OF CALIFORNIA

Adolph Knopf

Reprinted from *U.S. Geol. Survey Prof. Paper 157*, 1929, pp. 45-48.

ORIGIN OF THE GOLD DEPOSITS

The Mother Lode veins, as now long recognized and as amply confirmed by the present investigation, occupy fissures that were formed by reverse faulting—that is, by faulting in which the hanging wall has moved up relatively to the footwall. It is interesting to recall, however, that in 1913 the Kennedy Extension Gold Mining Co. attempted to wrest the title to the Argonaut vein from the Argonaut Mining Co., which had been operating 20 years on the vein, on the plea that the Mother Lode vein occupies a normal fault. The court rejected the contentions of the experts on both sides and confirmed the title of the Argonaut Mining Co. to the vein, on the ground that it is manifestly unjust to take away a property, on the basis of a geologic theory, after 20 years of undisturbed possession (Min. and Sci. Press, vol. 109, pp. 61-64, 1914). The faults dip less steeply than the slates. The powerful compressive force that produced the faulting has flattened the slates against the veins and produced a belt of schistose material along the fault zone. Naturally, then, in any exposure in a drift on a given vein the vein appears to lie parallel to the structure of the inclosing country rock, and for this reason few of the operators on the Mother Lode realize that the veins as they are followed downward gradually cut through belts of rock of diverse character.

The displacement along some of the faults amounts to 375 feet, but this total displacement is the cumulative result of a considerable number of displacements that occurred over a considerable span of time. That the present structure of the veins proves that they were opened intermittently was recognized by Ransome (Ransome, F. L., U.S. Geol. Survey Geol. Atlas, Mother Lode District folio (No. 63), pp. 7-8, 1900). The observational evidence for this conclusion lies in the ribbon structure of the veins and the quartz veinlets cutting older quartz. That the compressive forces that formed the fissures acted during a long span of time, while the vein filling was being deposited, is proved by the rotation of the ankerite augen in the slate wall rocks.

From the fact that the dislocations produced during historic earthquakes are not known to exceed 49 feet and as a rule are much smaller Hogbom (Hogbom, A. G., Zur Mechanik der Spaltenverwerfungen; eine Studie über mittelschwedische Verwerfungensbreccien: Geol. Inst. Upsala Bull., vol. 13, pp. 391-408, 1916) concluded that faults of much larger magnitude than 49 feet are the results of successive movements. This conclusion appears to be verified geologically by the fact that fault breccias of the larger faults show that they have been repeatedly crushed and recemented. Incidentally, it should be mentioned that Hogbom (Idem, p. 398) believes that the completely isolated fragments of country rock that occur in the cement of fault breccias have become thus isolated as a result of repeated brecciation and recementation. Flett (Flett, J. S., Notes on some brecciated stanniferous veinstones from Cornwall: Geol. Survey England and Wales Summary of Progress, 1902, pp. 154-159, 1903) found that the filling of the Cornish tin lodes, which according to geologic conceptions must have formed within a very short interval of time, shows abundant evidence of repeated crushing and recementation. Moreover, the same conclusion—namely, that a large displacement is the summation of a considerable number of minor displacements—follows as a logical consequence of the elastic rebound theory of faulting developed by Lawson (Lawson, A. C., Report of the California Earthquake Commission, vol. 1, pt. 1, pp. 147-151, 1908) and by Reid (Reid, H. F., Elastic rebound theory of earthquakes: California Univ. Dept. Geology Bull., vol. 6, p. 413-444, 1911) from their studies of the San Andreas fault of California. According to this theory faults are due to an elastic rebound on a rupture plane in the crust on which strain is suddenly relieved. The strain is slowly generated during a period of years, and when the strength of the rocks or the frictional resistance to movement on an old rupture is exceeded the rocks rupture. At the time of rupture the displacement is limited to the amount necessary to relieve the accumulated strain. The displacement is therefore proportional to the length of the fault along which the strain accumulates, and to judge from the small amount of elastic distortion that rocks can undergo before they rupture the displacement must necessarily be small—a conclusion that harmonizes with the relatively small displacements known to occur at the time of historic earthquakes.

The great linear extent of the Mother Lode system appears less remarkable since the results of the investigations of the San Andreas fault of California have become known. Here is a fault that has been traced for 600 miles, that is known positively to have been ruptured for a distance of 190 miles and probably 270 miles in 1906, and that traverses rocks of all degrees of strength, from granite to weak sedimentary rocks. The position of the San Andreas fault manifestly was not determined by a belt of weak rocks but by a deeper-seated, fundamental cause. Auxiliary fractures occur in a zone adjacent to the San Andreas fault (Lawson, A. C., op. cit., pp. 53-57).

The location of the Mother Lode fissure system, like that of the San Andreas fault, appears to have been determined by a deeper-seated, fundamental cause than a weak belt of rocks. It traverses rocks as diverse as slate, greenstone, amphibolite schist, and serpentine. In places there is evidence that the veins occupy auxiliary fractures in a zone parallel to a great reverse fault. This evidence is best shown near Plymouth, where amphibolite schist has been thrust up over Mariposa slate. Some mineralization has taken place along the master fault, but apparently it is nowhere of major importance. At Jackson the structural conditions appear to be like those at Plymouth, whereby the amphibolite schist has been thrust up over the Mariposa rocks. The fault zone is marked by a wide silicified belt, as at Jackson Gate, and 1½ miles farther south an important mine, the Zeila, was located on it. At the Eagle Shawmut mine, as at the Zeila, the main ore bodies were formed in or closely adjacent to the main reverse fault.

The movement along the fissures, which are markedly sinuous, would, according to the time-honored explanation of Werner, produce open cavities. It is argued by some that cavities would not stay open in slates, but drifts and crosscuts driven through normal Mariposa slate that has not been reduced to gouge remain open indefinitely. However, if the open cavity were large, some adjustment by gravity faulting would most likely take place. This possibility is strongly suggested by what is happening throughout the Argonaut mine as the result of stoping out large shoots of ore. Drill holes show normal faulting; horizontal fissures as much as 3 inches across have opened along joint planes; and highly suggestive of the way in which the ribbon structure of the veins was formed, some of the fissures have opened along horizontal quartz veinlets, the quartz with more or less black slate adhering to its upper surface forming the footwall of the open fissure. Granted that an open cavity existed before the main vein was filled with quartz, subsidence and the resulting opening of joints accounts perfectly for the horizontal quartz veins that in many places cut across the slates in the stringer halo of the main vein.

In recent years many explanations have been advanced to account for the quartz filling of veins of the Mother Lode type. The idea advocated by Lindgren (Lindgren, W., Characteristic features of California gold quartz veins: Geol. Soc. America Bull., vol. 6, p. 229, 1895) that the clean quartz in a vein is the result of the filling of cavities has proved unacceptable to many. Injection of a quartz magma in the form of a dike is an idea that has been revived, replacement is an alternative idea, and the pushing apart of the walls of a fissure by the force of crystallization of the growing quartz is still another. It has also been suggested that the original filling of the fissure, especially the gouge, may have been more nearly removed where the ascending currents of the ore-forming solutions were flowing most swiftly (Quiring, H., Thermenaufstieg und Gangeinschieben: Zeitschr. prakt. Geologie, Jahrg. 32, p. 166, 1924). Some, like Tronquoy (Tronquoy, R., Contributions à l'étude des gîtes d'étain: Soc. min. France Bull. vol. 35, pp. 456-465, 1912) in explaining the tin-bearing quartz veins of Villeder, France, have gone so far as to maintain that the quartz was injected as a jelly and then slowly became coarsely crystalline. Although this hypothesis appears to account satisfactorily for the "floating" or isolated "unsupported" inclusions of wall rock in the veins, yet it seems not applicable to the Mother Lode veins, because nowhere do they show any colloform structures; vugs are common, but they invariably have sharply angular surfaces instead of the cauliflower-like surfaces common in gel minerals that have become crystalline. Irregular shrinkage cavities might be produced by the loss of water during the crystallization of the gel, but they would be lined with chalcedony or fine-grained quartz and not with large crystals of quartz.

The hypothesis that the quartz veins are frozen magmatic injections—dikes, in short, or vein-dikes, as Spurr (Spurr, J. E., The ore magmas, vol. 1, pp. 76-85, 1923) calls them—is believed to be improbable because, first, the quartz veins nowhere show any evidence of the chilling in the form of fine-grained border facies, even if only narrow, that should be found in dikes that were injected into rocks relatively so cold as those of the Mother Lode belt; and, second, the enormous amount of replacement effected in the wall rocks is out of all proportion to the size of the veins and points to the long-continued flow of solutions through the fissures.

The relative importance of cavity filling, replacement, and the force of crystallization as factors in forming the quartz veins is an unsolved problem. The positive evidence, as pointed out on page 45, favors the predominance of cavity filling. Nevertheless the gross effect of the appearance of certain veins as viewed in the stopes or crosscuts is that they are the results of replacement. This strong suggestion of replacement origin is well shown by the quartz ore of some of the stopes in the Argonaut mine, as illustrated in Plate 5, A, which suggests quartz containing innumerable residuals of unreplaced slate. But the evidence as seen under the microscope is firmly against the idea of replacement of the slate by quartz.

Becker (Becker, G. F. and Day, A. L. The linear force of growing crystals: Washington Acad. Sci. Proc., vol. 7, p. 283, 1905) in 1905 suggested that the Mother Lode veins were opened by the force of crystallization of the growing quartz. Although it is highly improbable that the feeble force of crystallization could have crowded apart the walls of the veins to make room for the growing quartz, it is easily conceivable that after breccias had been formed in the fault fissure, or belts of schistose slate, or open spaces the force of crystallization thus working against no great resistance might pry apart the leaves of the slate and produce such closely spaced ribboned ore. But no criteria are known by which it can be established that the force of crystallization has been at work.

On the whole the theory that the Mother Lode quartz veins were developed by successive enlargements appears to fit the facts best. A later movement along the fissure might crush the earlier-formed quartz, and this would account for the fact that some of the gold occurs in crushed quartz. It is an interesting question whether such crushed quartz might not recrystallize under the influence of the later solutions and thus obscure or obliterate some of the evidence of successive movements along the fissure.

The great vertical extent of the Mother Lode veins indicates that they were formed by hot ascending solutions, as is shown also by their mineralogic character. They belong to the mesothermal group, formed under conditions of intermediate temperature, and the sporadic occurrence of pyrrhotite, tourmaline, and magnetite show that they are transitional to the hypothermal group. Probably the ore-depositing agency was hot water that carried in solution gold, silver, lead, zinc, and other heavy metals, also potassium, sulphur, arsenic, and carbon dioxide. The effect of the carbon dioxide and the potassium vastly exceeded that of the other constituents. The carbon dioxide liberated immense quantities of silica from the wall rocks, and this silica was delivered to the vein channels, where it was in part precipitated as quartz. The potassium was utilized in effecting the extensive sericitization of the wall rocks.

A question of much practical as well as scientific interest is what determined the very different behavior of the gold in the ore-forming solutions at different localities. In places the gold migrated freely into the wall rocks, forming the gray ore and the mineralized amphibolite schist ore. In places pyritized greenstone is valuable ore; in others, similarly pyritized and hydrothermally altered greenstone is valueless. Higher temperature of ore-forming solutions at some places than at others, the higher temperature giving the gold greater mobility, suggests itself as an explanation; but this suggestion appears to be negatived by a comparison with the mode of occurrence of the gold in the stringer lode of the Alaska Juneau mine at Juneau, Alaska; the gold there was deposited at a higher temperature than in the Mother Lode belt, but it is nevertheless confined to the quartz in the veinlets that ramify through the slates. It is this restriction of the gold to the quartz veinlets that makes possible a rough sorting of the ore and permits the successful working of an ore body that averages 63 cents to the ton.

We come now to the source of the ore-forming solutions. Long ago Richthofen, (Richthofen, F. von, Ueber das Alter der goldführenden Gange und der von ihnen durchsetzten Gesteine: Deutsche geol. Gesell. Zeitschr., Band 21, pp. 723-740, 1869) in a brilliant paper that still repays reading, maintained that the gold deposits of the Sierra Nevada were genetically related to the intrusive granite and were produced by emanations that issued from the deeper-lying portion of the magma, which was still fluid when that part of the granite now exposed to view had already solidified.

Ransome (Ransome, F. L., U.S. Geol. Survey Geol. Atlas, Mother Lode District folio (No. 63), p. 7, 1900) thought that "in all probability the waters which carried the Mother Lode ores in solution were originally meteoric waters, which after gathering up their mineral freight in the course of downward and lateral movement through the rocks were converged in the fissures as upward-moving mineral-bearing solutions." In the light of present ideas this explanation may be modified as follows. The carbon dioxide, sulphur, arsenic, gold, and certain other constituents were probably supplied by exhalations from a deep-seated consolidating magma, as was also a part of the heat. These substances and the heat were added to the meteoric circulation, which supplied the necessary motive power to lift the liquid solution to the earth's surface (Day, A. L., and Allen, E. T., The temperature of hot springs: Jour. Geology, vol. 32, pp. 184-185, 1924).

Which magma or magmas supplied the gold and other constituents is unknown. The attempt to ascribe their origin to gabbro, peridotite, and albite aplite dikes (Lindgren, W., Mineral deposits, 2d ed., p. 575, 1919) finds no support from the mode of occurrence of the gold along the Mother Lode belt; in fact, the belt is particularly poor in gold where these intrusions are most abundant. It appears to be a safe conclusion, however, that the ore-forming emanations were given off during the final stage of the epoch of plutonic intrusion at or near the end of Jurassic time. For wherever the age of the gold deposits in the Sierra Nevada can be even approximately dated it proves to be younger than the granodiorite intrusions, and the community of characters shown by the gold deposits over the whole region is so great that there can be little doubt that they are all of essentially the same age. Even the copper deposits, most of which are west of the Mother Lode belt, are of postgranodioritic origin. Reid's ascription of the hornblendite at Copperopolis as the ore bringer of the great copper deposits there (Reid, J. A., Econ. Geology, vol. 2, p. 414, 1907) has been proved erroneous by the subsequent discovery in the mine of large bodies of 4 per cent copper ore in the granodiorite, which is younger than the hornblendite. Lest the idea of a zonal arrangement of the gold and copper deposits of the Sierra Nevada be seriously entertained, it is well to point out that copper deposits—for example, in the Noonday mine, in Eldorado County, whose ore consists of massive pyrrhotite intergrown with chalcopyrite—occur within the Mother Lode belt itself. Obviously the occurrence of this pyrrhotitic copper ore, so unlike the geographically closely associated gold deposits, was not determined by so simple a principle as zonal arrangement around a center from which ore-forming solutions emanated. The true explanation is part of the larger problem of the Sierra Nevada considered as a metallogenic province, but this problem is one that awaits field study.

The Tertiary represents one of the major periods of epigenetic gold mineralization in the geological history of the earth. Likewise, this period represents a time when many of the great gold placers of the world were concentrated. Auriferous deposits of Tertiary age occur in the Cordillera of North and South America; in the Alpides of Europe and Asia; and in the Tertiary orogenic belts of the far eastern USSR and of New Zealand, Fiji, New Guinea, Papua, Indonesia, China, Philippines, Japan, and elswhere along the Pacific Rim.

The hypogene Tertiary deposits of the western United States have been extensively mined and studied for many years. Most are gold-quartz and allied types, often with an abundance of silver and base metal sulfides and sulfosalts. Here belong the deposits of Republic in Washington; Tonopah, Rawhide, Goldfield, Comstock Lode, Tuscarora, and Jarbidge in Nevada; Bodie in California; the Black Mountains in Arizona; De Lamar in Idaho; the San Juan region (Telluride, Ouray, Silverton, Lake City, Rico, Needle Mountains, La Plata and Creede) and Cripple Creek in Colorado; Pachuca, El Oro, and Guanajuato (Veta Madre) in Mexico; and a number of deposits in Central America. The disseminated Carlin, Pinson, Jerritt Canyon, Alligator Ridge, Cortez, Getchell and Gold Acres deposits in Nevada also belong in the listing, according to the latest dating investigations.

The host rocks of the Tertiary gold deposits of the Cordillera of the western United States are varied and include andesites, trachytes, rhyolites, and various porphyries of Tertiary age; Tertiary shales, limestones, and sandstones; and older Mesozoic, Paleozoic, and Precambrian rocks of sedimentary, extrusive volcanic, and intrusive igneous origin. The alteration processes are, likewise, varied depending on the rock types; characteristic, however, is the marked propylitization (chloritization, carbonatization, and pyritization) and sericitization manifest in the Tertiary andesites. In other rocks silicification, pyritizaton, sercitization, and alunitization are common. The gangue of the vein-type deposits is invariably quartz, often chalcedonic, banded or crustified, with abundant calcite in some places, rhodochrosite in others, and rhodonite, adularia, fluorite, and barite in a few deposits. The metallic minerals include pyrite, chalcopyrite, galena, sphalerite, argentite, polybasite, pyrargyrite, proustite, tetrahedrite, pearceite, silver selenides (naummanite), stibnite, molybdenite, bismuthinite, and wolframite. Arsenopyrite and pyrrhotite are relatively uncommon. The gold is generally free, although some is intimately associated with the sulfosalts. Much of the gold is extremely fine grained, giving a yellow hue to the quartz in some deposits. Tellurides are common as at Cripple Creek in Colorado. The Au/Ag ratio is relatively high in some deposits ranging from 5 to 40 with an average of about 10 at Cripple Creek; elsewhere and more generally the ratio is 1:1 or as low as 1:250 (Tonopah, Pachuca, etc.).

Many papers and memoirs have appeared on the vein-type and replacement deposits in this category since 1880, some advocating a lateral secretion origin (e.g., Becker, 1882), but most opting for a (magmatic) hydrothermal genesis. The classic memoir by Lindgren and Ransome (1906) on the Cripple Creek district, Colorado, is an example of the latter.

Cripple Creek, Colorado, is the best example of the gold telluride deposits of Tertiary age. The country rocks are Precambrian granites and gneisses broken through by a great mass of Tertiary volcanic rocks representing a large volcano. The core of this volcano is composed of tuffs and breccia of latite-phonolite, cut by dykes and masses of phonolite, syenite, monchiquite, and vogesite. The deposits are veins, mineralized sheeted zones, replacements in breccia and along fissures, and irregular pipes in mineralized breccia, cutting both the Precambrian and Tertiary rocks, but best developed in the latter. The principal ore mineral is calaverite, $AuTe_2$, with a silver content generally less than 4%. There is practically no hypogene native gold. Associated with the calaverite are small amounts of sylvanite, petzite, pyrite, sphalerite, galena, tetrahedrite, stibnite, cinnabar, molybdenite, and minor amounts of wolframite (huebnerite). Coloradoite has been identified in the primary ore, and acanthite, native gold, and jarosite in the oxidized zone. The gangue is quartz, fluorite, carbonate, and roscoelite (the vanadium mica). The vein structure is drusy. The alteration is pyritization, carbonatization, and propylitization. Adularia is developed in some veins.

Concerning the origin of the Cripple Creek ores Lindgren and Ransome (1906 p. 225) concluded:

> The waters which deposited the Cripple Creek veins were alkaline solutions containing the following compounds and ions, either free or in various combinations: SiO_2, CO_2, H_2S, CO_3, SO_4, S, Cl, F, Fe, Sb, Mo, V, W, Te, Au, Ag, Cu, Zn, Pb, Ba, Sr, Ca, Mg, Na, and K. We believe that at least some of the SiO_2, SO_4, Cl, Fe, Ba, Sr, Ca, Mg, Na, and K, are derived from the volcanic rocks by leaching of waters, while the remaining metals, as well as CO_2, H_2S, S, and some SiO_2, Cl, and K were more probably separated from intrusive cooling magmas at considerable depth, and brought up as solutions in magmatic water given off in the same manner.

The time of deposition of the ores was discussed as follows (p. 226):

> The ores were formed later than the latest actual eruptions; that is, later than the basic dikes. These dikes had solidified and had cooled at least to such degree that carbonates could form in them. Although the basic dikes to some degree followed the prevailing directions of the fissure system, the latter was not formed until after their intrusion. As the paths were opened they were filled by depositing solutions. The filling being in many cases only partial, we may infer that the solutions circulated for a limited time only.
>
> That the veins are not recent is indicated by the formation of considerable placer deposits and by the depth and extent of subsequent oxidation. If we assign a late Tertiary age to the close of direct volcanic activity, there is some reason for believing that the ore-forming epoch belonged to the close of that period.

Lindgren and Ransome (1906) also dealt in some detail with the source of the mineralizing solutions. Their discussion is of interest in the light of recent research based on hydrogen and oxygen isotopes (pp. 227-228).

> The ore-depositing water was derived either from the atmosphere or from magmas under diminishing pressure and temperature or from both of these sources. In other words, it may have been a part of the

ground water descending through the pores and fissures of the rocks from the surface on which it once fell as rain or snow, and possibly ascending, charged with dissolved material, from the lowest levels reached under the driving force of the volcanic heat encountered there. Or it may have formed part of the original molten phonolitic rocks and may have escaped from its bond during the ascent of this molten rock to levels of less temperature and pressure. Both hypotheses are plausible, though at first glance the former view seems much more natural and simple. Either may be difficult to prove, but it may be profitable to consider the probabilities involved.

It will be shown in Chapter XII that the conditions of underground drainage are unusual. The porous, shattered volcanic mass is deeply sunk in much more massive and impervious granite and metamorphic rocks. It therefore holds water much as would a sponge in a cup. The circulation of the ground water in this volcanic plug is exceedingly slow; in fact, the water is practically stagnant. The cold dilute sulphate solutions which constitute the ground water are evidently wholly impotent to deposit ores like those in the veins or to cause abundant pyritization of the rocks. They fill many open fissures in the rocks, but nowhere have they given the least indication of depositing telluride ores.

If surface waters were present they must have constituted currents under the influence of the heat of the volcanic rocks. This circulation was only a temporary phase, ceasing when the rocks had cooled sufficiently.

Such waters ascending vigorously throughout the volcanic mass could not reasonably have been derived from the very limited surface area of that mass itself. They must have been derived chiefly from the surrounding granitic plateau. They must have percolated through the granite to great depths near the volcanic mass, and finally have been driven up by the volcanic heat still existing in it. Considering, however, the almost impermeable character of the granite, as demonstrated by the mining operations, it becomes very difficult, if not impossible, to conceive how a sufficient amount of water could penetrate the porous volcanic mass from the surrounding granite to give rise to the strong ascending current which evidently streamed upward in every available fissure in this old volcano.

The second hypothesis of the derivation of the vein-forming waters is that they were originally an integral part of the intrusive phonolitic magmas and were given off by release of pressure or by cooling and crystallization after the magma had ascended to higher levels.

According to the general laws of solutions, pressure increases the solubility of water in magmas, and conversely, if all magmas contain more or less water which is just as much a part of them as is the silica, for instance, it follows that a portion of the water will be given off during the eruption.

The volcanic rocks of Cripple Creek are rich in combined water. The average of the analyses shows 1.62 per cent combined water given off above +110 C., and they range from 0.69 to 2.09 per cent.

> To a large extent this is contained in analcite, the primary nature of which is proved. Some of the phonolites contain 15 per cent of this analcite and the latite-phonolites average 4.6 per cent. To a smaller extent the water is present in kaolin or other secondary hydrous minerals. It is assuredly not an exaggeration to say that the rocks contain an average of 1 per cent of combined primary water. The presence of primary water being firmly established, it follows, if the statement in the preceeding paragraph is true, that the magma contained much more of it at greater depths. The water lost as steam by the intrusive bodies under our present range of observation doubtless partly permeated the rocks and was partly dissipated in the air at the time of the eruption. But unquestionably there are large intrusive masses which did not attain the level of those now visible, though they rose to much higher levels than they originally occupied. Their water was probably partly expelled from the cooling mass, but was held in its confines under strong pressure, having abundant opportunity to dissolve the other substances which may have emanated from the magma. As the volcanic mountain settled down deep fissures were created, which reached to the levels of these stored hot waters and afforded them means of escaping toward the surface. Such is the explanation of ore deposition by "magmatic" or "juvenile" waters, to follow Suess's terminology, and to these we are inclined to attribute the largest share of ore deposition, possibly the whole. The storage reservoir was limited, and the supply of these strange solutions was soon exhausted. Surface waters followed the retreating juvenile waters and filled the "sponge in the cup" until equilibrium was established. At the present time the volcano appears extinct, and yet a few hundred or a thousand feet below the surface the faint exhalation of carbon dioxide and nitrogen are met—the last volatile products of the phonolitic magmas.

The disseminated replacement-type gold deposits occur mainly in carbonate-bearing rocks such as limestone, dolomite, calcareous quartzites, and calcareous shales. They may be of any age, but the most productive to date are of Tertiary age. The orebodies are commonly irregular, but some are tabular following specific beds or series of beds; some occur in extensively brecciated fault or crush zones. Most of the deposits occur in highly faulted terranes, and the orebodies generally spread out into beds laterally from faults or fractured zones. The replaced beds or series of beds are often disturbed but not generally highly brecciated or contorted. The mineralization is variable, but the development of quartz, pyrite, arsenopyrite, and small amounts of other sulfides and sulfosalts is universal. Silicification is the main process in their formation. The principal gold mineral in most of these deposits is the native metal, commonly low in silver. The native gold is usually widely disseminated through the deposits and is generally microscopic, in the range 0.5-5 μ; in some ores the electron microscope reveals that the gold particles range down to 0.005 μ in size. Ore shoots usually have to be outlined by assay.

Numerous examples of this type of deposit occur throughout the world. Some of the replacement orebodies in dolomitic limestone and phyllitic quartzite of the Salsigne gold mine in France belong in this category. The protores of the Kuranakh-type residual deposits in southern Yakutia, USSR are also of this type. The best examples are in north-central Nevada, where a number of deposits (Carlin, Pinson, Jerritt Canyon, Alligator Ridge, Getchell, Cortez, and Gold Acres) have been extensively investigated.

Numerous papers describing the geology and origin of the Nevada disseminated replacement-type gold deposits have appeared in recent years. Two of those on Carlin, the most investigated of these deposits, have been chosen for discussion.

The gold orebodies of the Carlin deposit are in the upper part of the lower Silurian Roberts Mountains formation, several hundred feet below the Roberts Mountains thrust. The orebodies are diffuse irregular sheets that follow the stratigraphy of the carbonate beds but also cut across the bedding. In places a relationship to faults is evident. The mineralization comprises fine-grained quartz, barite, pyrite, stibnite, cinnabar, realgar, orpiment, sphalerite, and galena in variable amounts. The gold is present as the native metal in a very finely divided state (0.5-5 μ and in some cases <0.005 μ). The elements introduced during mineralization include Si, S, As, Sb, Hg, Tl, Ba, Pb, Zn, Cu, Mo, B, W, Se, Te, Au, and Ag. Silicification, during which silica (quartz) has replaced carbonates, is the principal alteration process. The grade of the Carlin orebodies ranges from 0.24 oz to 0.31 oz Au/ton according to the published accounts. A Tertiary (Oligocene?) age is ascribed to the mineralization.

Hausen and Kerr (1968 pp. 910–911) have studied and described many of the features of the Carlin deposit. Their informative abstract states:

> Fine colloidal gold near Carlin, Nevada is disseminated in leached carbonated strata of the Roberts Mountains Formation in the Lynn "window" of the Roberts Mountains thrust fault. The ore body is generally stratiform and is more or less conformable to altered beds near the top of the formation, underlying Devonian limestones.
>
> Two sequences of mineralization are recognized: (1) an early base metal-barite assemblage related to early Cretaceous intrusives (121 $\pm$ 5 m.y.), and (2) a later low temperature Au-As-Hg-Sb assemblage of near surface emplacement. The earlier sequence consisting of sparse galena and sphalerite in barite with anomalous amounts of zinc, lead, nickel and copper, associated with dikes of dacitic composition, is of little economic importance. The later sequence of gold, realgar, cinnabar, and stibnite associated with extensive silicification and argillic alteration of limestone beds, has resulted in important deposits of gold.
>
> Argillic alteration of hydrothermally leached carbonate strata has provided the environment in which the most prominent gold deposition took place. Carbonate minerals in the limestone host rock have been replaced by microcrystalline quartz and chalcedony to form stratiform silica masses and recrystallized lenses of euhedral quartz. Zones of porous silicification are light gray to white, elliptical in shape, and more or less follow bedding. Silicification is bordered by argillic alteration and pyritization. Deposition of gold usually lies in a zonal pattern that encircles chimney areas of silicification.
>
> Ore textures, mineral assemblages, and alteration criteria in the Carlin ore body all favor late stage epithermal mineralization. Gold

> introduction is attributed to late hydrothermal solutions rising along elliptical conduits controlled by permeability of select horizons in the Roberts Mountains Formation. Precipitation of gold has occurred mostly in illitic clays, organic matter, pyrite, and microcrystalline quartz.

In the conclusions to their study of the Carlin deposit Hausen and Kerr (1968 p. 939) state:

> The cumulative evidence at Carlin points strongly to the epigenetic origin of the gold. Among other features, argillic replacement of dolomitic host rock, major deposition of colloidal-sized gold in clay aggregates, chemical changes from rock to ore, nearby igneous activity, and elliptical chimneys of recrystallized silicification, all point to low-temperature epithermal origin. Major stages in the creation of the gold deposits as now found include:
>
> 1. Deposition of the Vinini, Roberts Mountains and "Popovich" formations.
> 2. Overthrusting as represented by the Roberts Mountains fault.
> 3. Cretaceous igneous invasion and subsequent deformation.
> 4. Base-metal emplacement in veins and replacements.
> 5. Hydrothermal alteration and gold deposition.
> 6. Epithermal silicification.
> 7. Supergene action, weathering, and oxidation.
>
> Distribution of gold depends largely upon the permeability of decalcified strata of the Roberts Mountains Formation to ore-forming fluids and precipitation in favorable hosts such as illitic clays, carbonaceous matter, and pyrite. Quite near surface emplacement by the release of H_2S may have contributed to the precipitation of gold, realgar, cinnabar, stibnite, and pyrite.
>
> The fine gold occurrence at Carlin may be grouped along with other deposits at Getchell, Bootstrap, and Manhattan in Nevada, and at Mércur in Utah, as a shallow low-temperature epithermal deposit where vein development is hardly recognizable and ore minerals are finely disseminated in replacement bodies in carbonate sediments.

Radtke and Dickson (1976 pp. 74-77) have also studied the Carlin-type deposits in considerable detail. An abbreviated form of their discussion of the conditions of ore formation of these deposits follows.

> Carlin-type gold deposits clearly resulted from the action of hydrothermal solutions on pre-existing sedimentary rocks. The ore-bearing solution migrated upward along steeply-dipping normal faults into near-surface zones, probably within 400 m of the surface, where they began interacting with wallrocks by dissolving calcite and precipitating hydrothermal minerals. They probably met and mixed with aqueous fluids from other sources, such as connate waters and ground waters. The hydrothermal solutions also moved petroliferous compounds and probably many other components from deeper sedimentary units that were undergoing heating, compaction and expulsion of fluids.
>
> According to a chemical model that best fits the data, the ore

solutions were relatively simple in composition. Studies of liquid inclusions in quartz from Cortez, Gold Acres and Carlin indicate that quartz associated with gold was deposited over a range of about 175-200°C from solutions of low ionic strength. Experimental studies on the capability of various kinds of solutions to transport gold suggest that under conditions sufficiently reducing for sulphide sulfur to be stable, and with excess sulfur present, gold forms a soluble and stable complex, $Au(HS)_2^-$. Slightly alkaline HS^- solutions are also capable of transporting SiO_2 and sulphides of As and Sb; if the solutions are sufficiently alkaline (pH ~8.5), they will also dissolve HgS.

As the ore solutions neared the surface, they were subjected to increasingly sharp changes in temperature, pressure and chemistry. Solutions initially saturated with gold, quartz and pyrite would become supersaturated owing to decreases in temperature and pressure. On penetrating favourable sedimentary beds, the solutions initially dissolved calcite and deposited SiO_2 simultaneously with gold and pyrite. Complexes of Au and Hg were removed from the ore solution through interaction with carbonaceous materials; Au, Hg, As and Sb were adsorbed on fine-grained materials, such as pyrite and clays. Subsequently, after the main period of hydrothermal gold deposition late sulphides of As, Sb and Hg were deposited in open fractures.

The hydrothermal episode reached a climatic stage during which the hydrothermal fluids penetrated to the surface. At this stage mineralization happened at and immediately below the surface. In addition, at these shallow depths and low pressures, boiling took place, and H_2O-H_2S vapours were given off. As the hydrothermal activity waned, the boundary between the boiling water and gas receded to greater depths, and the previously mineralized rocks were exposed to the action of H_2SO_4 and atmospherically derived oxygen. Because oxidation of both original and introduced hydrothermal pyrite in the upper part of the deposit would not produce enough H_2SO_4 to dissolve the total amount of carbonate removed, it is likely that most of the acid responsible for the strong leaching was formed by the oxidation of H_2S vapour given off during boiling of the solution.

The sulphate and carbonate produced in the oxidized zone, carried in solutions moving downward along fractures, reacted with components of the thermal waters to produce the late barite and calcite veinlets. Descending acid water that mixed with the thermal waters promoted a series of reactions, including the deposition of late As, Sb and Hg sulphides admixed with barite and calcite in veins of deposits such as Carlin, Getchell, White Caps and Mercur.

The temperatures and pressures of ore deposition ranged from surface hotsprings conditions (100°C, 1 bar) to at least 225°C and 30 bars. If we assume that a column of water at one time extended from the surface down to 400 m, and that the pressures and temperatures were adjusted to the boiling curve of water, a maximum temperature of 290°C and a pressure of 40 bars existed at the 400 m level. Sparse

> data from fluid inclusion and sulfur isotopes in minerals collected about 100 m below the surface at Carlin indicate that at one stage a temperature of about 175°C existed there. Some erosion has taken place at Carlin and the depth at the time the samples were formed is not known.
>
> Characteristic alteration effects are observed in all the deposits. Near-surface rocks have been intensely altered to assemblages of clay minerals. Limestones have been silicified along and near faults and along bedding, to form jasperoid bodies. Other types of alteration that affected rocks at greater depths and in outer zones of the ore deposits, include decarbonatization (removal of calcite and dolomite without addition of silica) and introduction of petroliferous compounds into shattered zones close to ore.
>
> The deepest ore found to date is at the Carlin deposit at about 400 m below the surface. No information is available as to the maximum depth of either mineralized or hydrothermally altered rocks in any of the deposits. On the basis of the little evidence available, vertical mineral zonation does not appear to be present.

The intimate relationship of hot springs and certain hydrothermal deposits was recognized by Élie de Beaumont and other early workers who laid the foundations of the hydrothermal theory. Later workers such as Lindgren likewise noted and commented on the relationship. In his textbook *Mineral Deposits,* Lindgren (1933 pp. 123-124) concluded:

> The quicksilver and stibnite deposits are satifactorily proved to have been formed by ascending hot alkaline springs, as at Sulphur Bank, California, and at Steamboat Springs, Nevada. The deposition proceeds almost under our eyes; small amounts of gold, silver, copper, lead, and zinc are also found in the sinter. Becker proved that the mercury and the antimony were contained in the waters as double alkaline sulphides; and he suggested that most of the other metals, perhaps excepting silver, were in similar combination.
>
> Regarding other deposits inference and speculation must be used. The epithermal deposits are so closely connected with hot-spring action and with deposition of colloidal silica that it is not possible to doubt that they were formed by similar ascending springs.

In recent years D. E. White of the United States Geological Survey has studied active geothermal systems and their relationship to hydrothermal mineral deposits in many parts of the world. The results of his studies are of fundamental importance in the science of hydrothermal gold, silver, and other types of mineral deposits. Paper 10-5 summarizes his general observations.

(Text continues on page 296.)

10-5: ACTIVE GEOTHERMAL SYSTEMS AND HYDROTHERMAL ORE DEPOSITS

Donald E. White

Abstract

During the past 25 years our understanding of hydrothermal ore deposits has progressed remarkably because of combined approaches through detailed study of actual deposits, laboratory experimental study of ore and gangue minerals and fluid inclusions, and study of active geothermal systems. My review emphasizes the active systems, which have recently become a focus of interest in a worldwide search for alternative energy sources.

Sulphur Bank, California, and Ngawha, New Zealand, have provided several keys for understanding the generation of many mercury deposits. Major requirements probably are: deep source regions of fluids and Hg at temperatures >200°C; metamorphic environments above subduction zones on continental margins; a through-going (rather than local) vapor phase enriched in CO_2 or other gases, migrating along with liquid water; and instability of HgS at high temperatures, decomposing to Hg^0 and S^0, with the migrating vapor required for major transport of Hg at temperatures <200°C; a coexisting liquid phase is generally required to transport SiO_2 and other nonvolatile constituents. This two-phase mechanism best explains the general absence of other significant ore metals. Vapor-phase transport of the Hg associated with other metals at higher temperatures is probably not essential.

Epithermal precious metal ore deposits are probably the fossil equivalents of high-temperature geothermal systems like Broadlands, New Zealand, and Steamboat Springs, Nevada. The evidence suggests that the fossil and active systems are similar in their rare chemical elements, ranges in temperature, pressure, compositions of fluids, isotope relationships, and mineralogy of ore, gangue, and alteration minerals. Broadlands and Steamboat Springs show a depth zoning of the "epithermal" chemical elements, Au, As, Sb, Hg, Tl, B, and some Ag, that selectively concentrate near the surface. Much Ag, base metals, and probably Se, Te, and Bi precipitate at somewhat greater depths and higher temperatures.

Nolan (1933) divided the epithermal precious metal deposits into a gold-rich group (Au > Ag by weight) and a silver-rich group. The concepts of depth zoning in active geothermal systems, if applied to epithermal deposits, suggest that some gold-rich deposits (including the recently recognized Carlin-type) form at relatively shallow depths and low temperatures. These may grade down into deposits enriched in Ag and base metals, perhaps in places separated by a relatively barren zone resulting from changes in the dominant complexing agent, Cl vs. S. This possibility, even if remote, justifies close examination.

Active systems that might form base metal ore deposits were virtually unknown 25 years ago. Discovery of the Salton Sea, Red Sea, and Cheleken thermal chloride brines in the early 1960s focused on Cl as the probable critical agent, permitting transport of base metals as metal-chloride complexes. Also, some oil field waters were found to have Pb and Zn contents in the range of a few parts per million (ppm) to many tens of ppm. The low-temperature brines have no sulfide within detection limits; only at temperatures >200°C can small quantities of sulfide coexist with the base metals in solution. All of these metal-bearing brines are deficient in sulfide; most of their metals can precipitate as ore deposits only where supplemental sulfide can be provided by any one of several proposed mechanisms. Comparable brines in the past probably formed low-temperature epigenetic deposits like those of the Mississippi Valley, as well as many marine sediment-hosted syngenetic and early diagenetic ore deposits.

Ore fluids rich in both base metals and reduced sulfide species probably require very high salinity, high temperature, and rock-water reactions buffered at low pH (thus, with little free S^{-2} immediately available). Hostile environments of extreme temperature and salinity, such as those indicated in generating porphyry copper deposits, cannot be drilled by present methods, even if we knew where to drill. Visual observation of comparable environments seemed unlikely until early in 1979, when Cu and Zn sulfides were found to be precipitating from spring vents on the spreading axis of the East Pacific Rise at temperatures exceeding 350°C. Low-temperature discharges on the sea floor had been known for a few years, but the activity in this hydrothermal area, known as Twenty-One North, is the first that bears directly on the origin of volcanogenic massive sulfide deposits (and indirectly on other deposits formed at extreme temperatures and salinities).

Brines of many origins can form base metal deposits; origin of the water may be less important than the physical and chemical environments of the brines and source rocks. Ocean water alone, ocean and fresh waters plus evaporites, evolved connate waters of marine sedimentary rocks, and magmatic waters are all effective solvents of base metals in suitable environments. Precipitation of metals from these brines can occur by decreasing temperature, mixing with low-salinity water, access of supplemental sulfide, and neutralizing reactions with wall rocks, as well as various combinations of these.

Suggestions for exploration for concealed deposits of the major groups considered here are offered, resulting from improved understanding of various genetic models.

To summarize, it is worth pointing out that, according to the original magmatic hydrothermal theory, the source of the mineralizing solutions with their contained gold, other ore elements, and gangue elements was ascribed to nearby igneous bodies (granodiorites, granites, quartz-feldspar porphyries, etc.). With time the theory has changed, the precise identification of the source igneous bodies has become vague, and the origin of the mineralizing solutions (water) has become a matter of conjecture as presaged in a number ofthe papers cited in this chapter, some investigators suggesting a meteoric origin for the water, others a metamorphic origin, and so on. At the present time (1985) even the source of the gold, ore elements, and gangue elements in the so-called magmatic hyrothermal gold deposits is controversial, as we shall see in the next two chapters.

REFERENCES AND SELECTED BIBLIOGRAPHY

Barnes, H. L., ed., 1967. *Geochemistry of Hydrothermal Ore Deposits,* Holt, Rinehart & Winston, New York, 670p.

Barnes, H. L., ed., 1979. *Geochemistry of Hydrothermal Ore Deposits,* 2nd ed., John Wiley & Sons, New York, 798p.

Bateman, A. M., 1954. *Economic Mineral Deposits,* 2nd ed., John Wiley & Sons, New York, 916p.

Beck, R., 1905. *The Nature of Ore Deposits,* W. H. Weed, trans., 2 vols., *Eng. and Mining Jour.* New York.

Becker, G. F., 1882. Geology of the Comstock Lode and the Washoe District, *U.S. Geol. Survey Monographs,* vol. 3, 422p.

Belt, T., 1861. *Mineral veins: An Enquiry into their Origin, Founded on a Study of the Auriferous Quartz Veins of Australia,* John Weale, London, 52p.

Berger, B. R., and P. M. Bethke, ed., 1985. *Geology and Geochemistry of Epithermal Systems, Rev. Economic Geology,* vol. **2,** Society of Economic Geologists.

Beyschlag, F., J. H. L. Vogt, and P. Krusch, 1914-1919. *The Deposits of the Useful Minerals and Rocks,* S. J. Truscott, trans., 3 vols., Macmillan, London.

Boué, A., 1820. *Essai géologique sur l'Ecosse,* Paris.

Boué, A., 1822. Mémoire géologique sur l'Allemagne, *Jour. Phys.* **94:**297-312, 345-379; **95:**31-48, 88-112.

Bowen, N. L., 1933. The broader story of magmatic differentiation, briefly told, in *Ore Deposits of the Western States* (Lindgren Volume), Am. Inst. Min. Metall. Eng. New York, pp. 106-128.

Boydell, H. C., 1924. The role of colloidal solutions in the formation of mineral deposits, *Inst. Min. Metall. Trans.* **34** (pt. 1):145-337.

Boyle, R. W., 1979. The geochemistry of gold and its deposits, *Canada Geol. Survey Bull. 280,* 584p.

Breislak, S., 1811. *Introduzione alla geologia,* Milan.

Clark, W. B., 1970. Gold districts of California, *Calif. Div. Mines Geol. Bull. 193,* 186p.

Darwin, C., 1846. *Geological Observations in South America,* Smith, Elder & Co., London.

Daubrée, G. A., 1879. *Etudes synthétiques de géologie expérimentale,* Paris.

Daubrée, G. A., 1887*a*. *Les eaux souterraines à l'epoque actuelle,* Paris.

Daubrée, G. A., 1887*b*. *Les eaux souterraines aux époques anciennes,* Paris.

de Launay, L., 1913. *Traité de métallogénie: Gîtes minéraux et métallifères,* 3 vol., Librairie Polytechnique, C. H. Béranger, ed., Paris.

Dolomieu, D., 1783. *Voyage au Îles de Lipari fait en 1781, ou notices sur les Îles Aeoliennes,* Paris.

Élie de Beaumont, J. B., 1847. Note sur les émanations volcaniques et métallifères, *Soc. Géol. France Bull.,* 2nd ser. **4** (pt. 2):1249-1334.

Emmons, W. H., 1937. *Gold Deposits of the World,* McGraw-Hill, New York, 562p.

Emmons, W. H., 1940. *The Principles of Economic Geology,* McGraw-Hill, New York 529p.

Gilpin, E., 1888. Notes on the Nova Scotia veins, *Royal Soc. Canada Trans.* **6** (sec. 4):63-70.

Graton, L. C., 1940. Nature of the ore-forming fluid; *Econ. Geology* **35** (suppl. to no. 2):197-358. See also discussions on this paper by C. N. Fenner, *Econ. Geology* **35:**883-904, and E. Ingerson, and G. W. Morey, *Econ. Geology* **35:**772-785.

Hausen, D. M., and P. F. Kerr, 1968. Fine gold occurrence at Carlin, Nevada, in *Ore Deposits of the United States, 1933-1967,* J. D. Ridge, ed., vol. 1, Am. Inst. Min. Metall. Petrol. Eng., New York, pp. 908-940.

Henley, R. W., A. H. Truesdell, P. B. Barton, and J. A. Whitney, 1984. Fluid-mineral equilibria in hydrothermal systems, *rev. Econ. Geol.* **1:**267p.

Ingerson, E., 1954. Nature of the ore-forming fluids at various stages—a suggested approach, *Econ. Geology* **49:**727-733.

Jensen, M. L., and A. M. Bateman, 1979. *Economic Mineral Deposits,* 3rd ed., John Wiley & Sons, New York, 593p.

Knopf, A., 1929. The Mother Lode system of California, *U.S. Geol. Survey Prof Paper 157,* 88p.

Korzhinskiy, D. S., 1964. A theory of the processes of mineral formation, *Internat. Geol. Rev.* **6** (3):387-399.

Lindgren, W., 1924. The colloid chemistry of minerals and ore deposits, in *Theory and Application of Colloidal Behaviour,* R. H. Bogue, ed., McGraw-Hill, New York, pp. 445-465.

Lindgren, W., 1933. *Mineral Deposits,* 4th ed., McGraw-Hill, New York, 930p.

Lindgren, W., and F. L. Ransome, 1906. Geology and gold deposits of the Cripple Creek District, Colorado, *U.S. Geol. Survey Prof. Paper 54,* 516p.

Lindgren, W., et al., 1927. Magmas, dikes and veins, *Am. Inst. Min. Eng. Trans.* **74:**71-126.

McLaughlin, D. H., 1931. The Homestake enterprise—ore genesis and structure, *Eng. and Mining Jour.* **132** (7):324-329.

Malcolm, W., 1976. Gold fields of Nova Scotia, *Canada Geol. Survey Mem. 385,* 253p.

Moore, E. S., 1940. Genetic relations of gold deposits and igneous rocks in the Canadian Shield, *Econ. Geology* **35:**127-139.

Murchison, R. I., M. E. de Verneuil, and A. von Keyserling, 1842. On the geological structure of the Ural Mountains, *Geol. Soc. London Proc.* **3:**742-753.

Necker, A. L., 1832. An attempt to bring under general geological laws the relative position of metalliferous deposits. . . , *Geol. Soc. London Proc.* **1:**392-394.

Niggli, P., 1929. *Ore Deposits of Magmatic Origin,* H. C. Boydell, trans., Thos. Murby & Co., London, 93p.

Park, C. F., and R. A. MacDiarmid, 1970. *Ore Deposits,* W. H. Freeman & Co., San Francisco, 522p.

Petrovskaya, N. V., 1973. *Native Gold* (in Russian), Izd. "Nauka", Moscow, 347p.

Posepny, F., 1894. The genesis of ore-deposits; *Am. Inst. Min. Eng. Trans.* **23:**197-369.

Radtke, A. S., and F. W. Dickson, 1976. Genesis and vertical position of fine-grained disseminated replacement-type gold deposits in Nevada and Utah, U.S.A., in *Problems of Ore Deposition,* B. Bogdanov, ed., vol. 1, 4th IAGOD Symposium, Varna, Bulgaria, pp. 71-78.

Ridge, J. D., ed., 1968. *Ore Deposits of the United States, 1933-1967* (The Graton-Sales Volume), 2 vols., Am. Inst. Min. Metall. Petrol. Eng. New York, 1880p.

Roedder, 1965. Evidence from fluid inclusions as to the nature of the ore-forming fluid, in *Symposium, Problems of Postmagmatic Ore Deposition,* vol. 2, Prague, pp. 375-384.

Roedder, E., ed., 1968-1974. Fluid inclusion research, 7 vols., *Commission on Ore-forming Fluids in Inclusions Proc.* Univ. Michigan Press, Ann Arbor, Mich.

Rye, D. M., B. R. Doe, and M. H. Delevaux, 1974. Homestake Gold Mine, South Dakota: II Lead isotopes, mineralization ages, and source of lead in ores of the Northern Black Hills, *Econ. Geology* **69:**814-822.

Scheerer, T., 1847. Discussion sur la nature plutonique du granite et des silicates cristallins que s'y rallient, *Géol. Soc. France Bull.,* 2nd ser., **4** (pt. 1):468-498.

Scrope, G. P., 1825. *Considerations on Volcanoes. . . ,* London.

Slaughter, A. L., 1968. The Homestake Mine, in *Ore Deposits of the United States,* J. D. Ridge, ed., 1933-1967, vol. 2, Am. Inst. Min. Metall. Petrol. Eng., New York, pp. 1436-1459.

Smirnov, V. I., 1976. *Geology of Mineral Deposits,* Mir Publishers, Moscow, 520p.

Smith, F. G., 1943. The alkali sulphide theory of gold deposition, *Econ. Geology* **38:**561-590.

Stelzner, A., 1879. Die über die Bildung der Erzgänge, u.s.w., *Deutsch. geol. Gesell. Zeit.* **31:**646-648.

Stelzner, A., 1881. Die uber die Bildung der Erzgange aufgestellten Theorien, *Neues Jahrb. Mineral. Geol. Palaeont.* **2** (ref. A):208-210.

Von Cotta, B., 1870. *A Treatise on Ore Deposits,* F. Prime, trans., D. Van Nostrand, New York, 574p.

Werner, A. G., 1791. *Neue Theorie von der Entstehung der Gänge,* Freiberg. (Trans. by C. Anderson as *New Theory of the Formation of Veins,* Edinburgh, 1809.)

White, D. E., 1967. Mercury and base-metal deposits associated with thermal and mineral waters, in *Geochemistry of Hydrothermal Ore Deposits,* H. L. Barnes, ed., Holt, Rinehart & Winston, New York, 670p.

White, D. E., 1981. Active geothermal systems and hydrothermal ore deposits, *Econ. Geology,* 75th anniv. vol., pp. 392-423.

CHAPTER

11

The Origin of Epigenetic Gold Deposits – The Granitization Theory

There are granites and granites . . . some formed in one way and some in another.
A. H. Green, 1882

The question of ore and granite has reached a position where, the more the mapping, the more instances are found of ore bodies that seem to be derived, just as so much of the granite does, from metamorphic rock.
A. Locke, 1941

The origin of granite and allied rocks has been a controversial topic among geologists and petrologists since 1785. Initially the disputation raged between those supporting A. G. Werner (1750-1817) in his Neptunist theory of the sedimentary origin of granitic rocks and those who followed J. Hutton (1726-1797) in his Plutonist theory of the igneous (magmatic) origin of granites. During the early part of the nineteenth century the controversy was partly resolved in favor of Hutton's magmatic theory, but during the early years and the last half of the nineteenth century another suggested origin for granitic rocks, the transformist theory, reactivated this controversial subject. Early ideas on high-temperature transformation of rocks (granitization) can be seen in the works of Ami Boué (1794-1881), J. Fournet (1801-1869), and H. Sainte-Claire Deville (1816-1881).

Advocates of the igneous (magmatic) theory consider granitic rocks to be the result of intratelluric crystallization from basaltic magma, derived presumably from the melting of a part of the mantle of the earth. The transformist school, on the other hand, has diverse opinions on the production of granitic magmas and granites from preexisting sediments and volcanics composing geosynclinal piles. Some transformists call upon solid state diffusion mechanisms to produce granite without the intervention of magma; others consider that granites are the result of metasomatic processes, elements being added or subtracted by ichors, emanations, migrating solutions, or diffusion currents to or from piles of geosynclinal sediments and volcanics, the end result being granodiorites and granites; and still others call upon selective and differential fusion (anatexis) of sediments and other rocks to produce the granitic rocks (the "per migma ad magma" school).

All of the fields of inquiry concerning these various processes have been extensively tilled, but the fruits of these labors have not brought unanimity among geologists respecting the precise origin of granite. As an aside, in the granite controversy the transformists have fixed their gaze principally on the vast migmatitic and granitic zones of the earth where deep erosion has prevailed, such as in the Precambrian shields; the magmatists, on the other hand, have been mesmerized by what is generally known as "high level" granites in shallow eroded terranes such as those of Tertiary age. When one has worked in both terranes one soon recognizes that both schools of thought are essentially correct, and that in reality no granite controversy exists. Thus, it seems to me that the so-called high level granitic bodies simply represent the mobilized (rheomorphic) parts of transformed (granitized) piles of sediments and volcanics in the deep depressed parts of geosynclines. There are, therefore, "granites and granites" as A. H. Green (1882) succinctly said and as reiterated at some length in the polemics of Rastall (1945), Backlund (1946), Reynolds (1946, 1947*a*, 1947*b*), Gilluly et al. (1948), Perrin and Roubault (1949), Guimaraes (1947, 1949), Raguin (1965), Mehnert (1968), and Winkler (1976), and in the meditations of H. H. Read (1957).

MINERALIZATION RELATED TO GRANITIZATION

The origin of granitic rocks bears on the genesis of gold and other metallic deposits in many respects. For instance, if differentiating magmas that yield granites are the source of the gold and gangue elements in auriferous deposits, one mechanism of concentration, namely the magmatic hydrothermal process, applies; if, on the other hand, the gold and accompanying elements in auriferous deposits are derived as the result of the metamorphic transformation of sediments and volcanics to granitic rocks, other mechanisms apply, principally diffusion and metamorphic secretion, plus chemical fronts and others. T. Sterry Hunt was the first geologist to recognize this fact. He held the opinion that granitic bodies were essentially alteration products of sediments, a view evidently conditioned by his work in the highly metamorphosed Laurentian (Grenville) rocks of Canada. Thus, in his paper *The Geognostic History of the Metals* he concluded (Hunt, 1873, p. 341):

> If the view which I hold, in common with many other geologists, that most, if not all, of our known eruptive (intrusive) rocks are but displaced and altered sediments, be true, then it may be fairly affirmed, not that eruptive rocks are the agents which impregnate sedimentary deposits with metals, but on the contrary, that in such deposits is to be sought the origin of metalliferous eruptive rocks, and that all our metallic ores are thus to be traced to aqueous solutions.

In his summary paper *The Origin of Metalliferous Deposits* Hunt (1872) emphasized the importance of an initial concentration of metals during sedimentary processes and the later influence of migrating solutions (metamorphic water) in concentrating the metals in veins. Speaking of gold and silver, which he recognized as being widely dispersed in sediments and other rocks but only in very small amounts, he wrote (p. 424): "But in the course of ages these sediments, deeply buried, are lixiviated by permeating solutions, which dissolve the silver diffused through a vast mass of rock, and subsequently deposit it in some fissure, it may be in strata far above, as a rich silver ore. This is nature's process of concentration."

Many years were to pass before the issue of the relationship of mineral deposits to granitization was addressed again, stimulated this time principally by the prolonged discussions on the origin of granitic rocks that took place in the 1940s and early 1950s (Gilluly et al., 1948; Read, 1957). One of the first geologists to enter the debate was Locke (1941). In the abstract to his paper *Granite and Ore* he set the problem and asked the question that led to much discussion in the years to follow.

> The accepted alliance of ore with intrusive igneous rock may be discussed in terms of granite, for granite is regarded as the predominant rock of this kind. However, much granite has been observed in recent years which seems to be, not "magmatic"—that is, not intrusive and igneous—but "metamorphic." This second kind of granite is thought to be at the end of a metamorphic series that started in great part with marine sediments; it is, essentially, recrystallized mud.
>
> How is the theory of ore genesis affected by what we know of this metamorphic granite? (p. 448)

His conclusions on the relationships merit consideration.

> The question of ore and granite has reached a position where, the more the mapping, the more instances are found of ore bodies that seem to be derived, just as so much of the granite does, from metamorphic rock. Even "cupola" ore is of this kind, for often cupolas have been proved to be drag folds in which, not only the internal structures of the granite, but also the surrounding structures of the wall rocks participate. While not venturing to predict how great a proportion of the ores will prove to have such an origin, the writer is convinced that none of the phenomena usually taken to denote magmatic origin are incompatible with the concept of metamorphic origin. (p. 454)

The paper by Locke was followed by one authored by Dunn (1942 p. 231), who recognized that "for some years now the replacement origin of certain granites has been in the foreground of discussion and, as a necessary corollary, certain accompaniments of magmation, such as ore liquids, will require reconsideration." After reviewing the data for granitization in the Archean of India, Dunn sought to explain the present day distribution of metallogenetic provinces as a result of early crustal processes. This important concept merits quotation.

> Ore minerals deposited in the early crust may become part of the magma formed from any part of the crust, or may be removed to a higher zone by solutions during diabrochomorphism (recrystallization induced by the action of solutions soaking through the rocks) before magma is actually formed. The initial distribution of metals in various parts of the primordial crust has presumably been largely responsible for the present day distribution of metallogenetic provinces in various parts of the world, in rocks of later age. Yet, it is not unlikely that regenerated magmas, forming at the base of the crust, may tap sub-crustal sources of metals in the sima. (p. 237)

In a series of papers on the origin of mineral deposits Guimaraes (1947, 1949) laid great stress on the fact that metallogenesis is but a chapter in the geochemistry of the cyclation (migration) of the elements. Considering the formation of granite-gneissic rocks and their relationship to (gold) veins in Brazil and elsewhere he (Guimaraes, 1947 p. 734) concluded:

> 1. That such rocks were formed by gradual metamorphism;
> 2. That all known Brazilian granites (batholitic) are palingenetic;
> 3. That granitization took place by the replacement of plagioclase and ferrous-magnesian minerals within pre-existing rocks, by microcline, quartz and muscovite;
> 4. That the agents of the transformation are magmatic emanations;
> 5. That magmatic emanations are not necessarily rich in metallic and metalloid deposit-forming elements;
> 6. That magmatic emanations promoting the migration of constituent elements that enter in the composition of percolated rocks, take along other elements either in the status of combination and spread over, or fixed in accessory and rare minerals within, the same rock;
> 7. That thus enriched with metallic and metalloid elements, magmatic emanations in fluid gaseous status infiltrate in the zone of ruptural deformation of the earth's crust, making possible the metasomatic processes yielding mineral vein-deposits.

The origin of the magmatic emanations referred to is rather vague in Guimaraes's papers. Presumably they derive from metamorphic processes, deep within geosynclines.

The paper by C. J. Sullivan (1948) is a landmark in the granitization theory because it gives many pertinent data in its support. His conclusions merit close study by those interested in metallogenesis.

> De Beaumont (De Beaumont, Élie, Note sur les émanations volcaniques et métallifères: Soc. géol. France, pp. 1290-1291, 1847) remarked that the study of the origin of granitic rocks is an indispensable counterpart of the study of the origin of ore deposits. The conception of "ortho-magmatic differentiation" led to certain hypotheses regarding the origin and behavior of ore deposits but these hypotheses must be reconsidered if the possibility is admitted that granitic rocks may have been formed mainly from crustal material by a process of "granitization" such as that described by Backlund (Backlund, H. G., The granitization problem: Geol. Mag., vol. 83, pp. 105-117, 1946).
>
> As a contribution to this reconsideration, outlines of five main concepts are put forward in this paper:
> 1. During the first cooling of the earth the elements were distributed according to geochemical laws outlined by Goldschmidt (Goldschmidt, V. M., The principles of distribution of the chemical elements in minerals and rocks: Chem. Soc. London, Jour. pp. 655-673, 1937). Geochemical equilibrium evidently required the formation of a granitic crust. The later addition to the crust of basic volcanic rocks set up a chemical gradient. Granitization represents

> the neutralization of this gradient—the approximate return to the geochemical equilibrium attained at the first earth cooling.
> 2. Magmatic differentiation and erosion do not adequately explain the distribution of granitic rocks and of ore deposits in geological time. The granitization hypothesis, when the effects of "volcanic" activity are considered, supplies a possible explanation.
> 3. The associations of specific ores with specific types of granitic rocks do not necessarily imply the correctness of the theory of "magmatic differentiation." During the formation of a granitic rock, those elements are concentrated which are not accepted into the crystal lattices of the common rock-forming minerals. The concentration of particular elements by particular rocks will, presumably, occur whether the rock crystallizes from a melt, or whether the crystallization results from the re-arrangement of atoms by some other processes—e.g., by diffusion in the solid (Reynolds, D. L., The granite controversy: Geol. Mag., vol. 84, pp. 216-218, 1947).
> 4. The differences between synchronous and subsequent batholiths become intelligible in the light of the granitization hypothesis. Synchronous granites show a close spatial relationship to contact metasomatism and to ore deposition; the two latter processes are intimately connected with each other. In areas where synchronous granites occur, these facts may be used to guide prospecting; in particular they may be used to localize ore search over non-outcropping granite cupolas.
> 5. The granitization hypothesis postulates a fundamental difference between granitic and "volcanic" rocks: similarly, ore deposits originating during a cycle of granitization are basically different from those of "volcanic," or abyssal, origin. Among these differences are the relative scarcity of lithophile ore elements in "volcanic" deposits; the relatively high sulphur content of many of the "volcanic" deposits; and the lack of zonal arrangement in "volcanic" deposits as compared with deposits associated with granitization.

In a later paper G. E. Goodspeed (1952 pp. 165-166) emphasized the hypothesis that the water constituting hydrothermal fluids arises from sediments undergoing granitization. He concluded:

> The most spectacular occurrences of granitic rocks formed by granitization are to be found in folded geosynclines (Misch, Peter, Metasomatic granitization of batholithic dimensions, Part 1: Amer. Jour. Sci., vol. 247, pp. 209-245, 1949). Many thousands of cubic miles of material are deposited in a thick geosynclinal prism; and the material has the widest range of chemical composition. The transformation of this material into granitic rock involves not only a vast chemical rearrangement (Sullivan, C. J., Ore and granitization: Econ. Geol., vol. 43, no. 6, pp. 471-498, 1948) but also a radical physical change. In a thick mass of sediments, the amount of water both connate and combined, may well be measured in cubic miles. The mobilization and release of this water by granitization forms an abundant source for hydrothermal solutions.
>
> Locally within the geosynclinal prisms argillaceous material would

> on recrystallization yield a greater amount of water than a more anhydrous rock like limestone. It is therefore possible that this is one factor which may have a bearing on the question of why some parts of a granitized mass show little mineralization. However, probably the most important factors are the presence of granitized satelitic stocks and favorable structures. It is also possible that the type of geosyncline—for example, ensialic or ensimatic (Wells, F. G., Ensimatic and ensialic geosynclines: Geol. Soc. Amer, Bull. vol. 60, p. 1927, 1949)—may have an important influence on the related later mineralization. Mineralized zones are commonly the loci for later metallization and, although in some cases metallization is directly related to granitization, in others a source beneath the geosyncline may be indicated.
>
> Ore deposits directly related to igneous intrusions are formed at the final stages of cooling of the magmatic body. The elements forming the deposits as well as the vehicle for transporting these elements came essentially from the magma. Hence in an igneous petrographic province, one might expect a certain regularity in the zones of mineralization. On the other hand, in a large area of granitization, the relatively gradual transformation of many rock types of diverse composition involves geochemical rearrangements which, combined with the varying amounts of volatiles, could give rise to many kinds of related mineralization.
>
> It is possible, however, that either an intrusive body or a granitized mass may be merely a structural control for later metallization. In this respect relatively simple structures of an igneous intrusion are in marked contrast to the complex structural features of large or small granitized masses with their irregular contacts and included relict material.
>
> In either case, the later and perhaps much later metallization may be due to the presence of profound fractures which provide access for emanations from deeper magmas, potential magmas, or even from subcrustal sources.

Read (1954 p. 97) also emphasized the role of the *water-front* in granitization as follows:

> One of the country-rock components not fixed in granitisation is of great interest—it is water. Its destiny may have a bearing, as discussed by Goodspeed, on the origin of hydrothermal veins and on the classical theory of ores related to igneous intrusion. The amount of water in magmas is small, but a great deal is required in metamorphism, mineralisation and wall-rock alteration generally. It can be suggested that this water comes from the country-rocks undergoing granitisation; it constitutes a *water-front* driven in advance of the granitisation-front just as, when we toast one side of a slice of bread, we produce a water-front on the other side. Goodspeed emphasises two characters of a thick geosynclinal pile, the wide range of chemical composition and an abundance of water, and concludes that the mobilisation and release of this water by granitisation forms

> an abundant source for hydrothermal solutions. "Such solutions, coming from depth and traversing heterogeneous sedimentary and volcanic rocks, have great opportunity for solution and transport of selected components which may be deposited under structural controls in the higher levels of the crust." During granitisation, a concentration of originally widely dispersed elements may take place; ore-minerals may not be directly derived from granitic magmas or emanations. But, if this were to be true, the classical theory of the relation of ore-deposition to magmatic differentiation requires consideration.

Taupitz (1954) offers a comprehensive theory for the origin of various types of mineral deposits based in part on the geochemical cycle of the elements. Here we are interested only in his views on the relationship of ore deposits to granitization processes. Taupitz distinguishes two types of magmas—mafic magmas originating in the eclogitic shell of the earth that give rise mainly to ore deposits containing one or more of Cr, Pt, Ni, Co, Cu, Fe, Ti, and V, and palingenetic (anatectic) magmas representing the melted products of deeply subsided geosynclinal rocks, principally sediments, that produce mineral deposits enriched in Fe, Mo, Sn, W, As, Au, Cu, Zn, Pb, Fe, Sb, Hg, F, and Ba. All of these elements are considered to have concentrated in the last magmatic fractions of the crystallizing magmas, giving deposits of pegmatitic, pneumatolytic, and hydrothermal character. In some places the mineral-laden exhalations and thermal waters from both types of magmas, but predominantly the mafic type, may discharge into the ocean, thus producing the so-called exhalite deposits discussed in the next chapter.

Moiseenko and co-workers attach considerable importance to the processes of granitization and metamorphism in the formation of gold deposits. Their general conclusion is that the primary sources of gold in the earth's crust are basic and ultrabasic magmatic rocks in which the element is present in a dispersed form (Moiseenko, 1965; Moiseenko, Nechkin, and Fat'yanov, 1970; Shcheka and Moiseenko, 1970; Moiseenko and Fat'yanov, 1972; Moiseenko and Neronskii, 1973. During granitization and metamorphism gold is released, is transferred to melts or solutions, migrates, and is deposited in local tectonic structures. Repeated mobilization and deposition of gold is frequently necessary for the formation of economic epigenetic gold deposits, according to these investigators.

CONCLUSION

There is a general consensus among granitizationists that two types of magmas are evolved in the earth, one a mafic variety giving rise to deposits containing Cr, Pt, Ni, Co, Cu, Fe, Ti, and V, another a granite-granodiorite variety that is the source of elements such as Au, Ag, Pb, and Zn. The first variety is thought to have its origin in the melting of the mantle or some other deep shell of the earth; the second is anatectic and originates from granitization of geosyncline piles of rocks, mainly sediments.

Few can doubt that the granitization of sediments and volcanics in a geosynclinal pile would yield enormous amounts of metals, gold included. My own calculations (Boyle, 1976) show that 1 cu km of granitized geosynclinal sediments would yield some 27×10^6 g Au and 450×10^6 g Ag; similarly 1 cu km of granitized volcanics would produce 12×10^6 g Au and 210×10^6 g Ag. These amounts alone are sufficient to form good-sized gold orebodies; when one considers that hundreds, and in places thousands, of cubic kilometers are involved in the granitization of a geosyncline one wonders where all the gold and silver have gone, because they are not present in any of the known gold deposits associated with the granitization centers.

Many questions remain to be resolved in the granitization theory. Chief among these is the question of timing, particularly at what stage mobilization and concentration of the elements present in gold and other deposits took place. For instance, were the elements mobilized and driven out of the zones undergoing granitization, advancing with the metamorphic (water) front, or did they remain in the anatectic granitic magma to be concentrated and expelled at its end stages of crystallization? The first possibility appears to be more probable on thermodynamic grounds, but we require much more field and laboratory data to demonstrate that it is indeed the case.

REFERENCES

Backlund, H. G., 1946. The granitization problem, *Geol. Mag.* **83:**105-117.
Boué, A. For a review of the many works of Ami Boué see E. W. Benecke, *Neues Jahrb. Mineral. Geol. Palaeont.* **1:**334-335, 1882; see also references in chapters 6 and 10.
Boyle, R. W., 1976. Mineralization processes in Archean greenstone and sedimentary belts, *Canada Geol. Survey Paper 75-15,* 45p.
Dunn, J. A., 1942. Granite and magmatism and metamorphism, *Econ. Geol.* **37:**231-238.
Fournet, J. F., For a review of the many works of Fournet see A. Caillaux, *Soc. Géol. France Bull.* ser. 2, v. **27:**521-539, 1870; see also the references in chap. 10.
Gilluly, J., et al., 1948. Origin of granite, *Geol. Soc. Am. Mem. 28,* 139p.
Goodspeed, G. E., 1952. Mineralization related to granitization, *Econ. Geol.* **47:**146-168.
Green, A. H., 1882. *Physical geology,* London, 728p.
Guimaraes, D., 1947. Mineral deposits of magmatic origin, *Econ. Geol.* **42:**721-736.
Guimaraes, D., 1949. Geoquimismo magmatico e origem dos batolitos graniticos, Estado de Minas Gerais, *Inst. Tecnol. Industr. Bol. 9,* 124p.
Hunt, T. S., 1872. The origin of metalliferous deposits, *Am. Inst. Min. Eng. Trans.* **1:**413-426.
Hunt, T. S., 1873. The geognostical history of the metals, *Am. Inst. Min. Eng. Trans.* **1:**331-346.
Locke, A., 1941. Granite and ore, *Econ. Geol.* **36:**448-454.
Mehnert, K. R., 1968. *Migmatites and the Origin of Granitic Rocks,* Elsevier, Amsterdam, 393p.
Moiseenko, V. G., 1965. Metamorfizm Zolota Mestorozhdenii Priamur'ya (Metamorphism of gold in deposits of the Amur River region); Khabarovsk: Kn. Izd., 127p.
Moiseenko, V. G., and I. I. Fat'yanov, 1972. Geochemistry of gold, 24th Int. Geol. Congr., Montreal, Sec. 10, *Geochemistry,* pp. 159-165.
Moiseenko, V. G., G. S. Nechkin, and I. I. Fat'yanov, 1970. Geochemical conditions of redistribution of gold associated with magmatic rocks, in *Problems of Hydrothermal Ore Deposition,* Z. Pouba and M. Stemprok, eds., Int. Union. Geol. Sci., ser. A, no. 2, pp. 115-120.
Moiseenko, V. G., and G. I. Neronskii, 1973. Confinement of gold mineralization to metamorphic complexes, Metamorf. Kompleksy Vostoka SSSR, A. M. Smirnov, ed., Akad. Nauk, SSSR, Dal'nevost Nauch. Vladivostock, pp. 219-225.
Perrin, R., and M. Roubault, 1949. On the granite problem, *Jour. Geol.* **57:**357-379.
Raguin, E., 1965. *Geology of granite,* Interscience, New York, 314p.
Rastall, R. H., 1945. The granite problem, *Geol. Mag.* **82:**19-30.

Read, H. H., 1954. Granitization and mineral deposits, *Geol. en Mijnbouw,* no. 4, pp. 95-99.
Read, H. H., 1957. *The Granite Controversy,* Thos. Murby & Co., London, 430p.
Reynolds, D. L., 1946. The sequence of geochemical changes leading to granitization, *Geol. Soc. London Quart. Jour.* **102:**389-446.
Reynolds, D. L., 1947*a.* On the relationship between "Fronts" of regional metamorphism and "Fronts" of granitization, *Geol. Mag.* **84:**106-109.
Reynolds, D. L., 1947*b.* The granite controversy, *Geol. Mag.* **84:**209-223.
Sainte-Claire-Deville, H. For a review of the works of Sainte-Claire-Deville see C. H. Friedel, *Soc. Minéral France Bull.* **3** (7):187-190, 1880.
Shcheka, S. A., and V. G. Moiseenko, 1970. Regular characteristics of gold distribution in basic and ultrabasic rocks, *Izv. Tomsk. Politek. Inst.* **239:**37-38.
Sullivan, C. J., 1948. Ore and granitization, *Econ. Geol.* **43:**471-498.
Taupitz, K. C., 1954. Uber Sedimentation, Diagenese, Metamorphose, Magmatismus und die Entstehung der Erzlagerstätten, *Chemie Erde* **17** (2):104-164.
Winkler, H. G. F., 1976. *Petrogenesis of Metamorphic Rocks,* 4th ed., Springer-Verlag, New York, 334p.

CHAPTER

12

The Origin of Gold Deposits—The Exhalite Theory

The theory of exhalative-sedimentary ores . . . suggested in this paper is a volcanologist's approach to ore geology.
C. Oftedahl, 1958

The volcanic exhalite theory has had a long history. Initiated during the early studies of volcanism in the first quarter of the nineteenth century, the theory has been revived from time to time. Its most recent rebirth, concerning mainly the origin of massive sulfide deposits and iron-formations (both of which may be auriferous), is due to Oftedahl (1958), followed later by Stanton (1959, 1960), Sangster (1972), Hutchinson (1973), and numerous others. There are a number of variants of the exhalite theory; as conceived by most advocates of the theory the massive sulfide bodies are coeval with their enclosing rocks (generally volcanics or volcaniclastic sediments), and their constituent sulfides were laid down in a sedimentary manner as a result of exhalative volcanic processes, or as an alternative, the masses were deposited as a sulfide replacement of unconsolidated or slightly consolidated sediments. Later the deposits are supposed to have been metamorphosed and sheared or otherwise deformed, and the resulting textures in both the enclosing rocks and deposits are said to reflect these processes. Many thorny issues have been stirred up by this theory, most of which have not yet been adequately explained either by the syngeneticists or hydrothermalists.

APPLICATION OF THE THEORY TO GOLD DEPOSITS

Few gold deposits have been directly attributed to volcanic exhalative processes although recently certain stratiform auriferous bodies, such as occur in iron-formations and tuffaceous beds, have been assigned such an origin. More generally such bodies and associated auriferous carbonate zones have been explained by a syngenetic-metamorphic mechanism.

The first investigator to explain certain gold deposits by such a mechanism was R. H. Ridler. Concerning the Kirkland Lake area of Ontario where some gold orebodies (e.g., Lake Shore, Macassa) occur in a thrust fault cutting syenites and others occur in schistose carbonate zones (e.g., Kerr Addison), Ridler (1970 pp. 40–41) concluded that:

> The Archaean stratigraphy of the Kirkland Lake area is characterized by three sequential volcanic cycles. Each is made up of an older effusive basalt platform followed by a younger salic pyroclastic edifice, and interbedded volcanogenic sediments.
>
> The volcanics of the Kirkland Lake area are anomalously potash rich. In particular they trend from older, saturated sub-alkalic tholeiites, andesites, and dacites to undersaturated, hyper-alkaline, trachytes, feldspathoidal lavas, and syenites. Noticeably alkaline volcanism preceded and presaged Timiskaming volcanism.
>
> The Timiskaming synclinorium is symmetrical about an arcuate axis, convex to the north and bisecting the Lebel syenite. One period of folding preceded the Timiskaming. At least two followed. In addition, there have been several periods of crustal fracturing, the most obvious being post-Archaean.
>
> The tectonic framework appears to be comprised of three elements: a mobile belt coincident with the Timiskaming synclinorium, a relatively stable shelf to the south, and a basin of mafic volcanism to the north.
>
> The Boston iron formation is upper Timiskaming. The phenomenon known as the Kirkland-Larder Lake *Break* appears to be largely the carbonate facies of the Boston iron formation. The carbonate facies is anomalously auriferous and has significant quantities of iron sulfides associated with it in the Larder Lake area. The highly folded nature of the carbonate facies suggests that many exploration targets remain.
>
> Important concentrations of gold and iron are spatially and genetically related to the youngest known, most differentiated salic unit, the Timiskaming volcanic centre enveloping the Lebel syenite. Vein gold-lode formation is considered a lateral secretion, dilational process, which occurred during regional relaxation at a time when the crust overlying the area was thick enough to create a mesothermal environment, perhaps in middle Proterozoic time.
>
> The relation of mineralization to volcano-sedimentary stratigraphy in the Kirkland Lake area is a classic example of the cyclic model of Archaean volcanism, sedimentation, and mineralization, proposed by A. M. Goodwin (*Canadian Inst. Min. Metall. Trans.* **68:**94-104; *Econ. Geol.* **60:**995-971; *Canadian Mining Jour.* **87** (5):57-60).

Some of Ridler's views, especially the origin of the Kirkland-Larder Lake *Break,* which he attributes to exhalative volcanic sedimentary processes, are at considerable variance with the observations of earlier (and many later) geologists who have interpreted this *break* as a carbonated fault zone (see Thomson, 1948).

The auriferous, pyritic silicified bodies and gold-quartz stockworks comprising the deposits of the "golden mile" at Kalgoorlie, western Australia, occur in shear zones, fractures, and pipes in Archean greenstones intruded by porphyries and granodiorite. They have generally been interpreted as hydrothermal (replacement) deposits (Travis, Woodall, and Bartram, 1971; Travis and Woodall, 1975), but recently a novel origin for these deposits has been suggested by Tomich (1974, 1976). He maintains that the deposits are essentially of volcanic origin and that some are exhalites. He admits that the deposits as they now occur may represent reconstituted (remobilized) parts of the volcanogenic-exhalite assemblage. The arguments for and against the two theories are much too complex to be considered in the space available here; they are outlined in the four publications just mentioned.

The latest summary of the exhalite theory concerning the origin of gold deposits in greenstone belts is by Hutchinson and Burlington (1984). The abstract of their detailed paper follows.

> Lode gold deposits are important and numerous in all Archaean greenstone belts, rare in similar early Proterozoic successions, absent in mid-Proterozoic rocks, but again important, although less numerous, in eugeosynclinal Phanerozoic volcano-sedimentary successions. This time-distribution resembles that of volcanogenic massive base metal sulphide deposits, which are often co-areal with the gold ores. The rarity of both types of deposit in Proterozoic rocks may reflect Earth expansion during mid-Proterozoic time.
>
> In addition to their geological setting in greenstone belts, lode gold deposits have many common characteristics, notably prominent association with carbonaceous strata, with abundant ferroan dolomite and a distinctive minor element/mineral suite, including significant Ag, Cr-muscovite, scheelite (W) and accessory Fe-Cu-Zn-Ni-(Pt) sulfides. However, individual deposits exhibit widely differing morphology. Some are concordant schists, stratiform beds or bedding-plane veins. Others are discordant ladder veins within single beds, or veins or stockworks that transect a number of beds. Some deposits are associated with felsic porphyries but others are not. This diversity has prompted conflicting genetic hypotheses.
>
> Recent studies indicate that the concordant deposits are exhalites, that is, sea-floor chemical precipitates from discharged hydrothermal fluids, mixed with varying proportions of volcaniclastic detritus. These primary strata have undergone a complex, prolonged, continuous spectrum of metamorphism. This commenced with diagenetic compaction and lithification; it included the effects of superimposed volcanism-subvolcanic plutonism during greenstone-belt development, and culminated in regional metamorphism, deformation and granitic plutonism. Successive metamorphic events formed distinctive secondary veins, vein systems and stockworks by remobilization of vein components from the primary Au-enriched strata. Polyphase metamorphism explains the morphological diversity of the deposits whilst primary exhalative processes explain their common characteristics.
>
> Sub-sea floor reduction of convectively circulated sea water, involving oxidation-reduction reactions between Fe^{2+} in silicate minerals and CO_2-H_2O, may have generated carbonyl (CO), cyanide (CN), thiocyanide (SCN) and hydrocarbons (CH_4, C_2H_6) in the exhalative

> fluids. Soluble carbonyl, cyanide or thiocyanide complexes may have transported and deposited the distinctive elements in the deposits. Discharged carbonyl, cyanide, thiocyanide and hydrocarbons may have caused the extensive carbonatization of the associated rocks and deposited the carbonaceous strata. (p. 339)

Quartz vein systems, saddle reefs, and other Bendigo-type auriferous bodies in Nova Scotia (see Paper 10-3), commonly hosted by turbidite sequences have been ascribed a sedimentary and/or exhalite origin by a number of early and more recent investigators.

T. Sterry Hunt (1868) writing on the gold region of Nova Scotia considered the auriferous quartz veins (of the Meguma group) to be essentially interstratified beds later folded and contorted into their present form. The gold of the Nova Scotia deposits and those of similar nature in North Carolina, according to Hunt, "was brought to the surface in a state of solution, and that the watery solvent held also alike the elements of the accompanying metallic sulphurets and the silica which now forms the quartz layers." H. Y. Hind (1870), discussing the origin of the Nova Scotia deposits, agreed in principle with the views of Hunt, and McBride (1978) has suggested that the gold in the veins of the Ecum Secum area in Halifax and Guysborough counties of Nova Scotia is syngenetic with the enclosing turbidites.

More recently Haynes (1986) advocates an exhalite origin for the turbidite-hosted gold deposits of eastern Nova Scotia. His views are outlined in his abstract on the geology and chemistry of turbidite-hosted gold deposits, greenschist facies, eastern Nova Scotia.

> Studies of the gold deposits of the Meguma Terrane have revealed that gold is present in both quartz veins and host rock metalliferous slates. Structural evaluation of these occurrences indicates most auriferous veins predate Acadian dynamothermal metamorphism and deformation, and Devonian granite intrusion and regional static metamorphism.
>
> Many gold districts are located near early monoclinal flexures on the flanks of Acadian major folds. It is proposed that these flexures represent tectonic activation of fault zones that acted as feeders for hydrothermal hot-spring systems. These precipitated silica, carbonates, gold, arsenic, mercury and other metals from low-density buoyant plumes ("white smokers") as laminated siliceous exhalites, exhibiting structures similar to modern geyserites. In contrast, silica in the underlying hydrothermal system was precipitated as non-laminated quartz-potassium silicates ± carbonate veins in the vertical conduits and sub-horizontal hydrothermal aquifers; accompanied by silicification, sericitization and arsenopyritization of their adjacent wall rocks.
>
> Mineral chemistry, fluid inclusion and isotope studies indicate: crystallization of the exhalative assemblages at below 8×10^4kPa and 320°C; and a range of temperatures for the feeder veins. These

values indicate crystallization of the exhalative silica gel under high sub-seafloor temperature gradients during turbidite burial.

Quartz vein arrays that formed during Acadian deformation and pressure solution display low gold values. Also, cleavages associated with this deformation cut the auriferous vein arrays. Thus, genetic theories for the gold, and associated quartz veins, involving Acadian folding and/or pressure solution are in question.

COMMENT

The exhalite theory as applied to lode gold deposits differs little from the metamorphic secretion theory suggested by a number of authors in the next chapter.

There is ample geochemical evidence to suggest that small amounts of gold were deposited in sediments associated with volcanism during many periods in the earth's history. Later metamorphic events may have mobilized this gold and concentrated it in lodes in various types of discordant structures. This latter process, however, remains to be proven. Certainly, many of the stratabound auriferous deposits in iron-formations and tuffaceous sediments seem to have had an initial exhalative origin; the distribution of the present auriferous ore shoots, however, appears to be related to local redistribution of the gold into structures such as drag folds, shear zones, disrupted beds, and chemically favorable (replaceable) beds within the iron-formations and tuffs.

REFERENCES

Haynes, S. J. (1986). Geology and chemistry of turbidite-hosted gold deposits, greenschist facies, eastern Nova Scotia, in *Turbidite-hosted gold deposits,* Geol. Assoc. Canada Spec. Paper 32.

Hind, H. Y., 1870. On two gneissoid series in Nova Scotia and New Brunswick, . . . *Geol. Soc. London Quart. Jour.* **26:**468–479.

Hunt, T. Sterry, 1868. *The Gold Region of Nova Scotia,* Canada Geol. Survey, 48p.

Hutchinson, R. W., 1973. Volcanogenic sulfide deposits and their metallogenic significance, *Econ. Geology* **68:**1223–1246.

Hutchinson, R. W., and J. L. Burlington, 1984. Some broad characteristics of greenstone gold lodes, in *Gold '82,* R. P. Foster, ed., A. A. Balkema, Rotterdam, pp. 339-371.

McBride, D.E., 1978. Geology of the Ecum Secum area, Halifax and Guysborough Counties, Nova Scotia, *Nova Scotia Dept. Mines Paper 78-1,* 12p.

Oftedahl, C., 1958. A theory of exhalative-sedimentary ores, *Geol. Foren. Stockholm Forh.* **80** (pt. 1, no. 492):1-19.

Ridler, R. H., 1970. Relationship of mineralization to volcanic stratigraphy in the Kirkland-Larder Lakes Area, Ontario, *Geol. Assoc. Canada Proc.* **21:**33-42.

Sangster, D. F., 1972. Precambrian volcanogenic massive sulphide deposits in Canada: A review, *Canada Geol. Survey Paper 72-22,* 44p.

Stanton, R. L., 1959. Mineralogical features and possible mode of emplacement of the Brunswick Mining and Smelting orebodies, Gloucester County, New Brunswick, *Canadian Inst. Min. Metall. Bull.* **52** (570):631-643.

Stanton, R. L., 1960. General features of the conformable pyritic orebodies, Part I—Field association, *Canadian Inst. Min. Metall. Bull.* **53** (573):24-29; Part II—Mineralogy, *Canadian Inst. Min. Metall. Bull.* **53** (574):66-74.

Thomson, J. E., 1948. Regional structure of the Kirkland Lake-Larder Lake area, in *Structural Geology of Canadian Ore Deposits,* Canadian Inst. Min. Metall., Montreal, pp. 627-632.

Travis, G. A., R. Woodall, and G. D. Bartram, 1971. The geology of the Kalgoorlie goldfield, J. E. Glover, ed., *Geol. Soc. Australia Spec. Pub. 3,* pp. 175-190.

Travis, G. A., and R. Woodall, 1975. A new look at Kalgoorlie Golden Mile geology, *Australasian Inst. Min. Metall. Proc.* **256:**33-40.

Tomich, S. A., 1974. A new look at Kalgoorlie Golden Mile geology, *Australasian Inst. Min. Metall. Proc.* **251:**35.

Tomich, S. A., 1976. Further thoughts on the application of the volcanogenic theory to the Golden Mile ores at Kalgoorlie, *Australasian Inst. Min. Metall. Proc.* **258:**19-29.

CHAPTER

13

Origin of Epigenetic Gold Deposits—Secretion Theories

I had long ago come to the conclusion, that most, if not all, the gold in the quartz reefs was derived from the rocks in which these reefs occur.
R. Daintree, 1866

Secretion theories have had a long history; the cardinal idea that the constituents of mineral veins may have their source in the host rocks of the veins began with Agricola in the sixteenth century. Since then secretion theories have undergone considerable modification and embellishment. Two general concepts implicit in secretion theories are extant: lateral secretion (*sensu stricto*), which postulates that the ore and gangue constituents were derived directly from the adjacent wall rocks of the veins, and secretion (*sensu lato*), which holds that the ore and gangue matter had their origin in a large volume of host rocks, generally in the piles of sediments and/or volcanics and their associated intrusives, all of which commonly host a mineral belt. The mechanisms of secretion are varied; the flow of ore and gangue constituents toward the veins may be lateral, vertical (ascending or descending), or inclined. The mode of transport may be by meteoric (ground) and/or stratal (connate) water, by metamorphic (hydrothermal) waters,* by diffusion currents, or by various combinations of these modes. Where diffusion is the principal mechanism of transport, the elements may migrate by surface diffusion along grain boundaries and other discontinuities or by diffusion through a standing fluid (water) that pervades the pores, grain boundaries, fractures, and other discontinuities in the rocks. Further details are given in Boyle (1963, 1979).

Agricola in his *De natura eorum quae effluunt ex terra* (1546*a*) and *De ortu et causis subterraneorum* (1546*b*) considered that veins were deposited from surface (meteoric) waters that had penetrated deeply into the earth, become heated, vaporized in part, and rose into canales (fissures) where they precipitated the gangue and ore constituents imbibed as the waters percolated through

*These waters are derived during metamorphism and granitization of piles of sediments and volcanics essentially as a result of dehydration processes. For example, during the metamorphism of shale to gneiss the loss of water is of the order of 4% or more.

the rocks. This statement, here greatly simplified and shorn of the many complications that Agricola appended in his great work on waters, stands as a landmark in geological thought. It is remarkable that it has stood the test of four centuries, and as regards the waters from which gold deposits (of mainly Tertiary age) have been deposited, it has been confirmed by recent isotopic work (see the papers noted in the last part of this chapter).

Numerous geologists writing in the seventeenth, eighteenth, and early nineteenth centuries advocated variants of the secretion theory. John Woodward (1695) in his *Essay towards a Natural History of the Earth and Terrestrial Bodies, Especially Minerals* called on the universal Deluge" (the Biblical Flood) for assistance in the lateral secretion formation of veins. He concluded:

> That the metallick and mineral matter, which is now found in the perpendicular intervalls (fissures) of the strata, was all of it originally, and at the time of the Deluge, lodged in the bodies of those strata . . . that it was educed thence and transmitted into these intervalls since that time, the intervalls themselves not existing till the strata were formed . . . that there do still happen, transitions and removes of it, in the solid strata, from one part of the same stratum to another part of it, occasioned by the motion of the vapour towards the perpendicular intervalls of these. (p. 188)

Later C. F. Zimmerman (1746) and C. T. Delius (1770) respectively called on transformation (metasomatism) of the enclosing rocks and the concentrating action of meteoric water in their lateral secretion theories. Likewise, W. Pryce (1778) in his classic work on the Cornwall deposits, *Mineralogia Cornubiensis,* concluded:

> We may reasonably infer that water, in its passage through the earth to the principal fissures, imbibes, together with the natural salts and acids, the mineral and metallic particles with which the strata are impregnated. [The metals and gangue elements thus dissolved were precipitated in the fissures] to form different ores more or less homogeneous, and more or less rich according to the different mixtures which the acid had held dissolved, and the nidus in which it is deposited. (p. 6)

The same view was taken by H. T. De la Beche (1839) in his classic *Report on the Geology of Cornwall, Devon and West Somerset.* Commenting on the origin of the tin lodes and cross-courses (sulfide veins) he opined:

> When we take a general view of the filling of the dislocations in the district, whether termed common faults, lodes, or cross courses, we see that it has depended upon conditions among which the mineral matter of the adjacent rocks holds a prominent place. Upon this character seems to have greatly depended the nature of the chief mineral substances in them. Among the limestones we find carbonate of lime abundant, and among the siliceous rocks, quartz. We

therefore infer either that water charged with matter derived from the adjacent rocks has infiltered into the fissures, or that their liquid contents have acted on the adjoining rocks and dissolved a portion of them.

Upon the mineral characters of the adjacent rocks seems also to have greatly depended the accumulation of the ores of the useful metals, it generally happening that when two or three dissimilar rocks are traversed by the same fissure they are most abundant. (p. 392)

Secretion theories received considerable impetus as the result of the analytical work of Forchhammer (1855), who demonstrated that ordinary rocks contain trace amounts of base metals and the other elements often found in veins. Bischof (1847-55), the renowned German geochemist, supported a lateral secretion theory, and similar ideas were held by Daubrée (1860, 1879, 1887), Hunt (1873*a*, 1873*b*, 1897), Phillips (1884), Sandberger (1882-1885), Emmons (1886*a*, 1886*b*), Becker (1882, 1888) and Van Hise (1901, 1904) during the nineteenth century and the early part of the present century. All these geologists ascribed the source of the metals to the host rocks of the vein deposits but differed in their opinions as to whether the metals were deposited by descending or ascending meteoric water. Some investigators like Daubrée thought that meteoric water penetrated deeply into the crust where it effected widespread metamorphism, dissolved the metals and gangue elements, and rose again to deposit them as vein minerals in fissures. Daubrée also gave us the important concept that the formation of mineral veins is essentially a special event in the hydrothermal activity apparent in metamorphism. Charles Darwin (1846), as noted in chapter 10, had earlier observed the similar relationship of metallic (gold) veins to metamorphic terranes in Chile. Other investigators, of whom S. F. Emmons was one, claimed that the veins were deposited from descending meteoric water that had leached metalliferous and gangue material from the neighboring rocks. Wallace (1861) held yet another opinion explaining the origin of the lead veins in the carboniferous limestones of northern England by deposition from descending meteoric solutions that had picked up mineral matter from exposed land surfaces.

It is to Forchhammer, Bischof, and especially Sandberger that we owe the clear statement that metals are widely disseminated in the sedimentary, metamorphic, and igneous rocks of the earth's crust, and that these metals are indigenous in these rocks and not introduced, except locally. Sandberger was too extreme in his view of the lateral secretion idea in that he advocated the genesis of the metals and gangue by secretion from the country rock immediately adjacent to the veins. It should be noted, however, that regarding certain gold veins he was essentially correct with respect to silica and a number of other compounds and elements, as will be discussed in other papers. Sandberger's quantitative analyses of country rocks and the minerals of these rocks established beyond all doubt that such metals as nickel, copper, cobalt, lead, zinc, and arsenic were widespread constituents of minerals such as olivine, micas, and augite. He also established the fact that slates and shales contain appreciable quantities of copper, lead, and zinc. His work has stood the test of time and has been amply confirmed by the researches of twentieth century geochemists.

SECRETION THEORIES OF THE ORIGIN OF EPIGENETIC GOLD DEPOSITS

Speculations on the origin of gold-quartz veins during the second half of the nineteenth century included variants of the secretion theory. Thus, Sir William Logan writing in 1863 in the *Geology of Canada* remarked:

> The observation among the gold-bearing rocks of the Southern States seems to show that the precious metal was originally deposited in the beds of various sedimentary rocks, such as slates, quartzites and limestones, and by subsequent process, it has been, in some instances accumulated in the veins which intersect these rocks. (p. 519)

Logan considered that the gold in the quartz veins in the Paleozoic schists of the eastern townships of Quebec probably had a similar origin.

Likewise, R. Daintree adopted a quite modern approach to the origin of the quartz reefs of Victoria, Australia. In *Geology of the District of Ballan* in 1866 he concluded:

> I had long ago come to the conclusion, that most, if not all, the gold in the quartz reefs was derived from the rocks in which these reefs occur. That the strata themselves received their supply of gold at the period of their deposition from the ocean in which they were deposited. That organic matter, and the gases generated therefrom on decomposition, sulphureted hydrogen, etc., was the cause of the precipitation; and that the amount of metallic deposit was in proportion to the amount of organic matter deposited with the oceanic sediment.
>
> That subsequent plication and desiccation of the sediment caused fissures, into which the mineral waters percolating the boundry rocks flowed and were decomposed, and their mineral contents were precipitated, possibly by magnetic currents, thus causing mineral veins. (p. 8)

It is of interest to note that A. R. C. Selwyn, then director of the geological survey of Victoria, in introducing Daintree's report wrote the following:

> I have carefully perused it (i.e. the report), and now beg to submit it, as requested by Mr. Daintree.
>
> In doing so, however, I wish to state that, though it contains a short summary of geological observations made during the survey, it can scarcely be regarded as a Geological Report of the district surveyed. The greater portion of the paper, as will be seen, is occupied with a somewhat imperfect statement of certain cosmical theories, first advanced by Sterry Hunt, on the possible relation of sedimentary deposits, generally, to upheavals and depressions of portions of the earth's crust, and their probable connection with the metamorphism and mineralisation of rock masses; together with

> other theories respecting the age, origin, and mode of occurrence of gold, quartz reefs, etc., in Victoria and other countries.
>
> Mr. Daintree has, I think, somewhat rashly applied these theories to the Ballan District, and has formed opinions on its geology deduced from them, which, in my opinion, are not at all borne out by the facts observed, if studied and reasoned on, as of course they should be, not by themselves, but in connection with the general geological features of the colony.

It should perhaps be mentioned that Selwyn's strong views were to reappear again in the acrimonious controversy between him and Sterry Hunt when they became co-directors of the Geological Survey of Canada (Boyle, 1971).

Gilpin (1888), speculating on the origin of the gold-quartz saddle reefs and veins of the Meguma group (slates and graywackes) in Nova Scotia, also advocated a secretion origin. He concluded from the evidence available to him (particularly the low gold content of the slates) that "the source of the gold in the Nova Scotia gold veins should be sought in the immediate enclosing strata," and that "so far as the subject has received attention the slates appear to be the source of the gold." Commenting on the origin of these veins and saddle reefs, I have come to the same conclusion (Boyle, 1966), basing my opinion on the low gold, arsenic, and other trace element contents of the shales, which taken as a whole provide an enormous reservoir for the vein elements (Au, Ag, As, S, etc.).

Becker (1882) was an enthusiastic advocate of the lateral secretion theory. Concluding his study of the Comstock Lode in Nevada, he observed:

> The diabase of the hanging wall when fresh was argentiferous and auriferous, and the precious metals of the lode are traced to this rock with much probability, the lateral-secretion theory being thus affirmed. It is further supported by the dependence of the other ore bodies of the district on the character of the inclosing rock. (p. xv)

Likewise, commenting on the auriferous quicksilver and other gold deposits of California, Becker (1888) opined:

> The evidence is overwhelmingly in favour of the supposition that the cinnabar, pyrite, and gold of the quicksilver mines of the Pacific slope reached their present positions in hot solutions of double sulphides, which were leached out from masses underlying the granite or from the granite itself. No one fact or locality absolutely demonstrates whether the metals were originally components of the granite or came from beneath it, but the tendency of the evidence at all points is to show that granite yielded the metals to solvents produced by volcanic agencies, and, when all the evidence is considered together, it is found that this hypothesis explains all the known circumstances very simply, while the supposition of an infragranitic origin leads to numerous difficulties. Though no one of these may be by itself fatal, when taken as a whole they appear to be so. As there is no known direct evidence pointing to an infragranitic origin

> of the quicksilver and the gold, I consider it tolerably well established that both were actually derived from the granite.
>
> I regard many of the gold veins of California as having an origin entirely similar to that of the quicksilver deposits. I also have some reason to suppose that some of the gold deposits were formed by the leaching of their walls by surface waters. (p. 449)

During the first half of the twentieth century secretion theories remained largely in abeyance especially as regards gold deposits, which were largely explained as originating from magmatic hydrothermal solutions as outlined in chapter 10. A number of geologists, however, noticed from detailed analyses that silica was lost from the altered zones of certain gold deposits and suggested that this silica was probably now present in the deposits as quartz. Strange as it may seem, the first observation of this phenomenon appears to have been by W. Lindgren, the great advocate of the magmatic hydrothermal theory. In a study of some gold deposits in amphibolites (greenstones) in western Australia, he (Lindgren, 1906, p. 539) noted that during alteration "the silica set free by the decomposition of the silicate has been deposited as quartz where it was not needed for the formation of the new silicates sericite and albite." Similarly, Moore (1912), another dedicated advocate of the hydrothermal theory, in a study of the alteration of a Precambrian granite to quartz-sericite schist at the St. Anthony gold mine, Sturgeon Lake, Ontario, found a decrease in the amount of silica as the quartz veins were approached and concluded that ". . . a comparison of the analyses shows a regular decrease in the silica of the granite indicating a transfer to the veins." Later McCann (1922) in his study of the gold-quartz veins in the Bridge River district, British Columbia, found a strong desilication of the wall rocks and postulated that "the silica which has been lost has probably contributed to the vein filling." And finally, Knopf (1929, p. 47) in his classic description of the Mother Lode system in California observed, after considering the alteration effects adjacent to the veins in greenstones, amphibolites, and serpentinites, that "the carbon dioxide liberated immense quantities of silica from the wall rocks, and this silica was delivered to the vein channels, where it was in part precipitated as quartz." Numerous other examples of the desilication phenomenon associated with the precipitation of gold-quartz veins are given in Boyle (1979).

The desilication phenomena that commonly accompanies gold deposition was examined in detail in the Yellowknife gold camp, Northwest Territories, Canada, by my own investigation during the 1950s. The results, pertaining particularly to the great shear zones of the greenstone belt, were summarized as follows (Boyle, 1955):

> The economic gold quartz veins and lenses at Yellowknife occur within shear zones in steeply dipping greenstones (amphibolites) of Precambrian age. The shear zones consist of chlorite schist that has been highly altered in the vicinity of the quartz veins and lenses. The zones of alteration enveloping the quartz bodies include an adjacent carbonate-sericite zone and an outer chlorite-carbonate zone that

> grades into the chlorite schist of the shear zone. Throughout the shear zones extensive chlorite-carbonate areas are present and these show no close relationship to the quartz bodies.
>
> Chemical analyses of samples taken across the strike and down the dip of the shear zones show that there is a marked loss of SiO_2 in the chlorite schist of the shear zones, when compared with the amphibolite of the wall rocks. This loss increases in the chlorite-carbonate zones and reaches a maximum in the carbonate-sericite zones. In one shear zone where the quartz lenses occur above extensive carbonate-chlorite schist portions, a marked decrease in SiO_2 with depth is evident.
>
> Graphs of the chemical analyses across the width of the alteration zones indicate that the loss of SiO_2 is compensated for by the addition of CO_2 and H_2O. The shape of the SiO_2 concentration line, exhibiting a low adjacent to quartz bodies, and increasing in value outward toward the wall rock, indicates that SiO_2 has migrated toward the present site of the quartz bodies.
>
> The chemical evidence strongly suggests that CO_2 and H_2O from mineralizing solutions displaced SiO_2 from the alteration zones and chlorite-carbonate areas of the shear zones and that this SiO_2 migrated toward dilation zones within the shear zones where it was deposited as quartz veins and lenses. (p. 51)

A study of the source and geochemistry of the volatiles (CO_2, H_2O, S, and B) that took part in the auriferous mineralization at Yellowknife was summarized as follows (Boyle, 1959):

> The gold deposits of the Yellowknife district occur in two distinct geological settings. The principal economic deposits occur in quartz-carbonate lenses in extensive chlorite schist zones (shear zones) cutting greenstone (amphibolite) rocks. The other deposits, of less economic importance, occur in quartz lenses in meta-sedimentary rocks.
>
> The deposits in the greenstones represent concentrations of silica, carbon dioxide, water, sulfur, arsenic, antimony, gold, and other metallic elements. Those in the sediments represent concentrations of silica, sulfur, boron, gold, and other metallic elements.
>
> For the deposits in the greenstones chemical evidence is presented to show that, under the influence of a strong thermal gradient, some of the carbon dioxide, water, sulfur, gold, silver, and other metallic elements in the original volcanic rocks was mobilized and migrated into the extensive shear zone systems. In the shear zones the chemical equilibrium was severely displaced, water and carbon dioxide reacted with the amphibolite rock producing extensive widths of chlorite and chlorite-carbonate schist, and silica, sulfur, gold, and numerous other elements present in the rock were mobilized. These mobilized constituents, together with those added by diffusion from

> the country rocks, migrated into dilatant zones, principally at shear zone junctions and flexures. In these sites they were precipitated as quartz, carbonates, sulfides, and gold.
>
> A similar process has operated to form the gold-quartz lenses in the meta-sediments. In these rocks silica, boron, sulfur, and various metallic elements were mobilized during the metamorphism of the sediments, and these migrated into and were precipitated in dilatant zones in faults, fractures, and drag folds in the rocks. (p. 1506)

Finally, a synthesis of the auriferous mineralization at Yellowknife, based on metamorphic and alteration secretion processes, was presented (Boyle, 1961 pp. 175-178) as follows:

> Throughout this memoir the thesis has been stressed that the elements present in the epigenetic deposits came from the country rocks in which the deposits occur and that they were concentrated by the interaction of metamorphic processes and the development of dilatant zones in the structures in which the deposits occur. The importance of the function of dilatancy cannot be too strongly emphasized. Its geological significance was first stressed by W.J. Mead (The geologic role of dilatancy, *Jour. Geol.* **33:**685-698), and in the author's opinion dilatancy is one of the most important factors in the genesis of epigenetic ore deposits. Dilatancy imposes a certain physicochemical effect on the surrounding rocks. Once a dilatant structure is formed, it constitutes a low-pressure zone as well as a zone of low-chemical potential for certain elements. To restore equilibrium such zones will be filled with the elements which are mobile within the rocks. During metamorphism the most mobile compounds and elements will be water, carbon dioxide, sulphur, arsenic, antimony, copper, lead, zinc, gold, silver, and in some cases silica. These elements and compounds probably migrate into dilatant structures by ionic diffusion through a nearly static flux of water vapour. Their route is mainly along crystal boundaries and minute cracks and fissures in the rock.
>
> The origin of the gold-quartz deposits at Yellowknife can best be illustrated by the following historical synthesis of the mineralization of the greenstone belt and sedimentary rocks. Figure 13-1 is a schematic representation of the synthesis as it applies to the greenstone belt.
>
> In early Precambrian time a thick series of marine sediments was laid down followed by extensive extrusions of lava flows which now form the greenstone belt. This was followed by further deposition of marine sediments (Yellowknife group).
>
> A period of great orogeny followed, during which the competent lavas were folded into a broad syncline and the less-competent sediments were complexly folded. Near the end of the orogenic period the sediments lying below the greenstones were granitized to form the western granitic mass. Large zones of the sediments lying above the greenstones (Yellowknife group) were also granitized to form the Prosperous Lake and southeastern granitic masses.

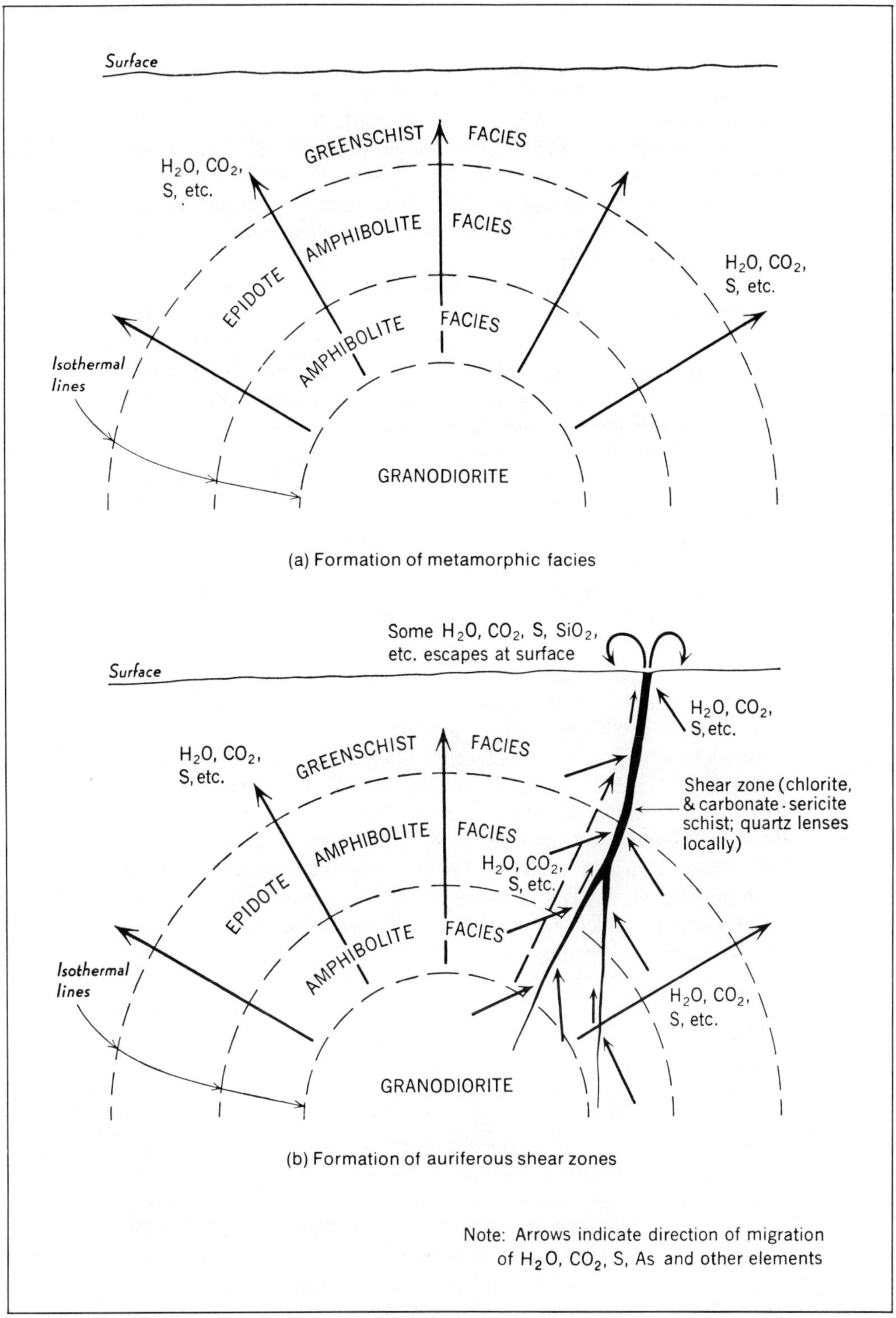

Figure 13-1. Schematic diagram illustrating the formation of metamorphic facies and auriferous shear zones.

During these orogenic processes, marked temperature gradients existed outward from the foci of granitization leading to the formation of the zone facies of metamorphism in the greenstone belt and sedimentary rocks. At this time much water, carbon dioxide, sulphur, and probably some of the chalcophile elements were mobilized and migrated toward the cooler parts, that is, down the temperature gradient at right angles to the isothermal lines as shown by the vectors in Figure 13-1*a*. If there had been no structural breaks in the rocks the result would have been a simple metamorphic halo about a granitized centre.

However, in the Yellowknife greenstone belt a great system of shear zones was formed near the end of the orogeny and formation of the granitic bodies. As a consequence the migration vectors were radically changed by the dilatancy of the shear zones as shown in Figure 13-1*b*. The shear zones literally sucked the mobile carbon dioxide, water, sulphur, and other elements from the country rocks and funnelled them toward the surface. Because of the great extent and depth of the shear zones the effects of dilatancy were undoubtedly felt over great volumes of the country rocks, and mobile elements were probably contributed from points thousands of feet horizontally and vertically from the shear-zone systems. Extensive structures which impose such marked dilatant effects on the country rocks may be termed 'first-degree' dilatant zones.

During the formation of the large shear zones, carbon dioxide, water, and sulphur reached a high concentration along them. As a consequence the chemical equilibrium was strongly displaced, and chloritization, carbonatization, pyritization etc., of enormous tonnages of volcanic rock took place. This led to the liberation of silica, potassium, calcium, iron, etc., in addition to gold, silver and other metallic elements present in the rock affected by alteration. These liberated compounds and elements, together with those added to the shear zones by diffusion from the country rocks, migrated laterally and vertically to 'second-degree' low-pressure dilatant zones at shear-zone junctions and other structural locales where contorted zones or openings were formed (Fig. 13-2). In these sites, secondary reactions, promoted by the low-pressure and lower-chemical potential, resulted in the precipitation of quartz, carbonates, and gold and silver bearing pyrite and arsenopyrite, forming quartz-sulphide lenses with adjacent alteration haloes.

During cooling of the deposits and further structural adjustments within the shear zones, much gold, silver, antimony, and other elements were exsolved from the early sulphides and found their way into local dilatant zones such as fractures in quartz lenses and local tension fractures and small faults in the shear zones. Some silica and the components of the late carbonates did likewise. This process gave rise to the successive later generations of native gold, sulphosalts, pyrite, quartz, and carbonates in the shear zones.

In the sedimentary rocks the quartz lenses containing sulphides and gold were concentrated by similar processes. During the severe structural deformation of the rocks many dilatant zones appeared in

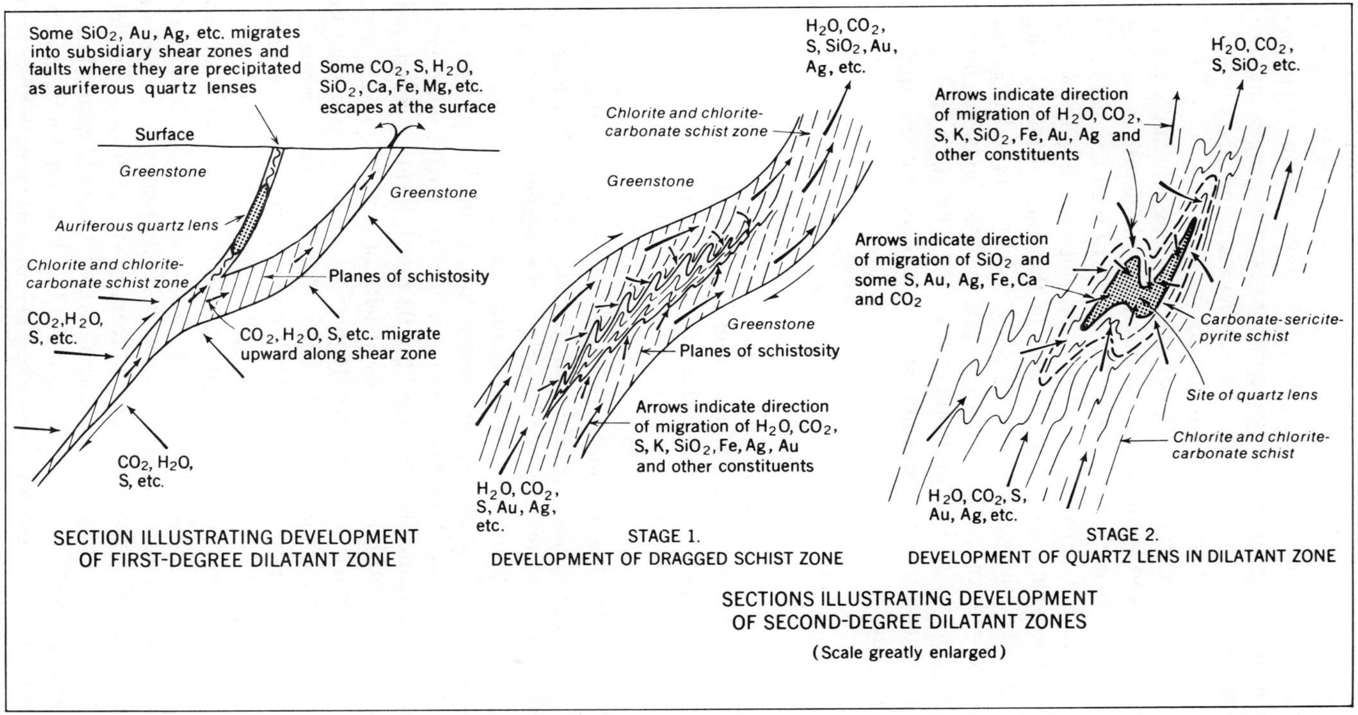

Figure 13-2. Diagrams illustrating development of first- and second-degree dilatant zones, alteration zones and auriferous quartz lenses.

> drag-folded parts of the beds, along faults, and along sheared axes of anticlines and synclines. To restore the chemical equilibrium within the rocks, silica, sulphur, boron, and various metallic elements migrated into these zones where they were precipitated, forming the gold-quartz-sulphide veins and lenses. The mobilization and migration of the compounds and elements forming the gangue and ore minerals was promoted by the high temperature and stress prevailing during the metamorphic processes.
>
> In late Precambrian time all rocks were extensively faulted, and this was followed by deposition of quartz and hematite at certain favourable sites in the late faults. The agents of regional metamorphism were not active at this time and only the highly mobile compound, water, took a major part in the mineralization. The water attacked the various minerals in the fault breccia and adjacent wall-rocks reducing them to chlorite and sericite. Silica was liberated together with iron and other elements. The silica and iron migrated into dilatant zones where they were precipitated as quartz and hematite.

In recent years numerous papers on the origin of epigenetic auriferous deposits in Archean terranes comprising greenstone and sedimentary belts have appeared, some advocating variants of the secretion theory.

Anhaeusser and co-workers (1969), Viljoen, Saager, and Viljoen (1969, 1970) and Anhaeusser (1976*a*, 1976*b*) consider that the gold in the deposits of the greenstone belts of southern Africa (Barberton Mountain land, Zimbabwe, etc.) and elsewhere was initially present in the mafic and ultramafic lavas and associated sediments and was mobilized and concentrated in dilatant zones during metamorphic events attendant upon granitic intrusion. In a more recent paper Saager (1973) advances a similar origin of epigenetic gold deposits in Precambrian rocks and suggests that the ultimate source of the gold in the vein deposits of the Swaziland greenstone belts is the "primitive" peridotitic and basaltic komatiites of the Onverwacht group rather than the intruding granites. The komatiites are relatively rich in gold (0.02 ppm). In more recent papers Anhaeusser (1976*a*, 1976*b*) and Anhaeusser with others (1975) note that the komatiites contain from 0.001ppm to 0.0015 ppm Au, a value substantially lower than the worldwide tholeiite average of 0.0025. They think two interpretations of these new data are possible. One is that these are primary concentrations, and the other is that the lavas contained significantly higher amounts of gold, but that later metamorphism and hydrothermal activity has erased all record of this fact. Belevtsev (1970) and Belevtsev with others (1972) consider that the source of the metals in a number of Precambrian deposits was the host rocks and think that they were mobilized and concentrated mainly by the action of metamorphism aided greatly by metamorphic water. The gold-quartz bodies in the greenstones of the Ramagiri goldfield in Andhra Pradesh, India, have been similarly explained by Ghosh and co-workers (1970), and Iyer (1970) considers these veins and others in the greenstone belts of southeast Karnataka (Mysore) to be of metamorphic secretion origin.

More recently numerous other deposits in Precambrian terranes have also been explained by metamorphic secretion theories. Here we may mention the Kalgoorlie goldfield in western Australia (Travis, Woodall, and Bartram, 1971); the Ashanti gold mine, Obuasi, Ghana (Wilson, 1971); the Kilo-Moto deposits, district of Mongbwalu, Zaire (Lavreau, 1973); the Homestake deposit, South Dakota (soon to be discussed in greater detail); Canadian vein and lode Archean gold deposits (Kerrich and Fryer, 1979, 1981; Hodder and Petruk, 1982, in which see especially Kerrich and Hodder, MacGeehan and Hodgson); a number of Precambrian and younger gold deposits in many parts of the world (Foster, 1984); and Precambrian gold deposits in general (Fyfe and Henley, 1973). Recent investigators in the People's Republic of China have explained the origin of the major Precambrian epigenetic deposits (veins, silicified bodies) by metamorphic secretion theories, the source of the gold being mainly remobilized from greenstones and associated sediments (Mu Ruishen, 1980; Liu Yingjun et al., 1984; Wang Konghai et al., 1984).

Knight (1957, p. 808) has postulated a source-bed concept of ore genesis of sulfide orebodies that can logically be applied to certain auriferous deposits. Briefly his idea is stated as follows:

> The *source bed concept* postulates that all sulfide orebodies of the majority of fields are derived from sulfides that were deposited syngenetically at one particular horizon of the sedimentary basin constituting the field, and that the sulfides subsequently migrated in varying degree under the influence of rise in temperature of the rock environment.

Among the examples that Knight includes in his discussion are the Witwatersrand gold field of South Africa, certain auriferous orebodies in western Tasmania, Australia, and presumably certain auriferous deposits in the Canadian Shield.

The origin of the famous Homestake gold deposit at Lead, South Dakota, has long been conjectural. Some investigators consider the deposit to be of Precambrian age; others are of the opinion that the gold mineralization is Tertiary, because deposits of that age occur in the vicinity. The gold-quartz replacement orebodies in the Homestake mine are, however, cut by Tertiary intrusives and faults, and the sulfur isotopic ratios of the sulfides in the Homestake are fundamentally different from those of the Tertiary deposits (Rye and Rye, 1974). Furthermore, on the basis of lead isotopes in galena from quartz veins, Rye, Doe and Delevaux (1974) concluded that the ore in the Homestake mine was emplaced by metamorphic secretion from a 2.5-billion-year-old source (i.e., the Homestake formation) during regional metamorphism and intrusion that occurred 1.6 billion years ago. The lead isotopes in Tertiary galenas from Paleozoic and Precambrian host rocks were found to have had different growth histories. The Tertiary galenas in Paleozoic rocks did not derive their lead entirely from Tertiary intrusive rocks in the area, but rather the lead may have been obtained mainly or in part from the Paleozoic rocks. Tertiary galenas in Precambrian rocks, on the other hand, seem to have obtained their lead largely from the transected metasedimentary host rocks or from the underlying basement, with little or no contribution from other sources. The abstracts of the two papers on the Homestake mine merit quotation in full because they bring unique methods to bear on the problem of the origin of this great deposit. The first by Rye and Rye (1974) follows.

> The Homestake gold mine is the largest producing gold mine in North America. The ore occurs as nearly conformable replacement bodies in dilatant zones of highly deformed Precambrian Homestake formation and is asssociated with arsenopyrite, pyrrhotite, and quartz in highly chloritized regions of sideroplesite or cummingtonite schist.
>
> The $\delta^{34}S$ values of pyrrhotite and arsenopyrite in the Homestake mine range from 5.6 to 9.8 permil in the gold-bearing Homestake formation, 2.7 to 5.1 permil in the underlying Poorman formation, and 4.1 to 29.8 permil in the overlying Ellison formation. The formational dependence of $\delta^{34}S$ values indicates a sedimentary origin for the sulfur in the deposit. Consideration of the influence of pH, f_{O_2}, temperature, and biogenetic activity on the $\delta^{34}S$ values of the primary sedimentary sulfides indicates that most of the sulfur in the Homestake mine was ultimately derived from sea-water sulfate.
>
> The $\delta^{34}S$ values of sulfides in individual ore bodies vary systematically with local structure and suggest that the original sedimentary sulfur migrated into dilatant zones during Precambrian metamorphism but not across formational boundaries.
>
> The $\delta^{18}O$ values of various type of quartz in the Homestake mine and regional segregation quartz veins show a strong dependence on the local wall rock ($>$13.8 permil in carbonate rock, $<$13.8 permil in silicate host rock) and indicate that the quartz in the Homestake mine was a normal part of the metamorphic sequence.
>
> Data on the δD of H_2O, $\delta^{13}C$ of CO_2, CO_2/H_2O mole ratios, and CH_4 content in fluid inclusions are consistent with a metamorphic origin for the ore body and are distinctly different from the data obtained on Tertiary ore deposits in the surrounding area. Furthermore, the Tertiary ore bodies in the area do not demonstrate any of the isotope systematics observed in the Homestake ore body.
>
> The isotopic and geologic data suggest that the gold and other constituents of the ore deposit were indigenous to the Homestake formation and were probably of syngenetic, "exhalative origin." The ore deposits were formed when the syngenetic components were concentrated in dilatant zones during metamorphism. (p. 293)

The abstract of the paper by Rye, Doe, and Delevaux (1974) details the lead isotopic data.

> A lead isotope study was carried out on galenas from vein quartz associated with gold mineralization in the Homestake and Clover Leaf mines, on galenas from ore deposits in the northern Black Hills of known Tertiary age, and on K-feldspar separates from Tertiary intrusive stocks from the same area. The leads from the Homestake and Clover Leaf mines have isotope ratios ($^{206}Pb/^{204}Pb$ = 15.653 to 16.517; $^{207}Pb/^{204}Pb$ = 15.375 to 15.595; $^{208}Pb/^{204}Pb$ = 35.415 to 35.551) that define a line with a slope of 0.2549 $\pm$ 0.0150 on a plot of

> $^{207}Pb/^{204}Pb$ versus $^{206}Pb/^{204}Pb$, which we interpret to represent lead derived from 2.5-b.y.-old source materials during metamorphism and intrusion at 1.6 b.y., thus indicating a Precambrian age for the Homestake gold deposit. Leads from deposits of known Tertiary age ($^{206}Pb/^{204}Pb$ = 18.434 to 20.858; $^{207}Pb/^{20}Pb$ = 15.666 to 15.992; $^{208}Pb/^{204}Pb$ = 38.196 to 38.744) fall into two distinct groups. Tertiary galenas found in Precambrian host rocks have lead isotope ratios that define a line with a slope of 0.1573 ± 0.0076, which again indicates an age of source materials of ~2.5 b.y. for a Tertiary mineralization age. The line defined by the Tertiary galenas in the Precambrian host rocks includes the K-feldspar data for the Tertiary intrusives. Galenas found in Paleozoic host rocks have lead isotope ratios that define a slope of 0.1392 ± 0.0113. From this set of data, the Tertiary veinlets in Precambrian host rocks and Tertiary intrusive rocks are interpreted to have been derived from Precambrian rocks. Lead in these galenas is interpreted to have been derived largely from the Paleozoic host rocks, and the high slope, indicative of an average source age of ~2.2 b.y., to partly reflect Precambrian source materials. (p. 814)

A number of epigenetic deposits of Paleozoic, Mesozoic, and Cenozoic age have been explained by metamorphic secretion, remobilization, or regeneration theories in recent years. Space precludes discussion here of more than a few of the papers on these deposits, but the interested reader may consult the monograph by Boyle (1979, p. 396) for more details. Of particular interest are the Tertiary deposits associated with volcanic centers, especially caldera.

It will be recalled from the first part of this chapter that Becker (1882) found the hard, fresh undecomposed diabase (andesite) hosting the Comstock Lode to contain twice as much gold and silver as the highly altered (propylitized) rocks in the immediate vicinity of the lode, and that from these data he drew the conclusion that the filling of the lode was by lateral secretion, the precious metals and the gangue being derived essentially from the propylitized rocks. The same idea was held by B. von Inkey, F. W. Hutton, J. M. Maclaren and others for the origin of the Hauraki gold-quartz veins in New Zealand (see Finlayson, 1909, and Williams, 1974).

Furthermore, it will be recalled that Agricola in *De ortu et causis subterraneorum* advocated a meteoric water secretion mechanism for the origin of auriferous and other veins in the Erzgebirge of central Europe, many of which are of Hercynian and Tertiary age.

Recent research tends to confirm these early views. Thus, investigations utilizing $^{18}O/^{16}O$, D/H and other isotopic ratios suggests that gold deposition in many of the Tertiary and younger deposits took place essentially from meteoric water (including in places sea water) set into convective circulation by interaction with rising igneous intrusions. Examples are widespread, including Tonopah, Goldfield, and the Comstock Lode, Nevada (Taylor, 1973, 1974); Bodie, California (O'Neil et al., 1973; O'Neil and Silberman, 1974); Creede, Colorado (Bethke, Barton, and Rye, 1973); Sunnyside Mine, Eureka mining district, San Juan County, Colorado (Casadevall, 1976); Tui Mine, Hauraki goldfield, New Zealand (Robinson, 1974); Kurokô deposits, Japan (Sakai and Matsubaya, 1974; Ohmoto and Rye, 1974); and epithermal Au-Ag districts in United States (White, 1955,

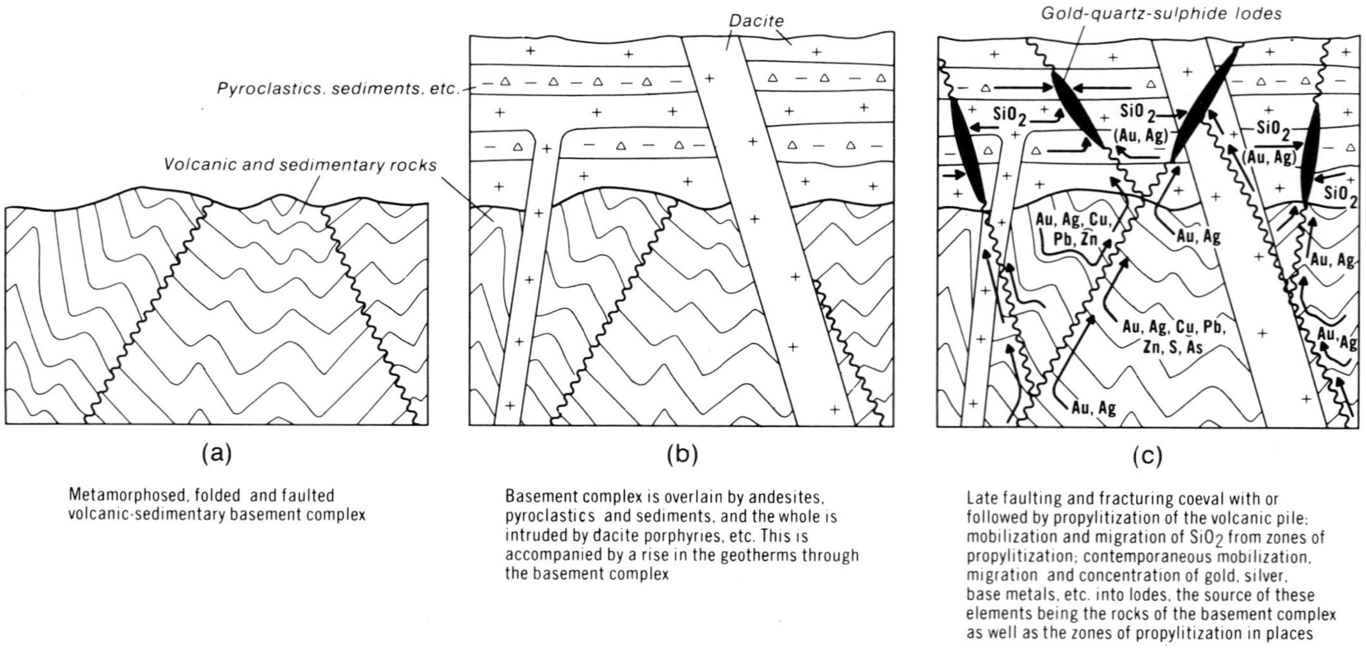

Figure 13-3. Schematic diagrams illustrating the sequence of events in the formation of certain gold-silver epithermal deposits.

1974). In general, therefore, it appears that meteoric waters have dominated during mineralization processes in epithermal gold-silver belts, but some investigators consider that magmatic water also played a part (Robinson, 1974; O'Neil and Silberman, 1974; White, 1955, 1974).

While the question of the source of the water seems at least partly answered for the so-called auriferous epithermal deposits, there still remain problems with respect to the source of the base metals, sulfur, gold, and silver. Some investigators have concluded that the magmas with which the meteoric water has interacted are the most reasonable source; others consider the volcanic sedimentary pile or the underlying basement rocks more likely sources. The last two appear more probable as suggested in Boyle (1979) and shown schematically in Figure 13-3.

This model appears to be particularly applicable to the origin of certain gold deposits in rocks overlying older geological terranes (e.g., Paleozoic rocks on Precambrian terranes and Tertiary andesites and pyroclastics lying on Mesozoic rocks). In this model the source rocks for much of the gold and its associated elements may be underlying pyritic graphitic schists, volcanics, and so forth, and the receptacle rocks for the deposits may be overlying faulted and fractured andesites, pyroclastics, limestone, and others. During igneous activity the heat wave associated with intrusive bodies or granitization fronts liberates water, carbon dioxide, and other volatiles from the basement rocks, promoting in turn the migration of silica, sulfur, base metals, silver, and gold, which rise into the available dilatant zones such as faults, fractures, and breccia pipes to be precipitated as quartz, sulfides, and native gold. Such an ore-genesis model appears to answer many of the problems associated with the formation of epigenetic Tertiary gold deposits. In this respect it is of interest to note that many of the basement rocks underlying Tertiary deposits in many parts of the world are Mesozoic or older graphitic, pyritic schists, graywackes, slate, and volcanic assemblages, rocks that are commonly enriched in gold, silver, sulfur, and the various base metals now present in the Tertiary gold deposits.

CONCLUSION

Secretion theories have a long history, having been in vogue at certain times and out of favor at others. As our geochemical knowledge of the distribution of gold in the earth grows, and as more is understood about the chemistry of the migration of gold during diagenetic and metamorphic processes, it seems certain that secretion theories will assume a dominant place in the genesis of auriferous (and other) mineral deposits.

REFERENCES AND SELECTED BIBLIOGRAPHY

Agricola, G., 1546*a*. *De natura eorum quae effluunt ex terra,* Froben, Basel.

Agricola, G., 1546*b*. *De ortu et causis subterraneorum,* Froben, Basel.

Anhaeusser, C. R., 1976*a,* The nature and distribution of Archaean gold mineralization in southern Africa, *Miner. Sci. Eng.* **8:**46–84.

Anhaeusser, C. R., 1976*b,* Archean metallogeny in southern Africa, *Econ. Geology,* **71:**16–43.

Anhaeusser, C. R., R. Mason, M. J. Viljoen, and R. P. Viljoen, 1969, A reappraisal of some aspects of Precambrian Shield geology, *Geol. Soc. America Bull.*, **80** (11):2175-2200.
Anhaeusser, C. R., K. Fritze, W. S. Fyfe, and R. C. O. Gill, 1975, Gold in "primitive" Archaean volcanics, *Chem. Geol.* **16:**129-135.
Becker, G. F., 1882. Geology of the Comstock Lode and the Washoe district, U.S. *Geol. Survey Monographs,* vol. 3, 422p.
Becker, G. F., 1888. The geology of the quicksilver deposits of the Pacific Slope, *U.S. Geol. Survey Monographs,* vol. 13, 486p.
Belevtsev, Ya. N., 1970. Sources of metals of metamorphic-hydrothermal deposits, in *Problems of Hydrothermal Ore Deposition,* Z. Pouba and M. Stemprok, eds., Int. Union Geol. Sci., ser. A, no. 2, Schweizerbart'sche, Stuttgart, pp. 30-35.
Belevtsev, Ya. N., V. Yu. Fomenko, V. N. Kucher, and S. V. Kuzenko, 1972. Mobilization of metals from sedimentary and metamorphic rocks by aqueous solutions, Geol. Zhur. **32** (3):42-51.
Bethke, P. M., P. B. Barton, and R. O. Rye, 1973. Hydrogen, oxygen and sulfur isotopic compositions of ore fluids in the Creede district, Mineral County, Colorado, *Econ. Geology* **68:**1205 (abstr.).
Bischof, K. G., 1847-1855. Lehrbuch der chemischen und physikalischen Geologie, vols. 1-3, suppl. by F. Zirkel, 1871. (Trans. by B. H. Paul and J. Drummond, 1854-1855, as *Elements of Chemical and Physical Geology,* 2 vols., Cavendish Society, London).
Boyle, R. W., 1955. The geochemistry and origin of the gold-bearing quartz veins and lenses of the Yellowknife Greenstone Belt, *Econ. Geology* **50:**51-66.
Boyle, R. W., 1959. The geochemistry, origin, and role of carbon dioxide, water, sulphur and boron in the Yellowknife gold deposits, Northwest Territories, Canada, *Econ. Geology* **54:**1506-1524.
Boyle, R. W., 1961. The geology, geochemistry, and origin of the gold deposits of the Yellowknife district, *Canada Geol. Survey Mem. 310,* 193p.
Boyle, R. W., 1963. Diffusion in vein genesis, *Problems of Post Magmatic Ore Deposition Symposium,* Prague, Czech., vol. 1, pp. 377-383.
Boyle, R. W., 1966. Origin of the gold and silver in the gold deposits of the Meguma Series, Nova Scotia (Abstract), *Canadian Mineralogist* **8:**622.
Boyle, R. W., 1971. Thomas Sterry Hunt (1826-1892)—Canada's first geochemist, *Geol. Assoc. Canada Proc.* **23:**15-18.
Boyle, R. W., 1979. The geochemistry of gold and its deposits, *Canada Geol. Survey Bull. 280,* 584p.
Casadevall, T., 1976. Sunnyside mine, Eureka mining district, San Juan County, Colorado: geochemistry of gold and base metal ore formation in the volcanic environment, *25th Int. Geol. Congr., Sydney, Australia, abstr.,* **3:**805.
Daintree, R., 1866. Geology of the District of Ballan . . . , *Geol. Survey Victoria (Australia)* 15, 11p.
Darwin, C., 1846. *Geological Observations in South America,* Smith, Elder & Co., London.
Daubrée, G. A., 1860. *Etudes et expériences synthétiques sur le métamorphisme et sur la formation des roches crystallines, Acad. Sci. Paris Mém.* pres. **17:**1-127.
Daubrée, G. A., 1879. *Etudes synthétiques de géologie expérimentale,* Paris.
Daubrée, G. A., 1887*a. Les eaux souterraines à l'époque actuelle,* Paris.
Daubrée, G. A., 1887*b. Les eaux souterraines aux époques anciennes,* Paris, 443p.
Daubrée, G. A. For a listing of Daubrée's voluminous works see De Lapparent, *Soc. Géol. France Bull.* 3rd ser. **25:**245-284, 1897.
De la Beche, H. T., 1839. Report on the Geology of Cornwall, Devon and West Somerset. London.
Delius, C. T., 1770. *Abhandlung von dem Ursprunge der Gebürge und der darinne befindlichen Eradern,* Leipzig.
Delius, C. T., 1773. *Anleitungen sur der Bergbaukunst nach ihrer Theorie und Ausübung,* Vienna.
Emmons, S. F., 1886*a*. Geology and mining industry of Leadville, Colorado, *U.S. Geol. Survey Monographs,* vol. 12, 770p.

Emmons, S. F., 1886*b*. The genesis of certain ore deposits; *Am. Inst. Min. Eng. Trans.* **15**:125-147.

Finlayson, A. M., 1909. Problems in the geology of the Hauraki gold fields, New Zealand, *Econ. Geology* **4**:632-645.

Forchhammer, J. G., 1855. Ueber den Einfluss des Kochsalzes auf die Bildung der Mineralien, *Poggendorf's Ann. Phys. Chem.* **95**:60-96.

Foster, R. P., ed., 1984. *Gold '82: The Geology, Geochemistry and Genesis of Gold Deposits,* A. A. Balkema, Rotterdam, 753p.

Fyfe, W. S., and R. W. Henley, 1973. Some thoughts on chemical transport processes, with particular reference to gold, *Miner. Sci. Eng.* **5**:295-303.

Ghosh, D. B., B. B. K. Sastry, A. J. Rao, and A. A Rahim, 1970. Ore environment and ore genesis in Ramagiri Gold Field, Andhra Pradesh, India, *Econ. Geology* **65**:801-814.

Gilpin, E., 1888. Notes on the Nova Scotia gold veins, *Royal Soc. Canada Trans.* sec. 4, **6**:63-70.

Hodder, R. W., and W. Petruk, ed., 1982. Geology of Canadian gold deposits, *Canadian Inst. Min. Metall. Spec. Vol. 24,* Montreal, 286p.

Hunt, T. Sterry, 1873*a*. The geognostical history of the metals, *Am. Inst. Min. Eng. Trans.* **1**:331-346.

Hunt, T. Sterry, 1873*b*. The origin of metalliferous deposits, *Am. Inst. Min. Eng. Trans.* **1**:413-426.

Hunt, T. Sterry, 1897. *Chemical and Geological Essays,* 5th ed.; Scientific Pub. Co., New York, 489p.

Iyer, G. V. A., 1970. *Geochemistry of the Alkali Feldspars and origin of Gold-Quartz Veins in the Precambrian of S.E. Mysore,* unpbl. Ph.D. thesis, Indian Inst. Sci., Bangalore, 224p.

Kerrich, R., and B. J. Fryer, 1979. Archaean precious metal hydrothermal systems, Dome mine, Abitibi greenstone belt. II. REE and oxygen isotope relations, *Canadian Jour. Earth Sci.* **16**:440-458.

Kerrich, R., and B. J. Fryer, 1981. The separation of rare elements from abundant base metals in Archean lode gold deposits: implications of low water/rock source regions, *Econ. Geology* **76**:160-166.

Knight, C. L., 1957. Ore genesis—The source-bed concept, *Econ. Geology* **52**:808-817.

Knopf, A., 1929. The Mother Lode system of California, *U.S. Geol. Survey Prof. Paper 157,* 88p.

Lavreau, J. J., 1973. New data about the Kilo-Moto gold deposits (Zaire), *Mineralium Deposita* (Berlin) **8** (1):1-6.

Lindgren, W., 1906. Metasomatic processes in the gold deposits of Western Australia, *Econ. Geology* **1**:530-544.

Liu Yingjun, Zhang Jingrong, Qiao Enguang, Mao Huixin, and Zhao Meifang, 1984. Geochemistry of gold deposits in west Hunan and east Guangxi, China, *Geochemistry* **3** (4):307-321.

Logan, W. E., A. Murray, T. Sterry Hunt, and E. Billings, 1863. *Geology of Canada,* Dawson Brothers, Montreal, 983p.

McCann, W. S., 1922. The gold-quartz veins of Bridge River District, B.C. and their relationship to similar ore-deposits in the Western Cordilleras, *Econ. Geology* **17**:350-369.

Moore, E. S., 1912. Hydrothermal alteration of granite and the source of vein-quartz at the St. Anthony Mine, *Econ. Geology* **7**:751-761.

Mu Ruishen, 1980. The main genetic types of Chinese gold deposits and their prospecting, *Chinese Acad. Geol. Sci. Bull.* ser. 5, **1** (1):20-40.

Ohmoto, H., and R. O. Rye, 1974. Hydrogen and oxygen isotopic compositions of fluid inclusions in the Kurokô deposits, Japan, *Econ. Geology* **69**:947-953.

O'Neil, J. R., and M. L. Silberman, 1974. Stable isotope relations in epithermal Au-Ag deposits, *Econ. Geology* **69**:902-909.

O'Neil, J. R., M. L. Silberman, B. P. Fabbi, and C. W. Chesterman, 1973. Stable isotope and chemical relations during mineralization in the Bodie mining district, Mono County, California, *Econ. Geology* **68**:765-784.

Phillips, J. A., 1884. *A Treatise on Ore Deposits,* MacMillan, London, 651p.

Pryce, W., 1778. *Mineralogia Cornubiensis,* London.
Robinson, B. W., 1974. The origin of mineralization at the Tui mine, Te Aroha, New Zealand, in the light of stable isotope studies, *Econ. Geology* **69:**910-925.
Rye, D. M., and R. O. Rye, 1974. Homestake gold mine, South Dakota: I. Stable isotope studies, *Econ. Geology* **69:**293-317.
Rye, D. M., B. R. Doe, and M. H. Delevaux, 1974. Homestake gold mine, South Dakota: II. Lead isotopes, mineralization ages, and source of lead in ores of the Northern Black Hills, *Econ. Geology* **69:**814-822.
Saager, R., 1973. Metallogenese präkambrischer Goldvorkommen in den vulkano-sedimentaren Gesteinskomplexen (greenstone belts) der Swaziland-Sequenz in Südafrika, *Gel. Rundschau* **62:**888-901.
Sakai, H., and O. Matsubaya, 1974. Isotopic geochemistry of the thermal waters of Japan and its bearing on the Kurokô ore solutions, *Econ. Geology* **69:**974-991.
Sandberger, F., 1882-1885. Untersuchungen über Erzgänge; 2 vols. Weisbaden.
Taylor, H. P., 1973. O^{18}/O^{16} evidence for meteoric-hydrothermal alteration and ore deposition in the Tonopah, Comstock Lode, and Goldfield Mining districts, Nevada, *Econ. Geology* **68:**747-764.
Taylor, H. P., 1974. The application of oxygen and hydrogen isotope studies to problems of hydrothermal alteration and ore deposition, *Econ. Geology* **69:**843-883.
Travis, G. A., R. Woodall, and G. D. Bartram, 1971. The geology of the Kalgoorlie goldfield, *Geol. Soc. Australia Spec. Publ. 3,* J. E. Glover, ed., pp. 175-190.
Van Hise, C. R., 1901. Some principles controlling the deposition of ores, *Am. Inst. Min. Eng. Trans.* **30:**27-177.
Van Hise, C. R., 1904. A treatise on metamorphism, *U.S. Geol. Survey Monographs,* vol. 47, 1286p.
Viljoen, R. P., R. Saager, and M. J. Viljoen, 1969. Metallogenesis and ore control in the Steynsdorp Goldfield, Barberton Mountain Land, South Africa, *Econ. Geology* **64:**778-797.
Viljoen, R. P., R. Saager, and M. J. Viljoen, 1970. Some thoughts on the origin and processes responsible for the concentration of gold in the Early Precambrian of southern Africa, *Mineralium Deposita* (Berlin) **5:**164-180.
Wallace, W., 1861. *The Laws Which Regulate the Deposition of Lead Ore in Veins, Illustrated by . . . the Mining District of Alston Moor,* London, 258p.
Wang Konhai, Qiou Youshou, Cui Keying and Han Shizheng, 1984. The control conditions of the gold deposits in Zaoyuan-Yexian Area, Shandong, Shenyang Inst. Geol. Min. Resources, *Chinese Acad. Geol. Sci. Bull. No. 9,* pp. 10-29.
White, D. E., 1955. Thermal springs and epithermal ore deposits, *Econ. Geology* **50th anniv. vol.** (pt. 1):99-154.
White, D. E., 1974. Diverse origins of hydrothermal ore fluids, *Econ. Geology* **69:**954-973.
Williams, G. J., 1974. Economic geology of New Zealand, *Australasian Inst. Min. Metall. Monog.* ser. 4, 2nd ed., 490p.
Wilson, I. R., 1971. Progress report on a study on geochemical variations in wall rock, Ashanti gold mine, Obuasi, Ghana, *15th Annual Rep., Res. Inst. Afr. Geol.,* p. 39-43.
Woodward, J., 1695. An essay towards a natural history of the earth and terrestrial bodies . . . , London.
Zimmerman, C. F., 1746. *Untererdischen Beschreibung der Meissnischen Erzgebirges,* Obersächische Bergakademie, Dresden.

CHAPTER

14

Gold Deposits—Quartz-Pebble Conglomerate and Quartzite Type

It seems to me that there are no valid objections to the theory of marine placer origin (for the Rand).
G. F. Becker, 1897

For my part, I find it simpler to regard the Rand deposits themselves as original hydrothermal products.
L. C. Graton, 1930

When the centenary of the Witwatersrand goldfields is celebrated in 1986, disputes will still be raging as to how and where the auriferous sediments formed.
D. A. Pretorius, 1975

Quartz-pebble conglomerates and associated quartzites host the largest known concentrations of gold, providing more than 50% of the world's annual gold production. Some of these deposits also contain economic concentrations of uranium, thorium, and rare earths, and minor amounts of recoverable platinoids. The largest economic gold deposits in quartz-pebble conglomerates are those of the Witwatersrand (Rand) in the Republic of South Africa; during a century of mining these deposits have produced 37 million kg (1200 million oz) of gold. Similar but smaller deposits occur in the Tarkwian system of Ghana and in the Jacobina series, Bahia, Brazil. Analogous deposits, but containing only economic quantities of uranium, thorium, and rare earths, occur in the Elliot Lake-Blind River district, Ontario, Canada.

The productive quartz-pebble conglomerate and associated quartzite deposits are all of Precambrian age. Briefly, the mineralized economic beds comprise oligomictic quartz-pebble conglomerates and relatively clean quartzites within which native gold is disseminated. The associated minerals in the Rand include principally pyrite, commonly spherical (buckshot pyrite), pyrrhotite, a large number of other sulfides and sulfosalts, thucholite and ucholite with associated uraninite and other Th-U minerals, and native gold. In the Tarkwian quartz-pebble conglomerates and quartzites the principal minerals are hematite, ilmenite, magnetite, rutile, and native gold. The Jacobina quartz-pebble conglomerates and quartzites contain essentially pyrite, a few other sulfides, uraninite (pitchblende), and native gold. Detailed descriptions of these deposits have been compiled by Boyle (1979).

There is a vast literature on the auriferous and uraniferous quartz-pebble conglomerates and associated quartzites. While the deposits are remarkably similar in many respects, in detail there are numerous differences, some small and some major, that serve to stir up controversies about their origin. In fact, perhaps no metallic deposit has exercised the minds of economic geologists as regards origin as much as the gold-bearing quartz-pebble conglomerates since their discovery on the Witwatersrand of South Africa in 1886. By the mid-1940s the protracted controversies between the placerists, modified placerists, and hydrothermalists had subsided, only to flare up again with renewed vigor and acrimony among a new generation when uranium was discovered or rediscovered in some of the deposits.

The principal theories advocated for the origin of the auriferous and uraniferous quartz-pebble conglomerates and quartzites are placer, modified placer (re-solution theory), and hydrothermal (infiltration theory). Many papers dealing with these three theories and their variants have been published; I have attempted only to include those of a definite landmark quality.

THE WITWATERSRAND

Early geologists on the Rand, including E. Cohen, J. Ballot, L. de Launay, D. A. Louis, and G. F. Becker, considered the gold to be strictly of placer origin without any significant modification. Of these, G. F. Becker (1897) was perhaps the most explicit. In his memoir on the Witwatersrand and other similar deposits he concluded:

> Thus, it seems to me that there are no valid objections to the theory of marine placer origin and no noteworthy features left unexplained by this theory, while so much can not be said of either the impregnation or the precipitation theory. The beach deposits of the Pacific form an excellent instance of such marine placers, although they are not so rich as those of the Rand. On the other hand, no case is known of an extensive gold deposit which is certainly an impregnation in sand or gravel, nor is there an established case of marine precipitation. (p. 176)

At the turn of the century J. W. Gregory (1907) held to the same view but thought that the black sand deposited with the gold was converted to pyrite and the gold was simultaneously dissolved and redeposited in situ. His extensive conclusions merit reproduction and are presented as Paper 14-1.

(Text continues on page 339.)

14-1: THE ORIGIN OF THE GOLD IN THE RAND BANKET

J. W. Gregory

Reprinted from pages 39-41 of *Inst. Mining and Metallurgy Trans.* **17**:2-41 (1907).

Summary of Conclusions.

I. The theory as to the origin of the banket in best agreement with the facts appears to be that which regards the banket as a marine placer, in which gold and black sand (magnetite with some titaniferous iron) were laid down in a series of shore deposits. The gold was in minute particles, and it was concentrated by the wash to and fro of the tide, sweeping away the light sand and silt, while the gold collected in the sheltered places between the larger pebbles. The black sand deposited with gold has been converted into pyrites and at the same time the gold was dissolved and re-deposited *in situ*.

II. The distribution of the gold agrees with that of placer deposits in the following respects:—

(*a*) The gold, as in a placer, is contained in the cement and not in the pebbles.

(*b*) The gold has a widespread horizontal, and narrow vertical range.

(*c*) The gold is spread through layers which are conformable to the sediments and is not deposited in verticals or fault planes across the bedding of the rocks, except in the case of a few unimportant secondary quartz veins.

(*d*) The gold is distributed in patches and not in shoots.

(*e*) The gold tends to occur on the footwall side of the conglomerate beds, or to rest on layers of quartzites which acted as false bottoms in the reef series. Owing to the redistribution of the gold during its solution this rule is not as general as in the case of a recent placer; but it appears to be the general experience through the Rand.

(*f*) The rich patches occur at varying horizons dependent upon the frequent local variations in the currents that necessarily occur during the deposition of a series of deposits upon a shore.

III. The objections to the infiltration theory include:—

(*a*) The absence of ore-shoots.

(*b*) The non-existence of the "verticals" up which the gold may have been introduced.

(*c*) The limitation of the gold to special seams of conglomerate, and its absence from beds of sand and bastard reef, which lie immediately below rich banket, and must have been equally open to percolating solutions.

IV. The essential difference in the distribution of the gold between the placer and infiltration theories is that, according to the former, the gold should originally have been deposited at the same time as the deposition of the conglomerates; whereas, according to the infiltration theory, the gold should have been introduced after the formation of the whole sedimentary series. That the gold was contemporary with the conglomerates is shown by—

(*a*) The beds of ore being always parallel to the bedding planes of the rocks.

(*b*) The absence of the "verticals" of South Dakota, and the copper slates of Thuringia.

(*c*) The presence of gold in the conglomerates before they were cut through by the contemporary erosion, which led to the formation of the "wash out" channel in the May Consolidated Mine.

V. The microscopic evidence shows—

(*a*) There is no evidence of infiltration, or the presence of the secondary minerals typical of infiltration processes.

(*b*) There has been no conversion of the banket into continuous sheets of vein quartz.

(*c*) The secondary minerals produced, such as the chloritoid, are typical of pressure-metamorphism, and have been developed alike in rich and barren rocks. Rich and poor banket, bastard reef and quartzite, are shown by the microscopic evidence to have all undergone the same changes, and the richest material sometimes shows less change than barren material.

VI. The banket differs from the gold ores due to infiltration in other fields, for example, those of South Dakota; and the best general agreement is with that band of modern beach placers which extends for 50 miles along the western coast of the South Island of New Zealand. The Kanowna lead is quoted as a case of the solution and re-deposition of gold in a modern placer.

VII. The absence of conclusive evidence of any considerable impoverishment in depth is consistent with the alluvial origin of the gold, and the placer theory is favourable to the further extension of the banket in depth.

A somewhat similar theory to that of the placer holds that the ores are syngenetic, but that the gold and/or uranium are the result of chemical precipitation processes rather than mechanical concentrations. Adherents of this theory include Penning (1888), de Launay (1913), Garlick (1953), Miholic (1954), and Reimer (1975). As regards gold, Penning's paper is a landmark in the elucidation of this theory, and his conclusions as reproduced below merit consideration.

> If the gold were alluvial, brought into the gravels in the same manner as the pebbles must have been—by the action of running water, it would certainly be found in "leads," and probably the larger portion of it accompanying the heavier pebbles. But thus far it appears to be equally distributed, for each banket, although it may differ in auriferous value from those above and below, maintains an equable richness for long distances. The conglomerate, as we have seen, was spread with remarkable regularity over a large area, and we are justified in concluding that, as in so many points within that area the conglomerate is rich in gold, so it will be in the intervening portions, because there is not evident reason why the agencies of gold distribution should not have been equally uniform. It is contended that, in the absence of any evidence to the contrary, it is right to assume that the banket is no less rich in the intervals between those points; that, in short, the opened ground is a fair indication of that which is undeveloped. Looking to the sharp form of the grains of this gold, to its freedom from the coating so often found on that which is really alluvial, and to the equality of its distribution, I am convinced that this was deposited—at the same time that the gravels were being accumulated—from water holding gold in solution. (p. 437)

A number of investigators of the Rand argue that the constituents of the ores (gold, uranium, thorium) are of hydrothermal origin, although some are hazy about the source of the gold and uranium. Presumably they consider that these metals were derived from a cooling magma. The supporters of this theory include Hatch and Corstorphine (1905), Horwood (1917), Graton (1930), and Davidson (1953, 1965).

The principal evidence advanced in favor of the hydrothermal theory by these writers follows.

1. The gold as it is now found is not in the form of nuggets or dust; on the contrary, it is similar in its various forms and aspects to epigenetic gold. The silver content is also high, a feature that is not generally found in placer deposits.
2. Much of the gold is not at the base of the conglomerate beds as is normal in placers; rather, the gold is commonly scattered throughout the bankets; some is even at the top of the bankets.
3. The gold is in shoots or pay streaks that are not identifiable with true pay streaks, leads, bars, channels, and so forth in modern placers.
4. The bulk of the pyrite is certainly not detrital; some pyrite is like that found in epigenetic deposits.

5. The bulk of the uraninite is probably not detrital; it is in veins and veinlets and scattered at random through thucholite (ucholite).
6. The bankets show a general paucity of detrital minerals, especially magnetite. The placerists claim, however, that the magnetite has been sulfidized.
7. Elements such as Cu, Pb, Zn, and Ag are present, bound in various sulfides and sulfosalts; these minerals, all investigators agree, exhibit epigenetic features.
8. Sericite, chlorite, carbonate, tourmaline, and other hydrothermal minerals are present in the deposits, showing in some places an intimate relationship to gold and seeming to have been precipitated with it.
9. Quartz veins carrying gold and a variety of epigenetic sulfides are found in faults, fractures, and gash fractures.
10. There is some similarity to copper deposits in conglomerates such as those in Michigan; the copper was obviously introduced into the conglomerates—why not the gold? The geological setting is somewhat similar as well. There are conglomerates and quartzites in a basin interbedded with volcanics and considerable evidence of igneous activity in both places (e.g., Duluth Batholith and Bushveld complex).

Hatch and Corstorphine (1905, pp. 145-146) in their book *The Geology of South Africa* reviewed the various theories of the origin of the gold and opted for the subsequent infiltration (hydrothermal) theory for the origin of the gold. Their summary concerning the infiltration theory follows.

> *The Infiltration Theory*—The infiltration theory satisfactorily accounts for most of the phenomena, without postulating any especially abnormal conditions. The fact that the cementing material of the conglomerates is so largely composed of secondary minerals shows that the beds must have been subject to much percolation by mineralising solutions before final consolidation. The slow and repeated passage of such solutions, even if carrying only a minute quantity of gold, would gradually enrich the matrix.
>
> The most striking feature of the gold contents of the Rand conglomerates is the limitation of the gold practically to the zone known as the Main Reef Series. That only certain beds should be the main gold carriers, seems to speak strongly for the theory of subsequent infiltration, for it is difficult to believe, as required by the placer theory, that only one series of conglomerate beds should be derived from pre-existing gold-bearers. The assumption necessary on the infiltration theory, that the entire series of conglomerates and quartzites was permeated by the auriferous waters, presents no serious difficulty. The limitation of the gold deposition to definite zones was probably governed by certain chemical conditions, such as, for instance, the presence of a reducing agent in these and not in the other beds. In this way alone does it seem possible to explain the existence of the gold in beds near the centre of a conformable series. At the same time, the theory of a general percolation of auriferous waters accounts for the occasional presence of payable gold in conglomerates above and below the Main Reef horizon, as well as for the existence of richer chutes or patches in the main gold-

> bearing series: for, wherever the necessary conditions prevailed, precipitation took place. What this reducing agent was it is difficult to say. The frequent association of the gold with pyrites suggests that the latter had something to do with precipitation. On the other hand, in some rich conglomerate reefs, carbonaceous matter is plentifully present: possibly this substance may have played a part in the precipitation.
>
> Summarising, it may be said that the theory of subsequent infiltration is preferable to the others for the following reasons:
>
> 1. The gold is practically confined to the matrix of the conglomerate, occurring there in association with other minerals of secondary origin: the rare cases in which gold occurs in the pebbles are obviously instances of infiltration along cracks, a fact which in itself lends support to the theory.
> 2. It occurs in crystalline particles often surrounding or lying in close association with pyrites crystals or marcasite concretions, which are of secondary origin.
> 3. It is uniformly distributed to a remarkable degree.
> 4. It is restricted to certain definite beds.

C. B. Horwood (1917), in a detailed and important memoir on the gold deposits of the Rand, reviewed all of the theories then extant and found all wanting except the hydrothermal theory. His concluding remarks represent a landmark in the controversies regarding the origin of gold in the Rand and are included in toto (Paper 14-2).

L. C. Graton, professor of geology at Harvard University, visited the Rand in the late 1920s and prepared a long treatise on the hydrothermal origin of the Rand gold deposits of which part 1 was titled *Testimony of the Conglomerates* (Graton, 1930). The publication of this treatise brought on such a torrent of criticism from the placerists that Graton hesitated, and part 2 of the treatise was never published. Most of the criticism is published as an annex to the *Transactions of the Geological Society of South Africa,* volume 34, 1931, pages 1-92. Many points noted by Graton are, however, critical in any discussion of the origin of the Rand; for this reason his conclusions are worthy of reproduction in full (Paper 14-3).

(Text continues on page 371.)

14-2: CONCLUDING REMARKS

C. Baring Horwood

A PERUSAL of the discussions, in the preceding chapter, on my original series of articles on the Rand bankets shows that there are still some who cling to the placer theory of the origin of the gold.

There are few British geologists who have had experience of lode mining in several parts of the world and who have made a special study of ore deposits. Further, the majority are more versed in stratigraphy than in petrology; and it is, therefore, not unnatural that many are fascinated by the fact that on the Rand the gold occurs in sedimentary conglomerates, and consequently they find difficulty in realizing the possibility, and still more the probability, of its origin being other than alluvial. If these conglomerates be cut in depth by fissures, it seems difficult for them to realize that from there upwards the routes chosen by any mineralizers were dependent largely on the relative porosity of the conglomerates, and of the upward extensions, if any, of these fissures. If the conglomerates were the more porous and so provided easier channels, they would naturally have been selected as channels of maximum circulation in lieu of any upward extensions of the fissures; being more open, the relief of pressure in them would have been greater and, with gradually decreasing temperature as distance was gained towards the surface, deposition of the mineral burden of the solutions would have occurred.

The superficial resemblance of the bankets to placers has apparently so obscured the main genetic features that, unfortunately, long argument has been necessary in order to demonstrate that in important *essentials* they differ little from ordinary lodes.

Consider, for example, the question of "shoots," which has given rise to a great deal of unnecessary discussion. Posepny very ably and clearly explained that the occurrence of shoots, or of rich ore alternating with poor or barren ground on the same vein-plane, is due to the fact that fissures do not, throughout their length, represent original spaces of uniform width; that, on the contrary, they were sometimes closed wholly or partially by movement of the walls and other causes; and that only the places remaining open permitted an active circulation of solutions and regular deposition from them, and that at obstructed points there would be either no circulation or only a very sluggish one, and that when accompanied by high pressure the solutions would doubtless penetrate the neighbouring rock.

The fact that beds so permeable as conglomerates have served in lieu of the upper extensions of fissures explains the widespread distribution, in the lode-plane, of gold in profitable quantity as compared with its occurrence in widely separated shoots in normal quartz veins. Whereas in the latter case the openings are irregular and frequently restricted, in the former the walls are kept apart practically throughout the whole extent, both along the strike and dip, of the lode-plane. Consequently, in the case of the bankets the wonder is not that well-defined shoots are relatively rare; but that when considering one or other individual sub-series of the Main Reef bankets the alternations between rich and poor ground are as marked as they are. The latter fact indicates the influence on the gold deposition of slight differences in permeability and explains why big pebbles and well-developed bankets usually are a sign of good ore, whilst small pebbles and ill-defined conglomerate usually mean poor ore, and bastard "reef" and quartzite are usually unprofitable and barren [1] respectively. Thus stretches of ground in which pebbles are absent, or only bastard reef occurs, correspond to the barren or poor zones that occur between the ore shoots in ordinary lodes.

Here might be emphasized what, perhaps, is not sufficiently appreciated, namely, that the stoping width, or width actually mined, usually contains seams of banket separated by bands of quartzite, it being the exception for this width to consist entirely of banket; consequently while the stope width may be 4·5 ft., and the stope value may be only, say, 7 or 8 dwt. per ton, the actual width of banket included in this may be only, say, 6 in., and its value 63 or 72 dwt. per ton. The Rand as a whole is now so generally considered as a low-grade proposition averaging

[1] The quartzite does, however, occasionally contain traces up to about 1 dwt. of gold per ton.

a little over 6 dwt. per ton milled, that in considering the genesis of the gold the fact is apt to be insufficiently appreciated that the gold is largely concentrated in narrow rich seams of banket.

Mr. T. A. Rickard pointed out that the non-concentration of the gold along the former bedrock (now the foot-wall), and the fact that the narrow seams of banket are richer than the wider ones, are points that are suggestively against an alluvial origin. He likened the Rand banket to the copper " banket " of the Lake Superior region, and dryly remarked that no American geologist had attempted to establish an alluvial origin for the copper.

Mr. David Draper [1] suggested that the gold in the dykes may be due to the latter having robbed the gold-bearing beds which they intersected. Previously Professor R. B. Young,[2] when admitting that I had proved the presence of carbon in the dykes, had suggested that the latter were not necessarily the source of the former as he suggested that the carbon may have been introduced into both dykes and reefs from some other source.

I have already dealt with this latter suggestion,[3] and it is sufficient here to remark that the distribution and tenor of both these elements in the dykes are sufficiently uniform to militate against the idea that they have been derived from any of the sediments cut by the intrusives. Further, at the time of these igneous intrusions the solutions were at a high temperature, and their circulation was consequently very active, and the diffusion of heat, according to the fundamental laws of thermodynamics, would have been from the hotter, *i.e.* the igneous, rocks to the cooler sediments. In spite of this, the fact that the present tenor of the gold in the igneous rock is less than that in the bankets,[4] where it may have been concentrated at the expense of the igneous rock and its associated magma, is apparently considered by some as an argument against the latter being the source of the gold, although they seem prepared to admit at least the possibility of such an origin for the vastly greater amount of pyrite that is present in the banket and which in any case is generally admitted to

[1] *See* his discussion in Chapter XI.

[2] *Proc. Geol. Soc. South Africa*, December 12, 1910.

[3] " The Rand Banket," by C. B. Horwood, *Min. and Scientific Press*, November 22, 1913, p. 812.

[4] As students of ore-deposits would naturally expect.

be a secondary mineral. Hence it is hard to understand why they find difficulty in admitting the possibility of a similar, or at all events a secondary, origin for the gold.

Dr. E. T. Mellor [1] maintains that enrichment of the banket when noticeable in the mines as an accompaniment of dyke intrusion and closely dependent on it is a purely secondary and extremely local result. In dealing with this same question, Professor S. H. Cox [2] in discussing Gregory's paper stated that from his experience he considered that "the fact that the proximity of dykes does not always exercise any influence on the grade of the ore" did not affect the argument. Personally, in actual mining practice on the Rand, I have always worked on the principle, which holds generally in lode-mining, that at and near the junction of the sites of fissures, now recognizable either by dykes or faults or both, with the lodes there may be, but there need not necessarily be, some change, either for better or worse, in the grade of the ore. On the Rand, just as secondary enrichment in the oxidized zone is local, so any adjacent or neighbouring variation in the grade of the ore due to dykes that is noticeable in any of the mines is admittedly a local effect, and its presence or absence is no argument against such sites marking a portion of the system of channels, whence those mineralizers ascended that are responsible for the contents of the lodes as a whole. There are many instances in lode mining of similar phenomena where there is ample proof that the dykes or faults do indeed furnish the present-day evidence of the former trunk channels of mineralization. In the case of the Rand the fissures which cut the conglomerates at greatest depths, depths to which mining has not yet reached, would naturally have formed the main trunk channels, for the reason that, owing to the greater length of their more open, porous, upper portions which consist of those conglomerate beds that have acted as their upper extensions, they would have supplied the easiest communications towards the surface.

Professor Gregory recognized that the gold in these deposits was epigenetic, and consequently in championing a placer origin realized that such a theory in its simple form, as ordinarily understood and accepted, was untenable. Consequently, he propounded

[1] *See* his discussion in Chapter XI.

[2] *Trans. Inst. Min. and Met.*, vol. xvii, 1907 to 1908, p. 43; *see also* my replies to Mellor's and to Gregory's discussions in Chapter XI.

the theory that the gold had since been dissolved and redeposited *in situ.*[1]

Dr. Hatch,[2] in discussing Gregory's contentions, expressed the opinion that the arguments brought forward for the placer, as against the infiltration, theory were not very convincing. He pointed out that Gregory was constrained to admit that the gold whatever its source must have been in solution before its final deposition in the conglomerate, and thought that Gregory could scarcely contend that the solutions from which the gold was precipitated were not in circulation. Hatch stated "that the gold in the banket had been precipitated from solutions circulating in the conglomerate was the only point that had been maintained by the upholders of the impregnation theory" who "had not ventured to speculate as to the original source of the gold." Thus, far from forming any opinion, Hatch admitted that he had not even speculated as to whether the gold in the bankets was due to deep-seated ascending solutions; to descending meteoric waters; or to lateral secretion. Consequently, Gregory,[3] in his reply, argued that the infiltration theory, as defended by Hatch, is not inconsistent with his (Gregory's) placer theory, as Hatch did not speculate as to whether the gold was originally alluvial or otherwise.

In the Introduction it has already been indicated that if Gregory's re-solution theory were correct, one would expect the gold to be very pure and remarkably free from silver, whereas the relatively high percentage of silver actually present is not explained on his assumption; nor is the decrease in the gold content of the banket and in the proportion of silver to gold with depth.

If the Rand gold be due to the degradation of pre-existing auriferous veins, the area denuded must surely, judging by the extent and richness of the bankets, have been many thousands of square miles in extent, in which case it is improbable that a metal of such high specific gravity as gold should have travelled great distances from many localities spread over wide areas, and have been concentrated to such an extent in this particular locality; and not in the sands, as is more usual with a heavy concentrate, but in the pebble-beds. To have been transported

[1] J. W. Gregory, *op. cit.*, pp. 2–41 and 77–85.

[2] *Trans. Inst. Min. and Met.*, vol. xvii (1907–1908), pp. 44–47.

[3] J. W. Gregory, *op. cit.*, p. 85.

such distances, much of it must surely have been worn so fine as to have become "floured," and concentration of this floured portion is unlikely; and any coarser remaining portion, more amenable to concentration, would be correspondingly less in amount, and have travelled a much shorter distance. Further, such an explanation does not account for the fact that, of the numerous conglomerate series in the Witwatersrand beds, the gold is practically confined to only one of them; nor does it explain the alternating character of its distribution, both along the strike and dip, as between the two principal groups of this series.

Dr. G. F. Becker,[1] in referring to the auriferous beach sands forming along the shores of extensive gold-bearing regions in New Zealand, and along the Pacific coast of North America, said it is hard enough to understand why they are not more abundant. He asked what had become of the immense quantities of gold swept away by erosion from the western slope of the Sierra Nevada and the auriferous territory of Northern California, and stated that it certainly had been carried to the sea, and, as a heavy metal, some of it had been left along the coast in beach sands. However, he recognized this difficulty of concentration and added that, although the aggregate quantity of this portion may be very great, he saw no probability that it represented more than a small fraction of the gold which had been denuded, and he considered the only explanation is that the gold had been triturated to such an extent that it had been "floured" and swept away into the abyss of the Pacific, and perhaps partly dissolved in the waters of the ocean. After such reasoning the weakness of his further argument is apparent, namely, that at the time of the formation of the conglomerate the Witwatersrand lay along the shore of an extensive auriferous area, and that it might have been predicted that if littoral deposits could be found within a moderate distance to the southward of Mashonaland and Matabeleland, and the Northern Transvaal, they would be nearly certain to show *some* alluvial gold. The italics are mine; but the very fact of his use of this word "some" suggests that he was aware of the weakness of his reasoning. The gold occurrences, as revealed to-day, of the three latter regions mentioned certainly do not warrant the

[1] "The Witwatersrand Banket, etc.," by Geo. F. Becker, *U.S. Geol. Sur., 18th Ann. Rep.*, 1896–1897, Part V, pp. 174 and 175.

assumption that they were a sufficient source to account for the vast quantities of argentiferous-gold bullion and gold-bearing pyrite of which that on the Rand would, in such case, merely form the small more concentrated portion, even supposing one were prepared to admit the possibility of any such detrital gold-bearing pyrite, under such conditions, having been preserved. It may be argued that some of the original deposits have since entirely disappeared, and that the present lodes are only the stumps of much larger ones. Even if such an assumption be accepted it does not weaken our line of argument, since the Rand, which has already in about twenty-eight years yielded gold to the value of £425,000,000 sterling [1] and will probably in the future yield, in addition, about twice this amount, must have had such a vast store of the precious argentiferous-gold bullion from which to draw, that in the light of the foregoing it is extremely difficult to imagine one that can be explained by any placer theory; and one instinctively turns to the ultimate source of these metals, namely, the igneous rocks, which are so abundantly represented by the diabase dykes of the Rand, and to their underlying magma, as the most probable direct [2] source of the gold.

The persistence of the gold at right angles to the strike of the beds for distances, as already proved, up to some 6000 to 7000 ft. and over, and on the Far East Rand to as much as 20,000 ft., is in marked contrast to the narrow width of placers in general, and to the disposition of the gold in the case of such typical marine placers as those at Nome, in which, as Rickard pointed out, although the gold extends a distance of some thirty miles, its persistence at right angles to this, corresponding to dip in lodes, only measures at most about 100 ft. The details enumerated in Chapter VIII with reference to the distribution of the gold are totally at variance with the observed facts of placer deposits. Gregory admits that on the infiltration theory a fall in the grade of the ore with depth would be expected; and in some of the preceding pages it has been shown that there is a distinct progressive diminution in the value of the banket with increasing depth.

[1] That is up to the end of March 1915.

[2] The word "direct" is here used to mean excluding the intermediate stages of former auriferous lodes, the denudation and transportation of their gold, its dissolution, reconcentration, and reprecipitation.

Although Gregory, while contending that the gold is of placer origin, maintains that it has since been dissolved and reprecipitated, yet the fall in grade and the present distribution of the gold in the Rand banket carry the same weight as before, as Gregory argues that the gold was reprecipitated *in situ*. Had he not made use of the qualification that the redeposition of the gold had taken place *in situ* his argument certainly would have had increased weight, but it would still have been necessary to invoke the aid of the neighbouring basic intrusives to explain the juvenation and setting in active circulation of the auriferous solutions.

He likens the Rand conglomerates to the now-forming beach placers of New Zealand, with their layers of black iron-sand, and assumes, in addition to the black sands, the presence of organic matter, and that the latter through its decomposition gave rise to sulphuretted hydrogen that converted the black iron-sands into pyrite. He then continues his series of assumptions and suggests that the subsequent action of water, containing oxygen, on the pyrite would have produced a solution of ferric sulphate, which would have dissolved the gold. Then he concludes that the gold and pyrite were redeposited by coming in contact with fresh reducing matter.

Now in the case of the Rand the pyrite occurs principally in the matrix of the banket itself, and also, though in much smaller proportion, irregularly scattered through the quartzites. It is quite the exception to find it in layers somewhat resembling in arrangement what might occur in the case of black sands. However, it is very occasionally so found in what is termed "the pyritic band" occurring in quartzites a little below the horizon of the Main Reef, and Gregory describes such a pyritic band of quartzite 18 in. thick, containing seven layers of pyrite, that occurs at the New Goch Mine.

In replying to Gregory's criticism on my paper I referred to a similar occurrence in the Wolhuter mine, adjoining the New Goch on the west, and pointed out that this was undoubtedly an instance of later infiltration as distinguished from reprecipitation *in situ*.

This pyritic band does not occur throughout the Rand as a whole, and is only found in a few places. Its gold content is very erratic and the amount mined is relatively insignificant; whereas, if Gregory's analogy between the banket and the New Zealand beach placers were correct, then, as Professor S. H. Cox pointed out when discussing Gregory's paper, one would expect the gold of the Rand to be concentrated in such layers as these, instead of among the pebbles; further, Cox, who is acquainted with these New Zealand placers, declared that he failed to see any resemblance between them and the Rand deposits. In the case of the former the auriferous black sand leads are found in places along a distance of fifty miles or more. They are not continuous—only along certain bays do they contain gold in profitable quantity, and this is confined to strips of sand near low-water mark. Thus, as in the case of the known placers already cited by Rickard, the extension of the gold at right angles to the strike of the occurrence is so small that in spite of the great length of the occurrence they are totally different from the deposits of the Rand, which are characterized by the great persistence of the gold for several thousand feet as measured along the dip.

Mr. S. J. Speak, during the discussion on Gregory's paper, showed how highly problematic and unlikely were Gregory's assumptions relative to the source of the sulphur and the solution and redeposition of the gold. At the same time Dr. MacLaren pointed out that the magnetite and ilmenite of black sands were extremely stable minerals, that their direct conversion into pyrite was unlikely, and that such a transition would appear to be possible in nature only by way of an intermediate hæmatite stage.

Even Van Hise,[1] who is the champion of the work of ground-water as distinguished from that of deep-seated magmatic water, refers the source of the sulphur of sulphides to igneous rocks; and he explains that in the belt of weathering these sulphides are largely oxidized to sulphites and sulphates.

Moreover, Dr. G. Becker, the early champion of the placer theory, although he firmly believed that the majority of the pyrite present in the banket was of direct detrital origin and consisted of well-rounded rolled pebbles, varying in size downwards to microscopic dimensions, yet in order to account for the crystallization of the gold and some of the pyrite, he did not hesitate to invoke the aid of "eruptive activity such as is manifested in the dykes of the Rand"; and he explained that a portion of the pyrite might be due to the conversion of magnetite into the sulphide "by the sulphur-

[1] "Some Principles Controlling the Deposition of Ores," by C. R. Van Hise, in "The Genesis of Ore Deposits," 2nd ed., 1902, p. 348.

bearing solutions accompanying the intrusion of the dykes," the solvent fluids being "probably hot waters containing sulphides of the alkalies and carbonates."[1] Professor R. B. Young called attention to the frequent marked increase in the amount of pyrite and other metallic sulphides in the banket as they approach the neighbourhood of dykes. I pointed out that, in mining at considerable depth, when one of the big dykes is pierced, water under pressure is sometimes encountered, and then the smell of sulphuretted hydrogen can occasionally be detected[2]; and Lindgren, in discussing my original series of articles, admits that it is "certain that the conglomerates were penetrated by hot waters containing hydrogen sulphide, which were competent to recrystallize the gold and convert detrital magnetite into pyrite."

Elsewhere Lindgren[3] had already stated that detrital pyrite may occur in gravels, but that there should always be some magnetite and ilmenite present, and that their absence in the Rand bankets is a strong argument against the theory of a direct placer deposition[4]; and that if these be placer deposits there has been extensive crystallization and some migration; and that advocates of the placer theory are compelled to admit a recrystallization of the gold and probably also a transformation of magnetite and ilmenite into pyrite.

Now for Gregory's complicated re-solution placer theory the dykes are even more necessary in order to explain both the source of the sulphur and the juvenation of the solutions.

As regards these solutions there can be no doubt as to either their circulation or their great chemical activity, as this is abundantly clear from a careful consideration of the present distribu-

[1] Geo. F. Becker, *op. cit.*, p. 174.

[2] Posepny pointed out that hydrogen sulphide plays an important part in the ascending waters; and that its presence seems to be the cause of a greater abundance of dissolved substances; and stated that "the most important geological factor in ascending waters is undoubtedly carbonic acid; for it is chiefly this compound which in the deep region, under high temperature and pressure, develops a greater solvent power for most of the elements of the rocks." (Posepny, *op. cit.*, p. 44.) In the particular analysis of water from one of the mines of the Rand, given in Chapter X, it will be noticed that although no hydrogen sulphide is shown, carbon dioxide is present to the extent of 1·82 grains per gallon.

[3] "Mineral Deposits," by W. Lindgren (1913), p. 222.

[4] *See also* Mennell's remarks, with reference to the absence of magnetite and ilmenite, in his discussion in Chapter XI.

tion of the gold ; in fact, Lindgren in his discussion declared that my statements and diagrams illustrating the gold distribution are perhaps the most convincing arguments brought forward to support the infiltration theory. Thus if Gregory's re-solution theory be correct, it must at least be modified to this extent, namely, that the gold was not merely redissolved and reprecipitated *in situ,* but must have been in active circulation and thoroughly redistributed.

Mr. Foster Bain likens the Rand banket to the placers that are dredged along the Sierra foothills of California. Although these are river gravels, Mr. Bain contends they are comparable with delta deposits, such as the bankets of the Rand, inasmuch as he says they occur at the embouchures of rivers upon the plains. He speaks of the profitable beds as miles in extent, and asserts that the gold is distributed in them as on the Rand; but he admits that the Californian gravels are not so rich as the Rand bankets.

In my reply to his discussion I asked if he could give a single example from among them showing not only regular alternations of relatively rich and poor zones in one and the same bed, but also similar alternations of rich and poor zones as between an upper and an underlying parallel bed some 50 to 200 ft. beneath it, *i.e.* the upper bed being rich opposite poor zones in the underlying bed and the reverse—like the Rand examples illustrated by me in Figs. 42 and 43. In reply to this he merely states that "there is at Oroville a lower pay-streak than the one now being dredged ; but being beyond the depth possible at present to work, it has been little studied and it is impossible to say how the gold is distributed in it" ; and in reply to my question relative to the extent of the distribution of the gold at right angles to the strike of the beds he gives this distance as fully 6000 to 7000 ft., or more.[1] On the other hand, Dr. Geo. F. Becker [2] when referring to the Californian gravels says they would scarcely be called gravels elsewhere, the pebbles being for the most part as large as cobblestones, and a good proportion of them reaching the size of the largest water-melons. Speaking of the width of the great gold-bearing channels he says they are not usually over a mile in width, though occasionally they extend to two miles. However, the point which I wish to emphasize is that

[1] *Min. and Scientific Press,* December 26, 1914, p. 1000.

[2] Becker, *op. cit.,* p. 162.

he is careful to add that only a small part of such masses carries any large quantity of gold per ton; that they are worked by hydraulicking, which is cheapest, and also by drifting; that a drift mine cannot be run much under 2*s.* per ton; and that in these great river channels the only portion which will pay for drifting is almost invariably the centre of the channel on the bed-rock; *and, as a rule, the width that will pay for drifting is not over* 200 *ft.*,[1] although in exceptional cases it extends to 400 ft.; and further, that much of the gold in these channels is coarse and nuggets are frequent. Further, Becker, as a result of his visit to the Rand in 1896, asks: "Could anything be more unlike the conditions on the Rand, where there is no real bedrock above the Swazi schists, no single surface along which the gold has accumulated, no conglomerate in the Main Reef series with pebbles larger than a goose's egg, where there are no nuggets, and where the remunerative ore is known, by actual exploration, to be fairly uniform for a width of some 6000 ft.?"

Lindgren, in 1913, in his work on mineral deposits,[2] published in the same year but prior to my original series of articles on the Rand banket, when referring to the latter deposits states that against the infiltration theory stand:

(1) The absence of channels followed by the solutions.

(2) The regular distribution of the gold in the conglomerate.

(3) The practical confinement of the gold to conglomerates, though the quartzitic sandstones are at least equally permeable.

(4) The fact that some reefs with angular pebbles are barren (bastard reefs).

He adds that the following considerations also militate against the infiltration theory: the conglomerates are, according to Becker and Gregory, characteristic of shore shingle with flat bun-shaped pebbles; they accumulated in a sea which washed the Swaziland schists, with their gold-bearing lenticular veins and sheared zones, and the shore gravels would almost certainly contain some gold; the crystalline character of the gold is accounted for by the adherents of the placer theory on the assumption of recrystallization and pressing of the thin gold flakes between recrystallized grains; the sulphur for the pyrite might be supplied by the ordinary surface waters in a sedimentary

[1] The italics are mine, not Becker's.

[2] "Mineral Deposits," by Lindgren (McGraw-Hill Book Co., New York, 1913).

formation more or less rich in sulphates and organic matter; and, finally, similar gold-bearing conglomerates of considerable geological antiquity occur, at various places, in the South African sedimentary rocks, and the conglomerates successfully exploited in West Africa also contain minute particles of crystalline gold, but here associated with magnetite, ilmenite, and chloritoid,[1] and these are regarded by Beck as of syngenetic origin, with subsequent recrystallization of the gold.

The flat bun-shaped pebbles to which Lindgren refers are not characteristic of the Rand bankets, and are quite the exception; whereas well-rounded ones like that illustrated at A in Fig. 9 are typical of the banket.

With reference to Lindgren's last argument, Mr. Mennell, in his remarks quoted in the preceding chapter, has clearly shown that in the case of at least one of these South African gold-bearing conglomerates, namely, that at the Eldorado mine in Rhodesia, the gold, which had been described by Gregory as of alluvial origin, is due (as is also the mineralization generally) to ascending deep-seated mineralizers, after the manner of ordinary gold-quartz veins; and as regards the auriferous conglomerates of West Africa, when I was there fifteen years ago, sufficient work had not been done to form a definite conclusion as to the origin of the gold, and much research is still necessary before the question as to how, and when, the gold came into these bankets can be finally answered.

After the publication of my articles on the Rand banket, Lindgren, discussing them in 1914, maintained that, as the conglomerates are only slightly more permeable than the original sandstones, hot magmatic waters would have permeated the whole, and so he considered the concentration of the gold in the conglomerate to be still a problem; and argued that if the solutions were due to these agencies they were different in character from what one would have expected by analogy with other regions. He also brought forward the absence of fissures as the one still remaining argument against the gold having been precipitated with the pyrite from magmatic waters derived from the same magmas as the Ventersdorp series of effusives.

Thus as a consequence of my paper he abandoned his second and fourth objections of the preceding year; and modified his third, inasmuch as he previously stated that the conglomerates

[1] R. Beck, "Erzlagerstätten," 2, 1909, p. 200.

and sandstones are at least *equally* permeable, and now admits that the former are slightly more so than the latter.

The original sandstones were doubtless permeated, to some extent, by the magmatic solutions, as the quartzites carry traces up to about 1 dwt. of gold per ton; but the reasons why the conglomerates were more permeable have been elaborated in the present volume, and the truth of this must seem obvious to any one who has observed along a sea beach how at low tide the pools of water still remain in the sands, but not in the coarse shingle. As regards the character of the mineralizers, this subject was dealt with when replying to Mr. Foster Bain's discussion, and I pointed out that Dr. J. Malcolm MacLaren [1] states that gold in vein-waters may most reasonably be considered as ionized, and balanced either as (i) auro-silicanion, (ii) thio-auranion, or (iii) telluro-auranion; that in the first case the deposition products are silica and free gold; and in the second free gold and sulphides, or possibly sulphides alone. He expresses the opinion that the first combination is the more probable in all those cases in which clean gold is found studding clean quartz; and adds that even in many pyritic veins deposition of pyrite, quartz, and gold has been contemporaneous; and that in such cases the gold would appear to have been held by both the first mentioned ions, and to have been freed on deposition. Such explanations as (i) and (ii) would apply in the case of the Rand, as will be referred to again later.

As regards Lindgren's other objection, namely, the absence of channels: throughout the Rand evidence of fracturing and fissuring is particularly abundant and is disclosed by the numerous occurrences of big strike faults, and by the longitudinal dykes that are usually accompanied by reversed faulting; and Dr. G. F. Becker, in criticizing the infiltration hypothesis, said that it implied the presence of deep fissures through which the auriferous solutions obtained access to the banket beds, and he freely admitted that "a system of fissures such as would be needed to convey metalliferous solutions certainly exists in the dykes and faults so abundant on the Rand." Further, Becker [2] admitted the probability of a genetic relationship between these dykes and

[1] *See* Chapter XI, where reference was made to this in my reply to Mr. Foster Bain's second discussion on my paper, "The Rand Banket."

[2] G. F. Becker, *op. cit.*, pp. 164, 170, and 174.

the Ventersdorp series of effusives, and also that eruptive activity such as is manifested by the dykes of the Rand is almost invariably accompanied by hot solvents of gold and pyrite in the form of sulphides of the alkalies and carbonates. Thus Lindgren's chief remaining difficulty disappears.

Dr. J. Malcolm MacLaren [1] emphasized the fact that in the investigation of auriferous deposits no feature stands out in greater relief than the constant association of the primary goldfields with igneous rocks,[2] and therefore the evidence available must be examined for the occurrence of gold as an original constituent of an igneous magma. He contended that the close connexion between igneous rocks and auriferous regions may have been brought about in either, or both, of two ways. The gold may have been brought near the surface and within the reach of meteoric waters by inclusion within an ascending magma; or auriferous solutions may have been introduced by ascending waters that have a connexion with igneous masses. In the absence of definite data, he assumes that the gold content of many waters, especially those set in circulation by intrusive igneous rocks, as by the pre-Cambrian diabases, is derived from emanations from intrusive magmas, the emanations being finally dissolved in percolating water, and by them carried into vein-fissures.[3] He regards primary auriferous deposits as phenomena dependent on the intrusion of igneous magmas, and, further, as having an origin indissolubly bound up with that of metalliferous sulphides or of the chemically related tellurides; and consequently, as he remarks, work that throws light on the origin of pyrite is to be welcomed as assisting inquiry into the genetic relations of gold.

These remarks of MacLaren's, including his ionization theory,

[1] "The Geological Occurrence and Geographical Distribution of Gold," by Dr. J. M. MacLaren (1908). *See also* the reference to this in my reply to Mr. Foster Bain's discussion in Chapter XI.

[2] This is also strongly emphasized by W. Lindgren, who states that almost all primary gold and silver deposits have been formed during or shortly after epochs of volcanic or intrusive activity; and he clearly shows that this holds good in the case of North and South America. ("Gold and Silver Deposits in North and South America," Bull. No. 112, *Amer. Inst. Min. Engrs.*)

[3] In this connexion a paper entitled "A Theory of Ore-Deposition," by J. E. Spurr (*Econ. Geol.*, vol. ii, No. 8, 1907, pp. 781–795), is of special interest. Spurr concludes that "metalliferous fluids, from which most ore-deposits are precipitated, are extreme differentiation phases of rock magmas."

to which reference has just been made, have a considerable bearing on the Witwatersrand deposits. The latter constitute a vast goldfield occurring in a very ancient formation, and associated with igneous rocks.

The oldest known sediments of the southern portion of the African continent are those comprising the Swaziland systems, their geological horizon being between the Witwatersrand bed and the basement Granite, and it is improbable that they were sufficiently extensive and auriferous to account for the vast amount of gold in the Rand banket, either on the theory of inclusion within an ascending magma ; or, as already indicated, on any placer theory ; although, possibly, some small proportion of the gold may be explained by both theories. His second alternative explanation that the gold was introduced by ascending solutions that have a connexion with igneous masses is, however, the view advanced in this work to account for the general mineralization, both auriferous and otherwise, of the Rand banket. The solutions may have been largely in the gaseous condition ; and the intrusion of the diabase dykes, that are doubtless genetically related to the Ventersdorp diabase, suggests the channels of ascent and also the agency by which the solutions were set in active circulation. The origin of the pyrite of the bankets was discussed in Chapters II and III and shown to be connected with neighbouring basic intrusions. The logical inference is that the mineralization of the solutions was due to emanations from the diabase magma. Further, the mineral burden of these solutions was deposited between the walls of those pebble-beds that presented easier upward prolongations of the passages in which the solutions in depth were circulating, rather than in the upward extensions of the fissures themselves, owing to the fact that the former, being the more porous, became channels of maximum circulation.

Tellurides do not figure in the Rand deposits ; but the latter are certainly indissolubly bound up with metalliferous sulphides in the form of iron pyrite. MacLaren's explanation of how the gold is carried in the mineralizers in an ionized condition and of the consequent deposition products would account for the gold and the pyrite ; and also for the secondary silica that, deposited between the pebbles in the matrix of the banket, corresponds to the clean quartz mentioned by him. His statement that in many pyritic veins the deposition of the pyrite, gold, and quartz has been contemporaneous is also significant, as in the preceding pages it has been shown that in the Rand bankets there certainly has been simultaneous crystallization of some of the gold and pyrite.

Now by the time Gregory's paper was read in 1907 the three original theories, assuming that the gold originated before, during, and after the deposition of the conglomerates respectively, had been reduced to two, the second one having been generally discarded as untenable. That is to say, opinion was then divided as to whether a placer or an infiltration origin should be assigned to the gold.

The result of Gregory's able paper brought the discussion within still narrower limits. Although previous to this several advocates of the placer theory, including G. F. Becker, had realized that it necessitated the admission of a recrystallization of the gold and of the alteration of magnetite and ilmenite into pyrite, yet it was not until Gregory's exposition of the subject had appeared that it was generally realized that the suggestion of a direct placer origin is untenable.

He made it quite clear that if the gold be of placer origin it must, since the bankets were laid down, have been redissolved and precipitated anew. He contended that redeposition had occurred *in situ.* However, in the preceding pages it has been shown that if the gold be of placer origin Gregory's re-solution theory must be still further modified, as the observed facts, including the present distribution of the gold, clearly indicate that it cannot merely have been redissolved and redeposited *in situ,* but that after dissolution the mineralizers must have been in active circulation, and the gold have been redistributed before it was again deposited.

The dryness of the mines in depth suggests that the gold is not due to meteoric waters, either descending directly or by means of a lateral secretion process ; whilst the observed facts indicate the agency of ascending mineralizers.

Therefore the problem is now confined within comparatively narrow limits : either the noble metal was placer gold which has been redissolved, redistributed, and reprecipitated ; or it is of lode origin due to ascending mineralizers. Whichever of these two alternative explanations more readily accounts for all the recorded phenomena, especially if at the same time it involves a simple process instead of a series of processes, can with confidence be accepted as correct.

Doubtless these ancient conglomerates, mainly consisting of well-rounded quartz pebbles derived from the denudation of the ancient Swaziland beds and from the archæan granite, would have originally contained some detrital minerals and gold; and these, if not already dissolved and carried away in solution, would through the agency of later intrusive activity have been dissolved and distributed afresh, possibly in more concentrated form. This is not disputed. The Rand gold of to-day may have been derived to some extent from such a source; and also from deep-seated emanations or mineralizers arising from the same common centre as the diabase dykes and the magma with which in depth they are connected.

The problem needing solution is to which of these two sources is the bulk of this gold due.

Now the re-solution theory, in the modified form which in the preceding pages I have shown is necessary for its acceptance, tacitly admits the deep region as the ultimate source of the metals, and assumes that the gold was conveyed upwards by the agency of deep-seated mineralizers, probably through the influence of igneous activity, and was then deposited in lodes in rocks older than the Witwatersrand system; that the latter during long geological time were worn down; that the gold was transported and, in some way hard to imagine, was deposited (probably originally in a widely dispersed condition through the sediments) on a vast scale over a comparatively small area; that, after the conglomerates had been formed, the gold went again into solution; and, finally, the mineralizing fluids were actively circulated and the gold was redistributed, concentrated, and reprecipitated in such a way as to produce a field containing the precious metal distributed in a workable form to the extent known to-day on the Rand.

As against this explanation being correct, we have seen how difficult it is to account for such a vast amount of placer gold concentrated within such a small area; especially is this so when it is realized how fabulously rich must have been the areas denuded for such rich concentration to have resulted, for, as Becker has pointed out, this concentration can only represent a very small fraction of the total amount of gold actually denuded and transported.

Moreover, the occasional association of gold with vein quartz ; the occurrence sometimes of such minerals as calcite, galena, blende, chalcopyrite, pyrrhotite, and tourmaline, the presence of the two latter being especially significant, as suggestive of high temperature conditions ; the structural geology of the district ; the presence of fissures or deep-seated communications, represented by big structural faults, sometimes accompanied by dyke intrusions ; the dryness of the mines in depth ; the great depth to which the profitable ore extends ; the pseudomorphic origin of the so-called pyrite pebbles, their shape and association with rich ore ; the close association of the gold with pyrite and carbon, of which the pyrite has been shown in the preceding chapters to be closely associated with dyke intrusions, and the carbon to be due to magmatic emanations, the globular form of the platinum metals that are present in minute proportion ; the alternating character of the distribution of the gold ; and the fact that, after the bankets were formed, coloration of the pebbles has occurred and is associated with rich ore—all these things, especially when considered collectively, are difficult to explain by any placer hypothesis, but are only the natural results if the mineralization be of deep-seated origin and associated with igneous intrusions.

Furthermore, the re-solution theory neither explains the relatively high proportion of silver in the gold, nor the decrease in this proportion with depth ; and it does not explain the decrease in the value of the ore in gold with increasing depth.

On the other hand, the alternate or lode theory is by comparison simple, involving instead of a series of processes the single well-known one of the mineralization of fissures by ascending solutions, liquid or gaseous. It accounts for the concentration within so small an area of vast metallic wealth, this being a recognized feature in lode deposits of the noble metals ; the Cobalt, Porcupine, Minas Geraes districts, and the Cordilleran regions of North and South America forming good examples.[1] Further, it explains the presence of the high

[1] Waldemar Lindgren, describing, recently, the primary gold and silver deposits of North and South America, laid considerable stress on how highly concentrated precious metals of deep-seated origin may be, resulting in great

proportion of silver; and the decrease in this proportion with depth [1]; and the decrease in the value of the ore in gold with depth.

Lindgren [2] remarked in 1913, before my original series of articles appeared, that in spite of long-continued discussion there was still no unanimity among geologists as to the genesis of the Rand gold, and so it would be necessary for a satisfactory discussion of the subject to go beyond the limits of the Johannesburg occurrences and consider the geological relations of the Transvaal and South Africa as a whole.

MacLaren [3] had already recognized this need, as he had pointed out the significance, as bearing on the genesis of the gold of the Rand, that at Barberton some of the numerous dykes occurring in the Barberton series (Swaziland formation) have exercised a notable influence on auriferous deposition, furnishing in the Barberton laminated quartzites well-marked shoots akin to those of Western Australia.

From the first I had realized that the Rand deposits must be studied in a broad way, having regard both to their relation to other South African gold occurrences and also to their analogy with ore-deposits in other parts of the world; and it is in this manner that I have investigated the problem; and the results are reviewed in Chapter X. The essential similarity of the various gold occurrences and of the paragenesis of the associated minerals were there noted; and the connexion between the Rand gold deposits and also those in the Black Reef and Dolomite formations in the Pilgrim's Rest district with the neighbouring basic igneous intrusions was emphasized.

There seems a general tendency to define too precisely the period of any particular igneous activity. The wide range of geological time over which it may extend before finally becoming extinct is not sufficiently appreciated; nor the fact that its declining, lingering activity may still be manifested, for example, after a sedimentary formation has, in whole or in part, been laid

richness within small areas. ("The Gold and Silver Deposits in North and South America," by W. Lindgren, Bull. 112, *Amer. Inst. Min. Engrs.*)

[1] Likewise, the proportion of silver in silver-lead deposits, and of galena to blende in lead-zinc deposits, usually diminishes with increasing depth. The phenomenon in each of these three cases has a similar explanation. (*See* foot-note 3 on p. 254.)

[2] W. Lindgren, *op. cit.*, p. 222.

[3] J. M. MacLaren (1908), *op. cit.*, p. 56.

down on the main products of its effusion ; and that consequently intrusions may be continued in such a formation.[1]

Nineteen years ago Becker [2] expressed the opinion that long after the Ventersdorp rocks had accumulated, expiring stages of the igneous activity to which they are due were manifested after the Black Reef beds, the Dolomite, and the Pretoria series had been deposited.

Seven years later, de Launay [3] pointed out that if the

[1] Some years ago I had occasion to emphasize very similar phenomena in connexion with the Old Granite of South Africa. I then stated: "In South and Central Africa there exists a Fundamental Granite-Gneiss formation, which formed the nucleus or core of the Continent, the base of the geological column on and around which the younger formations have been built up, and although later portions are intrusive in the Swaziland series, nevertheless it underlies and supports it." Also, "The Old Granites of South Africa, although all derived from similar, probably the same, magma, are not all of exactly the same age. Thus it would seem that the Old Granite is intrusive in the Swaziland series by after-protrusions due to the fracturing and crumpling of the old granitic crust produced by secular cooling. These intrusions have occurred at various intervals extending over vast geological time. For example, Dr. Molengraaff has shown that the Old Granite Boss of Vredefort is probably more recent than the Pretoria formation." ("The Old Granites of the Transvaal and South and Central Africa," *Geol. Mag.*, Decade V, vol. vi, October–December 1909, by C. B. Horwood and A. Wade. *See* Conclusions to my portion of this paper, pp. 548 and 549.)

[2] With reference to the Rand he wrote as follows: "The dislocations were accompanied by the injection of the country with large numbers of dykes, which appear to be diabase and related rocks. Between the Lower Cape and the Upper Cape is a sheet of amygdaloidal diabase, which probably escaped through some of these dykes, and both disturbances and intrusions seem to have occurred for the most part before the Upper Cape was deposited, though minor disturbances certainly occurred at a later period." (*See* Becker, *U.S. Geol. Sur.*, 18th Ann. Rep., 1896–97, Part V, p. 164.) Now at that time what was known as the Cape System comprised the rocks from and including the Hospital Hill series [Lower Witwatersrand beds] right up to and including the Pretoria series overlying the Dolomite. The division into Lower and Upper Cape was between the Witwatersrand beds and the Black Reef formation. (*See* Introduction to "The Geology of the Transvaal," by Dr. G. A. F. Molengraaff, 1901, translated by J. H. Ronaldson, T. & A. Constable, 1904; *see also* "The Geology of South Africa," by Hatch and Corstorphine, 2nd ed. (1909), p. 27.)

[3] "Observations on the Rand Conglomerate," by L. de Launay, *Eng. and Min. Jour.* (April 4, 1903).

Since de Launay wrote this, it is now known that the Witswatersrand beds are younger than the ancient granite, on which, and also on that ancient system of sediments known as the Swaziland series, they were laid down; and, therefore, it is not to this ancient granite that one looks for the source of these emanations.

gold was formed since the conglomerates, then the advent of the pyrite must also have been subsequent to their deposition; in which case the Rand ore deposits represent a lode formation of hydrothermal and deep origin, and are a simple but particular case of the type of deposit represented by the various formations of South Africa, which carry gold-bearing pyrite. Further, these deposits have been produced under pressure and in depth by mineralizing waters from an eruptive magma ; and during the intrusion of the ancient granite, or during the period marked by the Black Reef, or even much later, a far-reaching emanation of sulphide waters must have affected different geological formations throughout South Africa. Also at the same time he pointed out that there were those, on the Rand, who admitted a relationship between the gold and the basic dykes.

Two years later, in 1905, it was shown by Mr. A. L. du Toit, and also by Mr. George G. Holmes, and confirmed the following year by Dr. A. W. Rogers, Director of the Cape Geological Survey, that the Ventersdorp diabase is intrusive in the Black Reef and Dolomite formations.[1]

In this particular case the main phase of igneous activity was characterized by the outpouring and accumulation of the rocks now classified under the heading of the Ventersdorp system ; while the later and less active phase extended into Black Reef and Dolomite times. " Period " is the chronological equivalent of the stratigraphical term " system " ; and some may prefer to regard the mineralization as belonging to two distinct periods. However, when for the sake of convenience a series of igneous rocks is classified as a separate system, it cannot be expected that the conditions prevalent in the period responsible for them have ended so abruptly that its declining activities should not be manifested during the time, subsequent to its main outpourings, when those overlying sediments were forming which are grouped in the succeeding system. Especially is this so when, as in the present case, there is no great unconformity between the latter and the underlying igneous rocks.[2]

[1] *See* the latter part of Chapter X.

[2] Thus Draper in his discussion, in Chapter XI, points out that in the Vredefort area, on the southern side of the Vaal River, there is no sign of unconformity between the Black Reef beds and the underlying Ventersdorp diabase. Also Hatch and Corstorphine (" The Geology of South Africa," 2nd ed., pp. 173

I have, therefore, preferred, for the special purpose of this argument, to use the term "period" to designate the time during which the main and the later phases, or epochs,[1] of igneous activity occurred, the former being responsible for the Ventersdorp series of rocks, and the mineralization of the Rand; and the latter for intrusions into the Black Reef and Dolomite series of the Potchefstroom system, and for the mineralization of these two series.

Professor Grenville A. Cole has called my attention to a certain amount of similarity between this case and that of Rossland, British Columbia. In the latter the sulphide ores, containing gold, copper, and a little silver, entered during the Jurassic mountain-building epoch, when granodiorite and monzonite, and also dykes of "diorite porphyrite," accompanied by vein-formation, were intruded. In Cretaceous times denudation, followed by uplift and folding, occurred. Then in Miocene times the district was invaded by a new batholite and by lamprophyre dykes accompanied by secondary enrichment of the mineral deposits and the addition of free gold. In each case there were two mineralization epochs associated with intrusions from the same deep underlying cauldron, with an intervening unconformity.[2]

The two types of gold occurrences in the Black Reef formation are of special interest in helping one to realize better the relationship existing between the bankets of the Witwatersrand beds and the bedded gold-quartz seams in the Dolomite. The Black Reef beds may, more properly, be considered as forming the basal beds of the Dolomite formation rather than as a separate formation; and they are of no great thickness, varying from a few feet up to, in the Lydenburg district, about 1000 feet.

and 174), referring to the Black Reef series, state: "In the north-eastern districts of Cape Colony the series appears to be intimately connected with the lavas of the underlying Ventersdorp System. In the vicinity of Vryburg, the Cape Survey describes the two formations as always conformable in dip where they occur together, while in addition, lavas quite resembling those of the Pniel series, *i.e.* the upper division of the Ventersdorp rocks, occur interstratified with Black Reef quartzites. Messrs. Rogers and Du Toit incline to the view that where the Black Reef series rests on rocks older than the Ventersdorp System, this is due to the thinning out of the latter, and does not indicate an unconformity due to the elevation of the lavas prior to the deposition of the former."

[1] "Epoch" is the chronological equivalent of the stratigraphical term "series."

[2] *Canadian Geol. Surv.*, Mem. 77, *see* pp. 38–53.

In the Pilgrim's Rest district gold-quartz veins, similar to those in the overlying Dolomite, occur in them along well-marked horizons in the bedding; whilst in the Southern Transvaal, where conglomerates are developed, the mineralizers have followed the conglomerate horizons as offering the easier channels, with the result that, in that locality, bankets,[1] similar to those in the underlying Witwatersrand beds, have resulted. There is certainly no reason to ascribe a separate and different genesis to each of these two forms of occurrences: whether auriferous quartz seams or bankets resulted depended merely on the character of the local conditions in these Black Reef beds in the different districts. Though, as a consequence of the variation in these conditions, they differ considerably in external appearance, they resemble each other, and also the underlying Rand bankets and the overlying gold-quartz seams of the Dolomite, inasmuch as they all follow the stratification along well-defined horizons in bedded deposits; and also in the fact that the presence of shale has had a marked influence on the mineral deposition. Further, they are characterized by the presence of metallic sulphides, and in particular by gold-bearing pyrite (decomposed or otherwise); by their quartzose matrix; by considerable metasomatism; and by their intimate association with basic intrusions.

The bedded quartz veins in the Black Reef and Dolomite beds constitute quite a normal type of lode deposit, and their genesis is admittedly intimately connected with basic intrusions.

As might, however, be expected since they are products of a later and less active stage of that igneous period to which the Rand gold is due, the mineralization of the Black Reef conglomerates and of the Black Reef and Dolomite veins is less intense than that of the Rand, and the gold occurrences are of a more patchy nature, the economic result being that less reliance can be placed on their gold content, and in mining a greater amount of development work is required to keep the ore reserves sufficiently ahead of the capacity of the mills.

Consequently, the genesis of the gold, pyrite, carbon, and associated minerals of the Witwatersrand banket and of the Black Reef banket of the Southern Transvaal, and also of the bedded quartz veins in the Black Reef and Dolomite formations of the Northern Transvaal, can be satisfactorily explained by one general period

[1] *I.e.* auriferous conglomerates.

of igneous activity: the mineralization of the Rand on the one hand and of the Black Reef and Dolomite beds on the other hand being economic results of its most active and of its later, declining stages respectively.

The mineralizers were set in active circulation by the eruptive activity, and in their upward passage through the resulting fissures they found in certain of the Witwatersrand conglomerates ready-made, and more porous, extensions of these channels, and so in these conglomerates maximum circulation occurred, accompanied by gradually lessened temperature and lowered pressure and consequent deposition of their mineral burden.

Thus the simpler of the two remaining hypotheses and the one more in accord with observed facts ascribes the gold to mineralizers originating from the same deep-seated centre, or cauldron, as the magma from which the basic dykes of the Rand, the main mass of the Ventersdorp igneous rocks and the later basic intrusions associated with them, are derived.

This explanation, therefore, seems to be the true solution of the riddle of the genesis of the Rand gold.

14-3: HYDROTHERMAL ORIGIN OF THE RAND GOLD DEPOSITS

L. C. Graton

CHAPTER V. CONCLUSION OF PART I.

No attempt will here be made to review the many points touched upon in the preceding parts of this paper. On pages 40–42 is given a summary of the evidence of Chapter II. as to the origin of the conglomerate reef; this leads to the conclusion that syngenetic accumulation of detrital gold with the reef gravels is so highly improbable and so unlike any of the determining conditions under which gold placer deposits have been formed that the detrital hypothesis for the Rand deposits cannot be accepted. On pages 116–118 of Chapter III. the evidence afforded by the conglomerates as possible and actual channelways has been summarized and leads to the firm conclusion that the gold is and must be of hydrothermal introduction. Chapter IV. deals summarily with a variety of topics which, though incomplete both in number and in treatment, embrace. I believe, most of the considerations that have hitherto been taken into account as bearing upon the origin of the gold. Analysis of these several topics leads similarly to the unavoidable conclusion that the gold is of hydrothermal origin.

In his forceful and persuasive paper on the origin of the Rand gold, Dr. Mellor makes final disposition of infiltration in these words:

> The theory of infiltration from any outside source of supply demands so extended a series of purely hypothetical steps uninterrupted by any solid ground of observation and of demonstrable fact, that any one may be pardoned who comes to regard the further pursuit of an explanation in this direction as demanding a greater exercise of the imagination than is usually allowed in geological speculation.[166]

Perhaps it was only foolhardy to take up so ominous a challenge. But this is the way the question appealed to me: Dr. Mellor's argument assigns dominating value to simplicity. The advantage of simplicity would, at first thought, seem to belong to

[166] Mellor, 2, pp. 272–273.

the straight placer theory as contrasted with that of hydrothermal introduction. But mere apparent simplicity cannot be allowed to determine which hypothesis of origin is correct; determination must rest on which hypothesis best fits the facts. The notion that malaria results from damp night air seems simpler than the rather complicated chain of events connected with the sting of a certain variety of female *anopheles;* but the latter idea better accords with the facts now known. And so it is really " simpler " to acquire malaria by being mosquito-bitten than by breathing the night air. Also, it is far " simpler " to believe this now than it used to be before there was such general understanding of the germ theory.

Just so it seems to me the placer theory of deposition for the Rand gold holds out a promise of simplicity that proves to be spurious. Like other sedimentary accumulations, a placer is obviously a derived, not a fundamental or primary deposit. In the long run, the gold of placer deposits has come almost wholly from hydrothermal deposits; therefore the placer involves all those speculative elements of origin to which objection is raised. For my part, I find it simpler to regard the Rand deposits themselves as *original* hydrothermal products than as derived from some unknown, unlocated hydrothermal deposits. The conception that the Rand gold was hydrothermally deposited is simple in the sense that it finds innumerable analogs in other hydrothermal deposits in which placer action is absolutely ruled out; and on the whole it encounters no major feature or fact of occurrence which does not find approximate counterpart and explanation in some known deposit of undoubted hydrothermal origin. Harmonious thus with geological observation and theory, the hydrothermal hypothesis is in harmony also with physico-chemical fact and principle. Such are the tests of true simplicity.

The idea of placer accumulation for the Rand gold is a link in a chain of events. It is a relatively simple link. The other links in the chain are less simple. The whole of the chain, not the simplest link, must be considered in deciding on the probable origin of the actual deposits. Not only does the preceding link of source of the gold stand as a vague uncertainty; not only does the idea of placer accumulation itself, under the conditions obviously existent at the Rand, require a deal of making special assumptions unless vital facts are to be ignored; but even more fatal is that final link of assumed recrystallization of the gold, which alone can bridge the gap between observed fact and the idea of detrital deposition. Against the difficulty of solving and confirming that strange conception of recrystallization, which has no effective analog or support in the whole known range of ore deposits, the "complexity" of direct hydrothermal deposition of the Rand gold, in the light of all the facts, dwindles to insignificance.

Most serious and surprising of all is the failure of the placerists to recognize, appreciate or correctly interpret countless outstanding manifestations of hydrothermal activity which are intimately bound up with the occurrence of the Rand gold, features which in other regions are universally recognized as genetic associates of gold deposition. To claim advantage for the placer theory on the ground of simplicity when such strong evidences of a rival theory are missed or slightingly passed by seems hardly consistent, yet may well have contributed to the very general acceptance of the placer theory evident among many who have never had personal opportunity to examine the region in detail.

The method followed throughout this paper has been, first, to show what the placer theory would demand, to show how the placerists have seen the facts in a way to meet those demands, and to show that in reality the facts do not meet those demands but instead are incompatible with them; and second, to show that the facts are compatible from beginning to end with the view that the gold has been introduced from a deep-seated magmatic source by hydrothermal solutions of the same kind as have produced many of the other great gold deposits of the world.

It will be seen also that the treatment herein adopted is general rather than local. I have sought to catch and to employ the broad and characteristic facts, relations and tendencies rather than to risk losing true perspective by giving prime attention to the conditions to be found at individual, isolated points. Such treatment has an obvious weakness. It cannot hope so to fit every conceivable detail as to forestall the probability of innumerable allusions to specific occurrences where, it will be insisted, only conclusions that are the opposite of mine can possibly apply. That in so great a district many local features may be found that will puzzle the supporter of *any* hypothesis is only to be expected. Perhaps it is not too much to trust that a *barrage* of such local missiles which this paper may draw will not be mistaken for the heavy fire of attack on general facts and broad principles which one may earnestly hope the present contribution will stimulate.

If, in my faith and zeal, I shall be found to have fallen victim of the same errors of bias and distortion which I have attributed to the placerists, I shall have but poorly attained my objective of helping to shed more light on a problem of unexcelled importance, and I can only hope that those with whom I have differed will feel toward me as certain as I feel regarding them that these errors are unconscious and unintentional. One can only follow the light as he sees it, realizing that he cannot expect to make converts of those who strongly hold opposite views, but encouraged somewhat by the sense of doing what he cannot fairly shirk, and hopeful that his own efforts may, in the course of time, be found to have contributed in some measure toward illumination and recognition of that which is true.

CAMBRIDGE, MASSACHUSETTS,
March 31, 1930.

Professor Graton complained of the failure of the placerists to recognize, appreciate, or correctly interpret manifestations of hydrothermal activity, among which he includes in the text of his paper the occurrence of gold in ore shoots, size and purity of gold, relation of gold to the pyrite and hydrocarbons, obvious hydrothermal textures and mineral sequences, influence of dykes and faults, presence of abundant auriferous quartz veins, and so on. In recent years some of these features have been interpreted by the modified placerists as being due to metamorphism, remobilization, secretion, and other processes, but so far as I know there has been no detailed work to show that such are in fact the causes of these features.

C. F. Davidson was the last to advocate an infiltration (hydrothermal) theory for the origin of the bankets in the Rand and elsewhere. In his first of many discourses on the origin of the quartz-pebble conglomerate ores he (Davidson, 1953) pointed out the following considerations.

> In the gold-uranium ores of the Witwatersrand the two pay metals vary sympathetically and are clearly derived from the same mineralization. It is widely held in South Africa that the gold is of syngenetic origin; but since the uranium present (as uraninite) is certainly not of placer deposition this view cannot be maintained. The banket reefs differ from modern auriferous placers in possessing a much higher radioactivity than the latter, due to the ubiquitous presence of uraninite, a mineral which has never been recorded as a detrital constituent of any modern sediment. Conversely the refractory uranium and thorium minerals which account for the radioactivity of placers are absent from the South African ores. The distribution pattern of radioactivity throughout the Witwatersrand Series, as determined by radioactivity logging of bore-holes, is wholly dissimilar from that found in normal sediments. Lead isotope studies suggest that the uraninite together with galena (and, by inference, the gold and other sulfides) were introduced into the conglomerates by hydrothermal action about 1,700 million years ago. (p. 84)

Later he (Davidson, 1965) suggested that the gold and uranium were leached from overlying volcanics by heated saline waters. His theory is explained in the following succinct synopsis.

> In an attempt to resolve long-standing differences of opinion on the origin of banket-type mineralization, a new hypothesis of ore genesis is advocated. It is shown that the occurrence of uraniferous and auriferous conglomerates close to major unconformities within or bottoming deep confined basins of Proterozoic molasse sediments is compatible with the view that the metals have been leached from the overlying stratigraphical sequence. In the mature end-stage of a prolonged sequence of intrastratal migrations which the metals have undergone, mineralized groundwaters have sunk to the lowest permeable horizons; and moving outwards from a depth-source of heat, principally along the relatively open conglomerate channels, have deposited their load towards the cooler marginal zones. In all Proterozoic occurrences deposition of the uranium has preceded

> one or more periods of regional metamorphism in which the uraninite was rejuvenated with loss of radiogenic lead. Auriferous uranium deposits have been derived from the leaching of acid-intermediate volcanic-pyroclastic rocks, and uranium deposits devoid of workable gold from the leaching of granitic debris. The view that the Dominion Reef and Upper Witwatersrand ore deposits were generated in this way, from Dominion Reef and Ventersdorp volcanics respectively, is supported by a reinterpretation of isotopic data. Numerous Mesozoic and Caenozoic analogues of the Proterozoic formations are on record, but the distribution of mineralization in these has not always reached the mature basal-conglomerate end-stage seen in the more ancient fields. (p. 319)

This hypothesis brought forth a veritable blizzard of criticism, much of which was unfavorable to the mechanism presented. Nevertheless, the idea is novel and worthy of consideration in the pantheon of ideas concerning the origin of the auriferous and uraniferous quartz-pebble conglomerates.

Most South African geologists writing concerning the Rand postulate a modified placer theory. This theory, apparently initiated by J. W. Gregory (1907) (see Paper 14-1), has been further elaborated by Mellor (1916), Liebenberg (1955), and Pretorius (1975), among others.

For the Rand a considerable number of features are invoked to support the modified placer theory, among which the following are the most important:

1. The great persistence of the mineralization in individual beds or sequences of beds of conglomerate and quartzite
2. A ready source of placer gold in the Swaziland system known to contain numerous gold-quartz deposits
3. The presence of abundant quartz pebbles, a feature of many placers including the Klondike (White Channel gravels) and the Victoria placers in Australia (White Leads)
4. The presence of undoubted detrital minerals that commonly accompany gold in placers, examples being the platinoid metal minerals, chromite, and garnet (Uraninite is, likewise, said by some investigators to be detrital, but this suggestion has been disputed.)

E. T. Mellor examined and surveyed the Witwatersrand for many years; hence his views merit close study. The results of his work were published in great detail, but he summarized his views in a classic paper in 1916. The part on the origin and distribution of the gold is reproduced as Paper 14-4.

W. R. Liebengerg, among the newer generation of investigators, carried out a very detailed study of the distribution and origin of gold and radioactive minerals in the Witwatersrand system, the Dominion Reef, the Ventersdorp Contact Reef, and the Black Reef. The summary and conclusions from his comprehensive paper (Liebenberg, 1955) merit close consideration and are reproduced as Paper 14-5.

D. A. Pretorius, professor of geology in the University of the Witwatersrand, has studied and written on the Rand for many years, and his most recent summary (1975) of the history of the theories of origin of these famous gold deposits (Paper 14-6) is a classic and a landmark.

(Text continues on page 437.)

14-4: THE CONGLOMERATES OF THE WITWATERSRAND

E. T. Mellor

II.—Origin and Distribution of the Gold.

In the preceding portion of this paper considerable attention has been given to the sedimentary features of the conglomerates, their probable mode of origin and their relationships to the other portions of the Witwatersrand system. This has been done because, in the writer's opinion, these features have never received adequate consideration in connection with the question of the origin of the gold, attention having been largely focused on the evidence available from within the conglomerates themselves studied more or less in the form of petrological specimens or as mineral 'lodes.'

It should be stated at the outset, that before making a detailed study of the whole system and having an opportunity of becoming personally acquainted with the conglomerates over the whole extent of the Rand, no reason was seen to reject the 'infiltration theory' which was then, and is still, probably the most widely favoured of the various explanations suggested for the origin of the Rand gold. It would perhaps be more correct to say that my ideas had received a very distinct bias in favour of infiltrationist views.

Closer acquaintance with many aspects of the conglomerates, however, soon raised so many questions, which appeared to be answerable only when the gold was regarded as originally of placer or detrital origin, that I found it necessary for a long time to regard the matter as an open question upon which a decision could only be arrived at after much closer personal investigation of the evidence available.

As a result of such further study, I am strongly of opinion that the chief reason why the origin of the gold has so long remained a debatable question, especially with those who must rely for their data on the descriptions given by others, lies mainly in the fact that such descriptions have been largely based upon studies made in those portions of the Rand which were earliest opened up, namely the Central and Western portions, and also that in descriptions of the conglomerates of the Main Reef group no very clear distinction has been made between its various members. Had I been acquainted only with the Main Reef and with the Central and Western portions of the Rand, I might never have materially modified my inclination towards the infiltration theory. An extended study, however, of the Main Reef Leader, and of the conditions in the Eastern portion of the Rand area, appeared to place the problem in a very different light and has led me to a strong belief in the detrital origin of the gold.

Relationship of the occurrence of gold to sedimentary features.—From the point of view of the geologist, one of the most striking points in connection with the distribution of the Rand gold is the remarkable correspondence everywhere apparent between its occurrence and purely sedimentary features in the containing rocks. It will be convenient to deal with this correspondence first from the more general point of view of distribution of gold through the Witwatersrand system and then more in particular with regard to its distribution within individual 'reefs' and groups of conglomerates.

Distribution of gold throughout the Witwatersrand system.—The occurrence of gold in the Witwatersrand system is far more widespread than is generally recognized. The outstanding position occupied by certain reefs of the Main Reef group from a mining point of view is apt to give the impression that the distribution of gold is much more limited than it really is.

In the first part of this paper a number of conglomerate groups have been briefly described, whose distribution throughout the Witwatersrand system is shown in the section in Fig. 45. Some of these, like the Elsburg and Kimberley groups, may include in any particular section a hundred or more individual 'reefs' or conglomerate bands. After examining many detailed sections, in which assays have been made of many of these individuals, it would require some boldness to assert that any of the conglomerate bands throughout the system did not in some portion or other contain an appreciable quantity of gold, and quite encouraging assay results are frequently met with in groups like the Kimberley, which have furnished little in the way of actual mining propositions. Nearly all the known conglomerates in the Lower Witwatersrand system also occasionally show encouraging quantities of gold, and samples, which include a very small thickness, frequently give very high assays.

In one instance at least which has come under my notice, several thousand pounds worth of gold was very profitably obtained from a very small patch of 'reef' only a pebble or so in thickness, and the extent to which these indications encouraged extensive exploitation of these Lower Witwatersrand reefs has been already referred to.

Above the horizon of the Main Reef, the Livingstone, Bird and Kimberley groups have all furnished reefs which have been actually mined to a considerable extent, and which may assume greater importance at some later stage in the history of the field, when the more attractive propositions now available have ceased to absorb the available resources in capital and labour. In some portions of the groups just referred to, the difficulty appears to be not so much to find a reef which shows encouraging values, as one which maintains its character over a considerable lateral extent.

It might perhaps be said of some of the reefs in the Lower Witwatersrand system that it is the absence of any large body of conglomerate, rather than the nature of the values found in that which does occur, which has prevented such reefs from becoming important from a mining point of view, whereas in the upper division of the system it is the distribution of the gold which is present, through such large bodies and such very numerous bands, that has led to the same result, for it seems quite possible from what we know of them that had the gold distributed through the Kimberley or the Bird Reefs been confined to a more limited zone as in the Main Reef group, we might have had something equally important from a mining standpoint.

It is not so much the actual amount of gold present in the Main Reef group which appears exceptional, as its concentration within certain particularly well-defined and continuous beds; and, as will be presently shown, this concentration coincides with, and is probably the result of, special conditions of sedimentation.

From the above it will be seen that gold is of very general distribution throughout the whole Witwatersrand system, wherever conglomerates occur, and, as far as we know, it is practically limited to such beds or to the quartzites in their immediate neighbourhood, and, further, there appears to be a very close correspondence between the sedimentary character of the various groups of conglomerates and the extent to which gold is found associated with them.

Distribution of the gold within the conglomerate beds.—Similarly, if we study the distribution of the gold within the limits of the individual conglomerate beds themselves, a close relationship is usually seen between the occurrence of gold and purely sedimentary features.

Perhaps the most striking example of this is seen in connection with the Main Reef Leader in the Far Eastern Rand. The mode of distribution of the conglomerate in the Nigel neighbourhood in well-defined 'patches' or 'shoots' has been already described. The occurrence of gold in any notable quantity is practically limited to these patches of conglomerate, and in them gold is rarely absent, and this applies not only to the larger patches, but also to the numerous smaller ones which occur in the intervening spaces as shown in Fig. 48.

The explanation given by the infiltrationists of the presence of gold in the conglomerates as due to their acting as channels for the mineralizing solutions, find many difficulties of application in this case—for even granting the disputed point that the conglomerates offer easy channels as compared with the associated beds, it is difficult to see how this principle can apply in the Far Eastern Rand. The patches are separated in many instances by wide intervals occupied by beds for which no greater suitability for the passage of solutions can be claimed than for the general bulk of the formation.

They are practically isolated from each other by intervening stretches of the same quartzites which are supposed in other cases to have restricted the mineralizing solutions to definite channels offered by the conglomerates. In a building, a number of spacious halls would be of little use for the passage of a moving crowd if they were practically cut off from one another or connected only by narrow and difficult passages. Where the conditions are such as are met with in the Central Rand or are similar to those described for the Van Ryn where the 'shoots' are connected by intermediate areas also occupied by conglomerate, this intermediate material can always be requisitioned on the infiltrationist view to supply the necessary channels of transmission, but under the conditions prevailing on the Nigel area the infiltrationist idea of the conglomerates as widespread channels of communication does not appear in any way to meet the case.

The irregular disposition of values in the Main Reef, particularly with regard to their vertical distribution, corresponds with the impersistence of the various bands in that reef and with the irregular distribution of such larger pebbles as may occur, and this is in contrast to the much greater localization of values frequently evident in the Main Reef Leader corresponding to the grading of the pebbles in that bed, and especially to the frequently strongly marked preference of the gold for the foot of the reef, particularly when the foot is characterized by the presence of numerous pebbles above the average size.

The great frequency with which unusual values are associated with large pebbles, particularly on the foot of the Leader, is remarkable, and it is especially worthy of note that this association is not limited to bands of larger pebbles extending over some considerable distance, but is equally or perhaps more characteristic of isolated groups of larger pebbles and even of single individuals. I have frequently examined portions of the foot of the Leader which included such nests of large pebbles or single individuals, and in reef of moderate quality have rarely failed to find that specimens obtained from such a situation exhibited 'visible' gold which, as is well known, is not of very general occurrence in the conglomerates.

On the assumption that the gold is an original constituent this is a most natural and easily explained arrangement, for which we find innumerable parallels in typical placer deposits. On the infiltration hypothesis, however, it is difficult to find any feasible explanation of this association, particularly in the case of the large pebbles which occur in nests or groups or as isolated individuals.

Some definite explanation appears to be due from the infiltrationists as to how an auriferous solution traversing a sheet of conglomerate, hundreds of square miles in extent, should pick out small groups and single individual large pebbles in this extraordinary way. One could understand how a continuous tract of such larger pebbles might possibly allow larger quantities of the gold-bearing solution to pass and thus permit of a heavier deposit of gold in such an area, but when, as is often the case, such larger pebbles occur within patches of conglomerate entirely isolated from the main body, any such influence on deposition appears very difficult of satisfactory explanation.

Still more striking perhaps are the conditions presented by the well-known 'Bastard Reef,' of the Glencairn, May Consolidated, and neighbouring mines in the near East. Typically this 'Bastard' consists of pebbles of Main Reef type sparsely scattered through a dark fine-grained matrix which originally must have been a mixture of muddy and very fine sandy material, similar to that constituting the 'Black Bar.' In places the Bastard can be seen to grade into the Main Reef, in others into the Black Bar, and there can be little doubt that the Bastard originated by the mixture of Main Reef pebbles with finer material of the nature of the Black

Bar, probably in many cases along the margins and the beds of channels cut through the pre-existing pebbly deposits of the Main Reef horizon.

The Black Bar in itself is practically everywhere devoid of any notable amount of gold, yet in such admixtures with Main Reef material it acquires values such as might be expected from this mode of origin, and it has been largely mined. Its value appears to be proportionate to the quantity and grade of Main Reef material incorporated with it.

Typical 'Bastard' with its widely-scattered pebbles in a compact matrix appears an extraordinarily unsuitable medium for the passage of mineralizing solutions required by the infiltrationist theory, but on the other hand the distribution and character of the values found in it present no difficulty on the supposition that the Main Reef gravels, as appears probable on so many other grounds, already contained gold before their admixture with the Black Bar material.

The above are but a few among many points in connection with the occurrence and distribution of the Rand gold which appear to receive a much simpler explanation on the placer view than in any other way, and it seems likely that these will be considerably augmented as the exploration of the Far Eastern Rand progresses. It is, however, not possible to deal more fully with this question within the limits of the present paper except in so far as is necessary in considering some of the objections which have been raised to the placer theory.

Objections which have been raised to the placer theory of the origin of the Rand gold.—An objection which has frequently been raised to a placer origin for the Rand gold is that the deposits with which it is associated are not similar to those in which the majority of known placer deposits occur, and especially those which have been most conspicuous as gold producers.

This objection appears to be largely the result of ordinary stream-placers having been mainly used as a basis of comparison on the one hand, and, on the other, of attention having been largely directed to a consideration of such admittedly exceptional 'reefs' as the Main Reef Leader and the South Reef, thus leaving out of account the hundreds of other conglomerate bands which occur throughout the Witwatersrand formation, and which probably afford a much safer basis for comparison.

In the character of the sedimentation and in the quantity and distribution of the contained gold such conglomerates as those of the Bird and Kimberley groups appear to differ in no essential features from many well-known placer deposits, and to these might be added many members of the Main Reef group, including the Main Reef itself and numerous impersistent bands of conglomerate frequently found above and below it. For while it is true that these deposits have little in common with the more ordinary type of stream-placer, they show a very close correspondence with certain placer deposits of coastal origin.

The descriptions of the Alaskan 'gravel-plain placers' given by Brooks and others, abound with points in which these deposits show close similarity to the Rand conglomerates and their associated beds.

Comparison with the coastal-plain placers of Nome, Alaska.—The coastal gravel-plains of Alaska appear to have been formed by material of the nature of small delta deposits brought down by local streams, and more widely distributed by coastal currents and other agencies. If allowance be made for the vast difference in the scale of deposition in the two cases, and for the fact that the Nome deposits have been laid down close to a coast line of older works on a comparatively uneven floor, whereas the Rand conglomerates were distributed at a distance from the actual shore line over a widespread level expanse composed of previously deposited conformable beds, the similarities between the two occurrences are many and close.

There are few features of the more general types of Rand conglomerates that do not find a parallel in the Nome coastal gravel-plains, and in the beaches which have been locally formed in them from time to time. Of the various types of sedimentation exhibited by the Rand conglomerates, such beds as those below the Main Reef (Main Reef footwall quartzites), with their impersistent bands of conglomerate, sometimes lying on a bed of clayey character, may be compared to the gravel of the coastal-plain deposits, with which also such groups as the Bird and Kimberley Reefs have much in common.

The Main Reef itself appears frequently to correspond more nearly in type, but on a very much larger scale, with the Nome 'beach-placers,' and its origin may be, as suggested for these, due to concentration of gravel-plain deposits by wave action, with richer patches determined by the action of local streams crossing the surface of the beach.

When, however, we come to such uniform and extensive sheets of conglomerate as the Main Reef Leader and the South Reef of the Rand, which in the Upper Witwatersrand system are the only two known occurrences of the kind amongst hundreds of others for which parallels can be found in the Nome deposits, it is not surprising that an analogue is not so easily discovered.

It is, however, not difficult to see how in the Nome area an exceptional discharge from one of the rivers which traverse the Nome coastal-plain could give rise on a small scale to a deposit of essentially the same character as the Main Reef Leader, and the production of such a bed would be much more likely to occur under the widely-spread uniform conditions assumed to prevail in the Rand area when the Leader was laid down, and as a result of the redistribution of very much larger quantities of material than would be possible at Nome. It appears then that there is nothing in the sedimentary characters of the Rand gold-bearing conglomerates which is incompatible with the idea that the gold was an original constituent of the beds. Nor does there appear to be much in support of the further objections which have been made to the acceptance of a detrital origin for the Rand gold: that the values found in the conglomerates as a whole are of a different order from those characteristic of placer deposits, and that the Rand gold is too fine in character to be of placer origin, the rare occurrence of anything like a nugget being quoted in support of the same opinion.

With regard to the former objection the tendency to make comparisons with the exceptional values found in certain reefs and in special portions of the Rand rather than with the average must be remembered, for if the extensive low-grade areas of the Rand and also reefs outside the Main Reef group were more often taken into account as a basis of comparison greater similarity would certainly be found. Where concentration of the Nome coastal-plain gravels has taken place, as in some of the beaches which have been formed in them by wave action, values of a high order are also met with. In one of these beach-deposits values as high as $500 to the pan are recorded, and from one area 100 ft. by 15 ft. in extent more than $330,000 was taken, 90 % coming from the bottom 3 in. of the pay-streak.*

With regard to the second objection, the fineness of the gold in the Nome deposits is frequently referred to in the descriptions given of them. Brooks† states that anything like a nugget up to the value of one dollar is very exceptional, and that the average grains as saved by panning run from 70 to 80 to the cent, but that much

* 'Geology of the Nome and Grand Central Quadrangle, Alaska,' F. H. Moffit, *U.S. Bull.*, 533, 1913, p. 117.

† 'The Gold Placers of parts of Seward Peninsula, Alaska,' A. H. Brooks, *U.S. Bull.*, 328, 1908, pp. 153, 154.

of the fine gold is lost in separation, thus increasing the average size of the colors that are saved. He further mentions that beach gold from Randolph, Oregon, averages 110 colors to the cent, that from the Sixes Mine, Denmark, Oregon, about 600 to the cent, and the river-bar gold from Snake River, Idaho, from 900 to 1000 to the cent.

It will thus be seen that there is nothing remarkable in the absence from the Rand conglomerates of coarse gold, or of nuggets, for considering that such beds as the Main Reef Leader were probably laid down many times farther from the actual shore line than those of Nome, the gold in them might naturally be expected to be of still finer character.

It is perhaps worthy of note that in the Nome gravels, as in the reefs of the Main Reef group, the pebbles are all of comparatively small size, individuals of more than an inch in diameter being rare.

The Nome deposits are regarded as the redistributed material of small deltas, and the small range in the size of their pebbles and the rarity of larger constituents is no doubt due to the elimination of larger pebbles and boulders which this mode of origin entails, a process which must have been operative to a very much greater extent under the conditions governing the deposition of the Main Reef conglomerates.

In connection with the question of the fine character of the Rand gold and the absence of anything of the nature of nuggets it is interesting to consider what would be the result if the Rand deposits as they exist to-day could be subjected to rapid denudation, removal and re-deposition.

The pebbles from the existing conglomerates would probably form part of new beds possibly of a similar character and with them might conceivably be associated a large part of the existing gold, and this would presumably be in the same finely divided state in which it now exists. Anything in the nature of a nugget would be of extremest rarity, even if the new deposits were laid down at no great distance. Would therefore future generations of geologists be justified in saying that the gold could not be of detrital origin because of its uniform fineness?

Moreover, under such circumstances, large quantities of new quartz-pebbles formed from the debris of the innumerable small veins occurring on the Rand would be added to those derived from the present conglomerates, yet few of this multitude of new pebbles would be in the least likely to contain any gold, since the quartz-reefs from which they would be derived are rarely found to show any values. Yet the absence of gold in these pebbles might be used as an argument against the detrital origin of the gold then associated with them.

The Infiltration Hypothesis.—Difficulty of application in the Far East Rand.—The conception of the conglomerate beds, on the infiltration theory, as easy routes for the passage of ascending gold-bearing solutions, by which the conglomerates were converted into something bearing a close analogy to quartz-veins, probably owes its origin, and still more probably the continued advocacy it has received, to the study of sections, either in nature or on paper, in which these reefs appear in a vertical or highly-inclined position as is the case with sections taken from most outcrop mines in the Central and extreme Western portions of the Rand.

As usually represented in such sections, the conglomerate beds suggest analogy with a highly-inclined fissure, filled in this case by a pebble-bed whose interstices afford an easy passage for ascending solutions. The suggestion is so strong and the explanation of the deposits thus afforded appears so simple and attractive that it is embraced with readiness and it is one which in the earlier stages of my acquaintance with the Rand appealed to me as forcibly as it has done to others.

In sections and other illustrations of the Rand reefs it is, of course, usually quite impossible to represent them in full detail, or on a correct relative scale to the remainder of the section, and as a result such sections are usually very diagrammatic in character. This will be readily understood by reference to the figures in such well-known and standard accounts as those by Hatch and Chalmers* and by Truscott.

In viewing such necessarily diagrammatic figures one is apt to forget that the conspicuous spaces which appear between the pebbles as represented are really very fully occupied by a matrix, which may be practically as little pervious to the passage of solutions, as the beds outside the limits of the conglomerate are usually presumed to be.

In sections illustrating the more general relationships of the conglomerates, as for example those given with Hatch's map of the Southern Transvaal and similar figures, the band of conglomerate representing the Main Reef is necessarily shown much larger than it really is. In the example just mentioned, at the Main Reef horizon there is usually represented a band of conglomerate about 200 ft. in thickness, taken on the scale of the map, whereas, of course, 20 ft. for the aggregate thickness of all the pebble bands in

* e.g. Hatch & Chalmers' 'Gold Mines of the Rand,' Figs. 44, 50, etc.

this zone would usually be a liberal measurement, and this 20 ft. would not be in a single body but distributed in several layers through a thickness of quartzite ranging up to 200 ft. or more.

If now, instead of the Central and Western Rand with sections like these, so suggestive of easily-traversed channels leading downwards in the direction or whatever extraneous source of gold seemed most desirable as the origin of the mineralization, our first acquaintance with the gold-bearing conglomerates had been, say, in the central portion of the wide and comparatively shallow basin of the Far East Rand, where the conglomerate beds lie almost horizontally, and can be followed laterally through mines and bore-hole sections for many miles in all directions, the inclination to view them as comparable to ordinary lodes would probably have been very much less.

The suggestion that they themselves had supplied the channels by which their auriferous contents had arrived at their present position would not have been so obvious nor indeed so plausible, especially when in addition it was remembered that in the Far East the conglomerate frequently occurs in isolated patches and not as a continuous reef. We should then probably have been less inclined to accept the conglomerates themselves as channels for auriferous solutions, and should have sought other sources of supply.

These might have been looked for in the form of 'verticals' bringing up the solutions from below to be distributed far and wide along the slightly inclined conglomerate beds. It is, however, a very significant fact, which tells strongly against such extraneous origin for the gold, that in the great extent of country which can be said to have been subjected to close examination, and which has been traversed in all directions by a network of mine-workings, the total length of which must run into hundreds of miles and is increasing every day, nothing in the nature of such channels of supply has yet been discovered.

And when we consider that over very large areas in the Far East Rand the 'reef' lies directly on a 'slate' footwall, from which it is very carefully stripped, and in which any definite channel of supply could hardly escape being fairly conspicuous, the failure to detect any such channels is the more remarkable, especially as the richest patches, which one might expect to be most closely connected with any possible 'verticals,' are among those which have been most completely explored, worked, and otherwise subjected to the closest scrutiny.

And if we waive the necessity for evidence for the existence of channels of supply, we are still confronted with some very awkward questions, for we have either to assert that a certain special horizon a few inches thick, out of several thousand feet of associated strata, has some special quality which leads to its selection by the gold-bearing solutions, or we have to explain why numerous other bands of pebbles apparently offering far easier channels of circulation have been passed over or only favoured to a much less degree.

If we invoke the aid of the 'slates' underlying the Main Reef Leader in the Eastern Rand as a determining influence, we must offer some explanation as to why similar bodies occurring at no great distance both above and below the Main Reef horizon and also frequently associated with extensive bands of conglomerates, did not equally affect the deposition of the gold.

Some influence might more readily have been conceded to the slates had the conspicuous deposit of gold been along their *under* side, but as it occurs *above* them it is not easy to imagine the source or the course of the solution which deposited it, and it would be extremely interesting if some advocate of the infiltrationist hypothesis would give us some sort of diagrammatic representation, drawn approximately to scale, of the Far Eastern Rand, showing the probable route of the gold-bearing solutions from an outside source to its present wide distribution over practically the whole of the conglomerate sheet which forms the principal reef in that district, including the isolated patches, large and small, which are characteristic of the Nigel area.

In constructing such a chart or section it would have to be borne in mind that on a liberal allowance the conglomerate band forming the most favoured horizon for the deposition of the gold would bear to the remainder of the Witwatersrand succession about the same proportion and relationship as a single leaf in a book of 6000 pages does to the remainder of the volume.

When the favoured conglomerate sheet affords so much evidence of having resulted from a continuous act of deposition as does the Main Reef Leader on the Eastern Rand, the limitation of placer gold to the bed so formed is a possible and natural consequence of the mode of deposition. On the other hand for circulating solutions to follow a single horizon so faithfully over hundreds of square miles is a much more difficult operation to conceive.

In fact, when we come to deal with conditions such as are met with in the Far Eastern Rand, the theory of infiltration from any outside source of supply demands so extended a series of purely hypothetical steps uninterrupted by any solid ground of observation and of demonstrable fact, that any one may be pardoned who comes to regard the further pursuit of an explanation in this direction as demanding a greater exercise of the imagination than is usually allowed in geological speculation.

If the section represented in Fig. 45 be imagined to be reduced somewhat in thickness to correspond with the decrease in thickness of the formation in the Eastern Rand and then extended laterally to a length of about 38 in., corresponding to a distance of 25 miles, or about 130,000 ft., it would then represent on an approximately natural scale a section across the basin of the Eastern Rand from north to south. Assuming the reef in that area to average three feet in thickness it would be represented in the section by a line less than one-thousandth of an inch thick which would lie immediately above the band representing the 'footwall slate' of the East Rand.

On the infiltration theory we have now to suppose that solutions from some extraneous source travelling upwards through the many great bands of slaty rocks represented in the section, select a particular one which is not known to possess any special character and is not even the uppermost, and travelling laterally, follow faithfully the upper surface of that particular body of slate over the whole of the section, representing in nature an area of hundreds of square miles.

As an alternative we may suppose that the solutions, entering from above, pass downwards through a multitude of conglomerate bands in the Elsburg and Kimberley groups which to all appearances offer greater facilities for lateral movement, pass also through the big group of Kimberley slates and again unerringly pick out for the deposition of the major portion of their gold a single bed of conglomerate which over a large portion of the area is broken up into isolated patches with no apparent lateral connection. When viewed on a proper scale the difficulties in the way of such a course of events will appear to most enquirers so great as to render this mode of introducing the gold into the reef of the Eastern Rand still more difficult of application than it is in the Central area.

Possible influence of dykes.—Before leaving the question of the introduction of the gold into the conglomerates along definite channels of supply, the possible action of dykes in this direction may be briefly considered, since they have frequently been suggested as the possible source of mineralization.

In connection with visits to nearly all the principal mines of the Rand, I have made a point of enquiring whether any instance has been found of a distinct connection between a dyke and the general distribution of gold in the conglomerates. Up to the present I have not been able to find a single case in which such a connection was clearly shown.

In almost every instance the experienced mining men of whom my enquiry has been made have been quite definite in stating that they were unable to point out any example of clear connection between dykes and the values shown by the reef, except for extremely limited and local effects which belong to the secondary movements of the gold such as will be referred to later.

In one or two cases which have been suggested as possible instances, the probability of a different explanation being more applicable has been recognized and pointed out to me.

In this connection the Far Eastern Rand, with its simpler conditions, again appears to afford us the best field for judging of the possibility of dykes being concerned in the introduction of gold. And again, the Nigel Mine in particular appears to offer us a better means of testing this hypothesis than is available anywhere else. As already pointed out, the completeness with which this mine has been explored and worked makes it a particularly useful field of investigation.

The very marked distribution of the conglomerate in well-defined patches to which the gold is entirely confined has been described; a practically complete plan of these patches appears in Fig. 47 (B), and with them are shown the courses of all the principal dykes. It will be seen that these dykes traverse both auriferous patches and barren intervals in a manner which might have been specially devised to afford an answer to the question of their possible connection with the gold content of the reef. Yet no case is known to the officials on the mine in which any such connection was evident. The dykes passed through rich and poor conglomerate and barren ground indifferently.

It will be noted that in some cases the course of a dyke may coincide for some distance with the margin of a patch of conglomerate. In such a case, before the conditions were fully known, it might have been supposed that the dyke was the cause of the difference between the conditions met with on either side of it. This would have seemed still more possible if the patch of better ground had been surrounded by an area of inferior reef, as is usually the case nearer to the Central Rand.

In several such cases, however, when at first a connection has been supposed to exist between a dyke and the richer or poorer reef found on either side of it, more extended working of the area has shown that the dyke, when followed farther, had in other portions of its course richer or poorer ground on *both* sides of it, and that there was no real connection between the dyke and the gold distribution.

The coincidence of a dyke with a fault-plane along which richer and poorer ground have been brought together by lateral movement is, of course, in some instances the means of bringing about an apparent connection between the dyke and the character of the ground in its vicinity.

Against the experience of the great majority of mining engineers that it is exceedingly difficult to trace any connection between the numerous dykes which traverse all parts of the Rand and the distribution of the gold, we have the advocacy of such a connection by Mr. Horwood, who cites, as especially notable examples, dykes at Randfontein and Rietfontein.

The assays from the dykes given in the appendices to Mr. Horwood's paper do not appear very encouraging, especially when one compares them with the values shown by the conglomerates which they are supposed to have supplied with gold. The derivation of the gold in the dykes from the bankets near or through which they pass would appear to many a more probable deduction. Moreover, evidence is forthcoming which to most enquirers would appear to require a good deal of explanation if the dykes are the source of the gold.

A number of sections from the Randfontein area have recently come under my notice, of boreholes which passed through a large number of conglomerate bands, all of which were assayed. The evidence from all the examples is similar, and a few details of one may be given as typical. This hole intersected a numerous series of reefs lying about midway between the Randfontein Leader and the Monarch Reef (both well-known gold-bearing horizons). With the reef a number of dykes were also intersected.

In one portion of the hole a dyke 250 ft. thick was met with, having a reef immediately above it with a smaller dyke above that again—yet this reef, sandwiched between two dykes and, from the description, eminently suitable for the deposition of gold, gave only a trace.

In another portion of the section a dyke occurs which is 90 ft. thick with 65 in. of reef lying directly on it which also gave only a trace of gold, while 30 in. of reef directly below the dyke gave the same result. Yet in a third portion of the section, a reef only 6 in. thick which is 200 ft. from the nearest dyke gave 2 dwt., and one of 7 in. far distant from any dyke gave 3·5 dwt.

In endeavouring to supply a universal source for the gold and associated minerals not only of the Rand bankets but also of the

Black Reef and Dolomite in far distant parts of the Transvaal, Mr. Horwood seeks to connect all these, through the agency of dykes, with the period of volcanic activity associated with the Ventersdorp Amygdaloid.

The suggestion, as far as I am aware, is entirely novel, and that it has been made so late, among the multitude of alternatives, is probably due to the vast amount of field evidence which appears to be so directly opposed to it, and which has probably eliminated it very early from the minds of most observers as a possible explanation of the origin of the Rand gold. Some of the views on which Mr. Horwood appears to base his suggestion, as, for example, that the shales normally interbedded with the Black Reef quartzites in the Randfontein district are to be interpreted as intrusive sheets of Ventersdorp Amygdaloid,* certainly require revision, as they appear to be directly contrary to field evidence of the clearest kind.

It is also a noteworthy fact in this connection that, although so vast a mass of the Ventersdorp Amygdaloid lies immediately to the south of the Rand and once overlaid that district, the Elsburg group of conglomerates, which is nearest to the igneous mass, is probably the poorest in gold of all those associated with the Witwatersrand system.

Secondary changes in the conglomerates.—One of the main reasons for the divergencies of opinion which exist as to the origin of the Rand gold, is undoubtedly the condition in which the gold now exists, and its association with many minerals regarded as characteristic of deposits of the nature of veins or lodes.

These considerations attain greater importance when studies of the reef are largely directed to the examination of specimens from a mineralogical or petrographical standpoint, rather than to the geological stratigraphical characteristics of the beds. And in the case of many who have not been able to make an extended personal acquaintance with the Witwatersrand, the former class of evidence is likely to have the more weight because the papers in which it has been embodied are perhaps much more exact and scientific than those dealing with the latter aspects of the question, these being mostly of early date or confined to a very limited portion of the system. For that reason stratigraphical features have been given greater prominence in the present paper, and it is not proposed to add to the excellent descriptions of the mineralogical

* *Trans. Geol. Soc. S.A.*, vol. xv, 1912, pp. 77-80.

characters of the conglomerates which are available* except to enquire to what extent the conclusions which have been drawn from mineralogical features are opposed to the possible placer origin of the gold.

Prominent among the arguments which have been used against the placer origin of the Rand gold are those connected with the fine state of division in which the gold occurs, its crystalline character, and the absence of waterworn scales or fragments and of nuggets.

If the conglomerates originated as suggested in the earlier portion of this paper, either as simple deltaic deposits as in the case of the Main Reef Leader, or in modified form, redistributed by coastal currents, then the fineness of the gold and the absence of anything of a nuggety nature is only natural and in agreement with such deposits as those of Nome, and many other marine placers where the conditions are not even so favourable for the elimination of the coarser gold as in the Rand deposits, and instead of the crystalline character of the gold being an unexpected phenomenon, surely it would have been a matter for surprise had the original gold escaped solution and re-crystallization when we reflect that the Rand conglomerates are almost certainly of pre-Cambrian age, that they have undoubtedly more than once occupied a position many thousands of feet below the surface, and have been subjected to all the vicissitudes which this submergence entails, and, further, that apart from such a simple change as that involved in the re-crystallization of the gold, the conglomerates in common with the whole formation abound in evidences of profound and prolonged metamorphic action. In connection with the probable solution of the minerals originally present in the conglomerates and their re-precipitation in new forms in the same bed, and in close proximity to their original positions, it is somewhat remarkable that reference has not been frequently made to some very interesting observations on certain deposits in the old Roman baths connected with the thermal springs of Bourbonne-les-Bains recorded by Daubrée (Geologie experimentale, pp. 71–90).

The thermal waters have a temperature of about 150° F. and contain alkaline chlorides and sulphates. In carrying on certain operations in connection with these baths in 1874, there were found

* 'The Petrography of the Witwatersrand Conglomerates, etc.,' Hatch & Corstorphine. *Trans. Geol. Soc. S.A.*, vol. vii, 1904, pp. 140–5.

'Notes on the Auriferous Conglomerates of the Witwatersrand,' R. B. Young. *Trans. Geol. Soc. S.A.*, vol. x, 1907, pp. 17–30.

'Further Notes on the Auriferous Conglomerates, etc.,' R. B. Young. *Trans. Geol. Soc. S.A.*, vol. xii, 1909, pp. 82–101.

embedded in the mud and sand below the old Roman works and near the point of issue of the spring, a large collection of articles including rings, pins, statuettes, frames, etc., together with some thousands of silver, bronze and pewter coins. Four gold coins were also found.

One of the lowest layers of the deposits, which consisted of rock fragments and of coarse sandy material, contained hundreds of coins which showed all stages of solution by the thermal waters, in some cases an imprint of the coin alone remaining. The material derived from the solution of the metals had been re-deposited in the enclosing sandy matrix, which was thus for the most part cemented together by various minerals which frequently exhibited well-developed crystal forms. In some cases these new minerals had bound together the sandy matrix immediately surrounding an inclusion so as to form a flattened spheroidal mass.

Among the new minerals formed by the solution of the bronze, silver, lead, iron, etc., were black powdery oxide of copper, well-defined crystals of red copper, grey copper and copper-pyrites, comparable to the minerals in the lodes of Redruth, also well-defined crystals of galena. Crystals of pyrites were also found deposited upon fragments of sandstone and upon worked flints associated with the other ancient objects.

A feature of these deposits, which is specially notable in the present connection, is that although the circumstances were apparently very favourable to the removal of the minerals formed as a result of the solution of the coins and other objects, these minerals were largely re-deposited in the *immediate vicinity* and in some cases actually in contact with the coins from which the minerals in solution had been derived. Under the very much slower and more restricted circulation which might be expected to have occurred in the conglomerate beds of the Rand, the re-deposition of minerals in close proximity to their place of origin might be much more reasonably expected. If such changes as those described can be brought about by such insignificant agencies in comparatively so brief a time, the re-crystallization of the Rand gold seems scarcely a matter for surprise.

The association of the gold on the Rand with pyrites is frequently quoted in favour of the theory of the introduction of both these minerals into the conglomerates from external sources. While it is no doubt true that some of the pyrites has been so introduced, an adequate study of its distribution in the reefs, and especially in some of the quartzites associated with them, would probably convince most observers that much of the pyrites represents a modification of some of the original constituents of the rock. For frequently this distribution of the pyrites in relation to bedding-planes, current bedding and other sedimentary features of the conglomerates, and especially of the associated quartzites, as for example in the Main Reef Leader, the South Reef and the 'pyritic bands' associated with them and in particular with the Main Reef, is precisely similar to that of iron oxides such as ilmenite and magnetite in sediments of recent origin.

The natural inference, as suggested by Gregory and others, is that much of the pyrites we see so distributed, represents original oxides of iron. From the chemical point of view there seems no difficulty in the way of accepting such a natural interpretation.

The conversion of oxides into sulphides has been artificially effected by several experimenters and by different methods.* Clarke† mentions that Doelter prepared pyrites by heating hematite, magnetite or siderite with hydrogen-sulphide and water for 72 hours to 80° or 90°. If so little time and such a moderate temperature suffices for the conversion of magnetite into pyrites, the assumption that such a change has taken place in rocks which are most probably pre-Cambrian in age and which have certainly for long periods been subjected to the conditions prevailing thousands of feet below the surface appears quite a modest one, and surely demands far less indulgence than the supposition that such pyrites has been introduced from outside sources, and has disposed itself entirely in conformity with the usual mode of distribution of such constituents as the oxides of iron in sedimentary rocks.

In this connection, it may be recalled that iron oxides and also pyrites itself are among the most usual and abundant companions of placer gold. Pyrites figures constantly as a constituent of placer deposits in descriptions of the Nome gravels and of the Tertiary gold-bearing gravels of California.

The evidence for the derivation of much of the pyrites associated with the Rand gold from constituents originally present in the conglomerates is very strong. There can be no doubt, however, that considerable quantities of pyrites have also been introduced at some period subsequent to the deposition of the conglomerates. The existence of two different generations has been emphasized by Young,‡ and indeed it seems very probable that in more than one phase in the history of the conglomerates, pyrites was added to

* 'Data of Geo-chemistry,' F. W. Clarke, *U.S. Geol. Survey Bull.* 491, 1911, p. 316.

† Loc. cit. p. 317.

‡ *Trans. Geol. Soc. S.A.*, vol. xii, 1909, p. 95.

that derived from the original constituents of the rocks. The accession of some of this pyrites to the reefs was probably associated with the formation of numerous mineral veins met with in the Witwatersrand formation, which are of comparatively late origin.

In these veins, in addition to the coarse vein-quartz, which usually forms the bulk of the filling, galena, zinc-blende, copper-pyrites and other minerals usually associated with vein deposits are frequently met with. Quartz veins occur very abundantly in various parts of the Rand, and their outcrops strew the surface with abundant quartz-debris. In the earlier days of the goldfield such veins frequently attracted the attention of prospectors, and a large number of them were opened up and tested.

Some of the very earliest mining on the Rand was done by Struben on a vein of this character prior to the recognition of the importance of the conglomerates as a gold-bearing formation. The greater number of these veins show very little indication of containing much of interest; the great majority appear to be entirely barren, and they have long ceased to attract the smallest attention from mining men. It may therefore be said that the numerous quartz-veins which occur in the Witwatersrand rocks on the Rand are characterized as a whole by marked poverty in minerals of economic value, including gold.

Remarkable exceptions, however, are frequently found where such quartz-veins intersect the gold-bearing conglomerates. In these cases the veins in the immediate neighbourhood of the conglomerate may carry gold in phenomenally large quantities and some of the most remarkable gold specimens from the Rand have been obtained from such occurrences. In one of these, which I had an opportunity of examining in detail before it was removed in the course of mining, the relationship of the vein to its surroundings, the secondary nature of the deposit as a whole, and the derivation of the gold from the adjoining conglomerate were exceedingly clear.

The occurrence was met with on the New Goch Mine at a point where, in close proximity to a fault, the reef had been thrown into a very perfect fold about 50 ft. in width. In the neighbourhood of this fold the rock was traversed by large numbers of small irregular veins and lenticular masses of coarse quartz, which had filled the small fissures which originated as a result of the folding. It was in one of these which intersected the foot of the Main Reef Leader and the upper portion of the underlying 'Black Bar' that the coarse gold was found.

The relationship of the quartz to the Leader will be best

understood from Fig. 50. The gold occurred enclosed in the hungry white quartz as shown in the figure. Some of these patches of coarse gold obtained a diameter of a third of an inch or more. They were entirely confined in the immediate neighbourhood of the foot of the Leader, the more distant portions of the quartz mass being barren, as were also numerous other similar masses which occurred in association with it. The Leader in this particular locality showed very good values, which, as so frequently is the case, were associated more particularly with the actual foot of the reef along its junction with the underlying Black Bar.

The obvious secondary nature of the deposit of coarse gold, its deposition within a few inches of its most probable source, and the absence of even small quantities of gold from the same body of

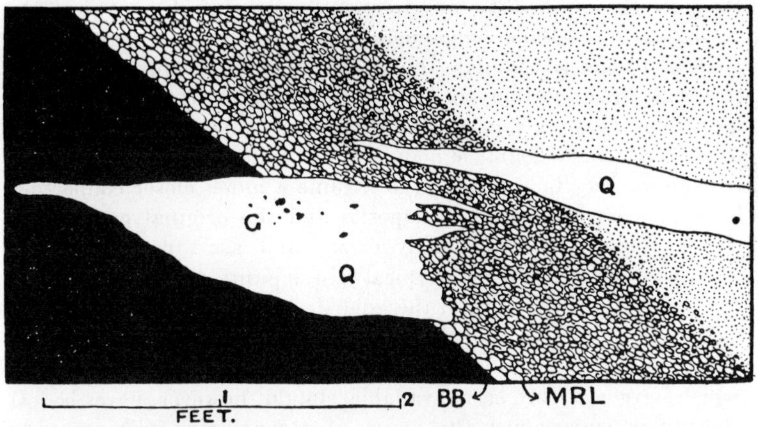

FIG. 50.—Secondary vein-quartz, Q, intersecting the Main Reef Leader, MRL, and the Black Bar, BB, and showing coarse gold at G.

quartz at any greater distance are all points of the greatest interest in connection with the question of the re-crystallization of the gold in the conglomerates, and its re-deposition after solution within a very short distance of its original position. Processes closely similar to those which took place locally in the formation of these small veins, would, on a more extended scale, account for the changes necessary to convert the original placer deposits into their present condition.

Similar occurrences of coarse gold associated with veins and masses of quartz are known from many mines. An example has been brought to my notice within the past few days from the Robinson Deep. In this case a quartz-vein of a more continuous character than those mentioned from the New Goch is found in one locality, where it traverses the South Reef, to contain an abundance of very coarse gold which in some cases occupies a conspicuous proportion of the surface of the specimen.

The quartz in this case frequently shows clearly its individual crystals, with occasional small drusy cavities. The coarse gold occurs between the quartz crystals and is moulded upon them in such a way as itself to appear very coarsely crystalline. Galena is also present in considerable quantities, with other minerals not yet determined.

Followed above and below the conglomerate of the South Reef, the vein ceases to carry gold, which appears to be limited to its intersection with the auriferous conglomerate. It is the coarse masses of gold found in similar veins, or in some cases perhaps recovered from mortar-boxes after the passage of such vein material through the mill, that are frequently spoken of as 'nuggets' occurring in the Rand conglomerates.

Descriptions of similar vein deposits and of their contained minerals, in connection with the conglomerates, without any clear indication being given of their secondary and entirely different character, have no doubt led many, who have not been able personally to examine the Rand deposits, to assume a much closer connection between such later mineral deposits and the original contents of the conglomerates than really exists, and as a consequence to regard the presence of such typical vein deposits as evidence against the possible placer origin of the gold.

The great age of the Rand conglomerates and the unlimited opportunities for modification and additions to the original deposits which have certainly been available should, however, never be lost sight of in endeavouring to arrive at a conclusion with regard to their original nature and contents, and the prominence frequently given to such widely separated and comparatively rare occurrences as those just described must not be allowed to obscure the evidence afforded by the main mass of the conglomerates over practically the whole of the field. In this connection it is noteworthy that Young, who has made the most extended studies of the petrological and mineralogical characters of the conglomerates, although strongly inclined to regard the gold as of outside origin, was evidently impressed by the fact that its distribution is frequently similar to that in placer deposits. This, however, he attributes to the distribution of the gold and pyrites having been influenced by that of some heavy mineral of placer origin already present in the conglomerate.*

* *Trans. Geol. Soc. S.A.*, vol. xii. 1909, p. 98.

Especially interesting in connection with the possible modifications of an ancient placer deposit, including the re-crystallization of the gold and its association with pyrites, are the auriferous conglomerates of the Black Hills of Dakota first described by W. B. Devereux* and afterwards examined and mapped by members of the United States Geological Survey, including S. F. Emmons and J. D. Irving.†

We have thus the advantage of independent observations by a practical mining-engineer and by experienced geologists specially connected with the examination of ore deposits, and the agreement shown on all essential points makes the data obtained from the study of the Black Hill deposits extremely valuable for purposes of comparison with the auriferous conglomerates of the Witwatersrand, with which they have many points in common.

The description of the Black Hill conglomerates should be read by all interested in the question of the origin of the Rand gold. It may here be stated that these deposits form part of a series of sedimentary rocks, which, on account of their contained marine fossils, can be definitely referred to the Potsdam formation, which is of Cambrian age.

The basal beds of the series are usually conglomerates which attain in places a thickness of about 100 ft. and which rest unconformably upon an ancient floor of schists (Algonkian). The schists are traversed by the famous Homestake vein, the source of the gold in the conglomerates.

It should be noted therefore that the gold-bearing deposits are thus basal conglomerates, and in this respect differ from the principal gold-bearing conglomerates of the Rand, including those of the Main Reef group, which were deposited as a part of a continuous conformable sequence of great thickness, and were composed of material, including the gold, which had travelled far from its source.

In the Black Hills conglomerate, on the other hand, the gold was derived from a closely adjacent quartz-vein. Its coarser character and the occurrence of small nuggets, of which the largest recorded attained a weight of 3 dwt., are thus accounted for. The Black Hill conglomerate is cemented together by material which in the gold-bearing portions is largely pyrites or iron-oxide, while the non-gold-bearing portions are cemented by a quartzitic matrix. Irving and Devereux agree that besides the gold which still retains its

* 'The Occurrence of Gold in the Potsdam Formation, Black Hills, Dakota,' *Trans. Am. Inst. Min. Eng.*, vol. x, 1882, pp. 465-473.

† 'Economic Resources of the Northern Black Hills,' J. D. Irving, *U. S. Geol. Survey.* Prof. Paper. No. 26, 1904.

detrital character, there is a portion which has been dissolved and re-deposited and this is found in a finely-divided condition. The values obtained from the Black Hills conglomerate are interesting, and are of the same order as those found in many cases in the Rand reefs, while the occurrence of very rich gold immediately below large pebbles recalls many similar occurrences met with in the Main Reef Leader.

From the above-mentioned descriptions of the Black Hills conglomerates, certain points of great interest bearing on the Rand deposits comes out very clearly, e.g.: (1) By far the greater portion of the gold is undoubtedly of detrital origin; (2) There have been extensive solution and re-precipitation of the detrital gold; (3) The re-precipitated gold was always met with in a finely-divided condition. The improbability of the solution and re-precipitation of gold in the Rand conglomerates and its finely-divided condition, have been regarded as among the strongest arguments against the placer theory of its origin, yet here, in one of the only occurrences known which are of similar age, these same conditions are conspicuously displayed and most clearly established.

A further noteworthy feature of the Black Hills conglomerate is that the gold is everywhere closely associated with pyrites, or with iron-oxides, derived from pyrites, and that where the gold is most abundant, so also pyrites or its derivatives is extremely plentiful. On the Rand where a similar association of large quantities of pyrites with rich gold is frequently observable, the deduction has been constantly drawn that the pyrites was intimately connected with the introduction of gold into the conglomerate. In the Black Hills example, the detrital origin of the gold is so clear that none of the observers who have studied it suggests that the pyrites has anything to do with the introduction of the greater portion of the gold.*

Hence the tendency to attribute to pyrites an important part in the *introduction* of gold into the Rand conglomerates, because it is frequently found in large quantities in the richer areas, is not supported by the very definite facts supplied by the Black Hills occurrence, where such a relationship is obviously absent.

Both Devereux and Irving regard the abundant iron minerals of the Black Hills conglomerate as having been introduced after the deposition of the rock. In reference both to these deposits and to those of the Rand, the question naturally arises as to why the close

* Irving considers that the introduction of a small but indeterminable quantity of gold with pyrites subsequently to the original deposition of the conglomerate seems probable, but hardly susceptible of proof.

association of the richer gold with abundant ferruginous material should not be largely due to the same sedimentary causes as bring about a similar association in so many well-known placer deposits, without, however, necessarily excluding a subsequent introduction of additional pyritic material which has certainly taken place, not only in connection with these gold-bearing beds, but also in very many other sedimentary rocks, where its presence has attracted less attention.

Summary.—It remains to summarize the main points dealt with in the preceding pages, and to consider briefly the ways in which the deductions which have been made bear upon economic questions connected with the conglomerates, particularly in the Eastern Rand, a district which in the near future, and for some time to come, appears destined to attract a very large share of the attention of mining-men as being the most important gold-field at present awaiting development.

The following points appear to be worthy of special attention from those interested in the question of the origin of the Rand gold, and in the future of the field :

(1) Gold occurs much more widely in the Witwatersrand system than is frequently assumed to be the case, being found in practically all the well-marked conglomerates, which are distributed throughout the system as shown in Fig. 45.

(2) Whereas certain of the conglomerates, like those of the Elsburg group, conform more or less to the generally-held idea that the pebble-beds were originally laid down as ordinary marine shore-deposits, there are other types, including the most conspicuous gold-bearing beds such as the Main Reef Leader and South Reef, which are of a very different character, and which are similar to certain persistent conglomerates in the Lower Witwatersrand system—e.g. the Government Reef. The characteristic features of these special types are their astonishing persistence as individual beds over very large areas, and in some cases the way in which they abruptly succeed a considerable series of fine-grained sediments.

(3) These features appear to be due to these particular pebble-beds having been laid down under conditions similar to those obtaining in an extensive deltaic area either upon its outer fringe or within the delta itself, and as a result of episodes of exceptional activity in sedimentation such as are known to occur in existing deltas.

(4) The character of the whole succession of strata of the Lower Witwatersrand system, whose deposition preceded that of the Main Reef group, is consistent with the explanation here suggested, there

being evidence that the shore-line of the land surface from which the sediments were derived was at first at some considerable distance from the present locality of the Rand, but occupied successively nearer positions until, at the time of the deposition of the Elsburg beds, it was in close proximity to it.

(5) The two most important gold-bearing conglomerate beds of the Rand—the Main Reef Leader and the South Reef—are characterized by the comparative uniformity of the material of which they are composed, and by the continuity of the whole deposit. These characters are such as would result from the rapid redistribution of large quantities of gravelly material already graded and accumulated in the lower portions of a great continental river and in the upper portions of its delta. The pre-eminence of the Main Reef group from a mining point of view is not exclusively due to an outstanding content of gold as compared, say, with the Bird or Kimberley Reefs, but to a large extent is a result of the degree of concentration of the gold within such beds as the Main Reef Leader. Had this bed, after its deposition, been subjected to the action of waves and coastal currents the result would have been a deposit more like the Main Reef or like portions of the Kimberley and Bird series, in which the distribution of the gold is very similar to that found in many recent marine placers. The concentration of the gold in such beds as the Main Reef Leader and South Reef is closely connected with their sedimentary features and is such as might be expected to result from the conditions of deposition, and there is a similar correspondence in other groups and in their individual pebble bands, between the distribution of gold and sedimentary features.

(6) The conditions found in the Far Eastern Rand, including the presence of only one principal reef and the possibility of following it in many different directions, afford much better opportunities for the investigation of the problem of the origin of the gold than the more complicated conditions of the Central Rand. In the former area the close connection between the occurrence of gold and of conglomerate, and between the richer deposits and certain types of conglomerate, even when these occur in isolated patches entirely separated from the main body, is so universal and striking as to render it exceedingly difficult to conceive how such associations could be due to other than a common cause. The continuity shown by the Main Reef Leader in the Far Eastern Rand and the occurrence of considerable quantities of gold in that particular bed over an area of hundreds of square miles, which, in this wide extent of horizontal or but slightly inclined beds, is more than ever difficult

to explain on the infiltration theory, receives a natural and adequate explanation when considered in connection with the stratigraphical features of the area and the probable mode of origin of the conglomerate bed.

(7) Difficulties in the way of accepting a placer origin for the gold have arisen from confining attention largely to a few particular reefs which are exceptional in character compared with a large number of associated conglomerates, and making comparisons between these and stream-placers or ordinary shore-deposits with which they have as little in common in sedimentary features as in regard to their gold contents. When a comparison is made between the more ordinary members of the Rand conglomerates and such deposits as those of Nome, a very great similarity is apparent, and the exceptional characters of such auriferous beds as the Main Reef Leader are not surprising now that fuller data concerning its stratigraphical relationships and the conditions leading up to its deposition are available.

(8) The solution and re-precipitation of the gold, which is generally accepted as having taken place by those who accept for it a detrital origin, has been shown to have similarly happened in the Black Hills conglomerate, one of the only known auriferous conglomerates of an age comparable to that of the Rand deposits. In this case, as well as in the secondary quartz-veins of the New Goch Mine, and also in the example given from Bourbonne-les-Bains, a very notable feature is the re-deposition of the gold and other minerals in the immediate neighbourhood of the spot in which they had been originally deposited.

It may further be noted that the features presented by the Rand conglomerates with regard to the distribution of the gold are no more exceptional than are the sedimentary features of the conglomerates themselves when compared with other known occurrences, and further, that whereas the evidence in favour of the gold being an original constituent of the conglomerates is direct, and becomes more and more complete with the fuller investigation of the sedimentary characters of the formation, the arguments for an extraneous origin are highly speculative, and are frequently of such a nature that, though apparently satisfactory when applied to one portion of the field, they are entirely inadequate to explain the conditions met with elsewhere.

The theoretical and highly speculative nature of much of the evidence adduced in favour of the infiltration hypothesis will be noticeable to anyone examining the fullest and most recent exposition of it by Horwood, and the fact that it is thought necessary to bring in so many subsidiary lines of argument indicates the absence of such direct evidence as would be considered in itself of a sufficiently convincing character.

The question of the nature of the association of gold with the Rand conglomerates is, of course, of the very greatest importance, both from the point of view of the student of ore-deposits and of those interested in economic and practical mining matters. Whether the conglomerates are to be regarded as 'fossil placers' or as 'lodes' has much to do with the future possibilities of the Rand, and of conglomerate occurrences in other localities in South Africa and in other parts of the world.

To my mind, after an extensive study of the whole formation, the evidence in favour of the former view is convincing and is increasing continually with the extension of our opportunities for the collection of information. The even balance in the weight of evidence in favour of opposing theories which obtained a few years ago has been materially disturbed by the data now available from the Far East Rand, and from a more extended study of the whole Witwatersrand system, and it seems likely that we must be prepared to admit a new and interesting variety into the family of known types of placer deposits.

If the Rand conglomerates are really placer deposits, then *the question as to whether mineral lodes decrease in value in depth has no application to them, since they belong to an entirely different class of occurrences.* The character and value of the conglomerates will be related to sedimentary causes, and though they may be subject to the vicissitudes of sedimentary deposits, they are not likely to exhibit the pronounced vagaries common in mineral veins, and, moreover, as our knowledge of the district increases, we shall be able to predict with a less degree of uncertainty future possibilities.

The much debated question of the possible impoverishment of the Rand ores in depth is one beset with many difficulties. It is almost impossible to give due weight to all the considerations which enter into that problem, or even to obtain the data regarding the earlier operations in this goldfield, which are necessary for a solution. That the question has remained so long in doubt shows, however, that the evidence in either direction is not of a very decisive character. Whatever the final result may be, it will probably be fully recognized at some later time that such diminution in value as very possibly occurs is not at all a function of *depth*, but one of direction, and depended not upon the flow of subterranean mineralizing solutions, but of surface streams.

In connection with this matter Rickard refers with ridicule* to the expectation of Rand engineers that profitable banket would extend to 18,000 ft. on the dip. It is only fair to assume that this attitude was adopted without knowledge of the amount of information we possess with regard to such a chain of mines as the Van Ryn, Kleinfontein, Van Ryn Deep, Brakpan and Springs. Here we have already the essentials of the expectations referred to more than realized.

Further, from a geological point of view, the Nigel has practically the same claim as Modderfontein or Kleinfontein to be regarded as the outcrop of the Springs Reef, and with very slight changes in the geological history of the district, the Nigel might very easily have been the *only* outcrop of the whole of the reef mined on the Far East Rand. This is an aspect of the matter which suggests a number of very interesting speculations and is worth consideration by those interested in the question under discussion.

In connecting the possible extension of gold-bearing conglomerates below the Witwatersrand basin with the destruction of auriferous quartz-veins *in the same area*,† Gregory appears to attribute to conglomerates like the Main Reef Leader and South Reef relationships similar to those of a basal conglomerate. Such considerations might apply to a deposit at or near the base of the system like that sometimes present below the Orange Grove quartzites (*see* Fig. 45). When, however, it is remembered that many thousands of feet of conformable beds separate the Main Reef Group from the older rocks at the base of the Witwatersrand System, it will be seen that the character and distribution of the quartz-veins in these rocks anywhere below the present basin of the Rand can have little to do with the contents of the conglomerates at the Main Reef horizon.

In a field like the Rand where there exist enormous quantities of auriferous material, whose payability or unpayability depends upon slight reductions in the cost of mining or of treatment, the question of the degree to which natural causes contribute to the cheapness of working is a very interesting one. In this connection, the degree to which the gold distributed through certain beds has been concentrated either laterally or vertically, is noteworthy. In the Central Rand the comparative degree of separation or approximation of the different reefs is often a very important factor, as for example, where two thin leaders are near enough together to be mined in the same stope, or where a thin but comparatively rich reef, as the

* 'The Persistence of Ore in Depth,' *Trans.* xxiv, p. 12.

† J. W. Gregory, loc. cit., p. 58.

Main Reef Leader may be in places, which cannot itself be profitably mined, lies sufficiently close to such thicker but poorer bodies as the Main Reef or the Bastard usually are, to enable the combination to be advantageously dealt with.

In the Far Eastern Rand, where usually a single reef has alone to be considered, the question of the lateral distribution of the gold becomes of greater importance, and it is a fortunate feature, throughout the greater portion of that area with which we are acquainted, that the conglomerate of the principal reef shows a marked tendency to concentration laterally into the well-defined patches described above. The corresponding concentration of the gold distributed over any particular area, which has so far been found to be usual, will considerably simplify the work of the miner in the Far East Rand.

Another fortunate circumstance, probably only fully appreciated by those directly concerned in actual mining, is the presence of the slate footwall. Exploratory work directed to finding reefs, temporarily or permanently lost by faulting or other causes, has probably formed a much bigger item in the cost of mining in the Central Rand than is generally recognized and might form an interesting, if not exhilarating, subject of research in statistics.

In the Far East Rand, with a flat dip and a single reef, absent altogether in places, such exploratory work might have been a still more formidable item were it not for the unusually good 'marker' afforded by the slate footwall in that area—another advantage of the close relationship everywhere existing between the distribution of gold and sedimentary features in the Far East Rand. In the earlier stages of exploration and development on the West Rand, the scanty and hurried attention bestowed on geological considerations resulted in useless expenditure, the extent of which has probably never even been remotely realized. The East Rand may possibly be more fortunate.

Of an abundance of points in connection with the Rand conglomerates, which are of the greatest interest, both from a purely scientific and from a practical mining point of view, no mention has been made in the present paper, which has, nevertheless, extended considerably beyond the length originally intended. In connection with some of these points, a considerable amount of information has already been collected; others remain for investigation. As stated at the outset, no attempt has been made to deal with the large body of literature which has already accumulated around the subject of the Rand conglomerates. The necessity for limiting consideration to the line of investigation which has fallen more especially within the scope of recent survey work also prevents adequate reference to the vast amount of work which has been done by previous writers in connection with aspects of the conglomerates not specially discussed in the present paper.

14-5: THE OCCURRENCE AND ORIGIN OF GOLD AND RADIOACTIVE MINERALS IN THE WITWATERSRAND SYSTEM, THE DOMINION REEF, THE VENTERSDORP CONTACT REEF AND THE BLACK REEF

W. R. Liebenberg

SUMMARY AND CONCLUSIONS

In addition to gold, uraniferous constituents are present in the conglomerates and pyritic quartzites of the pre-Cambrian Witwatersrand system, as well as in the Dominion, Ventersdorp Contact and Black Reefs.

The conglomerates are of the same general type and consist of various proportions of pebbles of quartz cemented together by a mosaic of fine-grained quartz, sericitic materials and chlorite. The valuable constituents of these conglomerates are confined to their matrices except for occasional veins and stringers of gold and uraniferous constituents occupying fissures in the pebbles. In the pyritic quartzites, the gold and uraniferous constituents are confined to the pyritic bands, i.e. the mineralized portions of these quartzites.

The gold in the various reefs occurs as native gold, i.e. a gold-silver alloy of a relatively uniform composition. The principal uraniferous constituents are detrital uraninite, secondary uraninite, thucholite (a heterogeneous mixture of hydrocarbon, detrital uraninite, secondary uraninite and some non-radioactive constituents). Oxidation products of uraninite are rare.

Detailed mineragraphic work on specimens from various localities shows that uraninite is common in the various bands of conglomerate and pyritic quartzites occupying the different stratigraphic horizons. This uraninite is one of the allogenic minerals. It occurs as detrital grains which are rounded to various extents ; oval grains predominate. Minute specks of " radiogenic galena " consisting of isotopic lead in combination with sulphur are ubiquitous in the uraninite. Veinlets of galena are common in the uraninite and some grains of uraninite are almost completely replaced by this sulphide. Ordinary galena is present in the banket and some of the galena associated with the uraninite may thus be of this variety. The grains of uraninite are frequently severely fissured ; the tiny fissures are healed to varying extents by gold and/or sulphides which sometimes also replace the uraninite.

X-ray powder diffraction data show that the uraninite is crystalline and has a composition close to UO_2. This uraninite is only slightly " metamict."

The grains of uraninite are generally less than 80 microns across and are remarkably uniform in size. This mineral is concentrated with other detrital minerals such as chromite, zircon and iridosmine, and rarely with cassiterite. The banket from one of the localities examined contains detrital grains of cyrtolite (uraniferous zircon). Detrital grains of monazite, garnet and cassiterite are relatively common in the Dominion Reef.

The most reliable age obtained on a uraninite concentrate from the banket is 1,913-2,020 million years. This high age represents the age of the source of the detrital grains of uraninite and indicates that the uraninite was, like the other components of the banket, derived from the waste materials of the " old granites," " gneisses " and pegmatites of the Basement Complex. The parent rocks were apparently situated to the north and north-west of the Witwatersrand. Rocks of the same apparent age as the Witwatersrand uraninite are found in the provenance suggested above.

The grains of uraninite are altered to various degrees as distinct from the changes caused by metamorphic agencies ; this alteration appears to have commenced during recent times—probably after exposure of the conglomerates by mining operations—and the uraninite in the freshly exposed conglomerate, e.g. in the Orange Free State mines, is not visibly altered. Large concentrations of the alteration products have not been observed and it must be stressed that they are found mainly on stope faces. Gummite, schoepite, uranophane and schroekingerite appear to be the main oxidation products. Some of these alteration products are intimately associated with the parent grains of uraninite, but the largest concentrations occur as stains on the matrix materials of the conglomerates and as films on granules of thucholite. Mixtures of schoepite and morenosite were encountered as friable crusts on portions of the banket exposed on a stope face in one of the mines on the Far East Rand and appear to have been precipitated from leach solutions which also contained Au, Ag and Pt-group metals. The morenosite is an alteration product of nickel-bearing sulphides such as pentlandite.

Secondary uraninite is often a prominent constituent of the conglomerates of the Bird Reef series, in the West and Far West Rand and Klerksdorp area ; it has also been found in the Basal Reef in the Orange Free State goldfields and in the Dominion Reef. It is relatively rare in the other conglomerate horizons. This uraninite is an authigenic constituent of the conglomerates (and pyritic quartzites) and occurs as small specks, veinlets and irregular patches, up to 1·0 mm. across, in the matrix and also in the thucholite. The secondary uraninite has the same composition as the uraninite, but differs slightly from it in appearance when viewed in polished section. Most of this uraninite represents localized redistribution of uranium derived from the detrital uraninite by leaching and redeposition of the uranium from the leach solutions; small deposits of secondary uraninite are often associated with the parent grains of uraninite and relatively large patches of this secondary mineral sometimes envelop and partly replace detrital grains of uraninite. Some of this secondary uraninite was probably formed by reconstitution of detrital uraninite during metamorphism and metasomatism. The secondary uraninite contains fewer inclusions of galena and gold than the detrital uraninite.

Radioactive leucoxene of the rutile type, often associated with partly altered detrital grains of a titanium-calcium-uranium mineral considered to be titanite, is a relatively common constituent of the Vaal Reef.

The thucholite, commonly known as "carbon," is an authigenic constituent of the banket. This hydrocarbon-uraninite complex is widely distributed throughout the conglomerate bands of the Witwatersrand system, Ventersdorp Contact Reef and Black Reef. It is rare in the Dominion Reef and pyritic quartzites. Where common it is macroscopically visible and occurs as black warty granules, small irregular patches and also as thin bands having a columnar structure. The bands of thucholite often mark the lower bedding planes of the conglomerate bands. The granules of thucholite, some of which are visible only under the microscope, are sporadically distributed throughout the matrix of the banket. They frequently show a tendency to be concentrated around the larger pebbles of quartz along the bedding planes of the conglomerate bands. In the pyritic quartzites, the granules are concentrated with heavy detrital minerals in the mineralized portions, i.e. the pyritic bands. The seemingly anomalous association of these comparatively light granules with the heavy detrital minerals is explained by the fact that the thucholite was formed by the replacement of detrital grains of uraninite by hydrocarbon.

Various degrees of replacement of the detrital grains of uraninite by hydrocarbon to form thucholite are evident. The granules of thucholite consist of remnants of detrital grains of uraninite, sometimes accompanied by specks and veinlets of secondary uraninite, in hydrocarbon. The shapes of the detrital uraninite grains in the granular and columnar thucholite are often preserved, but irregular remnants of uraninite are the most common. Most of the detrital uraninite grains are entirely free from microscopically visible hydrocarbon, but the intimate association of the hydrocarbon with uraninite and the widespread occurrence of this association in the conglomerates of the Witwatersrand system, indicate that the thucholite originated by the irradiation of gaseous and/or liquid hydrocarbons by the detrital uraninite. X-ray powder diffraction studies show that the uraninite in the thucholite is crystalline and structurally similar to the hydrocarbon-free detrital grains of uraninite. Further support for the mode of formation of the thucholite is provided by the age of about 2,000-2,200 million years determined for the uraninite in the thucholite, this age being comparable to that of the hydrocarbon-free uraninite. The thucholitization of the uraninite preceded the deposition of most of the secondary uraninite and a close association between this material and hydrocarbon was thus not established. Minute specks and veinlets of this secondary uraninite are, however, relatively common in the thucholite and appear to represent redeposition of uranium liberated during the replacement process. Secondary uraninite is sometimes relatively common in the thucholite from the Ventersdorp Contact Reef where it occupies fissures and cavities in the thucholite.

Authigenic gold and sulphides such as pyrrhotite, pentlandite, chalcopyrite, sphalerite and galena are often intimately associated with the thucholite; they are often macroscopically visible and mainly occupy fissures and cavities in the thucholite. The sulphides are frequently intergrown with the gold which was the last constituent to be precipitated from localized metalliferous solutions. Mineragraphic work shows that some of the detrital grains of uraninite are partly replaced by gold (and sulphides) which occupied available space in this mineral; this also applies to the secondary uraninite. It appears that the radioactive constituents exerted a precipitating effect on the metalliferous constituents of solutions. The associations of gold and sulphides with the uraniferous constituents are, however, sporadic and do not suggest simultaneous deposition from the same solution.

By far the greatest proportion of the gold in the conglomerates is extremely fine and is not directly associated with the radioactive constituents. The gold is confined to the matrix and the particles vary from less than 1·0 micron to about 100 microns across. Some of this gold appears to be an allogenic constituent, being present as grains which are more or less rounded. Most of the gold is, however, authigenic in its present form. Jagged grains of gold are the most common. The gold replaced matrix constituents to various extents and also occupied fissures and cavities in these constituents. Some of this gold is intimately associated with secondary sulphides such as pyrrhotite, pentlandite, chalcopyrite, sphalerite, galena, arsenopyrite, linnaeite and cobaltite. These gold-sulphide associations have been interpreted by other workers as evidence that the gold was derived from hydrothermal solutions of magmatic origin. On the other hand, they can also represent deposits from localized leach solutions or grains of primary gold and sulphides that were recrystallized during metamorphism. Some gold is associated with grains of buckshot pyrite. Some of it replaced the pyrite and also occupied available space in these grains; gold is rarely associated with pyrite representing a later generation. In many ways, the gold not intimately associated with metalliferous constituents, behaved like the detrital minerals. This indicates recrystallization, and/or solution and redeposition, of detrital grains of gold, e.g. during metamorphism and metasomatism, rather than deposition of gold from hydrothermal solutions of magmatic origin.

Confirmation of many of the views expressed on the placer origin of the gold and uraniferous products in the Witwatersrand banket is provided by information obtained from the pyritic quartzites in the Upper Witwatersrand beds and conglomerate bands known as the Ventersdorp Contact Reef and Black Reef. The bands of pyritic quartzites occur mainly in erosion channels in the footwall of the Main Reef and Main Reef Leader and derived their gold and uraninite by erosion of these conglomerate bands. The Ventersdorp Contact Reef and Black Reef do not belong to the Witwatersrand system proper, but derived their gold, uraniferous constituents and the bulk of their other components from the waste materials of conglomerates belonging to the Upper division of the Witwatersrand system. The gold in the pyritic quartzites, Ventersdorp Contact Reef and Black Reef is thus expected to occur principally as detrital grains, whereas it is in fact secondary in its present form. The uraniferous constituents are the same as; and of similar occurrence to, those found in the Witwatersrand banket. The manner of occurrence and identity of the gold and uraniferous constituents in these " reefs " thus support the views, amongst others, that recrystallization and redistribution of detrital gold and reconstitution of detrital uraninite could have taken place in the Witwatersrand banket. Since thucholite could not have been transported, as such, from the Witwatersrand conglomerates to these " reefs," support is also provided for the views that thucholitization of the detrital uraninite was a continuous process and that the thucholite in the Witwatersrand banket originated *in situ* by the irradiation of hydrocarbons by detrital uraninite.

A sympathetic relationship exists between the uranium- and gold-cosntents of the banket. Exceptions are relatively common, but, due to the inconaistent behaviour of minerals during their transportation and concentration in a vast body of sediments such as the Witwatersrand system, they are not unexpcted. The general sympathetic relationship between the gold and uranium is, however, so persistent and striking that whatever explanation is offered for the origin of the one constituent must necessarily also apply to the other. The uranium in the banket is derived from detrital uraninite, a placer mineral, and its derivatives ; most of the gold, i.e. the gold other than the detrital grains, is therefore also of placer origin notwithstanding its present form. This fact, considered in conjunction with the mineragraphic and geological evidence, bears out the validity of the *modified placer theory* advanced to account for the origin of the gold (and uranium) in the banket. Limited secondary enrichment of the banket in both gold and uranium cannot be entirely ruled out. It might, therefore, be valid to conclude that the Witwatersrand conglomerates represent ancient sediments which contained small fractions of heavy detrital minerals including gold and uraninite and which were slightly enriched, in some areas, by hydrothermal and/or pseudo-hydrothermal processes.

14-6: THE DEPOSITIONAL ENVIRONMENT OF THE WITWATERSRAND GOLDFIELDS: A CHRONOLOGICAL REVIEW OF THE SPECULATIONS AND OBSERVATIONS

D. A. Pretorius

[*Editor's Note:* In the original Figures 2 and 3 appear in color. They have been omitted here.]

INTRODUCTION

It is written in one of the accounts of the beginnings of the Witwatersrand goldfields that the Main Reef group of gold-bearing conglomerates was stumbled across on a Sunday morning, that there was general agreement by the Wednesday afternoon as to the discovery's being of unprecedented significance, and that by the Saturday evening a violent controversy was well under way as to what had been found. To this day, eighty-nine years later, no one single explanation of what the conglomerates really are has received universal acceptance. In 1887, only a year after the gold rush started, Mathers[1], wrote: 'Every second geological or more frequently non-geological expert who has visited the Rand has had a different theory to expound as to the origin of the banket. Where so much diversity of opinion exists among self-styled specialists the ordinary layman can take comfort to himself that his ignorance is not a distressing calamity and that it – sometimes – provides him with a little stock of that modesty which is always a welcome feature when found in mining circles'.

In July, 1886, four months after George Harrison and George Walker made their historic discovery on the farm Langlaagte, one alleged authority reported that the auriferous reefs would not continue to a depth greater than ten or fifteen feet below the surface. To him, the conglomerates represented the pebbly bed of a stream that had once run parallel to the ridge of hills. Owing to earth movements, the stream bed had been turned on edge, so that the depth to which the gravels would persist could not be greater than the original width of the stream. One year after digging started, Gardner Williams, the eminent mining engineer from Kimberley stated categorically, in March, 1887, that the conglomerates were merely surface deposits along ancient beach lines, and, because beaches were always narrow, no confidence could be attached to any predictions that the goldfields would have a long life.

Ballot[2], under the pseudonym Iones Beta, published in 1888 the first lengthy account of the mine workings that had developed around the Johannesburg camp. In his introductory preface, he stated:

Ever since the discovery of gold in the conglomerate of the Heidelberg, Potchefstroom and Pretoria districts, and more especially along the range of hills known as the Witwatersrand, there has been a considerable difference of opinion about the probable origin and the present position of this, as people supposed, entirely new and unequalled formation.

Some predicted a speedy collapse, supposing the reefs to be mere alluvial deposits, certain to run out at no great depths from the surface. Others thought the reefs of volcanic origin, forced up from below and that the conglomerate would gradually turn into solid quartz with increase of depth.

Another gentleman of practical reputation asserted the conglomerate to be of aqueous origin, formed under the ocean, but that it could not continue gold bearing, or go down to any depth; though not giving any reasons for such views, beyond saying that the reefs were bound to become barren or run out on striking the Silurian formation below.

Again we heard it advanced that the conglomerate and sandstones were of aqueous origin, deposited at the bottom of an ocean or lake; but that the gold was a chemical deposit precipitated from the waters of the ocean; in which it existed in the form of chloride of gold held in solution.

And lastly it has been held by some scientists that the conglomerates and sandstones are successive deposits of sand and gravel spread over the lake or ocean floors. That at certain intervals of time the floor of this ocean was ruptured by volcanic forces from below, and large quantities of auriferous lava in a molten state poured out and spread over these gravel beds, cementing them together. At the same time forcing the vapours of gold into and along the poured out gravel beds already covered up by beds of sandstone . . .

The nature of the environment in which the conglomerates and other gold-bearing strata were deposited has obviously been a matter of considerable debate for a long period of time. Normally, three broad classifications of depositional environments are recognized: terrestrial or continental, marginal or transitional, and marine. Conditions in the terrestrial class range from fluvial, through lacustrine, paludal, and desert, to glacial; in the transitional category from deltaic, through lagoonal, to littoral; and in the marine grouping from neritic (low-tide to 200 metres depth), through bathyal (200 to 4500 metres), to abyssal (4500 to 7000 metres). All of these environments, with the exception of bathyal and abyssal conditions, have found their proponents among the substantial number of authors of publications on the Witwatersrand group of rocks and their contained gold and uranium mineralization.

The object of this paper is to present a critical review, in chronological order, of the observations that have been made by many of the authors and of the speculations offered by a much larger proportion of the writers on the subject. As is generally the case in the sphere of geology, fancies outweigh facts, by far. For this review, 350 books, papers, and articles were selected from an impressive volume of literature as possibly containing information relevant to the theme of the exercise. A total of 170 was found to include observations and ideas of worth. Only 90 publications, in the final analysis, could be considered as having made significant contributions to a better understanding of the nature of the depository and the conditions under which the sediments, both auriferous and barren, accumulated. The present work deals with these 90 books and papers. This figure leads to the conclusion that, on average, only one important publication has appeared each year since the discovery of the Main Reef in March, 1886.

The history of exploration and exploitation of the Witwatersrand mineralization appears to follow a 21-year cycle. The phases that have been recognized are as follows:

PHASE 1 (1886-1906):
- discovery of banket
- development of outcrop mines as small units to shallow depths
- Stage 1 of Central Rand goldfield
  - of West Rand goldfield
  - of East Rand goldfield
  - of Klerksdorp goldfield

PHASE 2 (1907-1927):
- amalgamation of small units for deep-level mining
- Stage 2 of Central Rand goldfield
  - of West Rand goldfield
  - of East Rand goldfield

PHASE 3 (1928-1948):
- discovery of extensive carbon seams
- South Africa goes off gold standard
- development of lower-grade mines
- introduction of geophysical prospecting in search for new goldfields

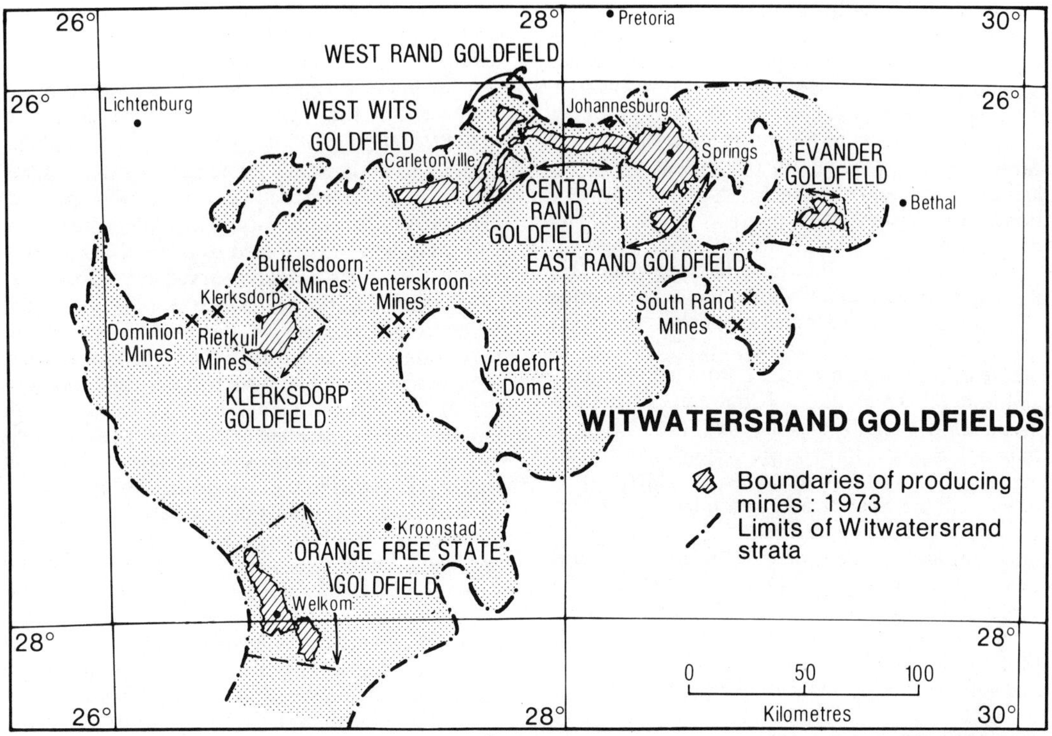

FIGURE 1 The location of the seven goldfields within the Witwatersrand Basin. The positions of the boundaries of producing mines at the end of 1973 are shown, as well as the presently known limits of Witwatersrand strata.

Stage 1 of West Wits goldfield
of Orange Free State goldfield
Stage 2 of Klerksdorp goldfield
Stage 3 of Central Rand goldfield
of West Rand goldfield
of East Rand goldfield

PHASE 4 (1949-1969):
recovery of uranium
devaluation of sterling and rise in price of gold
Stage 1 of Evander goldfield
Stage 2 of West Wits goldfield
of Orange Free State goldfield
Stage 3 of Klerksdorp goldfield
Terminal stage of Central Rand goldfield
of West Rand goldfield
of East Rand goldfield

PHASE 5 (1970-1990):
sale of gold on free market
Stage 1 of possible new goldfield(?)
Stage 2 of Evander goldfield
Stage 3 of West Wits goldfield
of Orange Free State goldfield
Terminal stage of Klerksdorp goldfield

The locations of the different goldfields mentioned above are shown in Figure 1, which also outlines the presently known limits of the Witwatersrand Basin.

The dates of publication of the more significant contributions on the depositional environment of the Witwatersrand strata have been grouped according to these five phases. In Phase 1, 12 papers merited serious consideration, in Phase 2, 19 contributions, in Phase 3, 12 publications, and in Phase 4, 42 books and papers. In the first five years of Phase 5, five publications are believed to have produced novel concepts of aspects of the mode of formation of the Witwatersrand conglomerate and carbon-seam auriferous horizons. The most fertile period for the generation of new ideas on the

FIGURE 2 (Top) Main Reef Leader from Crown Mines in the Central Rand goldfield. This is a specimen of conglomerate typical of that on which the Witwatersrand gold-mining industry was based for the first forty-odd years of its existence. The pebbles are exclusively of white vein-quartz, set in a matrix of quartz grains and flakes of phyllosilicates. The conglomerate rests with a sharp, planar contact on the Black Bar quartzitic argillite footwall. Detrital grains of gold are visible in the matrix and around the pebbles, a phenomenon which is rare in the banket. The specimen comes from depth in the vicinity of the discovery site at Langlaagte, southwest of Johannesburg, and portrays the type of banket that generally carries the richest concentrations of gold.
(Bottom) South Reef from the Robinson Deep Mine in the Central Rand goldfield. Two bands of conglomerate are separated by a narrow parting of coarse-grained pebbly quartzite. The pebbles are composed of vein quartz and several varieties of massive and banded chert, while the matrix consists of quartz, phyllosilicates, and abundant detrital pyrite. The foresets of cross-bedding within the upper conglomerate band are defined by oblique layers of pyrite granules. This type of banket is generally highly payable, but is not as rich as that exemplified by the specimen of Main Reef Leader.

problem was obviously the years between 1949 and 1969, when information became available in large volume for the important new goldfields beyond the limits of the original Witwatersrand. These good years saw geology, a rather poor relation of Witwatersrand mining practice in the three earlier phases, assume its rightful place in the hierarchy of research, development, and production in the greatest gold province the world has ever known.

In all the theories that have been advanced for the environment under which the accumulation and concentration of mineralization took place, five factors have had to be considered: (i) the nature of the conglomerates, both barren varieties and mineralized banket; (ii) the areal extent and distribution of the conglomerates; (iii) the variation in the composition of the sediments occurring above and below the mineralized horizons; (iv) the origin of the seams and granules of carbon, frequently extraordinarily rich in gold and uranium; and (v) the source of the gold, uranium, and other ore minerals in the banket and the carbon. The controversies, often acrimonious, which have been generated on the origin of the gold and uranium deserve a review to themselves, and will not be dealt with in this paper.

THE PERIOD 1886–1906

By 1888, the instantaneous impressions and immediate interpretations had run their courses, and serious thought was being given to the information that was rapidly accumulating from the mine development that was clearly showing that the Witwatersrand reefs were not the products of a shallow stream or of a narrow beach. In that year, Ballot[2] put forward the suggestion that older, auriferous quartz lodes had formed the rocky shores of a coast-line that was constantly undermined by waves and currents. The ocean advanced towards the land, destroying the coast-line as it transgressed, and on its floor the material from the shore was deposited. Periodic floods pouring off the adjoining highlands transported large quantities of gravel, grit, and mud over extensive areas of the ocean floor, adding to that gathered from the beaches. The farther the gravel was transported, the more waterworn and smaller the pebbles became, and the more gold was lost. He believed that the shales present in the Witwatersrand succession were laid down in deep, quiet water, well out into the ocean.

No further ideas of importance were presented until 1895 when publications by Hatch and Chalmers[3] and Kuntz[4] agreed with Ballot's conclusions that the conglomerates were marginal sea deposits. The latter regarded the shales as the products of an advancing coast-line and the conglomerates of a retreating shore. He thought that the coast-line was in a continual state of falling and rising caused by the folding of the crust. Kuntz was the first person to put forward the idea that the northern edge of the Witwatersrand basin occurred in the Johannesburg area and the southern margin in the northern Orange Free State.

In 1896, de Launay wrote that it was possible that the Witwatersrand strata could represent widespread fluviatile sediments deposited on torrential deltas distributed over an alluvial plain such as that which he had observed in Lombardy in Italy. However, his preference was for a marine beach origin because of the vast extent of the formations, the regularity of the beds, the frequent occurrence of shingle-type pebbles, and because the Witwatersrand conglomerates occupied an 'intermediate stratigraphic position between two formations known to contain marine fossils – the Bokkeveld beds and the Carboniferous limestones'. In the same year, Schmitz-Dumont voiced his disagreement with the generally prevailing ideas, and came out in favour of a fluviatile origin. The Main Reef he considered to represent the bed of an easterly-flowing river, which debouched onto a broad and large valley-plain. More than one river bed was formed in order to account for the great widths of the conglomerate horizons. The quartzites were thought to be the material formed on a delta or in an off-shore environment. As the continent subsided into the sea, the pebbles and sands that had formed in the river valleys gave way to river-borne sands and clays laid down in the sea.

The year 1897 saw the publication of an important contribution by Becker[7] of the United States Geological Survey, which became the most quoted study of the first phase of geological investigations. He could see no evidence for a lacustrine environment, because there were no large quantities of very fine-grained silt, because the coarseness of the conglomerates and sands showed the results of currents stronger than those that are developed in lakes, and because the areal extent of the gravels was too great to have been formed under lacustrine conditions. He viewed the presence of muffin-shaped pebbles, the absence of shingling, of regular imbrication of the pebbles, of river channels, of coarse nuggets, and of boulders, and the development of conformable stratification as convincing arguments for the depositional environment having been that of a marine beach. The currents flowed from east to west, in the same direction as the present currents around the southern African coast from the Indian Ocean to the Atlantic. The great thickness of the strata was considered to indicate substantial offshore subsidence. Becker saw similarities between the Rand deposits and the auriferous beaches of New Zealand and of the northwestern coast of the

United States. Denny[8], from his study of the Klerksdorp reefs, supported a beach environment where the action of waves was strongly felt. He added that '. . . the solidly pouring rains of Devonian times, when the heated earth was combining with the sun to evaporate ocean waters, from condensation in the upper rarified atmosphere, can easily be imagined of volume sufficient to create flowing seas of water, of which we can now have no possible illustration, and but faint conception'.

As the littoral origin became more and more the accepted theory, Carrick[9], in 1896, drew the attention of the various authors to the fact that Fred Struben, the original pioneer of the Witwatersrand area, at the very beginning of his mining of the conglomerates in the Lower Division of the Witwatersrand Sequence, considered the banket to be an ocean-beach deposit. At the same time, Garnier[10] put forward the first suggestion that the beaches had formed along a closed sea that was surrounded by high mountains, some of which formed a vast amphitheatre on the northern side. Streams converged in all directions towards the basin.

In the year 1898, Bleloch[11] described the Rand deposits as having some similarities to the Klondyke gravels. He saw the banket as a series of sub-shore deposits banked up by ocean currents and waves against a sloping shore. The alluvium was brought in by rivers and was spread over hundreds of miles on the offshore bottom. Bleloch introduced three important ideas. He was the first to question the generally held opinion that all the strata were completely conformable, and suggested that gradual subsidence at times produced estuarine deposits in a shallow tidal lake, and that, in such an environment, unconformity upon unconformity would have been produced. His was also the initial suggestion that, with recurrent subsidence and upheaval, the ocean currents caused previous deposits to be alternately overlapped and attacked, with later beds including some recycled material from earlier sediments. His third contribution was the suggestion that the original edge of the basin had not been that far removed from the positions of the present outcrops, so that little erosion had taken place of the Witwatersrand reefs. Prister[12], in 1898, also added his support to the shoreline concept. He thought that hot currents flowed along the coast, and that these were tidal, not ocean, currents. He visualized a large bay extending from the present Lebombo Mountains to Potchefstroom, with the water moving westwards because of the Earth's easterly rotation.

The last important contribution of Phase 1 was made by Hatch[13] in 1906. He subscribed to the theory that the material was derived from granites, gneisses, and schists of the Swaziland System, which were elevated to form a continental area. Rivers transported the debris to a shallow-water marine environment. The source-area lay to the west, and the sea to the east. The northern shore did not extend much farther north than the present site of Rustenburg. The clays and fine sands of the Lower Witwatersrand Division were laid down on a sinking bottom, whereas the conglomerates of the Upper Division accumulated on a rising bottom, with shallow water permitting the development of shingle and gravel.

In 1896, Schmitz-Dumont[6] wrote: 'Regarding the origin of the quartzitic sandstones, I have found some petrefacts in them like seaweeds'. However, it was Garnier[10], in 1896, who gave the first descriptions of the carbon, which had been observed associated with a number of the reefs in the East Rand, Central Rand, and Klerksdorp goldfields. He mentioned that the small carboniferous layer at the Buffelsdoorn Mine in the last-mentioned goldfield was a seam between one and five centimetres thick, which resembled lignite filled with gold. He suggested that 'organized beings' had flourished in the water, and that petrefaction of the decomposed organic matter had liberated gas, which precipitated gold present in the waters of the depository in the form of trichloride.

In summary, it can be said that Phase 1 of Witwatersrand mining history was a period in which speculation was rampant on the environment in which the gravels were laid down and the gold accumulated. Observations were almost exclusively qualitative, and no attempts were made to back up theories with quantitative data. The majority of investigators subscribed firmly to the idea that the conglomerates had been deposited along a marine beach and for a short distance off-shore. There were already suggestions that the source-area was a continental landmass to the north and west of the depository, that the sea was closed and surrounded by mountain ranges, and that later sediments contained contributions from reworked earlier material. The long-shore currents were generally believed to have flowed from east to west. Opinions were common that not much erosion of the Upper Witwatersrand Division had taken place, and that the original shore had not been too far removed from the present outcrops of conglomerate. There was one dissenting voice, which favoured a fluviatile origin for the gravels. Eastwards-flowing rivers deposited the pebbles on a large, broad valley-plain above a delta. Within the first ten years of mining, the importance of gold-bearing carbon had been appreciated, and it was thought that it had been derived from organisms that were present in the ocean waters.

THE PERIOD 1907–1927

Phase 2 opened with a notable contribution by

Gregory[14] in 1908, followed by a supporting statement in 1909[15]. This author saw the deposits as a long series of beach accumulations on a tide-swept shore where the material had been washed back and forth by the waves to yield a well-sorted end-product. The fines were removed to deeper water, leaving behind the pebbles to form small breakwaters, under the protection of which the fine-grained gold was concentrated. Channels were cut through the beach material by streams and tide races. Gregory concluded that '. . . the rich auriferous sheets of banket represent intervals during the formation of the Rand Series when by a temporary cessation of deposition the gold concentration took place by tidal scour on the shingly shore'. He saw a resemblance between these reefs and the black-iron-sand-bearing beaches of New Zealand, where the coarse gold had been left upstream in rivers and only the fine particles had been carried to the beaches along which they had been distributed by currents. However, he admitted that the restriction of the gold to certain definite beds was not inconsistent with alluvial placer deposits. Gregory also put forward the thought that the Witwatersrand beds were of Precambrian age, and were not as young as so many of the early investigators had assumed.

In 1909, Becker[16] confirmed his original ideas on the development of the gold-bearing reefs, saying that they had been formed on sand beaches along rocky coasts. Violent storms had thrown up banks of pebbles far beyond the highwater line. A heavier surf had then moved the pebbles below the highwater mark, and the quieter waters of tides had transported the sand, which infiltrated and covered the pebbles. That year also saw a most important contribution by Young[17] who drew attention to the presence of payshoots of richer reef, which had not been recognized previously because there had been no syntheses of the data from the many mine workings. The most obvious payshoot he found to occur on the eastern part of the Central Rand. This zone of enrichment had a distinctly south-westerly orientation. Young believed that shallow-water conditions prevailed throughout Upper Witwatersrand times, and that the strata had accumulated in an area of subsidence with a slowly encroaching sea. Two years later, Young[18] reviewed the state of knowledge and opinion on the Witwatersrand deposits, and stated that it was generally agreed that the banket was a littoral marine deposit, and that during Main Reef times, at least, there were wide stretches of shingly beach. The conglomerates and ripple-marked quartzites were formed during periods of pronounced shallow-water conditions, whereas the conformable sediments, several miles thick, below the Main Reef Series had been deposited many miles from the beach.

The year 1911 saw the first publications[19,20] on the subject of this review by one of the most distinguished of all the geologists who have been involved in the problems of the Witwatersrand goldfields – E. T. Mellor of the Geological Survey of South Africa. Mellor's work marked the beginnings of systematic observation and measurement of the stratigraphy of the depository, and his work of more than sixty years ago is still the basis for much of the present-day classification of rock-units within the succession. He was the first person after Schmitz-Dumont[6] to doubt the generally accepted hypothesis that the gold-bearing reefs had been laid down on a marine beach. As a result of his extensive mapping, Mellor felt that the sediments were representatives of a very large delta in which there were interrupted, but progressive, cycles of changes from deep-water in the lowest portions of the stratigraphy, through deltaic phases, generating the conglomerates, in the middle portions, to actual shore conditions in the uppermost portions. He concluded that the reefs had not all formed in the same environment, and that some were the products of intermediate conditions and others of proximal conditions, with respect to distance from the shoreline. The papers by Mellor provoked long and bitter arguments, during which he remained very much the lone voice in the wilderness. Doubt was cast on the accuracy of his observations and the integrity of his conclusions, and he was accused of deliberately ignoring the work of other investigators. He had committed the almost unpardonable crime of introducing fact, instead of fancy, into his pronouncements.

By 1915, Mellor[21] had added a very considerable volume of additional observations to his information. He became even more convinced that the sudden changes from fine-grained sediments to conglomerates, such as occurred in the Government Reef, the Main Reef Leader, and the Kimberley Reefs of the East Rand, could only have been produced in an extensive delta. The most spectacular of these sudden changes he considered to be the Speckled Bed in the lowermost assemblage of Witwatersrand rocks. For the most part, the finer-grained material of the Lower Division was deposited beyond the limits of the delta, while the Jeppestown and Kimberley shales he believed to be the equivalents of modern delta silts. The continual increase in coarseness upwards indicated the gradual elevation of high ground to the north of the depository. As a result, the points of discharge of the sediments slowly approached the position of the Central Rand. By Kimberley-Elsburg times, towards the close of the sedimentary history of the basin, the shoreline had reached as far south as the vicinity of Johannesburg. Mellor believed that vast quantities of material had already accumulated along the lower courses of the rivers and the upper portions of the deltas, and were available for sweeping out by floods. Persistent beds, such as the Main Reef Leader, were laid down in a single period of sedimentation, the duration of the period being a matter of only days. The limited amount of information available from the then-developing East Rand led Mellor to two erroneous conclusions. He thought that there was an absence of cross-bedding in the payshoot conglomerates, showing that the Main Reef Leader had been laid down rapidly and continously over the whole area. He also believed that there was no erosion of the underlying muddy deposits of the East Rand, which would have taken place had there been a gradual extension of the pebbly deposits over them. Mellor compared the size of the area of development of the Main Reef Leader to the much greater extent of the deltas of the Ganges, Hoang-Ho, and Yang-tse-Kiang rivers.

Certain of Mellor's ideas were amplified in 1916[22] in another important paper. He drew attention to the fact that most of the previous studies had been concerned with the Main Reef of the West Rand and the Central Rand. The opening up of the Main Reef Leader of the East Rand had disclosed a new dimension that made many of the previous arguments untenable. He saw the Main Reef as representing widespread coastal plain gravels that had been modified by wave-action. The Main Reef Leader and South Reef had been laid down in a pregraded river or the upper part of a delta, and had not been affected by subsequent wave-action and coastal currents. The Bird and Kimberley reefs, he thought, were further examples of coastal plain gravels. With gradual elevation in the northwest and subsidence in the southeast, denudation of the Lower Division sediments probably took place in the later stages of the Upper Division. One of the main criticisms levelled against Mellor's theory was that it considered only the East Rand, Central Rand, and West Rand, and ignored such important sections of the Witwatersrand depository as the Klerksdorp goldfield. If the latter were included, it was argued that the total area embraced by Witwatersrand strata would have been too large to be accommodated within a single delta. His opponents contended that only an ocean could distribute material similar to the original constituents of the banket over such a great distance. Wagner[23] postulated that Mellor's suggested mechanism of considerable floods sweeping out material in rapid pulses would not have led to the development of conglomerates as extensive as those of the Witwatersrand. If the pebbles had been laid down in a large delta, then they must have been reworked by waves and currents during episodic subsidence of the littoral zone, in order to produce the shapes, sorting, and imbrication that characterized the banket. Wagner stated that the lenticular nature of the conglomerates and the splittings and partings of the reefs were also convincing indicators of the presence of beach gravels.

Discussion on Mellor's papers produced a highly significant new comparison by Evans[24]. He viewed the Witwatersrand depository as similar to that presently existing in northern India, adjacent to the Himalayas. A range of mountains to the north of the Central Rand would have filled the role of the Himalayas, which were previously much farther north than they are today. Faulted ground bounds the present mountains, and has been moving progressively southwards, so that the earlier alluvial deposits on the upthrown side have been raised above the plains from time to time, and have been eroded and swept down on to the plains by torrents from the slopes of the Himalayas. The gradual rise of the mountains and the subsidence of the plains has led to the frequent recycling of material derived from the uplifted highlands to the north. Evans pointed to the fact that many smaller streams are developed along the slopes of the Himalayas, rather than one large river. The argillaceous sediments in the Witwatersrand depository would have been brought to the plains by gentle streams, while the pebbles would have been the product of sheetfloods associated with great torrents. Towards the end of Witwatersrand times, when the mountain-front was at its closest to the basin, the force of the streamflow swept the gold farther out on to the plain, to give larger volumes of conglomerates with smaller amounts of gold. Evans felt that there was no vegetation at the time the Witwatersrand rocks were formed, that there was no organic matter in the depository, and that there was no oxygen in the atmosphere.

It was only in 1920 that the first sedimentological observations were made in the Witwatersrand goldfields. It seems difficult of explanation that, with the economic mineralization being so intimately associated with well-defined aspects of the sedimentary history of the basin, no attempts had been made at studying the sediments in their own right during the previous 30-odd years of exploitation of the conglomerates. Pirow[25], consequent upon a systematic investigation of the pebbles in the reefs, added his support to Mellor's deltaic theory for the Main Reef Leader of the East Rand. He found that the distributary action of the streams and their velocities and carrying capacities diminished to the southeast, and that, in the same direction, there was a corresponding increase in wave- and tidal action. Three distinct types of ripple-marks were measured, as well as interference ripples, in both the hangingwall and the footwall of the East Rand. The pattern of footwall erosion indicated clearly to him that water-courses had existed below the Main Reef Leader, and that potholes had been scoured out by eddies in these channels. Pirow concluded that the laying down of the gravels on a delta was succeeded by a short period of subaerial exposure and local denudation by rivulets and flowing waters. These streams also carried in the gold, and Pirow was thus the first to propose that the gold might have been introduced into the conglomerates by a later pulse of sedimentation than that which laid down the gravels.

In 1926, Rogers[26] published the results of his detailed mapping of the Heidelberg area at the southeastern extremity of the East Rand goldfield and the northern limit of the South Rand mining area. He took the absence of calcareous beds as firm evidence against deposition in a marine environment. The presence of fresh feldspars in the rocks indicated a cold climate, and the absence of pseudomorphs after salt pointed against an arid climate. The remarkably regular, but repeated, succession of coarse sands and fine sediments over great areas could have been the product only of sedimentation in a delta, so that Mellor's conclusions received further strong support.

The first edition of '*The Geology of South Africa*' by Alex du Toit[27] appeared in 1926. In this was synthesized all the major investigations of the Witwatersrand Basin to that date. Du Toit only partially accepted the correctness of Mellor's ideas on deposition in a delta. The strata that had been laid down off the mouth of a large river were periodically submerged by waves, as the ocean floor intermittently subsided, so that shallow-water sands and gravels alternated with deep-water muds. The depository stood close to sea-level, with the result that fluviatile and estuarine conditions intermingled. The gradual transgression of the shoreline produced the irregular shingling and the prevailing muffin-shape of the larger pebbles. Du Toit thus saw Mellor's delta as having been subjected to littoral reworking, with the result that the conglomerates, in their final form, could be more satisfactorily compared with modern marine placers.

Nel[28], upon the completion of his mapping of the Vredefort area, concluded that the Witwatersrand sediments around the dome had accumulated in the same basin and under the same conditions as in the Witwatersrand proper and the Heidelberg locality. The presence of widely distributed beds and the sudden passage from fine-grained shales to coarse-grained sands over extensive areas, he took as evidence of a deltaic environment, with the Vredefort area being farther away from the points of discharge than the goldfields on the northern rim of the basin. Through the work of Mellor, Rogers, and Nel, the Geological Survey of South Africa was a powerful advocate of the theory that the Witwatersrand conglomerates and other sediments had been laid down in a delta.

The second phase of mining history came to a close in 1927 with the publication of Reinecke's[29] sedimentological and other studies of the East Rand and the Central Rand. This paper and the contributions of Mellor rank as the most important publications on the depositional environment of the Witwatersrand gold deposits that were presented in the first 40 years of geological investigations. Reinecke was the first person to prepare contoured plans of assay values, and, of the result, he said: 'This brings out the variation in value in a striking manner and enables one to grasp the mass of detailed information involved with the minimum of effort'. These plans clearly delineated the numerous paystreaks in the Main Reef Leader of the East Rand, and showed that they had a fan-shaped arrangement, spreading farther apart southwards. A series of elongated, discontinuous lenses of conglomerate was shown to have radiated out from near Benoni, with the currents having moved eastwards and southeastwards, and their strengths having declined in the same directions. The current patterns assumed a braided configuration. Among his conclusions, Reinecke stated that: 'The deposition of the lenselike conglomerate bodies of the East Rand, in the shape of a fan or a wedge, can, moreover, not be reconciled with parallelism to a strand line. The plan of these currents and of the conglomerate lenses does, on the other hand, resemble that laid down by a river of continental proportion on its flood-plain in times of exceptional high water'. On the Central Rand, the paystreaks were shown to have a more braided pattern than on the East Rand, and were smaller, less consistent, and poorer in gold. They tended to spread apart towards the southwest, and the areal percentage payability diminished in the same direction. Two interfering trends of paystreaks – one oriented southeastwards and the other southwestwards – were recognized in the central part of the Central Rand. The Main Reef Leader on the Central Rand and the East Rand was seen to have been deposited as the products of either two separate floods or of contemporary distributaries from the same main channel, one having broken to the east and the other to the west, from a point north of Benoni. The Lower Witwatersrand succession of alternating shales and grits was viewed by Reinecke as showing depositional environments varying from a large body of water to the flood-plain of a river. Upwards, there were more and more beds with true terrestrial characteristics. The thinning of the whole succession from northwest to southeast indicated that the currents had moved in that direction. Reinecke found that he could not support Mellor's deltaic theory. He favoured the depositional environment to have been that of a river flood-plain at some distance from the sea. He introduced for the first time a glacial concept into the environmental regime. The most likely agent for carrying the exceptionally coarse material of the banket was considered to have been an ice sheet that picked up alluvial gravels in the foothills of a mountain range and deposited the load far down the broad plain of a continental river. The pebbles were laid down partly as outwash gravel in front of the glacier and partly as ground moraine. The rapid change from cold to warm temperatures, accompanied by the sudden melting of large volumes of ice, furnished the necessary flooding conditions. Strong evidence of a glacial climate was believed to be provided by the tillite band in the Lower Witwatersrand Division.

The presence of uranium-bearing material in the auriferous reefs was first reported during Phase 2. Pirow's[25] work with the pebbles of the conglomerates brought him in contact with the diamonds that had been recovered from the reefs on the northern side of the East Rand goldfield. He suspected that the frequent green coloration of the diamonds was the result of radioactive bombardment. Black-sand concentrates from five mines in the East Rand, two mines in the Central Rand, and one mine in the West Rand were submitted for analysis to Ettlinger of the Physics Department of the University College, Johannesburg, but he reported in 1920 that there were positively no indications of radioactive minerals in the material from the reefs. However, in 1924, Cooper[30] identified uraninite in concentrates gathered from mines between Boksburg and Roodepoort. He found that the Central Rand samples were richer in uranium than those from the East Rand, a conclusion that was not to be borne out by the detailed work on uranium carried out twenty years later. Reinecke[29] also stated that uraninite occurred only west of Boksburg. Because, at that time, uranium was not of economic importance, no further attention was given to its presence in the banket, with the result that its distribution and mode of occurrence made no contribution, during Phase 2, to a better understanding of the conditions under which the conglomerates were deposited.

The presence of carbon was further commented upon in the period between 1907 and 1927, but its significance in unravelling the depositional environment was not appreciated. The most interesting remarks during this 21-year time-span were made in 1908 by Spilsbury[31], an American mining engineer, who compared some features of the Witwatersrand with the placer gold deposits of California. He wrote:

> *. . . that the origin of the banket deposits appears to be that of a marine placer seems well proved . . . I am, however, of the opinion that he [Gregory[14]] pays too little importance to the rôle that organic matter has played in the collection, or concentration, as well as in the deposition of gold in the gravel of the coastal rim . . . I am more and more impressed with the fact that the distribution of fine gold in the sedimentary deposits, as well as in some of our placers, is due mainly to organic agency . . . I think one of the best examples of the extent of this agency is in the recent growth of a new industry on some of our western rivers known as 'Moss Mining'. On the Trinity River in California, a stream which runs through some of the best gold-bearing ground in the State, and has rich placer ground on both banks nearly over its whole length, there occurs during the summer season of low water a heavy growth of algae in the low pools along the banks, and a species of dense moss covers the banks. These plants are constantly being submerged during local floods, and again on the water retiring subjected to the burning heat of the sun, so that partial decomposition is always going on. During the rainy season these growths are entirely covered by high water . . . Now it has been discovered that these plants contain considerable gold, and every spring, towards the end of the rainy season, the miners build light flat boats and float down on the last day of the flood, collecting these mosses and water plants, stocking them up and washing them for their gold contents . . . the function of the plant life in these cases is probably to a great extent mechanical, as the fine gold caught on their surfaces, while flaky and excessively fine, does not show sharp edges, and is generally dull on the surface; but in examining samples under the microscope we find quite a number of bright hair-shaped particles bent at different angles and having all the appearance of embryonic crystallization. This would tend to prove that the river waters carry gold in actual solution as well as in suspension, and that this gold is precipitated, probably on that part of the plants which is undergoing decomposition. Now, it seems to me that if this faculty of organic matter, or plant life, as a collector as well as depositor of gold is so active at the present time, and in fresh water, it is reasonable to assume that it was no less active on the shores of that great Devonian sea, during which epoch the algae grew in such remarkable exuberance. If instead of an exposed wave-lapped beach we should assume a coastal lagune, separated from the ocean by a sand bar, which from time to time might be broken over and inundated during violent storms, should we not have before us the ideal conditions for the distribution of gold in the banket under the exact conditions described by Professor Gregory . . .*

In 1909, Young[17] reported that carbon had been observed in reefs of seven gold mines in the Central Rand, East Rand, and Klerksdorp goldfields. In the following year, Horwood[32] presented the first systematic description of the carbon, and stated that it was present in most reefs, but was only well developed in the few mines which Young had mentioned. Horwood recognized that the presence of carbon indicated good values, even where no pebbles were present, and concluded that: 'There can hardly be any reasonable doubt that there is some clear, subtle connection between the presence of carbon and that of gold'. He did not think that the carbon was of organic origin, but saw it as being associated with an eruptive magma. There was a solfataric volcanic origin for a natural series of carbon compounds constituting petroleum, graphite and diamond being the end-products of this petroleum series. The carbon in the banket was thought to have been introduced by the longitudinal diabase dykes of Ventersdorp age that are abundantly present in the Witwatersrand strata. Young[18] considered the carbon to be essentially the same as anthracite, but doubted whether there was any abundance of organic matter in the original sediments, although he quoted other workers as saying that the '. . . beach was not the most salubrious, and the air was already tainted with mephitic exhalations from decaying seaweeds . . .

The second phase in the development of the Witwatersrand goldfields can be seen, in summary, to have been a period in which a number of significant changes were made in the ideas held about the nature of the depository in which the auriferous conglomerates accumulated. In addition, several new lines of investigation were initiated and concepts developed. Quantitative sedimentology was employed for the first time, in two separate investigations, but was not adopted as a standard technique, despite the fact that it yielded exceptionally useful new data. Contoured assay plans were developed, which revealed, in considerable detail, the distribution patterns of gold in the reefs and the limits of payshoots. It was put forward that all the reefs had not formed under the same conditions, and that different environments had led to the generation of different types of conglomerates with varying percentages of payability. The overall history of shrinking of the depository was emphasized more than once, and the importance was realized of the edge of the basin advancing closer and closer to the centre with time. The presence of uraninite was recognized, but it was not then of economic importance, and it contributed nothing towards a better appreciation of the depositional environment. The close relationship between high gold values and carbon emerged in several investigations. A suggestion that its origin might have been similar to that of the gold-extracting fresh-water algae of streams in the California placer goldfields was not deemed worthy of further consideration. In the early years of Phase 2, the majority opinion still saw the banket as a marine littoral accumulation. With the passage of time, this theory was seriously challenged, first, by the proposal that the conglomerates had accumulated in a large delta, and, second, by the argument that the banket was deposited in braided, fan-shaped channels on a fluvial floodplain at some distance from the sea. Differences of opinion existed as to whether the fluvial and deltaic material had been reworked later by littoral waves and currents. Up to virtually the end of Phase 2, all authors had agreed that rivers and streams had carried the pebbles and other sediments from a highland source-area to the river valley, or the delta, or the beach, or the off-shore ocean floor. Then, an entirely new theory was offered to the effect that the vast quantities of pebbles and boulders had been brought into the depository by glaciers. A great deal more observation was beginning to challenge the ideas built on the speculations of the first phase.

THE PERIOD 1928 – 1948

Phase 3 was hallmarked by the great controversy that arose about the placer *versus* hydrothermal theory for the mineralization of the banket. This was sparked off in 1930 by the publication of Graton[33], one of the foremost economic geologists in the world at that time, who lent all his authority to the hydrothermal school. The protests from South African geologists were long and voluble. Among the arguments that surged back and forth, there were several opinions voiced on the nature of the depository itself. Graton remarked that the conglomerates had a notable constancy of thickness and a remarkable areal continuity. The sediments were well sized and the sands clear and well sorted. The pebbles had a small size-range, were closely packed, well-rounded, ellipsoidal, and without marked imbrication. No cut-and-fill features were visible and lenticular bedding was strikingly absent. Some of these statements were wide of the truth, thus adding further to the indignation of the local geologists. He threw more fuel on the fire by claiming that: 'The accumulating confirmation of the essential conformity of the system from top to bottom, notwithstanding intensive search for unconformities, indicates a notable degree of uniformity and constancy in the dominant conditions and controls of sedimentation throughout the duration of this accumulation and over the whole of the area as now known'. All the above features, Graton believed, pointed firmly against the conglomerates being river-valley, deltaic, or beach deposits. Instead, he favoured neritic marine deposition on a nearly flat, even continental shelf, which underwent a gradual decrease in depth of water with the passing of time. Milling on a beach had produced the pebbles that were alternately carried away to deeper water by the undertow and returned to the beach for further milling. As soon as the beach mill had reduced the pebbles to the requisite degree of size and roundness, they were transported out to an off-shore position for the last time and there deposited. The pebbles were laid down without interstitial sand, which was washed in later. Contrary to almost all previous authors, he concluded that the East Rand had a shoreline that ran northwestwards, instead of northeastwards. He envisaged the greater part of the East Rand as having formed under deeper water, the Central Rand under shallow water, and the West Rand under deeper water, with the highlands being to the northeast and east-northeast of the East Rand, again the opposite to all the observations that had shown that the basin fill had moved from the northwest.

In 1930, Reinecke[34] expanded on his earlier contribution. Crossbedding was measured for the first time, and was found to be inclined in the direction of the paystreaks. He modified to a slight extent his earlier conclusions, stating that the Witwatersrand beds were laid down partly on a flood-plain of a river and partly over a piedmont area on the foothill slopes of a mountain range. A series of uplifts to the northwest caused the mountainland to approach gradually nearer the Witwatersrand proper. As it moved southeastwards, proximal facies were stacked upon distal facies. The earlier sediments in the northwest were progressively uplifted on to the mountain flanks from where they were removed by erosion and were fed back into the depository. The uniform, thin conglomerates were laid down between the piedmont and the lower flood-plain. Sudden increases in river volumes picked up the alluvial fans on the piedmont, and swept their contents over the plains below, with the result that the reworked sediments covered a much wider area than the alluvial fans, and were transported over flood-plain deposits, over channel deposits, over lagoons, and over the mudflats of the lower plains. The Main-Bird sediments were formed near to the mountain area, with the delta-plain much farther southeast towards the sea. Reinecke doubted whether the gravel sheet had ever extended as far as the seashore. In his attempts to find an analogue of the Witwatersrand environment, Reinecke revived Evans's[24] suggestion that the region straddling the southern foot of the Himalayas and the upper valley of the Indus River, in the northern portion of India, might hold the key to understanding the manner in which the banket was laid down. The Miocene-Pliocene Siwalik System of fluviatile origin was thought to bear strong similarities to the Witwatersrand succession.

Du Toit[35] disagreed with Reinecke's interpretations, maintaining his argument that only marine conditions could have given rise to such extensive pebble deposits. A shoreline advanced northwards, and immediately south of it sand and pebbles spread out at just below tide-level. He believed the Main group of conglomerates to be the outcome of the waning and cessation of such marine planation, of the establishment and consequent destruction of estuarine conditions, of the reworking of vast sheets of marine gravel by braided streams, and of the swamping of these deposits by sands and pebbles brought down from a rapidly rising interior with its mountain ranges advancing towards the depository. The Kimberley Shales, he thought, represented good examples of the establishment of estuarine or lacustrine conditions. In 1931, also, Gregory[36] restated his conclusions of more than 20 years earlier. The banket had been deposited on the southern slope of a mountainous coast-line where tidal action and backwash had produced the inconsistent pebble imbrication. The Main Reef Leader and the South Reef of the Central Rand could be viewed in no other light, but Gregory was prepared to concede that the Elsburg Reefs, formed in the closing stages of the Witwatersrand Sequence, could have been fluviatile, or perhaps deltaic, accumulations.

In criticizing Graton, Mellor[37] also re-affirmed his thoughts on the Main Reef Leader of the East Rand as having been formed on a deltaic fan. He stated that developments in the East Rand had served to produce only further evidence in support of the conclusions that he had drawn 15 years earlier. His comments on Reinecke's views emphasized that the main question that had to be settled was whether the conglomerates were the result of the actions of a number of rivers discharging from a mountainland on to an adjacent alluvial plain, or of some larger unit depositing material much farther from its source. On the completion of his mapping of the Klerksdorp area, Nel[39], in 1933, found his observations inclining towards either Mellor's delta theory or Reinecke's fluvial hypothesis. His earlier work in the Vredefort area had led him to believe that only the delta concept had merit. The Geological Survey[40] published, in 1936, its handbook on the mineral resources of South Africa, and gave its support to both Mellor's and Reinecke's ideas. It also quoted a verbal communication from Professor Douglas Johnson of Columbia University, who had ventured the opinion that the Main Reef Leader was similar to a gigantic piedmont alluvial fan deposit laid down on a nearly flat plain in front of a mountain range. He had observed such fan deposits to be made up of conglomerate lenses with a braided fan-shaped pattern. The texture and arrangement of bars in the Main Reef Leader did not appear to resemble marine or flood-plain deposits, while the variations in texture between the centre of the payshoots and the inter-payshoot areas were the same as those that could be seen on an alluvial fan.

In 1938, Roberts and Kransdorff[41] provided the first substantial evidence to show that the major conglomerate zones were in contact with distinct disconformities. The significance of this conclusion was to assume ever greater importance as more and more detailed studies were undertaken of the Witwatersrand reefs. Their work in the West Rand led them to concur with Du Toit's[27] ideas on the existence of shallow marine conditions in the Main Reef series of conglomerates. Oscillation ripples in intercalated shales were interpreted as marine current ripples produced by water moving from west to east. The lenticular nature and wide variations in character of the Bird Reefs were ascribed to their having formed under continental conditions. The presence of dolomitic shales and of an oolitic bed in the Kimberley Shales pointed to deep-water marine conditions, while a continental environment again prevailed during the formation of the Kimberley Reefs.

In the second edition of his textbook, published in 1939, Du Toit[42] said that the recognition of one tillite horizon in the West Rand and two at Klerksdorp indicated that glacial and fluvio-glacial conditions might have been more widespread than previously thought. The presence of wind-faceted dreikanters in the Promise Reef of the Lower Witwatersrand Division was also mentioned. Du Toit felt that there was no necessity to change the views he had held in earlier years. The worn-down southern part of the Transvaal had been invaded from the south by a shallow sea in which were deposited sands and muds during the middle period of Witwatersrand history. The emergence of the sea-floor produced extensive beaches, and ultimately led to the development of great delta-flats on which river systems from the northwest discharged sands and pebbles. Auriferous black sands were associated with the gravels. There was an abundance of gravel supply from accumulations partly of glacial origin in the interior mountains. Ultimate elevation in the north and northwest uplifted and denuded the Lower Division sediments, to give abundant coarse material for the Elsburg conglomerates. The planes of unconformity, the irregular shingling of the banket, and the muffin shapes of the larger pebbles were all still considered to be indicators of the gradual transgression of the shoreline.

The discovery of the new West Wits goldfield, which was to change many of the concepts previously held about the conditions under which the Witwatersrand reefs formed, led to De Kock's[43] paper in 1940 on a new reef of major economic importance – the Ventersdorp Contact Reef, present at the very top of the sedimentary assemblage. De Kock interpreted this horizon as being constituted by eluvial scree that had accumulated on an old land surface, with thicker sheets in the depressions, valleys, and stream beds. The nature of the reef, its mode of occurrence, and the gold content were described as varying with the morphology of the pre-Ventersdorp landscape. The detritus had not been subjected to prolonged or continuous wave-action. Thicker reef bodies were thought to be due to seasonal accruals of layers upon layers. The Ventersdorp Contact Reef was seen by De Kock as being a partly eluvial and partly alluvial placer deposit, and as representing the erosion product of the immediately underlying Witwatersrand formations, from which were derived the pebbles, the quartz particles, and the gold.

Nothing of any significance was written about uranium during this period, and its presence in the gold-bearing strata was all but forgotten. A minimum of new information was provided on the carbon deposits, despite the discovery of the exceptionally rich Carbon Leader of the West Wits goldfield. In 1931, Macadam[44] reported that very little visible gold was ever found without carbon being present. This substance was plentiful, not only in the banket, but also in bedded pyritic seams where no pebbles were present. Mellor[37] stated that, in the N.A. Leaders of the East Rand and the South Reef of the West Rand, very rich gold-bearing carbon was often present as a thin 'pencil line' on an obscure bedding plane. The only comment made during this period on the possible origin of the carbon was that by Du Toit[42], in 1939, to the effect that liquid and gaseous hydrocarbons had formed from organic matter.

To summarize the contributions made during Phase 3 to the arguments concerning the manner in which the auriferous horizons in the Witwatersrand Sequence developed, it is probably correct to say that, except for the first three years, it was a relatively non-productive period. It was a time of quiescence before the flood of papers that were to be presented during Phase 4. No new approaches were adopted towards settling the differences of opinion. Nothing was written about the uranium or the carbon in regard to their possible use as indicators of the depositional environment. Three main schools of thought prevailed – those supporting the previously postulated littoral, deltaic, and fluvial theories. However, there were signs of more observers beginning to subscribe to the idea that all of these had prevailed at one time or another, and that different reefs had not necessarily been formed in the same environment. There were reefs and reefs. At the beginning of the period, a new hypothesis was put forward to the effect that the gravels had been laid down in a neritic environment where the water was of substantial but not excessive depth. In the closing stages of Phase 4, the opening up of the West Wits goldfield encouraged a further theory that the terminal reef of the Witwatersrand Sequence was an eluvial and alluvial accumulation.

THE PERIOD 1949–1969

Time might well show that the fourth phase in the history of Rand mining was the golden period of Witwatersrand geology. In the twenty-one years between 1949 and 1969, as many significant contributions were made to an understanding of the depositional environment of the auriferous strata as were published in the previous 63 years. The primary reason for this was the opening up of the extensions of the Witwatersrand basin and the addition of voluminous new data from the Orange

Free State, Klerksdorp, West Wits, and Evander goldfields to supplement that which has formed the basis of the arguments concerning the origin of the reefs in the West Rand, Central Rand, and East Rand goldfields. The numbers of geologists employed in the gold mining industry increased exponentially. The complexity of the problems brought to light in the new fields demanded that quantitative studies be carried out, so that information became more accurate and less subjective. The uranium industry came into being, requiring that the uraninite component of the reefs be studied in immeasurably greater detail than before. The newly discovered reefs did not conform to the conclusions that had been drawn in regard to the classic conglomerate horizons, and it became ever more apparent that the origin of the carbon would have to be understood before the depositional environment of such new reefs as the Basal Reef of the Orange Free State goldfield and the Vaal Reef of the Klerksdorp goldfield could be determined. In 1959, the Economic Geology Research Unit was founded for the express purpose of working on the geological problems of the gold mining industry, and one of its first projects was the re-introduction of sedimentological methods into the gathering of data. The Geological Society of South Africa brought out, in 1964, a two-volume collection of papers on the geology of some ore deposits in Southern Africa, one volume being devoted in its entirety to the Witwatersrand basin. Throughout the whole period, geologists were encouraged to return from industry to universities for higher degrees, with the result that an appreciable amount of research was carried out on aspects of the stratigraphy, structure, sedimentology, and mineralization.

Sharpe[45] produced a valuable contribution in 1949, in which he stressed the cyclic nature of the sedimentation throughout the whole of the Witwatersrand succession. He recognized primary oscillations on a regional scale and secondary ones on a local scale, and attributed these to regional movements of uplift and subsidence, which progressively decreased in intensity stratigraphically upwards, so that, by the end of Witwatersrand times, conditions had become almost static. He viewed the argillaceous sediments and erosion channels filled with coarse debris in the footwall of the Main Reef Leader of the East Rand as typical lagoonal deposits. The Main Reef, Main Reef Leader, and Kimberley reefs were considered to be shoreline pebble deposits that settled on a wave-cut plane of erosion when the surrounding land was of low relief. The concentration of the gold took place during a gradual but relentless advance of an encroaching sea-front. He also emphasized the point that the payable reefs represented the initial deposits formed after a considerable break in sedimentation.

The idea of a neritic environment, put forward by Graton[33], was revived by De Jager[46,47,48] in a series of publications dating from 1949 onwards. His interpretations dealt specifically with the May Reef of the Kimberley group of conglomerates, which he saw as both depositional and lag gravels formed by wave-induced erosion of the underlying sediments. The associated chloritoid shales he thought were typical neritic accumulations on a shelf bordering an open sea. De Jager also subscribed to the idea that ice had played a major role in transporting material from a source-area to the marine shelf. The Elsburg Reefs, in particular, were seen as the representatives of ice-rafted coarse material, which had been dropped in the sea on to finger-grained material in the neritic environment.

The first detailed account of the Orange Free State goldfield was presented by Borchers[49] in 1950. He believed that the terrain in which the Witwatersrand depository formed contained distant, high, cold mountains above arid wastes of Archean basement rocks and disintegration products. Crustal movement associated with the outpouring of the Dominion Reef lavas had trapped a body of water. A flood-plain developed either in an inland valley-flat or on the landward side of a delta. Climatic fluctuations and extremes of temperature led to rapid and drastic break-down of the rocks. Cataclysmic floods transported vast quantities of detritus into enclosed basins where the rates of evaporation were very high. On occasions, glaciers reached the basins to deposit fluvio-glacial grits and tillite. Wave-action eroded the edges of the uplifted unconsolidated sediments, and a shoreline was formed, on which the coarser and heavier materials were concentrated. Because the basin was a shrinking depository, there was a progressive advance of the shoreline into the basin, the distance of the total advance being of the order of 50 km from the top to the bottom of the Witwatersrand Sequence. The numerous marginal unconformities which were apparent in the Orange Free State goldfield suggested that the present outcrops of the Upper Division lay close to the original limits of deposition. In fact, the Elsburg Reefs might well have been formed right on the edge of the basin, since they were constituted by detrital cones below faulted elevated ground, the fault being one of the bounding structures of the original basin.

From his studies of the Livingstone Reef in the West Rand goldfield, Pegg[50] came to the conclusion in 1950 that the conglomerates had resulted from a river discharging course pebble debris into an estuary. The gravels were washed in during flood periods. While the floor of the estuary was deep, little or no sorting took place; when it was raised by the accumulation of sediments, sorting occurred through the medium of wave-action, the pebbles remaining in the estuary and the argillaceous material being washed out to sea.

A notable synthesis of the geology of the basin was presented by Antrobus[51,52] in 1954 and 1956. The depository was seen as a continental feature, surrounded by land on all sides, the margins being close to the limits of the present extent of the Witwatersrand strata. Faulting, warping, and erosion were conspicuous on the periphery. The optimum conditions for reef formation were set during periods of stillstand and reduced sedimentation, accompanied by some erosion of the footwall. The economic reefs were located on disconformities, and represented the basal beds of a group of overlapping strata. Antrobus concluded that the reefs had been formed from detritus left on a pediment after a long period of subaerial weathering and concentration. These were exposed pediments and not bajadas of coalescing alluvial fans. The reefs accumulated on an erosion surface along the margin of the basin, the surface having been either completely dry or occasionally covered by thin sheets of water as a result of streams from the hinterland debouching on to it. The economic horizons represented material left upon the surface as a product of the disintegrational effects of attrition and weathering and of the washing effects of flood-waters. The debris took the form of a loose, unconsolidated, thin layer of sediment derived from the underlying rocks and from occasional inwashings. The less durable components were removed by sheetwash, floods, and winds. The paystreaks he considered to be braided streams produced during occasional floods when the material on the exposed surface was picked up, transported, and redeposited along the lines of strongest flow of flood-waters. In the Orange Free State and Klerksdorp goldfields, especially, there were many marginal faults that formed escarpments. Repeated uplift along the periphery caused the upthrown sides to be eroded and the debris to be deposited with the younger members of the succession. Antrobus provided the first detailed description of the facies changes along a particular horizon. Near the basin-edge, the reef was thicker, the pebbles more variegated, the conglomerates more lenticular, channels present, and the gold content low. Farther in, where washing and weathering were most effective, large pebbles were few, the reef thin, and the richest concentration of gold present. Towards the centre of the basin, the area was more often covered by water, with the result that the pebbles were smaller, the percentage of fine material greater, and the quantities of gold and other heavy minerals less. During the formation of the Witwatersrand rocks, there was no vegetation and little rainfall, and the climate was cold, as indicated by the presence of tillites and the absence of red coloration and saline deposits. Antrobus saw an analogy to the Witwatersrand environment in the Basin and Range Province of the western United States, where pediments form during periods of stillstand and are then eroded laterally by ephemeral and braided desert streams.

Brock[53] also presented a broad overview of the Witwatersrand Basin in 1954. He was convinced that it was a continental feature, with no encroaching ocean, ringed by granite mountains. These were lowered by erosion, and then re-elevated by vertical uplift, thereby maintaining the rim of the basin for a long period. There was no vegetation and no signs of life, the conditions were desert-like, and the climate was marked by heavy rainfall. Torrents swept gravel into the basin from all sides. Subsidence was gradual and sedimentation kept pace, so that shallow-water conditions, at least in the Upper Division, prevailed. Differential movement on the peripheral faults led to the development of several fan-like bodies of conglomerate at three points around the basin, at least – north of the Central Rand, west of the Klerksdorp, and west of the Orange Free State goldfields.

In the third edition of his textbook, which appeared in 1954, Du Toit[54] reiterated his continuing belief in the marine placer origin of the banket, the reefs being drowned pebble beaches that had extended progressively landward during regional subsidence. Reef formation was always intimately connected with crustal adjustments. An invading sea from the southeast caused marine planation, and pebble bars, graded by shore currents and pulsated by waves, advanced northeastwards over a subsiding coastal plain composed of loose deltaic materials and blown sands. In the same year, Miholic[55] wrote that deltas had formed in fairly deep water along a steep coast, and that the reefs had developed from the marine gravels accumulated in such an environment. He envisaged that the conditions must have been similar to those prevailing in Norwegian fjords at the present day, or along the bottom of the Black Sea where streams off the Caucasus mountains deposit gravel in the sea at considerable depths.

Bain[56], in 1955, and later in 1960[57], considered the conglomerates to be the equivalents of a parallel, linear pattern of gravels on a piedmont plain in which stream channels had been entrenched only slightly. There was a recurrent fill of the channels, followed by scour of the more mobile particles. He thought that the paystreaks showed an arrangement similar to that of tributaries converging downstream towards a major river flood-plain in the centre of an alluvial basin.

Instead of a radiating geometry to the lenses of banket, he saw a converging pattern. As a result, he envisaged the uplifted side of the basin to have been in the east and southeast and the movement of material towards the northwest, the opposite of almost all previous conclusions. He even alleged that the cross-bedding directions in the Main Reef Leader showed a northwesterly orientation, which was also contrary to all earlier observations.

The concept of transportation of debris by ice was revived by Wiebols[58] in 1955. The depositional area was considered to have been a vast peneplain bordering on a shallow inland sea, which was landlocked and of similar size to the Baltic or the Mediterranean seas. Sedimentation took place on the edge of an inland ice-sheet, which was comparable in size to the Gondwanaland ice sheet of Carboniferous age. The ice was the dominant medium of transportation, especially for the major reef horizons, which were seen as being composed of ground moraine or glacial boulder clay that was subsequently subjected to wave-action in a transgressing sea, the sand and pebbles being left behind on a peneplain as the ice-sheets retreated. Ice-flows from widely separated areas brought their own particular petrological mix, accounting for the conspicuous differences in composition between some of the reefs. Wiebols maintained that boulder clay, which had escaped wave-action, could still be seen as footwall argillaceous material below the Main Reef Leader and the Kimberley Reefs. Varved shales were also to be found in these footwalls. The erosion channels in the Main Reef Leader footwall were seen as drainage features through which the melt-waters of the ice masses had escaped. The payshoots were interpreted as eskers, long ridges of pebbly sand formed underneath the inland ice-sheet by water-courses. The gold content of the eskers was subsequently enhanced through milling by the transgressing sea. Contrary to the opinions of most other investigators, Wiebols doubted whether there had been frequent movements of uplift and subsidence, and he believed that the variations in grain-size were essentially a function of the distance between the inland ice-sheet and the sea.

Brock, Nel, and Visser[59] presented a paper in 1957 on the uranium deposits in the Witwatersrand Basin which they felt had been an extensive interior or intracratonic basin fed by a number of streams that had travelled no great distance from the source-area. The streams, which had entered on the northern and northwestern sides of the basin, had left fan-shaped deltaic remnants about their mouths. Deposition was in shallow water or sub-aerial, and was distinctly cyclic in nature with the fine-grained material having accumulated during sinking of the basin, and the coarse-grained during elevation of adjacent land-areas. The richest conglomerates were developed on planes of intra-formational diastems, disconformities, and unconformities. In general, conditions were fairly cold and dry. In 1958, Nel[60] added an important observation to the effect that a portion of the basin rim coincided with the outer limb of a major anticline wrapped around the Vredefort dome, some 40 miles away. This was the first reference to the influence of regional folding on the limits and geometry of the basin.

On the basis of the very coarse average grain-size of little-rounded quartz in the Upper Witwatersrand quartzites, Fuller[61] wrote in 1960 that the mineral was probably derived from a local coarse-grained primary source of quartz. On the other hand, the mineralized conglomerates were well sorted and continuous, and pointed to long transportation under conditions of sustained equilibrium that permitted the development of extensive gravel sheets. The quartzites were poorly sorted, massively bedded, and commonly cross-stratified, suggesting that the basin filled rapidly from local sources. As a result, he proposed that the quartzites and the conglomerates had been derived from different sources. In Upper Witwatersrand times, the drainage pattern was effectively disintegrated by recurrent faulting, so that, only at intervals, were master streams established. These introduced the gold, uranium, and other minerals from sources outside the confines of the basin. In the same year, Koen[62], on the basis of the first quantitative mineralogical study carried out of the Witwatersrand reefs, concluded that they were formed in two pulses – the first produced an openwork gravel on wide alluvial plains, and the second led to the infiltration of sand into the pores between the pebbles. A transgressive beach subsequently spread out the gravel accumulations.

The first results of the major programme of sedimentological investigation, which was launched by the Economic Geology Research Unit, appeared in 1962 when Hargraves[63] published a paper on transportation directions in the East Rand. The bimodality of the cross-bedding azimuths he interpreted as indicating movement of material in a south-southeasterly direction and its subsequent redistribution by high tides flowing towards the northeast. The fanning of the cross-bedding directions was found to be of similar pattern to the radiating geometry of the paystreaks. In the following year, Steyn[64] reported on the first detailed sedimentological study carried out since the pioneering work of Pirow[25] in 1920 and of Reinecke[29] in 1927. The Livingstone Reef of the West Rand goldfield was found to be fan-shaped, with an apex in the northeast, the quartzite ratio being highest towards this apex, the pebble-size larger, and the roundness less pronounced. The

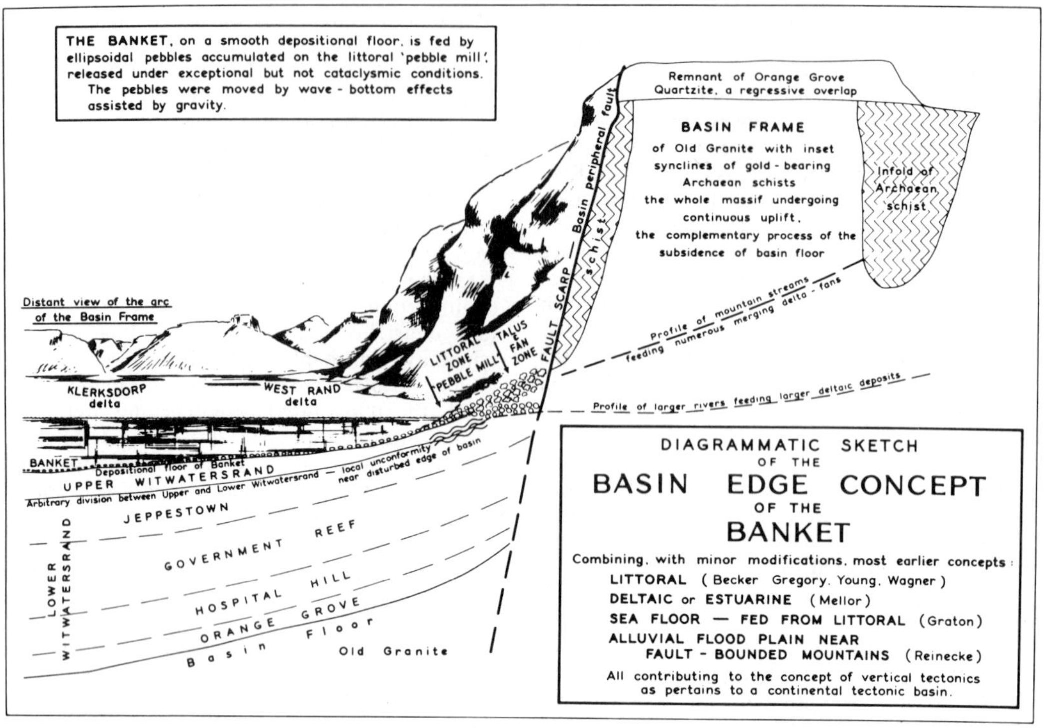

FIGURE 4 A pictorial conceptual model of the depositional environment of the Witwatersrand banket, as devised by Brock and Pretorius[66] in 1964. [reproduced by permission of the Geological Society of South Africa].

central main channel contained the higher gold values, and its location coincided with the axial direction of the major synclinal fold of the area. Cross-bedding in the channel was found to be parallel to the fold's direction of elongation. Transverse and longitudinal sandwaves and ripple-marks indicated both current-flow and wave-action, with an estuarine current moving from north to south and a weak longshore current from west to east. The sorting coefficients of the sands were found to be intermediate between those for a beach and a river, with a distinct tendency towards the latter. The type of cross-bedding pointed to a fluvial or deltaic environment and not to a beach. The paystreaks were observed to be arranged in a fan pattern typical of alluvial fans or marine deltas. Steyn concluded that the weight of the sedimentological evidence was in favour of the environment's having been that of a small delta formed where a river flowed directly from a highland into a sea or a lake. From the rates of change of certain of the properties of the sediments, he calculated that the apex of the delta had not been more than 7 to 9 km north of the position of the present outcrops.

Viljoen[65] published in 1963 what is considered to be the most extensive quantitative mineralogical study of the Witwatersrand arenaceous sediments that has yet been attempted. He examined many conglomerate and quartzite horizons in the East Rand, Central Rand, West Rand, and West Wits goldfields, and came to the conclusion that the micro-sedimentology gave strong support to the environment's having been that of a delta, the lower portion of which had been reworked by waves and currents under beach conditions. The submature matrix of the banket, with relatively large amounts of clay constituents, gave indications of a fluviatile origin. The degree of sorting suggested a fluviatile or beach situation. Viljoen identified separate deltas for the East Rand, Central Rand, and West Rand. Adjacent to the basin, elevated land occurred, from which gravels moved to the deltas by means of fast-flowing rivers. A systematic decrease in grain-size was apparent radially away each entry-point of material into the delta. The farther away from these entry-points, the more the material was reworked by littoral agencies. He found that the leucoxene, pyrite, arsenopyrite, zircon, and chromite were in hydraulic equilibrium, indicating that all of these minerals were of detrital origin. The hydraulic equivalent numbers showed that the pyrite had been laid down as pyrite, and not as black sands. The complete lack of oxidation of the detrital pyrite was taken as evidence of an atmosphere free of oxygen and of a cold climate.

The publication of the special volume on the Witwatersrand basin in 1964 by the Geological Society of South Africa saw a comprehensive synthesis and an attempt at reconciliation of the various theories by Brock and Pretorius[66]. Four distinct entry-points of material into the basin near Benoni, Krugersdorp, Klerksdorp, and Welkom were considered to be deltas formed on the basin-edge in structurally depressed portions of the rim between rising granite domes. Because of the major development of peripheral faults, the northwestern flank of the depository was a much more active source of sediments. The basin was bounded on all sides by land, and shallow-water conditions prevailed throughout the history of sedimentation. The skewness of the delta-shapes was ascribed to the effects of a longshore current flowing in a clockwise direction round the basin. In addition to the four main entry-points, it was thought that numerous small water-courses, talus slopes, and small merging delta-fans existed on the edge of the depository. Isopach maps were presented for the first time of the main groupings of strata in the Witwatersrand succession, and these unequivocally showed that the basin had been a shrinking feature with time. The Lower Division had occupied a larger area than the Upper Division. The basin frame underwent continuous uplift, causing a progressive advance of the zero isopachs towards the depositional axis. On the uplifted hinterland, remnants of Lower Division rocks were largely eroded to feed the Upper Division accumulations. Brock and Pretorius commented that: 'Seven or eight different hypotheses have been enunciated to explain this origin [of the banket]. Geological observations (including a preponderance of good observations) over a period of three-quarters of a century have not resulted in agreement'. The reconciliation that best fitted the facts was the one that they diagrammatically portrayed in Figure 4. The hinterland was separated from the depository by a fault-scarp of impressive dimensions along the peripheral fault zone of the basin. Immediately adjacent to the fault-scarp, talus and fan zones were formed at the mouths of major rivers debouching from canyons.

A pebble mill developed in the littoral zone at the base of the fans, and waves and longshore currents contributed to the distribution of pebbles over considerable distances parallel to the shore. The banket was laid down on a smooth floor fed by the littoral pebble-mill, the pebbles having been moved downslope from the shore by wave-bottom effects assisted by gravity. High rainfall and no vegetation characterized the overall environment. In two additional papers in the 1964 volume, Pretorius[67,68] stressed the observation that maximum gold concentrations occurred at a certain optimum distance from the shoreline. The Lower Division sediments, as presently exposed, had been laid down too far into the basin from the shore, while the Elsburg conglomerates, at the top of the stratigraphic column, represented too proximal a facies. He also drew attention to the fact that near-shore conditions on the southeastern side of the basin had not acted as host to the same assemblage of sediments as had been deposited on the more active, northwestern edge. Far less material had been contributed from the southeast, and very little gold had been associated with it, so that the economic potential of this limit of the depository was at a minimum.

In his two contributions to the volume, Winter[69,70] dealt with the main gold-bearing horizons of the northern and southeastern sections of the Orange Free State goldfield. He came to regard the general shape of the sedimentary units, the regular stratification, the variations in facies, and the relationships between facies as indicative of a shallow inland sea where the imprints alternated between deltaic, estuarine, beach, and even fluviatile environments. The elevated ground to the west moved eastwards with time, leading to the removal and redeposition of sediments. Material entered the basin from localized areas in the west and southwest, and there was a progressive northwards advance of the entry-points. Submergence kept pace with the influx of detritus. Owing to the periodic elevation of the source-area, cyclic sedimentation characterized the assemblage of strata. Slow upwarping to the south and east maintained a wave-cut surface, on which only the coarsest, heaviest, and most resistant particles were not swept away. As the shoreline receded from the land, a marginal, terminal, or regressive conglomerate was formed, as, for example, the Basal Reef. The smooth surfaces marked marginal unconformities, with their graded profiles typical of an epineritic environment. The payshoots in this reef were considered to be the products of longshore currents in the inland sea. Winter made the first comments on the pattern of coarse buckshot pyrite being closest to the shore, gold somewhat farther out, and uranium still farther. The displacement between the zones of maximum gold and uranium mineralization later came to be recognized in most of the other goldfields. A long period of sorting was necessary to produce the Basal Reef. Where a basal conglomerate developed on a rapidly transgressing shoreline, as in the case of the Leader Reef, the time of sorting was much shorter, so that such reefs assumed a lower economic potential than terminal conglomerates. The Kimberley Reefs were classified as typically fluviatile. Channels denoting transitional marine and continental environments were observed to occur only in the Kimberley strata. The Elsburg conglomerates were considered by Winter to be subsidiary deltaic deposits, because the reefs were clustered into vertically superimposed groups. The regular and sheetlike erosion of the marginal sediments was taken as indicating a shallow, aqueous environment. Deltaic and fluviatile conditions were concluded to have prevailed when the basin was at its smallest. For the rest of the time, a neritic environment existed, in which winnowing and sorting agents were optimally operative.

In the following year, Winter[71] looked at the Vaal Reef of the Klerksdorp goldfield, and found that the adbundance of trough cross-bedding was highly suggestive of a fluvial or deltaic environment. However, the absence of discordant channels, and the presence of mudstones, mudcracks, clay pellets, and slumped trough foresets mitigated against the fluvial interpretation. In addition, regular fluted scouring, regular current pattern, slight divergence in current directions, widespread narrow sheets of small-pebbled gravels, longitudinal elongation of sand-bodies, and a shallow-water regime, all favoured a water-covered deltaic environment. Mud galls in the upper portion of the Vaal Reef showed that no winnowing by wave-action had taken place. The Kimberley Shales, stratigraphically above the Vaal Reef, were stated to be of marine origin, having been formed on the basinward side of the delta. The still higher Elsburg reefs were seen as the products of fluvial or fluvio-deltaic conditions. Winter recognized distinct similarities between the Vaal Reef and the Pennsylvanian-age Sharon conglomerate of Ohio, which has been interpreted as of alluvial or deltaic origin. The Vaal Reef represents a facies somewhat more basinward than that of the Sharon formation.

In 1965, the results appeared of another outstanding sedimentological study. This was of the Kimberley Reef in the East Rand goldfield, and was undertaken by Armstrong[72]. He found abundant evidence for a continental and fluvial origin of this reef, despite the frequency with which arguments had been advanced in the past to the effect that a neritic environment had been in existence. An extensive subsiding pediplain developed, with features similar to that of a valley-flat environment. The floor of this valley was structurally deformed, and a distinct relationship developed between the orientation of the reef bodies and the paleotopography of the floor. The bulk of the reef was derived from the erosion of the floor. The conglomerate-filled channels ran parallel to the depositional slope. Oscillation ripplemarks showed the prevalence of shallow-water conditions, mudcracks pointed to subaerial exposure at times, and the imbrication of the pebbles was typically that of a fluviatile environment.

Knowles[73] published in 1966 the results of another detailed sedimentological examination, this one being of the Ventersdorp Contact Reef in the West Wits goldfield. The reef thickness, percentage conglomerate, pebble composition, pebble roundness, maximum pebble-size, size distribution, cross-bedding, pebble imbrication, and grain orientation led him to believe that this banket, at the very termination of Witwatersrand sedimentation, was the product of an alluvial fan that extended from a fault scarp in the northeast into an inland sea. The currents flowed south-south-westwards down the fan. Penecontemporaneous folding of the paleoslope controlled the distribution and deposition of the coarser material. Depositional gravels formed in the depressions and lag gravels on the elevations of the floor. The cross-bedding measured was typical of a fluviatile environment, as was the style of pebble imbrication. Observations were made of both a transporting current down the fan and a longshore current operative in the inland sea.

In 1966, Pretorius[74] produced a synthesis of the detailed sedimentological work that had taken place during the previous five years. He stated that it was generally agreed by all the investigators that the source of the sediments and the gold lay to the northwest of the basin, that the material was transported by rivers from the source-area to the depository, that the conglomerates, gold, and uranium were laid down close to the edge of the basin, and that the sediments had there been reworked and concentrated by wave-action and longshore currents in a closed basin, which was either a shallow inland sea or a lake. Reconciliation of the various interpretations produced a picture of a goldfield as most likely having been a wet alluvial fan or a fluvial fan that had developed at the point of discharge of a major river into the basin, the long axis of which trended east-north-east. The sediments were deposited on the fan by the river, and were later reworked by waters from both the river and the lake or inland sea. The fans all had varying degrees of asymmetry caused by the elongation of one lobe by longshore currents flowing in a clockwise direction round the basin. An uplifted highland surrounded the intermontane basin, and the interface between the two terrains was formed by a series of peripheral faults of major dimensions. Movement along the bounding faults caused recycling of the apices of earlier fans into the sediments of later periods of sedimentation. Four rivers were identified – one feeding the Orange Free State goldfield, a second the Klerksdorp goldfield, a third the West Wits, West Rand and western portion of the Central Rand goldfields, and a fourth the eastern section of the Central Rand, the East Rand, and the Evander goldfields. The strata of each goldfield represented the aggregate of a number of superimposed fluvial-fans formed at the mouth of the same river. As the basin grew smaller with time, so a regressive relationship developed between the mean positions of successive fans in the Upper Division. Present-day analogues of such an environment appeared to be more the alluvial fans and delta-fans of the foothills of the Alpine and Himalayan mountain fronts, with their much higher rainfall, than the fans of the arid Basin and Range Province. In this paper, Pretorius introduced two new concepts – identification of local depositional conditions according to fan facies, and employment of energy-levels as controlling factors in the nature of the sedimentary responses to the varying depositional processes. The conglomerates, grits, and quartzites had formed preferentially in the central and inner intermediate portions of the fanhead and the upper midfan, and the fine-grained quartzites, argillaceous quartzites, shales, and carbonaceous material in the outer intermediate and marginal portions of the lower midfan and fanbase. The sediments of the Lower Division were essentially products of fanbase and distal lacustrine environments, and those of the Upper Division of lower fanhead, upper midfan, and proximal lacustrine conditions. Goldfields developed only in high-energy regimes about the entry-points of the fluvial systems into the basin, while paystreaks were high-energy major stream channels on the fluvial fans.

In the only major publication that has so far appeared on the Evander goldfield, Tweedie[75] wrote in 1968 that the evidence favoured marine rather than fluvial conditions for the laying down of the Kimberley Reef. Uplift occurred to the east of this area. Material moved from the east-northeast, while longshore currents flowed from the west-southwest. Shoaling conditions and barrier beaches developed in the west and south, separating the Evander basin from the East Rand depository, which two goldfields were not connected from late Main-Bird times to the time of formation of the Kimberley Reef. The two areas might have been reconnected after the accumulation of the Kimberley Reef, when longshore currents swept sands eastwards from the East Rand entry-point.

In his book on the geological history of Southern Africa, published in 1969, Haughton[76] classed the Witwatersrand depository as an intracratonic closed basin, comparable in size to Lake Victoria. Rivers deposited the debris from surrounding highlands in deltas, at the edges of which waves and currents caused partial redistribution, particularly near the shore. In the Upper Division, differential tectonic movements produced local disconformities. Transgressions and regressions were partly due to climatic variations. Channels were cut and filled near the major feeding streams. Greater volumes of material were transported to the northwestern side of the basin, but by Upper Division times the main points of entry lay at only a few places around the periphery.

In 1969, Knowles[77] concluded that the Basal Reef of the Orange Free State goldfield had been deposited by fluvial agencies, in the form of channels and stream floods, on the central and lower portions of an alluvial fan extending into an inland sea. The sediments spread over the paleoslope by the lateral accretion of gravel point-bars. Later in the same year, on the completion of a comprehensive and valuable sedimentological investigation of most of the reefs in the same goldfield, Sims[78] came to the conclusion that all the products could be seen of the interaction between fluvial, valley-flat-pediment, delta-fan, littoral, and quasi-neritic lacustrine environments proximal to the entry-point of a fluvial system into a landlocked continental sea. Throughout the depositional history, constant adjustments, involving vertical tectonics along continuously active marginal faults, influenced and controlled sedimentation, gradually reducing the size of the basin and producing unconformities. The reef horizons were formed under essentially degrading conditions in the wake of the tectonic episodes, while the normal members accumulated in aggrading regimes with equilibrium conditions when the rate of subsidence was equal to the rate of sediment feed. The auriferous bodies were restricted to the margin of the basin, and always developed on unconformities. The Basal Reef was introduced from the southwest by a river which deposited on a delta-fan under shallow-water and intermittently subaerial conditions. It became the product of prolonged degrading and incising of the footwall by meandering streams which simultaneously introduced the gold-bearing detritus. Many distributaries migrated to and fro across the depository on a planar surface of erosion. Winnowing currents trended north-northwestwards. Sims indicated that the Leader quartzites were channel deposits, the Middle Reef fluvial channel accumulations, the lower Kimberley Reefs fluvio-deltaic sediments, the upper Kimberley Reefs fluvial deposits with deep erosion channels, and the Elsburg reefs the products of a rapidly aggrading delta-fan adjacent to a very active boundary fault.

The most significant study of the uranium present in the Witwatersrand reefs was carried out by Liebenberg[79] and published in 1955. He showed that it occurred in two forms, as thucholite associated with carbon and as detrital uraninite grains. The latter variety was gradually concentrated with other black sands in streams that discharged into marine placers. For the uraninite to survive during subaqueous transportation and deposition, Liebenberg concluded that a non-oxidizing atmosphere must have prevailed. Another notable contribution was made by Koen[62] in 1961, in which he stated that the well-rounded appearance and close sizing of the uraninite grains could be taken as strong indicators of reworking by vigorous wave-action. He believed that compact nodules of the mineral formed on the muddy floors of extensive marshes or shallow lakes, and that these nodules were subsequently churned up by the waves during the initiation of a new cycle of sedimentation whereby the uraninite became redistributed, was mixed with the normal sediments, was sorted, and then was deposited with the sand and gravel.

In 1949, Sharpe[45] reported that carbon was common on the peneplaned depositional floors beneath the Main Reef, Main Reef Leader, Hangingwall Leaders, May Reef, and Black Reef of the East Rand goldfield. It was also present in channel and lagoonal deposits. From this, he concluded that the carbon was most apparent in those horizons that formed along shorelines and between tides, in stream channels, and in muddy lagoonal deposits, the precise horizons that would favour the development of bacteria and possibly algae. He put forward that the carbon granules and other forms of carbon found in the banket might be the remains of some of the earliest forms of life. However, he felt that the association between gold and carbon was purely coincidental. In the following year, Fletcher[80] drew attention to the fact that certain elemental marine plants can concentrate radioactive substances in their systems, and suggested that the uranium-bearing Carbon Leader of the West Wits goldfield might have formed from such organic matter in a marine environment where the gold was concentrated under water. In 1951, Davidson and Bowie[81] disagreed with these ideas, and wrote that the evidence conclusively proved that gaseous hydrocarbons of hydrothermal origin had undergone polymerization around pitchblende nuclei. As a result, there was a consolidation of hydrocarbon gel, with an accompanying coagulation and flocculation of colloidally dispersed pitchblende into blebs. Two years later, MacGregor[82] suggested that the Witwatersrand basin had been fringed with algal peat bogs. The waters entering these swamps carried nitrates, chlorides, and sulphuric acid, which acted as solvents of the gold and uranium. Reducing conditions induced by dead organic matter in the stagnant waters precipitated the gold and uranium. Torrential floods washed out the peat, and carried it far out into the lake where it absorbed still more uranium and sank to the bottom. Miholic[55], in 1954, supported Sharpe's[45] earlier contention that the carbon was the product of organisms belonging to the oldest, if not the first, forms of life that appeared on Earth late in the Archean. A rich growth of vegetation concentrated the uranium under anaerobic conditions.

Columnar thucholitic carbon was reported by Liebenberg[79], in 1955, not to be as common as the carbon granules. His belief was that organic remains, probably of algae, in various stages of decomposition in marine bottom-muds generated a variety of hydrocarbonaceous materials, including methane, which were polymerized by radioactive emanations. He noted that the amount of hydrocarbon increased upwards from the Dominion Reef to the Black Reef, suggesting that there was a gradual proliferation of biological activity. Viljoen[65] subscribed to Liebenberg's concepts, and added, in 1963, that this enhanced activity probably took place on fluvial or delta-flats after the main influx of reef material. Brock and Pretorius[66], in the following year, concluded that the pattern of interlaced gold in carbon seams was reminiscent of algal structures. They considered that the biogenic material had accumulated in a lagoon or marsh, as part of an isolated embayment on a delta.

The year 1965 saw the first of the publications dealing with detailed microscopic and chemical studies of the carbon, which investigations led to more and more definite conclusions that the carbon was of biological origin. Snyman's[83] work revealed structures in the thucholite very similar to certain algal and fungal features in sapropelic coals, and he concluded that the thucholite represented highly coalified algae. He saw the uraninite as being precipitated from solution as a result of algal activity. In 1966, Pretorius[74] put forward the proposal for the first time that the carbon-bearing conglomerates, with varying quantities of fly-speck carbon granules in the matrix, were the product of the intermixing of algal mats laid down at the end of a preceding cycle of sedimentation and of gravels formed in the initial stages of the succeeding cycle. The first carbon isotope studies were reported on by Hoefs and Schidlowski[84] in 1967. Their results indicated that the Witwatersrand carbon was similar to sedimentary organic carbon of known biogenic derivation, mainly of the bituminous type. It was thought that the carbonaceous material was genetically related to oil rather than coal. They concluded that the waters of the Witwatersrand depository must have harboured a rich development of bacterial and algal life. In the same year, Prashnowsky and Schidlowski[85] added that amino acids and monosaccharides had been detected in the carbon of the Basal and B reefs of the Orange Free State goldfield. The only explanation that could be offered for their presence was that life processes must have been operative in the basin. Electron microscope studies also indicated the development of globose aggregates resembling cell colonies.

The years between 1949 and 1969 can be summarized as representing the most productive period in the history of Witwatersrand geology. The large number of objective investigations during Phase 4 removed many of the subjective elements from discussions on the nature of the depositional environment of the Witwatersrand sediments. Theories supporting wet alluvial fan, pediment, fluvial fan, valley-flat flood-plain piedmont-plain, delta-fan, estuary, beach, and neritic off-shore environments were all advanced during the period. However, an ever-increasing number of investigators came to appreciate that many of the features observed showed conditions transitional between continental and marine, and between fluvial and littoral. At the time of deposition of different reefs, different conditions prevailed, so that no one particular explanation could be offered that would satisfy all the features of all the reefs. The overall environment could best be described as fluvio-deltaic-littoral. The most satisfactory present-day analogies were believed to be represented by conditions prevailing in the valley-flats below the Alpine and Himalayan mountain-fronts and in the Basin and Range Province of the western United States.

During these twenty-one years, sedimentology assumed a justifiably important role in all serious attempts to unravel the problems of the depositional history. Cyclic sedimentation came to be recognized as the normal manner of accumulation of material. Reworking by wave-action and longshore currents was also seen to be an essential process in the formation of the reefs. The asymmetry of the delta-fans pointed to a clockwise motion of the currents in the intracratonic, closed basin, which took the form of a landlocked lake or shallow-water inland sea. Although glaciers were considered as a possible transporting medium for bringing the material from the source-area to the depository, the weight of the sedimentological evidence pointed strongly to rivers filling this role. The conglomerates were interpreted as having accumulated in two separate pulses of sedimentation – the first, of a higher energy-level, forming an openwork gravel, and the second, of an intermediate energy-level, distributing sand that infiltrated between, and covered, the pebbles to form the matrix of the banket and the quartzitic hangingwall. The restriction of payable gold- and uranium-bearing horizons to unconformities and breaks in sedimentation became an accepted fact. It was also determined that the maximum concentrations of gold occurred at a certain distance down the delta-fan and the greater accumulations of uranium somewhat farther into the basin. The importance of peripheral faults, particularly along the northwestern edge of the depository was clearly established, differential movement between the uplifted source-area and the subsiding basin taking place along these dislocations. It also was put forward that regional folding had contributed to the relative elevation and depression of hinterland and depository. At least four major points of entry of material into the lake or inland sea were identified, at each of which a large delta-fan developed at the mouth of a river. The rivers flowed from northwest to southeast.

The presence of uranium and carbon was taken into consideration in deciphering the depositional environment. Evidence from uraninite confirmed the conclusions drawn for the conditions under which the gold was concentrated. Sedimentological and chemical work removed the element of speculation about the origin of the carbon. By the end of Phase 4 there remained little doubt that it had formed from bacteria and algae which had flourished in the waters at certain times and which had precipitated or absorbed substantial amounts of gold and uranium that were thought to have entered the basin in solution, as well as in the form of the detrital particles that had concentrated in the conglomerates.

THE PERIOD BEGINNING IN 1970

The results of a very detailed and extensive sedimentological study of the Vaal Reef in the Klerksdorp goldfield were presented by Minter[86] in 1972. It was shown that a structurally determined entry-point had existed to the northwest of the area, and that a shoreline feature at a relatively short distance into the basin had created the conditions favourable for the formation of the Vaal Reef. The footwall of this stratum built out as a delta with a regressing shoreline down the paleoslope towards the southeast. As the regression became more extensive, so a vast, subaerial, wind-blasted sand-flat came into being. The shoreline then transgressed, truncating the deltaic plain to a very regular erosion surface, and thus formed an angular unconformity of wide extent. At the same time, estuarine braided channels were incised in a meandering pattern in a generally southeasterly direction. A protracted period of deposition of the Vaal Reef was marked by a mature pebbly quartz arenite and basal algal growth. The material filled irregularities in the erosion surface and spread between the channels to form a dendroidal belt of sediments carrying gold and uranium. The southeastwards migration of lunate dunes over a non-accumulating surface for a long time produced the patterns of mineral distribution. Continued transgression deepened the environment and caused the shoreline to advance, during which the pebbly arenites were cleaned up. The clay galls associated with the reef point to fluviatile conditions that existed in a de-oxygenated lagoon that was ultimately destroyed by the Vaal Reef transgression. The presence of slightly waterworn dreikanters suggested proximal, wind-blasted environments before the reef was laid down. Two separate entry-points were present, within several kilometres of each other, and differing pebble assemblages and gold contents developed about these discrete localities. The one to the west brought in very little gold. An offshore current moving to the northeast buried the Vaal Reef beneath a wedge of sediments prograding from the southwest. Whereas the Vaal Reef was seen as the product of a transgression, Minter considered the Kimberley Reefs to have been formed as fluvial gravels in a regressive environment and the Ventersdorp Contact Reef as an alluvial fan, also in a regressive circumstance.

In 1973, Antrobus[87] revised his ideas of twenty years earlier, coming to the conclusion that the large mass of information that had been gathered in that time no longer supported his original idea of the reefs representing residual detritus that accumulated on a pediment. Instead, he believed that they were formed of fluviatile gravels, which had been laid down on a smooth erosion surface that had been bevelled to the base-level of a river. Cross-bedding showed a radial distribution of the gravels from an entry-point on to the erosion surface. He saw an analogy in the coastal plains of the eastern United States where the Cretaceous-Tertiary Brandywine formation had formed under fluviatile conditions after spreading out from a hinge-line between the piedmont and the coastal plain in the area now covered by the Delaware and Potomac rivers.

Steyn[88] reported, in 1974, on the investigations that had been undertaken of the Bird Reefs of the West Rand goldfield. The White Reef was interpreted as a compound coalescing dispersal fan on a piedmont-plain near the head of a delta, which formed as the valley-fill of a major river prograding into the Witwatersrand Basin. During dry periods in this fluvial environment, wind-action produced pebble pavements. The portion of the original conglomerates that are now preserved belonged to the fanbase environment that merged into the upper valley flood-plain. A high concentration of heavy minerals accumulated as lag channel deposits in the more permanent distributary channels on the fan, while the rest of the gravels were deposited by shifting braided streams. The poorly mineralized conglomerates were laid down high upslope, near the apex of the compound fan, as channel bar deposits of braided streams oversupplied with sediments. Farther down the slope, where the currents were undersupplied with bedload materials, water continuously washed over thin sheets of sediments causing extensive winnowing and concentration over floor-highs in the channels. The payable reefs, therefore, represented residual lag gravels in the beds of the larger streams.

Convincing evidence in favour of the biogenic origin of the carbon was presented in 1973 by Hallbauer[89]. Through electron-microscope studies, he was able to observe that each column of carbon was surrounded by a thin membrane, and contained inside an irregular framework of hyphae-like filaments. Gold encrustations were abundantly developed on these filaments. He found the fly-speck carbon to be similar in size and shape to living fungus spores, and these also had inclusions of gold. He interpreted the carbon seams as the fossilized remains of lichen-like plants and the carbon granules as the reproductive form of the plant.

SUMMARY

From an analysis of the 90 significant contributions to the geological literature on the depositional environment of the Witwatersrand conglomerates, carbon seams, and other sediments, which have been reviewed, the following events and developments are considered to be the ones which contributed most towards helping to understand the conditions under which the gold- and uranium-bearing reefs were formed:

1896 suggestion that carbon might have been formed by 'organized beings' (Garnier[10])

1897 comparison of conglomerates with auriferous beaches of New Zealand and the northwestern United States (Becker[7])

1908 recognition that payable reefs are associated with breaks in sedimentation (Gregory[14])
comparison of carbon with present-day gold-bearing fresh-water algae in streams in California (Spilsbury[31])

1909 recognition of paystreaks of richer conglomerates (Young[17])

1911 report on first systematic and comprehensive investigation of geology of Central Rand and West Rand; recognition that not all reefs were formed under the same depositional conditions; first proposal of deltaic environment (Mellor[19,20])

1915 conclusion that general coarsening of sediments stratigraphically upwards due to hinterland advancing progressively into basin (Mellor[21])

1916 suggestion of reworking of reefs by wave-action; of recycling of Lower Division sediments into Upper Division (Mellor[22])
suggestion that conglomerates might have formed in alluvial valley-flat, such as that occurring in plains below fault-bounded Himalayan mountain-front (Evans[24])

1920 report on first sedimentological studies; recognition of two pulses (first openwork gravel, second infiltrating sand) to form conglomerates; first suspicion of radioactive material in reefs (Pirow[25])

1924 presence of uraninite proved in conglomerates (Cooper[30])

1926 proposal that elements of fluviatile, estuarine, deltaic, and littoral environments can be recognized in different reefs (Du Toit[27])

1927 first use of contoured assay plans to delineate paystreaks; recognition of fan-shaped pattern of paystreaks in East Rand, of two interfering directions of paystreaks at right-angles to each other on Central Rand, of braided stream patterns, of lobes on fans, of entry-point of material north of Benoni; suggestion of ice-sheets as being transporter of sediments rather than rivers; first substantive arguments for valley-flat, fluvial flood-plain environment (Reinecke[29])

1930 first report on use of cross-bedding to determine paleo-current directions; recognition that proximal facies stacked on distal facies stratigraphically upwards (Reinecke[34])

1931 introduction of concept of delta-fan instead of previously assumed oceanic delta (Mellor[37])

1936 suggestion of conglomerates having formed on alluvial fans in piedmont environment (Johnson[40])

1938 recognition that major conglomerate horizons occur on disconformities (Roberts and Kransdorff[41])

1940 proposal that certain conglomerates were formed from eluvial scree on arid pediment (De Kock[43])

1949 first description of ubiquitous cyclic sedimentation in Witwatersrand Sequence; proposal that carbon occurs on those stratigraphic horizons where conditions could have favoured presence of bacteria and algae (Sharpe[45])

1954 proposal that basin completely surrounded by land; identification of facies changes in reef horizons with respect to distance from basin-edge; suggestion that Witwatersrand be compared with Basin and Range environment of western United States (Antrobus[51,52])

recognition of separate, discrete entry-points of material for different goldfields; suggestion that depository is intracratonic basin (Brock[53])

1955 first comprehensive description of nature of uranium mineralization (Liebenberg[79])

1958 observation that geometry of basin influenced by folding (Nel[60])

1962 first report on re-introduction of sedimentological methods of investigation (Hargraves[63])

1963 first comprehensive macro-sedimentological investigation (Steyn[64])

first comprehensive micro-sedimentological investigation (Viljoen[65])

1964 first attempt at overall synthesis of previous work; first compilation of isopachs to outline geometry of basin; recognition of clockwise direction of movement of longshore currents; identification of confinement of locations of goldfields to synclinal downwarps between rising granite domes (Brock and Pretorius[66])

recognition that southeastern edge of basin more passive than northwestern, that minimum amount of sediments introduced from this side, that no significant amounts of gold brought in from southeast (Pretorius[68])

recognition of optimum zone of uranium mineralization as lying somewhat basinward of zone of maximum gold mineralization (Winter[69,70])

1965 first microscopic identification of possible algal origin of carbon (Snyman[83])

1966 first use of conceptual process-response models in synthesizing Witwatersrand data; employment of principle of energy-levels as measures of variations in depositional environments; proposal that goldfield represents higher-energy fluvial fan rather than lower-energy conventional delta-fan; design of standardized sedimentological investigation to produce maximum information; suggestion that carbon-bearing conglomerate result of breaking up of algal mat formed at end of preceding cycles by pebbles at beginning of succeeding cycle (Pretorius[74])

1967 carbon-isotope studies used to determine origin of carbon (Hoefs and Schidlowski[84])

chemical identification of biogenic derivation of carbon (Prashnowsky and Schidlowski[85])

1972 report on first comprehensive sedimentological study of whole goldfield as a single entity (Minter[86])

1973 detailed electron microscope investigation of carbon definitely establishes algal origin; suggests lichens might have fixed gold and uranium (Hallbauer[89])

1974 second attempt at overall synthesis of previous work (Pretorius[90,91])

The progressive changes in ideas concerning the nature of the depositional environment and the various conditions that prevailed in it have been summarized in Table 1.

TABLE 1

THE PERCENTAGES OF AUTHORS WHO FAVOURED DIFFERENT TYPES OF DEPOSITIONAL ENVIRONMENTS FOR BANKET DURING VARIOUS PHASES OF MINING HISTORY

Environment	Percentage of Authors					
	Phase 1 1886–1906	Phase 2 1907–1927	Phase 3 1928–1948	Phase 4 1949–1969	Phase 5 1970–	Average 1886–1974
neritic	19	8	4	2	—	6
littoral	63	43	26	28	30	38
shoreline delta	12	33	29	16	—	18
shoreline fluvial fan	—	—	—	17	37	11
coastal fluvial	—	2	—	4	9	3
interior fluvial	6	14	32	22	11	17
interior desert	—	—	9	11	13	7

The trends in speculations and observations over the years are readily discernible in the table. The neritic environment was popular only during Phase 1, thereafter fading rapidly until there are no proponents of it at the present time. Elements of a littoral regime have been conspicuously present at all times, and the reworking of the conglomerates by wave-action and currents has generally been considered to have taken place on a beach, whether on an open ocean, an inland sea, or an intermontane lake. This environment received most support during Phase 1 and then drew progressively less adherents up to Phase 3. Thereafter, the littoral component was again incorporated into more and more hypotheses. The conventional delta gained its largest number of proponents in Phase 2. As it became less accepted, so it was replaced by the fluvial fan environment, which, today, has the largest percentage of supporters. Fluvial deposits on a coastal plain have never attracted many advocates. Interior fluvial processes in a flood-plain or valley-flat system have always had strong supporters, particularly during Phase 3. The number of investigators has been growing steadily since Phase 3, who have given preference to an inland desert environment embracing alluvial fans, bajadas, and pediments. Opinions over the past 89 years, in general, have favoured the littoral environment, with the deltaic and interior fluvial regimes next in popularity.

Only in Phase 1 was there a marked tendency to interpret the depositional conditions attending the laying down of the conglomerates as being restricted to one environment only. Geologists have usually regarded the Witwatersrand sediments as the products of at least two interactive regimes. The preferred combinations over the years have been the following:

Phase 1 (1886-1906): littoral + neritic
Phase 2 (1907-1927): shoreline delta + littoral
Phase 3 (1928-1948): interior fluvial + shoreline delta + littoral
Phase 4 (1949-1969): interior fluvial + shoreline fluvial fan + littoral
Phase 5 (1970-1974): interior desert + shoreline fluvial fan + littoral

TOWARDS A POSSIBLE RESOLUTION

Ten years after the *magnum opus* of the Geological Society of South Africa, on the geology of the Witwatersrand Basin and its contained deposits of gold and uranium, the time seemed appropriate to attempt a synthesis of all the work that had taken place since the appearance of the publication. The substantial amounts of data that had been gathered during extensive and systematic studies of the quantitative sedimentology, mineralogy, and geochemistry of the banket and carbon seams had either reinforced or undermined the theories that had been put forward in 1964 concerning the nature of the depositional environment. The generally accepted model required certain modifications. These were coordinated in 1974 in two papers by Pretorius[90,91] who attempted to resolve the differences of opinion that were becoming lesser in number and smaller in magnitude. After almost 90 years, there were signs that the beginnings of a consensus were being reached by the majority of geologists working on the problems of the Witwatersrand, although there still remained a small group of dissenters whose voices could not be ignored.

It would now appear that the Witwatersrand depository was a intermontane, intracratonic, yoked basin with a fault-bounded northwestern edge and a gently downwarping more passive southeastern boundary. The enclosed basin was at least 350 km long in an east-northeasterly direction, and 200 km wide in a north-northwesterly direction. The structural environment resembled that of the Basin and Range Province, but a far wetter climate prevailed. The basin was a shallow-water lake or inland sea, no connexion to an open ocean having yet been found. The depository became structurally more unstable with time, and a pattern of interference folding produced structural depressions and culminations both on the rim and within the depository. The various goldfields developed in downwarps between basement domes. The northwestern side was episodically but continuously rising, causing the basin-edge to advance progressively farther towards the depositional axis. The final depository was smaller than the original, so that, overall, the sediments were laid down in a shrinking basin. Conditions were generally transgressive in the Lower Witwatersrand Division and generally regressive in the Upper Division. Second-order transgressions and regressions were superimposed on these primary trends. Between the base of the Lower Division and the base of the Upper Division, the edge of the basin moved southeastwards by 60 km, and the depositional axis by about 10 km.

A high-energy transfer system from the source to the depository took the form of a relatively short, linear fluvial array. From the areal geometry of the different stratigraphic horizons, the patterns of facies variations, the trends in the changes of grain-sizes of sediments, the directions and patterns of paleoflow, the nature of the environmental indicators, and the distribution of heavy minerals, it would appear that a goldfield is a fluvial fan or fan-delta that was formed where the river system debouched into the lake via a canyon cut through the high ground to the northwest of the peripheral faults. In this type of environment, there were far

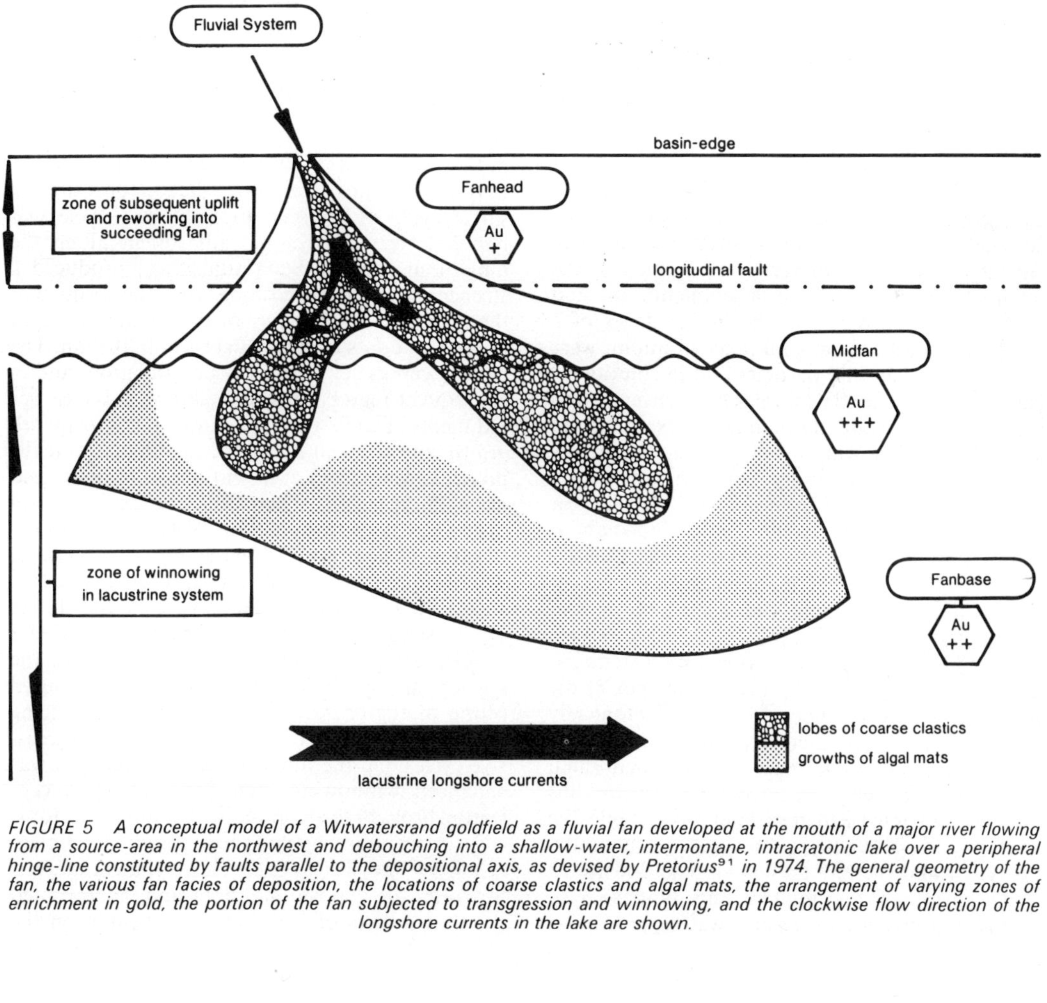

FIGURE 5 A conceptual model of a Witwatersrand goldfield as a fluvial fan developed at the mouth of a major river flowing from a source-area in the northwest and debouching into a shallow-water, intermontane, intracratonic lake over a peripheral hinge-line constituted by faults parallel to the depositional axis, as devised by Pretorius[91] in 1974. The general geometry of the fan, the various fan facies of deposition, the locations of coarse clastics and algal mats, the arrangement of varying zones of enrichment in gold, the portion of the fan subjected to transgression and winnowing, and the clockwise flow direction of the longshore currents in the lake are shown.

greater amounts of water than on a typical alluvial fan and the energy-level was higher than that on an oceanic delta. After emerging from the canyons, the rivers flowed short distances over a piedmont plain and then dispersed through a braided-stream pattern into the basin. The fluvial fans were restricted to the northwestern margin of the depository, and some of them coalesced in their more distal parts, leading to the impression of extensive sheets of uniform gravel. The largest of the fans was that constituting the East Rand goldfield, which measured 40 km in length down the central section from the apex to the fanbase, 50 km in width in the midfan section, and 90 km in width in the fanbase section. The western margin of this fan was 45 km long and the eastern flank 60 km. Six fluvial fans have so far been discovered – those constituting the Orange Free State, Klerksdorp, West Wits, West Rand, East Rand, and Evander goldfields. The original Central Rand, on which the Main Reef conglomerates were discovered, would seem to be a geographical entity only, since, geologically, it represents the coalescence of the eastern part of the West Rand fan and the western part of the East Rand fan.

The gold and uranium mineralization has been recognized to occur in five forms: (i) in the matrices of the conglomerates, (ii) in heavily pyritic sands filling erosion channels, (iii) on quartzites along a plane of unconformity between successive cycles of sedimentation, (iv) on shales along planes of unconformity, and (v) in carbon seams on, or adjacent to, planes of unconformity. The three last-mentioned types of reef were formed in the terminal stages of one cycle of sedimentation, and the first-mentioned two in the initial stages of a succeeding cycle. The gold and uranium were transported as detrital particles and in solution as chloride- and cyanide-complexes. Concentration took place physically, through gravity settling and subsequent winnowing by wave- and current-action, and biochemically through interaction between the gold and uranium and the algal or lichen colonies that preferentially developed about the mouths of the major rivers, in the quieter-water conditions on the margins and fanbases of the fluvial fans, and at the end of certain cycles of sedimentation.

The conceptual model of a typical Witwatersrand goldfield is portrayed in Figure 3. The apex of the fluvial fan was located along the tectonically unstable basin-edge where repeated uplift of the source-area side took place along longitudinal faults. The fanheads of earlier fans were thus uplifted and reworked into later fans, while the midfan and fanbase sections were structurally depressed and thereby preserved. The downward displacement of the midfan and fanbase also caused transgression of the lake waters, producing winnowing of the fines and lag concentrations of the heavier minerals. Longshore currents moved the finer sediments farther away from the entry-points to form asymmetrical fans owing to the clockwise movement of the water in the depository. The typical fluvial fan had two main lobes in which were located a larger number of braided-stream channels, thicker and coarser clastic sediments, and higher concentrations of detrital gold and uranium. The material that was laid down between the lobes took the form of sands, silts, and muds, similar to that which accumulated on the fan margins and base. Conditions under these lower-energy regimes at times provided the optimum environment for the growth of algae or lichens, which took the form of thin algal mats. The gold was of too fine a grain to settle in the fanhead facies. The highest concentrations took place in the midfan lobes, with the peak of uranium mineralization a little farther down the slope than the peak of the gold. The energy-level dropped too low to permit the transportation of detrital gold to the fanbase section. However, the gold and uranium that were in solution did interact with the biogenic material that was present in the low-energy environments.

A fluvial fan was built up in a series of pulses of sedimentation, which started with progradation during regression, went through aggradation during transgression, and ended with degradation during stillstand. These three stages constituted a single cycle of sedimentation. A new cycle was initiated through tectonic adjustment along the longitudinal faults. Such adjustment produced a steepening of the paleoslope, with the result that the increased competency of the streams brought greater amounts of coarse debris onto the fan. The higher energy-level caused progradation and a consequent regressive relationship with the earlier sediments. The first pulse laid down an openwork gravel, and the next pulse, the sand matrix. Heavy minerals were brought in with the sand phase and not with the gravels. Thereafter, as the energy-level dropped, transgression took place, with the deposition of finer-grained material, until a state of equilibrium was reached and deposition came to a standstill. End-of-cycle winnowing by the waters of the streams and the lake produced a greater concentration of residual heavy minerals on the erosion surface. Incipient tectonic activity caused tilting of the erosion surface, thereby producing the unconformable relationships between successive cycles. On the tilted surface, degradation was enhanced, winnowing intensified, and lag concentration brought to an optimum. Continued tectonic adjustment culminated in the prograding sedimentation of the next cycle. The turbulent gravels broke up the depositional floor and incorporated the thin streaks of lag gold and uranium on the unconformity. Thus, these minerals could have been introduced into the gravels in two processes – pick-up from the footwall sediments and downward infiltration during the sand pulse that succeeded the laying down of the pebbles.

That is the 1974 model. It will no doubt undergo still many more refinements before gold mining comes to an end in the Witwatersrand Basin. There are no complete and final solutions to problems in geology.

REFERENCES

1. MATHERS, E. P. *The gold fields revisited.* Durban, Davis and Sons, 1887. 352 pp.
2. BALLOT, J. *The banket formation: its probable origin and present position.* Johannesburg, Mendelssohn and Scott, 1888. 18 pp.
3. HATCH, F. H., and CHALMERS, J. A. *The gold mines of the Rand.* London, Macmillan, 1895. 306 pp.
4. KUNTZ, J. The Rand conglomerates: how they were formed. *Trans. geol. Soc. S. Afr.*, vol. 1, 1895. pp. 113-122.
5. DE LAUNAY, L. *Les mines d'or du Transvaal.* Paris, Baudry, 1896. 540 pp.
6. SCHMITZ-DUMONT, P. The river-bed theory of the Witwatersrand conglomerates. *Trans. geol. Soc. S. Afr.*, vol. 2, 1896. pp. 141-142.
7. BECKER, G. F. The Witwatersrand banket, with notes on other gold-bearing pudding stones. U.S. Geol. Surv., *18th Ann. Rept.*, Part V, 1897. pp. 153-184.
8. DENNY, G. A. *The Klerksdorp goldfields.* London, Macmillan, 1897. 251 pp.
9. CARRICK, J. T. On faulting along the Main reef line. *Trans. geol. Soc. S. Afr.*, vol. 2, 1896. pp. 39-41.
10. GARNIER, J. Gold and diamonds in the Transvaal and the Cape. *Trans. geol. Soc. S. Afr.*, vol. 2, 1896. pp. 91-103, 109-120.
11. BLELOCH, W. Rand conglomerates. *Trans. geol. Soc. S. Afr.*, vol. 4, 1898. pp. 175-181.
12. PRISTER, A. Notes on the origin and formation of the Witwatersrand auriferous deposits. *Trans. geol. Soc. S. Afr.*, vol. 4, 1898. pp. 19-36.
13. HATCH, F. H. The geological history of the South African formations. *Proc. geol. Soc. S. Afr.*, vol. 9, 1906. pp. xxi-xxxiv.
14. GREGORY, J. W. The origin of the gold in the Rand banket. *Trans. Inst. Min. Metall.*, vol. 17, 1908. pp. 2-41.
15. GREGORY, J. W. The origin of the gold of the Rand goldfield. *Econ. Geol.*, vol. 4, 1909. pp. 118-129.
16. BECKER, G. F. Discussion on 'The origin of the gold of the Rand goldfield' by J. W. Gregory. *Econ. Geol.*, vol. 4, 1909. pp. 363-384.
17. YOUNG, R. B. Further notes on the auriferous conglomerates of the Witwatersrand with a discussion on the origin of the gold. *Trans. geol. Soc. S. Afr.*, vol. 12, 1909. pp. 82-101.
18. YOUNG, R. B. The problem of the Rand banket. *Proc. geol. Soc. S. Afr.*, vol. 14, 1911. p. xxi-xxix.
19. MELLOR, E. T. The normal section of the lower Witwatersrand system on the Central Rand and its connection with West Rand sections. *Trans. geol. Soc. S. Afr.*, vol. 14, 1911. pp. 99-131.
20. MELLOR, E. T. Some structural features of the Witwatersrand system on the Central Rand, with a note on the Rietfontein series. *Trans. geol. Soc. S. Afr.*, vol. 14, 1911. pp. 24-42.
21. MELLOR, E. T. The upper Witwatersrand system. *Trans. geol. Soc. S. Afr.*, vol. 18, 1915. pp. 11-56.
22. MELLOR, E. T. The conglomerates of the Witwatersrand. *Trans. Inst. Min. Metall.*, vol. 25, 1916. pp. 226-291.
23. WAGNER, P. A. Some problems in South African geology. *Proc. geol. Soc. S. Afr.*, vol. 20, 1917. pp. xix-xxxix.
24. EVANS, J. W. Discussion on 'The conglomerates of the Witwatersrand' by E. T. Mellor. *Trans. Inst. Min. Metall.*, vol. 25, 1916. pp. 291-294.
25. PIROW, H. Distribution of the pebbles in the Rand banket and other features of the rock. *Trans. geol. Soc. S. Afr.*, vol. 23, 1920. pp. 64-97.
26. ROGERS, A. W. The geology of the country around Heidelberg. Geol. Surv. S. Afr., *Explanation Geological Map*, 1922. 80 pp.
27. DU TOIT, A. L. *The geology of South Africa*, 1st edition. Edinburgh, Oliver and Boyd, 1926. 463 pp.
28. NEL, L. T. The geology around Vredefort. Geol. Surv. S. Afr., *Explanation Geological Map*, 1927. 130 pp.
29. REINECKE, L. The location of payable ore-bodies in the gold-bearing reefs of the Witwatersrand. *Trans. geol. Soc. S. Afr.*, vol. 30, 1927. pp. 89-119.
30. COOPER, R. A. Mineral constituents of Rand conglomerates. *J. Chem. Metall. Min. Soc. S. Afr.*, vol. 24, 1924. pp. 90-93.
31. SPILSBURY, E. G. Discussion on 'The origin of the gold in the Rand banket' by J. W. Gregory. *Trans. Inst. Min. Metall.*, vol. 17, 1908. pp. 66-69.
32. HORWOOD, C. B. The mode of occurrence and genesis of the carbon in the Rand banket. *Trans. geol. Soc. S. Afr.*, vol. 13, 1910. pp. 65-92.
33. GRATON, L. C. Hydrothermal origin of the Rand gold deposits: part I - testimony of the conglomerates. *Econ. Geol.*, vol. 25, supplement to No. 3, 1930. 185 pp.
34. REINECKE, L. Origin of the Witwatersrand system. *Trans. geol. Soc. S. Afr.*, vol. 33, 1930. pp. 111-133.
35. DU TOIT, A. L. Discussion on 'Origin of the Witwatersrand system' by L. Reinecke. *Proc. geol. Soc. S. Afr.*, vol. 34, 1931. pp. xxxvi-xxxviii.
36. GREGORY, J. W. Professor Graton on the Rand banket. *Annex. Trans. geol. Soc. S. Afr.*, vol. 34, 1931. pp. 23-36
37. MELLOR, E. T. The origin of the gold in the Rand banket: discussion on Professor Graton's paper. *Annex. Trans. geol. Soc. S. Afr.*, vol. 34, 1931. pp. 55-69.
38. MELLOR, E. T. Discussion on 'Origin of the Witwatersrand system' by L. Reinecke. *Proc. geol. Soc. S. Afr.*, vol. 34, 1931. pp. xxxviii-xl.
39. NEL, L. T. The Witwatersrand system outside the Rand. *Proc. geol. Soc. S. Afr.*, vol. 36, 1933. pp. xxiii-xlviii.
40. GEOLOGICAL SURVEY OF SOUTH AFRICA. The mineral resources of the Union of South Africa. Geol. Surv. S. Afr., *Handbook*, 1936. 454 pp.
41. ROBERTS, E. R., and KRANSDORFF, D. The upper Witwatersrand system at Randfontein Estates. *Trans. geol. Soc. S. Afr.*, vol. 41, 1938. pp. 225-247.
42. DU TOIT, A. L. *The geology of South Africa*, 2nd edition. Edinburgh, Oliver and Boyd, 1939. 539 pp.
43. DE KOCK, W. P. The Ventersdorp Contact reef: its nature, mode of occurrence, and economic significance, with special reference to the Far West Rand. *Trans. geol. Soc. S. Afr.*, vol. 43, 1940. pp. 85-107.
44. MACADAM, P. The distribution of gold and carbon in the Witwatersrand bankets. *Annex. Trans. geol. Soc. S. Afr.*, vol. 34, 1931. pp. 81-88.
45. SHARPE, J. W. N. The economic auriferous bankets of the upper Witwatersrand beds and their relationship to sedimentation features. *Trans. geol. Soc. S. Afr.*, vol. 52, 1949. pp.265-288.
46. DE JAGER, F. S. J. Discussion on 'The economic auriferous bankets of the upper Witwatersrand beds and their relationship to sedimentation features' by J. W. N. Sharpe. *Trans. geol. Soc. S. Afr.*, vol. 52, 1949. pp. 290-296.
47. DE JAGER, F. S. J. Morphological reconstruction of the Kimberley-Elsburg series, with special reference to the Kimberley group of sediments in the East Rand basin. *Ann. Univ. Stellenbosch*, vol. 33, section A, nos. 1-11, 1957. pp. 125-190.
48. DE JAGER, F. S. J. The Witwatersrand system in the Springs-Nigel-Heidelberg sector of the East Rand basin. *The geology of some ore deposits in southern Africa.* S. H. Haughton, ed. Johannesburg, *Geol. Soc. S. Afr.*, 1964. vol. 1. pp. 161-190.
49. BORCHERS, R. The Odendaalsrus-Virginia goldfield and its relation to the Witwatersrand. *Ph.D. thesis*, Univ. S. Afr., 1950. 120 pp.
50. PEGG, C. W. A contribution to the geology of the West Rand area. *Trans. geol. Soc. S. Afr.*, vol. 53, 1950. pp. 209-224.
51. ANTROBUS, E. S. A. A study of the Witwatersrand system. *Ph.D thesis*, McGill Univ., 1954. 78 pp.
52. ANTROBUS, E. S. A. The origin of the auriferous reefs of the Witwatersrand system. *Trans. geol. Soc. S. Afr.*, vol. 59, 1956. pp. 1-15.

53. BROCK, B. B. A view of faulting in the Orange Free State. *Optima*, vol. 4, 1954. pp. 5-17.
54. Du TOIT, A. L. *The geology of South Africa*, 3rd edition. Edinburgh, Oliver and Boyd, 1954. 611 pp.
55. MIHOLIC, S. Genesis of the Witwatersrand gold-uranium deposits, *Econ. Geol.*, vol. 49, 1954. pp. 537-540.
56. BAIN, G. W. Discussion on 'The occurrence and origin of gold and radioactive minerals in the Witwatersrand system, the Dominion reef, the Ventersdorp Contact reef, and the Black reef, by W. R. Liebenberg. *Trans. geol. Soc. S. Afr.*, vol. 58, 1955. pp. 236-240.
57. BAIN, G. W. Patterns to ores in layered rocks. *Econ. Geol.*, vol. 55, 1960. pp. 695-731.
58. WIEBOLS, J. H. A suggested glacial origin for the Witwatersrand conglomerates. *Trans. geol. Soc. S. Afr.*, vol. 58, 1955. pp. 367-382.
59. BROCK, B. B., NEL, L. T., and VISSER, D. J. L. The geological background of the uranium industry. *Symposium on Uranium*, Johannesburg, 1957. Assoc. Sci. Tech. Soc. S. Afr., 1957. vol. 1. pp. 275-305.
60. NEL, L. T. The occurrence of uranium in the Union of South Africa. *Proc. 2nd U.N. Int. Conf. Peaceful Uses Atomic Energy*. Geneva, United Nations, 1958. pp. 54-86.
61. FULLER, A. O. Discussion on 'Further observations on uraniferous conglomerates' by C. F. Davidson, *Econ. Geol.*, vol. 55, 1960. pp. 842-843.
62. KOEN, G. M. The genetic significance of the size distribution of uraninite in Witwatersrand bankets. *Trans. geol. Soc. S. Afr.*, vol. 64, 1961. pp. 23-46.
63. HARGRAVES, R. B. Cross-bedding and ripple-marking in the Main-Bird series of the Witwatersrand system in the East Rand area. *Trans. geol. Soc. S. Afr.*, vol. 65, 1962. pp. 263-275.
64. STEYN, L. S. The sedimentology and gold distribution pattern of the Livingstone reefs on the West Rand. *M.Sc. thesis*, Univ. Witwatersrand, 1963. 132 pp.
65. VILJOEN, R. P. Petrographic and mineragraphic aspects of the Main Reef and Main Reef Leader of the Main-Bird series, Witwatersrand system. *M.Sc. thesis*, Univ. Witwatersrand, 1963. 193 pp.
66. BROCK, B. B., and PRETORIUS, D. A. Rand basin sedimentation and tectonics. *The geology of some ore deposits in southern Africa*. S. H. Haughton, ed. Johannesburg, Geol. Soc. S. Afr., 1964. vol. 1. pp. 549-599.
67. PRETORIUS, D. A. The geology of the Central Rand goldfield. *The geology of some ore deposits in southern Africa*. S. H. Haughton, ed. Johannesburg, Geol. Soc. S. Afr., 1964. vol. 1. pp. 63-108.
68. PRETORIUS, D. A. The geology of the South Rand goldfield. *The geology of some ore deposits in southern Africa*. S. H. Haughton, ed. Johannesburg, Geol. Soc. S. Afr., 1964. vol. 1. pp. 219-282.
69. WINTER, H. D. The geology of the northern section of the Orange Free State goldfield. *The geology of some ore deposits in southern Africa*. S. H. Haughton, ed, Johannesburg, Geol. Soc. S. Afr., 1964. vol. 1. pp. 417-448.
70. WINTER, H. D. The geology of the Virginia section of the Orange Free State goldfield. *The geology of some ore deposits in southern Africa*. S. H. Haughton, ed. Johannesburg, Geol. Soc. S. Afr., 1964. vol. 1. pp. 507-548.
71. WINTER, H. D. Trough cross-stratification in the upper division of the Witwatersrand system at Hartebeesfontein Mine, Klerksdorp. Unpub. rept., Econ. Geol. Res. Unit, Univ. Witwatersrand, 1965. 88 pp.
72. ARMSTRONG, G. C. A sedimentological study of the U.K.9 Kimberley reefs of the East Rand. *M.Sc. thesis*, Univ. Witwatersrand, 1965. 65 pp.
73. KNOWLES, A. G. A paleocurrent study of the Ventersdorp Contact reef at Western Deep Levels Limited on the Far West Rand. *M.Sc. thesis*, Univ. Witwatersrand, 1966. 125 pp.
74. PRETORIUS, D. A. Conceptual geological models in the exploration for gold mineralization in the Witwatersrand basin. *Symposium on mathematical statistics and computer application in ore valuation*, Johannesburg, 1966. S. Afr. Inst. Min. Metall., 1966, pp. 225-275.
75. TWEEDIE, K. A. The stratigraphy and sedimentary structures of the Kimberley shales in the Evander goldfield, eastern Transvaal, South Africa. *Trans. geol. Soc. S. Afr.*, vol. 71, 1968. pp. 235-254.
76. HAUGHTON, S. H. *Geological history of southern Africa*. Johannesburg, Geol. Soc. S. Afr., 1969. 535 pp.
77. KNOWLES, A. G. A sedimentological investigation of borehole intersections of the Basal Reef in the southern part of the central area of the Orange Free State goldfields. Unpub. rept., Anglo American Corpn. S. Afr., 1969. 10 pp.
78. SIMS, J. F. M. The stratigraphy and paleocurrent history of the upper division of the Witwatersrand system on President Steyn mine and adjacent areas in the Orange Free State goldfield, with special reference to the origin of the auriferous reefs. *Ph. D. thesis*, Univ. Witwatersrand, 1969. 181 pp.
79. LIEBENBERG, W. R. The occurrence and origin of gold and radioactive minerals in the Witwatersrand system, the Dominion reef, the Ventersdorp Contact reef and the Black reef. *Trans. geol. Soc. S. Africa.*, vol. 58, 1955. pp. 101-227.
80. FLETCHER, H. Discussion on 'Radioactivity logging' by D. J. Simpson and R. F. Bouwer. *Trans. geol. Soc. S. Afr.*, vol. 53, 1960. p. 11.
81. DAVIDSON, C. F., and BOWIE, S. H. On thucholite and related hydrocarbon-uraninite complexes. *Bull. geol. Surv. Gt. Brit.*, no. 3, 1961. pp. 1-18.
82. MACGREGOR, A. M. Discussion on 'The gold-uranium ores of the Witwatersrand' by C. F. Davidson. *Min. Mag.*, vol. 88, 1953. pp. 281-282.
83. SNYMAN, C. P. Possible biogenetic structures in Witwatersrand thucholite. *Trans. geol. Soc. S. Afr.*, vol. 68, 1965. pp. 225-235.
84. HOEFS, J., and SCHIDLOWSKI, M. Carbon isotope composition of carbonaceous matter from the Precambrian of the Witwatersrand system. *Science*, vol. 155, no. 3766, 1967. pp. 1096-1097.
85. PRASHNOWSKY, A. A., and SCHIDLOWSKI, M. Investigation of Precambrian thucholite. *Nature*, vol. 216, no. 5115, 1967. pp. 560-563.
86. MINTER, W. E. L. The sedimentology of the Vaal reef in the Klerksdorp area. *Ph.D. thesis*, Univ. Witwatersrand, 1972. 170 pp.
87. ANTROBUS, E. S. A. Presidential address. Unpub. rept., Congress, Bloemfontein, 1973, Geol. Soc. S. Afr., 1973.
88. STEYN, L. S. Sedimentological studies of the Bird reefs of the West Rand goldfield. Unpub. rept., Econ. Geol. Res. Unit, Univ. Witwatersrand, 1974, 32 pp.
89. HALLBAUER, D. K. The biological nature of the carbon in some Witwatersrand reefs and its association with gold. *Research Rept.*, Chamb. Mines S. Afr., no. 9173, 1973. 14 pp.
90. PRETORIUS, D. A. The nature of the Witwatersrand gold-uranium deposits. *Inform. Circ.*, Econ. Geol. Res. Unit, no. 86, 1974. 50 pp.
91. PRETORIUS, D. A. Gold in the Proterozoic sediments of South Africa: systems, paradigms, and models. *Inform. Circ.*, Econ. Geol. Res. Unit, no. 87, 1974. 22 pp.

One of the interesting and important features of the Rand is the presence of "fly speck and columnar carbon" with which gold and uranium are commonly associated. This so-called carbon of the Rand conglomerates and other types of closely associated deposits is a hydrocarbon-uraninite mixture probably best referred to as thucholite or uraniferous carbon. It was observed by the early workers, has been studied in detail by Liebenberg (1955), and is referred to at some length by De Kock (1964) and by others in the book edited by Haughton (1964). The mineraloid occurs in a granular, columnar, and massive form and is jet black to dull black in color. Occasional thin sheets, pencil lines, stringers and veinlets are also encountered in some reefs. The granular variety (fly speck carbon) occurs in the matrix of most reefs as small (0.1-4 mm) black spheroidal or nodular forms with a warty outer appearance. The ovoid granules frequently extend beyond the ends of the conglomerate lenses and are the only indication of the position of conglomerate horizons. The granules are generally distributed throughout the reefs, but in places there are often concentrations along the parting planes between the hanging wall and footwall of the reefs; in some areas the concentrations of granules are localized along the parting planes on the footwalls of the reefs. The massive thucholite resembles vitrinite and the columnar variety occurs as bundles of vitreous fibers normal to the bedding. The thickness of the seams of columnar and massive thucholite is generally only a few millimeters, rarely more than 1 cm. Both varieties are commonly associated with one another, or they may occur alone. Both may occur in the conglomerates, although they are more commonly found along the bedding planes of conglomerate bands and along the basal contacts of the reefs. In the latter situation the carbon is extensively developed in the Carbon Leader reef of the West Wits Line (De Kock, 1964).

The replacement, displacement, and other growth features of the thucholite indicate that it is an authigenic mineraloid. Associated minerals include uraninite, gold, platinoids, and sulfides such as pyrrhotite, chalcopyrite, pentlandite, sphalerite, cobaltite, linnaeite, galena, and arsenopyrite. Microscope examination of the thucholite indicates that the uraninite commonly occurs as a large number of tiny inclusions in the mineraloid. Gold occurs as veinlets, specks, and small irregular patches within the body of the thucholite; bands of gold also surround the granules of thucholite, and thin films and veinlets cut across the columns or occur with phyllosilicates between the columns of the columnar variety of the mineraloid. The various sulfides noted above commonly occur in the interstitial spaces between granules of thucholite and also appear in veinlets and patches in cracks and cavities in the mineraloid. Inclusions of sulfides in thucholite are also observed in places.

The intimate association of gold and thucholite in the Rand has been known and commented on for nearly fifty years. In and near the mineraloid, visible and occasionally coarse gold can be frequently seen, a feature that is unusual in the Rand ores. Thucholite is also rich in other metals in the Carbon Leader; besides gold there is native silver, a variety of platinoid minerals (sperrylite, braggite, cooperite, platiniridium, platinum, and osmiridium), chalcopyrite, a number of Ni-Co arsenides and sulfides, sphalerite, and galena (De Kock, 1964).

The origin of the thucholite is uncertain. Davidson (1965) and Davidson and Bowie (1950) consider the mineraloid a carburan polymerized radiogenically from the methane and other hydrocarbon gases and fluids so prevalent in the South African mines. De Kock (1964) thinks the thucholite is possibly the organic remains of a primitive form of algae that existed in the waters of the basin in which the reefs were deposited. Actually, the methane and other hydrocarbon compounds may be similarly derived, although they may also have come from a very deep source as the result of abiogenic processes. Schidlowski (1966) considered that the carbonaceous material (thucholite) in the Rand formed by radiolytic polymerization of originally mobile (preferably gaseous) hydrocarbons percolating through the conglomerates, the prerequisite for its formation being the presence of detrital uraninite within the heavy fraction of the individual reefs.

The origin of the thucholite has been variously related to biogenic sources. Snyman (1965) observed structures analogous to those of algae and fungi in the thucholite, Schidlowski (1965) found cell-like structures in the Witwatersrand rocks, and Prashnowsky and Schidlowski (1967) identified amino acids and monosaccharides in the carbonaceous substances (thucholite). Hoefs and Schidlowski (1967) concluded from carbon isotope data of the thucholite that its constituents were of biogenic origin. The sulfur isotopic data on the allogenic and authigenic pyrite presented by Hoefs, Neilsen, and Schidlowski (1968), however, showed only a small enrichment in ^{32}S in the recrystallized authigenic pyrite, suggesting only slight, if any, activity of organisms operating on the sulfur cycle in the conglomerates.

In a recent contribution to the origin of the carbon of the Rand, Hallbauer and van Warmelo (1974) have examined both the columnar and fly speck types of thucholite from the Carbon Leader Reef and other auriferous reefs by scanning electron microscope. In the columnar type internal structures morphologically resembling filamentous, branched, and apparently septate cells of obvious biological origin partly encrusted with gold were identified. In addition silicified structures identified as primitive fungi were noted. From these observations they concluded that the organism was some kind of a fossilized symbiotic lichenlike plant with no modern equivalent that formed carpetlike colonies of up to several square meters in extent. These organisms were capable of extracting gold and uranium from the environment and depositing them inter- and intracellularly. The fly speck carbon constitutes spherical nodules with irregular and pitted surfaces showing a remarkable morphological resemblance to living fungal *sclerotia* (reproductive hyphae). The amorphous coal-like thucholite is envisaged by Hallbauer and van Warmelo (1974) as due to exogenous plant debris washed into its present position and there attacked by a fungal organism.

The paper by Hallbauer and van Warmelo (1974) marks a major advance in our understanding of the carbon of the Rand, meriting reproduction in full as Paper 14-7.

(Text continues on page 453.)

14-7: FOSSILIZED PLANTS IN THUCHOLITE FROM PRECAMBRIAN ROCKS OF THE WITWATERSRAND, SOUTH AFRICA

D. K. Hallbauer and K. T. Van Warmelo

ABSTRACT

Hallbauer, D.K. and van Warmelo, K.T., 1974. Fossilized plants in thucholite from Precambrian rocks of the Witwatersrand, South Africa. Precambrian Res., 1: 199—212.

Examination of columnar carbonaceous material or thucholite from gold-bearing conglomerate revealed internal structures morphologically resembling filamentous, branched and apparently septate cells of obvious biological origin partially encrusted with gold, as well as silicified structures which could be identified as primitive fungi. The columnar structures could be part of a differentiated, apparently symbiotic organism which formed carpet-like colonies of up to several square metres in extent. On the basis of the chemical composition and the apparent ability of the suggested organism to assimilate gold and other inorganic material, the apparent presence of a symbiotic alga and the assumed nature of the Precambrian environment, an organism is proposed that has many morphological similarities with lichens but has otherwise no known living equivalent.

INTRODUCTION

The earliest evidence of life on earth is found in unicellular microorganisms occurring in Early Precambrian carbonaceous sediments of the Fig Tree Group of the Swaziland Supergroup near Barberton in the Eastern Transvaal, South Africa (Pflug, 1966). These microorganisms, almost certainly of prokaryotic affinities, were relatively advanced, which consequently suggests that life began even earlier. The presence of eukaryotic microorganisms in the Late Precambrian, that is 0.9 b.y. ago, is firmly established. The occurrence of some older microfossils, however, has suggested the first appearance of eukaryotic organisation near the beginning of the Late Precambrian, about 1.7 b.y. ago (Schopf, 1972).

The rocks of the Witwatersrand Group in South Africa, which are approximately 2.3—2.7 b.y. old (Allsopp, 1964; Van Niekerk and Burger, 1965) yielded only microfossils of prokaryotic affinities and "chemical" fossils (Schopf, 1972). The carbonaceous matter, or thucholite, which is commonly found in a number of gold-bearing reef deposits of the Witwatersrand group, at or close to the peneplaned deposition floor, has been recognized as being of organic origin (Liebenberg, 1955; Snyman, 1965; Hoefs and Schidlowski, 1967; Prashnowski and Schidlowski, 1967; Oberlies and Prashnowski, 1968; Schidlowski, 1968; Plumstead, 1969) but no definite cellular structures have been proved.

It is thought that the geological environment was that of a composite environment at the edge of a tectonic basin including lagoonal facies and the shallow interface between an arid land surface and the basin (Sharpe, 1949; Brock and Pretorius, 1964).

This paper describes studies of structures found in carbonaceous material which occurs in gold-bearing reefs and an attempt is made to determine its origin and nature.

Studies have recently been made of the shape and size of gold particles in the Carbon Leader Basal Reefs (Hallbauer and Joughin, 1972). In these studies the gold was separated from the reef by dissolution of the rock in hydrofluoric acid.

Residues from this treatment were occasionally found to contain large amounts of apparently undistorted carbonaceous material. The internal constitution of these particles was studied extensively by means of a scanning electron microscope.

Although some surface structures could be developed by treatment of the carbonaceous material in hot concentrated perchloric acid, better results have been obtained by complete oxidation of the combustible matter at about 500° C, which left behind only the inorganic substructure. This was examined by optical microscopy as well as in the scanning electron microscope.

Gold particles of predominantly fibrous shape which were found to occur within this inorganic substructure were extracted from the oxidized mineral matrix by repeated washing with hydrochloric acid and examined in the scanning electron microscope for structural details.

Apart from the carbonaceous material itself, three other of its constituents, uranium oxide, silica and gold, were found to have preserved biogenic structures in varying degrees of preservation.

STRUCTURE OF THE CARBONACEOUS MATERIAL

Most of the carbonaceous material in the Witwatersrand reefs, which is the source of all structures discussed in this paper, occurs as individual seams which range in thickness from one to several millimetres and in length from a few centimetres to a few metres, and very rarely with a thickness of a few centimetres.

In most instances the seams have been found to consist of aggregates of vertical columns of coal-like material about 0.5—5 mm long and 0.2 mm in diameter. After having been cleaned in hydrofluoric acid some better-preserved specimens show a distinct base of carbonaceous material from which the columns appear to have grown (Fig.1). Close examination of the surfaces reveals occasional individual fibres or bundles of fibres parallel to the long axis of the column. This feature was, however, often masked by small slickensides caused by small movements within the rock. The lower portions of the columns appear to be slightly tapered towards the base. The rounded upper ends have pitted surfaces and, sometimes, a slightly frayed appearance. Fracture surfaces are pitch-like in appearance and, at magnifications of up to 10,000, show no internal structural details.

The second type of carbonaceous material, known as "fly-speck" carbon, consists of spherical nodules ranging in diameter from 0.2 to 1 mm. These nodules occur separately in gold-bearing conglomerate above the layers of columnar carbon or in association with them. Their surfaces are irregular and pitted (Figs.2 and 3) and they show a remarkable morphological resemblance to living fungal *sclerotia* (Fig.4). Some carbon nodules after oxidation have been found to contain particles of gold having diameters ranging from 0.5 to 1 μm.

It was principally the columnar carbon which was examined for its contents of amino acids (Prashnowski and Schidlowski, 1967; Oberlies and Prashnowski, 1968). The latter report a distribution of carbohydrates which, together with a Pentose/Hexose ratio of 1:1, points to algae. The biogenic nature of the columnar carbon is underlined by its internal structure revealed in X-ray photographs and slow oxidation of specimens at about 500°C.

Oxidation left the outer surfaces of the individual columns virtually intact and rendered further details of the inner and outer structures discernible, so that it has been possible to distinguish between a thin membrane-like structure surrounding each column and an irregular supporting framework of fibres within it (Fig.5). This structure was not a local phenomenon but could be observed in specimens from widely separated localities. Chemical analysis has shown that the material obtained after oxidation by heat consists mainly of uranium and thorium oxide, silica, titanium oxide and traces of rare earths. Samples from different localities showed only insignificant fluctuations.

The membrane-like material enclosing each columnar particle, apparently less than 1 μm thick, is perforated at irregular intervals by holes of about 1 μm in diameter. Occasionally a number of holes are connected to form longer chain-like series.

Examination by optical microscopy of fragments from oxidized specimens usually revealed membrane-like material without much detail, and bundles of more or less parallel, often septate, fibres (Fig.6). Detailed examination of these fibres sometimes revealed cell-like structures about 1.5 μm in diameter. The cells range in shape from almost spherical to elongated. Of particular interest is what appears to be true branching of the fibres (Fig.6). The colour of these fibres is a greenish-brown, suggesting, considering the radioactivity of the specimens and their composition, a replacement of the original biogenic structures by uranium oxide.

Small spherical and fibre-like gold particles, some of which formed aggre-

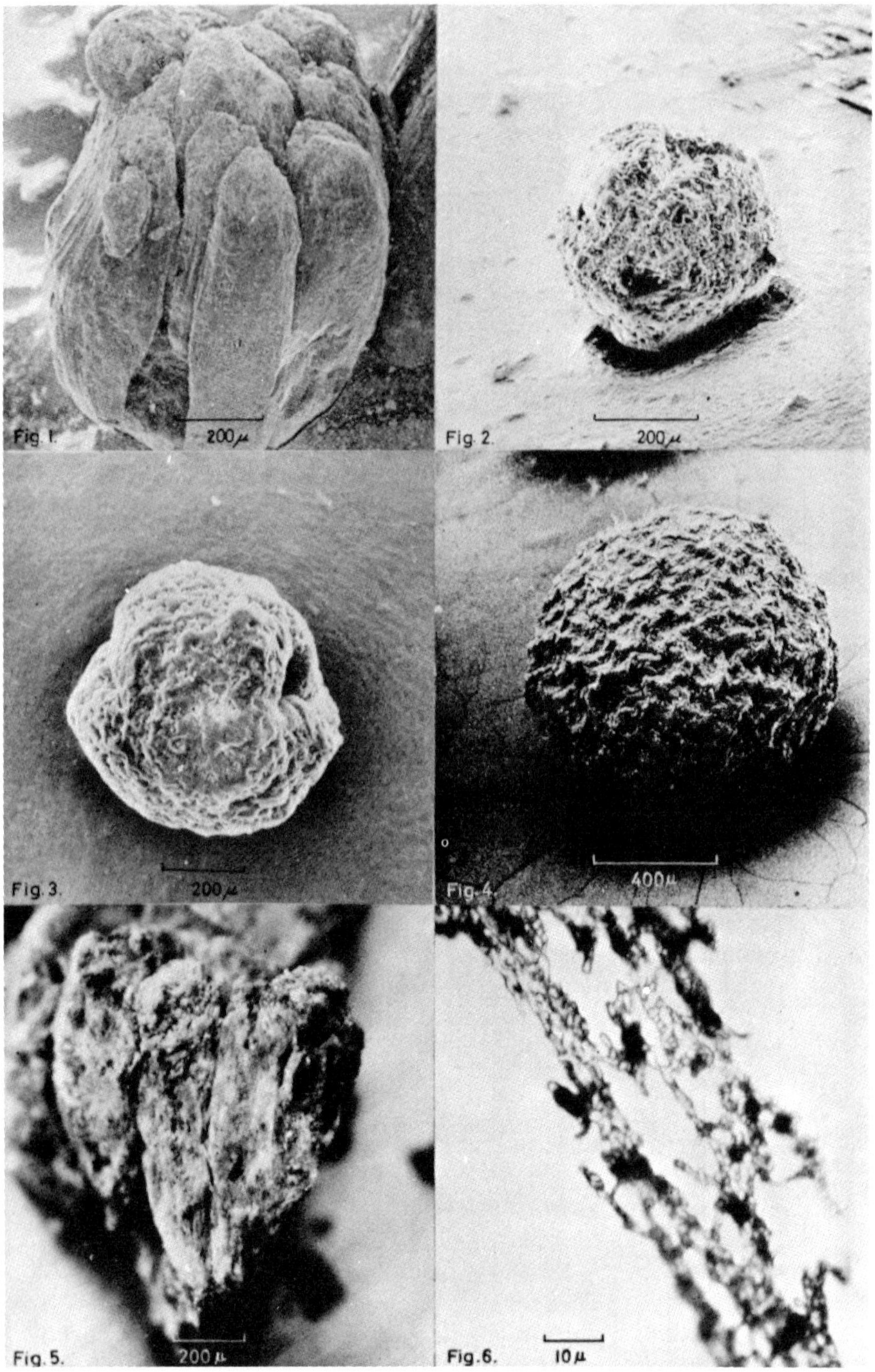
Fig. 1.
200μ
Fig. 2.
200μ
Fig. 3.
200μ
Fig. 4.
400μ
Fig. 5.
200μ
Fig. 6.
10μ

gates, have been observed on both the outer and inner sides of the columns.

Stereo X-ray photographs (Hamblin, 1971) of rock slabs about 2–3 cm thick containing columnar thucholite bands were made to determine the relationship between the quartzite rock and the carbonaceous material. A typical example given in Fig.7 shows the columnar thucholite, outlined by numerous vertical gold fibres, to be positioned on a rough footwall. A layer of small rounded pyrite grains, zircon and patches of gold grains can be observed microscopically in the immediate footwall contact, suggesting a heavy mineral sand on which the assumed plant life was growing and is now preserved in situ. The structure of the fibrous gold particles inside the columnar carbonaceous material is totally different from gold particles found in the footwall contact or in typical sedimentary reef deposits, as will be discussed later.

Silicified structures

Apart from the columnar material occasional carbonaceous seams of up to 5 cm thickness are found which have the typical amorphous appearance of coal. The oxidation of specimens appears to confirm the first impression. The grainy, irregularly structured ash rather looks like debris washed from elsewhere into the present location. Fibrous gold particles oriented at random occur in varying amounts. As a new component white silica fibres can occasionally be observed in nest-like agglomerations in the ash which, so far, has been observed only in samples of Basal Reef from the St. Helena Gold Mine. Fresh samples as well as older samples from collections show the same phenomenon.

Generally these white fibres have the appearance of fungal growth. Agglomeration of fibres, as shown in Fig.8, can frequently be found inside the ashed material, while single fibres normally abound throughout the specimens.

Microscopic examination of individual, as well as "nests" of fibres, shows structures which have all the characteristics of hyphae and were identified as

---

Fig.1. Scanning electron photomicrograph of a group of columnar, carbonaceous particles from Carbon Leader Reef, Western Deep Levels Gold Mine, Carletonville, Transvaal.

Fig.2. Spherical nodule of carbonaceous matter extracted from a sample of Basal Reef, St. Helena Gold Mine, Welkom, Orange Free State. Side view in scanning electron microscope.

Fig.3. Top view of a spherical, carbonaceous nodule showing detail of surface structure. Origin as in Fig.2.

Fig.4. Scanning electron photomicrograph of *Sclerotium rolfsii.*

Fig.5. Photomicrograph of a group of columnar, carbonaceous particles after oxidizing at 500° C. Carbon Leader Reef, Western Deep Levels Gold Mine, Carletonville, Transvaal.

Fig.6. Photomicrograph of fibrous material from oxidized carbonaceous matter showing septate fibres and true branching. Western Deep Levels Gold Mine, Carletonville, Transvaal.

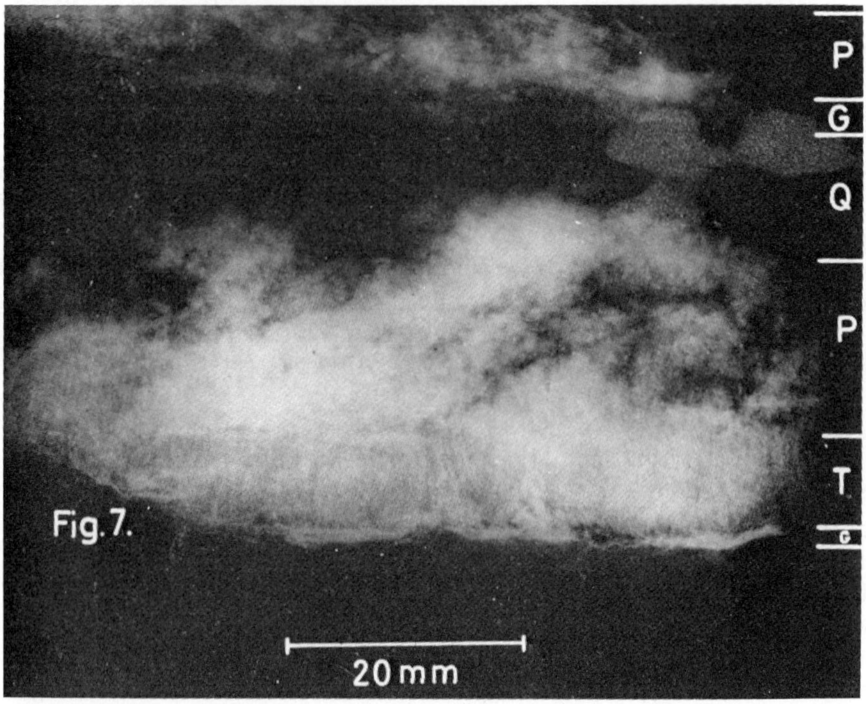

Fig.7. X-ray photograph of a specimen of Basal Reef, St. Helena Gold Mine, showing a deposit of heavy minerals rich in gold (G, white) on a rough footwall, followed by a layer of columnar thucholite (T) containing a mass of vertical, thin gold fibres. This is followed by successive layers of quartz and pyrite pebbles (P), quartzite (Q), heavy minerals with gold grains (G) and pyrite pebbles (P).

such*. Fig.9 shows the typical appearance of an agglomerate of fibres. The average diameter ranges from 1—2 μm while that of individual fibres with a particularly thick silica encrustation can reach 4 μm across. Most of the fibres appear to consist of solid silica. Occasionally, however, hollow structures could be observed (Fig.10). The detailed structure of the silica often resembles scales, easily observed in microscopic preparations under polarized light. A feature strongly suggestive of an affinity with extant Fungi Imperfecti, are conidium-like structures. No attempt at closer classification has yet been made. The conidium-like bodies are mostly elongated with a width to length ratio of 1:3 (Fig.11). Sometimes structures suggesting complete conidiophores were found (Fig.12). The cells of the silicified fungi were best observed by optical microscopy. In order to avoid fracturing of the very brittle and fragile material, microscope slides with a thin layer of "lakeside cement" were prepared

*Prof. D. Gottlieb, Dept. of Plant Pathology, University of Illinois. Verbal opinions expressed after examination of prepared material in the SEM.

and the fibres placed carefully on the cement. Gentle heating of the slides caused the fibres to sink into the cement without disturbing their structure. After covering in the normal way they were ready for examination.

The hyphae were clearly visible under bright-field illumination (Figs.13,14) showing a variety of elongated and short cells. The internal structure of the individual cells could sometimes be vaguely observed, suggesting round to oval sub-structures without, however, much detail.

Gold particles showing biogenic structures

Most observations discussed in this paper have been made on gold particles separated from the oxidized columnar particles by gentle washing with hydrochloric acid. In many instances structures similar to those in the silicified fibres have been found. Gold particles separated from normal conglomerate reef by hydrofluoric acid extraction (Hallbauer and Joughin, 1972) had forms typical of those found in sedimentary deposits, i.e., plate-like, round, and nugget-like or irregular recrystallized grains but never the filamentous forms observed in the inner structure of the thucholite. Occasionally plate-like or nugget-like gold particles were found to be attached to the outside of columnar thucholite and between columns, but were never part of the internal structure.

Filamentous gold particles can occur naturally. During the investigation of a hydrothermal nickel deposit in Eastern Turkey (D.K. Hallbauer, unpubl.), needle-like gold particles were observed as inclusions in niccolite. After dissolution of the niccolite in nitric acid samples were examined in the scanning electron microscope. The particles were 1—2 μm in diameter with a width to length ratio of sometimes 1:20, had smooth faces and appeared to be distorted and elongated octahedrons.

Gold from the internal structure of the carbonaceous material has been observed as single fibres protruding from a comparatively undifferentiated surface with an actual outer cell structure replicated in gold (Figs.15,16). The diameters of such apparently septate fibres ranged from 1.5 to 2 μm. Branching of the fibres is a common feature. The surfaces of some fibrous gold particles at higher magnification have a grained appearance (Fig.17), not unlike the inorganic crusts on hyphae in lichens (Fig.18). The ends of all fibres protruding from the more massive gold particles normally are closed, suggesting a total replacement of the fibre by gold.

While the abovementioned structures are characteristic of samples of the Carbon Leader Reef, structures which show the negatives of fibres as well as replicas have been observed in samples of Basal Reef and Intermediate Reef. Irregular fibrous and massive structures of gold are perforated by round to slightly oval holes ranging in size from 0.5 to 1 μm (Figs.19 to 21). Bundles of apparently septate fibres which often display true branching have similar dimensions.

The characteristic structural features of gold particles extracted from the carbonaceous matter, where the gold appears to have been deposited in the

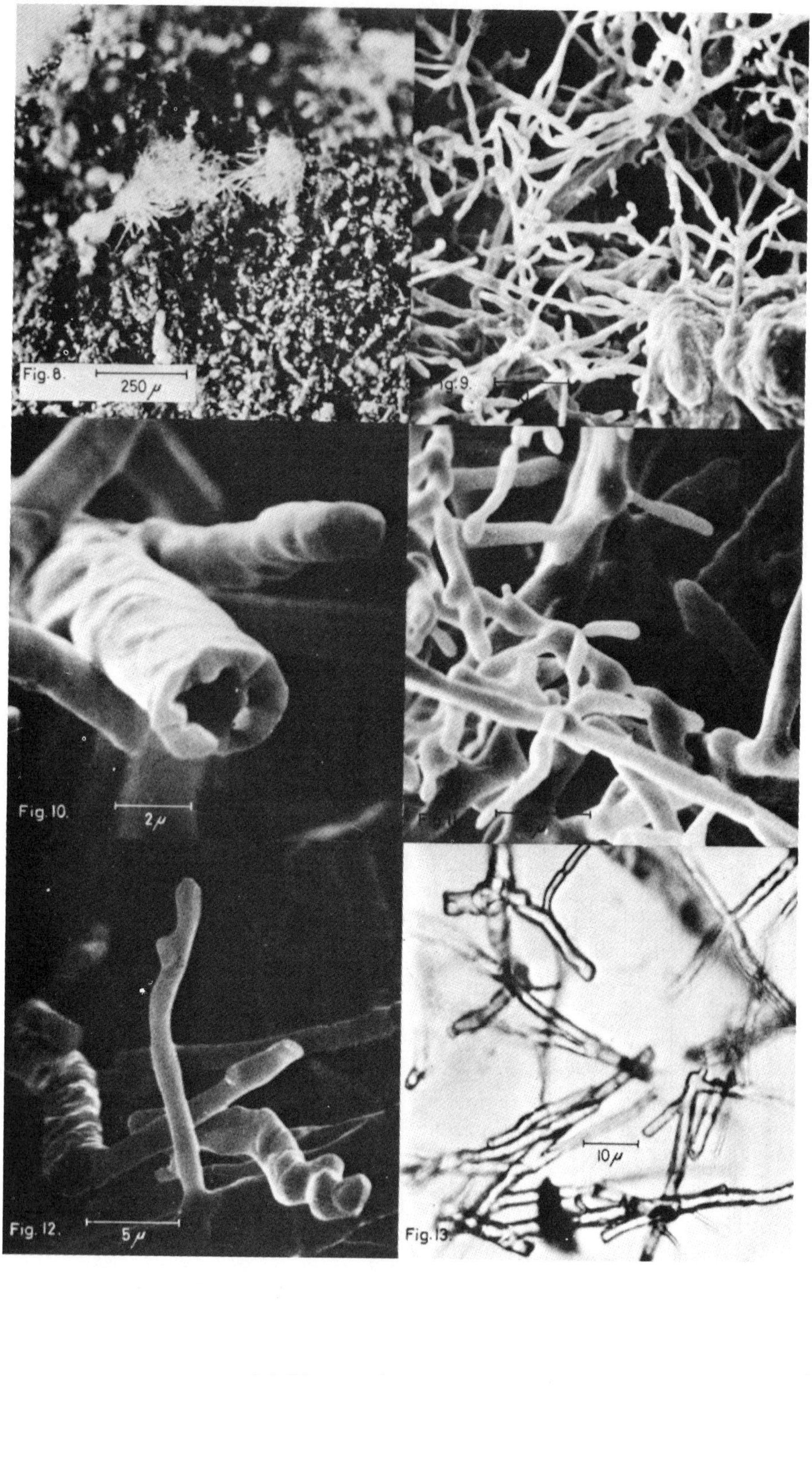
Fig. 8.
250 μ
Fig. 10.
2 μ
10 μ
Fig. 12.
5 μ
Fig. 13.

interstitial spaces in a fibrous framework, or to have replaced or coated single fibres, was very prominent in samples of Intermediate Reef from the St. Helena Gold Mine. In some cases hollow remnants of filaments protruding from the more or less amorphous gold mass (Fig.21) or parts of filamentous tubes (Fig.20) have been observed.

A reference specimen of a crustose lichen growing on rocks in the Namib Desert showed similar structures, especially in those parts of the thalli which are close to the rock surface. Here the hyphae are embedded in an amorphous, gel-like material with a very high silica content (Fig.22). A similar ability of a lichen thallus to accumulate inorganic material, especially radioactive and heavy metals, has been reported by Hale (1973).

Still another morphological similarity to lichens can be observed occasionally on the surface of some gold particles. These are short, rounded stubs, sometimes with basal constrictions, protruding from the irregularly formed gold particles (Figs.23,24), similar to those found on the surface of the cortex of a crustose lichen (Fig.25).

DISCUSSION

When considering the chemical composition of the carbonaceous material or thucholite little doubt remains about its biogenic nature (Prashnowski and Schidlowski, 1967; Oberlies and Prashnowski, 1968; Schidlowski, 1968). Taking the morphology and internal structure into consideration, we can visualize the columnar particles representing the fossilized remains of a Precambrian plant. As the columnar form is so consistent over large distances and is usually in an upright position an exogenous origin can be ruled out. Columnar thucholite in a horizontal position has not been observed, except for odd particles.

The upright position of the columnar particles after diagenesis points to a tough, leathery plant texture which appears to rule out structures normally encountered in algae. In reconstructing the original organism a carpet-like colony of columnar individuals can be envisaged, each about 0.2—0.5 mm in diameter and up to 7 mm in length, with a membranous outer covering and irregularly arranged fibres within. Gold and other materials were apparently

---

(Figs.9—12 scanning electron photomicrographs)

Fig.8. Photomicrograph of oxidized "amorphous" thucholite showing a nest of white silicified fibres. Basal Reef, St. Helena Gold Mine, Welkom, Orange Free State.

Fig.9. Agglomerate of silicified fibres showing the characteristic appearance of mycelium. Extracted from Basal Reef, St. Helena Gold Mine.

Fig.10. Single silicified, hollow filament showing details of the internal structure and surface. Basal Reef, St. Helena Gold Mine.

Fig.11. Silicified filament with conidium-like structures, Basal Reef, St. Helena Gold Mine.

Fig.12. Silicified filaments with conidiaphore-like structures. Basal Reef, St. Helena Gold Mine.

Fig.13. Photomicrograph of silicified filaments showing septate fibres and branching. Basal Reef, St. Helena Gold Mine.

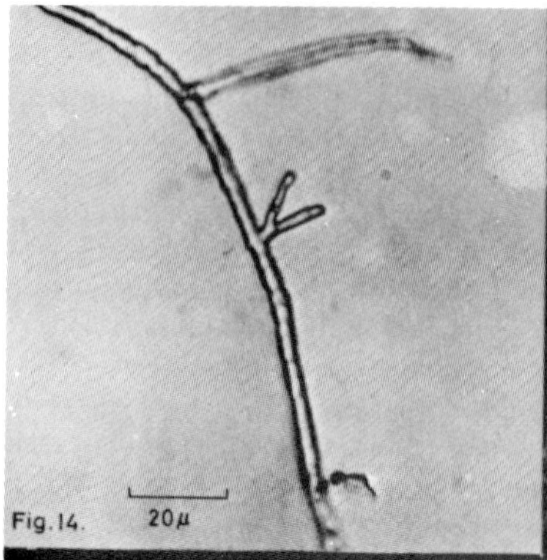

Fig. 14.

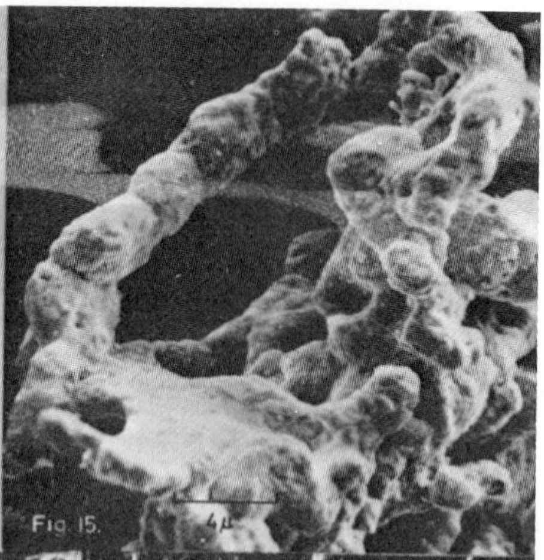

Fig. 15.

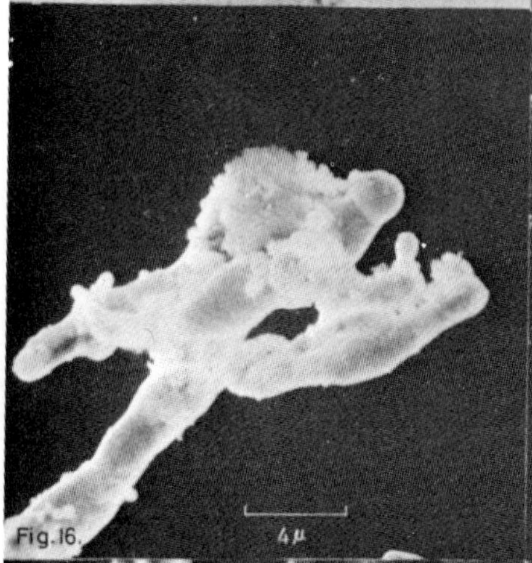

Fig. 16.

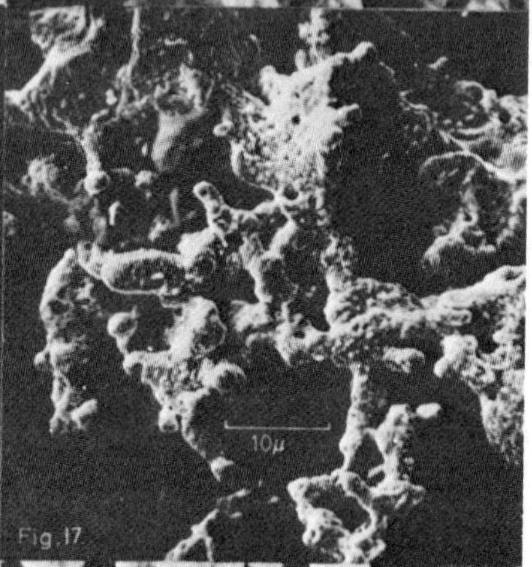

Fig. 17.

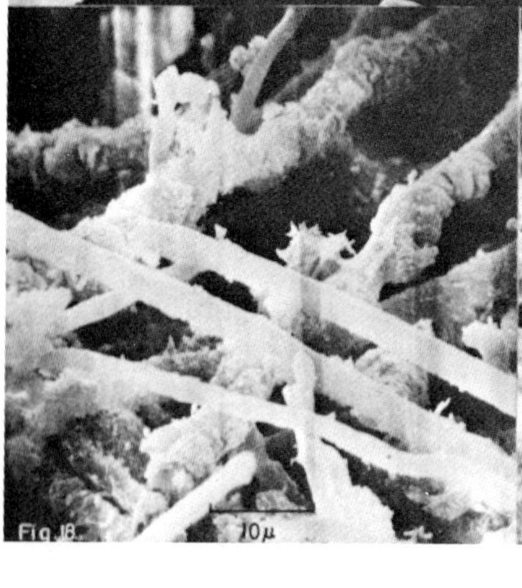

Fig. 18.

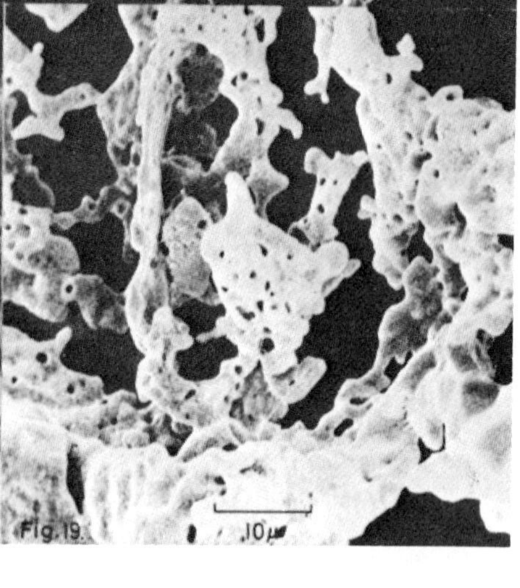

Fig. 19.

extracted from the environment by the organisms and deposited inter- and intracellularly. This process, as well as the structural details, points to fungal activity while chemical evidence appears to favour a photosynthesizing plant. Fungi have been reported to extract gold from colloidal solutions with the metal stored in the cell walls and spores (Williams, 1918). The mechanism of fungal attack on rock and minerals appears to be one in which the organism produces acid which then attacks these materials (Silverman and Minoz, 1970). The existence of fungi in Precambrian times appears to be confirmed, as shown earlier.

These considerations lead to the postulation of a symbiotic association which had evolved at Precambrian times, similar to existing lichens. The concept of an association of photosynthetic algae with a non-photosynthetic filamentous organism in Precambrian times has been advanced previously (Echlin, 1966). Photosynthesis of the algal component makes the organism almost independent of an external oxygen supply. This would be of importance in an atmosphere of low oxygen content as generally assumed for the Precambrian and supported by the presence of water-transported pyrite in the mineral assembly of the Witwatersrand reefs, as observed by the authors. Lichens, as the nearest living example of such an organism are also known to absorb large quantities of mineral matter which are then redeposited around the hyphae. Hale (1967) mentioned three factors in connection with ion absorption and deposition of insoluble inorganic material in the thallus, namely (a) unspecified cytoplasmic resistance to metallic ions inherent to lichenized fungi and algae, (b) immobilization of the ions within the cytoplasm by means of chelators or other metalbinding substances, and (c) active and passive transport of the ions to regions external to the plasma and cell wall. Hale further describes extracellular deposition in *Acarospora sinopica* (lichen) as a thick surface layer of iron salts.

The redepositioning of metallic salts appears to be confined to the fungal

---

(Figs.15—19 scanning electron photomicrographs)

Fig.14. Photomicrograph of single silicified filament showing faint cell-like divisions and branching. Basal Reef, St. Helena Gold Mine, Welkom, Orange Free State.

Fig.15. Part of a gold particle extracted from oxidized carbonaceous matter, showing a single fibre with septae-like constrictions protruding from a mass of irregular fibres. Basal Reef, St. Helena Gold Mine, Welkom, Orange Free State.

Fig.16. A gold particle extracted from oxidized carbonaceous matter, showing septae-like constrictions and unidentified spherical particles on surface of gold fibre. West Driefontein Gold Mine, Carletonville, Transvaal.

Fig.17. Part of a gold particle extracted from carbonaceous material showing septae-like constrictions and branching of fibres. Basal Reef, St. Helena Gold Mine, Welkom, Orange Free State.

Fig.18. Hyphae from the medulla of a crustose lichen showing thick encrustations of inorganic material.

Fig.19. Gold extracted from oxidized carbonaceous material showing fibres exhibiting branching and septae-like constrictions as well as negatives of fibres in irregularly formed gold particle. Western Holdings Gold Mine, Welkom, Orange Free State.

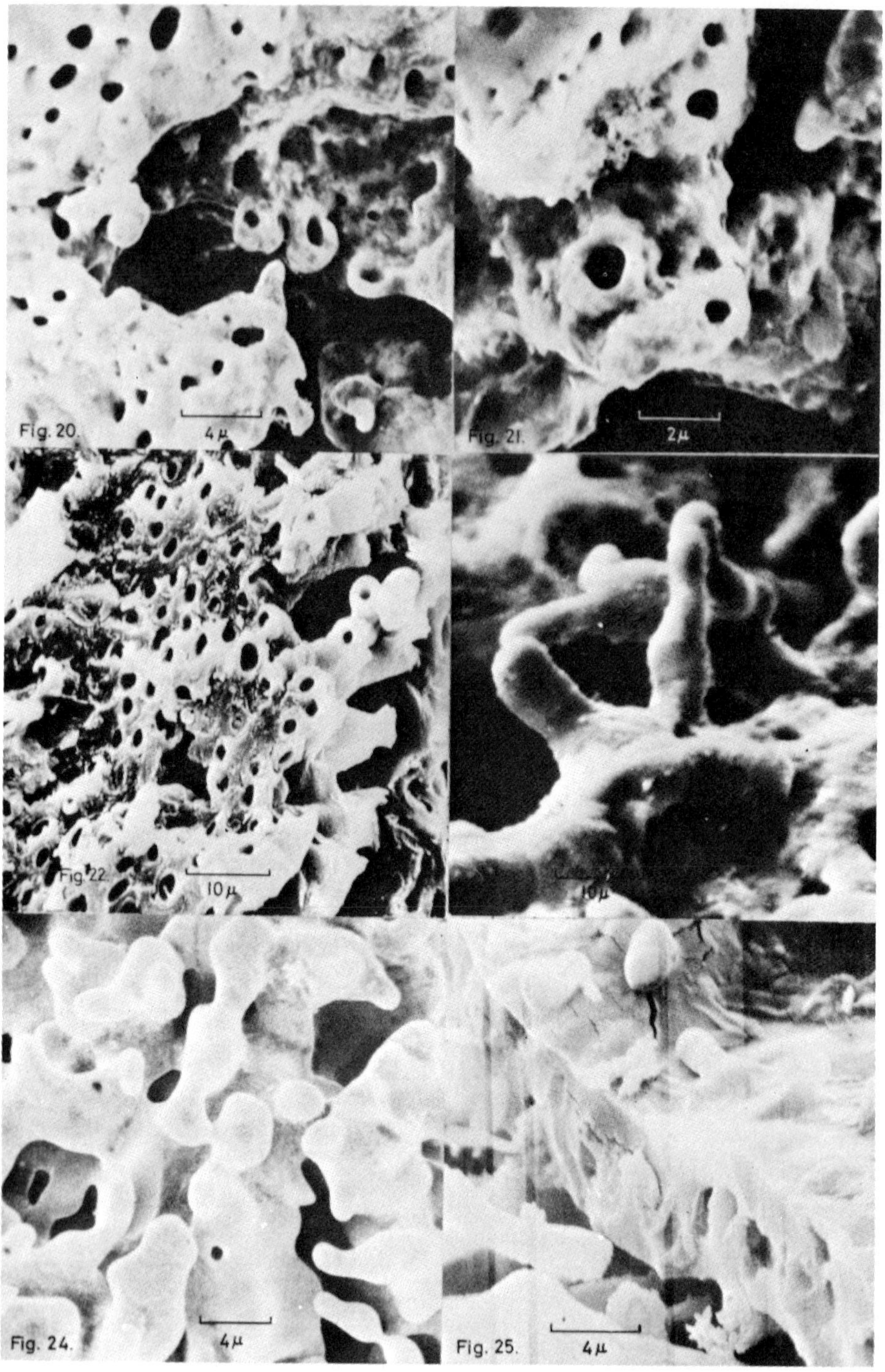
Fig. 20.
4µ
Fig. 21.
2µ
Fig. 22.
10µ
10µ
Fig. 24.
4µ
Fig. 25.
4µ

symbiont only. One would, therefore, be unlikely to find remnants of the algal partner in the type of fossilized material as described above. Even in recent lichen material the algal symbiont is quite inconspicuous when viewed in the scanning electron microscope. Structures related morphologically to the fossilized material, however, can be observed frequently (Hale, 1973).

The membranous covering in the fossilized material can be seen as being similar to the outer cortical layer in existing lichens which serves as a protective covering and consists of compressed, sometimes parallel, heavily gelatinized hyphae cemented firmly together. The bulk of a lichen thallus consists of hyphal threads which are interwoven irregularly in a loose, often fibrous or cottony mass with little compression. When living lichens are heated to about 500°C, a thin membrane remains, representing the outer cortical layer similar to the membrane-like material which can be observed after oxidation of the combustible material of the columnar carbonaceous material.

It has been observed in polished sections that the carbonaceous material apparently "digests" detrital uraninite (Liebenberg, 1955). The high gold and uranium content of the fossilized material therefore points to a process of active assimilation of these substances, a process similar to the dissolution and accumulation from the substrate of radioactive and other minerals by lichens (Hale, 1967).

In conclusion, we advance the theory that the columnar carbonaceous material is the fossilized residue of a Precambrian symbiotic association, which probably consisted of an algal partner and a fungal organism, and occupied areas ranging from a few square centimetres to a few square metres on bare rock or gold-bearing heavy mineral sand. The *sclerotium*-like "fly-speck" carbon could have been a reproductive mode. The immediate surroundings of these plants could have been partially submerged in a lagoonal facies, at the interface between an arid land surface and a shallow basin or on the dry parts in a braided river system. This supposition is supported by geological evidence (Sharpe, 1949; Brock and Pretorius, 1964).

The amorphous coal-like thucholite on the other hand can be seen as exogenous plant debris washed into the present location and there attacked by a fungal organism. The distance of transport can be taken as small as this type of carbonaceous matter is normally separated by not more than a few

---

(Figs.20—25 scanning electron photomicrographs)

Fig.20. Gold particle extracted from a carbon layer showing short hyphae-like tubes and faintly marked structures suggesting tubes enclosed in the amorphous matrix. Intermediate Reef, St. Helena Gold Mine, Welkom, Orange Free State.

Fig.21. Tubular features protruding from the gold matrix. Same sample as above.

Fig.22. Medulla of a crustose lichen showing hyphae in a gel-like matrix of organic material and amorphous silica. Namib Desert, South West Africa.

Fig.23. Surface of a gold particle from carbonaceous material showing short, apparently septate stubs protruding from outer surface of pore-like cavity. Basal Reef, St. Helena Gold Mine, Welkom, Orange Free State.

Fig.24. Surface of gold particle with short, rounded stubs occasionally showing septae-like constrictions. Western Holdings Gold Mine, Welkom, Orange Free State.

Fig.25. Cortex and medulla of a crustose lichen showing short stubs of hyphae protruding from the cortex.

metres from the columnar material. Only the "fly-speck carbon" is often found by itself in otherwise pure pebble reefs. The age of the carbonaceous material examined has important implications on the dating of the origin of life. The proposal that symbiotic associations were formed in the Precambrian means that the more differentiated organisms must have formed much earlier than has hitherto been believed. Although the organism appears to have many morphological and perhaps physiological similarities with lichens no known living equivalent can be proposed until more detailed investigations have been made.

REFERENCES

Allsopp, H.L., 1964. Rubidium/strontium ages from the Western Transvaal. Nature, 204 (4956): 361–363.

Brock, B.B. and Pretorius, D.A., 1964. The Geology of Some Ore Deposits in Southern Africa. Geol. Soc. S. Afr. (Publ.), pp.544–599.

Echlin, P., 1966. Origins of photosynthesis. Science Journal, 2: 42–47.

Hale, M.E., 1967. The Biology of Lichens. Edward Arnold, London, pp.59–61, 86.

Hale, M.E. 1973. Fine structure of the cortex in the lichen family *Parmeliaceae* viewed with the scanning-electron microscope. Smithson. Contr. Bot., 10: 15, 16, 54.

Hallbauer, D.K. and Joughin, N.C., 1972. Distribution and size of gold particles in the Witwatersrand reefs and their effects on sampling procedures. Trans. IMM, Lond., 81 (A): 113–143.

Hamblin, W.K., 1971. X-ray photography. In: R.E. Carver (Editor), Procedures in Sedimentary Petrology. Wiley Interscience, New York, N.Y., 653pp.

Hoefs, J. and Schidlowski, M., 1967. Carbon isotope composition of carbonaceous matter from the Precambrian of the Witwatersrand System. Science, 155: 1096–1097.

Liebenberg, W.R., 1955. The occurrence and origin of gold and radioactive minerals in the Witwatersrand system, the Dominion Reef, the Ventersdorp Contact Reef and the Black Reef. Trans. Geol. Soc. S. Afr., 58: 101–254.

Oberlies, F. and Prashnowski, A.A., 1968. Biogeochemische und elektronenmikroskopische Untersuchung präkambrischer Gesteine. Naturwissenschaften, 55: 25–28.

Pflug, H.D., 1966. Structural organic remains from the Fig Tree series of the Barberton Mountain Land. Econ. Geol. Res. Unit Witwatersrand Univ., Johannesburg, Inform. Circ., 28,1.

Plumstead, E.P., 1969. Three thousand million years of plant life in Africa. Trans. Geol. Soc. S. Afr., Annex., 72, A.L. du Toit Memorial Lectures, 11.

Prashnowski, A.A. and Schidlowski, M., 1967. Investigation of Pre-Cambrian Thucholite. Nature, 216: 560–563.

Schidlowski, M., 1968. Critical remarks on a postulated genetic relationship between Precambrian thucholite and boghead coal. In: P.A. Schenck and I. Havenaar (Editors), Advances in Organic Geochemistry. Proc. 4th Int. Meet., Amsterdam. Pergamon, Oxford, pp.579–592.

Schopf, J.W., 1972. In: C. Ponnamperuma (Editor), Exobiology. North Holland, Amsterdam, pp.16–61.

Sharpe, J.W.N., 1949. The economic auriferous bankets of the Upper Witwatersrand Beds and their relationship to sedimentation features. Trans. Geol. Soc. S. Afr., 52: 265–288.

Silverman, M.P. and Minoz, E.F., 1970. Fungal attack on rock: solubilization and altered infrared spectra. Science, 169: 985–987.

Snyman, C.P., 1965. Possible biogenetic structures in Witwatersrand thucholite. Trans. Geol. Soc. S. Afr., 63: 225–235.

Van Niekerk, C.B. and Burger, A.J., 1964. The age of the Ventersdorp System. Ann. Geol. Survey S. Afr., 3: 75–86.

Williams, M., 1918. Absorption of gold from colloidal solutions by fungi. Ann. Bot., 32: 531–534.

OTHER DEPOSITS

One would be remiss in not mentioning the large amount of work that has been done on the auriferous and uraniferous quartz-pebble conglomerates and quartzites other than those of the Witwatersrand. Here can be mentioned the auriferous conglomerates of the Tarkwaian system of Ghana, the auriferous and uraniferous conglomerates of the Jacobina series, Bahia, Brazil, and the uraniferous conglomerates of the Elliot Lake-Blind River district, Ontario, Canada. The origin of all these deposits has, likewise, been a bone of much contention since their discovery. Both modified placer and hydrothermal processes have been vociferously advocated for their origin, and the pros and cons supporting each are essentially similar to those stated for the Rand. The interested reader is referred to the detailed reports by Bray (1928) and Junner, Hirst, and Service (1942) on the Tarkwaian of Ghana; White (1961), Cox (1967), and Gross (1968) on the Jacobina series of Bahia; and Davidson (1965) and Roscoe (1969) on the Elliot Lake deposits. Their views and those of numerous others have been summarized by Boyle (1979).

REFERENCES

Becker, G. F., 1897. The Witwatersrand Banket, with notes on other gold-bearing pudding stones, *U.S. Geol. Survey 18th Ann. Rept.*, pt. 5, pp. 153-184.

Boyle, R. W., 1979. The geochemistry of gold and its deposits, *Canada Geol. Survey Bull. 280,* 584p.

Bray, A., 1928. Notes on the banket reefs of the Gold Coast Colony, *Inst. Min. Metall. Trans.* **38:**21-69.

Cox, D. P., 1967. Regional environment of the Jacobina auriferous conglomerates, Brazil, *Econ. Geology* **62:**773-780.

Davidson, C. F., 1953. The gold-uranium ores of the Witwatersrand, Mining Mag., **88:**73-85.

Davidson, C. F., 1965. The mode of origin of banket orebodies, *Inst. Min. Metall. Trans.* **74:**319-338.

Davidson, C. F., and S. H. U. Bowie, 1950. On thucholite and related hydrocarbon-uraninite complexes, *Great Britain Geol. Survey Bull. 3,* p. 1-19.

De Kock, W. P., 1964. The geology and economic significance of the West Wits line, in *The Geology of Some Ore Deposits in Southern Africa,* S. H. Haughton, ed., vol. 1, Geol. Soc. South Africa, Johannesburg, pp. 323-386.

de Launay, L., 1913. *Gîtes mineraux et métallifères;* ts. 1, 2, 3; C. H. Béranger, Editeur, Paris.

Garlick, W. G., 1953. Reflections on prospecting and ore genesis in northern Rhodesia; *Inst. Min. Metall. Trans.* **63** (563):9-20.

Graton, L. C., 1930. Hydrothermal origin of the Rand gold deposits: Part 1. Testimony of the conglomerates, *Econ. Geology* **25** (suppl. to no. 3):185p.

Gregory, J. W., 1907. The origin of the gold in the Rand Banket, *Inst. Min. Metall. Trans.* **17:**2-41.

Gross, W. H., 1968. Evidence for a modified placer origin for auriferous conglomerates, Canavieiras mine, Jacobina, Brazil, *Econ. Geology* **63:**271-276.

Hallbauer, D. K., and K. T. van Warmelo, 1974. Fossilized plants in thucholite from Precambrian rocks of the Witwatersrand, South Africa, *Precambrian Res.* **1:**199-212.

Hatch, F. H., and G. S. Corstorphine, 1905. *The Geology of South Africa,* Macmillan, London, 348p.

Haughton, S. H., ed., 1964. *The Geology of Some Ore Deposits in Southern Africa,* vol. 1, Gold deposits of the Witwatersrand Basin; Geol. Soc. South Africa, Johannesburg, 625p.

Hoefs, J., and M. Schidlowski, 1967. Carbon isotope composition of carbonaceous matter from the Precambrian of the Witwatersrand System, *Science* **155:**1096-1097.

Hoefs, J., H. Neilsen, and M. Schidlowski, 1968. Sulfur isotope abundances in pyrite from the Witwatersrand conglomerates, *Econ. Geology* **63:**975-977.

Horwood, C. B., 1917. *The Gold Deposits of the Rand,* Charles Griffin & Co., London, 400p.

Junner, N. R., H. Service and T. Hirst, 1942. The Tarkwa Goldfield, *Gold Coast Geol. Survey Mem. 6,* 75p.

Liebenberg, W. R., 1955. The occurrence and origin of gold and radioactive minerals in the Witwatersrand System, the Dominion Reef, the Ventersdorp Contact Reef and the Black Reef, *Geol. Soc. South Africa Trans.* **58:**101-227.

Mellor, E. T., 1916. The conglomerates of the Witwatersrand, *Inst. Min. Metall. Trans.* **25:**226-291.

Miholic, S., 1954. Genesis of the Witwatersrand gold-uranium deposits, *Econ. Geology* **49:**537-540.

Penning, W. H., 1888. The South African goldfields, *Royal Soc. Arts Jour.* **36:**433-444.

Prashnowsky, A. A., and M. Schidlowski, 1967. Investigation of Precambrian thucholite, *Nature* **216:**560-563.

Pretorius, D. A., 1975. The depositional environment of the Witwatersrand goldfields: A chronological review of speculations and observations, *Minerals Sci. Eng.* **7**(1):18-47.

Reimer, T. O., 1975. The age of the Witwatersrand system and other gold-uranium placers: Implications on the origin of the mineralization, *Neues Jahrb. Mineral. Monatsh.* **2:**79-98.

Roscoe, S. M., 1969. Huronian rocks and uraniferous conglomerates in the Canadian Shield, *Canada Geol. Survey Paper 68-40,* 205p.

Schidlowski, M., 1965. Probable life forms from the Precambrian of the Witwatersrand-System (South Africa), *Nature* **205:**895-896.

Schidlowski, M., 1966. Mineralbestand und Gefügebilder in Faseraggregaten von kohliger Substanz (Thucholith) aus den Witwatersrand-Konglomeraten, *Contr. Mineralogy and Petrology* **12:**365-380.

Snyman, C. P., 1965. Possible biogenetic structures in Witwatersrand thucholite, *Geol. Soc. South Africa Trans. and Proc.* **68:**225-235.

White, M. G., 1961. Origin of uranium and gold in the quartzite-conglomerate of the Serra de Jacobina, Brazil, *U.S. Geol. Survey Prof. Paper 424-B,* pp. 88-89.

CHAPTER

15

Gold Deposits—Placers

But gold which is formed in sands, as a kind of grains, larger or smaller, is formed from a hot and very subtle vapour, concentrated and digested in the midst of the sandy material, and afterwards hardened into gold.
Albertus Magnus, 1260

The large nuggets found in the drift are simply the reliquiae of the chief masses of gold which once occupied the uppermost parts of the reefs, and that like the blocks of many an ancient conglomerate, they have been swept from the hilltops into adjacent valleys by former great rushes of water.
Murchison, 1872

Placer deposits provided early man with the first samples of gold and thereafter have accounted for a large production of the metal. If we include the Witwatersrand and other quartz-pebble conglomerates as fossil placers or modified placers, the placer type of auriferous deposit has provided more than two thirds of the world's store of gold, about 80×10^9 grams (2.6×10^9 oz).

A number of treatises on auriferous and other types of placers have appeared occasionally since about 1890; of general interest are the works by Raeburn and Milner (1927), Gardner and Johnson (1934-1935), Wells (1973), Shilo (1981), and Macdonald (1983).

Auriferous placers are of three types—eluvial, alluvial, and aeolian. The first is formed in the weathered detritus (eluvium) at or near the outcrop of auriferous deposits; alluvial placers are formed in the sands and gravels of streams, rivers, beaches, and deltas; and the rare and unimportant aeolian placers accumulate in windblown sands and detritus usually near the outcrops of auriferous deposits. Fossil (lithified and poorly lithified) equivalents of the three types of placers are known. They are not considered further here; the interested reader can refer to Boyle (1979) for descriptions of these placers.

The terminology of the zone or stratum containing an economic concentration of gold in eluvial and alluvial placers is called the *pay streak*. The economic mineral in auriferous pay streaks is invariably native gold; some gold placers may contain recoverable amounts of cassiterite, tantalite-columbite, and gem stones. The pay streaks of gold placers usually rest on or near bedrock (the true bottom). When the streaks rest on a well-defined stratum of sand, gravel, shingle, or clay above the bedrock they are said to be on a "false bottom."

The gold in auriferous placers may come from one or more of the following sources:

1. Auriferous quartz veins and other types of gold-bearing deposits (e.g., skarn deposits)
2. Auriferous sulfide impregnation zones, porphyry copper deposits, and so on
3. Auriferous polymetallic deposits; massive sulfide deposits
4. Slightly auriferous quartz stringers, blows and veins in schists, gneisses and various other rocks
5. Various slightly auriferous minerals such as pyrite and other sulfides in graphitic schists and other rocks
6. Slightly auriferous conglomerates, quartzites, and other rocks
7. Auriferous laterites and other similar weathered products of weakly gold-bearing schists, shales, sandstones, greenstone terrains, and so on
8. Old (former) placers

For the development of placers of any type, four requisites are necessary:

1. The occurrence of gold in deposits, in widespread quartz veins and blows, or in a disseminated form in pyritic shales, other country rocks, laterites, and so on
2. A fairly long period of deep secular chemical and mechanical weathering on a surface of submature-to-mature topography, during which time the gold is set free from the deposits or country rocks
3. Concentration of the gold by some agency, generally gravity and water
4. Absence of extensive glaciation: Glaciation does not entirely preclude the occurrence of placers because both eluvial and alluvial types of placers may be overridden and little disturbed by the glaciers in some cases and buried by their deposits of till, clay, and so on in others. Such placers occur in Alaska, British Columbia, Yukon, Quebec, and the Lena district of USSR.

Country kindly to the occurrence of extensive placers is readily recognized. The topography is subdued and marked by broad, often terraced, entrenched valleys and rounded, deeply weathered hills commonly with nearly accordant summit levels. Few extensive placers are found in terrains marked by sharp alpine features and high-gradient V-shaped valleys; similarly, excessively flat terrains far from mountain systems and their foothills yield few productive placers. Like all generalities in geomorphology and geology there are some exceptions to these observations.

ELUVIAL PLACERS

Originally eluvial gold placers were considered to be represented by decomposed and disintegrated vein material overlying or only slightly removed from underlying auriferous deposits. Some slumping or downslope creep was often assumed, but in general the auriferous material was considered not to have been transported far from its source. In recent years eluvial placers have been further classified as eluvial, deluvial, and proluvial. By this classification eluvial placers are those whose outlines coincide more or less with those of the primary deposits. Deluvial (scree or talus) placers are those whose upper limit is at or near the primary source and whose downhill front lies at the foot of a slope. Proluvial placers form in the disintegrated debris and sediment in deltaiclike deposits at the foot of hills or mountains.

Despite their frequency few auriferous eluvial placers have been studied in detail. This circumstance probably stems from the general view that eluvial deposits are simple. Such is far from the truth. While the source of the gold in eluvial placers is seldom in doubt, there are a host of problems concerned with the mode of concentration of the gold and the nature and origin of the pay streaks.

The source of gold in eluvial placers is most commonly auriferous quartz veins, shear zones, polymetallic deposits, massive sulfide bodies, and porphyry copper-molybdenum deposits, and only rarely, lithified placers. Three types of gold are commonly present in eluvial placers: residual gold, chemically precipitated gold, and nuggets (pepites) composed of the other two types. The residual gold is formed essentially in the zones of oxidation of auriferous deposits; it is usually splendent, hackly, unworn, and derived directly by the disintegration of the gangue enclosing the primary gold particles. The silver (and other metal) content of residual gold is commonly high (similar to that present in the primary ore), giving a relatively low fineness (commonly less than 900). Residual gold is usually associated with gold-quartz veins and lenses sparsely mineralized with sulfides. Chemically precipitated gold is formed both in the zones of oxidation of auriferous deposits and in the residuum of the eluvial, deluvial, and proluvial deposits; some varieties are dull in appearance and pulverulent (mustard gold), occurring as dust and in small aggregates; other varieties occur as reticulated, dendritic, arborescent, filiform, and spongy groups, aggregates, and masses; only rarely does this variety occur as foil and crystals. All varieties exhibit little if any mechanical abrasion. The silver (and other metal) content of chemically precipitated gold is low, and hence the fineness is high (generally much greater than 900 in fineness). Chemically precipitated gold is often difficult to win by panning and other mechanical modes of concentration because of the minute grain size of the gold. With time, however, this type of gold may be aggregated into small flakes and nuggets, some very large. Chemically precipitated gold appears to be most frequently associated with sulfide deposits or with gold-quartz deposits greatly enriched in sulfides.

Nuggets (pepites) in eluvial placers occur as small rounded fragments, flattened grains, scales, plates, and irregular gnarled masses. The nuggets commonly show evidence of chemical accretion, and some present a worn and abraded appearance. This type of nuggety gold is characteristic of deluvial and proluvial placers and appears to represent residual gold to which chemical gold has been accreted; the whole nuggets in some cases have been extensively recrystallized. The fineness of the nuggety gold is variable, depending essentially on the contributions of residual and chemical gold; normally the fineness is greater than the particles of primary gold. This nuggety eluvial gold appears to derive most commonly from gold-quartz deposits greatly enriched in sulfides, from deeply weathered auriferous massive sulfide bodies, and from slightly auriferous porphyry copper-molybdenum deposits.

The literature is replete with general descriptions of eluvial placers in many countries, but particularly in those where deep secular weathering of rocks and their auriferous deposits has resulted in the formation of lateritic soils and residuum containing workable placers overlying or downhill from outcrops of gold-bearing deposits. Here we may mention a few of these descriptions, beginning with one by Becker (1895) of the saprolitic eluvials in the southern Appalachians.

Gold is found in loose material of two very distinct kinds in the Southern Appalachians. True stream gravels carrying gold are not wanting, but much more common are auriferous accumulations of rotten rock in place. As is well known, decomposition of the bed-rock in the unglaciated South often extends to a depth of from 50 to 100 feet from the surface. Where the mass was originally intersected by gold-quartz seams, perhaps accompanied by impregnation of the wall rock, the decay of the mass to soft earth takes place without sensible loss of the precious metal. Such deposits can be worked with pick and shovel or, when they are rich enough, by the hydraulic process. In such deposits, as a rule, the original structure of the rock is perceptible to within a couple of feet of the surface. The rock in decomposing may have undergone some change of volume and a trifling amount of movement, but the material is substantially in place.

There is no term in general use to designate this decomposed rock in place, although it is found almost universally in unglaciated regions, even within the arctic circle. It is by no means often possible to name a given occurrence of this kind from the original rock, because, when there are no exposures of unchanged material, it is usually doubtful which of several allied rock species is really present. The word "Geest" was long ago proposed as a general term for such material. This is a provincial German word meaning dry land as distinguished from marshy land. The name does not seem aptly chosen and has not been adopted by many writers. The German term "Gruss" has sometimes been made to serve, but this word denotes a mass consisting of angular fragments, as distinguished from the rounded pebbles of gravel, and it is constantly in use for transported material, as for example "Grusskohle," equivalent to slack-coal.

I propose the term *saprolite* as a general name for thoroughly decomposed, earthy, but untransported rock. When the exact character of the original rock is known it is easy to qualify this term and to speak of "granitic saprolite," and the like.

The deposits referred to above, then, are gold-bearing saprolites. In these the original quartz veins are usually but little decomposed, and can be followed at small expense. Near Brindletown the hills are scored with deep trenches thus excavated. In the Dahlonega region a system often adopted is to hydraulic the saprolite, the fine earth usually being allowed to escape after passing through sluices, while the fragments of vein quartz are thrown by grizzlies, or equivalent devices, and passed through the stamp mill. The losses in this process, however, are very great, owing, it would seem, to the inclosure of the gold particles by films of iron oxides or other substances which prevent amalgamation.

Geologically the saprolites are of course identical with the more solid masses beneath them, yet they sometimes permit the observation of relations less well seen elsewhere. Thus in some saprolites an almost endless number of the thinnest possible quartz seams and the most perfect quartz lenses are accessible not merely in section but in

> three dimensions. At the Kin Mori mine one can pull out of the saprolite pieces of vein of several square inches area and scarcely thicker than writing paper. But this is not always the case; some auriferous saprolites show almost no quartz. There are instances in which the observer can scarcely believe the results of assays from carefully taken samples, so little indication is there of any gold. Now, the investigations of late years make it certain that quartz is often attacked and dissolved, or is replaced by mica, iron oxides, etc. It seems to me probable that the deceptively rich saprolites are those in which the quartz has been thus removed.
>
> The gold found in the saprolite deposits is naturally very rough, and in some cases, as at the Loud mine, north of Dahlonega, masses of wire gold are met with. If anyone still doubts the origin of the gold in stream gravels he may readily convince himself in the South that the origin of the metal is in veins, for gold can be seen there in all stages from the roughest to the smoothest as the saprolites are followed into the waterworn gravels. (pp. 289-290)

In proposing the term *saprolite* Becker noted that the derivation is from the Greek 'rotten.' In recent years the more inclusive term *laterite* has been used, being thought more appropriate.

Becker considered that most of the gold is derived from the weathering of veins; he allowed for no chemically accreted gold from pyritic schists and other slightly auriferous rocks.

Much of the placer gold of the Guianas has been won from eluvials developed on auriferous Archean greenstone and sedimentary belts. The eluvials of British Guiana have been described in great detail by J. B. Harrison (1905, 1908). One of the deposits is of considerable interest in that the auriferous eluvium was developed on a gold-bearing pyritiferous aplite dike, a circumstance that is of frequent occurrence in a number of deeply weathered auriferous belts in many parts of the world. Maclaren (1908) has summarized most of the data provided by Harrison and others on this deposit as follows:

> An interesting occurrence is that of Omai, Essequibo river. The surface rock here is a diabase which is associated with aplite and granitite. At a depth of 964 feet borings through the acid rocks came upon epidiorite. The Archaean rocks of the country are apparently intruded by this mass or stock of aplitic granite. After its intrusion there was a succession of outbursts of diabase, and the latter rock is now developed both above and below the aplite. The interest of the occurrence lies in the fact that the aplite is auriferous, selected specimens assaying as high as 15 dwts. per ton. The aplite further carries in depth a great deal of pyrites, and the gold found is probably to be associated with that mineral. Small quartz veins, which are exceedingly numerous in the aplite, are slightly auriferous. During 1904-5, ninety per cent. of the total yield of the company, or more than 22,600 ounces gold, was obtained by sluicing the highly decomposed aplite. It was worked out in benches to 150 feet, to which depth the aplite was freely decomposed. The gold was free and often well crystallized, and there was no pyritous residue in the

> wash. The acidic rock is therefore the primary source of much of the alluvial gold of Omai. The only occurrences readily comparable with that of Omai are those of Berezovsk, in the Urals, and of Gallinazo, Colombia. (p. 638)

An alluvial deposit, somewhat similar to that at Omai, occurs on Serra Pelada, Para, Brazil. There, rich streaks and irregular pods of gold grains and nuggets, some large, are concentrated at the noses of folds and in intensely faulted and shattered zones in deeply weathered, friable gray and red siltstones and manganiferous breccia of the Precambrian Rio Fresco formation. Since its discovery in 1980 *garimperos* (prospectors) have won more than 12 tons of gold from the Serra Pelada deposit. Some details of the deposit are given by Meireles and Teixeira (1982).

Considerable amounts of eluvial gold have been won from the lateritized rocks and auriferous deposits of Brazil. In addition to the normal eluvial placers there are unusual types that merit attention, particularly those in the Quadrilatero Ferrifero, Minas Gerais. All of these are associated with gold-bearing iron-formations (itabirites). General descriptions of these deposits are given by Derby (1884, 1903), Bensusan (1929), De Oliveira (1932), Gair (1962), and Dorr (1969). One type of secondary deposit, not necessarily strictly eluvial but often grading into eluvial deposits, is locally known as *jacutinga* (De Oliveira, 1932). The jacutinga occurs as thin (centimeters to a meter) lines or bands in itabirite and is a decomposition, or more accurately a surficial chemical disintegration product of the iron-formation. It is composed essentially of loosely consolidated to powdery ferric oxide (limonite and hematite), manganese oxides, clay minerals, and talc in which nuggets, plates, and threads of native gold are present. Evidently the gold originated by chemical solution and reprecipitation of the metal from low grade gold-bearing iron-formation. From the descriptions, some of the jacutinga deposits were rich, up to 0.5 oz Au/ton or more, and some were mined to depths of 230 m. Some of the gold was rich in palladium (up to 5% Pd).

De Oliveira (1932) states six conclusions concerning the origin of jacutinga deposits and emphasizes the role of sulfuric acid, hydrochloric acid, manganese dioxide, and nascent chlorine in the solution of gold (see also chap. 16).

> *Conclusions*—In conclusion, it may be stated that:
>
> 1. The gold of the jacutinga is of secondary origin, that is, it is derived from solution and precipitation of gold from pyrite quartz veins.
> 2. The manganese dioxide, lithomarge, and limonite result from the above mentioned reactions.
> 3. The absence of pyrite in these deposits is now explained.
> 4. At depths below the water table, the tenor of gold of these deposits must diminish rapidly.
> 5. The best guide for detecting gold deposits in the itabirite formation is the pulverulent manganese dioxide.
> 6. It is not possible for gold to exist in jacutinga where there are no auriferous pyrite quartz veins, that is, primary deposits. (p. 748)

The other type of deposit associated with iron-formations in Brazil is the gold-bearing *tapanhoancanga* (*canga* for short), an irregular layer or blanket up to 10 ft (3 m) or more thick of limonite-cemented fragments of iron-formation (itabirite). It is commonly developed on all iron-formations throughout the world that are extensively oxidized. Much of the limonite appears to derive from the oxidation of iron silicates and pyrite in the iron-formations. Where the iron-formations are also auriferous the canga is commonly enriched in gold, the metal occurring in small flakes, wires, specks, and also in a submicroscopic form associated in some manner or other with the limonite. De Oliveira's six conclusions just described probably apply also to the origin of gold in the canga deposits. In the past, deposits of this type were worked in a small way in many parts of the ferriferous quadrangle in Minas Gerais.

Eluvial gold was relatively common in Australia before the turn of the century, but most of these placers are now exhausted. Fabulous nuggets were found in some districts in the eluvium or weathered rubble of the oxidized deposits. At Big Nugget Hill in the Hargraves goldfield of New South Wales a nugget weighing 106 lb was found in the eluvium by an aboriginal shepherd. At Lucknow, also in New South Wales, the eluvium was of great richness at and near the outcrop of the veins. In Victoria even larger nuggets were found in the eluvium and disintegrated country rock near the veins of Ballarat, Tarnagulla, and other places, including the Welcome Stranger (2,516 oz), Welcome (2,195 oz), Blanch Barkley (1,743 oz), Canadian (1,319 oz), Dunolly (1,364 oz), and Sarah Sands (775 oz). In western Australia the eluvial placers were small, rich, and soon worked out in the Pilbara, Kimberly, Coolgardie, Kalgoorlie, and other Yilgarn fields.

Eluvial placers have been worked extensively in USSR In addition to the normal eluvial placers there are two types of unusual formation that merit brief mention: the Kuranakh type and the talus type.

The Kuranakh gold deposits found in 1959-1962 are mainly in Lower Cambrian bituminous limestones and dolomites in the Aldan anteclise of southern Yakutia (Razin and Rozhkov, 1966; Borodaevskaya and Rozhkov, 1974). There is considerable similarity between the primary gold occurrences and the disseminated deposits of north-central Nevada (Carlin, Cortez, etc.) Summarizing from the landmark papers previously noted it can be said that the primary occurrences in the Kuranakh district are mainly potash feldspar (adularia) and quartz metasomatites with about 6 ppm gold occurring as irregular replacements of the limestones and dolomites in places in stratiform bodies up to 3 km in length, 300-500 m in breadth, and 5-10 m in thickness. The gold in the adularia zones is very finely divided (2-5 μ) and mainly in pyrite. Thin kersantite dykes and sills of late Mesozoic age are also accompanied by adularization along their contacts both in the Cambrian strata and in Lower Jurassic sandstones, and these igneous rocks have a comparable tenor of gold within the ore zones. Oxidation of the primary gold-bearing zones and karstification of the Lower Cambrian limestones in post-Jurassic and recent times has given rise to economic eluvial deposits consisting essentially of an aggregate of highly weathered material composed mainly of oxidized ore, limonite, clay and sandy materials in the karst cavities. The gold in the eluvial material occurs as very finely divided flakes ranging in size from 5 μ to 20 μ. The ratio of concentration is not given, but presumably it is about 2, making the grade of the deposits about 12 ppm (0.35 oz Au/ton).

A typical example of the talus (deluvial) type of placer occurs near the

Belaya Gora deposit in the Lower Amur region of USSR (Borodaevskaya and Rozhkov, 1974). In summary these authors state that the bedrock deposit occurs near the summit of Belaya Gora in a strongly fractured, kaolinized, and silicified Oligocene trachyte volcanic neck. The gold is mainly free and accompanied by about 0.5% combined pyrite, arsenopyrite, sphalerite, and argentiferous sulfosalts. The talus placer associated with the primary deposit lies below the deposit on a gentle slope and averages some 5-6 m in thickness. In composition the talus material is largely variegated clay and sandy clay and oxidized rubble with a heavy mineral suite composed mainly of limonite, magnetite, ilmenite, chromite, epidote, and small amounts of zircon, sulfides, manganese oxides, and other minerals. The gold is disseminated throughout the talus material and also forms short streaks and enriched lenses. The gold in the talus is not much different from that in the primary deposit. The gold grains range in size from 0.05 mm to 1.5 mm. Some absorbed and very finely divided gold also occurs in concentrations of kaolinite and quartz. Downhill from the talus placer there are proluvial and alluvial placer systems associated respectively with springs and the river fed by the springs. The gold in the proluvial placers is of interest in that it differs markedly from that in the deluvial placers in color and character. Nuggets weighing up to 190 g have been found in the proluvial placers in addition to abundant monocrystals and crystal intergrowths of gold. The fineness of the gold in the deluvial placers is 630-670; in the proluvial placers, 663-694.

ALLUVIAL PLACERS

Alluvial placers were the type of gold deposits worked earliest, as described in chapters 2, 3, and 4. By the time of the rediscovery of America by Columbus in 1492 nearly all of the old placers of the Mediterranean area, western and eastern Asia, and western Europe were exhausted.

When Columbus made his first landfall in the New World he found the natives in possession of gold, and while exploring in Hispaniola (Haiti) in 1493 on his second voyage he observed crude placer operations in the streams and rivers of the interior. In 1494 Alonso de Ojeda discovered the rich Cibao placers, and a year later Pablo Belvis arrived from Spain with a large quantity of mercury for amalgamation purposes. The first gold won was sent immediately to the King of Spain, who donated it to Pope Alexander VI in Rome, where it was dedicated to the service of Christianity in the gilding of a cathedral dome. Thus began the *auri sacra fames* of the Spaniards in America that was to wreak such havoc on the natives and was also to introduce the Negro slave trade to the New World. The sordid story has been told many times and need not be repeated here.

A more pleasant story can be written about the great gold rushes of the 1880s and 1890s. Alluvial gold gravels were worked in the Altai of Siberia as early as 1820, but some nine years later extensive alluvial deposits were found in the Lena and Amur basins, probably the largest alluvial gold deposits known. These discoveries led ultimately to the colonization of southern Siberia. These and other deposits in eastern Siberia are still worked extensively today. The great gold rush of 1849 to California can be said to have opened up the American west. The California gold rush was followed in 1851 by the Australia rushes to New South Wales and Victoria, an impetus to mining that has kept Australia in the forefront of world mineral production ever since. The golden gravels of the Fraser River in British Columbia were known as far back as 1852, but it was not until 1858 that the great stampede up the Fraser began, ultimately in 1861 reaching Williams and Lightning creeks in the Cariboo, the most celebrated of all gold creeks in the province. Gold placers in the Yukon Basin were worked as early as 1880, but it was not until 1896, and perhaps two years earlier according to some accounts, that gold was discovered on the tributaries of the Klondike River. The great rush took place in 1897-1898 to Dawson—one of the greatest gold rushes in history and certainly the most colorful, made immortal by Robert W. Service in his novel *The Trail of '98* and in his poems *Songs of a Sourdough* and *Ballads of a Cheechako.*

Alluvial gold placers in streams and rivers have been the richest and of most interest to the prospector, but some beach placers have also been extensively worked. Gold was discovered in Alaska as far back as 1865-1866, and alluvial prospecting in the streams and rivers near Nome was in full swing in 1899, the year that gold was discovered on the beaches by a soldier and a prospector. There followed a frenzied digging and grubbing along the coast for many miles, and more than one million dollars were won by hand rockers in less than two months (Collier et al., 1908).

The methods of the alluvial placer miner have changed much since the heavy splendent golden nuggets were plucked by hand from the gravels of the streams. This is the stage at which Columbus and his mining engineers found some of the natives working the placers of Haiti. Later the gravels of the streams were stirred up by workers using crude booming (hushing) operations and sluiced over the fleeces of sheep and goat pelts, the gold remaining mainly trapped within the wool and goat's hair. This method is apparently still employed in some of the placer streams of Asia Minor, Afghanistan and Mongolia. Panning is an old technique certainly known to the ancients of the Old World and the natives of Africa, who used the calabash (gourd), and the natives of Central and South America, who employed the batea. The rocker and the sluice were known to the Greeks and Romans, and they were adept at booming (hushing) operations. The dry washer or dry blower has long been known to those who have sought gold under desert conditions. The water monitor-, bulldozer- and dragline-sluice operations and the great bucket-line dredges are modern mechanical adaptations of age-old techniques of separating the gold from the dross.

Two general types of alluvial placers are recognized: those formed in streams and rivers and those formed along lake or ocean beaches. Another type, deltaic, is of little importance economically and is not discussed further.

The literature contains such a vast number of entries on alluvial placers that it is virtually impossible to deal adequately with landmark papers on the subject. My choice of some of these papers is strictly arbitrary and based almost entirely on field areas with which I am familiar.

The location of the pay streaks in alluvial placers is of prime interest to the placer miner and also of scientific interest to those studying placers. Given an adequate primary gold source, pay streaks in general are fairly uniform and have considerable continuity in moderately hilly country where uniform rainfall has prevailed and where deep secular decay has been followed by a gradual restricted uplift. Any aberrations in this ideal pattern invariably cause marked variations in the tenor and continuity of the pay streaks. The richest pay streaks are the ones produced by reworking of preexisting auriferous gravels.

The law of the pay streak in placer deposits is variable depending on whether the placer is formed in gulches, in river channels, on flood plains, or in deltaic deposits. Referring to the deposits of the Sierra Nevada of California, some of which are buried beneath Tertiary lava flows, Lindgren (1911, p. 66) commented on such variability.

> It has become almost an axiom among miners that the gold is concentrated on the bedrock and all efforts in placer mining are generally directed toward finding the bedrock in order to pursue mining operations there. It is well known to all drift miners, however, that the gold is not equally distributed on the bedrock in the channels. The richest part forms a streak of irregular width referred to in the English colonies as the "run of gold" and in the United States as the "pay streak" or "pay lead." This does not always occupy the deepest depression in the channel and sometimes winds irregularly from one side to the other. It often happens that the values rapidly diminish at the outside of the pay lead, but again the transition to poorer gravel may be gradual. An exact explanation of the eccentricities of the pay lead may be very difficult to furnish. Its course depends evidently on the prevailing conditions as to velocity of current and quantity of material at the time of concentration.

Tyrrell (1912, p. 596) while admitting the fickle nature of the pay streak, concluded that "the pay-streak is a feature in the structure and growth of the valley in which it occurs, its formation is governed by certain geological laws, and those laws should be recognizable without great difficulty if the growth of the valley can be traced with reasonable accuracy." His paper, and the discussion that followed, is a landmark in the study of auriferous placers and merits reproduction in full as Paper 15-1.

(Text continues on page 486.)

15-1: THE LAW OF THE PAY-STREAK IN PLACER DEPOSITS

J. B. Tyrrell

TWELVE years ago I had the pleasure of reading a paper before this Institution* on "The Gold-bearing Alluvial Deposits of the Klondike District," in which the topographic features of the country were briefly outlined, and the general character of the gravels and the underlying rocks were indicated. At the same time it was pointed out that the two sources from which to obtain an adequate water supply for the efficient mining of Bonanza Creek were the Rocky Mountains to the north and the conservation of the water of the creek itself. It is interesting to record that both these projects, first laid before the public through this Institution, have now been completed by the building of a great ditch and flume from the Twelve Mile River, at the foot of the Rocky Mountains, and by the building of a dam across the upper part of Bonanza Creek.

This evening it is my intention to present to you, very briefly, some of the results of a study of the placer deposits of that northern country, especially with regard to any light that they may throw on the laws governing the deposition of placers and the formation of the run of coarse gold which is usually found in the bottoms of the larger valleys, and which is known as the "pay-streak" or "pay-lead." It is believed that the laws or principles here enunciated not only explain the occurrence and characteristics of "pay-streaks" in the Klondike district, but that they have general application to the concentration of heavy metals or minerals in alluvial deposits.

Placer deposits may be defined as "detrital deposits of heavy metals or minerals mechanically concentrated by natural agencies."

Prof. James Geikie defines a placer as "an alluvial deposit derived from the disintegration of metalliferous rocks and ore-bodies of various origin."

* *Trans.*, viii, 217–229.

Richard Beck says: *

"By detrital deposits we understand accumulations of ore formed by the destruction and re-deposition of primary deposits. These two results have been accomplished, in the main, in a mechanical, but in part, also, in a chemical way. In both cases water was the main agent used by nature for the purpose. Such a destruction and re-deposition of primary deposits may have taken place in remote geologic periods, but only in comparatively rare cases have the products of such periods been transmitted to us in a recognizable condition. On the other hand, the Tertiary and Pleistocene formations of the earth's surface contain a great number of such detrital deposits, as they are commonly called. It is customary to use the term *placer gravels* for the Pleistocene and Tertiary alluvial gravel deposits."

And again: †

"Placer gravels are deposits of loose, more or less rolled, material derived from the destruction of older deposits, lying on the earth's surface, or at least very close to it, and containing paying amounts of ore or precious stones.

As the material composing placer gravels has been exposed to all the influences of the atmospheric air and of the water seeping through the upper strata of the soil, placers will be found to contain, in the main, only relatively insoluble, and, in general, refractory metallic compounds, which, moreover, are protected by their great specific gravity against easy removal by water.

These placer gravels are usually grouped into two classes, according to their position with reference to the deposit from which they are derived, and in part, also, according to the manner of the original process in which they are derived from the primary ore deposit:

1. Residual gravels, *i.e.*, of local-origin (eluvial gravels).
2. Alluvial gravels, *i.e.*, formed by washing. These may again be subdivided, according to age, into recent, Pleistocene and Tertiary gravels.

Residual gravels, the rarer of the two groups and certainly the less extensive, are found in the immediate vicinity of the original ore deposits, and quite independent of water-courses, viz., on mountain slopes, plateaus, and sometimes even on mountain summits.

On the contrary, the gravels formed by the transporting and washing action of water are found only in the channels of brooks and rivers, in fresh-water lakes or along the sea-coast. They lie for the most part within the present valleys or along the present shore, but are also often found in stretches of fluviatile sediments, sometimes intersecting the present direction of the valley on old river terraces, or in sheets covering plateaus (California, Ohlapian in Transylvania), and, finally, in old shore terraces above the present level of the sea. Their material is always much rolled, and for the most part is assorted, according to the size of the ingredients, into shingle, gravel, sand, clay, mud, etc."

* "The Nature of Ore Deposits," by Richard Beck. Translated by W. H. Weed, 1905, p. 611.

† *Op. cit.*, p. 617–618.

Residual gravels occur on many of the higher slopes in the Klondike, but only in few cases do they form workable placers. The best illustration of such placers which came to my notice was on the upper portion of Victoria Gulch, one of the tributaries of Bonanza Creek, where some beautiful sharp "spinel twins" of gold were found, just in the condition in which they had been washed out of a vein that outcropped higher up towards the summit of the ridge.

Most of the placers in the country are such as are designated above "alluvial gravels" and belong to the class of alluvial gravels found "in the channels of brooks and rivers."

In many of these alluvial gravels that occur throughout the Klondyke some gold can be found, but in the gravel deposits in the bottoms of most of the wider valleys, whenever gold is present, it is not evenly distributed, for most of the coarser particles are found in a band of restricted width which lies on or close to bedrock, and wherever the bedrock is fissured these particles descend into it for varying distances. This band or run of coarse gold is known as the "pay-streak," and the discovery of it beneath the gravel of the alluvial plain is the constant desire of the prospector.

The existence of this pay-streak has been recognised by placer miners from time immemorial.

A. G. Lock* refers to it as the "gutter," which he defines as the lowest portion of a lead, which contains the most highly auriferous dirt."

Posepny states † "The gold occurs concentrated in the deepest portion of the weather-detritus, that is to say, on the contact with bedrock, and has penetrated all the open, loosely-filled fissures in the latter."

Beck states: ‡ "It would, however, be an error to assume that in a cross-section of a river valley the lowest layers of shingle, gravel or sand are throughout the richest. On the contrary, the values in this horizon are variable and pay gravel is ordinarily limited to streaks of greater or less width, which are found in one place in the centre of the valley, in another along one side, now nearer, now further away, from the present water-course."

W. Lindgren writes of the pay-streak as follows: § "It is well known to all drift miners, however, that the gold is not equally

* "Gold," by A. G. Lock, 1882, p. 1181.

† "Genesis of Ore Deposits," by Prof. F. Posepny, New York, 1902, p. 153.

‡ "The Nature of Ore Deposits," by R. Beck, translated by W. H. Weed, 1885, p. 620.

§ "The Tertiary Gravels of the Sierra Nevada of California," by W. Lindgren, U.S. Geol. Surv. Prof. Paper 73, 1911, p. 66.

distributed on the bedrock in the channels. The richest part forms a streak of irregular width referred to in the English colonies as the 'run of gold' and in the United States as the 'pay-streak' or 'pay-lead.' This does not always occupy the deepest depression in the channel and sometimes winds irregularly from one side to the other. It often happens that the values rapidly diminish at the outside of the pay-lead, but again the transition to poorer gravel may be very gradual. An exact explanation of the eccentricities of the pay-lead may be very difficult to furnish."

It is true that the pay-streak very often seems to be one of the most elusive of phenomena, and time and again the prospector is inclined to say that there has been no advance in the knowledge of the laws which govern the deposition of placer gold since the days of Job, 35 centuries ago, and that all that can be said now, as then, is that "There is a vein for silver and a place for gold."

But the pay-streak is a feature in the structure and growth of the valley in which it occurs, its formation is governed by certain geological laws, and those laws should be recognizable without great difficulty if the growth of the valley can be traced with reasonable accuracy.

In what we now know as the Klondike district, marine sediments were laid down at various periods up to the beginning of Tertiary times, and after their deposition they were raised, crushed and bent into their present form and position.

The country was then worn down to base level, and a peneplain, the remains of which can now be seen at an elevation of about 3500 ft. above the sea, was formed. This peneplain may be called the "dome peneplain," as portions of it are distinctly recognizable in the vicinity of the mountain known as "The Dome." For our purpose the period of its formation may be designated as the "first cycle of erosion," since the history of the gold-bearing gravels would appear to begin with it and no gravel deposits have yet been recognized on it.

After the dome peneplain was formed the "first period of elevation" began, and the country was raised to a considerable height above the sea. The Yukon River, which had probably been outlined at an earlier period, immediately began to erode its channel, while the water, which fell as rain on the elevated Klondike land, carved out smaller valleys to carry the drainage from it to the larger river. As the Yukon River was a powerful eroding agent it deepened its valley rapidly, and at the same time the smaller streams radiating from The Dome, such as Bonanza, Hunker, Dominion, Sulphur Creeks, etc., kept excavating their channels to keep pace with the lowering of the bottom of the valley of the Yukon River, which was the master-stream into which they flowed.

During all this time the valleys of these smaller streams maintained the general character of gulches or young valleys, with V-shaped cross-sections. But little gravel or loose material remained on the rock which formed the bottoms of their channels, for it was being constantly moved downward by the current towards the Yukon River, and, on the way, was helping to cut deeper and deeper into the rock over which it travelled.

While this process of deepening the valleys was in progress, detrital material was being constantly brought into them by wash from their sides and by smaller streams from the ridges between them, and, as the rocks from which this material was derived were gold-bearing, the detritus contained a small quantity of gold. Thus gold and particles or masses of rock were fed gently into the main streams.

Now, a stream with a certain velocity is able to carry pebbles of a certain size and specific gravity. If the specific gravity is constant the diameter of the pebbles which it can carry will vary according to the square of the velocity, and if the velocity remains constant, the size of the pebbles will vary according to the specific weight of the substance composing them weighed in water. For instance, if the velocity of a stream is doubled it is able to carry pebbles of quartz four times the diameter, or 64 times the weight, of those which it could carry before. If, on the other hand, one pebble is of quartz and the other is of gold, which is 11 times as heavy as quartz weighed in water, the volume of a pebble of quartz which can be carried by the current will be 121 times (11^2) as great as that of a pebble of gold, or, in other words, the diameter of the pebble of quartz will be about five times the diameter of the pebble of gold.

Again, if particles of quartz and gold of equal size are dropped into water the gold will sink to the bottom with more than three times the velocity of the quartz.

Where the fragments of rock, consisting of quartz, schist, granite, etc., and gold, are fed into the stream, they are caught by it and carried along the bottom until they lodge in some crevice or opening, from which they cannot be dislodged except by upward currents, and these upward currents will lift any pebbles of quartz or similar rock which are less than five times the diameter of nuggets, or grains of gold occurring with them, before they will lift the gold, even if the quartz and gold are equally accessible. This makes the removal of the gold exceedingly difficult as long as the crevice remains, for the upward currents will constantly carry away the finer and lighter rock, and undermine the grains of gold and allow them to sink. When the finer and lighter material is carried away, the coarser and lighter pebbles are exposed to the force of the current, and the smaller and heavier grains of gold are able to obtain lodgment beneath and between them so as to be almost inaccessible. In fact, under normal conditions, the spaces between the lighter pebbles are large enough to hold any grains of gold which could be carried by the current flowing over them.

It is thus shown that gold will remain permanently in a fissure of the rock in the bottom of a stream as long as that fissure remains in existence, and also that it will remain between or beneath larger pebbles and boulders as long as these remain unmoved.

Now, the small streams of the first period of elevation, which developed into, or was succeeded by, the second cycle of erosion, continued to cut down their channels as long as the Yukon River continued to deepen its valley. During all this the bottoms of their valleys continued to act as sluices, which were more or less efficient agents in collecting and retaining gold according to the character of rock of which they were formed. If the rock where the gold was discharged into the stream was a fissile schist standing on edge the gold would be caught at once, while, if it was a massive granite or other similar rock, without joints or fissures, or a smooth horizontal schist, the gold would be carried down-stream over it until it would be caught by some more favourable rock. In this way there would be rich places, poor places and blanks in the streak of gold deposited in the bottom of the valley.

As the stream would continue to deepen its valley very gradually, almost imperceptibly, by downward erosion, those places which were underlain by schists standing in a vertical or highly inclined attitude would continue to hold the gold which they had already caught, and to accumulate more, for fissures would open as fast as the surface was worn away, and the gold would sink into them as they opened. On the other hand, those places which were underlain by a harder bedrock, and which had probably also a steeper grade, would remain barren. If, again, the character of the bedrock should change from " open " to " tight," the gold which it had held might be undermined by the continual downward erosion, and so be brought again within the influence of the transporting power of the running water, by which it would be carried along to find some new resting-place farther down the stream.

When the Yukon River had eroded its valley down to base-level, the smaller inflowing streams were no longer obliged to continue to deepen their respective valleys to keep pace with it, but were able to

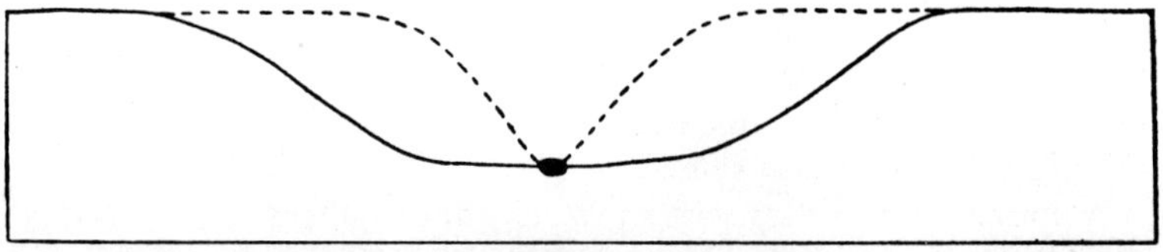

Fig. 96.—Diagrammatic representation of Pay-streak in the bottom of a simple valley.

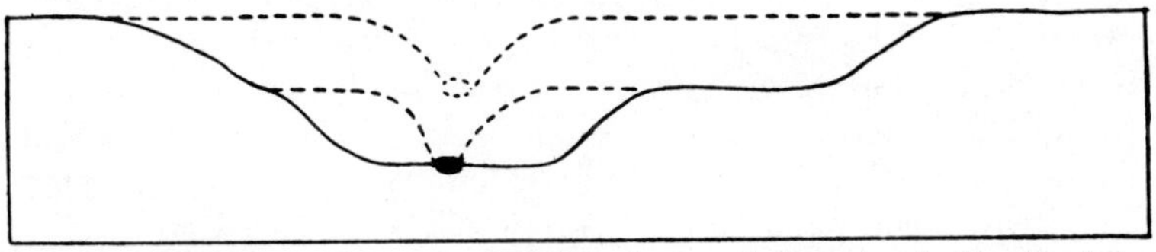

Fig. 97.—Diagrammatic representation of second Pay-streak directly below the first.

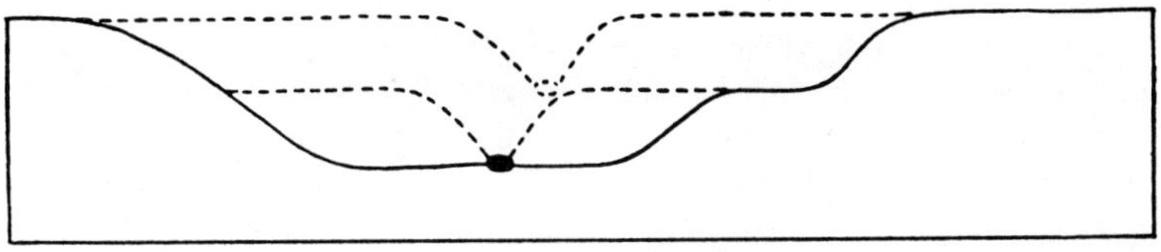

Fig. 98. —Diagrammatic representation of second Pay-streak obliquely below the first.

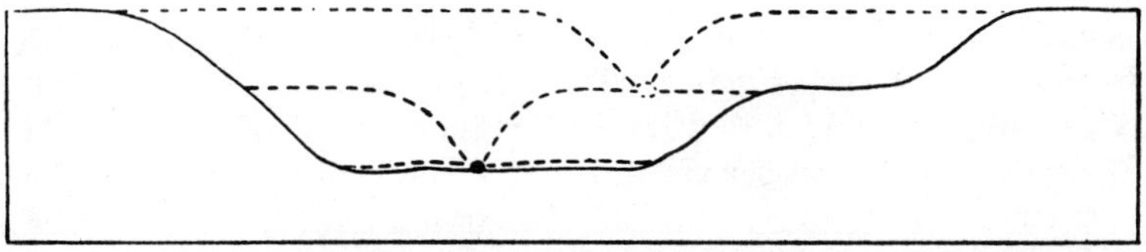

Fig. 99.—Diagram showing how first Pay-streak may be distributed in second valley.

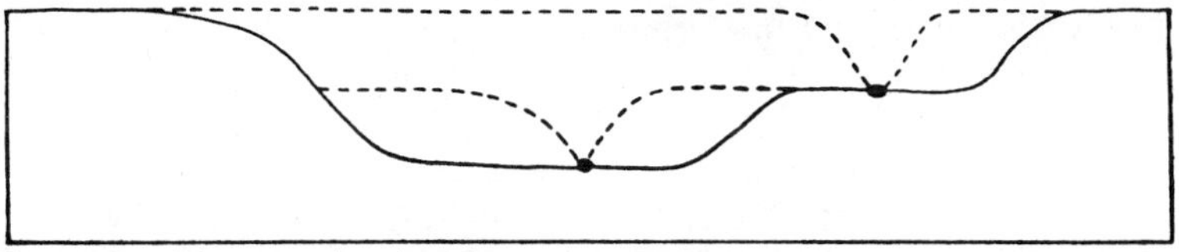

Fig. 100.—Diagram showing first Pay-streak on terrace and second lighter Pay-streak in second valley.

cut them down to grade, and then to widen and form flood plains in them, thus changing the V-shaped valleys into U-shaped ones, floored by alluvial plains through which the rivers and brooks meandered from side to side.

A normal stream decreases in velocity and gradient as it descends its valley and reaches grade near its mouth before it has cut down the rest of its valley to grade. So, when each of these streams had cut down the lower portion of its valley to correspond with the base-level established by the Yukon River, it would begin to meander and extend the width of its floor. At the same time, with the decrease in gradient the velocity of the current would decrease, and its transporting power would be diminished. Consequently, part of the detrital material which would be brought down by the upper and swifter portion of the stream would be dropped where the current was retarded by the decreased gradient, and would lodge in the bottom of the valley and form a "flood plain" or "alluvial plain." This alluvial plain would be first formed where the V-shaped valley changed into a U-shaped one.

Most of the gold which had previously been discharged into the stream with the detritus from the adjoining hills and ridges would have already lodged in the bottom of the V-shaped valley, and would have settled down almost vertically as the bottom was lowered by the downward erosion of the stream. If any gold was carried down to the mouth of the V it would have a very strong tendency to settle just where the velocity of the current was diminished, or at the head of the flood plain, and the weaker current would have no power to pick it up again, or to release that gold which was already present beneath it on account of having been previously caught in the bottom of the V-shaped valley. Thus the pay-streak would be formed. Afterwards the gravel, sand and alluvium of the flood plain would be deposited over and beyond it, but it would continue to mark the position of the bottom of the old V-shaped valley, no matter how wide the bottom of the mature valley might afterwards be extended.

After a flood plain had been formed at the mouth of a valley the river farther up stream would still continue the downward erosion of its channel until it reached the grade of that below it, when lateral plantation and the formation of the flood plain would begin. Thus the flood plain was formed gradually up the valley from its mouth, and always, where the old V-shaped valley changed into a U-shaped valley, there was left a trail of gold beneath it.

The gold which was collected and stored in the bottom of the V-shaped valley had been derived from the rocks of the adjoining country. At the same time the lighter material derived from the disintegration of these rocks had been carried through the valley and out beyond its limits, for the stream was then cutting down and enlarging it, and not filling it up, and there was very little room beside the stream for the accommodation of loose rock material. At the head of the flood plain this gold, which had been concentrated from the rocks of the surrounding country through previous ages, was gradually covered, and hemmed in on both sides, by gravel and alluvial material brought down by the stream at a later date. Therefore the gold in the pay-streak was derived from its home in rocks at a date which preceded that of the formation and deposition of the gravel which overlies and surrounds it.

The gravel of the flood plain may itself contain some gold which had been washed down the stream with it, or which had been washed into the valley from the sides, but this gold is usually very fine, such as might be carried readily by the stream for long distances.

If, after the flood plain was once formed, the stream should continue to deposit gravel to considerable thickness in the bottom of the valley through which it meanders, the source of supply for the gold would, on account of the general wearing down of the country, become more and more remote, and the average gold contents of the gravel would gradually decrease from below upwards.

The laws governing the formation and position of the pay-streak in an alluvial plain in the bottom of a valley may therefore be stated as follows :—

1. It was formed in the bottom and at the mouth of the V-shaped valley, which was the young representative of the present valley.

2. It marks the position formerly occupied by the bottom of that V-shaped valley.

3. The gold contained in it was washed out of the surrounding country and collected into approximately its present position before the gravel of the flood plain (or terrace) was deposited over and around it.

It has been assumed, for purposes of illustration, that the growth of the valleys in the Klondike district, which empty into the Yukon River, was continuous and regular throughout the second cycle of erosion, and in view of their symmetrical character, and the regularity of the pay-streak, which has been shown to have existed in them, it is probable that this assumption is not very far from correct; but nevertheless there were doubtless interruptions and cessations, both in the regular course of erosion and sedimentation.

After the Yukon River had cut its valley down to base level in this White Channel period, or second cycle of erosion, the tributary streams flowing from the Klondike district also widened their valleys and formed flood plains, as has just been described.

Then there was a long period of quiescence, during which the base-level of the country was raised, permitting heavy accumulation of gravel in the valleys, while at the same time the hills and ridges were worn down to mature forms. At the mouth of the valley of Bonanza Creek the local gravels, derived from the watershed of the creek itself, accumulated to a thickness of more than 200 ft. These gravels can still be recognised forming terraces at many places on the hills several hundred feet above the bottom of the valley, and Mr. McConnell,* who has carefully measured them, has shown on a map accompanying his report a pay-streak running in a very straight line through and beneath them. According to the laws here formulated, this pay-streak was formed in the bottom of the old V-shaped valley, which represented the valley of Bonanza Creek at the White Channel period in its youthful stages, and it now tells us the original position of the bottom of that V-shaped valley.

Just before, or at the termination of, the second cycle of erosion, the Klondike River brought a heavy load of sediment down from the mountains to the east, and covered the bottom of its own valley, and the mouths of its tributary valleys, with a bed of gravel, which, opposite the mouth of Bonanza Creek, has a thickness of 150 ft. The influx of this gravel caused the lower portion of the latter stream to move westward, almost to the limit of its own flood plain, and to be ready to begin a new rock valley with the advent of the next erosion cycle.

After the deposition of this upper gravel in the valley of the Klondike River a period of elevation† set in and the third cycle of erosion was inaugurated, which has continued down to the present time.

With the advent of this cycle of erosion the Yukon River was rejuvenated and again began to actively deepen its channel, and at the same time the tributary streams also began to deepen their old channels, or to cut out new ones, in order to keep pace with the master-stream. The Klondike River, the largest affluent of the Yukon in this district, probably did not lag very far behind it in the work of downward erosion, but its tributaries, such as Bonanza and Hunker Creeks, undoubtedly continued to flow in narrow, V-shaped

* Report on "The Gold Values in the Klondike High Level Gravels." By R. G. McConnell, Ottawa Government, 1907.

† In a paper published in the *Scottish Geographical Magazine* for June, 1900, entitled "The Basin of the Yukon River in Canada," I stated that this elevation was probably in the nature of a tilting from the south-west towards the north-east, but in view of the fact pointed out by Mr. McConnell that the rock terrace of the White Channel period, on the sides of the valley of the Yukon River, now rises steadily northward from the mouth of Stewart River to the mouth of Forty Mile River, this opinion is no longer tenable.

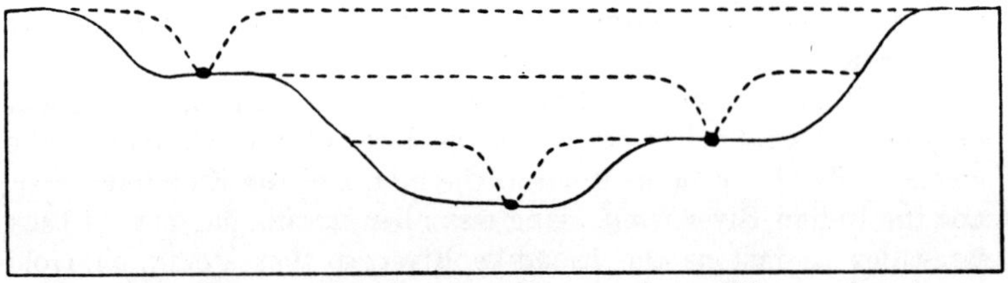

FIG. 101.—Diagram showing three Pay-streaks at different elevations.

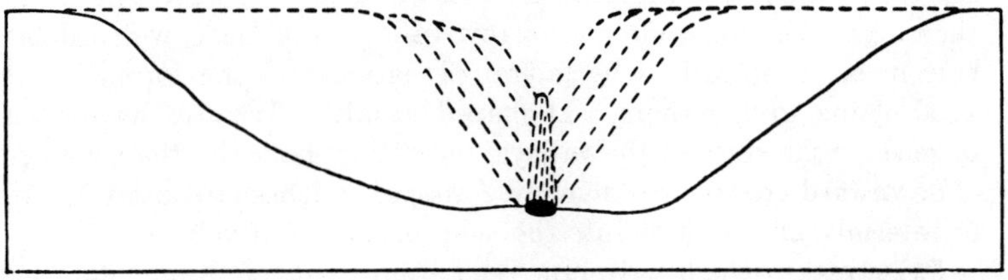

FIG. 102.—Diagram showing formation and downward growth of a Pay-streak in a wide valley.

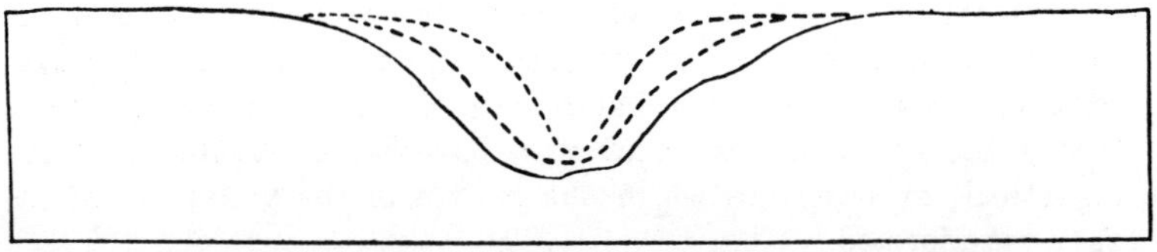

FIG. 103.—Diagram illustrating the transformation of a V-shaped valley into a U-shaped valley. (After Chamberlin and Salisbury.)

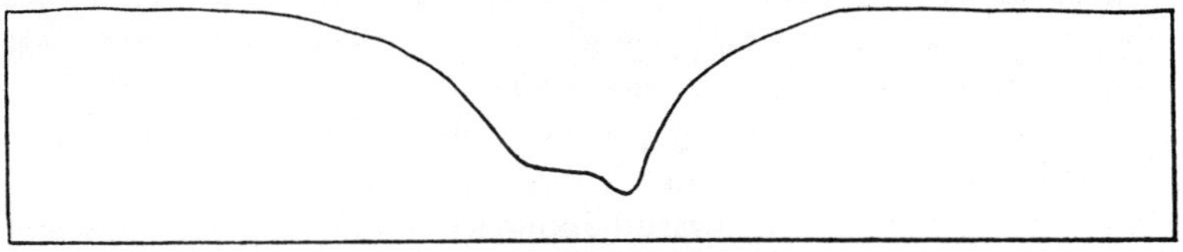

FIG. 104.—Diagram to illustrate the widening of a valley flat by erosion. (After Chamberlin and Salisbury.)

valleys as long as the main stream was actively engaged in deepening its channel.

Opposite the mouth of Indian River the Yukon River has not deepened its channel as far below the level of the channel of the second cycle of erosion as it has at the mouth of the Klondike River, and the Indian River itself, being a smaller stream, has not cut back its valley as fast as the Klondike River, so that Dominion, Gold Run, Sulphur, and the other tributaries of Indian River have not had the same opportunity to deepen their channels as the tributaries of the Klondike River.

During the third cycle of erosion the smaller streams, and especially those flowing into the Klondike River, have cut down their channels to grade in narrow valleys, and have widened the bottom of those valleys by lateral planation and the formation of flood plains, giving them a U-shaped profile. Terraces have been formed on the sides of the valleys, indicating halts in the progress of downward erosion, and narrow V-shaped gulches still carry small, or intermittent, streams into the sides of the main valleys.

Pay-streaks, which have now been almost entirely mined out, ran beneath the flood plains down the bottoms of these valleys, or crossed the terraces on their sides, and other pay-streaks were in process of formation in the gulches until that process was arrested by the work of the miner.

It is not necessary for our present purpose to follow the growth of these younger valleys in detail, or to trace the formation of the pay-streak in them, for that was clearly governed by the laws which we have already enunciated, but it will be interesting to indicate a few of the eccentricities which may have been introduced in the pay-streak by irregularities in the growth of the valleys in which they were formed.

We have already seen that difference in the character of the bed-rock will produce a marked difference in the quantity of the gold in the pay-streak.

A variation in the supply will also influence the richness of the deposit, as may be clearly seen in many of the small lateral streams which flow into the main creeks. Some of these cut across the old pay-streak of the second cycle erosion, and where this occurs the gravels in the bottoms of these streams are enormously enriched.

Temporary cessation of downward erosion, with the corresponding formation of flood-plains at successive levels, would appear, however, to exert the most powerful influence in affecting the nature of the pay-streak and introducing irregularities into it.

Let us suppose that a valley has been eroded down to the first level, and that a flood plain has been formed at that level. The pay-streak will occupy its normal position in this flood plain on the line of the bottom of the old V-shaped valley, as shown in Fig. 96.

If the stream is rejuvenated and again begins to deepen its valley a number of other conditions may occur.

1st.—It may cut down its channel directly beneath Pay-streak No. 1, in which case the pay-streak will simply be lowered, and will contain practically all the gold from the older pay-streak, as well as any gold that may have been collected into the channel since the time of its formation, as shown on Fig. 97.

2nd.—It may cut down its channel to one side of Pay-streak No. 1, and while still actively engaged in downward erosion may undercut the pay-streak, and allow the gold to slide down the side of the valley into the stream, where it will be carried downwards until it finds a new resting-place. In this case, too, the second pay-streak will contain most of the gold that was in the first, but it will have undergone a decided movement down the stream. (*See* Fig. 98.)

3rd.—The stream may cut out its second V-shaped valley entirely to one side of the first pay-streak, but when it again begins the process of lateral planation, and forms its second flood plain, it may undercut the pay-streak and allow it to fall into the meandering stream, where part of it may quickly sink and form a pocket off the line of the true second pay-streak altogether (though it will give an indication of the former position of the first pay-streak) while part of it may be carried down by the stream and distributed in its winding channel. The true second pay-streak itself will, in this case, probably be very weak. (*See* Fig. 99.)

4th.—The second channel may be formed altogether to one side of the first pay-streak, in which case the first pay-streak will be on a terrace and the second pay-streak will probably be weak. (*See* Fig. 100.)

Any of these conditions may occur in different parts of the same valley, and their relative intensity, or rapid changes from one to another, may cause great variations in the character of the pay-streak.

A greater number of stages in the deepening of a valley would allow for a still greater complexity in the character of the one or more pay-streaks which might be found in it, and these might be still further added to by a filling of the valley with detritus and partial re-excavation at one or more different times. But, for the period in which it was formed, the pay-streak represents the bottom of the young V-shaped valley, which formerly occupied part of the present valley.

DISCUSSION.

Mr. Newton B. Knox said the paper was on a subject of the greatest economic importance. It was surprising how little had been written, and how little was really known, on the laws of alluvial pay-streaks. Any law which would afford the slightest assistance in the tracing of the position of a pay-streak was to be most heartily welcomed.

The laws set forth in the paper, while most ingenious and interesting, dealt with the origin of the deposition of pay-streaks. They explained how pay-streaks in valleys were formed, but unfortunately held out no clue as to their practical application in the event of attempting to trace the position or define the width of virgin pay-streaks. To do that recourse must still be had to lines of pits or boreholes across the valley.

It seemed as if the author had attempted to give his laws too wide a scope. He should have called his paper "The Laws of Pay-streaks in the Placer Deposits of Klondyke," and have limited their application to that district or to districts possessing the same natural conditions.

As it was, the author had derived laws from one set of unusual conditions and offered them as applicable generally to all placer deposits "in the bottom of a valley." Now the valleys of the Klondyke were ideal for the concentration of gold, possessing as they did natural conditions approaching those of a long sluice, that was, rapid flowing waters, a regular gradient, rough bottoms, and, most important, few or no floods.

Under those conditions, the regular pay-streak for which the district had been noted would naturally occur, and the laws stated would apply. If, however, the conditions were changed to those of other districts—for instance, if a few sudden floods were introduced—the laws would break down, and it would no longer hold that the pay-streak was a marker of the former bottom of V-shaped valleys.

The destructive effect of mountain torrents swollen by floods, or by what was known in the West as a "cloud burst," was enormous.

He did not know whether "cloud burst" was an English term or not, but it was used in the Western States. Cloud bursts occurred with great frequency in Nevada and Colorado and some of the dry valleys in California, and they did enormous damage. He had seen a wall of water 10 ft. deep rushing down a valley, carrying everything before it.

During its transit the order of arrangement of the materials laid down during the more placid movement of the stream was upset, and sudden rearrangement followed. For instance, the transporting power of a current, or the weight of the largest fragment it would carry, varied with the sixth power of the velocity. A current running 3 ft. a second, or about two miles an hour, would move, according to Le Conte, fragments of stone the size of a hen's egg, or about 3 oz. in weight. From the law just quoted it followed that if the current were increased during flood to ten miles an hour the river would carry boulders of 1½ tons, and a torrent of 20 miles an hour would carry boulders 100 tons in weight. The placer miner recognised that law and its effect when he employed the method of mining known as "booming."

The author on p. 600 stated that the pay-streak is formed at the mouth of the V valley where the velocity of the current was diminished, at the head of the flood plain where the velocity of the weaker current would have no power to pick up the gold again. Let them imagine the effect of a cloud burst, or the sudden flood which at times became a wall of water of considerable height, rushing like an advancing cataract down the valley. That would sluice the valley free of most of its fine gold, and scatter it, together with the accumulated and as yet uncovered gold at the mouth of the valley, broadcast over the flood plain. Thus would be destroyed and dissipated the trail of gold marking the position of the bottom of the receding V-shaped valley.

Referring to Fig. 100, the effect of a flood rushing over the bench would be to sweep away the first pay-streak on the terrace and destroy it as limiting the path of a V valley.

He did not doubt, however, that the laws set forth in the paper applied to the formation of pay-streaks in the Klondyke districts, and it was interesting to note that they also seemed to fit in with the deposition of the platinum pay-streak as mentioned by Mr. Perret on p. 661 of his paper. Mr. Perret noted that the platinum deposits were characterized by straight lines. Thus those deposits *might* mark the trail of a receding V valley.

On the other hand it would be difficult to explain by means of that law the origin of the pay-streak in the gravelly deposit at Oroville. The deposit there represented the flood plain of an ancient stream of the Sierra Nevada, near the point where it debouched in the gulf which occupied the present great valley of California.

It had now been determined that the pay-streaks there occurred as irregular deposits with islands of sandy material between. At times there were several pay-streaks one above the other. Floods, with their burden of sand and silts, were evidently important factors in the deposition of those celebrated placers, and would largely explain the irregularity of the pay-streaks.

He agreed with the author's remark on p. 598 that the rôle played by bedrock in retaining the gold was most important, and the richness of bedrock depended on the character of the rock.

He could not, however, bring himself to agree with the author when he stated that gold rested permanently in a crevice until that crevice had gone. He thought the gold migrated more than the author would give them to understand, especially in districts given to floods and torrents.

There were two factors which played a part on the gold on its journey down the river. One of those forces was the current of the water; the other force, which was not so well recognised, was the creep of the whole river bed. He thought that that creep, which was something like an ice creep and explained how the gold got down from hillsides to the river in regions without much rain, might have something to do with moving the particles of gold along bedrock.

On the other hand, attention should be called to the remarkable depth to which the particles of gold were sometimes driven into bedrock. He remembered a case in point. Some years ago he had had occasion to examine a gold quartz prospect in Butte, California, about 50 miles above Oroville. On the property was a vein about 6 ft. wide, of shattered white quartz, lying in the upturned schists. The croppings of the vein were crossed by the Feather River, one of the rich placer rivers of the State. Where the river crossed the croppings the veins assayed about 1 oz. of gold per ton, and had every indication of being the top of an ore-shoot. A shaft had been sunk to a depth of about 50 ft., from which a drift had been run under the river. Only barren white quartz without a vestige of an ore-shoot was encountered. It was afterwards proved that the apparent shoot on the surface was in reality only placer gold driven, by the river's incessant hammering, into the cropping to a depth of about 15 ft. Otherwise the vein was quite barren.

Before closing, he wished to take up arms on behalf of one whose statements on the occurrence of the precious metals had been unconsciously misinterpreted, and whose reputation as a mining geologist had in consequence suffered. He referred to poor, patient, long-suffering Job. The poor gentleman was always misquoted; everybody was doing it. Job never meant that there was a place for gold, as was quoted by the author on p. 596 of his paper. What he did say (and he had traced it down only after a very long search) was: "Surely there is a vein for the silver and a place for the gold where they fine it," that was, where they refined it, *i.e.* there was a refinery for gold.

Job stated further: "He cutteth out rivers among the rocks; He bindeth the floods from overflowing." Surely the old gentleman who could make such observations on rivers and floods was quite incapable of issuing such vague and somewhat idiotic geologic laws as those with which the author had credited him.

There was another thing which he would like to mention, and that was the law of deposition of flour gold. Flour gold was that very fine gold which was continually moved about, which migrated by means of floods, and which was never found concentrated below the surface; although it was by itself of no economic importance, yet it added appreciably to the clean-ups. It was found on bars and the heads and tails of islands, and in little eddies caused by boulders. Such deposits could not be dignified by the name of "pay-streak," and the author had not mentioned them; but he thought that they played an important part in the aggregate clean-up, and should therefore have been mentioned.

Mr. H. W. Turner said that Mr. Knox must pardon him if he misunderstood what he said, but he thought he had objected to the author's statement that gold in a crevice remained there till the crevice was gone, and then almost immediately afterwards he emphasized the way in which gold was driven down in schistose rocks. In some of the slate bedrock of California the clay slate was taken up for about 1½ ft., and was washed for the placer gold in it because it was profitable. He thought that was commonly understood, and Mr. Knox had brought it out himself. He supported the author in his statement.

Mr. C. W. Purington thought the paper extremely interesting, and it dealt with a subject which had not been much commented upon. Nobody was more competent to pass a judgment on the matter than Mr. Tyrrell, so far as the conditions in Klondyke were concerned, because he was in the Klondyke for many years, and had certainly studied that district as fully and scientifically as anyone could do.

He remembered meeting Mr. Tyrrell in Klondyke, and they had had a good many talks about the local occurrence of the gold. The author had called his attention to the remarkable crystals of gold which were found in Victoria Gulch.

He would like to offer a similar criticism to that which Mr. Knox had offered, namely, that the author's remarks were perhaps of local application to the Klondyke. They certainly would not apply, or at any rate must be greatly modified when applied, to torrential streams such as must have existed in California, and which in fact did exist in California to-day, and in other regions where the erosion appeared to have gone on much more quickly than it had in the Klondyke.

With regard to another phase of the author's conclusions, he gave the impression that those pay-streaks were of a narrow character. In the Fairbanks district, a few hundred miles from the Klondyke, the pay-streaks were equally rich with the Klondyke, and much wider. He remembered that the pay-channels in the Fairbanks district varied from 300 to 500 ft. in width.

He would also like to call attention to the fact that the deposits on the Lena River in East Siberia were covered with from 100 to 50 ft. of overburden, and the pay-gravels were from 7 to 10 ft. in thickness underneath the overburden. At the junction of the famous Akanak-Nakatami Creek and the Bodaibo River the entire width of the pay-streak was 770 ft., and it continued its way for over two miles in length, gradually narrowing to 500 ft. below. Further than that, on the so-called bench, claims had been located on a lateral extension of the pay-channel, and it had been ascertained that the entire width of the pay-gravel, including the so-called bench, was over 1000 ft. There really was no proper bench gravel at all, but the sheet of gold-bearing gravel which it paid to mine was continuous practically at the same horizon, covering the entire width of the valley under the varying thickness of overburden.

The apparent bench was caused by a greater thickness of overburden at the sides than in the centre.

He did not think the author's hypothesis could possibly be applied to such extremely wide pay-channels as were attained in those two districts, and he must simply limit his conclusions to certain districts which had suffered comparatively gentle and long-continued erosion, to say nothing of their non-application to torrential districts in regions like California, where the gradient was much heavier and where the erosion had been faster.

Mr. T. A. Rickard ventured to suggest that the author, or the honorary editors, should add a foot-note to the paper defining the term "pay-streak," and that having defined it they should delete it. His friend Mr. Purington had spoken of a pay-streak 350 ft. wide. Now the first characteristic of a streak was narrowness. He thought that members would agree with him that the term "pay-streak" was borrowed by the placer miner from the vein miner. A "streak" as used by an Englishman was the same thing as the word *hilo* as used by a Mexican, namely, something that was particularly narrow and rich, and usually more nearly vertical than horizontal.

There was no need for the word "pay-streak," which was used in vein mining, and which when used in placer mining was only confusing. He thought that any intelligent man who had never heard the word before would assume that a pay-streak was something extremely narrow and ribbony. It need not be used, because there were other terms. The term "gold-bearing channel" described all that a pay-streak suggested, and also correctly suggested a great many things, while "pay-streak" made incorrect suggestions. Australians spoke of the "gutter." It was a miner's term, but was preferable because it suggested the bottom of an alluvial channel. A pay-streak, of course, was the gold-bearing bottommost portion of the gravel lying usually on bedrock. "Gold-bearing channel" said all that was necessary.

It was a great pity to introduce terms that were mainly local.

Mr. G. W. Blakemore said that what appealed to him about the deposition of gold was that it followed the same law as the sand in any river. If one went to the concave side of a river, one would find the coarse gravel; in the coarse gravel one would find the coarse gold, and in the fine gravel the fine gold.

That condition of affairs would be upset by a flood, as a previous speaker had said.

It seemed to him, with regard to the deposition in a river, whether ancient or modern, that the same laws of gravity applied, no matter how long ago it was that that particular lot of gold was laid down. A river might be covered over with a lava bed and the conditions arrested, but while that river was in being it seemed to him that the gold shifted according to the volume of water that travelled down it.

He agreed with Mr. Rickard as to the use of the word "pay-streak." He could never quite see the point of the expression, whether applied to a lode or a vein or anything else.

In Australia they had got in the habit of calling a deposition "wash-dirt." In the case of tin, for instance, in one place he knew, one could go across a deposit of tin sands for nearly a quarter of a mile and put the bores down in almost any place and find an almost equal distribution of the tin. A condition of that kind to his mind simply arose because the waters which flowed down at that particular time were of a quiet nature; there was nothing to disturb the sand or create channels of any kind. But if one got a fast-flowing river like some he knew, gold was found first on the one side and then on the other. There seemed to be no law which would define how or where one might positively expect to find gold. In a deposition of modern date what one had to do was to find the latest channel. In ancient gravels it was quite possible to work "wash-dirt" for a certain distance and find it suddenly cut off, the sand having nothing in it that would pay. After going across that for a certain distance, it would be found that the "wash-dirt" continued, showing that the payable sand had been cut through by a modern change in the stream which had disturbed the whole condition of affairs.

Mr. H. W. Turner said he would like to ask if the term "pay-streak" did not come into use from the way mining was conducted by the prospector. The prospector sank a hole in the bar or the deposit and when he came across a thin layer containing most of the gold he called that a pay-streak, and he was perfectly right, as it was a streak in the vertical section; he could not know anything about the breadth of it until later. Of course it might prove to be a sheet or layer. From the miner's standpoint it was a pay-streak.

Mr. E. T. McCarthy thought that the use of the word "pay-streak" might be defended from the analogy of streaky bacon. The layer of lean which was found in streaky bacon represented the pay-streak of the miner. Whether it was a mile wide or only 3 in. wide it was still a pay-streak, for the pay-streak was not governed by the width alone, but by the proportion between the thickness and the width, *i.e.* the pay-streak meant to the ordinary prospector or miner something narrow in thickness as compared with the width or depth.

The President said that, notwithstanding the disparaging remarks which had been made in reference to the word "pay-streak," the paper would be of practical value to those concerned in the exploitation of placer deposits. As the author had remarked, "the pay-streak was very often one of the most elusive of phenomena," although its formation was governed by certain geological laws.

CONTRIBUTED REMARKS.

Mr. Edward Halse: I quite agree with Mr. Rickard that it is inadvisable to apply the term "pay-streak" to the pay-channel of placer deposits, but this or "pay-lead" appears to be of local signification in the Klondyke district, and localisms are proverbially hard to kill.

Dr. R. W. Raymond* defines pay-streak as "the zone in a vein which carries the profitable or pay-ore." For alluvial deposits we already have the terms "pay-dirt," "pay-gravel" or (U.S.A.) "pay-channel," therefore why introduce "pay-streak"? Of the first three "pay-channel" is perhaps the best, as "dirt" is too vague, and, as regards the second, the material of the pay-channel is not necessarily gravel, it may consist of rock-fragments, pebbles, gravel, sand or clay, or a mixture of some or all of these; *e.g.* at El Caratal, Venezuela, it is composed of a yellow ferruginous clay covering *cascajo* or decomposed schist (bedrock), and is called *greda* by the natives.

"Pay-lead" in California, U.S.A., means the richer portion of the deep placers, and corresponds with the "gutter" of Australian miners. R. Brough Smyth† defines "gutter" as "the lower portion of a lead. A *gutter* is filled with auriferous drift, or *wash-dirt*, which rests on the palæozoic bed-rock," and "lead" as "a deep alluvial auriferous deposit or gutter. A lead, correctly defined, is an auriferous gully or creek, or river, the course of which cannot be determined by the trend of the surface, in consequence of the drainage having been altered either by the eruption of basalt or lava, or the deposition of newer layers of sand or gravel." Hence the term "pay-lead" seems as inapplicable to the Klondyke placers as "pay-streak"—only more so.

In Spanish-America various terms are used to signify the pay-streak of veins, such as *zona, banda;* (Mexico) *cinta* (ribbon), *cordón,* (Portuguese, *cordão*, cord or string), *faja* (Portuguese *faixa*, band), *hilo* (thread). In Colombia the term *cinta* is confined to the pay-channel of placer deposits; if of unusual thickness it is termed a *cintarrón.* Rafael Uribe U.‡ defines *cinta* as the "productive layer of an alluvial mine, or the band (*faja*) of mineral earth located above the bedrock (*peña*), and below the overburden (*barros*) or *guache.*" The *guache*, it is necessary to explain, is a rather hard mineral layer lying between the overburden and the *cinta*, and may be described as a false pay-channel.

In Bolivia and Peru, generally, the pay-channel of recent placers is called *venero;* this word comes from *vena*, a seam or bed. In French Guiana the richest part of the pay-channel is called *vena*, and in Bolivia, when it lies on a false bedrock of rounded pebbles, it is termed *venerillo* (or little *venero*) to distinguish it from the true pay-channel or *venero* proper.

* "Glossary of Mining and Metallurgical Terms." New York, 1881.

† "The Gold Fields of the Mineral Districts of Victoria." Melbourne, 1869.

‡ "Diccionario Ábreviado de Galicismos, Provincialismos y Correcciones de Lenguaje." Medellin (Colombia), 1887.

The following extracts from the voluminous literature on the great gold placer districts of the world have been chosen mainly to illustrate landmark principles of placer formation rather than to give detailed descriptions of the placers. For such descriptions the interested reader should consult the original works.

The fabulous gold placers of Victoria, Australia, were first discovered at Clunes in 1850 and were mined for nearly three quarters of a century thereafter, yielding more than 10 million oz. The gold was derived from myriad saddle reefs, legs, and other veins developed in the Ordovician and Silurian slate-graywacke-quartzite series of the state. The placers of Victoria are described in a classic memoir by Hunter (1909) and more recently by Bowen and Whiting (1975). Hunter classifies the Victoria placers as normal alluvial and deep leads. The latter are covered in places by Tertiary basalts; in an abbreviated form their formation and classification are detailed by Hunter (pp. 8-10) as follows:

Formation of Deep Leads

> The same laws that now control the formation of modern river beds are responsible for the origin of the deep leads. The same denuding agencies, the same steady earth movements slowly uplifting or lowering the surface, the same catastrophic convulsions suddenly altering the contour of the land, have all played their parts from the remote past down to the present hour in the making, destroying, and remaking of our deep leads. The vastly greater number of these are unaltered from the time they were cradled and covered in the ancient valleys, but there are some, classed as "elevated deep leads," which are in the old age of their geological life, and only the isolated patches of auriferous gravels covered by a protecting armour of hard basalt represent the once swift rivers flowing onward to "the ultimate slime of the world."
>
> These rivers would rise in flood time to a formidable width and strength, and would be able to transport untold millions of tons of disintegrated rock oceanwards from the highlands. This denuded material consisted during the early life of the river largely of quartz broken down from the gold bearing reefs. The gold-studded quartz fragments were rolled along in the flood waters, and worn, broken and shattered, by attrition and the concussion of boulder with boulder, and the gold was freed in "nuggets," "slugs," and "flour gold." After some time, the supply of auriferous quartz becoming limited, other fragmentary material from the granites, slates, and sandstones took its place, and covered up the auriferous gravel deposits with silt, drift, or clay. It must not be supposed that the whole of the gravel or wash deposit in a deep lead is payably auriferous; those portions immediately down stream from a known auriferous quartz zone are nearly always so, but as the lead trends away from such a zone it becomes poorer until another auriferous quartz zone is crossed from which it gathered fresh supplies of gold-bearing gravels. This crossing and recrossing alternate areas of rich and poor country has many times caused the unexpected to happen in alluvial mining, a poor mine following a highly payable one and *vice versa.*

Classification of Deep Leads

Our deep leads may, on geological grounds, be described in detail under the following headings—

Alluvium Covered Deep Leads.

Here the auriferous wash is covered by drifts, gravels, and clays to the surface. Here and there in boring operations the older stream bed can be located at various depths, where it has left traces in its meanders across each succeeding and rising valley surface. The terracing occasionally observed towards the lower end of coastal valley surfaces indicates that the present river has almost ceased its work of erosion and is now engaged in widening the valley on the outer bank of each bend.

Sub-basaltic Deep Leads.

The drifts, gravels, and clays covering the auriferous wash have in this case been covered by a flow of lava (basalt). Various cross-flows of basalt have occurred, as successive eruptions took place. In many localities we find that a considerable period of time elapsed between the volcanic outbreaks, sufficient in fact for the surface of the basalt to decompose and support a growth of vegetation or for small lakes to form wherein lived animals of which we have only the fossil remains to indicate their existence. It is noticeable that a stream invariably exists on either side of the basalt flow which has filled an old valley and the streams follow closely the contact line between the edge of the basalt flow and the older rocks.

This physical peculiarity is due partly to the fact that the adjoining rocks are, in the majority of instances, more easily decomposed and disintegrated than the basalt, which is a tough slowly weathering rock; but chiefly because the sides of the valley were much higher than the surface of the newly formed volcanic plain at the time of consolidation and hardening of the basalt flow. The rain waters, therefore, flowing down the sides of the valley and meeting with the hard edge of the basalt soon commenced to erode a channel along the surface line of junction of the two formations.

Elevated Sub-basaltic Deep Leads

There are many typical examples of this class of deep lead in Victoria. The geological conditions were at one time identical with those described under the heading of sub-basaltic deep leads, but owing to slow regional elevation of the surface and the denudation of the softer sedimentary bedrock, extensive isolated portions of the leads originally formed at much lower altitudes, and covered with a protecting sheet of basalt, are now left stranded on high mountain tops at heights varying from 1,000 to 4,500 feet above sea-level and some 2,000 feet above the streams, which formed (as explained under the previous heading) along the junction of the basalt and the older rocks.

Isolated Deep Leads

> There are some rather puzzling auriferous alluvial deposits in various parts of Victoria, which may be termed "isolated" deep leads. They consist of large well-rounded and water-worn boulders, gravel, and clay resting on a bottom which is usually almost level and has no rising banks on either side to indicate the remains of a small valley. The total length rarely exceeds a mile and a half and the width may vary from 30 to 100 feet, and the depth of sinking occasionally reaches 70 feet. They have the appearance of being remnants of a wide spread littoral deposit. Mention is here made of them although they can hardly be classed from the mining point of view as "deep leads."
>
> In addition to this form there are the auriferous portions of deep and widespread gravels exposed at the surface. In this instance the auriferous area differs apparently nowise in appearance from the surrounding deposit but in many parts of Victoria, notably Stawell, Pitfield, Ballarat, St. Arnaud, etc., narrow auriferous tracts have been worked while the surrounding material similar in general character is proved not to be payably auriferous. The sinking rarely exceeds 40 feet.

Deep leads covered by thick sequences of alluvium and volcanic flows are not unique to Victoria; they occur in California, in the Atlin district of British Columbia, in Africa, and elsewhere.

Indications of the rich and extensive placers of the Sierra Nevada of California were discovered by James W. Marshall at Coloma on the American River in 1848, after which followed the great rush of 1849. Since then some 55 million oz of gold have been won from the placers that lie mainly on the southwestern flank of the Sierra Nevada along a distance of some 250 m.

The productive gravels of the Sierra Nevada are mainly of Tertiary and Quaternary age and are described at length in a classic memoir by W. Lindgren (1911); more recent descriptions include those by Clark (1970) and Yeend (1974). The most productive placers were those in the drainage basins of the Feather, Yuba, Bear, and American rivers.

Chapter 4 of Lindgren's classic memoir is a landmark in the study and description of alluvial placers. The parts of chapter 4 dealing with the geology and origin of the placers are reproduced in full as Paper 15-2.

(Text continues on page 501.)

15-2: GOLD OF THE TERTIARY GRAVELS

Waldemar Lindgren

Reprinted from The Tertiary gravels of the Sierra Nevada of California, *U.S. Geol. Survey Prof. Paper 73,* 1911, pp. 65-76.

GEOGRAPHIC DISTRIBUTION.

The occurrence of gold in paying quantities in the Tertiary gravels of the Sierra Nevada is limited almost entirely to the gravels in which quartz and metamorphic rocks form the principal components. This is natural because the gold is derived wholly from veins occurring in the metamorphic rocks of the range. In places gold-bearing deposits of primary character are also found in granitic rocks adjacent to the metamorphic area, but these granitic rocks rarely furnish material for pebbles and cobbles on account of their rapid disintegration. A little fine or flour gold is found in the sands and clays which cover the gravels. Gravel beds embedded in the volcanic series and consisting chiefly of andesitic pebbles contain gold only when during their deposition adjacent beds of older gravels or parts of the "Bed-rock series" happened to be exposed to erosion. The Tertiary rocks of the western slope of the range, almost without exception, are barren of precious-metal deposits.

The distribution of detrital gold is strictly dependent on the distribution of primary deposits in the pre-Tertiary rocks of the range. These are confined almost exclusively to the Paleozoic and Mesozoic sedimentary rocks and to the igneous rocks which are associated with them and which were erupted and metamorphosed before the principal intrusion of the great granitic masses of the Sierra took place. The primary gold deposits were formed shortly after these granitic intrusions. It is a remarkable fact that the large areas of granite in the Sierra are almost wholly barren except close to the contact with the metamorphic series, where smaller veins may begin to appear. About the same time as the main intrusion minor masses of granitic and dioritic rocks were forced into the adjacent older series. In and surrounding these smaller intrusions gold-bearing deposits are particularly abundant. In their distribution the gold-bearing gravels reflect these conditions in a most accurate manner. The Tertiary and recent rivers traversing the large granite area of the upper part of the range are in general entirely barren, but after entering the metamorphic areas they speedily become charged with auriferous detritus. The amount of gold contained in the streams changes within short distances. Adjacent to the main granite contact, in Eldorado, Amador, and Calaveras counties, are considerable areas of the Calaveras formation—monotonous clay slates or siliceous slates without many areas of igneous rocks. Here the Tertiary channels are poor as a rule, but lower down they become greatly enriched on reaching the areas in which sedimentary and igneous rocks are intimately mingled.

In Sierra, Yuba, and Butte counties the Tertiary channels are rich in gold almost up to the divide of the range, the conditions corresponding to those outlined above. In Nevada County they are barren in the extreme eastern part, but soon after entering the metamorphic area they become greatly enriched, first by the Washington belt of quartz veins and second after crossing the long complex dike known as the Serpentine belt. In Placer County the channels are almost barren in the eastern part but become tremendously enriched in crossing the continuation of the Washington belt of quartz veins, here appearing in the vicinity of the Hidden Treasure mine. The Serpentine belt is crossed near Forest Hill and here the result is again a great enrichment. In Eldorado County the upper channels to points within a few miles of Placerville are generally poor, but at those points, where they cross the Mother Lode, coarse gold appears in enormous quantities and the enrichment continues for a considerable distance below this line. In the counties farther south similar conditions prevail. Wherever the channels cross areas rich in gold-bearing quartz veins they become heavily charged with gold. Above such deposits the channels grow rapidly poorer; below them the decrease in tenor is gradual. In Amador, Calaveras, and Tuolumne counties it so happens that most of the ancient river deposits below the great Mother Lode are either eroded or so heavily covered that they can not be mined. This great source of enrichment being absent, the general grade of the gravels in these counties is lower than in those farther north. In Tuolumne County, just previous to the Table Mountain flow, a drainage channel was established for a short time across the Mother Lode and this water-course, covered by a basaltic flow of great resistance, has escaped subsequent erosion. The gravels deposited in it have been mined underneath the Table Mountain west of the Mother Lode. They were rich in places, but the channel existed for too short a time to become heavily enriched. Smaller patches of gravel preserved in the same position west of the Mother Lode, as near Chinese Camp, have proved very rich. South of Tuolumne County few Tertiary gravels have been preserved from erosion.

The geographic relations sketched above in merest outline prove conclusively the dependence of the gravels for their enrichment on the distribution of the primary vein deposits, and it may be safely asserted that the gold in the channels is almost exclusively derived from such deposits.

DISTRIBUTION OF THE GOLD IN THE GRAVELS.

It has become almost an axiom among miners that the gold is concentrated on the bedrock and all efforts in placer mining are generally directed toward finding the bedrock in order to pursue mining operations there. It is well known to all drift miners, however, that the gold is not equally distributed on the bedrock in the channels. The richest part forms a streak of irregular width referred to in the English colonies as the "run of gold" and in the United States as the "pay streak" or "pay lead." This does not always occupy the deepest depression in the channel and sometimes winds irregularly from one side to the other. It often happens that the values rapidly diminish at the outside of the pay lead, but again the transition to poorer gravel may be very gradual. An exact explanation of the eccentricities of the pay lead may be very difficult to furnish. Its course depends evidently on the prevailing conditions as to velocity of current and quantity of material at the time of concentration. The gravel outside of the "pay streak" would ordinarily be regarded as extremely rich by the hydraulic miner, who would be content with a yield of 10 cents a cubic yard; but the drift miner is obliged to leave as unpayable gravel containing from 75 cents to $2 a cubic yard. Figure 12 (p. 151) illustrates the position of the pay lead in the Mayflower channel, according to Ross E. Browne.

SIZE OF THE GOLD.

Although the larger part of the gold in the channels is fine or moderately fine, large nuggets are sometimes found and much speculation has been indulged in as to their origin. It has been repeatedly stated in the literature that large nuggets occur more commonly in the gravels than in the veins. It is difficult to trace the origin of this tradition; it certainly has little foundation in fact. The largest masses of gold found in California are said to be that from Carson Hill, which weighed 195 pounds troy, and that from the Monumental quartz mine, in Sierra County, which weighed about 100 pounds troy. The mass at Carson Hill, if not directly in a quartz vein, was at any rate immediately below the croppings and not in any well-defined alluvial channel. The well-known heavy nuggets obtained near Columbia, Tuolumne County, were found in a vicinity of rich pocket veins where decay of rocks has proceeded without much interference since Tertiary time, and in which assuredly there has been little transportation. Heavy masses of gold are exceedingly common in the so-called pocket veins of Sonora. Many of the veins near Alleghany and Minnesota, in Sierra County, contain remarkably heavy masses of gold. Hanks,[1] in his list of nuggets found in California, states that a slab of gold quartz extracted from the Rainbow mine, near the locality just mentioned, was calculated to contain gold to the value of $20,468. The total yield from a single pocket of this mine was $116,337.

The Ballarat nuggets, some of which weighed from 100 to 200 pounds, found near the town of Ballarat, in Victoria, Australia, are often quoted as conspicuous examples of masses of gold

[1] Hanks, H. G., Second Rept. State Mineralogist California, 1882, p. 49.

found in streams for which an explanation is difficult. It is true that these nuggets were recovered by mining channels underneath the basalt, but it is not ordinarily noted that these channels were simply small gullies or ravines heading a short distance from the place where they were mined and traversing the decomposed outcrops of an exceedingly rich system of gold-bearing quartz veins. Of the direct derivation of these nuggets from such veins by processes of erosion and rock decay there can be no doubt.

The gold in the larger channels of the Sierra Nevada is usually fine to medium fine. Grains of the size of wheat kernels are considered as being very coarse gold, and in most places the size of the average grain corresponds more nearly to that of a mustard seed. In form most of the grains are flattened, a natural result of the continual pounding of the particles by the cobbles in the moving gravel. A certain proportion of the gold is extremely fine, and this part constitutes the so-called flour gold, which may be so fine that one or two thousand particles must be obtained to get the value of a cent.

Few systematic investigations are available regarding the proportion of coarse and fine gold in the channels, and the various localities show indeed great divergence. The data given by Hanks [1] and Blake [2] regarding the occurrence of nuggets show that in the main channels large masses of gold are on the whole rare. Most of the masses noted are from gulches or minor streams close to croppings. Very coarse gold was found in the tributary channel extending from Minnesota to Forest. In the Live Yankee claim, at Forest, 12 nuggets were found weighing from 30 to 170 ounces. At Remington Hill and Lowell Hill, in Nevada County, both of which are on a tributary to the main river, pieces weighing from 58 to 186 ounces are recorded. The gold is rarely found in the quartz pebbles and bowlders of the channels; however, Blake records the discovery at the Polar Star mine of a white quartz bowlder which yielded gold to the value of $5,760. This is in the gravel of a principal tributary to the Tertiary Yuba River, at a point where the contact of slates with the "Serpentine belt" is crossed. The White channel, mined by the Hidden Treasure mine, contains rather unusually coarse gold. It is a broad gravel deposit, 800 feet wide in places, accumulated on a tributary to the main river descending by way of Long Canyon, Michigan Bluff, and Forest Hill. The coarse gold is explained by the fact that the stream followed a belt of clay slate rich in auriferous quartz veins. Some of the gold occurs in rounded grains, many of which have pitted surfaces, but most of the pieces are flat. Small nuggets of a value of 10 to 50 cents are common, and larger pieces worth from $10 to $400 are occasionally encountered. At the celebrated Morning Star and Big Dipper drift mines, at Iowa Hill, the gold is also decidedly coarse, some pieces of a value up to $20 being found, but at other places along the same main branch of the Tertiary Yuba River much finer gold prevails, and a small part of it, which is difficult to recover, can even be classed as flour gold. Blake states that in the deep channels at You Bet, in Placer County, the gravel is in some places literally packed with small scale gold. He found that in a sample from American River the scales averaged less than 1 millimeter in diameter. The thickness is usually from one-third to one-fifth of the diameter.

Hoffman [3] has furnished a valuable description of the Red Point channel, which forms a tributary or upper extension of the White channel, on the Forest Hill divide. The gold obtained in this drift mine was classified by him as follows: Coarse, 15.78 per cent; medium, 48 per cent; fine, 36 per cent; powder, 0.32 per cent. Coarse gold is defined as that which will not pass a sieve of 10 meshes to the inch. Medium gold is defined as that which will not pass a 20-mesh sieve; this is more scaly and uniform in size, averaging 2,200 colors to the ounce. Fine gold is defined as that which will not pass a 40-mesh sieve; this averages 12,000 colors to the ounce. The remaining part, or powder, passed through a 40-mesh sieve and averages 40,000 colors or more to the ounce.

On the whole, it may be said that flour gold, such as is found in the beaches of the California and Oregon coasts or in the sands of Snake River, is not abundant in the Tertiary

[1] Hanks, H. G., Second Rept. State Mineralogist California, 1882, pp. 148-150.
[2] Blake, W. P., The various forms in which gold occurs in nature: Repts. Dir. Mint on Precious Metals, 1885, pp. 573-597.
[3] Hoffman, C. F., The Red Point drift gravel mines: Trans. Tech. Soc. Pacific Coast, vol. 10, No. 12, 1894, pp. 291-307.

or present gravels of the Sierra Nevada. During both Tertiary and present time the grades of the rivers have been such as to prevent its accumulation, and the largest part of such material has undoubtedly been swept out among the sediments which now fill the Great Valley of California.

RELATIVE VALUE OF QUARTZ GOLD AND PLACER GOLD.

Observations in all parts of the world have shown that placer gold is always finer than the gold in the quartz veins from which the placers were derived. The explanation, as has been shown in a most convincing manner by Ross E. Browne,[1] among others, is that the silver alloyed with the gold is dissolved by the action of surface waters. The purity of the gold becomes greater as the size of the grains diminishes, the explanation being, of course, that the proportionate amount of surface exposed to the action of solutions is greater in the finer gold. An interesting confirmation of this view is recorded by McConnell,[2] who states that examination of nuggets from the Klondike shows that their surfaces consist of gold of greater fineness than their insides. Some interesting data on the fineness of California gold have been contributed by F. A. Leach and C. G. Yale.[3] A few of these data, which were obtained from mint returns for a period embracing several months of 1898, are mentioned below. The average fineness of the gold of Nevada County is given as 855; of Placer County, 792; of Plumas County, 851; of Sierra County, 858; of Calaveras County, 835; of Tuolumne County, 804. This includes both placers and quartz mines. The finest gold produced in California is that from the San Giuseppe quartz mine, near Sonora, Tuolumne County. This gold runs from 982 to 998, or $20.63 an ounce. On the whole, however, the gold from quartz veins is decidedly lower in grade than that from placers. The highest average of fineness in California is that of the gold from the placers at Folsom, Sacramento County, which runs from 974 to 978. The gold from the dredging areas of Butte County, near Oroville, is also of high grade, averaging about 922. At the localities cited the gold is obtained mainly from Quaternary deposits in the present rivers.

In Plumas County the listings of quartz gold run from 627 to about 850 and the placer gold from 800 to 950.

In Sierra County gold from quartz mines varies from 622 to 883; gold from the hydraulic mines at Port Wine (a Tertiary channel) is 948 in fineness. At Gibsonville similar deposits show a fineness of about 900.

In Nevada County the quartz veins produce gold ranging from 645 to 890. The Tertiary gravels of the Harmony channel show the lowest grade of placer gold; it is 790 fine, but is derived from a small channel immediately crossing a number of rich veins so as to offer little chance of enrichment. In the main channel, at the Manzanita mine, at Nevada City, the fineness is 830. Gold from the deep channels of North Bloomfield and Relief has a fineness of 906 to 935; at the Alpha hydraulic mine, 940 to 950; at American Hill and French Corral, all in main channels, 930 to 950.

In Placer County the quartz veins carry gold of a fineness from 580 to 921. In the main channels of Tertiary gravels may be noted the Morning Star mine, at Iowa Hill, where the gold is 900 fine; the Big Dipper, on the same channel, 884; Michigan Bluff, 940 to 970; the Red Point drift mine, 927; and the Hidden Treasure mine, on the White channel, 924 to 941.

In Eldorado County the quartz gold varies from 570 to 901 in fineness. At the Excelsior claim, at Placerville, on one of the principal channels, the gold on the bedrock had a fineness of 925, while that in an upper stratum at the same place, on "false bedrock," reached 975. The gold in the Snow mine, above Placerville, a gravel deposit in the main Tertiary river, runs 948 fine. A drift mine at Grizzly Flat, on a small Tertiary stream near the headwaters and near some quartz veins, runs 871 fine.

In Calaveras County the quartz veins yielded gold ranging in fineness from 627 to 885; one exceptional quartz mine near Angels Camp shows a fineness of 960 to 975. The gold in the

[1] California placer gold: Eng. and Min. Jour., vol. 59, 1905, pp. 101–102.
[2] McConnell, R. G., Report on gold values in the Klondike high-level gravels. Geol. Survey Canada (pub. No. 979), 1907, p. 14.
[3] California mines and minerals, San Francisco, 1899, pp. 175–187. See also Bowie, A. J., jr., Hydraulic mining in California, 1885, p. 289; Hittell, J. S., Fourth Ann. Rept. State Mineralogist California, pp. 219–223.

gravels of the main Tertiary river draining this county yielded at Vallecito gold from 940 to 987 fine. In the Green Mountain hydraulic mine, at Mokelumne Hill, the fineness was 919.

The figures quoted show very clearly that in the main Tertiary streams a considerable refining of the gold has been going on, so that the average grade is now decidedly above 900. It is difficult to compare accurately the tenor of the gold in the present streams with that in the Tertiary channels, for it must be remembered that the former contain a mixture of detrital gold derived from Tertiary channels with much new gold set free during the erosion of the present canyon system.

DEPOSITION OF PLACER GOLD FROM SOLUTIONS.

In spite of the fact that the geologic conditions indicate so clearly a direct derivation of the gold from quartz veins, there have always been a number of adherents of the view that placer gold is formed by chemical deposition in the gravels. This view was held extensively among the Australian geologists of earlier years and was also earnestly advocated by Prof. Egleston, of Columbia University, New York. In recent years A. Liversidge [1] has made an extensive examination of nuggets from various sources in order to ascertain whether they bear any evidence of segregation in water. He concludes that they are derived entirely and directly from veins and that "any small addition they may have derived from meteoric water" is quite immaterial and may be neglected. Only in two nuggets from New Guinea were concentric lines of accretion observed. All other nuggets examined on etching developed signs of crystalline structure such as is entirely natural to find in vein gold. Maclaren [2] regards this structure as an argument in favor of growth in place, but it is difficult to understand his reason for this opinion.

The long exposure during gradual accumulation and the long rest of the gravels in channels exposed to percolation of atmospheric waters since Tertiary time has evidently produced a great enrichment in the fineness of the gold. The average grade in the main Tertiary channels is clearly much over 900. The highest grades of the fine-sized gold in Quaternary deposits where the canyons open into the valley are from 922 to 978. In this connection it is interesting to recall the statement by C. F. Hoffmann that the gold in the Red Point channel was of the highest grade wherever the gravels appeared to be particularly exposed to the percolation of water.

It is a well-known fact that solutions containing ferric chloride have the power to dissolve metallic gold to some extent and it is believed that in most places where solution and precipitation can be proved such solutions have been active. It was thought for a long time that ferric sulphate had the same property, but the investigations of Stokes [3] indicate that this is the case only where ferric chloride is also present. Pearce, Rickard, and lately W. H. Emmons [4] have shown that nascent chlorine is the really important agent in the solution of gold, while ferrous sulphate probably is the main precipitating agent. The action of manganese dioxide on acid solutions of sodium chloride would produce the necessary nascent chlorine. Undoubtedly some such action has taken place in the gravels of the Sierra, but there is little evidence that it is quantitatively important. The circulating waters are of great purity and probably contain extremely little sodium chloride and free acid. The gravels contain little recognizable manganese.

The evidence of secondary precipitation of gold in the California gravels is exceedingly meager and appears to be confined to two modes of occurrence. As noted above, pyrite and marcasite deposited in the gravels are in some places auriferous, though they have not been found to contain large amounts of the precious metals and there is usually difficulty in proving that no detrital gold was included. The second mode of occurrence consists in the deposition of gold on particles of magnetite or ilmenite associated with the gold. Microscopic preparations clearly showing that such a deposition had taken place were shown to the writer by Mr. J. A. Edman, of Meadow Valley, who has made a special study of the black sands of California. The particles referred to came from the Tertiary gravels of Providence Hill, in Plumas County, and the black grains are partly covered with a thin crust of gold. Mr. Edman admits that

---

1 Jour. Roy. Soc. New South Wales, vol. 40, 1906, p. 161.

2 Maclaren, J. M., Gold, London, 1908, p. 83.

3 Stokes, H. N., Experiments on the solution, transportation, and deposition of copper, silver, and gold: Econ. Geology, vol. 1. 1906, p. 650.

4 Bull. Am. Inst. Min. Eng., No. 46, October, 1910.

such occurrences are very rare. Several instances of the occurrence of precipitated gold in the roots of grasses near the surface and also in compact clays have been noted in the literature. Maclaren [1] cites an occurrence of this kind in the crystallized gold of Kanowna, Western Australia, where tiny yet bright and sharply defined octahedral crystals occur in the so-called "pug" or ancient clay gravel. McConnell [2] cites from the Klondike a quartz pebble carrying numerous thin specks and scales of crystallized gold dendritically arranged. Maclaren in the place referred to lays much emphasis on the occurrence of crystallized gold from alluvial mines. Such occurrences are certainly extremely rare in California. A number of specimens of this kind were found at Byrds Valley, near Michigan Bluff, in Placer County, but they were very near their source in local pocket veins and were partly rounded. The writer believes it very improbable that large crystals of gold have ever been formed under the conditions prevailing in the Tertiary gravels of California.

In all placer mines it is exceedingly common to find that the gold works downward into the softened bedrock immediately underlying the gravel and places are known where it has descended to a depth of 2 or 3 feet. On limestone bedrock extremely deep and irregular cavities of dissolution are often formed, and in these placer gold may be carried down to depths of 30 or 40 feet or even more; this was frequently observed in the rich placers of Columbia, Tuolumne County (Pl. XI, *A*, p. 72). In a recent textbook on mining geology this gold which is mechanically admixed with the bedrock is asserted to be due to chemical precipitation. The phenomenon and its true explanation are perfectly well known to every practical placer miner. At all drift mines the bedrock is removed to a depth of at least a foot below the gravel and washed with that in subsequent operations.

So far as the Tertiary gravels of California are concerned, the conclusion of the writer is that solution and precipitation of gold have played an absolutely insignificant part.

TENOR OF THE GRAVELS.

The productiveness of a channel is best measured by its yield per linear foot. A distinction must of course be made as to whether the hydraulic method is employed and thus the whole amount of gravel is washed or whether only the rich bottom layer is mined by drifting. In general it may be said that the channels yield from $70 to $500 to the linear foot; in drifting operations rarely more than half of the total amount of gold contained in the gravel is obtained, for besides the inaccessible gold in the upper gravels there are usually considerable bodies of gravel on the bedrock of too poor a grade to pay for extraction.

Pettee [3] mentions the instance of American Hill, near San Juan, in Nevada County, which was mined by the hydraulic method many years ago. The mass was 3,000 feet long, 1,000 feet wide, and approximately 150 feet high; it yielded $1,241,000, equivalent to $414 a running foot. Among channels mined by drifting the gravels at Red Point described by Hoffmann averaged 200 feet in width and yielded for a distance of somewhat over a mile at the rate of $71.65 a foot.

The Mayflower channel, also in Placer County, in which the average width of breasted gravel was 75 feet, yielded for 3,900 feet at the rate of $150 a foot, according to Ross E. Browne.[4] The Paragon channel, which is the upper continuation of the Mayflower, yielded $125 a running foot. In the same mine an upper channel 225 feet wide on tuff bedrock produced $300 a foot. At the Hidden Treasure mine the width of pay gravel averaged for a long time 250 feet and the average yield was $150 a running foot. The Ruby Gravel mine, in Sierra County, in which the channel was from 50 to 300 feet in width, was worked for a distance of 3,850 feet and yielded at the rate of $465 a linear foot. The cost is stated to have averaged $240 a foot. A number of data regarding the yield of gravels in drift and hydraulic mines are found in the reports of the State mineralogist of California.[5]

---

[1] Maclaren, J. M., Gold, London, 1908, p. 83.
[2] McConnell, R. G., Ann. Rept. Geol. Survey Canada, vol. 14, 1901, p. 64B.
[3] Report by J. D. Hague in Whitney, J. D., Auriferous gravels, 1879, p. 206.
[4] Tenth Ann. Rept. State Mineralogist, 1890, pp. 435–466.
[5] See especially Hammond, J. H., The auriferous gravels of California: Ninth Ann. Rept. State Mineralogist, 1890, pp. 105-138; Browne, R. E., The ancient river beds of the Forest Hill divide: Tenth Ann. Rept. State Mineralogist, 1890, pp. 435–466.

For comparison it may be mentioned that according to A. H. Brooks the average value of the principal creeks at Nome, Alaska, is approximately $100 a foot. The White channel in the Klondike yielded $380 a foot. In the Berry drift mines in Victoria, Australia, the yield per foot ranged from $440 to $1,293, the width of gravel mined being from 330 to 1,000 feet.

The amount of gold contained in the gravels is usually measured by the cubic yard of gravel, more rarely by the ton or the "carload." The latter is, of course, an indefinite quantity, but usually about equivalent to or a little less than 1 ton. One ton of broken gravel is assumed to contain about 18 cubic feet.

The gold content of the gravel varies, of course, enormously. In general it may be said that the upper gravels, sands, and clays are very poor; and although more gold is contained in the lower gravels it is only within a few feet of the bedrock that the rich material begins to appear. By far the greatest part of the gold is ordinarily contained in the gravel within 3 feet of the bedrock, and in many places within the last foot above the bedrock. In drifting operations only a few feet of gravel above bedrock are extracted; in many hydraulic operations the whole mass is washed, including parts of the almost barren overburden of fine gravels, sand, clay, or rhyolitic tuff. Wherever possible the overlying andesitic breccia is excluded from the bank wash, for besides being barren the tenacious masses of this material are difficult to handle.

Though it is difficult to give exact figures, it may be said that within the productive region the hydraulic washing of deep banks varying perhaps from 50 to 300 feet in height yielded, including the top and bottom gravels, between 10 and 40 cents a cubic yard. The top gravels alone will vary between 2 and 10 cents a cubic yard and the drifting ground on the bedrock from 50 cents to $15 or more.

In the following paragraphs a few data regarding the grade of the gravels are assembled from the detailed descriptions in Part II of this report.

At Morris Ravine, near Oroville, the best drifted ground yielded from $4 to $9 a cubic yard; this is a minor channel. A main branch of the Tertiary Yuba River at Poverty Hill, Sierra County, yields $2 a cubic yard within 5 or 6 feet from bedrock; the lower 2 feet contains most of the gold, but much is also derived from the upper part of the section. At La Porte, Plumas County, the deep gravels yielded from $2 to $20 a cubic yard. According to W. H. Pettee, one bank of gravel, covering an area of 250 by 100 feet and 30 feet high, yielded at the rate of $21 a ton. Most of the gold is said to be within 2 feet of the bedrock. At North Bloomfield, Nevada County, the gravels average between 200 and 300 feet in depth. The upper 120 feet of fine gravel contains small values but a considerable number of pieces of scaly gold of little value or weight. The lower 87 feet contains most of the gold, and the last 8 feet above the bedrock yields high values, averaging about $1.50 a ton. A large amount of the upper gravel at North Bloomfield, washed from 1870 to 1874, yielded, according to A. J. Bowie, jr.,[1] 2.9 cents a cubic yard. This work afforded practically no profits. In 1877 the bottom gravel, 65 feet deep, was found to yield 32.9 cents a cubic yard, but the top gravel, which was up to 200 feet deep, yielded only 3.8 cents. The Derbec drift mine, near North Bloomfield, working on a branch of the same channel, mined gravel from which $2.47 a ton was recovered.

The thick gravels between Dutch Flat and Indiana Hill are stated to yield 11 cents a cubic yard; the cemented gravel on the bedrock at Indiana Hill contained up to $9 a yard.

Rich hydraulic ground was found in the gravel areas extending from North San Juan, in Nevada County, to Smartsville, in Yuba County; large masses of them, with banks up to 150 feet in height, are said to have yielded from 30 to 37 cents a cubic yard. Between Cherokee and North Columbia, below North Bloomfield, in Nevada County, the gravels are up to 600 feet in thickness and very extensive; these top gravels are said to contain from 10 to 15 cents a cubic yard. At Omega, Nevada County, where the hydraulic banks reach 150 feet in height, the yield is said to be 13½ cents. At Blue Tent, Nevada County, the upper gravels and sands are practically barren, but the lower gravels are said to contain 15 cents a cubic yard. At this place the channel is 1,000 feet wide, and 5 feet of gravel next to the bedrock is stated to contain 50 cents a ton.

---

[1] Hydraulic mining in California, 1885, p. 74.

The Forest Hill divide, in Placer County, is particularly rich in drift mines. South of Iowa Hill the Morning Star and Big Dipper drift mines worked the main channel for a length of 10,000 feet. From 6 to 7 feet of gravel was extracted, and the contents ranged from $10 to $14 a cubic yard; gravel containing less than $2 was not considered payable. At the Dardanelles mine, at Forest Hill, the channel was 75 feet wide, and gravel to the depth of 5 feet was extracted; in one part of the mine gravel was extracted in floors to the height of 38 feet.[1] A number of smaller channels at Forest Hill were extremely rich. Pettee mentions a piece of ground on the New Jersey claim, 800 by 300 feet, from which $1,500,000 is believed to have been taken by drifting.

The Mayflower channel was mined for a distance of 3 miles and had an average width of 75 feet. A thickness of 2 to 14 feet of gravel was extracted and is said to have yielded $7 a ton; 66 per cent of the bottom gravel was found to pay for mining. At the Paragon mine, where the same channel was mined, an upper lead, 150 feet above bedrock, was discovered. The false bedrock consisted of rhyolite tuff; the channel was 225 feet wide and the lower 5 feet of gravel is said to have yielded $4.50 a ton.

The Hidden Treasure, northeast of Forest Hill, has worked for more than 8,000 feet a channel containing loose quartz gravel and ranging in width from 250 to 800 feet; from 4 to 7 feet of gravel is extracted, together with 1 foot of decomposed bedrock, and the yield is stated to range up to $1.75 a ton; at this mine the costs are unusually low.

Rich gravels were also mined near Placerville. At the Excelsior claim a considerable mass of gravel 100 feet in thickness is stated on reliable authority to have averaged $1 a cubic yard, worked by the hydraulic method. At this place two upper pay streaks occurred, one 25 feet and the other 60 feet above the bedrock. The deep gravel of the so-called Blue lead at Placerville, averaging about 100 feet in width, yielded cemented gravel containing from $2 to $3.50 a cubic yard. From Smiths Flat to White Rock many rich benches produced gravel containing as much as $19 a carload, which is somewhat less than a ton.

In Calaveras County the gravels are ordinarily of somewhat lower grade, although, of course, many rich localities were found. The gravel in the Deep Blue lead of Mokelumne Hill, at the North Star mine, is said to average $1.95 a ton in drifting operations. At the Banner drift mine, on the Fort Mountain channel, the bottom gravel is said to contain $3 a ton, and the costs of mining and milling are given as $1 to $1.25 a ton.

THE BEDROCK.

The bedrock of the Tertiary channels may, of course, consist of any of the great variety of rocks of Jurassic or earlier age which make up the Sierra Nevada. The channels occupy flat trough-shaped depressions in these rocks; the form of one of the larger channels is well illustrated by Plate V, *B* (p. 30), showing the Dardanelles mine in Placer County, the bottom of the channel being laid wholly bare by hydraulic work. On either side of the channel the rising rims may flatten into smaller benches. The bottom is, like any river channel of to-day, of irregular form, ridges alternating with depressions or potholes. The general trough shape is likely to be broken by a deeper gutter of varying width, but this feature is not always present. The ups and downs of such an old river bottom are well illustrated in figure 12 (p. 151), showing the bedrock of the Mayflower channel of the Forest Hill divide.

The surface of the bedrock is usually hard, contrasting strongly in this respect with the soft and clayey slate bedrock found in the drift mines of Victoria, Australia. In some places, however, as in the slates of the Hidden Treasure mine, the bedrock has been greatly softened and bleached so that there is no difficulty in removing it with the pick. In the few places where the gravels rest on granitic bedrock the bedrock is greatly softened and usually possesses the disagreeable quality of swelling. This was particularly well illustrated in the mines on the Harmony channel, near Nevada City, where the swelling took place so rapidly that drifts not attended to would be closed within a few months.

---

[1] Browne, R. E., Tenth Ann. Rept. State Mineralogist of California, 1892, p. 447.

In limestone areas the bedrock is extremely irregular (Pl. XI, *A*) and solution has produced holes which in places may be 50 or 75 feet in depth. Accumulation of rich gravels often takes place in these cavities.

A soft bedrock is considered advantageous because of its property of catching the gold driven across its surface in the moving gravels. Sometimes the gold will work down into the soft mass for a depth of 1 to 2 feet. On the other hand, a hard and smooth bedrock is less efficacious as a gold catcher, and serpentine is said to be especially unfavorable in this respect. The steeply dipping ridges made by alternate strata of slate serve to catch the gold, but at many places it is held to be more advantageous if the strike of the slates runs parallel to the channel than if they cross it.

In many parts of the United States gold-bearing gravels rest on clays or tuffs above the true bedrock, and this means, as a rule, several epochs of gold concentration. In the Tertiary rivers of California such secondary pay streaks and false bedrock are of comparatively rare occurrence. Gold is not retained on the surface of sand and gravel, and during the deposition of the gold-bearing gravels proper such thick clay beds were not ordinarily formed on account of the generally steep grade of the watercourses in a region of accentuated topography. Later, during the epoch of the rhyolitic eruption, such tuffs and clays were frequently deposited, but at that time there was little opportunity for the accumulation and concentration of gold in the wide flood plains. Some notable occurrences of false bedrock are mentioned in the detailed descriptions. An excellent example is that of the upper channel 150 feet above bedrock between Mayflower and Bath, on the Forest Hill divide. This channel was 225 feet wide and 5 feet deep and yielded $4.50 a ton. The lower channel, only 75 feet wide, was richer, averaging in the drifting ground $7 a ton. Another excellent example is found in the three pay streaks of the Excelsior mine near Placerville, which has been mentioned above.

MINERALS ACCOMPANYING GOLD IN THE TERTIARY GRAVELS.

Comparatively few useful minerals are found with the gold in the Tertiary gravels, but naturally the concentration which sorted out the gold from the bedrock also accumulated in the sands and gravels such heavy minerals as may be contained in the rocks. In the sluice boxes which are used for the washing of gravels these heavy minerals accumulate, and from the prevalence among them of magnetite and ilmenite they are usually referred to as "black sands."

The minerals occurring in the gravels may be divided into those of detrital origin and those which have been formed by chemical action within the gravels themselves.

DETRITAL MINERALS.

As stated above, magnetite and ilmenite are the most common of the minerals which accompany the gold, and their derivation is easily found in the basic rocks, like diabase, gabbro, and allied greenstones, which occupy so much space in the gold-bearing region. The granodiorites also furnish a considerable amount of magnetite. Most of the ilmenite is doubtless derived from the basic rocks mentioned. The Tertiary volcanic rocks are also rich in these constituents and channels traversing them are likely to contain an exceptional amount of black sand. A number of detailed determinations of the quantity of these minerals present were made in the examination of the black sands by D. T. Day at Portland in 1905,[1] and the mineralogical classification was carried out by Charles H. Warren, of the Massachusetts Institute of Technology. From the results it appears that magnetite largely prevails, but that chromite is also present in considerable quantity, as was indeed to be expected from the occurrence of large areas of serpentine in the gold belt. The black sand of Oroville contains, for instance, 1,400 pounds of magnetite, 250 pounds of chromite, and 150 pounds of ilmenite to the ton; this is the average black sand from dredging operations. At Cherokee, Butte County,

[1] Day, D. T., and Richards, R. H., Useful minerals in black sands of the Pacific slope: Mineral Resources U. S. for 1905, U. S. Geol. Survey, 1906, pp. 1175–1258.

pannings from old dumps yielded 16 pounds of magnetite and 356 pounds of chromite to the ton. In Calaveras County, black sand from a point near Murphy yielded 1,416 pounds of magnetite and 200 of ilmenite to the ton. Samples from Placerville, in Eldorado County, yielded 32 pounds of magnetite and 1,500 of ilmenite. Sands from North Bloomfield, Nevada County, yielded 8 pounds of magnetite, 200 of chromite, and 200 of ilmenite. From Nevada City, where the bedrock is granodiorite, no magnetite and chromite are recorded, but one sample showed 1,024 pounds of ilmenite to the ton. Gravels from Spanish Ranch, Plumas County, yielded black sand containing 1,760 pounds of magnetite and 218 pounds of ilmenite to the ton. Concentrates from dredges near Marysville contained 1,256 pounds of magnetite and 267 pounds of ilmenite to the ton.

The claim is often made that the black sands contain gold, but as a rule it is safe to assert that this is simply admixed detrital gold.

Platinum is of widespread occurrence in the Sierra Nevada and is always associated with the gold. Its origin is, however, entirely different, for its distribution shows clearly that it is derived from serpentine, peridotite, or gabbro, of which it is a constituent of primary origin, like the magnetite in igneous rocks. The platinum is always accompanied by small quantities of iridosmine and probably also iridium. Bright scales of iridosmine are locally present in considerable quantities. Though of widespread occurrence platinum is not recovered on a commercial scale except at Oroville and Folsom, where it is obtained by panning from the black sand after the gold has been extracted by amalgamation. A few hundred ounces represented the total yield of California in 1908, and of this amount the larger part came from the dredges at Oroville. The examination by Day, referred to above, has, however, shown that the metal is widely distributed in the Tertiary gravels. Its presence was proved in sands from the following places, besides the localities mentioned: In Butte County, Oroville, Butte Creek (Nimshew), Cherokee, Brush Creek, and Buchanan Hill; in Calaveras County, at Douglas Flat; in Nevada County, Rough and Ready and Relief Hill; in Placer County, Butcher Ranch, North Fork of American River, East Auburn, and Blue Canyon; in Plumas County, Genesee, La Porte, Nelson Point, and Rock Island Hill; in Sacramento County, Folsom; in Yuba County, Brownsville and Indian Hill. It is safe to say that platinum is universally present in the gravels of the Sierra Nevada, wherever these have been derived from the erosion of serpentine areas.

Small flakes of metallic copper are observed occasionally.

Detrital pyrite is not uncommon in the gravels. The mineral is derived from rocks of the "Bedrock series," such as amphibolite schist or clay slate; the latter especially is likely to contain well-developed crystals of pyrite. Pyrite may also be derived from the disintegration of quartz veins but is probably preserved from oxidation only where immediately covered by sand or gravel shortly after disintegration. Pyrite in large amount was noted in the White channel of the Hidden Treasure mine, in Placer County; at this place it is in part doubtless derived from the bedrock, of which about 1 foot is extracted with the overlying gravel; but there are also present here waterworn grains of pyrite which indicate the mechanical action of the streams. In the Harmony channel at Nevada City the gravel contains some pyrite derived from quartz veins crossed by the ancient watercourse.

Monazite, a phosphate of the rare metals, which so constantly accompanies the gold in some districts—as in the South Mountains of North Carolina and Idaho—is rather conspicuously absent in the gravels of the Sierra Nevada. Small amounts were identified by Warren from Rough and Ready, Nevada County. Rutile is found occasionally. Cassiterite, or oxide of tin, has been reported by Edman from Plumas County.

Zircon, on the other hand, is universally present and locally in considerable quantities. Black sand from Placerville, according to Day's report, contained 176 pounds of zircon to the ton, and similar material from the North Fork of American River in Placer County contained 340 pounds to the ton. The greatest relative quantity was found in black sand from a channel in granodiorite at Nevada City; it yielded 928 pounds to the ton.

Garnet is another mineral of wide distribution, especially in the vicinity of granitic contacts. It is usually found in small rounded grains of red to purplish color. It is nowhere very abundant, the largest quantity recorded in the black sand being from Rough and Ready, Nevada County; this material contains 446 pounds to the ton.

Of other materials there is little of interest to record. Olivine, epidote, and pyroxene occur here and there in small grains; the larger part of these minerals have been destroyed by oxidation before the accumulation of the gravels. In sand from the Hidden Treasure mine a number of small pale-reddish grains were found which were identified with some doubt as ruby or corundum. Cinnabar and amalgam have been found at the Odin drift mine, Nevada City. It is unnecessary to state that the gravels also contain a large amount of quartz sand.

One of the most interesting of the minerals occurring with the gold and one which requires some special consideration is the diamond. Its occasional occurrence with the gold was known at an early date and was discussed in some detail by Whitney.[1] The occurrences have been summarized by Turner.[2] The principal localities are at Cherokee Flat, in Butte County, and in the gravels at Placerville, Eldorado County. It is said that at Cherokee 56 specimens have been found; at other localities they have been less abundant. The diamonds found have generally been of small size and yellowish color; the largest size reported is about 1½ karats. Turner points out that at all except one of the localities where diamonds have been found in California areas of serpentine occur in the vicinity and he infers that the diamond, like platinum, once formed an original constituent of the peridotites, which were later altered into serpentine. This view is probably correct. On account of the occurrence of numerous diamonds in Butte County some mining operations have recently been undertaken near Oroville to search for the rock from which the diamonds were derived. It is probable, however, that the occurrence is too scattered to make mining operations like those in South Africa profitable. The following list of diamond localities in the Sierra Nevada is taken from Turner's paper:

Eldorado County: Placerville, south side of Webber Hill, White Rock Canyon, Dirty Flat, Smith's Flat.
Amador County: Jackass Gulch, near Volcano; Rancheria, near Volcano; Loafer Hill, near Oleta.
Nevada County: French Corral.
Butte County: Cherokee Flat, Yankee Hill.
Plumas County: Gopher Hill, upper Spanish Creek.

Turner states that a number of diamonds have in recent years been found at Placerville and that his informant, G. W. Kimble, is of the opinion that many diamonds have been crushed in the gravel mills. Undoubtedly the recovered specimens represent but a small fraction of the gems originally present in the gravels.

AUTHIGENETIC MINERALS.

Since their deposition the gravels have for long ages been exposed to percolating surface waters, and at many places the andesitic flows must have furnished a considerable amount of heat, so that it would be safe to infer that for a while at least after being covered by the flows these percolating waters were moderately warm. At the present time the waters are entirely cold. The basins of the Tertiary rivers form in fact reservoirs that gather the descending surface water, which has percolated through the overlying gravels, sands, clays, and volcanic tuffs and breccias. This abundant stored water finds its way to the rivers from a number of springs which in places very clearly mark the line between bedrock and gravel. This percolating action has continued since the deposition of the gravels in Tertiary time and it might be supposed that a great number of minerals would have been formed by it. As a matter of fact, however, the minerals formed in the gravel—the authigenetic minerals—are remarkably few. Carbonates, like calcite and dolomite, are not plentiful. Silica is the substance which has been most generally deposited and the commonly observed cementing of the deeper gravels is probably to be ascribed to a deposition of this substance, most likely in the form of opal or chalcedony. Newly formed quartz has not been observed cementing the pebbles, though small crystals of this mineral have been noted in silicified wood, which in places occurs in large amounts in the gravels. Similar minute crystals, according to C. F. Hoffmann, cover pebbles here and there in the

[1] Whitney, J. D., The auriferous gravels of the Sierra Nevada: Mem. Mus. Comp. Zool. Harvard Coll., vol. 6, 1879, pp. 362–364.
[2] Turner, H. W., The occurrence and origin of diamonds in California: Am. Geologist, vol. 23, 1899, pp. 182–191.

Red Point drift mine, in Placer County. Partly or wholly carbonized fragments of trunks and roots of trees are frequently converted to opaline masses, usually of gray, white, or black color. Pyrite is of common occurrence and has doubtless been formed through the reducing action of organic matter on sulphates contained in the waters. The pyrite is always most abundant near masses of vegetable remains, though in places it coats the surface of pebbles. Assays of such material frequently show small quantities of gold.[1] It is difficult to prove that this gold was originally present in the solution, for minute quantities of gold occur almost everywhere in the gravels. It is believed, however, that such a solution and precipitation of gold may actually have taken place on a small scale. When the gravels become exposed to the air the pyrite or marcasite oxidizes rapidly to limonite, and in many freshly exposed banks the distinction between the upper red and the lower blue parts is prominent. The blue gravel is simply that in which the pyrite or the ferrous silicates have not yet been decomposed to limonite.

As already stated, the placer gold of California is, as a rule, of a high degree of fineness. If it is assumed that this fineness, in the main Tertiary channels, is 920, this means, of course, that 92 per cent by weight is composed of pure gold. The remaining 8 per cent consists almost entirely of silver. Here and there are small quantities of lead and copper, amounting at most to about 0.25 per cent; there is also in places a little platinum and associated platinum metals, but this, of course, is really only a mechanical admixture.

Most of the gold shows a bright surface and deep yellow color, but locally, especially in the fine or scaly gold, each flattened grain is coated by some foreign substance that renders it difficult or impossible of amalgamation. Under the microscope the surface of such scales appears brown, gray, or black. In many cases at least the coating disappears on treatment with acid or by rubbing, and it is inferred that the substance is limonite, silica, or peroxide of manganese.

[1] J. A. Edman (Min. and Sci. Press, Dec. 15, 1894) reports auriferous pyrite from French Corral, Nevada County, and states that nodular and granular pyrite from a gravel mine in Butte County yielded gold at the rate of $173 to the ton.

The Klondike placer district in Yukon was discovered in 1896, after which the great rush recorded in poem and prose by Robert Service and others took place in 1898. Since then more than 10 million oz of gold have been won from the placers of the Klondike. Their heyday is past, although some small hydraulic and bulldozer operations continue on some of the streams. Extensive descriptions of the placers of the Klondike are given by McConnell (1905, 1907) and by Gleeson (1970). The history of placer mining in the Klondike is covered in an interesting book by Green (1977).

The most productive creeks were Bonanza, Hunker, Dominion, Gold Run, Sulphur, and Quartz. The underlying bedrock is mainly the Klondike Schist (Klondike series) comprising folded and faulted sedimentary quartz-mica schists, chlorite schists, sericite schists, quartzites, phyllites, pyritic graphitic schist, and highly sheared quartz porphyry sills or flows sandwiched between the sediments. The age of these rocks is not precisely known; they are probably Precambrian.

McConnell (1905) classified the auriferous Klondike gravels into three categories as follows:

- High-level gravels
  - River gravels
  - White Channel gravels
    - Yellow gravels
    - White gravels
- Gravels at intermediate levels—Terrace gravels
- Low-level gravels
  - Gulch gravels
  - Creek gravels
  - River and stream gravels

The high-level (terrace) White Channel gravels are of Tertiary age derived from deep secular decay of the Klondike Schist terrane. With downcutting of the streams they gave rise to the rich low level gravels.

The gold in the placers appears to have been principally derived from quartz boudins, blows, stringers and bedded veins sandwiched between the sedimentary beds of the Klondike Schist; quartz-barite veins containing minor amounts of galena; quartz veins in the chloritic phase of the Klondike Schist; stringers, veins, and irregular lenses of quartz in extensive, wide, northwest-trending shear zones that exhibit marked sericitization and pyritization; and the feebly auriferous pyrite of the graphitic schists and phyllites of the Klondike Schist.

McConnell (1907) was one of the first, if not the first, to observe the rind effect on nuggets marked by lower silver contents in the surface layer. His observations are a landmark in the study of nuggets.

> The variation in grade of the placer gold appears to depend mostly on original differences in grade of the vein gold from which it was derived. Creeks draining certain areas in the district carry low grade gold, while other areas supply high grade. An important centre of dispersion for low grade gold occurs west of the lower portion of Hunker creek. Hester and Last Chance creeks, Henry gulch and

Bear creek all head in the same ridge within a comparatively short distance of each other and all carry low grade gold. Big Shookum creek, a tributary of Bonanza creek, heads in a low grade area and the gold brought down by it lowers appreciably the general grade of the Bonanza Creek gold for several claims. The Dome and surrounding region furnishes a good example of a high grade area. The streams flowing outwards from this centre, including Upper Dominion, Upper Hunker, Sulphur and Gold Bottom creeks, all carry high grade gold although the values differ considerably.

While the grade of the placer gold is supposed to conform in a general way with that of the original vein gold some changes are evidently produced by leaching out of a portion of the silver contents.

Mr. M. Carey Lea in a series of articles in the American Journal of Science, commencing in Vol. XXXVII, p. 491, has shown that silver passes readily when treated with certain re-agents into an allotropic form, one of the distinguishing characters of which is its easy solubility, and the same process may go on in nature.

Evidence of loss of silver is afforded by the fact that fine gold which would necessarily be affected more by leaching than the accompanying coarse gold invariably carries a smaller percentage of silver.

Nuggets also assay higher as a rule on the surface than in the centre. Five assays of selected nuggets made by Mr. Connor in the laboratory of the Survey gave the following results:

		Center of nugget	Surface	
1.	Silver	35.8	29.4	Trail Hill, Bonanza creek.
	Gold	64.2	70.6	
2.	Silver	39.9	33.5	Chechaco Hill, Bonanza creek.
	Gold	60.1	66.5	
3.	Silver	37.3	30.3	Bonanza Creek, No. 12 below.
	Gold	62.7	69.7	
4.	Silver	46.1	41.0	Treasure Hill, Last Chance creek.
	Gold	53.9	59.0	
5.	Silver	33.0	33.5	Bonanza Creek, No. 3 below.
	Gold	67.0	66.5	

All the nuggets with the exception of No. 5 show losses in silver of from five to seven per cent on the surface, assuming that the composition was originally uniform. No. 5 was a large nugget filled with quartz and its exceptional character is probably due to its being much younger than the others. (pp. 13-14)

The study of the placers of the Chaudière River basin (Beauceville area) in Quebec, Canada, by MacKay (1921) resulted in a classic memoir in which experiment served to elucidate the general distribution of gold pay streaks in placer streams. The figure reproduced in the reprint below in which galena (as the heavy mineral) was utilized as a surrogate for gold is believed to represent the first extensive experiment in placer geology. Those who have observed the features of pay streaks in gold placers will recognize the analogy between field and experiment immediately. The conclusions drawn by MacKay from his experiments are given in chapter 6 of his memoir (Paper 15-3).

(Text continues on page 511.)

15-3: ECONOMIC GEOLOGY: PLACER GOLD

B. R. MacKay

Reprinted from Beauceville Map-area, Quebec, *Canada Geol. Survey Memoir 127*, 1921, pp. 63-69.

GENERAL DISCUSSION

ORIGIN OF PLACER GOLD

The unequal distribution of the gold values in placer deposits, the large size of many of the nuggets, the mammillary form of the nuggets, together with the increase in fineness of placer gold over vein gold, and the occurrence of crystallized gold in numerous placer deposits, has led many observers to the conclusion that the gold found in placers is a chemical precipitate rather than a product of the weathering and disintegration of primary deposits. Although most of these characteristics have been satisfactorily explained without recourse to the precipitation theory, there are still a number of advocates of that view. Experiments[1] have shown that gold may be deposited from chlorine solutions by organic matter such as peat, leather, leaves, cork, petroleum, and wood, and also by metallic compounds and metals. The difficulty, however, does not lie in the precipitating of the gold from solution, but in getting the gold into solution; the presence of MnO_2 being considered favourable if not essential to the solution of gold.[2] It is now generally conceded, however, that precipitation of gold has played an insignificant part in the formation of placer deposits and that gold placers may be considered almost wholly a product of mechanical concentration. This has been firmly established by the fact that the primary deposit has been frequently located by tracing the trail of gold from fine colours through deposits in which it progressively increased in coarseness and quantity until the parent source was crossed, when the values ceased abruptly.

DISTRIBUTION OF GOLD VALUES IN RELATION TO PRIMARY DEPOSIT

The coarse gold as a rule has not travelled far from its original source, but the trail of fine gold may be traced for great distances along stream courses. Unlike the fine gold, which is generally concentrated on the surface of bars or rarely more than a few inches beneath the surface, the coarse gold, once in the stream channel, with each movement that takes place in the gravels works rapidly downward, and unless arrested by a clay stratum soon reaches bedrock. This is so widely recognized that placer gold prospectors always search either the gravels lying directly over the bedrock or the bedrock itself. This enriched zone has been termed the "paystreak" or "paylead". If large nuggets are present they do not travel far after reaching bedrock and as far as known, no large nuggets have been found at any great distance from their primary deposit, unless transported by some other agency than water.

[1]Egleston, T., "The formation of gold nuggets and placer deposits," Trans. A.I.M.E., vol. IX, p. 633
[2]Emmons, W. H., Min. and Sc. Press, XCIX, 1909, pp. 751–754; 783–787. Trans. A.I.M.E., XLII, 1912, pp. 1–73, and Brokow, A. D., "Secondary precipitation of gold," Jour. of Geol., vol. XXI, 1913, pp. 251-267.

Coarse gold is not necessarily stopped on reaching bedrock, for it moves slowly with the water-soaked gravel mass until finally caught by projections or crevices in the bedrock, or until it finds a lodging place in some protected part of the channel. The slowness with which coarse gold travels in actively eroding streams has been pointed out by McConnell[1] in the case of Hunker and Bonanza creeks of the Klondike district. "The paystreak of the elevated White Channel gravels has been destroyed in places along both these streams. Whenever this occurs the creek bottoms directly opposite the destroyed portions are immediately enriched, showing that the gold, or a large part of it at least, has remained almost stationary during all the time the creeks were employed in deepening their channels from 150 to 300 feet. The horizontal movement in some instances scarcely exceeds the vertical movement. The complementary relationship existing between the creek and the hill pay-gravels has been recognized by the miners, and whenever the creek gravels are lean, pay is confidently expected on the hills, and in the productive portion of the creeks is usually found."

It is not to be inferred that the enriched bottom zone is continuous or uniformly rich. The gold, especially if coarse, occurs in rich pockets, separated often by gravels which are either barren or very poor. The richest zone does not necessarily occupy the centre or deepest part of the channel, but often crosses and recrosses the channel. In some places the values diminish gradually, whereas at others the transition from rich to poor gravels is abrupt. The auriferous zone may be wide or narrow. Owing to the configuration of the channel and the differences in texture and structure of the bedrock, certain parts of the stream undoubtedly favour more than other parts the concentration of gold, but so many factors influence the deposition that each case must be judged on its own merits.

Factors in the Distribution of Values in Placer Deposits

In studying the general problem of the distribution of values in placers, use was made of a laboratory stream, 20 feet in length, having the configuration shown in Figure 5. In addition to the bends, islands, gorges, tributary streams, depressions in bedrock, etc., introduced, riffles representing rock ledges were placed across the stream at different angles from the course and dipping in different directions. The currents set up by these various factors are shown in the upper part of the diagram. In the experiments, galena, garnet sand, and quartz sand of the same mesh were used and the concentration effected was as shown in the lower part of the diagram. Galena, having the greatest specific gravity, was taken as indicating the values in placer deposits, and the concentration effected in the laboratory stream was checked with numerous field observations on the distribution of values in placers.

Texture of Bedrock. The texture of the bedrock is one of the most important factors affecting the localization of the values. If the bedrock wears to a smooth surface, as is the case in granite, serpentine, and soft schists, or if it is covered by a clay stratum, very little of the gold is apt

[1]McConnell, R. G., "Gold values of Klondike high level gravels," Geol. Surv., Can., 1907, p. 15.

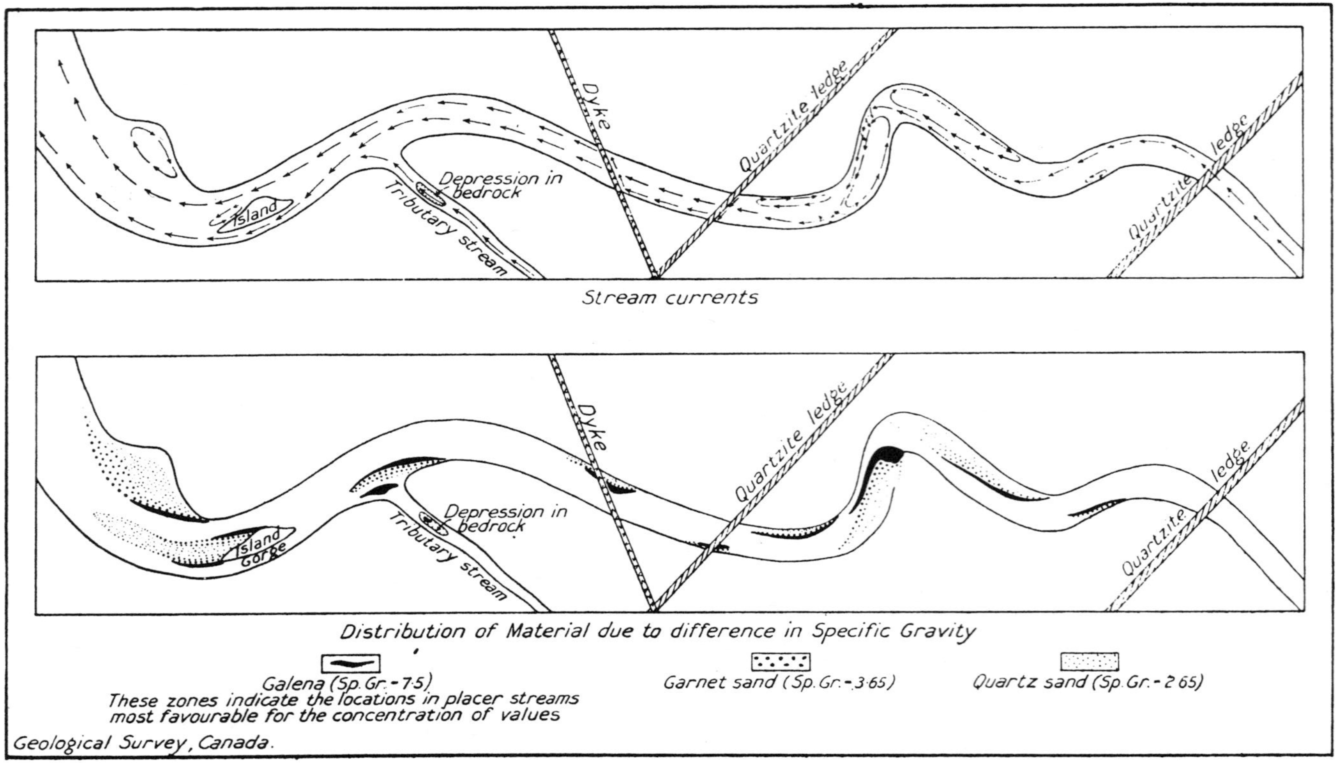

Figure 5. Diagram of laboratory stream showing currents and distribution of material of different specific gravity.

to be caught, but most of it will probably be carried downstream until a more favourable collecting rock is encountered. A rough bedrock, on the other hand, not only makes a better retaining surface, but upward currents are produced which give a jig-like movement to the gravels and result in increased concentration of the values on bedrock. Other factors, however, such as curvature of channel, decrease in gradient of stream bed, etc., may affect the localization of values, and in certain placer streams it was found that the gravels overlying rough bedrock were very poor in values, whereas the gravels overlying smoother bedrock retained their values.

Structure of Bedrock. The structure of the bedrock is also an important factor in gold concentration. Should the bedrock consist of fractured diorite or similar igneous rock, or of tilted beds of slates or quartzites, excellent pockets and riffles exist for collecting and retaining gold. This is especially so if the bedding or structure planes crossing the channel dip with the stream. The gold, being moved along the bedrock, becomes wedged in the riffles and is held there until the rock is weathered, when it works its way gradually down the cleavage, or bedding planes of the rock. In some deposits the crevices of the decomposed slate or schist are so rich that the bedrock has been mined to a depth of 8 feet. This retention of values is so effective that the statement is frequently made that in placer areas characterized by such steeply dipping bedding or schistose planes, the vertical distance the values have travelled exceeds the horizontal distance.

Should the riffles dip upstream they are not so effective. Moreover, values once caught are frequently dislodged through weathering of the rock and may travel a considerable distance downstream before they are again arrested.

When the beds or structure planes strike with the stream the values are moved along the troughs of the bedrock and are generally caught in the crevices.

Should the beds strike at an angle to the stream course, and especially if they project sufficiently above the bed of the channel to deflect the bedrock currents, the values moved along the bottom are deflected down along the bedding or schistose planes and accumulate in the acute angle formed by their intersection with the side of the channel, or if not retained at this place, are carried down along this side until again deflected. Numerous examples can be cited showing that the schistosity of the bedrock with reference to the direction of the stream was of great importance in determining the localization of values, but there are exceptions to this, as to every rule[1].

Cross-section of Valleys. The concentration is richer and, other things being equal, the paystreak is more sharply defined, the more V-shaped the valley is, owing to the restricted area in which the gold accumulates. The exploiting of many placer deposits has been abandoned because the width of the channel was too great to allow of sufficient concentration. The concentration of values in valleys of narrow channels is so common an occurrence that, where well-defined paystreaks in broad valleys have been encountered, they have been interpreted by some investigators as

[1]Spur, J., "Geology of the Yukon gold district, Alaska," 18th Ann. Rept., U.S. Geol. Surv., pt. 3, pp. 300–363.

indicating the position of the bed of a former V-shaped valley which was the younger representative of the present valley[1].

Gorges. Where canyons or gorges occur the effect is to increase the velocity of the stream currents and especially of the bed currents, so that values, having once entered the gorge, are likely to be carried through and deposited at the lower end where the channel widens out, unless on their journey they are arrested by projecting ledges or are caught in crevices on the bed. The formation of such placer deposits at the mouths of canyons is a very common occurrence. In some cases, however, the absence of values in the gorge and the occurrence of high values at its mouth are due to other causes.

Curvature of the Channel. The curvature of the channel is also important. In a straight channel the current is swifter in the middle than near the sides and if no other factors enter into the case, the concentrates would be laid down along or near the middle of the channel. In a curved channel, however, the action is greatly modified, as cross currents are set up, each contributing its part to the localization of the values. On reaching a bend in the channel the stream resists change of course and the central part, having the greater velocity, is least deflected and impinges against the outer bank. In so doing the water displaces the slow-flowing water near the bank which descends obliquely and displaces the slower-flowing water below. This lower water is crowded toward the opposite side, and the water previously near that bank moves toward the centre of the channel as an upper layer. This transfer gives to the current a twisting motion accompanied by the setting up of eddies in adjacent parts of the channel, as indicated by the arrows shown in Figure 5. As the current passes from the outer wall to the inner, the velocity is gradually decreased, and sorting of the material takes place. The result is that active erosion occurs on the outer or concave side of the channel and deposition in the inner side. In this way a gravel ridge is gradually built downstream from the spurs. The heavy material is deposited on the outer slope of this ridge, and the lighter material is carried up over the crest and dropped on the inner slope.

When the ridge reaches a certain height its upward growth is checked by surface currents and it builds toward the inner bank. In experiments with the laboratory stream using galena, garnet, and quartz sand of the same size, the zones of deposition of the different minerals were, at all times, sharply defined.

The greater the specific gravity and the more uniform the size and shape of the particles, the more sharply defined will be the zone occupied by the mineral. Thus, the mineralized zone ought to be much broader in gold placers than in platinum placers, for the gold particles are of a lesser specific gravity and often flaky, whereas the platinum has a greater specific gravity and the particles are of a more uniform size. That such is the case appears from Perret's[2] observations in Russia, in which he states " In broad valley sand especially in curves, platinum follows rich pay

[1]Tyrrell, J. B., "Law of the paystreak in placer deposits," Trans. Inst. of Min. and Met., 1912, pp. 593–613.
[2]Perret, Leon, "Gold and platinum alluvial deposits in Russia," Trans. Inst. of Min. and Met., 1912, p. 661.
Spur, L., "Geology of the Yukon gold district, Alaska, 18th Ann. Rept., U.S., Geol. Surv., pt. 3, pp. 360–363.
Pringle, L. M., "Gold placers of Forty-mile region, Alaska," Bull. 251, U.S. Geol. Surv.
Purington, C. W., Trans. A.I.M.E., vol. 29, p. 6.

streaks in a way similar to gold, but the platinum content taken in the transverse direction decreases more rapidly than does the gold content."

With continued deposition a ridge is finally built across the stream from the inner side of one curve to the inner side of the next curve. With later erosion the part of this ridge crossing the deep channel may be removed, but unless exceptional circumstances intervene, parts of the ridge remain as bars projecting downstream from the bends. These bars occur on opposite sides of the channel, each immediately below the bends where the current flows off at a tangent to the bank. The values in such bars are not distributed uniformly throughout the bar, but are generally concentrated on the outer and upstream borders. Deposits bearing similar relations to bends have been noted by Spur and Pringle in Alaska and by Purington in Russia. The same relation applies to deposits of both coarse and fine gold, except that the fine placer gold is concentrated on or close to the surface of the bar, whereas coarser gold is generally found on bedrock.

Islands. Where islands and gravel bars occur in auriferous channels the concentration of values is along the outer and upstream face of the island in positions comparable to those at bends in the stream channel.

Boulders. Large boulders in a channel set up eddies and protect any material which finds lodgment under or behind them, from further transportation. Consequently, in auriferous channels the areas underlying the large boulders have been favourable spots for the localization of values and are generally noted for their richness.

Pot-holes. Pot-holes seldom pay for the labour of cleaning them out. In certain places small channels leading into the pot-hole are rich, but the gravels of the pot-hole itself are practically barren. The reasons for these abnormal occurrences are not clear.

Change in Gradient of Channel. A decrease in the gradient of the channel causes a decrease in the velocity of the current and an increase in the friction of particles moving along it. This, together with the increase in the width of the channel, is the cause of deposition at the mouths of tributaries, and in placer districts search at these localities for auriferous deposits has frequently proved remunerative. On the other hand, as the gradient of the stream decreases the size of the gold particles transported becomes less and the concentration of values is not so marked. Along the larger streams very little coarse gold is met with, the gold occurring as minute flakes and particles, the average coarseness being about $\frac{1}{100}$-inch in diameter. Owing to the small size of the gold particles they do not concentrate on bedrock and all attempts made at locating a continuous paystreak have proved unsuccessful.

Depressions in Bedrock of Channel. Where depressions occur in the bedrock of auriferous channels the values in many places are concentrated on the downstream or backward slope of the depression, owing, doubtless, to the increase in frictional resistance of the particles moving along the reversed slope of the bedrock.

Flood Conditions. The increase in volume of water and the velocity of the current during flood conditions are apt to destroy completely the concentration which was formerly effected, or so alter the localization of values as to render it difficult, if not impossible, to determine the position of the paystreak. Frequently during flood conditions a river changes its channel completely. Moreover, if a river be greatly overloaded, sand and gravel are deposited too rapidly to permit of a concentration of the heavy minerals and the alluvium thus formed is generally low in values. When such deposits are re-sorted by stream action, however, the heavier material is gradually concentrated on bedrock unless arrested in its downward course by a stratum of clay.

Effect of Clay Strata. Where clay strata occur in auriferous gravel deposits the upper surfaces are invariably found to be rich in values concentrated from the overlying gravels. In some deposits, as many as seven such clay layers are known to occur, upon each of which lies a layer of auriferous gravel.

General Application of Foregoing Principles to Prospecting

Although too much confidence must not be placed in predictions as to the probable location of values along streams which have not been thoroughly prospected, yet in such prospecting much useless search may often be avoided by the application of the foregoing principles. This is especially so in recent placer deposits, of which numerous examples can be cited supporting the conclusions reached by the laboratory experiments.

The principles are also applicable to buried placers. In prospecting for buried pre-glacial placers it should be remembered that the former streams which deposited the values had characteristics similar to those of present streams, with their irregularities of channel and differences in structure and texture or bedrock. In the pre-glacial channels, as in the present stream channels, certain localities were much more favourable than others for the concentration of values. By taking into consideration, therefore, the configuration of the buried valley, the probable position of the stream within it, the location of spurs, bends, the position of tributary stream junctions, the texture and structure of the bedrock, and other factors above mentioned, the most likely localities for valuable placer deposits may be determined. In glaciated districts, however, the pre-glacial concentration so effected is apt to be completely destroyed by the erosive action of the ice-sheet. This is specially likely should the trend of the valleys be with the direction of ice movement. When the channels lie transverse to the direction of ice movement, the deposits, being protected from the sweep of the ice, are more apt to remain undisturbed, but in such cases the deposition of glacial drift is often so great as to obliterate the configuration of the pre-glacial rock valley, and thus make the position of the paystreak in the buried channel indeterminable. In studying the localization of values in the deep channels care must be taken not to confuse the pre-glacial form of the valley with that resulting from post-glacial stream cutting. Although in most cases the present and pre-glacial channels lie in the same rock valleys, their courses seldom coincide, but are generally found to cross and recross one another many times.

Modern beach placers carrying economic amounts of gold are relatively rare. The classic example is the beach placers at Nome in the Seward Peninsula, Alaska. Other examples are known along the Oregon and California coasts, in Nova Scotia, in British Columbia, along some of the beaches in Chile, along the coasts of Westland and Southland in New Zealand, and elsewhere. Those at Nome demonstrate most of the features of gold-bearing beach placers.

The beach placers at Nome and the geology of the area are described in detail in the publications by Collier and co-workers (1908) and by Nelson and Hopkins (1972). Briefly, the bedrock in the vicinity of Nome is composed of highly folded and faulted metamorphic rocks, mainly of Paleozoic age, consisting of various sediments now largely schists, graphitic quartzites, phyllites, limestones and graphitic slates, all cut in places by granitic rocks and greenstones. Veins and small lenticular masses of quartz and calcite are widely distributed in all of the metamorphic rocks but most abundantly in the schists of the Nome group near the localities rich in placer gold. Most of the veins, especially the larger ones that crosscut the foliation, carry free gold and sulfides. Silver-lead veins are also known from Seward Peninsula, and there are quartz veins bearing cassiterite and a little gold in the far western part.

The Nome area is characterized by a flat, crescent-shaped, alluvial coastal plain or tundra between the sea and the highlands that rise to 1,000 ft or more some 5 mi inland. The alluvium of the coastal plain is predominantly fine, usually sand and sandy clay, with some coarser gravel layers and frequent beds of clay. Most of it is slightly auriferous. These sediments represent late Tertiary and Quaternary deltaic deposits laid down in the sea by streams and rivers bringing material from the auriferous highlands. The gravels and sands are up to 100 ft or more thick in places and are covered with muck to varying depths. Inland, the alluvial deposits are river, stream, and lake gravels and sands. The Nome area was extensively glaciated during Pleistocene time, and parts are underlain by permafrost.

The classic description of the Nome placers is that by A. J. Collier and F. L. Hess from the bulletin on the gold placers of Seward Peninsula (Collier et al., 1908), classifying the various placers in the Seward Peninsula as follows:

> All the productive gold placers here to be considered occur in the gravels of alluvial origin and are deposits concentrated from the bed rock by stream or wave action. Elsewhere in this report is given a discussion of the underlying principles which govern the occurrence of the alluvial gold, and with it is discussed a scheme of classification based on genesis. For the purposes of the present description the writers will hold to a topographic grouping of the various types, which is almost identical with the classification of placers recently published by Purington (Purington, C. W. Methods and costs of gravel and placer mining in Alaska, *U.S. Geol. Survey Bull. 263,* p. 27, 1905). This group is as follows:
>
> *Classification of placers in Seward Peninsula.*
>
> Creek placers: Gravel deposits in the beds and intermediate flood plains of small streams.
>
> Bench placers: Gravel deposits in ancient stream channels and

> flood plains which stand from 50 to several hundred feet above the present streams.
>
> Hillside placers: A group of gravel deposits intermediate between the creek and bench placers. Their bed rock is slightly above the creek bed and the surface topography shows no indication of benching.
>
> River-bar placers: Placers on gravel flats in or adjacent to the beds of large streams.
>
> Gravel-plain placers: Placers found in the gravels of the coastal or other low-land plains.
>
> Sea-beach placers: Placers reconcentrated from the coastal-plain gravels by the waves along the seashore.
>
> Ancient beach placers: Deposits found on the coastal plain along a line of elevated beaches.
>
> These types of placers in occurrence and origin have close affinities with the present topographic forms with which they are associated. Genetically they fall into five groups. One embraces those which are found in the present stream channels, such as the creek, river-bar, and in part the hillside placers. In a second group fall the gravel-plain placers, also of fluvial origin, but laid down by streams that have since shifted their channels. The bench placers form a third group, which includes most of the hillside placers as well, and these were for the most part stream deposits that have been elevated and dissected. The present sea-beach placers constitute the fourth subdivision, and the elevated sea beaches the fifth.
>
> All these land forms are the result of extensive periods of denudation through which the peninsula has passed. This region has probably been exposed to subaerial erosion since late Tertiary time, and in this long interval there have been many upward and downward movements of the land masses relative to sea level. These movements have been irregular and broken by long intervals of stability during which erosion went on, and all parts of the peninsula have not been equally subjected to these influences, for the movements have been differential, so that while one part was elevated above sea level other parts were flooded. In other areas the land surface has been warped. This irregularity of uplift may make a coastal plain in one part of the region of identical age with a high bench in another part. The possible economic importance of this fact will be readily recognized.
>
> A logical discussion of the various types of placers demands that the oldest be taken up first, and this class comprises the high-bench deposits which formed before the development of the present topography. Next in order of genesis are the benches on the slopes of the present valleys; then the gravel-plain placers, together with the elevated beaches, and finally the placers in the beds of the existing watercourses as well as those of the present shore line. (pp. 142-143)

Concerning the beach placers they wrote:

> The beach placers are confined to the narrow strip of ground along the coast which is affected by the sea waves. As has been noted, the

> coastal plain is for the most part bounded on its seaward side by an escarpment from 10 to 20 feet high, which marks the inland edge of the beach. The waves are continually encroaching on the coastal plain, cutting back this escarpment and concentrating the gold from its face in the beach sands. Every year the streams probably bring down a small amount of fine gold, which is also caught up by the waves and added to the beach placer. This action has been continuous for the long period of time that has elapsed since any movement of land has taken place relative to the sea. The amount of gravel, therefore, thus affected must have been very great, but as the concentration occurs only in the strip subjected to wave action the resulting placer is confined within the same limits. Some fine gold is also found in the gently sloping floor of the sea, but since this is probably derived from the beach, it is more disseminated and finer than beach gold, and can not at present be regarded as forming a workable placer. (p. 145)

This brief summary is followed by long sections on the beach placers at Nome and on the various other types of placers in the Nome district.

Gold placers are widespread in USSR. Many are now largely exhausted, but those in the Lena, Aldan, Amur, Amgun, Kolyma, and other drainage systems of Siberia are particularly productive, are the largest river and bench placers known, and produce much of the gold of the Soviet Union. Recent descriptions of the placers of Siberia include those by Shakhov (1961), Kartashov (1971), Borodaevskaya and Rozhkov (1974), Shilo (1981), and Shilo and Shumilov (1976).

The placers of Siberia occur mainly near the headwaters of the Ob, Yenisey, Lena, Aldan, Kolyma, Amgun, and Amur rivers or near the headwaters of tributaries to these great systems. Nearly all are in deeply dissected terrains marked by ranges of low hills and an irregular network of streams. They owe their origin to the weathering and erosion of uplifted zones of considerable extent that are marked by gold-bearing deposits of many types in rocks of both sedimentary and volcanic origin. Most river and bench placers in Siberia carry one pay streak on or near the bedrock that commonly varies in width from 100 ft to 1,000 ft or more and ranges in thickness from 3 ft to 25 ft; some have multiple stacked pay streaks.

Most of the placers of Siberia are of late Tertiary and Quaternary age. Many are in the permafrost zone, and a number contain remains of the woolly mammoth *(Elephas primigenius)* and other animals. Some 20,000 woolly mammoths are said to have been found in the placers, many with the fleshy parts of the bodies intact, a feature suggestive of a catastrophic onset of the frigid climate of the glacial period.

N. Shilo has studied the placers of Siberia for many years and has published a number of detailed papers on all aspects of these placers; an extensive summary of his research is embodied in his book *Fundamentals of the Study of Placers* (Shilo, 1981). A succinct view of Shilo's ideas about the formation of placers is given in a recent abstract presented as Paper 15-4.

Another view of the origin of the placers of the northeastern USSR is given in the abstract by Zhelnin and Travin, presented as Paper 15-5.

(Text continues on page 518.)

15-4: MECHANISMS OF BEHAVIOUR OF GOLD DURING PLACER FORMATION PROCESSES IN THE NORTH-EAST OF THE USSR

N. A. Shilo and Yu. V. Shumilov, Academy of Sciences, Moscow, USSR

Mechanisms of gold behaviour during the processes of placer formation have been experimentally studied as follows:

1. release of metallic gold from lode matter;
2. gravity concentration of gold in alluvial environments;
3. corrosion of gold particles in stream channels and the formation of finely dispersed gold;
4. postsedimentary cryogenic subsidence of gold grains within placer deposits.

Laboratory modelling has revealed that the release of gold from ore sources proceeds within a polydispersion system in which prolonged static stress, thermal and glacial stress, and hydration of rock-forming minerals are important destructive processes. In a lithofacial environment (i.e. watershed, slope, channel) the volume of gold set free per unit of geological time is controlled by the rate at which dispersed material having the same (or smaller particle size) as that of the gold particles is released by weathering processes. This mechanism of gold release is well illustrated by the presence of well preserved dendrites and/or composite particles of gold even in the most hydrodynamically active sedimentary environments. The maximum possible amount of gold released from one environment can be graphically demonstrated by examination of the area of overlap of separate particle size distribution plots for weathering products and gold particles. This relationship has an important geological implication for estimating gold-productivity for various placer-generating formations of Pleistocene and older epochs during which formation of placers took place.

The gravity separation mechanism of gold concentration processes involves vertical differentiation of alluvial sediments into two parts: an upper part depleted of, and a lower part impregnated by gold. The differentiation occurs due to different mineral particles having varying hydraulic properties under conditions of turbulent stream flow. The efficiency of the gold concentration process increases under conditions of high stream-turbidity, irregular sediment particle size distribution and maximal differences between hydraulic values of high specific gravity "placer" minerals and those of the host alluvium. This causes high grade gold concentrations (as well as concentrations of the majority of other "placer" minerals) to be confined to the most coarse-grained beds of auriferous strata usually sand and conglomerates. The accumulating qualities of coarse-grained and detrital beds are re-emphasised during periods of reworking of sediments when particles of greatest hydraulic value are localized as a bed lying on bedrock below an actively downcutting river.

Gold comminution is caused by gradual abrasion of gold flakes by detrital material having sizes as great as, and smaller than those of the gold grains and hardnesses of more than 50-75 kg/mm. Many new gold particles are formed, the proportion of which may be as great as 0.06-0.2% of the whole metallic mass within the water-borne detritus. The "new-born" gold consists of particles finer than 0.1 m (75-95%). In a closed system some gold may become colloidal and concentrations can reach 130 milligrammes per ton of aqueous medium.

Postsedimentary "cryogenic" subsidence of gold particles in the body of placers subsequent to their formation is caused by the formation of microlamina and ice-filled crevices in gravel, sand and loamy materials of the auriferous bed. Owing to differences between the thermal conductivity of gold and the host material, gold particles either melt the underlying icy substratum, or slip down into the crevices formed during thawing. Either process will result in minute movements of the particles over a distance from parts of a millimetre to dozens of millimetres.

The experimentally studied mechanisms of behaviour of gold during placer formation are supported by field observations. The inferred mechanisms enable us to interpret the processes of placer formation more adequately for purposes of placer exploration and prospecting.

15-5: COMPARATIVE FEATURES OF PLACER GEOLOGY IN THE MODERN STRUCTURE OF THE NORTH-EASTERN U.S.S.R.

S. G. Zhelnin and Yu. A. Travin, Academy of Sciences, Moscow, USSR.

The northeastern U.S.S.R. lies within the permafrost zone. Subpolar climatic conditions categorise it as a region of periglacial lithogenesis. Despite this background the region is characterised by a diversity of placer formation conditions, ages and geomorphical arrangements. As a result of differentiated neotectonic movements of different amplitudes three main types of modern tectonic structure are evident in the region, each having varying amplitudes of vertical displacement relative to the palaeoplain of levelling. Each type is characterised by specific morphological relief. Three types in particular can be recognized:

1. superimposed depressions with framework zones which are tectonically stable or have been subjected to uplift of minor amplitude (relief—plains or hills);
2. zones of uplift of moderate amplitude (relief—low to medium height mountains);
3. large amplitude uplift (high mountain relief).

Placer gold deposits are connected with each type of recent structure though the deposits are qualitatively different.

Placer deposits of superimposed depressions and their framework zones are fairly widespread. They are characterized by the close genetic relationship of the chemical weathering products of the primary deposits and, to a lesser degree, with clastogene host rocks belonging to an intermediate stage and native sources related to the present erosion level. The placers also have different ages from Neogene through late Pleistocene and Holocene, the youngest being characteristic of the surfaces of depressions and framework zones.

The alluvial placer is dominant; eluvial and proluvial types being poorly developed. The gold placers typically have complex morphological features. Gold grains are not coarse but widespread and high clay content is characteristic of the productive beds with quartz abundant within the clastic sediments. The placers are commonly buried beneath unconsolidated sublithified overburden.

The most abundant placers are related to neostructures of moderate uplift. Direct or indirect relationships to the primary sources can usually be detected and in a few cases they occur as chemically weathered relict crusts on the primary ore. The majority of the placers are of late Pleistocene-Holocene Age and pre-Pleistocene placers are rare.

Alluvial placers occur within Recent valleys in terraces and in relict ancient isolated valleys. The morphological distribution of the gold is simple in the small valleys and complicated in the larger valleys. Gold fractions of medium size predominate. The grades of the ores and the volume of overburden present varies within wide limits.

The small group of placers localised within neostructures of greater uplift amplitude are related to thin Holocene alluvium. These placers lie in small valleys and are distinguished by their simple morphological setting and coarse particle gold. The differences between the placers in terms of geological structure and setting have been caused by prolonged evolution of depositional processes from the Pleiocene through the Holocene under conditions of changing relief. There has been a progression from a level palaeo-plain to a strongly differentiated relief today, together with a climatic change from moderately warm to sub-polar.

A comparative analysis of the geological features of the placers suggest that the formation conditions and nature of the relief may have been pre-determined prior to destruction of the palaeo-plain. In the course of the development of the three types of neo-structure, the previously formed placers have been buried and preserved. In the framework zones their complete or partial transformation took place, the nature of which depended on the time, character and direction of tectonic movements. Within the neostructures of moderate uplift (the latter being cyclic during the Pleistocene—10 to 12 levels of river terraces) there occurred a repeated transformation of the ancient placers accompanied by gradual dispersion of the metal content. New placers formed both from container rocks belonging to an intermediate stage (chemical weathering crusts, ancient placers) and directly from primary sources, the destruction of the latter being accompanied by the general development of physical weathering processes which were superposed upon the chemical ones. Physical weathering favoured placer formation processes and became the most important process due to a sharp fall of temperature during the Pleistocene. During that period the old placers in uplifted areas were destroyed under conditions of periglacial lithogenesis, and new placers were formed which differed from the pre-Pleistocene ones. Differences in the dynamics of streams created conditions favourable for the formation of placers within neostructures of moderate uplift and did not favour uplift areas of greater amplitude, where free gold dissemination processes prevail over concentration.

SPECIAL PROBLEMS OF PLACER GOLD

Despite the fact that the origin of gold in placers seems at a cursory glance to be relatively simple there still remain a number of mystifying features that have perplexed geologists for a century or more. Many outstanding investigations on gold in placers have been completed, yet some of the problems remain unsolved, including (1) the precise details of the source of the gold in placers, (2) the reasons for the common occurrence of gold on the bedrock or on false bottoms, (3) the origin of nuggets, and (4) the reasons for the formation of a large number of productive gold placers in the Cenozoic era, particularly during Tertiary time in all of the climatic zones of the earth. Each of these problems will now be discussed.

Source of Gold in Placers

The precise details of the source of the gold in placers has long been a problem in certain placer districts. In some districts one can point to gold-bearing deposits of various types as a source for the gold, but in others large auriferous deposits are lacking. In their place large quantities of widely distributed slightly auriferous quartz stringers, blows, boudins, and irregular quartz masses are commonly found and are said to have provided the gold to the placers. Examples of this feature are widespread, a striking one being the localization of the principal placers of the Klondike within the confines of the Klondike Schist. Another is found in the Bendeleben Quadrangle, Seward Peninsula, Alaska, where practically all of the placers of the quadrangle occur in areas underlain by the York Slate, which is marked by intrusive bodies of acid to intermediate composition, small base metal veins carrying gold, and major fault zones that have localized a plethora of auriferous quartz veinlets. Furthermore, in still other districts it is evident from analyses that the pyrite and arsenopyrite in graphitic schists and other pyritiferous rocks could have supplied large amounts of gold to placers during weathering of these rocks. Finally, large tracts of slightly auriferous laterites and blankets of other types of weathered residuum appear to have been the source of gold in some placers.

A number of possible origins for the gold in placers are therefore evident—auriferous deposits, widely distributed slightly auriferous quartz bodies, various types of rocks, especially those bearing abundant pyrite, tracts of laterites and other weathered residuum, and combinations of all four. Strange as it may seem, considering how long placers have been exploited, we have at present no certain idea of the contribution of each of the possible three principal sources of gold in the placers of any district. This problem merits intensive research, for its solution could be particularly useful in geochemical prospecting, as is evident to anyone searching for economic primary gold deposits. If for instance the gold in placers or stream sediments is wholly or at least partly derived from primary auriferous quartz or sulfide deposits, use of the element as an indicator of such deposits is valid. If, however, the sources are widely distributed quartz blows, stringers, and so on, or the country rocks or their weathered equivalents (laterites, etc.), the use of gold as an indicator for the location of economic gold-bearing primary quartz or sulfide deposits is invalid.

Origin of Pay Streaks in Gold Placers

The question of why pay streaks of gold occur on or near the bedrock or on false bottoms, considering the great quantities of sand and gravel that often overlie the pay streaks, has intrigued many students of placers. One would suspect on casual examination that the gold would be evenly distributed throughout the gravels and sands rather than concentrated in well-defined streaks.

In a classic paper discussing the distribution and origin of the gold placers of Alaska, J. B. Mertie (1940) included a long section on the formation of the fluviatile placers and commented at length on the problem of increasing fineness of gold in placers with increasing distance from the source of the gold. His views merit reproduction in full as Paper 15-6.

(Text continues on page 531.)

15-6: PLACER GOLD IN ALASKA

J. B. Mertie, Jr.

FORMATION OF PLACERS

Since most of the placers of Alaska are of fluviatile origin, and even the beach placers were in some measure concentrated as preexisting fluviatile deposits, the formation of placers deals mainly with the erosion of gold from bedrock sources, and its transportation and concentration by the action of streams.

Many data indicate that most of the placer gold of Alaska was liberated from its bedrock sources long before it was finally deposited in the placers that are now being mined. In other words, it is believed that much of this gold has been handled and rehandled by streams in many successive geomorphic cycles. In east-central Alaska, for example, the gold that originated in the Mesozoic granitic rocks south of the Yukon River, began to be freed from its bedrock sources when those rocks were first bared to erosion; and since a considerable part of the gold was deposited, and some of it still remains, in the early Tertiary conglomerates of this area, the long alluvial history of the gold in this area is not open to question. At most localities, however, this generalization is not directly provable, though it is strongly suggested by the presence of large quantities of placer gold in certain localities where few or no evidences of mineralized bedrock can be found; by great differences in fineness between placer gold and the gold of geographically contiguous lodes; and in fact by the mere presence of deeply truncated laccoliths and stocks of granitic rocks in the drainage basins where workable placers occur. Naturally, since the parent rocks range in age from Middle Jurassic to mid-Tertiary, no generalized statement is possible regarding the ancient climatological conditions under which the country rock was first disintegrated and made available to stream action. But during some epochs, notably just before the general epeirogenic uplift at the end of the Pliocene epoch, a large volume of residual and eluvial material is believed to have mantled much of Alaska. In interior Alaska, for example, there are placer camps, as at Poorman, where practically all the gravels of the placers are vein quartz and chert. In such localities, it seems certain that the concentration of siliceous rocks is due to the disintegration and destruction of the other rocks with which they were originally associated. Probably, therefore, a large part of these siliceous gravels were derived from bedrock sources long before the Quaternary period.

In streams having gradients of the same order as those obtaining in the medial courses of the usual placer streams of interior Alaska, the downstream vector of movement for gold appears to be small. Splendid examples of this feature are apparent in Fourth of July, Coal, and Woodchopper Creeks, in east-central Alaska, where important placers have been concentrated from the above-mentioned Tertiary bedrock in the present valleys during several erosional cycles. The Tertiary rocks cross these three valleys as a belt several miles in width. Upstream from these gold-bearing rocks no placers exist, and downstream from them the workable placers terminate in a very short distance. The same feature may also be noted in Hunter and Little Minook Creeks, in the Rampart district, where the present placers have been reconcentrated from a belt of unconsolidated auriferous Pliocene gravels, that crosses the valleys of these two streams. Some of the very fine gold, of course, travels many miles downstream, and lodges in the large trunk valleys; but the amount appears to be very small in comparison with that which is repeatedly handed by streams in successive erosional cycles, and still remains in the original valleys where it was first concentrated.

Since most of the placers of interior Alaska are classified as compound types, no simple exposition of their general geomorphic history can be attempted; but it is relevant to review the classical hypothesis of their formation during a single cycle of erosion. (See Fig. 6.) In most small streams much of the alluvial material is in course of progressive movement from the headwaters downstream. In the uppermost stretches all this alluvium, from the surface to bedrock, at times of flood is moved downstream, and redeposited. But in most small streams there is a zone in the valley downstream from which the al-

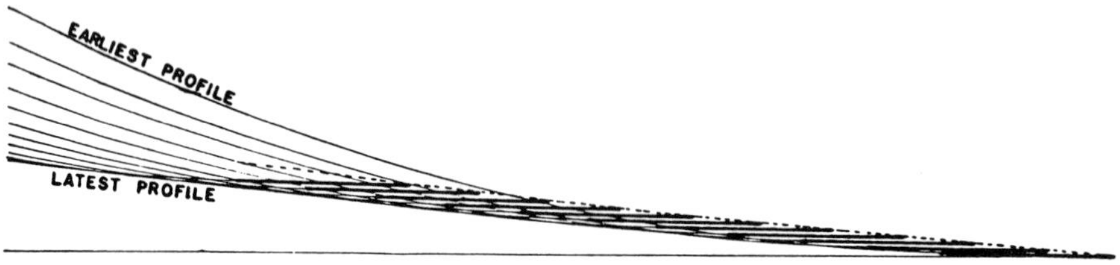

Fig. 6.—Successive longitudinal profiles of a valley, showing formation of a paystreak progressively upstream, with overlying alluvium omitted. (Vertical scale greatly exaggerated.)

luvial material on or near bedrock will not be further disturbed, even at the highest flood stages, unless the stream is rejuvenated by a lowering of its base level of erosion. The position and length of this critical zone varies with the strength of the current, the size and specific gravity of the alluvial materials, and with several other factors; yet its existence is fairly well substantiated. If a gold lode occurs at or near the head of a valley, the gold on being liberated by the process of weathering migrates downstream with the other stream detritus, gradually working its way toward bedrock. Somewhere in the critical zone, however, most of this gold, and all the coarse gold, finally comes to rest; and from this zone downstream the current of the stream is slower, and the detritus becomes thicker, so that the stream can no longer erode to bedrock. This critical zone, which lies between the headwater stretch of intermittent movement of all debris and the downstream stretch of no movement of the debris near bedrock, marks the downstream terminus of the paystreak; but the gold in process of downstream migration is also present upstream from the critical zone, and such gold may or may not constitute a paystreak, depending upon various factors. But stream erosion is a continuous process, in the course of which the valley is either extended backward into its divide; or, if another headwater stream is flowing in the opposite direction, the divide between the two streams will be lowered.

In either case, the net result is a change in the longitudinal profile of the stream bed, such that the headwater gradient is diminished and the critical zone of deposition migrates slowly upstream. Hence that stretch of gold placers, no longer subject to downstream movement, is lengthened, and a paystreak is deposited progressively upstream. The concept thus results that the part of a placer farthest downstream was deposited first and that the formation of the paystreak took place progressively upstream, as a series of overlapping wedges. This mode of paystreak formation is particularly applicable in areas where the local base level of erosion has remained sensibly constant over a long period of time; and such conditions have apparently obtained during certain erosional cycles in parts of interior Alaska.

If a paystreak was derived from a lode located in the headwater part of a valley, and if it was formed in the manner above outlined, it should be expected that most of the placer gold will be formed near, on, or in bedrock. This is actually true for most of the placer streams of Alaska, the gold being found in the lowermost few feet of gravel, on the surface of bedrock, and if the latter is greatly fractured to a depth of as much as 6 feet in bedrock. Some of the early dredge operators in Alaska failed to recognize the depth to which gold can penetrate in bedrock; and as a result of this, and also of inadequate washing in the trommel, some of the old dredging sites are now being reworked at a good profit. The absence of this localization of the gold becomes immediately a reason for searching a valley for uneroded lode sources not in the headwaters, or for postulating a rapidly changing local base-level of erosion during the deposition of the gold, or after a part of it had been deposited. One of the most striking examples of gold that is not concentrated near bedrock, is found in the placer streams which derived their gold from the Tertiary sedimentary reefs, above described.

It is obvious that this idealized mode of placer accumulation, which stresses the lack of geologic simultaneity in the formation of a paystreak, may be modified in many ways, not merely by a succession of erosional cycles, but also by conditions and events that may exist within a single cycle of erosion. The theory, as sketched, applies particularly to placers that accumulate from lodes that are localized in or near the headwater portion of a valley, and many examples of such conditions in interior Alaska could actually be cited. But the bedrock source of gold is not always thus localized, as for example where the locus of a lode system is more or less coincident with a valley, or where mineralized zones occur intermittently, crossing the valley at different places. Likewise there may be present proximate sources of gold, other than bedrock gold, such as auriferous bench gravels of any origin, which are distributed along the sides of a valley, or auriferous gravels or conglomerates which cut across valleys. All such conditions, and combinations of them, tend either to modify or to render entirely inapplicable the idealized concept; and many such examples are known in interior Alaska where the general hypothesis does not apply, even within a single erosional cycle.

Summarizing, it needs to be stressed that no general hypothesis of placer accumulation can be presented. The volume of available debris, derivable from bedrock, is affected by climatological conditions, past and present; and such conditions also control the volume of water formerly, and at the present time, existing in valleys. The character of the alluvial deposits is likewise affected by climatological conditions, as for example in interior Alaska where these deposits are frozen to great depths. The velocities and erosive power of streams, on the other hand, are functions that depend upon many variables, among which are the volume of water, the transported load and the valley gradients. And finally, the valley gradients, though locally influenced by the character of bedrock and other factors, are in large measure controlled by the duration, changes, and accelerations in local base levels of erosion. Some of these data can be deduced or inferred from geologic studies; others can not. Hence, the history and mode of formation of placer deposits in a region constitute a series of individual, yet related problems, which are seldom completely solvable.

THE PROBLEM OF FINENESS

The fineness of lode and placer gold is an economic factor of considerable significance. Gold that is 900 fine, for example, yields a profit 20 percent greater than gold having a fineness of 750; and since many examples could be cited of finenesses of this order, the illustration is by no means overdrawn. In general, therefore, the matter of fineness has been approached from a purely economic rather than a genetic point of view; and thousands of assays have been made of Alaska gold, with few attempts to correlate and to understand these significant data.

The genetic problem of fineness has several aspects, of which the most general has to do with the range and limits of the ratios of gold to silver in all the natural alloys of these metals. A second phase of the problem is concerned with the recognition and explanation of variations in the grades of lode and placer gold, and this veers into and is really a specialized part of the theory of placer genesis. Still another phase of the problem is what might be called a problem of fineness of lower magnitude, dealing with the variations in fineness within individual nuggets, grains, and crystallites of gold-silver alloys; and this in turn leads to a physical-chemical study of free gold.

It has already been shown that artificial alloys of gold and silver form a continuous series of solid solutions, with all possible compositions from pure gold to pure silver. The natural alloys of gold and silver, however, seldom contain more than 40 percent and never more than 60 percent of silver; whereas at the other end of the system, native silver is usually free of gold, and seldom contains more than traces of it. Therefore, it is possible that a miscibility gap exists in the solid state of the gold-silver system, as developed in nature. It has also been shown that the specific volumes of the natural alloys of gold and silver depart materially, and in a reverse manner, from the specific volumes of the corresponding artificial alloys. Hence, it appears doubtful that the natural gold-silver alloys are altogether solid solutions; or, if they are, the presence of a small percentage of base metals in the dross has produced some remarkable atomic readjustments. Moreover, since most geologists believe gold-quartz veins to be of hydrothermal origin, what basis is there for assuming that gold and silver will crystallize in the presence of silica, water, base metals, and mineralizers to produce a solid phase identical with that which solidifies from dry melts of gold and silver alone? And finally, is it not possible that allotrophic modifications of gold and silver may crystallize under the conditions of cooling formulated by geologists? These questions will be answered only by laboratory work on specimens of the natural gold-silver alloys; and such studies will probably have to include the investigation of many physical properties of such alloys, both on macroscopic and microscopic scales, in addition to chemical and thermal analyses.

High-grade placers can be developed by the partial erosion of high-grade lodes, thus making it possible to interpret preexisting bedrock conditions in the light of the present bedrock; yet this condition seldom obtains, because high-grade lodes are rare. On the other hand, high-grade placers may also be produced from low-grade lodes, but under such conditions the life history of the placers is so extended that the original lodes may be largely or completely removed by erosion. At some localities the roots or basal parts of the lodes may still remain, but the contained ores may be quite different from the medial and apical horizons that have been removed. Hence, it is seldom that observations on bedrock lead to a complete understanding of the character and genesis of the preexisting ore deposits that served as the sources of the placers.

Few data have been collected in Alaska that bear upon the character of the eroded lodes, but one generalization has been adduced, partly from theoretical reasons and partly from observed relationships, that seems to bear upon the distribution of fineness in placers. It is probably the usual, rather than the unusual, condition for the fineness of gold to vary in different parts of a lode. Pertinent data on this point are lacking, partly because the depth of many gold mines is less than the thickness of rocks eroded to produce the present placers, and partly because this topic has not received the consideration it could have received in our deep mines. But since the primary ores of the precious metals are known in some mines to give place at depth to base ores, it is a reasonable hypothesis that free gold itself, in the apical horizons of an original ore body, may be of higher fineness than that which occurs at considerably greater depths, as a result entirely of hypogene processes. The apical enrichment of lode gold by supergene processes can hardly be questioned at some localities.

Gold lodes have been either wholly or partly eroded to produce their derived placer deposits. Let it be assumed that the apical portions of lodes contained gold of higher grade than the lower horizons. Then if they have been wholly eroded, and if no enrichment of the gold has occurred during or after the transformation of the lode gold to placer gold, the average fineness of the placer gold should equal approximately the average fineness of the preexisting lode gold. But if the lode deposit has been only partly eroded, as is usually true, then the average grade of the gold in any one paystreak should be higher than that of the gold in the uneroded part of its antecedent lode. Moreover, and regardless of the degree of erosion, it follows that the average fineness of all the gold recovered from existing lodes should be less than that of the average fineness of all placer gold; and this is actually true. It should also follow from these considerations that gold of lower grade could be found in existing lodes than in placers; and this is likewise true, as little or no gold having a fineness of less than 600 has been found in placers, whereas lode gold has been found to have a fineness as low as 400.

In addition to variations in the primary fineness of gold within a preexisting lode system, it is probable, at least in interior Alaska, that a zone of oxidation and enrichment has existed continuously from the time that these lode systems were bared to erosion to the present day. If surficial enrichment within a lode system has operated to increase the fineness of lode gold, then in a country free from glaciation, this has been a continuous process to which all the lode gold was subjected, though perhaps in varying degrees, before it was liberated from its bedrock sources. This process of enrichment may, and in some areas certainly has, operated to produce marked differences in the grades of lode gold and placer gold derived therefrom; but unless this process varies greatly in intensity from one geologic epoch to another, it is improbable that it has been more than a minor factor in producing variations in the grade of the gold within any one paystreak.

One of the best examples known in Alaska of a great divergence between the lode gold and the placer gold derived from it, has been described by the writer[17] in the Nixon Fork district. Here occurs a quartz monzonite of Tertiary age, which has been the bedrock source for gold placers in the streams draining from it; yet the roots of the lodes are still preserved, and are being actively mined as gold lodes. From the records of thousands of ounces of this lode gold, its average fineness is known to be 735, with maximum and minimum values, respectively, of 781 and 715; yet in Hidden Creek, which drains out of the lode area, the maximum, minimum, and mean finenesses are, respectively, $961\frac{3}{4}$, $892\frac{1}{2}$, and 928 parts gold in a thousand. It should be stressed also that at this particular site the fineness does not increase progressively downstream; and although most of the paystreak that has been mined is underlain by a bedrock of limestone, nevertheless this paystreak extends upstream into the zone of quartz monzonite bedrock. Certainly no better example than this could possibly be found to prove that the gold eroded from the apical part of this lode system was of higher grade than that now being mined in the roots of the lode; but such conditions indicate an enrichment of the alloy in gold in the zone of weathering, before it was liberated from its bedrock source. In other words, as no marked variation of fineness is known within the paystreak, this locality may not be cited as an example of primary differences in the grade of the lode gold.

The general accepted theory has been stated, which pictures the paystreak as a series of overlapping wedges of gold-bearing alluvium, which are progressively deposited upstream during a single erosional cycle. According to this view the downstream end of a paystreak is

[17] Mertie, J. B., Jr. *Mineral deposits of the Ruby-Kuskokwim region, Alaska.* U. S. Geol. Surv. Bull. **864**: 193–194, 229–242. 1934.

its oldest, and the upstream end is its youngest part, if the deposition has taken place during a single cycle of erosion. A hypothesis has also been stated, which assumes that the apical part of a gold lode contains gold of higher grade than its lower horizons. Taken together, these two concepts lead directly to the conclusion that the gold of highest grade is likely to be found in the downstream end of a paystreak, and that the grade of the gold may diminish progressively upstream. This relationship actually exists in some placer paystreaks in Alaska, but in others the fineness has been observed to change erratically, or not at all.

The progressive increase in the fineness of placer gold, in going downstream, has heretofore been explained as due to the removal of silver from gold-silver alloys as a result of solution by cold surface waters, during or after the formation of a paystreak. According to this hypothesis, the gold farthest downstream has traveled a greater distance from its parent lode than the gold farther upstream. Therefore it has suffered the most handling by streams, has been to the greatest degree comminuted, and for both these reasons has been the most vulnerable to solution, Also, it has been longest separated from a bedrock source, and for this reason, too, should have been most affected by solution. This hypothesis is further supported by an experiment performed by McConnell,[18] on some of the placer gold of the Klondike district, in Yukon Territory, Canada. He found that gold shaved from the outer surface of a nugget assayed 60 to 70 parts per thousand finer than gold from the inside of the nugget; and this has been generally accepted as a proof that surface waters dissolve an appreciable amount of silver from gold-silver alloys in a paystreak. Let us look at the supporting data.

It has already been shown that the silver contained in artificial alloys of gold and silver can be completely dissolved by strong inorganic acids, only when the ratio of gold to silver is 1:3 or less; also that if this ratio is greater than 1:1.5, corresponding to a fineness of 400 parts gold in a thousand, no silver will be dissolved. It has also been shown that the solubility of silver in pure water is very slight. But practically all placer gold is of higher grade than 600 fine, and moreover we are dealing with ordinary cold surface waters, instead of strong inorganic acids. Finally, it must be remembered that any gold that has remained undisturbed in the placer paystreaks of interior Alaska since the beginning of Pleistocene time has probably

[18] McConnell, R. G. *Report on the gold values in the Klondike high-level gravels.* p. 979. Geological Survey of Canada, 1907.

been entirely untouched by water for the last million years, because these alluvial deposits were then, as now, largely frozen.

Still other facts have to be considered. The assayer knows that even strong inorganic acids do not readily remove all the silver from large pieces of a gold-silver alloy; and for this reason, a sample of the alloy, after quartation, is hammered flat on an anvil before treatment with nitric acid, in order to present a large surface to solution. Hence, the weak solvents postulated to exist in cold surface waters should affect only the outer layers of grains of placer gold; and for this reason the surficial volume of nuggets and coarse grains of gold, in which solution of silver might occur, is a smaller proportion of the total volume than in small grains of placer gold. It, therefore, follows, insofar as enrichment by solution of silver is concerned, that nuggets and coarse grains of placer gold should be of lower grade than the finely comminuted grains of placer gold. Actually, in any one paystreak, and at any one place in the paystreak, the reverse is usually true. Furthermore, when any appreciable amount of silver is dissolved in the laboratory from a gold-silver alloy, the sample becomes distinctly porous; and if much of or all the silver is removed, the sample becomes very fragile and may even crumble to a powder. Therefore, if solution of silver is actually accomplished by cold surface waters to an extent sufficient to change appreciably the fineness of placer alloys, this process should be reflected in a marked surficial porosity. But this also has not been observed.

As for McConnell's experiment, nothing is proved except that the outer surfaces of certain nuggets were of higher grade than the inner parts. It does not at all follow that this relationship is due to solution of silver by cold surface waters, for either it may have been an original characteristic of the primary lode gold, or, more probably, it was caused by surficial enrichment in the zone of oxidation, long before the gold was liberated from its bedrock source.

As a result of these considerations, the writer is not disposed unreservedly to accept the idea of any progressive change of fineness in a placer paystreak, as a result of solution of silver from gold-silver alloys by the action of cold surface waters. But it must be admitted that the alternate hypothesis is also unproved. In the first place it rests upon another hypothesis regarding the vertical distribution of different grades of gold in preexisting and present lodes. And secondly, it rests upon physical-chemical data that assume an essential identity, or at least the great similarity, of artificial and natural alloys of gold and silver, in so far as their chemical reactions are concerned. Hence, additional geologic, mining, and chemical data will be needed to prove or to disprove the hypothesis above outlined. If, for example, the emplacement of a granitic body long antedated the formation of its associated lode deposits, and if a period of diastrophism had intervened between these two processes, the apex of the intrusive mass and the apex of the mineralized zone would not necessarily correspond. And if diastrophism had occurred after either or both of these processes had occurred, the present cropping of an intrusive might correspond to neither of these antecedent apices.

Mining and chemical data are required, most of all to learn the three-dimensional variations in fineness that may exist in present gold lodes. Assays, both of bulk samples and of individual grains, should be made of uncontaminated samples of free gold taken from different horizons in lodes; and for this purpose assays of run-of-mine bullion may or may not suffice, depending upon the methods employed in recovering the gold. Assays, for example, of bullion recovered by cyanidization would certainly be useless. Complete chemical analyses should also be made, in order to learn the character and quantity of the metals in the dross; and for this purpose, even amalgamated free gold would not be serviceable.

Finally, the physical and physical-chemical properties of natural gold-silver alloys need to be studied, in order to learn how they differ from those of the corresponding artificial alloys. One of the most important problems of this work should be a complete physical and chemical examination of individual grains and nuggets of natural gold. According to Raydt's equilibrium diagram, it is possible, if the solidification of gold took place rapidly enough, that individual crystals of gold could be zonally grown, in the manner of the plagioclase feldspars. If this occurs, and if his equilibrium diagram applies to natural alloys, the outer zones of crystallites should be lower in gold than the cores. Such a condition could hardly influence materially the range of fineness in a placer paystreak, as it would be a microscopic phenomenon, of a lower order of magnitude. But the investigation of this and related phenomena is equally a part of the general problem of fineness.

Mertie thus pictures the pay streak as a series of overlapping wedges of gold-bearing alluvium progressively deposited upstream during a single erosional cycle. According to his view the downstream end of a pay streak is the oldest, and the upstream end the youngest, if the deposition has taken place during a single cycle of erosion. His idea that the apical part of a gold lode contains gold of higher fineness than its lower horizons is of interest, for if taken together his two concepts lead directly to the conclusion that the gold of highest fineness is likely to be found in the downstream end of a pay streak, and that the fineness of the gold may diminish progressively upstream.

Cheney and Patton (1967) mention that most economic geology text books either do not mention the origin of pay streaks or treat the problem rather vaguely. These authors consider two general processes that might produce pay streaks on bedrocks or false bottoms: downward chemical movement of gold, which they dismiss, or downward physical movement of particles of gold through unconsolidated sediments, which they also consider improbable. They view the streaks on the bedrock as the result of extensive scouring (sluicing) of the whole sediment column in stream and river channels during floods of unusually large magnitude at irregular intervals of decades or even centuries. Their short paper explains the mechanism in more detail (Paper 15-7).

Gunn (1968), commenting on the paper by Cheney and Patton, concluded that there can be no all-embracing explanation for bedrock placers. From his experience, however, he believes that there is a downward physical movement of gold through the unconsolidated materials in placers. The source of the agitation (or jigging) to produce this movement is microseisms, one activating agency of which is moving water. Gunn also noted that in some placers the barren gravels may have been deposited under quite different conditions than were the gravels containing the pay streaks. His views are further explained in Paper 15-8.

Tuck (1968), after mentioning that bedrock concentrations are not the sole occurrence of gold in placer deposits, noted that there are also concentrations on various types of false bottoms and disseminations throughout some gravels. He emphasized the general complexity of placers in Paper 15-9.

L. Krook (1968) maintains that climate is an important factor in the formation of pay streaks in gold placers, a point also stressed by Mertie (1940). Krook's short discussion is reproduced as Paper 15-10.

Finally, we should mention that Kolesov (1975*a,* 1975*b*), in recent efforts to answer the question of why placer gold tends to find the bedrock or false bottoms, maintains that the laws of hydrodynamic sorting in a flowing aqueous medium must be taken into consideration. He develops a number of equations indicating that large gold particles tend to accumulate in the bottom of the placer and that the finer particles are distributed relatively uniformly throughout the cross section of the placer. Details should be sought in his original papers.

(Text continues on page 539.)

15-7: ORIGIN OF THE BEDROCK VALUES OF PLACER DEPOSITS

Eric S. Cheney and Thomas C. Patton

The concentration of placer ore minerals within and near bedrock (or false bedrock) is axiomatic (6). Yet no thoroughly acceptable explanation is presented in modern text books (1, 7, 8) for the origin of ores overlain by several tens of feet of nearly barren alluvium.

Chemical redistribution occurs in hydrothermal situations such as the Witwatersrand and may be locally important in placers permeated with chloride-rich ground water, but in most placers it is insignificant for gold (6) and probably almost all other ore minerals as well. Nor can the heavier ore minerals sink appreciably through even unconsolidated sediments. The jigging action of swirls and eddies of water may be important to depths of a few feet. Abrasion of gangue minerals might conceivably reduce the sediment thickness somewhat, but this process cannot be important where the ore minerals themselves (such as diamond or cassiterite) are also brittle. Floods are capable of removing thin gravel deposits, thereby concentrating the contained heavy minerals on the bedrock (7), but the problem of bedrock values beneath thick alluvium persists.

A simple explanation for bedrock values under tens of feet of sediment is that they originate during infrequent floods of unusually large magnitude, floods that may not recur for decades or even centuries. The depth of river channels increases markedly during floods (5), especially during unusually large floods (3, 9). It is not unreasonable to assume that during such floods scouring sometimes extends to deeply buried bedrock.

An unusual confirmation of this hypothesis was discovered during the excavation for Hoover Dam (2) and has been eloquently summarized by Legget (4, p. 323); ". . . a most interesting feature revealed by excavation [of about 40 feet of sands and gravels] was the existence of an 'inner gorge' forming a narrow and tortuous channel roughly along the center of the main gorge at a depth of 75 to 80 feet below the rock benches on either side. The side rock benches were generally smooth and uniform (although showing some potholes), but the inner gorge was pitted and fluted, being generally very uneven in form and depth. This form suggested that the whole of the previously superincumbent mass of sediment had moved in great flood 'tides,' being subject to scour, whereas in the center there had been a considerable whirling action, setting up 'pothole erosion' at greater depth. Although this would seem hard to believe, it was proved beyond doubt by the essential continuity of river fill from top to bottom and also by the finding of a sawn plank of wood embedded in the gravel at the edge of the inner gorge, a position that it is believed could have been reached only by burial during a recent flood."

REFERENCES

1. Bateman, A. M., 1950, Economic Mineral Deposits (2nd ed.) : John Wiley and Sons, New York, 916 p.
2. Berkey, C. P., 1935, Geology of Boulder and Norris Dam sites: Civil Engineering, v. 5, p. 24–28.
3. Jahns, R. H., 1947, Geologic features of the Connecticut Valley, Mass. as related to recent floods: U.S.G.S. Water Supply Paper, 996 p.
4. Legget, R. F., 1939, Geology and Engineering (1st ed.) : New York, McGraw Hill Book Co., 650 p.
5. Leopold, L. B., and Maddock, T., Jr., 1953, The hydraulic geometry of stream channels and some physiographic implications: U. S. Geol. Survey, Prof. Paper 252, 57 p.
6. Lindgren, W., 1911, The Tertiary gravels of the Sierra Nevada of California: U. S. Geol. Survey, Prof. Paper 73, 226 p.
7. Lindgren, W., 1933, Mineral Deposits (4th ed.) : New York, McGraw-Hill Book Co., 930 p.
8. Park, C. F., and McDiarmid, R. A., 1964, Ore Deposits: San Francisco, W. H. Freeman & Co., 475 p.
9. Wolman, M. G., and Miller, J. P., 1960, Magnitude and frequency of forces in geomorphic processes: Jour. Geol., v. 68, p. 54–74.

15-8: ORIGIN OF THE BEDROCK VALUES OF PLACER DEPOSITS

Christopher B. Gunn

Sir: In a recent issue (1) Cheney and Patton state that although the concentration of placer ores on and in bedrock is axiomatic "no thoroughly acceptable explanation is presented in modern textbooks for the origin of ores overlain by several tens of feet of nearly barren alluvium." I suggest that there can be no all-embracing explanation for bedrock placers, just as there can be no all-embracing explanation for any other class of deposit. We must satisfactorily account for individual deposits first and, with luck, generalizations will be possible later.

Cheney and Patton state that heavier ore minerals cannot sink appreciably through thick unconsolidated sediments. They do not quote any studies that have led them to this conclusion and so we cannot tell upon what authority it rests, but my own experiences of alluvial prospecting in West Africa and North America (1966, 1967) have satisfied me that heavy minerals will very readily settle downwards when below water level and agitated in some way. The necessity of having water in an alluvial deposit at some time present or past should pose no difficulty. A source of agitation may at first sight be absent but, as anyone who has operated a seismograph well knows, the surface of the earth is constantly in motion. Moving water is one source of microseisms and it may well be that in river channels where placers are forming the magnitude and duration of the agitation are sufficient to effect a downward migration of heavy minerals.

Be that as it may, alluvial deposits necessarily take some time to form and one must not fall into the trap of regarding an alluvial section as the result of a single geological event. The "tens of feet of nearly barren alluvium" may have been deposited under quite different conditions from the basal gravels that contain the ore. Each case must be considered separately and on its own evidence.

Cheney and Patton suggest in connection with the settling of heavy minerals through unconsolidated sediment that "abrasion of gangue minerals might conceivably reduce the sediment thickness somewhat, but this process cannot be important where the ore minerals themselves (such as diamond or cassiterite) are also brittle." The present writer cannot draw any meaning from this suggestion and possibly the authors would care to enlarge on this point.

Finally, Cheney and Patton suggest that bedrock values below tens of feet of sediment might be explained by infrequent floods of unusually large magnitude, but they do not explain how this could occur. The evidence which they cite as "confirmation" of their hypothesis would seem to suggest that channel fill may on occasions move bodily "in great flood 'tides' " and this is surely the perfect antithesis of conditions which would be suitable for the separation of heavy minerals from light ones.

CHRISTOPHER B. GUNN

REFERENCES

1. Cheney, E. S., and Patton, T. C., 1967, Origin of the bedrock values of placer deposits: ECON. GEOL., v. 62, p. 852–853.
2. Gunn, C. B., 1966, An alternative method for the collection of material for heavy mineral studies: New York State Geol. Assn. 38th Ann. Meeting, Guidebook, (abstract), p. 86.
3. Gunn, C. B., 1967, Provenance of diamonds in the glacial drift of the Great Lakes region, North America: M.Sc. thesis, University of Western Ontario, London, Ontario.

15-9: ORIGIN OF THE BEDROCK VALUES OF PLACER DEPOSITS

Ralph Tuck

reprinted from *Econ. Geology* **63:**191-193 (1968).

Sir: I wish to amplify Cheney and Patton's (1967) explanation of bedrock concentrations with special reference to gold, which is the principal mineral of value having such an occurrence.

The writer has investigated many placers in Alaska, Canada, and the western United States. Particularly, he participated in detailed studies of the Fairbanks and Nome, Alaska, areas where excellent records of exploration and production had been kept for many years during the extensive operations of the United States Smelting Refining and Mining Company. Although these two major placer gold camps had different origins—that is, one by stream action and the other marine—they have many common characteristics, and at both the greater part of the values occurred as bedrock concentrations.

Bedrock concentrations are not the sole occurrence of gold in placer deposits. It also occurs as a concentration at or near the top of gravel, as definite horizons between bedrock and the top of gravel, as a dissemination throughout the gravel, and usually as a combination of several of these circumstances. However, the most productive of these have been the bedrock concentrations, which as the term is commonly used, are the concentrations directly on bedrock and in the gravel within a few feet of it. These occurrences are a reflection of the history of the area, whether it be a stream or marine deposit, either of which may be modified by glaciation.

Commercial bedrock concentrations, in the case of stream deposits, are generally the result of a long period of erosion during which the stream is dominantly downcutting. Assuming a source of gold, weathering and erosion are separating and concentrating the gold from the country rock into the bottom of the valley. All during this period, which may represent several thousand feet or more of vertical erosion, gold as it is liberated is on or near bedrock since the stream is essentially flowing on bedrock. The action has been linkened to that of a sluice box on a large scale, and there is no better analogy. Once gold reaches bedrock it generally remains there. All during this period the stream valley is essentially youthful. Eventually a baselevel is reached, maturity is attained, and dominantly the stream begins to widen its valley.

Characteristic of many of the important bedrock concentrations of gold is a narrow rich concentration meandering through a wider valley in which the bedrock of the bordering area may or may not have about the same elevation as that of the rich bedrock concentration. The bordering bedrock area may also have a concentration of gold, but it is of much lesser value than the narrow rich streak. In many cases, the narrow rich channel was high-grade enough to be hand-mined while later it and the bordering low-grade areas were dredged. The most logical explanation for this narrow rich channel is that it marks the bottom of the valley when the stream ceases dominantly to downcut and begins to cut laterally having reached a mature stage. The narrow rich channel is the concentration of gold from a volume of country rock very large relative to that which furnished the lower grade concentrations in the bordering area. It also logically follows that generally the narrow rich channel contains much coarser gold, and the particles have a higher gold-silver ratio, than the bordering area. Floods have their greatest importance in carrying gold to bedrock principally in the stage when the valley has reached maturity and is being widened with aggradation and degradation alternately occurring. At a still later stage aggradation becomes dominant, and the bedrock concentration formed during the youthful and early mature stages becomes covered with alluvium.

Few stream histories are as simple as stated above. Streams are usually aggrading in their lower reaches while degrading above. Temporary base levels are reached and then later destroyed. The availability of gold is reflected in its deposition. If all during the erosional history of a stream, gold is being freed in essentially constant amounts, then a distribution as described above, with a narrow rich channel bordered by lower-grade bedrock concentrations, is generally the result. If only a narrow high-grade channel is found in a broad valley, it would appear that there was a source of gold only during the youthful stage. Commonly the alluvium overlying the bedrock concentration contains disseminated values indicating a continuing source of gold. This aggradational period may be interrupted by a period of downcutting in which case the disseminated gold in the alluvium may be reconcentrated at or near the top of gravel. Subsequently, the stream may aggrade again, the gold reconcentration is covered, and a gold horizon above bedrock is effected. Generally the gold up in the alluvium is of a finer size than that on bedrock.

Marine gold deposits are commonly bedrock concentrations that are dominantly of two types. One, is a narrow beach concentration on or very near bedrock, and the other, when there is a very rich source of gold, is a concentration at bedrock on the abrasion platform. Generally the abrasion platform concentration is of coarser gold than that on the beach. Here again the erosional history is usually complex with changes in the land elevation relative to the sea resulting in the burial of some concentrations and in the formation of new concentrations above bedrock.

In addition to a favorable erosional history the bedrock source must contain gold in the free state and in particles of some appreciable size in order to form a commercial placer. Large sulfide deposits, such as the large copper deposits of the west, which in the aggregate may contain much gold and have suffered extensive erosion, have not resulted in commercial placers because of the fine state of subdivision of the gold when liberated from the enclosing sulfides. It is apparently scattered far and wide. Many of the rich Tertiary gold camps of the west did not form placers for similar reasons.

Gold particles of any appreciable size—say more than one milligram in weight which seems essential for a commercial placer—are not moved far from their source even through a long period of erosion. An exception to this is where glaciation has taken place and moved and destroyed previously formed concentrations. The horizontal movement of gold of any appreciable size from its source appears to be principally a result of the vertical distance that the gold travelled during erosion. Studies seem to indicate that during say 5,000 feet of erosion, coarse gold (+ 1 milligram in weight) liberated at the beginning of this cycle may not travel more than 10,000 to 15,000 feet horizontally. There are a number of examples of gold concentrations that have later been dissected by stream action and lowered from 100 to 400 feet below the old level where the bulk of the gold in the eroded portion has only moved a similar distance horizontally. This is true, although to a lesser extent, of marine deposits where surf action during storms is a much more powerful transporting agent than streams. At Nome, where the sources of placer gold have long been eroded, stream placer gold deposits had been formed in the valleys entering onto the old coast line. The sea advanced onto the land mass, bevelled it off, and destroyed these old stream concentrations. However, the bedrock concentrations of coarser gold found on the now four-mile wide coastal plain still roughly indicate the location of the old drainage lines. The finer gold was redistributed into beach concentrations by surf action during storms.

Although coarse gold does not travel far, there is a decrease in average particle size away from the source and they become smoother and more rounded. Very fine flake gold, of such a size that several thousand colors are necessary to have a value of one cent, is carried great distances by water as surface tension effects become a dominant factor. Such fine gold is characteristic of some localities along the Colorado and Snake Rivers in the western states but has never made commercial deposits.

Because of its physical and chemical characteristics gold survives although subjected to long periods of erosion. The only changes that take place are liberation from any adhering rock, rounded and smoothing of the particles, and an increase in the ratio of gold to silver. Away from the source the original silver with the gold is progressively slowly leached, thus purifying the gold to a small degree.

RALPH TUCK

REFERENCE

1. Cheney, E. S., and Patton, T. S., 1967, Origin of the bedrock values of placer deposits: ECON. GEOL., v. 62, p. 852–853.

15-10: ORIGIN OF BEDROCK VALUES OF PLACER DEPOSITS

Leendert Krook

reprinted from *Econ. Geology* **63**:844-846 (1968).

Sir: After several discussions on the problem of bedrock values I may draw the attention to an agency that so far has not been mentioned, viz. the climate.

Cheney and Patton (1967) explained the bedrock values by assuming that they originated during infrequent floods of unusually large magnitude. This apparently needs some further information. If a river deposit were rather homogeneous throughout the profile, an enormous flood might indeed remove a great part of the sediment which, after renewed deposition, would have the coarsest particles at the bottom and the finest at the top. The heavy ore minerals would be among the first to be deposited and thus appear in the gravel. This is what is often found, a gravel layer rich in ore minerals directly overlying the bedrock and rather barren layers of sand and clay on top. I do not agree with Gunn (1968) who states that Cheney's and Patton's hypothesis is the "perfect antithesis of conditions which would be suitable for the separation of heavy minerals from the light ones." Such a separation would occur in a greater or lesser degree. If the top layers attain depths of tens of feet, however, it seems improbable that the scouring would have reached that depth across a great part of the surface of the river bottom. Gunn is right, when he says that the "tens of feet of nearly barren alluvium" may have been deposited under quite different conditions from the basal gravels that contain the ore minerals. The contribution of Tuck (1968) is very elucidating. He explains the high bedrock concentrations of gold in fluvial deposits as a "result of a long period of erosion during which the stream is dominantly downcutting." The course of the stream where the downcutting ended usually shows the highest values. After this the stream "begins to cut laterally having reached a mature stage." This is pure davisian cyclic thinking and may often be correct. In many cases, however, the transition from vertical to lateral erosion and aggradation is caused by a change of climate. This has been the case during the Pleistocene where lateral erosion and aggradation took place during the glacial periods under periglacial conditions, while the warm interglacials showed downcutting to a high degree, thus in the course of time giving rise to a series of river terraces of wide extension.

In this case it is only to be expected that ore minerals like gold reach higher values in the lower terrace deposits, since with each phase of downcutting the gold is concentrated because the heavy components are transported only little horizontally as Tuck pointed out, whereas the bulk of the sediment that is being eroded is transported far away downstream.

In the environments of Benzdorp on the Lawa River in Surinam which for a long time has been the most important gold producing area of the country, several terraces occur along the creeks, each of which has a coarse layer of gravel with sands and clays on top. Gold is found mainly in the gravel and has its source in the quartz veins occurring in the crystalline basement rocks. The greatest concentrations are usually found in the youngest gravels underlying the present alluvial plains (Brinck, 1956). No clear concentration was found in the course of the stream where the original incision had ended before lateral erosion took place. On the contrary, the highest values were often encountered in the gravel along the borders of the valleys. Nowadays only some sand and clay is transported by the small creeks flowing through very wide, flat-floored valleys. Even with the heaviest cloudburst transport of gravel does not occur, since it is not available, due to the heavy cover of tropical vegetation. Hence conditions during the deposition of the gravel must have been different from the present ones.

The occurrence of gravels at the base of river- and creek-deposits, commonly containing placer minerals such as gold, cassiterite, and diamonds, has been mentioned by Tricart and Cailleux (1965). They occur for instance in Africa, S.E. Asia, and the Guianas. The gravels are mostly covered by sterile sands and clays, which are still deposited during floods. The deposits have been studied mostly in parts of West Africa, where recent tectonic deformation is minimal: Ivory Coast, Gambia, and Senegal. In areas with a dense tropical forest the gravels are generally covered with fine-grained deposits, in savanna areas they are mostly uncovered. The petrographical composition of the gravels shows great variations along the length of the rivers, which points to a mainly lateral supply.

Vogt (1959), quoted by Tricart and Cailleux, ascribes the deposition of the coarse sediments to a climate with occasional heavy showers falling on a soil scarcely protected by vegetation, a kind of savanna climate with a very irregular precipitation. During a downpour the coarse particles and the heavy ore minerals were only little transported and were deposited soon after the velocity of the stream decreased. The greater part of the sand and clay was carried away much further downstream. Later humid conditions caused the return of the tropical rain forest, which inhibited the supply of gravel. The sediments deposited now were sand and clay that contained very little ore minerals, firstly because these heavy particles remained in the soil in the source area and secondly because most sands and clays came from areas far upstream outside the source area of the ore.

In great parts of Africa during the Pleistocene climates characterized by dominating chemical weathering alternated with more arid climates where mechanical weathering and erosion played an important role due to the lack of a dense vegetational cover. During the first sands and clays were deposited or downcutting took place, whereas during the latter wide valleys with very coarse deposits developed. The more arid climates occurred contemporaneously with the glacial periods on the northern and southern hemisphere (Fairbridge, 1963; Derruau, 1962) when the equatorial forests were replaced on a great scale by savannas. In West Africa the tropical rain forest has only remained intact during the Quaternary in S.W. Ivory Coast and Liberia. In these countries coarse sediments are hardly found in the terrace deposits.

There are many indications that during the last glacial period (Würm) the climate of parts of South America, too, was much more arid than it is at present. A few will be mentioned here. Pollen of Würm deposits show a poor savanna vegetation in Northern Guyana (Van der Hammen, 1963). A fossil coral reef of Würm age along the edge of the continental shelf of the Guianas (Nota, 1958–1967) points to little clay supply by the Amazon River, causing a clear sea along the coast of the Guianas which allowed the growth of the reef. This may be an indication of less rain in the catchment area or less chemical weathering, or both. Derruau mentions the fact that the Amazon at present indeed contains much more water than the Amazon during the last glacial period.

At many places in Surinam very coarse gravels occur, which may have been deposited in glacial times. The youngest gravels in the above mentioned Lawa area, however, probably are of a more recent date as neolitic stone artifacts of Amerindians have been found in the gravels. This may indicate more recent arid conditions, the nature of which is not yet understood as no detailed studies have been made so far.

LEENDERT KROOK

REFERENCES

Brinck, J. W., 1955, Goudafzettingen in Suriname (Gold deposits in Surinam), Thesis Leiden: Leidse Geologische Mededelingen, Part XXI, pp. 1–246.

Cheney, E. S., and Patton, Th. C., 1967, Origin of the bedrock values of placer deposits: ECON. GEOL., v. 62, p. 852–853.

Derruau, M., 1962, Précis de Géomorphologie: Masson et Cie., Paris.

Fairbridge, R. W., 1963, African ice-age aridity: Nato Advanced Study Institute on Palaeoclimates, 7–12 January 1963, King's College, Newcastle upon Tyne, England.

Gunn, C. B., 1968, Origin of the bedrock values of placer deposits: ECON. GEOL., v. 63, p. 86.

Hammen, Th. van der, 1963, A palynological study of the Quaternary of British Guiana: Leidse Geologische Mededelingen, Part 29, pp. 125–180.

Nota, D. J. G., 1958, Sediments of the Western Guiana Shelf: Thesis Utrecht, 98 p.

——, 1967, Geomorphology and sediments of Western Surinam Shelf; a preliminary note. Scientific Investigations on the Shelf of Surinam, H.NL.M.S. Snellius 1966: Hydrographic Newsletter, Special Publication Number 5, 1967, pp. 55–59. Hydrographer of the Royal Netherlands Navy, The Hague.

Tricart, J., and Cailleux, A., 1965, Le modelé des régions chaudes: Traité de Géomorphologie, Tome V, Sedes, Paris, 322 pp.

Tuck, R., 1968, Origin of the bedrock values of placer deposits: ECON. GEOL., v. 63, p. 191–193.

Vogt, J., 1959, Observations nouvelles sur les alluvions inactuelles de Côte-d'Ivoire et de Guinée: Actes 84e Congr. Sav., Dijon, p. 205–210.

Origin of Gold Nuggets

The origin of gold nuggets has long been a subject of discussion among students of gold placers. Three general theories prevail: One holds that the nuggets are formed mainly by chemical accretion processes; another maintains that they are detrital in origin and had essentially the same primary weight as they now possess, but that their shape and features are due to the rolling and hammering received as they were moved along in the gravels. A third theory is a compromise holding that nuggets are partly detrital and partly chemical in origin.

There has been no consensus about the origin of nuggets for centuries, as indicated by the discussion and quotations from Albertus Magnus, Calbus, Biringuccio, and Agricola, given in some detail in chapters 4 and 5.

Egleston (1881) considered that nuggets, and indeed most placer gold, were of chemical origin, the gold having been deposited from surface waters and waters that circulated through the placers. He explained the gold values on bedrocks by stating that as the gold-bearing solutions reach bedrock they can go downward no farther and so precipitate their gold around nuclei because of the presence of reductants such as lignite, fossil wood, pyrite, and so on. Gold in the upper few feet of bedrock below placer deposits was similarly explained.

Liversidge (1893) reviewed the literature and theories on the origin of gold nuggets back to 1860. He mentions that A. R. C. Selwyn was the first to propound a chemical accretion theory for the origin of nuggets in the Victoria placers, Australia. Selwyn thought that the nuggets were chemically accreted around small nuclei of gold in the placers from large quantities of saline and acid thermal waters that had obtained gold by percolation through great thicknesses of gold-bearing gravels and sands. This process supposedly took place during the period of great volcanic activity that produced the basalts covering the deep leads in some districts of Victoria. It should be mentioned in passing that a similar theory was advocated by some investigators for the deep leads in California.

Liversidge (1893) carried out many experiments on the precipitation of gold on gold nuclei and conceded that gold nuggets could be formed in this way. However, he concluded that large nuggets have not been formed in situ in this manner. He stated that they had been set free from reefs, and that any small additions of chemically precipitated gold were immaterial. In the case of gold dust and gold grains he thought the situation might be different, for they expose a large surface area, and thus chemical accretion (he called it "electroplating") may have had an appreciable effect in increasing the amount of this type of gold.

Maclaren (1908) reviewed all the available evidence about the origin of gold nuggets up to the turn of the century and concluded that chemical accretion played a large part in the formation of the nuggets and other gold particles in placers. Lindgren (1933) doubted that chemical accretion played any part in the formation of placer gold. Fisher (1935), who carried out an extensive investigation of the nature and structure of alluvial gold in the Morobe goldfield, New Guinea, concluded that the nuggets and flakes were derived mechanically from denuded gold-quartz veins or lodes. McConnell (1905) and Johnston and Uglow (1926) advocated a similar origin for the placer gold of the Klondike and Cariboo respectively. Mustart (1965) also concluded that most of the placer gold he examined from the central Yukon was essentially detrital. Desborough (1970) and Desborough, Raymond, and Iagmin (1970) have inferred that the placer gold they investigated from the western United States is also largely detrital in origin.

Boyle (1979) has advocated a multiple hypothesis for the origin of placer gold that can be summarized in the following statements. In Yukon and elsewhere there is undoubted evidence that the gold in the oxidized zones of auriferous deposits is chemically accreted from very finely divided or lattice gold in pyrite and arsenopyrite. There is also positive evidence that the coarse gold in the upper parts of oxidized zones of auriferous deposits is finer than that in the primary deposits. What happens to the gold as it passes from the oxidized zones into the eluvium and from there into the alluvial placers and thence downstream is a problem that has not been adequately researched. There is certainly positive evidence that the soil solutions and stream waters contain both gold and silver in auriferous regions, and all of the chemical evidence points to the probability that gold will be precipitated on any available gold nuclei, provided reductants are present, which is nearly always the case in eluvium and alluvium. There is also considerable evidence that the gold of placer deposits has been extensively recrystallized, and the weight of the chemical evidence with respect to the relatively pure gold rinds on nuggets is that the rinds in most cases represent chemically precipitated gold and some silver rather than that the silver was leached from the surface layer. There may, however, be cases where the silver was leached from the surface of nuggets, but how this leaching could have proceeded throughout a nugget the size of a hen's egg, thus increasing its fineness considerably in excess of that found for the gold in the primary source, is difficult to understand.

These considerations suggest that the gold of placers is of both detrital and chemical origin. In some placers the detrital component may predominate; in others the chemical is more important. One should view the formation of gold nuggets in a dynamic sense, the agents forming them being both chemical and mechanical and their action being concomitant. As auriferous deposits weather down, particles of gold pass into the oxidized zones; these particles are partly of chemical origin, partly of mechanical origin, and partly of chemical-mechanical origin depending on the many factors discussed in chapter 16. When the particles pass into the eluvium, they slowly accrete dissolved or colloidal gold (and some silver) from the oxidizing solutions in the veins and from the soil solutions. As the weathering cycle proceeds, the material in eluvial deposits may be further decomposed, eroded, transported, and redeposited. This cycle can be effected by gravity, wind, streams, rivers, ice, or the surf of the sea, resulting in aeolian, alluvial, and beach placers. During these movements the gold-silver particles may be hammered, comminuted and reduced to a flour, or, as is more often the case, especially in an aqueous environment, the small particles may be partly dissolved and the gold reprecipitated on the larger particles that may then be cold-worked and recrystallized as they are moved along in the placers. Because silver is more mobile (more soluble) than gold the chemical processes of solution and reprecipitation generally lead to an increase in the fineness of the gold flakes and nuggets with increasing distance from the source.

A recent and important discovery concerning the origin of gold nuggets has been made by Watterson, Nishi, and Botinelly (1985). They have observed that the spores of *Bacillus cereus* precipitate gold and that laboratory-grown gold crystals around these spores are indistinguishable from dodecahedral gold crystals found in nature (placers).

There is a centuries-old tradition among placer miners that gold "grows" in placers. There is probably much truth in the adage.

Time Factor in the Formation of Productive Gold Placers

This subject has been discussed very little in the literature. Boyle (1979), however, has dealt with some aspects of the subject as follows:

> The reasons for the formation and preservation of large numbers of productive gold placers throughout the world during Cenozoic time, particularly in the Tertiary period, as compared with earlier periods, has received little mention in the literature. But, before continuing the discussion of why these placers are so prevalent some of the general features of the distribution and nature of gold placers with time should be briefly reviewed. Large economic placerlike deposits occur in the Precambrian (e.g., Witwatersrand), but they are of restricted occurrence, of controversial origin, and if originally placers have been highly modified. The sedimentary rocks of the Paleozoic and Mesozoic contain a few fossil placers, but all are low grade, the gold being finely divided and generally widely disseminated throughout the conglomerates and sandstones. In Cenozoic sediments, particularly those of Tertiary age, gold placers are widespread occurring in practically every country. Many are or were phenomenally rich. An interesting feature about gold placers is the relationship of the size of the gold particles with time. Precambrian placers are all characterized by very fine-grained gold, often in the range 50 to 100 μ. Paleozoic and Mesozoic gold placers are invariably also marked by fine-grained gold, but in places small nuggets a few millimetres in diameter can be found. They are, however, relatively rare. Tertiary and younger placers often contain coarse gold, nuggets weighing 1 oz or more being relatively common in places.
>
> The Tertiary was a period of great volcanic activity and gold deposition in many parts of the world. But this abundant primary source of gold cannot be the sole parameter, or indeed a vital parameter, in the production of the many placers in the Tertiary and younger periods, for older eras were also marked by an abundant primary gold deposition, particularly the Precambrian.
>
> The Tertiary was marked by extensive tectonic uplift in most parts of the world, and the paleoclimatologists tell us that the climate was generally hot and humid over almost the whole face of the earth, from the poles to the equator. Great forests of tropical to subtropical type also covered large areas of the land. The weathering, mainly lateritic, was deep and the downcutting by stream and river systems due to uplift was prolonged and extensive. Certainly these were optimum conditions for the formation of gold placers. But they are not unique in the geological history of the earth. Similar conditions

> probably existed in late Precambrian time, and certainly prevailed during parts of Paleozoic and Mesozoic time.
>
> On close scrutiny it appears that neither source, tectonics nor climatic conditions are the decisive factors which gave rise to the multitude of placers in Tertiary and later times. One factor alone seems to be responsible, namely the generation of the sediments in which the placers occur. In nearly all Tertiary and younger gold placers the sediments are first generation or in some placers locally second generation. In other words the natural milling process has been relatively limited, and the gold has not been comminuted and dispersed. In older rocks most of the sediments represent many generations (cycles), and the natural milling has been extensive and pervasive. The result has been the wholesale comminution of gold and its dispersal throughout a vast volume of sandstones, conglomerates, etc. This is apparently why there is a diminution in the size of gold particles in placers with time. Recognition of these features is important in prospecting for fossil placers. First or second generation sediments in the sedimentary column are most likely to contain placers; only rarely in certain basins, such as the Witwatersrand, has the dispersed gold been reconcentrated into payable quantities by secondary (diagenetic) processes. (pp. 385-386)

While some aspects of the time factor in the formation of productive gold placers are expressed in the preceding excerpt, I believe this subject requires much more research to reach unequivocal conclusions.

REFERENCES

Becker, G. F., 1895. Gold fields of the Southern Appalachians, *U.S. Geol. Survey 16th Ann. Rept.,* pt. 3, pp. 251-331.

Bensusan, A. J., 1929. Auriferous Jacutinga deposits, *Inst. Min. Metall. Trans.* **38:**450-474.

Borodaevskaya, M. E., and I. S. Rozhkov, 1974. Gold deposits, in *Ore Deposits of the U.S.S.R.,* V. I. Smirnov, ed., vol. 3, "Nedra" Moscow, pp. 3-81. (Trans. Pitman Publishing, London, 1977).

Bowen, K. G., and R. G. Whiting, 1975. Gold in the Tasman Geosyncline, Victoria, in *Economic Geology of Australia and Papua New Guinea,* 1-Metals, C. L. Knight, ed. *Australasian Inst. Min. Metall. Monograph Ser. 5,* Victoria, Australia, pp. 647-659.

Boyle, R. W., 1979. The geochemistry of gold and its deposits, *Canada Geol. Survey Bull. 280,* 584p.

Cheney, E. S., and T. C. Patton, 1967. Origin of the bedrock values of placer deposits, *Econ. Geology* **62:**852-853.

Clark, W. B., 1970. Gold districts of California, *Calif. Div. Mines Geol. Bull. 193,* 186p.

Collier, A. J., F. L. Hess, P. S. Smith, and A. H. Brooks, 1908. The gold placers of parts of Seward Peninsula, Alaska, *U.S. Geol. Survey Bull. 328,* 343p.

De Oliveira, E. P., 1932. Genesis of the deposits of auriferous jacutinga, *Econ. Geology* **27:**744-749.

Derby, O. A., 1884. Peculiar modes of occurrence of gold in Brazil, *Am. Jour. Sci.,* ser. 3 **28:**440-447.

Derby, O. A., 1903. Notes on Brazilian gold-ores, *Amer. Inst. Min. Eng. Trans.* **33:**282-287.

Desborough, G. A., 1970. Silver depletion indicated by microanalysis of gold from placer occurrences, western United States, *Econ. Geology* **65:**304-311.

Desborough, G. A., W. H. Raymond, and P. J. Iagmin, 1970. Distribution of silver and copper in placer gold derived from the northeastern part of the Colorado mineral belt, *Econ. Geology* **65:**937-944.
Dorr, J. V. N., 1969. Physiographic, stratigraphic, and structural development of the Quadrilatero Ferrifero, Minas Gerais, Brazil, *U.S. Geol. Survey Prof. Paper 641-A,* 110p.
Egleston, T., 1881. The formation of gold nuggets and placer deposits, *Amer. Inst. Min. Eng. Trans.* **9:**633-646.
Fisher, M. S., 1935. The origin and composition of alluvial gold, with special reference to the Morobe goldfield, New Guinea, *Inst. Min. Metall. Bull. 365,* pp. 1-46.
Gair, J. E., 1962. Geology and ore deposits of the Nova Lima and Rio Acima quadrangles, Minas Gerais, Brazil, *U.S. Geol. Survey Prof. Paper 341-A,* 67p.
Gardner, E. D., and C. H. Johnson, 1934-1935. Placer mining in the western United States, Pt. I, General information, hand shoveling, and ground sluicing, *U.S. Bur. Mines Inf. Cir. 6786,* 74p.; Pt. II, Hydraulicking, treatment of placer concentrates, and marketing of gold, *U.S. Bur. Mines Inf. Circ. 6787,* 89p.; Pt. III Dredging and other forms of mechanical handling of gravel, and drift mining; *U.S. Bur. Mines Inf. Circ. 6788,* 82p.
Gleeson, C. F., 1970. Heavy mineral studies in the Klondike area, Yukon Territory, *Canada Geol. Survey Bull. 173,* 63p.
Green, L., 1977. *The Gold Hustlers,* Alaska Northwest Pub. Co., Anchorage, Alaska, 339p.
Gunn, C. B., 1968. Origin of the bedrock values of placer deposits, *Econ. Geology* **63** (1):86.
Harrison, J. B., 1905. *Report on the petrography of the Cuyuni and Mazaruni districts, and of the rocks at Omai, Essequibo River, with some notes on the geology of part of the Berbice River,* Georgetown, British Guiana (Rep. Gov. Geol.), 71p.
Harrison, J. B., 1908. *The Geology of the Goldfields of British Guiana,* Dunlau & Co., London, 320p.
Hunter, S., 1909. The deep leads of Victoria, *Victoria (Australia) Geol. Survey Mem. 7,* 142p.
Johnston, W. A., and W. L. Uglow, 1926. Placer and vein gold deposits of Barkerville, Cariboo district, British Columbia, *Canada Geol. Survey Mem. 149,* 246p.
Kartashov, I. P., 1971. Geological features of alluvial placers, *Econ. Geology* **66:**879-885.
Kolesov, S. V., 1975*b*. Flattening and hydrodynamic sorting of placer gold, *Internat. Geol. Rev.* **17** (8):940-944.
Kolesov, S. V., 1975*a*. Subsidence of gold grains in allochthonous alluvial and coastal marine placers, *Internat. Geol. Rev.* **17** (8):915-953.
Krook, L., 1968. Origin of the bedrock values of placer deposits, *Econ. Geology* **63** (7):844-846.
Lindgren, W., 1911. The Tertiary gravels of the Sierra Nevada of California, *U.S. Geol. Survey Prof. Paper 73,* 226p.
Lindgren, W., 1933. *Mineral Deposits,* 4th ed., McGraw-Hill, New York, 930p.
Liversidge, A., 1893. On the origin of gold nuggets, *Royal Soc. New South Wales Jour.* **27:**303-343.
McConnell, R. G., 1905. Report on the Klondike gold fields, *Canada Geol. Survey Ann. Rept.,* pt. B, **14:**1-71.
McConnell, R. G., 1907. Report on gold values in the Klondike high level gravels, *Canada Geol. Survey Rept. 979,* 34p.
Macdonald, E. H., 1983. *Alluvial Mining,* Chapman & Hall, London, 508p.
MacKay, B. R., 1921. Beauceville Map-area, Quebec, *Canada Geol. Survey Mem. 127,* 105p.
Maclaren, J. M., 1908. Gold: Its geological occurrence and geographical distribution, Mining Jour., London, 687p.
Meireles, E. M., and J. T. Teixeira, 1982. Serra Pelada gold deposit, Int. Symp. Arch. Early Prot. Geol. Evol. Met.-ISAP, Abstracts/Excursions, pp. 72-75.
Mertie, J. B., Jr., 1940. Placer gold in Alaska, *Washington Acad. Sci. Jour.* **30:**93-124.
Mustart, D. A., 1965. A spectrographic and mineralographic investigation of alluvial gold from the central Yukon, B.S. thesis, Univ. British Columbia, Vancouver, 46p.
Nelson, C. H., and D. M. Hopkins, 1972. Sedimentary processes and distribution of particulate gold in the northern Bering Sea, *U.S. Geol. Survey Prof. Paper 689,* 27p.

Raeburn, C., and H. B. Milner, 1927. *Alluvial Prospecting,* Thos. Murby & Co., London, 478p.

Razin, L. V., and I. S. Rozhkov, 1966. *Geochemistry of Gold in the Crust of Weathering and the Biosphere of Gold-ore Deposits of the Kuranakh Type;* "Nauka" Moscow, 254p. (rev. in *Econ. Geology* **62:**437-438.

Shakhov, F. N., 1961. Major trends in scientific investigations in the auriferous districts of Siberia, *Geol. Geofiz.,* no. 10:89-101 (Eng. trans. in *Internat. Geol. Rev.,* 1964, **6** no. 2:202-211.)

Shilo, N. A., 1981. *Fundamentals of the Study of Placers,* "Nauka," Moscow, 383p.

Shilo, N. A., and Yu. V. Shumilov, 1976. Mechanisms of behaviour of gold during placer formation processes in the north-east of the U.S.S.R., *25th Internat. Geol. Cong., Australia, Abstracts,* vol. 1, p. 224.

Tuck, R., 1968. Origin of the bedrock values of placer deposits, *Econ. Geology* **63** (2):191-193.

Tyrrell, J. B., 1912. The law of the pay streak in placer deposits, *Inst. Min. Metall. Trans.* **21:**593-613.

Watterson, J. R., J. M. Nishi, and T. Botinelly, 1985. Evidence that gold crystals can nucleate on bacterial spores, *U.S. Geol. Survey Circ. 945,* pp. 1-4.

Wells, J. H., 1973. Placer examination; principles and practice, *U.S. Dept. Interior, Bureau of Land Management, Tech. Bull. 4,* 204p.

Yeend, W. E., 1974. Gold-bearing gravels of the Ancestral Yuba River, Sierra Nevada, California, *U.S. Geol. Survey Prof. Paper 772,* 44p.

Zhelnin, S. G. and Travin, Yu. A., 1976. Comparative features of placer geology in the modern structure of the north-eastern U.S.S.R., *25th Internat. Geol. Cong., Australia, Abstracts,* vol. 1, pp. 229-230.

CHAPTER

16

Oxidation and Secondary Enrichment of Gold Deposits

But iron veins (gossans)... should not be despised... you will generally find gold or some other valuable ore after the iron has been mined.
Calbus, 1497

The supergene enrichment of gold in deposits may take place by a simple mechanical process, by chemical means, or by a combination of these. First, the enrichment may be primarily the result of the chemical removal of a large part of the gangue minerals, in which case the gold migrates slowly downward by gravity or is left behind as a residual component. Such enrichments require no chemical movement of the gold; the lodes simply weather down and the gold collects mainly by gravity in the accumulated residuum. Second, the gold content may be enriched as a result of chemical migration of the element. In most deposits the movement of the gold is downward, although some exhibit chemical enrichments that indicate a degree of lateral movement of the auriferous solutions. In the oxidized zones of gold deposits both mechanisms have generally operated, and there are rarely sufficient criteria to distinguish which mechanism is dominant.

Furthermore, the supergene enrichment of gold in deposits is often highly irregular and not all deposits exhibit the same features. Practically every deposit is an individual study. There are, however, certain general regularities that can be mentioned, one of which is the fact that many auriferous deposits exhibit a vertical zonation manifest by a surface-oxidized zone or gossan in which there is frequently, but not always, an enrichment of gold often in the form of the native metal. Such enrichment passes imperceptibly into the primary gold-quartz or gold-sulfide ores that are leaner in gold. This common near-surface enrichment must be taken into account when examining surface prospects lest grave errors be made in estimating the grade of the primary ores. Where conditions were such as to produce supergene sulfides in the zone of reduction at depth, enrichments of gold may also be encountered in this zone. The ideal vertical zonation in an oxidized gold lode is, therefore, a moderately gold-enriched near-surface gossan or oxidized zone, a highly gold-enriched secondary sulfide zone at or near the present or former water table, and a lean gold-bearing primary zone that extends to depth. Considerable divergence from this ideal zoning arrangement is common.

The chemistry of the migration, dispersion, and concentration of gold in auriferous deposits undergoing oxidation is complex and depends on many extrinsic parameters, including type of gangue and wall rocks; degree of fracturing, crushing, or shearing of the deposit and its enclosing rocks; climatic factors; position of the water table; and presence of organic agencies. In detail the intrinsic chemical factors that play a part in the oxidation processes in auriferous deposits are intricate and depend essentially on the Eh and pH of the system. Colloidal and coprecipitation phenomena also play a large part. Iron and manganese minerals and carbonates in the gangue and ore greatly influence the reactions that lead to the secondary enrichment of the element. Gold is not easily solubilized in nature, and its soluble forms are readily reduced to the metal by a great variety of natural materials. The result of this behavior is that the only common gold mineral found in the oxidized zones of auriferous deposits is the native metal. In the zones of reduction below the oxidized zones of auriferous deposits native gold is the most stable of the gold minerals, occurring usually in a very finely divided form in secondary sulfides such as chalcocite, secondary pyrite, and so on.

Details of the redox chemistry of gold during the oxidation of auriferous deposits are given in Boyle (1979). More generalized accounts of oxidation processes in mineral deposits, but dealing only incidently with the behavior of gold, include those by Sato (1960*a,* 1960*b*) and Garrels and Christ (1965). (See also the papers on the geochemistry of gold in chapter 8).

Historically, the oxidation of mineral deposits did not occupy the minds of the ancient miners, naturalists, and early mining geologists to any great extent. Pliny, Albertus Magnus, and Calbus each mentioned gossans, as discussed in chapters 3, 4, and 5, respectively. The very old German mining proverb likewise contains much wisdom:

Es thut kein Gang so gut	There is no lode like that
Er hat einen eisernen Hut	Which has an iron hat

The veins mined by the ancients were largely oxidized, yet one finds relatively little detail on oxidation processes in their writings. The reason may be that when the old miners reached the hard unoxidized ore, operations ceased because of economics or their inability to cope with the hard materials. Hence, the hypogene ores were rarely examined in any detail. Another reason was undoubtedly the fact that chemistry had not developed to the point where the processes could be understood. In the old literature mining geologists rarely differentiated between oxidized and hypogene ores, a feature that colored their views on the hypogene origin of veins. In countries such as Canada and Sweden, where glaciation had largely removed the oxidized zones, a knowledge of supergene impoverishment and enrichment was unnecessary in the exploitation of the ores, and hence little mention of redox processes appears in their literature.

The discovery of the great ore deposits of the central and western United States and of Australia, South America, and elsewhere, especially the vast iron ore and porphyry copper deposits and the great gold and silver lodes, all of which are deeply oxidized and secondarily enriched, led to a detailed study of the processes of oxidation and reduction in the second half of the nineteenth century and the first quarter of the twentieth century. There followed then a gradual decline in the investigation of oxidation phenomena in mineral deposits until about 1945, when the advent of geochemical prospecting renewed interest in the problems of oxidation in most mineral-producing countries.

Early studies of the solution, migration, and concentration of gold in the surface and near-surface oxidizing environment were recorded by Wilkinson (1867) and Newbery (1868) and concerned mainly the formation of nuggets in the auriferous drift of Victoria, Australia. Later W. Skey (1871), analyst to the Geological Survey of New Zealand, applied his knowledge of chemistry to the problems of the migration of gold and its precipitation by a great variety of substances. The resulting paper is a landmark in the natural redox chemistry of gold.

Skey reviewed the results of Wilkinson and Newbery and carried out many experiments on the reaction of soluble gold salts (mainly gold chloride) with various natural sulfides and arsenides. He showed that pyrite, arsenopyrite, galena, and many other natural metallic sulfides and arsenides reduce gold to the native metal without the intervention of organic matter, as thought by Wilkinson. Skey also remarked on the observation that gold itself is nuclear to gold, or in more modern terms that there is a great tendency for soluble gold to be precipitated by and grow on nuclei of native gold, a feature common in nature. Finally, he concluded with the following observation, since verified by a multitude of investigators:

> The great deoxydizing power of sulphides generally upon most gold, silver, or platina salts, as manifested by the experiments just described, renders them so absorbent as it were of these metals, when presented to them as chlorides or oxy-salts (the forms usually contemplated for them when in solution), that any such solutions traversing even a very thin vein or reef of the common metallic sulphides, would in all probability be completely divested of these metals. (p. 229)

Skey's research was followed by a classic paper by Schuermann (1888), who established the series of sulfide precipitation that bears his name. There followed numerous papers on the oxidation of various types of mineral deposits, on the processes of sulfide and metallic (gold, silver) enrichment, and on the chemistry of these processes. W. H. Emmons, longtime professor of economic geology in the University of Minnesota, devoted many years to the study of secondary enrichment of metalliferous deposits. His classic *Enrichment of Ore Deposits* contains a long section on gold (Paper 16-1).

(Text continues on page 568.)

16-1: GOLD

William Harvey Emmons

Reprinted from The enrichment of ore deposits, *U.S. Geol. Survey Bull. 625*, 1917, pp. 305-324.

SOLUTION OF GOLD.

Although gold belongs to the same chemical family as copper and silver its activities differ in many respects from both these metals. It forms no insoluble compound in the oxidized zone and its sulphide is not precipitated by mineral waters. Unlike copper and silver it is insoluble in sulphuric acid. Wurtz[1] stated, in 1858, that ferric sulphate dissolves gold, and his statement has frequently been quoted in discussions of the processes of enrichment of gold deposits. It is, indeed, a common statement that ferric sulphate is the principal agent in the enrichment of gold deposits. Stokes[2] showed, however, that ferric sulphate will not dissolve gold at 200° C., except in the presence of a chloride, and its insolubility in ferric chloride and in ferric sulphate at ordinary temperatures has been verified by A. D. Brokaw.[3] W. J. McCaughey[4] found that gold is dissolved in a concentrated solution of hydrochloric acid and ferric sulphate and also in a concentrated solution of hydrochloric acid and cupric chloride, both experiments being carried on at temperatures from 38° to 43° C. Pearce,[5] Don,[6] and Rickard[7] performed experiments in which gold was dissolved in the presence of a chloride and manganese dioxide. Hydrochloric acid forms in the presence of sodium chloride and sulphuric acid, and in the presence of an oxidizing agent the hydrogen ion is removed to form water, leaving the chlorine in the so-called "nascent state." In this state it is in the uncombined or atomic condition and its attack is more vigorous. The reaction may be written:

$$MnO_2+2NaCl+3H_2SO_4 \rightarrow 2H_2O+2NaHSO_4+MnSO_4+2Cl.$$

At the beginning of the reaction manganese has a valence of 4; at the end a valence of 2. It is known that several oxides will release "nascent chlorine" at low temperature if the solutions are sufficiently concentrated, but in moderately dilute solutions manganese oxides

---

[1] Wurtz, Henry, Contributions to analytical chemistry: Am. Jour. Sci., 2d ser., vol. 26, p. 51, 1858.

[2] Stokes, H. N., Experiments on the solution, transportation, and deposition of copper, silver, and gold: Econ. Geology, vol. 1, p. 650, 1906.

[3] Brokaw, A. D., The solution of gold in the surface alterations of ore bodies: Jour. Geology, vol. 18, p. 322, 1910.

[4] McCaughey, W. J., The solvent effect of ferric and cupric salt solutions upon gold: Am. Chem. Soc. Jour., vol. 31, p. 1263, 1909.

[5] Pearce, Richard, Discussion of paper by T. A. Rickard on The origin of the gold-bearing quartz of the Bendigo Reefs, Australia: Am. Inst. Min. Eng. Trans., vol. 22, p. 739, 1894.

[6] Don, J. R., The genesis of certain auriferous lodes: Am. Inst. Min. Eng. Trans., vol. 27, p. 599, 1898.

[7] Rickard, T. A., The Enterprise mine, Rico, Colo.: Am. Inst. Min. Eng. Trans., vol. 26, pp. 978–979, 1897.

are probably the only common ones that are effective. Ferric chloride, ferric sulphate, and cupric salts are not more than a fraction of 1 per cent as effective as manganese salts and doubtless this fraction is exceedingly small.[1] Lead oxides are probably efficient, but they are rare in gold deposits.

A number of experiments on the solubility of gold in cold dilute solutions were made at my request by Mr. A. D. Brokaw.[2] The nature of these experiments is shown by the following statements, in which *a* and *b* represent duplicate tests:

(1) $Fe_2(SO_4)_3+H_2SO_4+Au$.
(*a*) No weighable loss. (34 days.)
(*b*) No weighable loss.

(2) $Fe_2(SO_4)_3+H_2SO_4+MnO_2+Au$.
(*a*) No weighable loss. (34 days.)
(*b*) 0.00017 gram loss.[3]

(3) $FeCl_3+HCl+Au$.
(*a*) No weighable loss. (34 days.)
(*b*) No weighable loss.

(4) $FeCl_3+HCl+MnO_2+Au$.
(*a*) 0.01640 gram loss. Area of plate, 383 square millimeters. (34 days.)
(*b*) 0.01502 gram loss. Area of plate, 348 square millimeters.

In each experiment the volume of the solution was 50 cubic centimeters. The solution was one-tenth normal with respect to ferric salt and to acid. In the second and fourth experiments 1 gram of powdered manganese dioxide also was added. The gold, assaying 0.999 fine, was rolled to a thickness of about 0.002 inch, cut into pieces of about 350 square millimeters area, and one piece, weighing about 0.15 gram, was used in each duplicate.

To approximate natural waters more closely, a solution was made one-tenth normal as to ferric sulphate and sulphuric acid and one-twenty-fifth normal as to sodium chloride. Then 1 gram of powdered manganese dioxide was added to 50 cubic centimeters of the solution, and the experiment was repeated. The time was 14 days.

(5) $Fe_2(SO_4)_3+H_2SO_4+NaCl+Au$.
No weighable loss.

(6) $Fe_2(SO_4)_3+H_2SO_4+NaCl+MnO_2+Au$.
Loss of gold, 0.00505 gram.

---

[1] Emmons, W. H., The agency of manganese in the superficial alteration and secondary enrichment of gold deposits in the United States: Am. Inst. Min. Eng. Trans., vol. 42, pp. 25–27, 1912.

[2] Brokaw, A. D., op. cit., pp. 321–326.

[3] This duplicate was foubd to contain a trace of Cl, which accounts for the loss.

The loss is comparable to that found in experiment 4, allowing for the shorter time and the greater dilution of the chloride.

Although the concentration of chlorine in most of these experiments is greater than that which is found in many mineral waters, it is noteworthy that solution of gold will take place with even a trace of chlorine (see experiment 2*b*), and without much doubt these reactions will go on also in the presence of only small quantities of manganese oxides.

SOURCES OF ACID IN GOLD DEPOSITS.

Gold can not be carried into solution under these conditions except in the presence of acid. If the solution is neutralized the gold is quickly precipitated. In sulphide ore deposits, as already shown (p. 91), acid is generated by the oxidation of sulphides, particularly by the oxidation of iron sulphides. The solution of gold is most important in the upper parts of the oxidized zones, where, in the presence of gold, pyrite is oxidized to ferric sulphate. The sulphuric acid which, under these conditions, is necessary for the reactions could easily be supplied, even above the zone where pyrite persists, by the leaching of basic iron sulphates, the formation of which tends to delay the downward migration of a part of the sulphuric acid that is released where ferric sulphate alters to limonite.

It is evident that the presence of an iron sulphide is an essential condition for the solution of gold. Ore bodies that do not contain pyrite or other iron minerals will show little evidence of transfer of gold, and in those that contain only a little pyrite the gold will not be carried far down, because the small amount of acid formed is quickly neutralized as it passes downward.

SOURCES OF CHLORINE IN GOLD DEPOSITS.

The sources of chlorine in ore deposits have been discussed elsewhere (p. 92), but, as it has been shown that chloride is the only natural solvent for gold in reactions causing superficial enrichment of gold deposits, the sources and distribution of chlorine are of special interest here. Other valuable metals are dissolved as sulphates with or without ferric ion but not gold, as has been shown. Chlorine has been found in all waters of gold mines where it has been sought. The amounts stated in analyses range from traces to 842.8 parts in a million. That a very small amount of chlorine is sufficient to dissolve gold in the presence of manganese is evident from experiment 2, above, where there is loss of gold through the introduction of a trace of chlorine without intent. Chlorine is carried by many sedimentary rocks and some minerals, such as apatite, scapolite, haüyne, and nosean, carry appreciable amounts. Salt cubes are present in fluid inclusions in some vein quartz. In general, however, these sources of chlorine are unimportant. Its chief source is from finely divided salt or salt water from the sea and from other bodies of salt water. The salt is carried by the wind and precipitated with rain.[1] The amount of chlorine in natural waters varies with remarkable constancy with the distance from the shore; several determinations very near the seashore show from 10 to 30 parts of chlorine per million; a few miles away it is generally about 6 parts per million; 75 miles from shore it is generally less than 1 part per million. (See p. 93.)

The amount of chlorine contributed from this source even near the seashore appears small (from 6 to 10 parts per million), but it may be further concentrated in the solutions by evaporation or by reactions with silver, lead, and other metals, forming chlorides, which in the superficial zone may subsequently be changed to other compounds. In arid countries, as suggested by C. R. Keyes,[2] dust containing salt doubtless contributes chlorine to the mine waters. Penrose,[3] discussing the distribution of the chloride ores, pointed out long ago that these minerals form most abundantly in undrained areas.

SOURCES OF MANGANESE IN GOLD DEPOSITS.

It has been shown that the solution of gold depends on the presence of chlorine in the mineral waters, and the distribution of chlorine has been discussed briefly above. But chlorine is an efficient solvent of gold only when in the "nascent" state, and under natural conditions nascent chlorine is released principally by manganese oxides. The efficiency of other compounds that occur in gold deposits to release nascent chlorine is so small that they can probably be disregarded.[4] The sources and distribution of manganese are, therefore, important in connection with a study of the concentration of gold. The manganese minerals include the oxides (pyrolusite, psilomelane, manganite, manganosite, and pyrochroite); the sulphates (szmikite, mallardite, and apjohnite); the sulphides (alabandite and hauerite); the carbonates (rhodochrosite, manganiferous calcite, and manganiferous siderite); and the silicate (rhodonite). Besides these many rock-making silicates carry manganese. (See p. 437.)

[1] Jackson, D. D., The normal distribution of chlorine in the natural waters of New York and New England: U. S. Geol. Survey Water-Supply Paper 144, 1905.

[2] Keyes, C. R., Origin of certain bonanza ores in the arid region: Am. Inst. Min. Eng. Trans., vol. 42, p. 500, 1911.

[3] Penrose, R. A. F., Jr., The superficial alteration of ore deposits: Jour. Geology, vol. 2, p. 314, 1894.

[4] Emmons, W. H., The agency of manganese in the superficial alteration and secondary enrichment of gold deposits in the United States: Am. Inst. Min. Eng. Trans., vol. 42, p. 25, 1912.

The commonest sources of manganese are probably the rock-making silicates of iron and magnesium, a little manganese being very commonly isomorphous with these elements. Manganese is more abundant in basic than in acidic rocks. The average of 1,155 determinations of igneous rocks[1] is 0.10 per cent manganese protoxide (MnO), though some contain more than 1 per cent.

Rhodonite is closely allied to the pyroxenes in crystallization, but, unlike them, it occurs as a gangue mineral in veins formed by ascending waters at moderate depths. It is found in ore of the Camp Bird, Tomboy, and other mines of the San Juan region, Colo., in the silver-bearing veins of Butte, Mont., at Philipsburg, Mont., and at many other places.

Rhodochrosite is more common in mines of the United States than rhodonite. It is present in many veins of the San Juan region, Colo., at Butte and Philipsburg, Mont., at Austin, Nev., and in many other western districts. Rhodonite is considered primary in practically all its occurrences and rhodochrosite is deposited in the main from ascending hot solutions also, but in some occurrences it has been regarded as a secondary deposit from cold solutions.

Manganiferous calcite, an isomorphous compound of rhodochrosite and calcite, is a source of abundant manganese in many later Tertiary deposits in Nevada, Montana, and Colorado. It is probably present in certain Mesozoic copper deposits of Shasta County, Cal., where, according to Graton,[2] wad occurs in the oxidized zone. Manganiferous siderite is abundant at Leadville, Colo., and according to Argall[3] it has supplied a large part of the manganese ore in that district.

Manganiferous ores of the precious metals are in general deposited at moderate depths. Consequently they are common in middle or late Tertiary deposits of western North America but very rare indeed in contact-metamorphic deposits, in the California gold veins, and in deposits formed at equal or greater depths.[4]

The sulphates of manganese are soluble and do not accumulate in veins to any important extent; all are secondary. The oxides and hydrous oxides, which are very numerous, are the products of weathering of all manganese compounds. None of them are known to be deposited by ascending hot waters except near the orifices of hot springs.

---

[1] Clarke, F. W., The data of geochemistry, 3d ed.: U. S. Geol. Survey Bull. 616, p. 27, 1916.

[2] Graton, L. C., The occurrence of copper in Shasta County, Cal.: U. S. Geol. Survey Bull. 430, p. 100, 1910.

[3] Argall, Philip, The zinc carbonate ores of Leadville: Min. Mag., vol. 10, p. 282, 1914.

[4] Emmons, W. H., The agency of manganese in the superficial alteration and secondary enrichment of gold deposits in the United States: Am. Inst. Min. Eng. Trans., vol. 42, pp. 47–49, 1912.

The sulphides of manganese are exceedingly rare. Alabandite has been found in the gold mines of Nagyag, in Transylvania; at Gersdorf, near Freiberg, Saxony; on Snake River, Summit County, Colo.; at Tombstone, Ariz., and at a few other places. Nishihara has shown, however,[1] that there are traces of a manganese sulphide, probably alabandite, in nearly all specimens of galena. Hauerite is a very rare mineral. I can find no record of its occurrence in the United States.

As to the amount of manganese required to accomplish appreciable enrichment of gold in its deposits, there are few data at hand. Doubtless only a little will permit some solution, but if ferrous sulphate is present gold will immediately be precipitated. Manganese dioxide, however, will convert ferrous sulphate at once to ferric sulphate, which does not precipitate gold. To convert the ferrous to ferric salt more manganese is required and appreciable amounts are necessary for the extensive downward migration of gold. In the deposits that show transportation of gold with which I am familiar manganese is conspicuously developed, particularly in the oxidized material. Its black or brown powder is easily recognized, as are the dendritic growths which some of it assumes.

PRECIPITATION OF GOLD.

The downward transportation of gold in chloride solutions obviously depends on the rate at which the solutions move downward, and the rate at which they react with ores and wall rock or with the reduced and neutralized solutions that have been shown (p. 149) to prevail in the deeper zones. The gold is precipitated from chloride solutions by ore and gangue minerals, by ferrous sulphate, by alkaline waters, and by hydrogen sulphide. Although gold sulphide is easily formed in the laboratory, it is noteworthy that this compound is unknown in nature. As shown by Brokaw[2] ferric sulphate converts the sulphide to native metal. The reaction is probably as follows:[3]

$$Au_2S_2+6Fe_2(SO_4)_3+8H_2O=2Au+12FeSO_4+8H_2SO_4.$$

As hydrogen sulphide does not accumulate in the superficial part of the zone of alteration and as other precipitates of gold are generally abundant it is probably of subordinate importance as a precipitant of gold.

---

[1] Nishihara, G. S., The rate of reduction of acidity of descending waters by certain ore and gangue minerals, and its bearing upon secondary sulphide enrichment: Econ. Geology, vol. 9, pp. 743–757, 1914.

[2] Brokaw, A. D., The secondary precipitation of gold in ore bodies: Jour. Geology, vol. 21, p. 254, 1913.

[3] Idem, p. 256.

The minerals that precipitate gold are legion. Copper, silver, mercury, and tellurium, all of which precede gold in the electromotive series (see p. 112), may displace it from gold-bearing solutions. Nearly all the sulphides of the metal[1] are fairly effective.

Brokaw[2] recently performed experiments precipitating gold, some of them earlier performed by Skey and Wilkinson. When gold chloride solution was in contact with sphalerite, after standing 24 days, the surface of the crystal was covered with shining flakes of gold. With pyrrhotite the action was more rapid, and the gold was precipitated in 3 days. The pyrrhotite was covered with a yellowish-brown coat, which was made up of minute crystals of gold partly embedded in material resembling limonite. A small fragment of a crystal of polybasite was placed in 5 cubic centimeters of gold chloride solution containing 0.5 per cent gold. After 10 days the gold had been completely precipitated, forming a dull coat over the crystal.[3] Palmer and Bastin[4] showed that galena, stibnite, copper-iron sulphides, and chalcocite will readily precipitate gold. Fused chalcocite is an exceedingly efficient precipitant of gold. According to Vautin[5] it will remove all the gold from a solution containing only 1 part in 5,000,000. The carbonates[3] also—calcite, siderite, and rhodochrosite—rapidly precipitate gold from the solutions in which it is held as chloride. Siderite is particularly efficient, probably because in acid abundant ferrous sulphate forms. Nepheline and leucite reduce acid solutions with great rapidity, as recently shown by Nishihara. Even comparatively stable minerals, like the feldspars and micas, give a distinctly alkaline reaction,[6] and, given time enough, an auriferous chloride solution would be neutralized, and gold would be precipitated by many minerals of the gangue and of the wall rock.

Where feldspar or rhyolite glass is attacked by sulphuric acid kaolin will form. Along with these changes, acidity is reduced, and the gold solution is neutralized. The association of rich gold ore

---

[1] Skey, W., On the reduction of certain metals from their solutions by metallic sulphides and the relation of this to the occurrence of such metals in their native state: New Zealand Inst. Trans. and Proc., vol. 3, 1871. Wilkinson, C., On the theory of the formation of gold nuggets by drift: Royal Soc. Victoria Trans. and Proc., vol. 8, 1867, pp. 11–15, 1867. Brokaw, A. D., The secondary precipitation of gold in ore bodies: Jour. Geology, vol. 21, p. 253, 1913. Palmer, Chase, and Bastin, E. S., Metallic minerals as precipitants of silver and gold: Econ. Geology, vol. 8, p. 140, 1913. Grout, F. F., On the behavior of cold acid sulphate solutions of copper, silver, and gold with alkaline extracts of metallic sulphides: Econ. Geology, vol. 8, p. 417, 1913.

[2] Brokaw, A. D., op. cit., p. 253.

[3] Idem, p. 256.

[4] Palmer, Chase, and Bastin, E. S., op. cit., p. 160.

[5] Vautin, Claude, The decomposition of auric chloride obtained in the chlorination of gold-bearing materials: Inst. Min. and Met. Trans., vol. 1, p. 274, 1893.

[6] Clarke, F. W., U. S. Geol. Survey Bull. 167, p. 156, 1900. Steiger, George, U. S. Geol. Survey Bull. 167, p. 159, 1900. Clarke, F. W., U. S. Geol. Survey Bull. 616, p. 480, 1916.

with kaolin at Bullfrog, Nev.,[1] particularly in the Montgomery Shoshone mine, and that in deposits at De Lamar, Idaho,[2] may have formed by some such reaction.

The minerals that precipitate gold as stated above include practically all the natural sulphides as well as the carbonates, many silicates, and organic matter. Among the common minerals, calcite, siderite, pyrrhotite, and chalcocite are noteworthy, however, for with these gold is precipitated very rapidly in zones of superficial

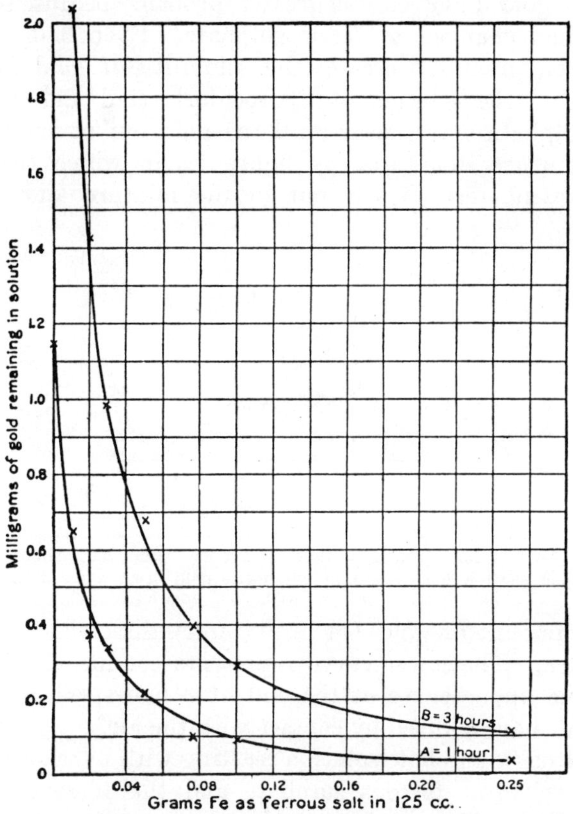

FIGURE 19.—Diagram illustrating the effect of ferrous sulphate in suppressing the solubility of gold in ferric sulphate solutions, where gold is dissolved as chloride. (Illustrating experiment by McCaughey.)

alteration. But any gold that may have been precipitated by these agents will be redissolved by chloride solutions in the presence of manganese oxides and ferric sulphate or sulphuric acid. It has been shown (p. 150) that ferric sulphate decreases and ferrous sulphate increases with increase of depth. Ferrous sulphate causes precipitation and inhibits solution of gold. Where it forms abundantly no gold is redissolved, and any in solution is reprecipitated.

[1] Ransome, F. L., Emmons, W. H., and Garrey, G. H., Geology and ore deposits of the Bullfrog district, Nevada: U. S. Geol. Survey Bull. 407, p. 104, 1910.

[1] Lindgren, Waldemar, The gold and silver veins of Silver City, De Lamar, and other mining districts in Idaho: U. S. Geol. Survey Twentieth Ann. Rept., pt. 3, p. 171, 1900.

Experiments of McCaughey[1] show the effect of very small amounts of ferrous sulphate on solutions of gold in ferric sulphate. To a solution, 125 cubic centimeters, containing 1 gram of iron as ferric sulphate and 25 cubic centimeters of hydrochloric acid, ferrous sulphate was added in quantities containing from 0.01 to 0.25 gram of ferrous iron. The solutions were immersed in boiling water and subsequently 250 milligrams of gold was added. The dissolved gold was determined at the end of 1 hour and 3 hours. At the end of 3 hours the gold dissolved was greater, probably because some ferrous sulphate had changed to ferric sulphate. Even 0.01 gram of the ferrous iron greatly decreases the solubility of gold in the ferric sulphate and the solution of hydrochloric acid and 0.25 gram of ferrous sulphate drives nearly all the gold out of solution. These experiments are illustrated by figure 19, in which the horizontal lines represent ferrous salt put in the mixture and the vertical

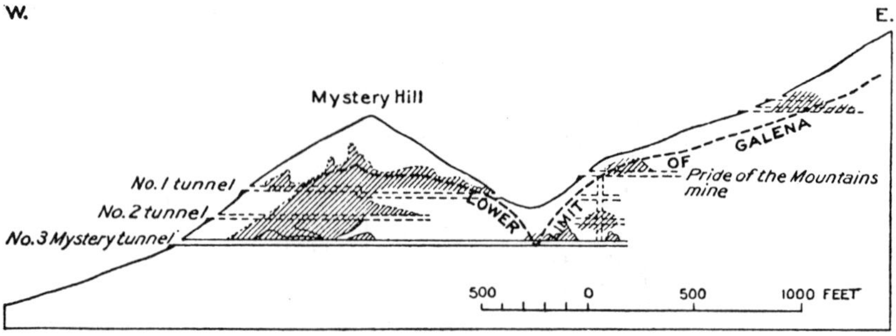

FIGURE 20.—Vertical longitudinal section of Mystery-Pride vein, Monte Cristo, Wash., showing shallow enrichment in pyrrhotite gold ore. After J. E. Spurr.

lines the amount of gold (in milligrams) dissolved by chlorine in the solution. The lower curve represents conditions at the end of 1 hour, the upper curve at the end of 3 hours, when some of the ferrous salt had oxidized by contact with the air.

An acid ferric sulphate solution reacting with pyrrhotite or siderite will readily form ferrous sulphate and the ferrous sulphate, as already stated, will precipitate gold and prevent its solution. Figure 20 is a longitudinal section of a pyrrhotite gold-bearing vein which has a shallow enriched zone. Calcite and other carbonates will decrease acidity of the solutions, and any deposit containing these minerals in abundance will carry only shallow secondary gold bonanzas, for in such deposits the gold can not migrate deeply. Nepheline and leucite also, as shown by Nishihara, are very effective. If inert minerals, like quartz and sericite, or minerals that act slowly, like pyrite, are present exclusively, gold will migrate to greater depths.

[1] McCaughey, W. J., The solvent effects of ferric and cupric salt solutions on gold: Am. Chem. Soc. Jour., vol. 31, pp. 1261–1270, 1909.

It is therefore evident that the depth to which gold will be carried by downward-moving waters in which it is held in solution depends not only on the permeability of the deposits but also upon the character of the ore and gangue minerals. If the gangue minerals precipitate gold rapidly it is not carried to great depths. If they reduce and neutralize the solution quickly no gold is redissolved. Secondary gold bonanzas are rare below the depths where such minerals prevail.

TRANSPORTATION OF GOLD AND DEPTHS TO WHICH IT MAY BE CARRIED.

The foregoing discussion of the conditions of the solution and precipitation of gold shows that it is carried downward in superficial alteration by stages. It is doubtless repeatedly dissolved and reprecipitated as weathering extends downward, and it will be fixed permanently only when it reaches an environment where the conditions become permanently reducing and alkaline. These conditions may exist in the zone of alteration above the water level, particularly where gold is associated with much calcite, siderite, or pyrrhotite, or where the deposits are in minutely fractured rocks containing olivine, nepheline, leucite, or other minerals that rapidly neutralize acid waters. (See p. 121.) At Creede, Colo.,[1] in the Amethyst vein, a fractured auriferous deposit containing mainly amethystine quartz, thuringite, barite, and pyrite, accompanied by galena, sphalerite, and chalcopyrite, some secondary gold has been carried downward to considerable depths. Rich bunches of ore consisting of the sulphides named, cut by stringers of gold-bearing manganese oxide, are found in this vein about 1,000 feet below the surface. Even in this deposit, however, where conditions appear to be favorable for the transportation of gold, the largest secondary gold deposits lie 200 to 700 feet below the surface in a zone where anglesite and cerusite are mingled with the manganese oxides and gold. At Philipsburg, Mont.,[2] the secondary ore in which gold is important is almost wholly in a zone between 200 and 800 feet deep, and in this the gold is concentrated in the upper part. At Bodie, Cal.,[3] the veins were not workable below a depth of 500 feet.

In a calcite gangue the secondary gold is deposited near the surface; at Marysville, Mont., according to Weed,[4] principally within 200 feet of the surface. At Cripple Creek, Colo.,[5] where the gold deposits carry a little manganese and are associated with nepheline

---

[1] Emmons, W. H., and Larsen, E. S., Geology and ore deposits of Creede, Colo.: U. S. Geol. Survey Bull. — (in preparation).

[2] Emmons, W. H., and Calkins, F. C., Geology and ore deposits of the Philipsburg quadrangle, Montana: U. S. Geol. Survey Prof. Paper 78, p. 178, 1913.

[3] McLaughlin, R. P., Geology of the Bodie district, California: Min. and Sci. Press, vol. 94, p. 796, 1907.

[4] Weed, W. H., Gold mines of the Marysville district, Montana: U. S. Geol. Survey Bull. 213, p. 70, 1903.

[5] Lindgren, Waldemar, and Ransome, F. L., Geology and ore deposits of the Cripple Creek district, Colorado: U. S. Geol. Survey Prof. Paper 54, pp. 167–168, 1906.

rocks, there is some evidence that gold has been dissolved and re-precipitated near the surface; but, as shown by Lindgren and Ransome, there has been little or no downward migration of gold. As long as alkali minerals, such as nepheline, are present to neutralize any acid formed by oxidation of pyrite, gold could not be dissolved, and any that may have been locally dissolved in an upper zone leached of alkalies would almost immediately be precipitated below. Here mass action due to the great abundance of precipitating agents would be most effective.

Several groups of calcitic gold veins have recently been described by Eddingfield,[1] Knopf,[2] and Ferguson.[3] The experimental work of Brokaw showed that calcite will rapidly precipitate gold from solutions in which it is dissolved in acid in the presence of chlorine and manganese. (See fig. 21.) It follows from this also that gold will not be dissolved in the presence of manganese and chlorine so long as calcite is effective. This was shown experimentally also by Eddingfield. Eddingfield states also that gold in deposits with calcitic gangue will not be transported, and he attributes the numerous rich stringers of gold and manganese in many Philippine deposits to the mechanical migration of gold. It is clearly unnecessary, however, to invoke mechanical migration, and it seems highly probable that iron and manganese oxide would be carried along mechanically by downward-moving waters quite as rapidly as gold and much more abundantly, and if so there would be little localization of gold. Since, as already stated, the calcite neutralizes or makes alkaline the acid gold-bearing solutions, as long as calcite is in contact with the solution gold will not be dissolved, but if any gold has gone in solution it will be precipitated. But the downward-moving acid, reacting on the calcite, uses it up and ultimately insulates passages through which acid solutions formed by oxidation of pyrite can move downward to greater depths, and they may carry gold downward until it comes into contact with fresh calcite. This is pointed

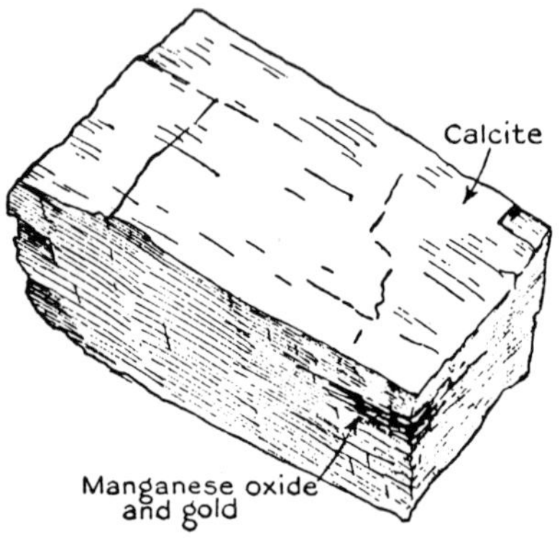

FIGURE 21.—Calcite that was immersed in a manganiferous acid gold solution. Gold and manganese have been precipitated together in cleavage cracks.

[1] Eddingfield, F. T., Alteration and enrichment in calcite-quartz-manganese gold deposits in the Philippine Islands: Philippine Jour. Sci., vol. 8, No. 2, sec. A, April, 1913.

[2] Knopf, Adolph, Ore deposits of the Helena mining region, Montana: U. S. Geol. Survey Bull. 527, p. 65, 1913.

[3] Ferguson, H. G., Gold lodes of the Weaverville quadrangle, California: U. S. Geol. Survey Bull. 540, p. 42, 1914.

out by Ferguson in the paper mentioned above, and by Bard,[1] in connection with a discussion of downward migration of copper in calcite gangue. Indeed, an experiment recorded by Eddingfield in the paper mentioned above illustrates this fact admirably. A solution of 10 per cent by weight sulphuric acid was allowed to percolate through a mixture containing 45 per cent powdered calcite. The filtrate was neutral to litmus after 48 hours and was still neutral to litmus after 72 hours, but after 75 hours it was acid. The solution had formed channels in the calcitic mass and was no longer reacting on the limestone, at least not rapidly enough to be neutralized. Briefly, in a gangue containing manganese with abundant calcite or other minerals that react rapidly with the acid waters, the solution of gold is delayed, but when the calcite or other active mineral has been dissolved and carried away, gold may then go into solution. It will be carried downward until it encounters ferrous sulphate, calcite, or some other precipitating material lower in the altering zone, when it will be deposited. In general calcite-rich veins are likely to carry any secondary gold bonanzas very near the surface, and the outcrops of such veins may be their richest parts. Siderite, nepheline, and pyrrhotite, as stated above, likewise neutralize acid solutions and precipitate gold very readily. Gold therefore descends tardily in deposits carrying appreciable amounts of these minerals.

PLACERS AND OUTCROPS.

Those deposits in which the transportation of gold is believed to have taken place are, probably without exception, manganiferous. Inasmuch as enrichment is produced by the downward migration of the gold instead of by its superficial removal and accumulation, it should follow that both gold placers and outcrops rich in gold would generally be found in connection with nonmanganiferous deposits; and this inference is confirmed by field observations. Placer deposits are in general associated with nonmanganiferous lodes, and such lodes are generally richer at the outcrops and in the oxidized zones than in depth, the enrichment being due, in the main, to a removal of the material associated with gold. Even under favorable conditions, however, gold is generally dissolved less readily and precipitated more readily than copper. Consequently its enriched ores are likely to be found nearer the surface.

As already stated, the rate of the transfer of gold from the surface downward depends on many factors, such as the fracturing of the deposit and its mineral composition. When erosion is rapid it overtakes solution and then auriferous outcrops or placer deposits may

[1] Bard, D. C., Absence of copper sulphide enrichment in calcite gangues: Econ. Geology, vol. 5, pp. 59–61, 1910.

be formed from manganiferous gold lodes. This is most likely when solution is slow, as it is where carbonates, particularly calcite and siderite, are abundant in the gangue, or where inclosing rocks contain nepheline, leucite, olivine, or other minerals that reduce acid solution rapidly. In a manganiferous calcite gangue gold may accumulate at the very outcrop, for the solutions could not long remain acid if passing through alkaline minerals. Some placer deposits are associated with gold lodes having a manganiferous calcite gangue. Examples of such deposits are found in the Philippine Islands and at Marysville, Mont. These are described by Eddingfield and by Knopf in the papers mentioned above.

CONCENTRATION IN THE OXIDIZED ZONE.

The concentration of gold in the oxidized zone near the surface, where the waters remove the valueless elements more rapidly than gold,[1] is an important process in lodes which do not contain manganese or in manganiferous lodes in areas where the waters do not contain appreciable chloride. In the oxidized zone in some mines it is difficult to distinguish the ore which has been enriched by this process from ore which has been enriched lower down by the solution and precipitation of gold and which, as a result of erosion, is now nearer the surface. It can not be denied that fine gold migrates downward in suspension, but this migration probably does not occur to an important extent in the deeper part of the oxidized zone. If the enrichment in gold is due simply to the removal of other constituents it is important to consider the volume relations and the mass relations before and after enrichment and to compare them with those now prevailing. It can be shown that the enriched ore in some lodes occupies about the same space that it occupied before oxidation. Let it be supposed that a pyritic gold ore has been altered to a limonite gold ore, and that gold has been neither removed nor added. Limonite (specific gravity 3.6 to 4), if it is pseudomorphic after pyrite (specific gravity 4.95 to 5.10) and if not more cellular, weighs about 75 per cent as much as the pyrite. In those specimens which I have broken cellular spaces occupy in general about 10 per cent of the volume of the pseudomorph. With no gold added, the secondary ore should not be more than twice as rich as the primary ore, even if a large factor is introduced to allow for silica and other gangue material removed and for cellular spaces developed.

Rich bunches of ore are much more common in the oxidized zones than in the primary zones of gold lodes. They are present in

[1] Rickard, T. A., The formation of bonanzas in the upper portions of gold veins: Am. Inst. Min. Eng. Trans., vol. 31, pp. 198–220, 1902.

some lodes which carry little or no manganese in the gangue and which below the water level show no deposition of gold by descending solutions. Some of them are doubtless residual pockets of rich ore that were richer than the main ore body when deposited as sulphides, but others are very probably ores to which gold has been added in the process of oxidation near the water table by the solution and precipitation of gold in the presence of the small amount of manganese contributed by the country rock. In view of the relations shown by chemical experiments it is probable that a very little manganese will accomplish the solution of gold, but it requires considerably more manganese to form appreciable amounts of the higher manganese compounds that delay the deposition of gold, suppressing its precipitation by ferrous sulphate. In the absence of larger amounts of the higher manganese compounds the gold would probably be precipitated almost as soon as the solutions encountered the zone where any considerable amount of pyrite or other reducing materials were exposed in the partly oxidized ore. From this it follows that deposits showing only traces of manganese, presumably supplied from the country rock, are not appreciably enriched by solution and downward migration, although they may show some solution and reprecipitation.

VERTICAL RELATION OF DEEP-SEATED ENRICHMENT IN GOLD TO CHALCOCITIZATION.

In several of the great copper districts of the West gold is a valuable by-product. In another group of deposits, mainly of Tertiary age and younger than the copper deposits, silver and gold are the principal metals, and copper, when present, is only a by-product. But in some of these precious-metal ores chalcocite is nevertheless one of the abundant metallic minerals and constitutes several per cent of the vein matter. In many ores it forms a coating over pyrite or other minerals. Some of this ore, appearing in general not far below the water table, is fractured spongy quartz coated with pulverulent chalcocite. A part of it contains a good deal of silver and more gold than the oxidized ore or the deeper-seated sulphide ore. Clearly the conditions that favor chalcocitization are favorable also to the precipitation of silver and gold.

The replacement of pyrite by chalcocite is, according to Stokes,[1] as follows:

$$5FeS_2+14CuSO_4+12H_2O=7Cu_2S+5FeSO_4+12H_2SO_4.$$

[1] Stokes, H. N., Experiments on the action of various solutions on pyrite and marcasite: Econ. Geology, vol. 2, p. 22, 1907.

This reaction is considered by Spencer[1] as comprising several stages, which may be indicated as follows:

$$2FeS_2+2CuSO_4=Cu_2S+2FeSO_4+S.$$
$$3S+2CuSO_4=Cu_2S+4SO_2.$$
$$6H_2O+5SO_2+2CuSO_4=Cu_2S+6H_2SO_4.$$

The reactions may differ as to details, but without any doubt ferrous sulphate is commonly present in zones where chalcocite is forming. The abundant ferrous sulphate must quickly drive the gold from solution. If the solutions that deposit secondary chalcocite are alkaline gold would quickly be precipitated. Hence it follows that there may be no appreciable enrichment of gold below the zone where chalcocitization is the prevailing process. Moreover, chalcocite will itself rapidly precipitate gold from acid solutions in which it is held as chloride. (See p. 311.) In deposits like disseminated chalcocite in porphyry, in which the chalcocite occurs in flat-lying zones related to the present topography, where the ore from which chalcocite was derived carried gold and where suitable solvents were provided, the gold, at least in the upper part of the chalcocite zone, should be rather evenly distributed and should increase and decrease in quantity with the chalcocite of the secondary ore. According to reports there is a fairly constant ratio between copper and gold in the disseminated deposits at Ely, Nev., and Bingham, Utah. In the copper deposits at Rio Tinto, Spain,[2] the secondary ores between the gossan and the primary sulphides carried high values in gold and silver. In view of the chemical relations it would appear that whatever gold and silver are present below a chalcocitized pyrite ore zone are, without doubt, primary.

Gold and silver are commonly associated in their deposits, and an exact knowledge of the solution and precipitation of the two metals in experiments where both are present would have great practical value. Although gold is dissolved in chloride solutions, silver chloride is but slightly soluble, and high concentration of the two metals could not exist in the same solution. There is so little exact information regarding the solubilities of gold chloride and silver chloride in solutions containing both metals that a discussion of their relations is little more than speculation, yet certain data should be considered in this connection. Silver chloride is slightly soluble in water, and silver may be held in small concentration in solutions in which gold also is dissolved. A mine water from the Comstock

[1] Spencer, A. C., Chalcocite deposition: Washington Acad. Sci. Jour., vol. 3, p. 73, 1913.

[2] Vogt, J. H. L., Problems in the geology of ore deposits, in Pošepný, Franz, The genesis of ore deposits, p. 676, 1902.

lode, according to Reid,[1] carried 188 milligrams of silver and 4.15 milligrams of gold in a ton of solution. Since ferrous sulphate and certain sulphides precipitate both gold and silver from acid solution, alloys of these metals might form as secondary minerals.

The rapidity with which chalcocite precipitates both silver and gold[2] would prevent migration of these metals through a zone where chalcocite prevailed.

PRECIPITATION OF GOLD WITH MANGANESE OXIDES.

Where it is held in solution as chloride, gold is readily precipitated by ferrous sulphate, which, as already stated, is formed by the action of acid on pyrite or other iron sulphides. As long as ferrous sulphate is present gold will not be redissolved. If much manganese oxide is present, however, the ferrous sulphate is immediately oxidized to ferric sulphate, which does not precipitate gold from solutions in which it is held as chloride. In the presence of manganese oxides, therefore, gold is not only dissolved in acid solution but the conditions under which it is precipitated may be delayed. Gold may be carried in acid solution so long as the higher oxides of manganese are present.

In many gold deposits manganese oxides and gold are intimately associated and without doubt have been precipitated together. This association is by no means uncommon. The deposits of the Camp Bird mine,[3] of the Tomboy mine,[4] of the Amethyst vein at Creede, Colo.,[5] of the Dahlonega mines in Georgia,[6] and of mines at Philipsburg, Mont.,[7] are noteworthy examples.[8]

These observations indicate a process by which gold is precipitated with manganese oxides in a reducing environment. Both gold and manganese[9] are held in the acid solution descending through the ore, but in depth by the continued reaction with alkaline minerals of the gangue and wall rock the solution loses its acidity when manganese and gold are precipitated together.

---

[1] Reid, J. A., The structure and genesis of the Comstock lode: California Univ. Dept. Geology Bull., vol. 4, p. 193, 1905.

[2] Palmer, Chase, and Bastin, E. S., Metallic minerals as precipitants of silver and gold: Econ. Geology, vol. 8, p. 140, 1913.

[3] Ransome, F. L., A report on the economic geology of the Silverton quadrangle, Colorado: U. S. Geol. Survey Bull. 182, p. 202, 1901.

[4] Purington, C. W., Preliminary report on the mining industries of the Telluride quadrangle, Colorado: U. S. Geol. Survey Eighteenth Ann. Rept., pt. 3, pp. 838–841, 1908.

[5] Emmons, W. H., and Larsen, E. S., A preliminary report on the geology and ore deposits of Creede, Colo.: U. S. Geol. Survey Bull. 530, p. 58, 1913.

[6] Laney, F. B., oral communication.

[7] Emmons, W. H., and Calkins, F. C., Geology and ore deposits of the Philipsburg quadrangle, Montana: U. S. Geol. Survey Prof. Paper 78, 1913.

[8] See also Tolman, C. F., Jr., Secondary sulphide enrichment of ores: Min. and Sci. Press, vol. 106, p. 41, 1913.

[9] Emmons, W. H., The agency of manganese in the superficial alteration and secondary enrichment of gold deposits in the United States: Am. Inst. Min. Eng. Trans., vol. 42, p. 28, 1912.

Experiments in the precipitation of gold have been made by A. D. Brokaw.[1] Into an acid solution in which gold was dissolved in the presence of manganese a crystal of calcite was introduced. On decreasing acidity of the solution with calcium carbonate gold was precipitated with manganese oxide on the surface and in the cleavage cracks of the calcite crystals. (See fig. 21, p. 315.) Brokaw considers the reactions to be as follows:

$$2AuCl_3+3MnCl_2+6H_2O \rightleftarrows 2Au+3MnO_2+12HCl.$$

Or

$$2AuCl_3+3MnSO_4+6H_2O \rightleftarrows 2Au+3MnO_2+3H_2SO_4+6HCl.$$

These reactions indicate processes by which gold held in acid solution in the presence of manganese salts may be precipitated in the deeper zones, together with manganese oxides, when the solutions reacting on alkaline minerals lose acidity. Because gold and manganese dioxide were precipitated before the solution became completely neutralized, Brokaw regarded the manganese salt as the precipitating agent, as is indicated in the reactions stated above, or more simply as follows:

$$2Au^{+++}+3Mn^{++}=2Au+3Mn^{++++}.$$

In general, however, the manganiferous gold ores formed in the lower zones of gold deposits carry more manganese than gold and a larger proportion of manganese than would be precipitated by the reaction above indicated.

Very recently Lehner[2] has discussed this reaction in considerable detail. He states that manganese dioxide and gold chloride will be reduced by the process of "autoreduction." When the two are brought in contact manganese dioxide loses some oxygen and gold loses chlorine and becomes metallic gold. The process he compares with the reaction of peroxide of hydrogen and gold oxide:

$$Au_2O_3+3H_2O_2=2Au+3H_2O+3O_2.$$

It is supposed that the atoms of oxygen in the gold oxide and in hydrogen peroxide become molecules of oxygen. In a similar way, as above stated, manganese dioxide will itself reduce gold chloride. But the reaction goes on only in alkaline, in neutral, or in very feebly acid solutions. With even moderate acidity, as is shown in experiments stated on page 306, manganese does not precipitate gold but by releasing nascent chlorine aids its solution. In the

[1] Brokaw, A. D., The secondary precipitation of gold in ore bodies: Jour. Geology, vol. 21, p. 251, 1913.

[2] Lehner, Victor, On the deposition of gold in nature: Econ. Geology, vol. 9, p. 527, 1914.

oxidizing acid zone, with moderate supply of pyrite undergoing alteration, "autoreduction" would rarely take place, if ever, and there is excellent geologic as well as experimental evidence that gold is dissolved in the presence of manganese and acid.

Lehner states also that when a large excess of manganese chloride or sulphate is introduced into an open vessel with calcite and the solution greatly diluted to cause deposition to take place slowly, the calcite is after several days covered with a thick layer of manganese dioxide which contains only a very small amount of gold. When this experiment is repeated with a pure manganese solution and calcite, manganese dioxide is formed without gold in the same manner.

But this reaction, precipitating manganese dioxide without gold, could not go on at depths where air is excluded, for obviously there must be a source of oxygen to form an oxide in which manganese has a valence of 4 from manganous sulphate or chloride, in which the valence is 2. In many deposits gold is precipitated with manganese oxides below the zone where air is present. At such places no source of oxygen is apparent and probably no more manganese dioxide would form than would correspond to oxygen released by the gold solution. The so-called autoreduction of gold solutions can not be appreciably effective in the zone of solution as long as the solutions have moderately high acidity. It is true that in some deposits where gold and manganese have been precipitated together more manganese is present than gold. Why manganese is in excess has not yet been explained.

Possibly some of the manganese was precipitated on neutralization as hydroxide, which oxidizes almost immediately on being exposed to air, or perhaps as some other manganese compound that is soluble in acid but insoluble in alkaline solutions.

In some deposits there is evidence that gold has been dissolved and reprecipitated in the surficial zone, yet the secondary gold ore carries no manganese, or at least not more than a trace of manganese compounds. Such relations, according to Ransome,[1] are indicated by the gold veins of Farncomb Hill, in the Breckenridge district, Colorado, where in primary ores that carry manganese the secondary gold ore is almost free from manganese. It has been stated that gold may be precipitated from solutions in which it is dissolved as chloride either by the neutralization of the solution or by ferrous sulphate generated by the action of the solution on pyrite. Ferrous sulphate will precipitate gold, even from strongly acid solutions in which manganese would still remain in solution. It follows that manganese might not be precipitated with gold from acid chloride waters, even where manganese dioxide has supplied conditions for its solution, but

[1] Ransome, F. L., Geology and ore deposits of the Breckenridge district, Colorado: U. S. Geol. Survey Prof. Paper 75, p. 170, 1911.

from neutralized solutions the gold and manganese could go down simultaneously. Gold is precipitated from chloride solution also by native metals, sulphides, organic matter, and many other materials. Experiments in the geological laboratory at the University of Minnesota show that many of these will precipitate gold with little or no manganese dioxide.

The size of the particles of gold has little value as a means of determining its origin. The gold deposited by the reduction or the neutralization of manganiferous compounds may be finely pulverulent or it may form nuggets of considerable size. Brokaw, in the experiments cited above, where gold and manganese were precipitated on calcite, obtained masses of gold distinctly visible without a hand lens.

In some deposits the purity or fineness of the gold may afford a means by which primary may be distinguished from secondary gold. In a deposit in northern Nevada the rich gold ore is electrum, a light-colored natural alloy of gold and silver. On this, deposited presumably from cold solutions, are small masses of dark-yellow gold which is apparently of great fineness. This criterion should be applied with caution, however, and the relations should be established independently for each particular deposit, for under some conditions doubtless secondary gold and silver are precipitated together.

GOLD AND SILVER TELLURIDES.

Petzite, sylvanite, krennerite, and calaverite are tellurides of gold and silver, the precious metals being present in varying proportions. All are primary. Hydrogen telluride (H_2Te) is made by the action of acids on metallic tellurides, but it decomposes readily even below 0° C. No clearly defined examples of secondary gold-silver tellurides are known to me.

SUMMARY.

Briefly stated, the chemical processes by which secondary deposits of gold are formed in nature are almost identical with those employed in barrel chlorination, a process that before the introduction of cyaniding was the chief wet method employed in recovering gold. The ore in the chlorination process is agitated with sulphuric acid, salt, and manganese dioxide; the gold solution is decanted or filtered off and treated with ferrous sulphate, which precipitates the gold. In nature the conditions that are favorable, if not essential, for gold enrichment are (1) chloride solutions, (2) iron sulphides, (3) manganese compounds. Where these conditions are supplied and where no very effective precipitant is at hand and erosion is not too rapid, gold placers are rarely formed, and outcrops of gold ores are likely to be less rich than the ores that lie deeper. Where these conditions exist and where the lodes are fractured, gold will migrate downward. Many minerals will precipitate gold, but in the presence of some its precipitation is particularly rapid. These minerals include calcite, siderite, rhodochrosite, pyrrhotite, chalcocite, nepheline, olivine, leucite. In deposits that carry appreciable quantities of these minerals in ore or wall rock the downward migration of gold is delayed and gold bonanzas are likely to form at or very near the surface. Under these conditions, also, placers may form, even from manganiferous lodes. In a gangue of adularia, sericite, and quartz, with pyrite, chalcopyrite, galena, and sphalerite, gold may be carried downward several hundred or even a thousand feet, but this depth should be regarded as nearly the maximum and exceptional. As gold is so readily precipitated by many common minerals, it will generally migrate slowly in ground water.

The section just quoted from Emmons, with its accompanying bibliography, summarizes most of the work done on the near-surface redox processes in auriferous deposits up to 1915. It is followed by a long section on examples of enrichment in gold deposits, mainly in North America but also in Australia (Mount Morgan).

Relatively little detailed knowledge based on modern investigations is available on the oxidation of gold telluride deposits. Many years ago Lindgren and Ransome (1906) carried out an investigation of the oxidation of the gold telluride lodes of Cripple Creek, Colorado. Their findings are summarized in the section on gold in chapter 32 of Lindgren's (1933) Mineral Deposits as follows:

> The gold-telluride lodes of Cripple Creek, Colorado, are mainly sheeted zones in which the seams are filled with quartz, fluorite, and calaverite [$Au(Ag)Te_2$]. These deposits oxidize to brownish clayey material in which the original vein structure is no longer apparent. As quartz is not abundant, the main product of the oxidation is kaolin, with some limonite. The fluorite is carried away, while the tellurides are very easily reduced to dark-brown powdery gold. The tellurium is partly carried away in solution but to some extent remains as colorless tellurite (TeO_2) and green ferric tellurites like durdenite and emmonsite.
>
> The oxidation extends to the water level, which is from 300 to 900 feet below the surface, and in places the ore is oxidized for some distance below the water level. There has been little or no enrichment of gold in the oxidized zone, but a decided leaching of the small amount of primary silver originally contained as telluride or tetrahedrite. No secondary silver sulphides were detected, nor is there evidence of secondary deposition of tellurides. (pp. 859-860)

Work in the Soviet Union on the redox processes in the surface and near-surface parts of mineral deposits was pursued mainly by academician S. S. Smirnov in the late 1940s. The results of his research on gold are contained in his book (Smirnov, 1951) *The Zone of Oxidation of Sulphidic Mineral Deposits.* Paper 16-2 is an abridged free translation of the section on gold.

(Text continues on page 576.)

16-2: GOLD

S. S. Smirnov

This article was translated expressly for this volume by the Translation Bureau, Department of the Secretary of State, Government of Canada from pages 198-207 of *The Zone of Oxidation of Sulphidic Mineral Deposits*, Izd. Akad. Nauk USSR, Moscow, 1951, 335p.

Hypogene Mineralogy:

The hypogene mineralogy of gold is extremely simple. If the comparatively rare tellurides of gold (petzite, calaverite, krennerite, nagyagite and sylvanite) are excluded there is only one gold mineral that has a universal occurrence, namely, native gold.

The forms of native gold in ores are varied, and it is necessary to consider them before discussing the behavior of gold under the conditions of the zone of oxidation. First, it must be emphasized that in addition to the so-called visible gold, which is apparent either to the naked eye, or what is more often the case, only under the microscope, there is also some gold that cannot be detected even with microscopes of the highest magnification. Furthermore, in some deposits (e.g., Boliden) a considerable part of the total content of gold is present as invisible gold. Another marked characteristic is the frequent specific association of gold with various sulphides, primarily with pyrite and arsenopyrite, partly with chalcopyrite and stibnite, and to a lesser extent, with other sulphides.

These two characteristics, the occurrence of invisible gold and the specific association of the metal with sulphides, have not been adequately explained.

Former hypotheses concerning the occurrence of gold sulphide or the presence of certain chemical combinations of gold within separate minerals are not convincing in the light of the latest data and are consequently rejected by most researchers. Most investigators are inclined to explain the combination of gold in minerals as the result of hypogene fixation (Buschendorf, 1928), since, as a rule, a study of the paragenesis of gold discloses its younger age compared with most sulphides. It is thought that the sulphides exerted a marked precipitating (reducing?) influence upon gold transported in thermal waters, and for this reason the metal became fixed primarily in intimate association with sulphides.

Regarding invisible gold it is generally assumed that this variety of gold is also present in the native form, but in an exceedingly dispersed state. Even the gold that is visible is represented mainly by very fine grains that at times approach the dimensions of colloidal particles. The data given by Head (1934) for certain auriferous pyritic ores can be used as an illustration. According to his data many of the particles of gold have dimensions of only a few microns.

It is, however, still not clear how the extremely fine (at times invisible) particles of gold in various sulphide grains originate.

In many cases the nature of the distribution of gold in sulphides definitely points to a successive deposition of the metal along the minutest fissures and cavities in sulphide grains. In others, the more or less dispersed nature of the impregnation, at times fairly regular, suggests exsolution of the gold from solid solution in various sulphides.

Without dealing further with these vague phenomena concerning the characteristics of gold in ores we will emphasize two facts that are of importance to us. The first fact is the development of gold primarily in the native form, and the second fact is the common minute size of gold grains, which, no doubt, at times have the dimensions of colloidal particles.

Solution of Gold in the Zone of Oxidation:

Without considering the alteration-processes of gold tellurides, in view of the rarity of these minerals,* we will deal here only with the alteration processes of native gold in the zone of oxidation. The high stability of gold under the various physical-chemical conditions, that exist in the surface zone of the earth's crust, is well known. An excellent illustration of this stability is the common occurrence of gold in alluvial deposits, often in the form of grains of an extremely small size (up to 2-3 μ).

Nevertheless, a number of facts gathered from deposits of placer gold, and also, in particular, from deposits of native gold indicate clearly that under certain conditions gold has considerable mobility in the zone of oxidation. The manner in which this migration of gold takes place has not yet been definitely ascertained.

Up to the present the problems of the chemical migration of gold are, according to the works of Hoover, "a convenient field for various speculations."

Actually, all the existing hypotheses concerning the modes of transport of gold in solutions in the zone of oxidation are not entirely convincing, and not one of the hypotheses explains the entire diversity of the observed facts.

The most authentic and developed hypothesis concerns the migration of gold in the form of chloride. This hypothesis, postulated some time ago, has acquired a special popularity following the work by Emmons and Brokaw (Emmons, 1917). The essence of this hypothesis, which is based on a well-known method of processing gold ores by means of chlorination, can be stated as follows: Gold chloride, which is soluble in water, forms readily when gold is exposed to the action of nascent chlorine. The occurrence of chlorine in the nascent state in the zone of oxidation is possible when H_2SO_4 and NaCl occur simultaneously in the waters, and the ores contain various higher manganese oxides, e.g., psilomelane or pyrolusite. The reaction can be written as:

$$MnO_2 + 2NaCl + 3H_2SO_4 = 2H_2O + 2NaHSO_4 + MnSO_4 + Cl_2.$$

Chlorine is formed, which in turn dissolves gold as the chloride.

Of the numerous experiments carried out concerning the dissolution of gold under these conditions we can mention those by Brokaw (Emmons, 1917). A piece of gold leaf (assaying 0.999 Au) weighing 0.15 g, with an area of 350 mm^2 and a thickness of 0.05 mm was placed in a solution containing 0.1 N. $Fe_2(SO_4)_3$, 0.1 N. H_2SO_4 and 0.04 N. NaCl. To this solution, which had a

*Lindgren (1919) characterizes the alteration of gold tellurides in the zone of oxidation at Cripple Creek, Colorado in the following way. Tellurides decompose very readily, and the gold precipitates in the form of a finely-dispersed powder with a dark brown color. The tellurium passes partly into solution, and part is fixed in the form of colorless tellurite and as greenish tellurites of iron, namely, emmonsite and durdenite.

volume of 50 cm^3 was added 1 g of powdered manganese dioxide. The experiment was continued for 14 days, after which it was observed that 0.00505 g of gold of the original 0.15 g has passed into solution.

Sulphuric acid is almost always present in the oxidation zones of sulphidic gold deposits, especially those containing pyrite, marcasite, pyrrhotite, and sphalerite. Sodium chloride is always detected in mine waters, and in some cases sodium chloride is present in considerable quantitites. Furthermore, the sulphides themselves, as shown by the investigations of Newhouse (1932), contain frequent inclusions of saturated NaCl solutions.

Manganese oxides frequently occur in the ores of a number of deposits, but even if the hypogene ores lack manganese compounds the latter may appear in oxidized material as a result of infiltration from wall rocks. In general it seems possible that all the enumerated manganese compounds occur in the zone of oxidation in many cases.

But the simple fact of the occurrence of these manganese compounds does not entirely suffice. It is necessary, as Boydell (1924) has pointed out, that the solutions containing H_2SO_4 and NaCl, manganese oxides and gold, occur in intimate contact. Chlorine, especially in the nascent form, is such an active element that one can hardly conceive of its extensive migration in the waters of the zone of oxidation, especially when occurring in an uncombined (nascent) condition. Concerning this fact considerable dissolution of gold in the manner just indicated is doubtful in general. Of course, in deposits with a high sulphide content, with an abundance of manganiferous minerals, and where the waters have a high chloride content, the dissolution of gold in the form of chloride can be extensive; such conditions are, however, rare.

Furthermore, it should be borne in mind that even in deposits where the conditions are ideal one does not always observe widespread migration of gold in solution. On the contrary, considerable mobility of gold is observed in deposits that do not meet the conditions of the "chloride theory."

Thus, the chloride theory of the transportation of gold, generally speaking, is indisputable, but it explains only a part, actually a very small part, of the phenomena.

The other hypothesis, which in its day was generally accepted, and is still adhered to by many, is the hypothesis of the solvent action of $Fe_2(SO_4)_3$ on gold in the presence of oxygen. Experiments concerning the action of $Fe_2(SO_4)_3$, carried out by Stokes and Brokaw (Emmons, 1917), disclosed the complete absence of the solvent action of $Fe_2(SO_4)_3$ on gold. Nevertheless, field observations which often indicate a relatively large-scale migration of gold where high concentrations of ferric sulphate occur in the waters, suggest that we cannot disregard the role of $Fe_2(SO_4)_3$ in the dissolution of gold.

A third hypothesis, strongly supported by a number of authors (among them Lindgren, 1919 and Boydell, 1924), concerns the solution of gold as a colloid under the conditions existing in the zone of oxidation. The extreme dispersive nature of gold in ores, mentioned frequently in the literature, strongly favors this hypothesis. It is possible that, when sulphides containing extremely finely divided colloid gold are dissolved, the gold passes directly into colloidal solution. Furthermore, considering the time-factor, one cannot discount the solvent action of water upon native gold, especially in view of the extremely dispersed nature of the particles of the gold.

Actually, there are many investigators who do not reject the possibility of dissolution of gold in the manner indicated. The main objections concern the extreme instability of colloidal gold solutions. This feature will be dealt with where the problems of migration are examined.

Of all the existing theories the colloidal theory of the dissolution of gold is the most tempting one in view of its universality. It is probable that the migration of gold in both primary and placer deposits occurs primarily in this manner.

A fourth hypothesis, of recent origin, concerns the solution of gold by waters containing humic substances, provided there is an absence of oxygen and electrolytes. In Brazil, Freise (1931), the author of this hypothesis, observed the astonishing fact of the rapid restoration of worked out placer gold deposits. Thus, according to Freise, one of the placer deposits in the eastern part of Rio state, contained 11.6 g/t Au when mined in 1912 (at which time 10.85 g/t were extracted); in 1926 this placer yielded 4.66 g/t near the bedrock in the worked out part. In the state of Minas Gerais, another gold placer which was worked out in 1908-1909, with a gold content of 8.5 g/t, yielded a gold content of 4.85 g/t in 1926. In both cases the mechanical transportation of gold was completely ruled out; furthermore, the newly deposited gold in the placers differed sharply from the original variety. The new gold was of a greenish color, its purity was extremely high (almost 1000), the specific gravity was somewhat less, the nature of the surface of the gold flakes differed, and the reaction to Hg and KCN was also different; as a rule the gold responded poorly to amalgamation and cyanidation.

The general environment of the placer gold deposits and the marked analogies in physical and chemical properties of the new gold with the so-called "black gold" (i.e., gold coated with a brownish-black substance, found by Freise to be an iron humate) suggested the idea of solution and transportation of gold by humic compounds. The experiments conducted in this respect confirmed the accuracy of this supposition. Thus, experimenting with finely granulated gold and an extract from lignite disclosed a considerable solubility of gold. The following experiment was of particular interest. Finely granulated gold was placed in a slightly inclined trough, 50 feet long, 6 feet wide, and 4 feet deep, filled in one case with sand and in another with sand containing clay and so on (the gold was placed in the elevated end of the trough); the contents of the trough were then sluiced (at a speed that ruled out mechanical transport) with nearby swamp waters having a high concentration of humic substances. The results showed that in every case the gold had migrated along the trough, but at varying rates, depending on the conditions of the experiment.

Freise assumes an active role for humic suspensions in the dissolution and transportation of gold, provided oxygen and electrolytes are absent. However, it is not entirely clear how accurate this interpretation is of the observed facts and whether certain other factors not yet taken into account played a more important role. In this regard, the experiments conducted by Fetzer (1934) on the dissolution and transportation of gold by humic suspensions produced negative results. Hence the reliability of the humic theory can be seriously questioned. Furthermore, regarding the conditions in the zone of oxidation where an abundance of oxygen and electrolytes are present, the humic theory, even if it is correct, cannot be of any great significance.

These four theories encompass all the known viewpoints regarding the principal modes of solution of native gold under the conditions existing in the zone of oxidation. It is plain that none (with perhaps the exception of the colloidal theory) are entirely satisfactory either because of doubtful reliability (the sulphate theory), or for some other reason (the humic theory).

Transportation and Deposition of Gold:

The question of the migration of gold in solution is as obscure as the problem of its dissolution. All the types of gold solutions, known to us as being possible under the conditions under investigation, are distinguished by an exceptional instability. The extent of our lack of knowledge of the manner in which gold is dissolved is compensated by the extent of our knowledge of the deposition of gold.

Thus, if one, for instance, examines the migration conditions of gold in the form of chloride it can be said that gold is precipitated from alkaline, neutral and even slightly acid solutions by an overwhelming majority of compounds usually found in the zone of oxidation. Almost all sulphides (especially sphalerite and pyrrhotite), native metals, silicates, and particularly carbonaceous materials precipitate gold in the native form from its chloride solutions. Furthermore, all the typical reducers such as the various organic compounds, hydrogen sulphide, and of course, $FeSO_4$, quickly terminate the migration of gold in the solutions being examined. The influence of $FeSO_4$, (a compound common in the zone of oxidation) upon the migration of gold merits special attention. A number of experiments, described by W. H. Emmons (1917), prove the marked precipitating effect of $FeSO_4$ upon gold solutions. This sulphate is justly regarded as one of the most universal precipitants of gold under the conditions existing in the zone of oxidation.

It can be said in general that because of the abundance of precipitants it is difficult to envisage any kind of perceptible migration of gold in alkaline, neutral, and even slightly acid solutions. One must agree, therefore, with Locke (1926) that "the presence of abundantly occurring feldspars makes the migration of gold unlikely and the occurrence of an abundance of calcite, impossible." The acidity of solutions, which neutralizes the effect of the previously enumerated precipitants, will, however, improve the conditions for the migration of gold, and therefore, as always, this factor facilitates or favors the transportation of gold. An abundance of $Fe_2(SO_4)_3$ also improves the conditions for the migration of the metal.

We now pass on to an examination of the conditions of migration of gold in colloidal solutions. Such colloidal solutions can originate either in the manner indicated or by reducing auriferous solutions, for example, when $FeSO_4$ acts upon a solution of gold chloride. Most researchers, however, regard the stability of colloidal gold solutions as extremely negligible. Thus, Maclaren (1908), who spent much time studying the problems of the transportation of gold in colloidal solutions, points out that such solutions are coagulated so readily by various electrolytes that they can exist only in pure water; even in pure water the colloidal solutions of gold disintegrate relatively rapidly. Therefore, Maclaren assumes that colloidal solutions play a relatively unimportant role in the migration of gold. However, it must be stated that the occurrence of protective colloids in the waters (such as SiO_2 and $Fe(OH)_3$) in the zone of oxidation introduces another factor in the views of the impossibility of a considerable migration of gold in colloidal solutions.

Boydell (1924), in particular, emphasizes the effect of protective colloids and assumes that the principal mechanism for the transportation of gold is by colloidal solutions.

When speaking of the migration of gold and its precipitation, mention must be made of a frequently interesting peculiarity related to various types of gold solutions. Quite commonly one observes a marked gold enrichment in various so-called glinki (clays) comprising kaolin and similar substances that often occur in the zone of oxidation. It is not yet clear what causes this kind of local concentration, but is seems possible that the phenomenon is the result of adsorption. Mention must also be made of the frequent association of gold with accumulations of manganese oxides and hydroxides. Some investigators (e.g., W. H. Emmons, 1917) are inclined to regard this association as the result of the reducing action of manganese oxides upon gold chloride (the so-called autoreduction phenomenon). Alternatively, it appears possible that the association results from adsorption. At any rate, the quantity of manganese oxides in accumulations of this kind greatly exceeds the quantity of native gold present in them.

Concentration of Gold in the Zones of Oxidation and Sulphide Enrichment:

When summarizing all that has been said about the dissolution, migration, and precipitation of gold, it must be admitted that the conditions for extensive migration of the element are generally unfavourable. However, when prolonged re-working by surface waters is a feature, gold may repeatedly pass into solution, each stage experiencing a small but significant primarily downward migration. Furthermore, the farther the gold passes downward the more unfavourable become the conditions for its further migration. And, finally, in the lower parts of the zone of oxidation, and especially in the upper horizons of the zone of sulphide enrichment, the conditions become such (presence of $FeSO_4$, an abundance of non-decomposed sulphides, lack of oxygen, etc.) as to entirely exclude the possibility for further migration. Here, as a result of the foregoing factors one can expect a secondary gold enrichment zone, which as a rule, is extremely poorly defined, due to the sporadic nature of the migration of gold; what is usually represented is a series of enrichment pods. Instances of the formation of such pods (or sections) are rare, but are, nevertheless, encountered and provide an excellent illustration of the fact that under certain conditions gold is capable of considerable migration.

What conditions can then be regarded as conducive to the formation of a zone or of separate pods (sections) of enrichment? Because of the general vagueness respecting the nature of the migration of gold it is difficult to give a definite answer to this question. Nevertheless, it is clear that the following conditions can be regarded as favourable: inert wall rock and vein rock, an abundance of pyrite and other sulphides that can supply H_2SO_4 and $Fe_2(SO_4)_3$; the occurrence of manganiferous minerals; widely dispersed fine-grained gold; good permeability of ore deposits and wall rocks that permits a relatively rapid migration of solutions; and a comparatively stationary groundwater level that produces a prolonged re-working of the ore materials by surface waters.

In most deposits the combination of circumstances that are conducive to extensive migration of gold do not occur, and no extensive zone of enrichment occurs. At best only separate local concentrations, sporadically dispersed in the various horizons of the zone of oxidation are formed. More frequently, such sporadically enriched pods (sections) are lacking, and one cannot speak of any degree of gold migration.

However, it must be borne in mind that gold by virtue of its marked immobility may produce an enrichment in the zone of oxidation simply as the result of the lixiviation (solution and removal) of the more mobile elements. This "relative" enrichment at the cost of the removal of the lode gangue is widespread and one must always consider it when evaluating gold deposits. And furthermore, when this "relative" enrichment of gold is present, the difference in the gold contents of the ores in the zone of oxidation as compared with those in the hypogene zone will be marked. As a consequence extreme caution must be used in interpretating the results (assays) of the surface parts of gold deposits. In most cases the hypogene ores will contain less gold than the corresponding ores in the zone of oxidation.

References

[As given in the original text]

Boydell, H. C. The rôle of colloidal solutions in the formation of mineral deposits. Bull. Inst. Min. Met., 1924. No. 243, 1-108.

Buschendorf, F. Betrachtungen über die Gangkomponenten, sowie über das Vorkommen und die Verteilung des Goldes in der Primärzone alter Goldquartzgänge. Zs. prakt. Geol., 1928, H, 1, 1-11.

Emmons, W. H. The enrichment of ore deposits. Bull. U.S. Geol. Survey, 1917, No. 625.

Fetzer, W. G. Transportation of gold by organic solutions. Econ. Geol., 1934, 29, No. 6, 599-604.

Freise, F. W. The transportation of gold by organic underground solutions. Econ. Geol., 1931, 26, No. 4, 421-431.

Head, R. E. Gold in pyrite. Canad. Min. Journ., 1934, No. 6, 275-277.

Lindgren, W. Mineral deposits. 1919, 882-883.

Locke, A. Experiment in ore-hunting geology. Min. Met., 1922, No. 184, 27-29.

Locke, A. Leached outcrops as guides to copper ore. Baltimore, 1926.

Maclaren, J. M. Gold, its geological occurrence and geographical distribution. London, 1908.

Newhouse, W. H. The composition of vein solutions as shown by liquid inclusions in minerals. Econ. Geol., 1932, 27, 421-436.

It will be noticed that Smirnov is skeptical of the migration of gold in humic waters, and indeed he also questions the chlorine-manganese dioxide and the ferric sulfate theories. He favors the colloidal theory for the transport of gold.

More recently, considerable research on the oxidation of auriferous deposits has been carried out in the Soviet Union by V. M. Kreiter and others. Much of this research is contained in their outstanding book *The Behavior of Gold in the Zone of Oxidation of Auriferous Sulphide Deposits* (1959), which has been succinctly reviewed by O. Zvyagintsev in *Geochimia* (no. 6, 1959, pp. 560-561). An abridged form of the English translation of this review is presented as Paper 16-3.

(Text continues on page 580.)

16-3: REVIEW: BEHAVIOR OF GOLD IN THE ZONE OF OXIDATION OF AURIFEROUS SULPHIDE DEPOSITS, by V. M. Kreiter, V. V. Aristov, I. S. Volynskii, A. B. Krestovnikov and V. V. Kuvichinskii (Gosgeoltekhizdat, Moscow, 1959, 268 pages)

O. Zvyagintsev

Reprinted from *Geochemistry,* no. 6, 1959, pp. 683-685.

There are many gaps in our knowledge of the geochemistry of gold which need filling. The problem of the circulation of gold in the earth's crust has not been solved: great amounts of gold dissolved by ground waters are carried out into seas and oceans, but its further history is not known. In general, the sedimentary rocks contain very little gold, and this means that it is preferentially precipitated somewhere from marine waters. Perhaps the deposits of the Rand in South Africa, colossal in reserves and richness, were formed from marine sediments.

There are many other unsolved problems in the geochemistry of gold. Are there two gold fractions deposited in hydrothermal veins, an early and a late one, or is there only one? Can gold be carried by hot solutions over large distances? What role is played by colloidal suspensions in the transfer of gold? And so on. The solution of some of these problems would have not only purely scientific but also practical value.

One of these still incompletely solved problems in the geochemistry of gold is the problem of the solution of gold in the oxidized zones of sulfide deposits and of precipitation of secondary gold in the underlying zone of cementation. The authors of the book reviewed here made a thorough study of these processes.

In the words of the authors--geologists, mineralogists and chemists united under the guidance of V. M. Kreiter--the object of their work was to "give the investigators a practical basis, following from the ideas concerning the supergene migration of gold, for determining the primary and secondary zones of gold deposits."

The first chapter of the book gives a summary of the modern concepts of hypogene and supergene gold and makes it evident that there are many different opinions about the origin of hydrothermal gold and its later history in the zone of weathering.

The second chapter presents the results of geological and mineralogical observations at the Maikain (Kazakhstan), Dzhugaly (Kazakhstan), Blyava and Novyi Sibai (Southern Urals) deposits. The observations include geology, mineralogy and geochemistry of gold in the oxidized zone. There are numerous sketches of polished sections of minerals, including secondary gold. From the material of this chapter, an entirely correct conclusion is drawn that secondary (supergene) gold can form in considerable quantities only if sulfides are present in the primary ores containing microscopic and submicroscopic gold.

The third chapter describes the formation of secondary auriferous zones. A general account is given of the formation of different minerals in the zone where the solutions descending from the oxidized zone enter a reducing environment. Here supergene gold is deposited in a variety of forms and compositions, together with argentite and, less commonly, native silver.

The process which interested the authors the most is that of the solution of gold in the oxidized zone. Experiments and calculations were made to discover the possible solvents of gold and the rates of solution.

The results of these experiments and calculations are given in chapter four. They show that gold may be dissolved in "geologically short time" in acid solutions of ferric sulfate, in iron chloride solution and in hydrochloric and nitric acids, provided it is finely dispersed. Under the same conditions, gold dissolves much more slowly in sulfuric acid.

Very interesting is the method used by the authors in the study of the solubility of gold. They prepared thin films of metallic gold on glass, which were dried and heated to the melting point of glass.

It is shown by calculations what volume of water or of some other solvent is required to dissolve a given weight of dispersed gold. For water, the volume is very large, but for iron chloride or ferric sulfate solutions, it is relatively small. From this, the time required to dissolve 10g of coarse gold dispersed through a block of ore ($0.4\ m^3$) is calculated and found to be from 2000 to 35000 years, depending on the concentration of the solution.

Chapter five contains a description of the conditions of solution and migration of gold in the oxidized zone. Here the authors review the literature on the subject and give their conclusions. They consider that solution of dispersed gold in $Fe_2(SO_4)_3$ and sulfuric acid is the most probably mode of transfer of gold. The solution of gold in iron chloride is regarded as less probable by the authors. The behavior of silver, which migrates with gold or a little ahead of it, speaks in favor of the first hypothesis. The authors discard as improbable and contradictory to observation and experiment the hypothesis of F. Freize of the transfer of gold in organic solvents, that of A. E. Fersman of its migration in the form of cyanides, F. V. Chukhrov's idea that gold is transported in the form of bromides and iodides and M. N. Al'bov's hypothesis of the transfer of gold in suspensions.

The authors document their opinion by observations made at the deposits.

The last two chapters are essentially appendices to the main text. In the sixth chapter there are notes on the practical methods of valuation of gold ore outcrops and in the seventh, some methodological questions of mineragraphy of gold ores are discussed.

Chapter five, which summarizes the findings of the authors, is of especial interest to geochemists. Can the authors' main conclusion that gold is dissolved in the sulfates in the oxidized zone be regarded as proved? Not quite, the reviewer believes. Undoubtedly the migration of gold from the oxidized zone into the underlying rocks is connected with the presence of sulfates and dispersed gold in it. There is no doubt that as a result of oxidation of sulfides free sulfuric acid and ferric sulfate are formed. The authors' experiments have proved that ferric sulfate in a sulfuric acid environment will dissolve gold.

But under natural conditions gold may be dissolved in hydrochloric acid formed from the action of sulfuric acid on sodium chloride. This process is definitely stimulated by the oxidizing action of $Fe_2(SO_4)_3$. The authors of the book consider solution in hydrochloric acid and ferric sulfate improbable and refer to the behavior of silver migrating together with gold. Gold chloride, in the authors' opinion, would remain in the oxidized zone. But they forget that silver chloride is itself easily dissolved by solutions of sodium chloride in hydrochloric acid, forms complex compounds and may migrate into the zone of cementation together with gold. The argument based on the behavior of silver is therefore not decisive.

The reviewer would accept the authors' view of the migration of gold in the form of sulfate solutions as true if they had proved the existence of compounds of gold with ferric sulfate and sulfuric acid and had studied its properties. They have not done this but merely remarked that the formation of complex compounds is probable. Their hypothesis remains a hypothesis. Speaking of the solution of gold in ferric sulfate-sulfuric acid solutions, the authors remark (p. 215) that the presence of chlorides hastens solution. This argues a greater probability of solution of gold in hydrochloric acid formed under the oxidizing action of $Fe_2(SO_4)_3$. Complex chloride compounds of gold which may form under these conditions, such as $Na[AuCl_4]$, are known, and others are possible, including those mentioned in the book on p. 207.

These comments indicate that some points have not been sufficiently developed in the book but are not intended to belittle its importance. The book is very interesting, gives abundant factual material, discusses a difficult problem from many points of view and contains practical and methodological suggestions usable in field and laboratory work.

The publication of the book should be welcomed and the authors encouraged to continue their investigations.

A number of field and accompanying laboratory studies in recent years have dealt with the migration and secondary enrichment of gold in its deposits. Most of these studies are referenced and reviewed in Boyle (1979). I have selected three representative studies, one by Kreiter and co-workers (1958), one by Kelly and Goddard (1969), and one by Lesure (1971).

The studies by Kreiter and his co-workers (1958) are detailed and touch upon nearly all aspects concerning the migration and secondary enrichment of gold in sulfidic deposits. Their book was reviewed in the preceding paragraphs.

The study by Kelly and Goddard (1969) deals with the oxidation of the telluride ores of Boulder County, Colorado. There, the primary telluride veins are composed of an interlacing network of pyritic or marcasitic horn quartz seams in which the ore minerals are quite sparse and irregularly distributed. The primary ore and gangue minerals are numerous (about 67), the most important being quartz, roscoelite, ankerite, calcite, fluorite, barite, pyrite, marcasite, galena, sphalerite, sylvanite, petzite, hessite, calaverite, krennerite, various other tellurides, native tellurium and native gold.

The telluride veins have not been deeply weathered and the residual enrichment of gold is correspondingly slight. Partially oxidized ore contains abundant jarosite, limonite, and tellurium oxides, and in places some supergene native tellurium, mercury, hessite, and the copper tellurides. Fine spongy gold in limonite ("rusty gold") is common in the outcrops and is in places associated with native silver and the silver halides. The geochemical behavior of the principal metals (gold, silver, tellurium, and iron) is discussed in terms of acidities, oxidation potentials, and chloride ion activities in the oxide zone. The mobility of gold has been generally low in the deposits; on the other hand silver has been extensively leached close to the surface, and only minor amounts were reprecipitated at depth. Tellurium was highly mobile in the oxide zone and has been extensively or completely leached from most of the vein outcrops. Eh-pH diagrams for the systems Te-H_2O, Ag-Te-S-H-O-Cl, and for the relations in the near-surface telluride ores are given, as is also an equilibrium oxidation potential-chloride ion activity diagram for the system Ag-Te-S-H-O-Cl. Kelly and Goddard (1969), Kelly and Cloke (1961), and Cloke and Kelly (1964) place considerable emphasis on the role of chloride in the various reactions involving the supergene history of gold, silver, and tellurium.

Lesure (1971) dealt with the enrichment and supergene transport of gold in the saprolites of the Dahlonega gold belt in Georgia. (See also the summary description of these deposits by Becker in chap. 15). Lesure's findings are summarized in his abstract (Paper 16-4).

16-4: RESIDUAL ENRICHMENT AND SUPERGENE TRANSPORT OF GOLD, CALHOUN MINE, LUMPKIN COUNTY, GEORGIA

Frank G. Lesure

Abstract

Gold is present in weathered wall rock and quartz veins of the opencut of the Calhoun mine, Lumpkin County, Georgia. Eleven samples of mica schist saprolite contain 0.02 to 0.1 parts per million (ppm) gold and three samples of weathered vein quartz, 0.04 to 0.6 ppm (limit of detection is 0.02 ppm). Gold was not found in any of the 18 samples of fresh mica schist and was found in only 3 of 16 samples of fresh vein quartz from underground workings 40–100 feet below the opencut.

Whole-rock chemical analyses of the unweathered schist and schist saprolite show large losses of minor elements and addition of water during weathering. The average specific gravity of the unweathered schist is 2.8 and of the schist saprolite, 1.65. The chemical analyses recalculated to weight per volume show a loss during weathering of large amounts of all major constituents including silica, alumina, and total iron. Of the minor elements, gold and arsenic show the most residual enrichment, but boron, beryllium, cobalt, lanthanum, niobium, nickel, lead, zinc, and zirconium show some enrichment in the saprolite as compared with the fresh rock. Most of the other minor elements have been at least partly removed, the greatest loss being shown by calcium, potassium, magnesium, sodium, and strontium. Fresh limonite coating the wall of the adit contains 2.9 ppm gold and appreciable amounts of these elements leached from the schist, particularly barium, calcium, and manganese.

Some layers of unweathered schist contain as much as 1 percent gold-bearing pyrite and may have a gold content of 0.01 and 0.02 ppm, or just at or below the limit of detection. Loss of 30 to 45 percent of the rock during weathering increases the gold content above the lower limit of detection. The gold in recently deposited limonite shows that some gold is also being transported with iron during weathering. If a concentration of limonite were found near the water table, it might be expected to show supergene enrichment in gold.

REFERENCES

Boyle, R. W., 1979. The geochemistry of gold and its deposits, *Canada Geol. Survey Bull. 280,* 584p.

Cloke, P. L., and W. C. Kelly, 1964. Solubility of gold under inorganic supergene conditions, *Econ. Geology* **59:**259-270.

Emmons, W. H., 1917. The enrichment of ore deposits, *U.S. Geol. Survey Bull. 625,* 530p.

Garrels, R. M., and C. L. Christ, 1965. *Solutions, Minerals, and Equilibria,* Harper & Row, New York, 450p.

Kelly, W. C., and P. L. Cloke, 1961. The solubility of gold in near-surface environments, *Michigan Acad. Sci. Arts and Letters Papers,* **46:**19-30.

Kelly, W. C., and E. N. Goddard, 1969. Telluride ores of Boulder County, Colorado, *Geol. Soc. America Mem. 109,* 237p.

Kreiter, V. M., V. V. Aristov, I. S. Volynskii, A. B. Krestovnikov, and V. V. Kuvichinskii, 1958. *Behavior of gold in the zone of oxidation of auriferous sulphide deposits,* Gosgeoltekhizdat, Moscow, 268p. (English rev. by O. Zvyagintsev, in *Geochemistry,* no. 6, pp. 683-685, and by C. F. Davidson in *Econ. Geology* **55:**1761).

Lesure, F. G., 1971. Residual enrichment and supergene transport of gold, Calhoun Mine, Lumpkin County, Georgia, *Econ. Geology* **66:**178-186.

Lindgren, W., 1933. *Mineral Deposits,* 4th ed., McGraw-Hill, New York, 930p.

Lindgren, W., and F. L. Ransome, 1906. Geology and gold deposits of the Cripple Creek District, Colorado, *U.S. Geol. Survey Prof. Paper 54,* 516p.

Newbery, J. C., 1868. Introduction of gold to, and the formation of nuggets in, the auriferous drifts, *Royal Soc. Victoria Trans.* **9:**52-60.

Sato, M., 1960*a*. Oxidation of sulfide orebodies: I. Geochemical environments in terms of Eh and pH, *Econ. Geology* **55:**928-961.

Sato, M., 1960*b*. Oxidation of sulfide ore bodies: II. Oxidation mechanisms of sulfide minerals at 25°C, *Econ. Geology* **55:**1202-1231.

Shuermann, E., 1888. Über die Verwandtschaft der Schwermetalle zum Schwefel, *Liebig's Annalen* **249:**326-350.

Skey, W., 1871. On the reduction of certain metals from their solutions by metallic sulphides, and the relation of this to the occurrence of such metals in a native state, *New Zealand Institute, 1870, Trans. and Proc.* **3:**225-231.

Smirnov, S. S., 1951. *The Zone of Oxidation of Sulphidic Mineral Deposits,* Izd. Akad. Nauk USSR, Moscow, 335p.

Wilkinson, C., 1867. On the theory of the formation of gold nuggets by drift, *Royal Soc. Victoria Trans. and Proc.* **8:**11-15.

CHAPTER

17

Gold Deposits – Special Topics

The presence and distribution of gold within veins is not a matter of chance, but is the logical result of a unique train of events occurring under certain favourable conditions
W. H. White, 1943

Gold deposits possess a number of special features that deserve mention in a landmark book, particularly including the associated minerals and elements in gold deposits, the fineness (Au/Ag ratio) in auriferous deposits, wall rock alteration effects, and the structural environment of deposition of gold in the various deposits in which it is a constituent. Because of space considerations no attempt is made to deal with these subjects in any detail, especially the problems of fineness (Au/Ag ratio) of gold and the chemistry of wall rock alteration in auriferous deposits. The interested reader can refer to Boyle (1979) for summaries of worldwide investigations on these subjects.

ASSOCIATED MINERALS AND ELEMENTS IN AURIFEROUS DEPOSITS

Boyle (1979) has described the mineral associates of gold in detail and has dealt with the elemental associates of the precious metal according to the groups of the periodic table. Briefly, the most common hypogene (endogene) mineral associates of gold include the following in their approximate order of frequency: quartz, carbonates, pyrite, arsenopyrite, sphalerite, galena, pyrrhotite, chalcopyrite, stibnite, sulfosalts, tellurides, molybdenite, scheelite, wolframite, tourmaline, and hematite; less common are fluorite, barite, alunite, magnetite, niccolite, cobaltite, argentite (acanthite), selenides, cinnabar, uraninite (pitchblende), and brannerite. The supergene mineral associates of gold include the following in approximate order of frequency: limonite, wad, jarosites, scorodite, azurite, malachite, anglesite, bindheimite, beudantite, antimony ochres, gypsum, tellurites and tellurates, clay minerals, calcite, and opaline quartz; supergene uranium minerals (e.g., autunite, carnotite, and zeunerite) characterize auriferous deposits enriched in pitchblende and other U-Th minerals.

The elements with which gold is most commonly associated in hypogene (endogene) and supergene deposits are mainly chalcophile and include S, Se, Te, As, Sb, Bi, Cu, Ag, Zn, Cd, Hg, Sn, Pb, Mo, W, Fe, Pt, Pd, Co, and Ni. To these elements must be added Si, which is almost universally enriched in gold deposits, and the constituents of the various other gangue and ore minerals, particularly Ca, Mg, Mn, Al, Ba, B, U, Th, V, Cr, F, and P.

The microscopic features of auriferous quartz, the most important gangue mineral in epigenetic gold deposits, have been described in a classic paper by White (Paper 17-1). He places great stress on cataclasis, a feature repeatedly observed by numerous other investigators of gold-quartz. The microscopic characters of vein carbonates have been described in great detail by Grout (1946) in a memorable paper. Space does not permit reproduction of the whole paper here, but some general observations may be noted. Concerning sulfides in auriferous deposits Grout found a general trend (not without exceptions) for the presulfide carbonate to be mixed and complex, whereas the late carbonate is commonly calcite. The early carbonates are commonly replaced by sulfides, whereas the late calcite fills minute fractures and only rarely replaces the sulfides. Grout noted the common occurrence of native gold with ankerite and along cleavage planes of calcite and other carbonates. In some deposits the carbonates associated with gold began forming early and continued even after the deposition of native gold. Although numerous complexities exist in auriferous deposits, a review of Grout's paper suggests that it is usual to find a paragenetic sequence of carbonates in gold-quartz deposits, beginning with ankerite, giving way to more calcic carbonates, and ending with calcite.

AU/AG RATIOS AND FINENESS OF NATIVE GOLD IN AURIFEROUS DEPOSITS

These two factors have been repeatedly discussed in the literature, resulting in a considerable disparity of opinion as to their genetic significance. It should be emphasized that a clear distinction should be made between fineness of gold and the gold-silver ratio of deposits. The latter refers to the total gold and silver in the deposits as determined by analyses (assays) of the ores expressed as the ratio Au/Ag; the former is calculated in parts per thousand and refers to the proportion of gold present in the naturally occurring metal in the deposits. In general the minor element in native gold is mainly silver, but other base and precious metals may also occur, (e.g., Cu, Fe, and Pt metals). Where the amount of base metals is high in the native gold or bullion the ratio $(Au/Au + Ag) \times 1000$ or the true fineness should be used as suggested by Fisher (1945).

The gold/silver ratios of most types of auriferous deposits have been calculated from a large spectrum of data and presented as Table 17-1.

The significance of these ratios in a metallogenic context is difficult to assess, but a few generalizations are apparent. Gold-quartz deposits and quartz-pebble congomerate deposits of Precambrian, Paleozoic, and Mesozoic age tend to have a high Au/Ag ratio, averaging about 4.2. Many Tertiary and younger gold-quartz deposits have a low Au/Ag ratio averaging about 0.05, but there are many exceptions to this generalization (e.g., Cripple Creek, Colorado: Au/Ag = 10). Most massive sulfide, polymetallic, and porphyry copper depos-

Table 17-1. Au/Ag ratios of a variety of mineral deposits

Type of deposit	*Au/Ag ratio (range)*	*Remarks*
Shale deposits (Kupferschiefer type)	0.006-0.025	
Disseminations, veins, etc, in sandstones (red-bed type)	0.003-0.01	Variable
Quartz-pebble conglomerates (Rand type)	5-20	Witwatersrand bullion ranges between (5.8-15.6) (Hargraves, 1963). On channel samples of the Ventersdorp Contact Reef and Main Reef the Au/Ag ratios range from 0.6-13.7 (Von Rahden, 1965)
Disseminated lead-zinc deposits in carbonate rocks (Mississippi Valley-Pine Point type)	0.001-0.1	Few data
Skarn-type deposits	0.005-10	Variable
Massive sulphides (Ni-Cu Sudbury type)	0.03-0.07	Shcherbina (1956) gives 0.07 for Norilsk, and Hawley (1962) gives 0.03 for Sudbury ores
Porphyry copper type	0.001-0.1	Few data
Polymetallic massive sulphides (Flin Flon-Noranda-Bathurst type)	0.006-1	Average about 0.025
Polymettalic massive sulphides (Kurokô type)	0.005-0.05	Average about 0.02
Polymetallic veins, mantos, etc. (Keno Hill-Sullivan-Coeur d'Alene type)	<0.0001-0.02	
Gold-quartz veins, lodes, etc. (Precambrian, Paleozoic and Mesozoic age)	1.37-12.5	Average about 4.2
Gold-quartz veins, lodes etc. (Tertiary age)	0.005-0.33	Average about 0.05
Native silver-Co-Ni-As-Bi-U veins, (Cobalt-Jachymov type)	<0.001-0.01	
Hot spring siliceous sinters	<1	
Gold placers (all types and ages)	>1	

Source: Boyle (1979), p. 202.

its have low Au/Ag ratios, ranging from 0.001 to 1 with an average of about 0.02. Skarn-type auriferous deposits vary widely in their Au/Ag ratios (0.005-10). Gold placers of all types and ages have a high Au/Ag ratio, generally much greater than 1, a reflection of the natural refining action that takes place in the oxidized zones of auriferous deposits, in eluvium, and finally in alluvium.

Concerning the fineness of native gold in deposits, a paraphrased account of the summary by Boyle (1979 p. 207) follows.

To summarize the data available on the variations in the fineness of gold in deposits one can say that the situation is generally complex, and in some cases each gold belt and commonly each deposit seems to have its own characteristics. In some auriferous deposits there is an indication that the early deposited gold is more silver-rich (lower in fineness) than that deposited later in the paragenetic sequence. At Kirkland Lake for instance the gold in early pyrite tends to be silver-rich (low in fineness) whereas the later generations of free gold are silver-poor (high fineness). A similar relationship in some of the gold mines in other Precambrian terranes has been observed, but many of these investigations were carried out by reflectometry measurements about which there is some controversy as to their accuracy. The author's investigations in this matter are equivocal, but on a statistical basis there is an indication that the early gold in sulphides is lower in fineness than the later generations of free gold. This probably only applies to old (Precambrian and Paleozoic) deposits, since there is undoubted evidence that the late generations of gold in certain Mesozoic-Cenozoic deposits are exceptionally rich in silver compared with early generations.

There appears to be considerable variation of fineness with depth from deposit to deposit. In some cases there is an increase in the fineness of native gold with depth, a feature explained by some as due to an approach to a hot centre; silver being more mobile migrates farther from the hot centre than gold. In other deposits there is an indication that the fineness decreases with depth, and in still others there is no particular relationship, the fineness remaining relatively constant.

The data are equivocal concerning lateral zonations of the fineness of native gold in deposits with respect to thermal centres of intrusion or granitization. There are indications that the fineness of native gold in deposits near these centres is higher than in more remote deposits. Numerous aberrations seem, however, to complicate this picture and more data are required before definite statements can be made.

Within individual deposits there are often major differences in the fineness of native gold from oreshoot to oreshoot. Most individual oreshoots, however, generally record a relatively uniform fineness when their dip and strike dimensions are short.

In some deposits as revealed by microprobe studies the gold grains have a relatively uniform silver composition; in others they are markedly heterogeneous being zoned or having silver-rich or silver-poor domains. Numerous deposits also seem to have marked intragrain compositional differences with respect to silver and other lattice constituents. These inhomogeneities obviously have a marked effect on systematic variations within deposits, features that must always be considered by those doing work on the fineness of gold in deposits.

There is obviously a wall-rock effect on the fineness of native gold, but the details are obscure and the whole problem requires detailed investigation.

The effect of temperature on the fineness of native gold in deposits should be carefully investigated, especially where deposits have a great vertical range and pass from low grade metamorphic facies into intrusives. Some of the data at hand suggest that there is a temperature effect, native gold near hot centres or fronts being finer than that in remote sites. This is most noticeable where intrusives such as dykes cut pre-existing auriferous oreshoots. Adjacent to the dykes the native gold is often much finer than that at a distance from the dyke.

All of this discussion has centred essentially on the fineness (i.e., the silver content) of native gold. It seems probable that the contents of Cu, Te, Hg and the various other elements in lattice sites in native gold would also vary in certain systematic ways with depth, wall rock and other geological and geochemical features of the deposits. This problem has not been investigated in any detail within gold belts or deposits, although some preliminary work has been done.

Oxidation processes have a marked effect on the fineness of gold in deposits. In many deposits the free gold in the oxidized zones is considerably finer than that in the hypogene zones. The range in fineness of hypogene gold is usually 700-925; that for native gold in oxidized zones is commonly 850-999. In placers the gold is normally finer than the hypogene gold in the original deposits, although there are situations where there is an equivalence in the fineness of the two types of gold.

WALL ROCK ALTERATION EFFECTS IN AURIFEROUS DEPOSITS

The literature on this subject is immense; beginning with the classic paper by Lindgren (1901) more than a hundred papers have appeared to date on the subject, a general review of which is given in Boyle (1979). Here we can do no more than deal briefly with the principal types of wall rock alteration associated with gold deposits.

The types of wall rock alteration associated with auriferous deposits depend essentially on two parameters: the type of host rock and the nature of the mineralizing medium (media). Under the first the rocks are generally subdivided into igneous and sedimentary, the former including ultramafic, mafic, intermediate, and felsic; the sedimentary category includes carbonate rocks, pelites (shales), arenites (sandstone and graywacke), and rudites (conglomerates). The metamorphic equivalents of these various rocks have essentially the same types of wall rock alteration effects as their parents. The nature of the mineralizing medium (media) is determined by its state—magmatic liquid, aqueous solution, gas, diffusion current, or composites of these phases—and by its pressure, temperature, pH, Eh, and concentration (fugacity) of dissolved or contained constituents.

Wall rock alteration effects associated with auriferous deposits generally have a marked zonal distribution that may be both lateral and vertical with respect to the vein systems, breccia zones, or other dilatant sites. Most of the research carried out on wall rock alteration in auriferous deposits has emphasized the lateral variations, but there are a few cases where studies of the vertical distributions have been done. Unfortunately, the latter are rarely extensive enough to draw any generalized conclusions.

The most pronounced visual regularity in the lateral zonation of wall rock alteration is the presence near the deposit of the most intense phase of the alteration, which fades outward through less intense phases into unaltered rock. This phenomenon is normal in most types of deposits and is manifest by intense mineralogical and chemical changes adjacent to the deposit and consistent decreases in these changes outward from the deposit.

The regular lateral zonation adjacent to gold-quartz veins and other similar types of epigenetic deposits has been interpreted in two ways. One school holds that the zonation is contemporaneous or nearly so with the deposition of the ore and gangue minerals, and that it is due to an attenuation of the altering medium (solution, gas, or diffusion current) as it migrated outward from the fracture or other dilatant feature. The other school holds that the lateral zones outward from deposits are of different ages and are due to successive waves or pulses of altering media. Thus, the farthest-out zones are the earliest, the intermediate zones next oldest, and the intensely altered phases adjacent to the deposits the very latest or the final stages of the mineralizing process. The example of gold-quartz veins attended by an adjacent pyrite-sericite-carbonate zone that fades laterally through carbonate-chlorite and chlorite zones into unaltered andesite will serve to illustrate the two divergent views.

According to the first school, the sequence is due to an attenuation of the mineralizing and altering medium as it migrated laterally from the dilatant zone (the site of the quartz veins). As this medium diffused outward, it reacted with the wall rock minerals, most actively with those near the vein and decreasingly so with those farther out as the concentration of its constituents, particularly K, CO_2, S and H_2O, declined through reaction. This school supports its argument mainly by the bilateral symmetry of the zonation with respect to the veins.

The other school interprets the zonation as due to successive waves or pulses of altering media. According to them, the chlorite zone formed first by extensive introduction of water. Then followed a wave containing some CO_2 which superimposed a weak carbonatization on a part of the chlorite zone, and finally a wave greatly enriched in CO_2, K, and S superimposed an intense sericitized, carbonatized, and pyritized zone on the earlier chlorite-carbonate phase adjacent to the vein. This school supports its argument by various paragenetic relationships and by the fact that patches, veinlets, and protuberances of the more intense phases are frequently seen in an apparent cutting relationship with the less-intense phases.

The few descriptions of vertical zonation of wall rock alteration in auriferous deposits are rarely extensive, mainly mineralogical, and, with a few exceptions, give no detailed chemical analyses, probably because most studies of wall rock alteration are rarely continued throughout the history of a gold deposit. Further research is therefore desirable on the features of vertical zonation, particularly from the chemical aspect, which should be aimed at material balance calculations with respect to additions and subtractions within the vertical extent of entire gold-quartz vein systems and other types of auriferous deposits.

Certain gold-quartz and polymetallic deposits of Tertiary age exhibit extensive near-surface argillization with increasing propylitization at depth, but these effects depend to a large extent on the type of host rocks. According to some investigators the intensity of biotitization and development of actinolite increases with depth in some gold-quartz deposits, a phenomenon that is often related to the presence of granitic intrusives at depth. This feature can be seen in the Madsen mine at Red Lake, Ontario. Other gold-quartz deposits show an increase in carbonatization and sericitization with depth, as at Yellowknife (Boyle, 1961). This extensive carbonatization has released large quantities of silica that probably migrated upward to form the quartz lenses at higher elevations. There are also brief references in the literature to increases and differences in argillic types of alteration with depth in a variety of vein-type deposits.

One of the features of wall rock alteration in many auriferous deposits that merits emphasis is the tendency of dissimilar rocks to converge to a uniform type of alteration. This feature has been noted by numerous investigators and is evidently caused by prolonged attack by the mineralizing medium (media). There are, for instance, numerous references in the literature of interbedded slates, graywackes, amphibolites, and other quite dissimilar rocks all being converted to carbonate-sericite schists, and of quartz porphyries, tuffs, and shales intensely altered to an assemblage of quartz, sericite, and clay minerals. Silicification of carbonate rocks, shales, and sandstones, likewise, yields a convergent type of alteration in some deposits that is indistinguishable with respect to the host rock except by very detailed studies.

An abbreviated version of the subject of wall rock alteration taken from Boyle (1979, p. 208) follows, discussing the principal types of hypogene wall rock alteration associated with auriferous deposits. Numerous examples of deposits with particular types of alteration are given in the original text.

Skarnification consists generally of the development of calcium-magnesium-manganese-iron silicate minerals, quartz, magnetite, and a host of other minerals in limestones, dolomites, or calcareous shales and schists. Gold deposition may or may not be associated with the process. The calcium-magnesium-manganese-iron silicates formed include olivine, wollastonite, garnet, pyroxenes, amphiboles, uralite, and scapolite. Quartz, plagioclase, and carbonates are generally present, and there may be a development of many minerals containing the mineralizers, S, H_2O, F, Cl, and B, such as pyrite, pyrrhotite, serpentine, tourmaline, axinite, scapolite, vesuvianite, topaz, micas, and fluorite. The process of skarnification may be essentially isochemical with removal of CO_2; more generally, however, it involves the introduction of SiO_2, Mg, Fe, and the volatiles S, As, B, F, Cl, and H_2O. Under these conditions removal of CO_2 is usually extensive.

In some skarn deposits the introduction of gold, economic sulfides, scheelite, and various oxides appears to have taken place simultaneously with the formation of the principal skarn minerals. In other skarn deposits the introduction of the economic minerals appears to be later than the formation of the skarn. In this case the skarn is altered, the main effects being replacement of pyroxene by tremolite and actinolite and the development of quartz, epidote, calcite, sericite, and chlorite or serpentine.

Amphibolitization and *pyroxenitization* involve the development of amphiboles and pyroxenes respectively in the wall rocks of auriferous deposits. Minerals accompanying these types of alteration include feldspar (commonly albite), epidote, tourmaline, biotite, magnetite, pyrite, pyrrhotite, garnet, biotite, and occasionally andalusite and cordierite. Both amphibolitization and pyroxenitization are characteristic of certain deep-seated (high temperature?) gold-quartz veins and impregnation deposits mainly in igneous rocks of intermediate and basic composition but also in certain gneissic (originally partly calcareous) sedimentary terranes. Some of these deposits seem to be transitional to skarn bodies.

Feldspathization consists of the development of secondary feldspars in the host rocks of epigenetic gold deposits. Albitization is perhaps the most common, but the development of potassium feldspar (microcline, orthoclase, or adularia) is common in certain types of deposits. Chemically the process may involve the introduction of K, Na, and Al, although in some deposits these elements are simply rearranged, and the system is essentially isochemical.

Tourmalinization consists of the development of tourmaline in the wall rocks of certain types of epigenetic gold deposits. Chemically it involves the introduction of boron and in some cases Mg, Fe, Na, Ca, Al, and Li. In zones of intense tourmalinization in shales and schists there is frequently a decrease in the SiO_2 and CO_2 content of the altered rock.

Fluoritization involves the introduction of fluorine into wall rocks with the development of fluorite. *Topazization* implies the development of topaz and an introduction of fluorine. *Greisenization* is an all-inclusive term involving the processes of fluoritization, topazization, sericitization, and tourmalinization. These combined processes entail the conversion of feldspar, and other constituents of granitic and pelitic rocks in places, to aggregates of quartz, topaz, tourmaline, fluorite, and commonly lepidolite or lithium-bearing muscovite. Judging from the available analyses there is a general introduction of fluorine, boron, lithium, tin, and iron, and small abstractions of silica, calcium, sodium, and potassium; alumina and magnesium generally remain relatively constant. All of these various fluoriferous alterations are generally associated with tin, tungsten, and beryllium mineralizations. They are relatively rare in auriferous deposits. Fluoritization is, however, common in some Tertiary gold deposits (e.g., Cripple Creek, Colorado); topazization is notable at the Brewer gold mine, Chesterfield County, South Carolina; and greisenization is marked in some auriferous deposits bearing tin and tungsten minerals in the Mesozoic-Cenozoic tin-tungsten-gold belts of the far eastern USSR.

Silicification consists of the development of secondary quartz, jasper, chalcedony, chert, opal, or other varieties of silica in the wall rocks of epigenetic gold deposits. By another terminology it also includes the development of various silicates in the host rocks, particularly in carbonate rocks or calcareous shales and schists. As already noted, the latter process is called skarnification.

The chemistry of silicification is varied and depends essentially on the type of host rock that is affected by the process. In carbonate rocks there is generally a major introduction of SiO_2 and a wholesale removal of Ca, Mg, Fe, CO_2, and other constituents. In silicate rocks SiO_2 may be redistributed within the host rocks, that is, leached in one place and added in another.

Biotitization is an uncommon type of alteration in gold deposits but occurs in certain places. It usually involves the development of biotite or hydrobiotite from various cafemic silicates. During the process H_2O and K are introduced, and some SiO_2 may be extracted. Less commonly the constituents of biotite, including K, Mg, Fe, Al, and SiO_2, are all introduced into the rock.

Chloritization is one of the most common types of alteration in gold deposits. In many cases it develops directly by the hydrous alteration of cafemic silicates. During this process H_2O is introduced and some SiO_2 may be removed. In other cases constituents of chlorite, including Mg, Fe, Al, and SiO_2, are introduced into the rock.

Sericitization is perhaps the most common of all of the types of wall rock alteration in gold deposits developed in acidic intermediate and basic igneous and metamorphic rocks, and also in certain pelitic and other sedimentary rocks. It consists of the development of potash mica, generally sericite or hydromuscovite, as a result of the hydration of feldspars or from the rearrangement of K, Al, and SiO_2 within intensely altered wall rocks. Chemically, sericitization generally involves the introduction of K and H_2O into the rocks affected, and there is usually a removal of some SiO_2, Fe, and Ca. The development of fuchsite and mariposite, the chromium-bearing micas, in basic and ultrabasic rocks adjacent to gold-quartz veins usually does not imply an introduction of chromium. Rather, this element appears to be redistributed during the chemical breakdown of magnetite, pyroxenes, and amphiboles, the principal chromium-bearing minerals in the rocks. The development of the mica, roscoelite, probably has a similar origin deriving its vanadium from magnetite and so on.

Granitic rocks, granite gneisses, quartz porphyries, and certain types of schists heavily sericitized, albitized in places, and impregnated with pyrite are called *beresites* by the USSR geologists. The term *beresitization* is commonly used in the Ural goldfields and in other gold-bearing districts of USSR. Numerous studies of the process have been carried out. According to a number of Russian geologists the wall rocks are enriched in K, CO_2, and Ca and exhibit depletions of Fe^{3+}, Al, and SiO_2. Sericite and Ca-Mg carbonates are the principal resultant minerals. According to the same investigators the deposition of gold occurs during the late beresitization—early silicification (quartzitization) stages. Beresites and greisens are apparently formed within a similar temperature interval of 270-400°C. According to some investigators fluorite is generally absent in beresites because the F concentration in the solutions from which gold was precipitated was lower than in those that gave rise to rare metal deposits (Be, Mo, W, Sn, Rb, Li). In the latter the F content range is often 940-8500 ppm, whereas in beresites the range is much smaller, 550-740 ppm. Most of the fluorine in beresites is in the micas (sericite).

Beresitization is particularly complex in certain deposits and may be zoned. Complex patterns with respect to Na and K as the veins are approached appear to be frequent, although in a general way the K_2O/Na_2O ratio commonly exhibits an increase towards the auriferous deposits. In addition a vertical zonation over 1 km has been observed in some deposits, the most outstanding feature being the development of seybertite (a high-temperature brittle mica) at deep horizons and lower-temperature chlorite in the upper horizons. This phenomenon is partly attributed to a temperature gradient along the direction of flow of solutions. Marked lateral zonations have also been noted in beresitized rocks in certain gold-sulfide deposits. The zoning usually comprises an external chlorite-albite-sericite-carbonate zone, an intermediate quartz-carbonate-sericite-pyrite zone, and an inner quartz-sericite-pyrite and quartz zones adjacent to the orebodies. During the beresitization process the content of Na, Mg, and other bases is usually decreased, and S and CO_2 are introduced. The Si and K contents are increased in the inner zone. In some deposits Al displays an inert behavior.

The development of the sodium mica, paragonite, is not common but occurs in some alteration zones where soda metasomatism is a marked feature of gold deposition. The process can be called *paragonitization;* it usually involves introduction or rearrangement of considerable amounts of Na and H_2O.

Carbonatization consists of the formation of secondary carbonates in the host rocks of epigenetic gold deposits. The phenomenon is particularly common in intermediate to basic rocks, and also in certain ultrabasic rocks such as serpentinites. *Dolomitization* consists of the formation of dolomite in the wall rocks of deposits, usually at the expense of calcium carbonates but sometimes at the expense of magnesium silicates and other minerals. The process is relatively uncommon in gold deposits, but an analogous process *ankeritization*, the formation of ankerite, $Ca(Fe, Mg, Mn)(CO_3)_2$, is especially common.

The chemical processes involved in carbonatization are complex and depend essentially on the type of host rock affected. The ankeritization of basic and intermediate intrusive rocks and volcanics, such as frequently occurs in gold-quartz deposits, appears to require only the introduction of CO_2, the other constituents of ankerite such as Ca, Fe, Mg, and Mn coming from the chemical breakdown of amphiboles, pyroxenes, and Ca-feldspar. During this process much SiO_2 is removed and is probably transferred to the veins where it crystallizes as quartz.

Sideritization and other types of carbonatization of sedimentary rocks, especially quartzites and sandstones, generally requires the introduction of nearly all of the components of the carbonates formed in the rocks.

Basic and ultrabasic rocks, heavily carbonated, sericitized and pyritized are called *listwanites* by the Soviet geologists. The term *listwanitization* is commonly used in the Ural goldfields and in other auriferous districts of the USSR. The main characteristic of listwanitization is the conversion of serpentine into talc and/or carbonates. The chemical composition of listwanite is variable and is controlled by zonal factors and the composition of the host rocks. In general there is an introduction of K, Ca, Al, CO_2 and H_2O and an abstraction of SiO_2.

Pyritization and *arsenopyritization* are two of the most common types of alteration in gold deposits. They consist of the development of disseminated pyrite and arsenopyrite in the host rocks of epigenetic deposits. In many types of deposits only sulfur and arsenic are introduced, the iron coming from the breakdown of iron silicates and oxides. In other types of deposits all of the components of pyrite and arsenopyrite are introduced.

Propylitization is a particularly common type of alteration associated with gold-quartz and polymetallic deposits in basic and intermediate intrusive rocks and volcanics. The principal secondary minerals developed are chlorite, epidote, zoisite, leucoxene, carbonates, sericite, feldspar, clay minerals, pyrite, and arsenopyrite. Propylites commonly form the fringe zone of highly altered rocks where sericitization, carbonitization, and pyritization are intense.

Chemically, much water is introduced during propylitization, and there may also be additions of CO_2, S, and As. Some SiO_2 is generally extracted during the process, and there may be losses of Na, K, and the alkaline earths in some districts; in others, K and Na may be enriched.

Propylitization tends to pervade large volumes of rock in many districts and therefore may not be directly connected with epigenetic gold deposits. In places it appears to grade imperceptibly into the greenschist facies of regional metamorphism.

Rocks that are heavily propylitized exhibit several types of alteration including chloritization, carbonatization, pyritization, argillization, and sericitization. The term *propylitization* is therefore a broad inclusive one.

Much has been written on the processes of propylitization in the past, especially with reference to the Mesozoic-Cenozoic gold deposits of Europe and the United States, and numerous papers have appeared in recent years on the subject, most dealing with Tertiary rocks and deposits in which propylitization is especially marked. There are many questions about the causes of propylitization, the source of the volatiles, such as CO_2 and H_2O, that take part in the regional phenomena, and the relationship of propylitization to mineral (gold) deposition. These problems lie more in the field of petrology and cannot be discussed here. About all that can be said with assurance respecting gold deposits is that many, especially those in Tertiary andesitic terranes, appear to be closely associated with extensive propylitization linked in some manner with a high mobility of potassium resulting in potash metasomatism of the propylitized rocks and the wall rock alteration zones of the gold-quartz veins.

Argillic alteration consists of the development of clay minerals in the wall rocks of epigenetic gold deposits. The common clay minerals found include kaolinite, montmorillonite, dickite, illite, and halloysite. The principal chemical effect appears to be the introduction of water, and there may be a removal of some SiO_2, Fe, K, Na, Ca, and Mg, depending on the type of rock affected.

Alunitization is common in some low-temperature (epithermal) gold deposits, principally in felsic and intermediate volcanic rocks. It generally consists of the development of alunite, $(K, Na) Al_3 (SO_4)_2(OH)_6$, and quartz in feldspathic rocks, the alunite being largely the result of alteration of the feldspars. Chemically, the alteration generally involves an introduction of S and H_2O and removal of some SiO_2, Na, Ca, Mg, and Fe.

According to most investigators a fairly high H_2SO_4 concentration is required in the mineralizing solutions before alunite can be produced in equilibrium with muscovite, kaolinite, or K-feldspar. Alunite-kaolinite (including dickite or pyrophyllite) are the most common products of strong hydrolytic (advanced argillic) types of alteration in either hot spring or slightly higher temperature environments, whereas muscovite-alunite is favored by higher temperature as well as by higher potassium concentrations.

In general, alunitization signifies acid attack by H_2SO_4 on the alkaline silicates of the wall rocks, particularly on the feldspars. In some deposits, however, the components of alunite, including K, Na, Al, and SO_4 seem to have been introduced, at least locally, possibly as soluble sulfates.

Alunitization may be hypogene and/or supergene in origin in some gold deposits.

Hematitization is a type of alteration that frequently accompanies gold deposits, particularly those containing uranium. It is also found in a variety of other deposits, often in syenitic rocks as at Kirkland Lake, Ontario. The presence of hematite generally indicates a low-partial pressure of sulfur in the mineralizing solutions or diffusion currents.

The iron producing the hematitization may or may not be introduced. In some cases there is abundant chemical evidence to show that the element has been introduced as ferric iron into the wall rocks; in other cases where pervasive hematitization is present the iron is probably only redistributed while undergoing oxidation from the ferrous to ferric state. In the latter case some iron may actually be leached.

Serpentinization and the closely allied development of talc *(talcification)* generally consists of the formation of serpentine and talc in ultrabasic rocks such as dunite, peridotite, and pyroxenite. Serpentinization and the development of talc may also occur in crystalline limestones and dolomites. The effect is marked in some gold-quartz and other types of deposits in ultrabasic rocks.

Serpentinization and the development of talc in ultrabasic rocks generally involves only the introduction of H_2O and the redistribution of the other constituents of the rocks. Where serpentine and talc are developed in limestones and dolomites there is generally an introduction of SiO_2 and H_2O and frequently some Mg.

Hydration is nearly universal in all types of wall rock alteration in gold deposits. It indicates an introduction of water that is fixed either as hydroxyl or as water of hydration in the various alteration minerals. *Dehydration* involves the removal of water, such as takes place in the formation of certain types of wall rock alteration zones.

Phosphatization is common in the alteration zones of some gold deposits. It involves the introduction or rearrangement of phosphate resulting mainly in the development of apatite. Some gold deposits exhibit this feature to a moderate degree.

Pyrophyllitization consists of the development of the hydrous aluminum silicate, pyrophyllite. In most types of epigenetic gold deposits this type of alteration is apparently rare. Pyrophyllite is, however, a very common mineral in the matrix of the auriferous quartz-pebble conglomerates of the Witwatersrand.

Zeolitization consists of the development of zeolites such as stilbite, natrolite, heulandite, chabazite, and so on. The widespread process is frequently not associated with mineral deposition. The process, however, commonly accompanies native copper deposits in amygdaloidal basalts and is also associated with certain types of gold and silver veins. The zeolites are often accompanied by calcite, prehnite, pectolite, apophyllite, and datolite.

Concerning the timing of wall rock alteration, open and closed systems, and the nature of ore-forming solutions giving rise to hypogene auriferous deposits, the abridged text continues.

The timing of the introduction and abstraction of the various constituents that participate in wall rock alteration is a most complex problem—one that has not yet been satisfactorily solved. According to one school the various exchanges are an integral part of one surge of mineralization, the various reactions and exchanges taking place in a complex diffusion system within a relatively short geological interval. The other school advocates a number of successive but discontinuous surges or pulses, each producing characteristic effects that depend on the nature of the mineralizing medium, the nature of the wall rocks, the sequence of fracturing and other factors. One stage of wall rock alteration may be superimposed on another, and the chemical and mineralogical relationships may be very complex indeed. The whole problem of the timing of wall rock alteration is intimately associated with mineral paragenesis and zoning, two subjects that are too extensive and complex to be discussed here. The student of ore genesis and wall rock alteration should always be aware of the two possibilities and should strive to obtain unequivocal criteria for the solution of the chemical and mineralogical relationships.

The problem of open and closed chemical systems in wall rock alteration processes associated with epigenetic gold deposits remains essentially unsolved and will remain so until detailed material balance calculations are carried out throughout all parts of the various types of auriferous deposits, including the ores and the lateral, sub- and supra-alteration haloes. Similarly, the problem of equilibrium in the various chemical reactions involved in wall rock alteration is unanswered. Probably most reactions are not equilibrium reactions in the commonly accepted chemical sense. It is more fruitful to approach wall rock alteration from a dynamic (kinetic) viewpoint, because diffusion seems to play such a large part in all of the processes.

It has long been recognized that the ore-forming medium may be gaseous, an aqueous solution, meltlike, or in the nature of a diffusion current of mobile ions or complexes. The features of wall rock alteration in auriferous deposits do not seem to give unequivocal criteria for any of these mechanism of transport, although there are certain trends that may be indicative.

The features of auriferous skarn zones suggest extensive diffusion of SiO_2 and other constituents into the carbonate rocks and CO_2 out of them. It also seems probable that diffusion of mineralizers, such as H_2O, S, As, F, Cl, and B, played a large part in the formation of auriferous skarns.

The formation of the various types of auriferous veins and lodes probably took place mainly by diffusion in the deep zones of the earth and by precipitation from aqueous solutions under near-surface conditions. In certain cases, such as in the slightly auriferous massive nickel-copper ores, the sulfides may have crystallized from melts, as many have little or no associated wall rock alteration. It seems impossible from studies of wall rock alteration to decide which process—diffusion or precipitation from aqueous solutions—was the important agent in the formation of auriferous veins and lodes. Other criteria (Boyle, 1963) must be brought to bear on this problem. Whatever the agent or mechanism, it is certain that water played a large part in the formation of auriferous veins. Other volatiles, such as CO_2, S, As, Se, Te, Sb, B, and F, have played a leading role in the formation of certain types of gold deposits. Wall rock alteration zones associated with auriferous deposits rarely show the effects of chlorine metasomatism, a feature that might lead one to suspect that chlorides are not as important in the transport of gold and other metals as some investigators would have us believe. On the other hand sodium chloride is particularly abundant in some hot springs and thermal waters and is found in liquid inclusions in auriferous quartz and other gangue minerals. The reason for the low abundance of chlorides in wall rock alteration zones and in auriferous veins may be the great solubility of most of these compounds. In other words, they are largely removed from the deposits at a late stage by the mineralizing solutions or diffusion currents.

Judging from the chemistry of wall rock alteration, the constituents of most auriferous veins and lodes were deposited from an aqueous medium that was highly fluxed with CO_2 and S in most cases and with As, Sb, Se, Te, Bi, B, F, and Cl in some other cases. The gold was probably transported mainly as sulfide, arsenide, antimonide, or telluride complexes, and most of the accompanying metals were carried as complex bicarbonates, sulfides, arsenides, antimonides, selenides, tellurides, bismuthides, fluorides, or chlorides. Quartz and many of the constituents of the other gangue minerals, such as Fe, Ca, Mg, and Mn, may have been largely derived from the adjacent wall rocks.

The pH of the ore forming medium is difficult to assess from wall rock alteration studies of auriferous deposits. In some cases there is evidence that the solutions or diffusion currents were acid and that they gradually became alkaline by reaction with the wall rocks. In other cases they evidently were neutral or slightly alkaline. In most cases where direct measurement of or calculation from hot spring data have been done the solutions are at or near the neutral point.

The reader will note that little or no reference has been made to alteration effects in the auriferous quartz-pebble conglomerates, the most productive of all gold deposits. The consensus of many geologists is that these deposits are essentially modified placers. If so, the alterations associated with these deposits are essentially the result of diagenetic and metamorphic agencies, including particularly recrystallization of the matrix to yield fine-grained quartz, sericite, chlorite, chloritoid, and pyrophyllite (in some deposits). The origin of the pyrite, pyrrhotite, and iron oxides (in some deposits) is problemical. Some contend they are (recrystallized) detrital minerals, others that they are the result of complex biogeochemical agencies, and still others that they originated by sulfidization of detrital iron oxides to form mainly pyrite.

STRUCTURAL ENVIRONMENT OF DEPOSITION OF EPIGENETIC GOLD DEPOSITS

The details of the environment of deposition of gold in epigenetic deposits and the responsible agencies are of fundamental importance in the understanding of auriferous deposits. First, these details involve certain types of major structures that are usually self-evident in most auriferous belts, although their detailed structural analysis may be complex. Second, minor structures have also played a major part in the localization of ore shoots in most epigenetic gold deposits.

Both of these subjects have been addressed by a number of geologists, resulting in four landmark papers (Bain, 1930; Mawdsley, 1938; White, 1943; and Ebbutt, 1948). The first two deal mainly with Precambrian deposits; the last two cover the age spectrum of gold deposits.

Bain (1930) studied both the major and minor structural features that localize gold in the Canadian Shield. His general conclusions are that the identification of gold-bearing quartz depends upon the occurrence of gold in secondary crush fissures in a primary quartz vein or silicified rock mineralized with iron pyrite. This theme is repeated in all of the four landmark papers reproduced here. In more detail Bain's general summary and conclusions in his paper *Structure of Gold-bearing Quartz in Northern Ontario and Quebec* are as follows:

Major Structures

1. Gold of the Laurentian Shield occurs almost entirely in quartz veins of medium grade and can be recovered by cyanidation but recovery is usually expedited by flotation or tabling.
2. The quartz veins occupy fracture systems in, around, or adjacent to a massive intrusive rock whose chief role seems to have been to localize regional strains.
3. Intrusives have sometimes effected a concentration of movement of solutions and produced rich shoots.

Microscopic Structures

1. Gold deposited above the zone of flowage fills minute fractures of a second or third generation in any mineral, but usually in quartz or pyrite.

2. Gold deposited within the zone of flowage occurs as a replacement of any mineral along a plane of movement but shows a preference for chlorite.
3. The form and pattern of the gold-bearing fractures is determined by the type of movement producing them. Simple shear is least effective; crushing is most effective.
4. Rock brittle enough to fracture and give the primary openings in which quartz is deposited and strong enough to transmit crushing stresses to produce fractures in the quartz or pyrite, is the only favorable wall rock. Chemical makeup is unimportant: physical behavior is all that has any effect.
5. Movement causing a single gold-bearing fracture has never been observed to exceed 1 mm. Aggregate movement along a system of shatter planes and the sheared walls of the Stabell vein amounted to 5 ft., or less than 3 per cent. of the total movement. Horizontal displacement causing the primary, secondary and tertiary fractures with metallic mineralization at Kirkland Lake is only 150 feet.
6. The average original width of the gold-bearing fractures is between $^{1}/_{50}$ and $^{1}/_{25}$ mm. This is frequently increased by solution by the ore-bearing solutions. (p. 43)

Mawdsley (1938) studied the details of the localization of gold in a number of gold deposits in the Canadian Shield (Ontario and Quebec) and concluded that gold is generally the last mineral to be deposited. The abstract of his paper *Late Gold and Some of Its Implications* gives his explanation for the occurrence of gold late in the paragenetic sequence.

> In many Canadian gold deposits the gold is one of the last minerals deposited. In some cases, considerable time has elapsed between its introduction and that of the other minerals; in others, a subordinate amount of gold was introduced with at least some of the other minerals, but the main gold enrichment was later.
>
> The tentative hypothesis is advanced that in such cases the bulk of the vein material left the magma before the gold. This early magmatic product left in the gaseous state but condensed to a liquid prior to its final deposition; the gold remained behind and was concentrated in the liquid end-product of the magma. The early gaseous phase when condensed to a liquid phase rich in superheated water, could dissolve, transport and deposit much silica and account for much of the vein quartz. The cases in which both early and late gold are present are explainable on the basis of the partial intermingling of the two magmatic products.
>
> Distinctly late gold raises the question of the soundness of classifying many gold deposits on the basis of their associated vein minerals. The occurrence of gold-bearing ore shoots is often easily explainable, assuming gold to be late, by inferring late stresses resulting in suitable structures. Weak members, such as vein quartz, are loci for such favorable channels. (p. 194)

The mechanism and environment of gold deposition in veins was studied in great detail by White (1943) and published in a classic paper that merits reproduction in its entirety (Paper 17-1). He stresses the role of cataclasis in the localization of gold in ore shoots.

Ebbutt (1948) stressed the role of minor structures in gold deposition, presenting his evidence in a number of remarkable color photographs that unfortunately cannot be reproduced here. The text of his paper is of interest because it is the distillation of a geologist's lifetime observation of the role of minor structures in the formation of auriferous ore shoots (Paper 17-2).

The reasons for the late appearance of gold in paragenetic sequences of gold-quartz deposits, as manifested by its common occurrence in late fractures, has not yet been adequately explained. I (Boyle, 1979, p. 423) attribute the cause mainly to the high mobility of gold in its deposits. Each time a fracture appears in gold deposits it tends to be filled with microcrystalline quartz and gold, both minerals representing remobilized gold and silica from preexisting minerals (e.g., quartz, finely dispersed gold, auriferous pyrite and arsenopyrite, sulfosalts, etc.) in the deposits.

REFERENCES

Bain, G. W., 1930. Structure of gold-bearing quartz in Northern Ontario and Quebec, *Am. Inst. Mining Metall. Eng. Tech. Pub. 327,* 44p.

Boyle, R. W., 1961. The geology, geochemistry, and origin of the gold deposits of the Yellowknife district, *Canada Geol. Survey Mem. 310,* 193p.

Boyle, R. W., 1963. Diffusion in vein genesis, *Problems of post-magmatic ore deposition Symp.,* Prague, Czech., vol. 1, p. 377-383.

Boyle, R. W., 1979. The geochemistry of gold and its deposits, *Canada Geol. Survey Bull. 280,* 584p.

Ebbutt, F., 1948. Relationships of minor structures to gold deposition in Canada, in *Structural Geology of Canadian Ore Deposits,* Canadian Inst. Min. Metall., Montreal, pp. 64-77.

Fisher, N. H., 1945. The fineness of gold with special reference to the Morobe goldfield, New Guinea, *Econ. Geology* **40:**449-495, 537-563.

Grout, F. F., 1946. Microscopic characters of vein carbonates, *Econ. Geology* **41:**475-502.

Hargraves, R. B., 1963. Silver-gold ratios in some Witwatersrand conglomerates, *Econ. Geology* **58:**952-970.

Hawley, J. E., 1962. The Sudbury ores: their mineralogy and origin, *Canadian Mineralogist* 7 (pt. 1):207p.

Lindgren, W., 1901. Metasomatic processes in fissure-veins, *Am. Inst. Min. Eng. Trans.* **30:**578-692.

Mawdsley, J. B., 1938. Late gold and some of its implications, *Econ. Geology* **33:**194-210.

Shcherbina, V. V., 1956. Geochemical significance of quantitative Au-Ag ratio, *Geokhimiya* **3:**65-73. (Also, *Geochemistry* **3:**301-311.)

Von Rahden, H. V. R., 1965. Apparent fineness values of gold from two Witwatersrand gold mines, *Econ. Geology* **60:**980-997.

White, W. H., 1943. The mechanism and environment of gold deposition in veins, *Econ. Geology* **38:**512-532.

17-1: THE MECHANISM AND ENVIRONMENT OF GOLD DEPOSITION IN VEINS

William H. White

ABSTRACT.

A study of ores from twenty-seven districts, representing the main types of gold deposits in Canada and the United States, shows that they have a number of characteristic features, some of which are common to nearly all deposits. The process of cataclasis, one of minute fracturing and almost contemporaneous recrystallization of quartz in gold veins, is believed to have been a very important factor in permitting the residual solutions carrying the gold in the veins to permeate the quartz and distribute the gold. Vacuoles are very common in vein quartz and they also play a part. There are some evidences of the temperature at which gold was deposited and it is concluded that in all cases it was relatively low.

CONTENTS.

INTRODUCTION.

THIS paper presents briefly the results of an inquiry into the mechanism and environment of vein-gold deposition. Since no conclusions of consistent validity can emerge from an investigation of the occurrence of gold in any one deposit, twenty-seven vein and lode gold deposits, which varied widely in form, mineralogy, and genetic classification, were studied in both field and laboratory.

These deposits represent most of the important gold camps on the continent (Fig. 1). The general geological setting and structural relations of each deposit were studied and the important mineral associations, textures, and micro-structures were critically ex-

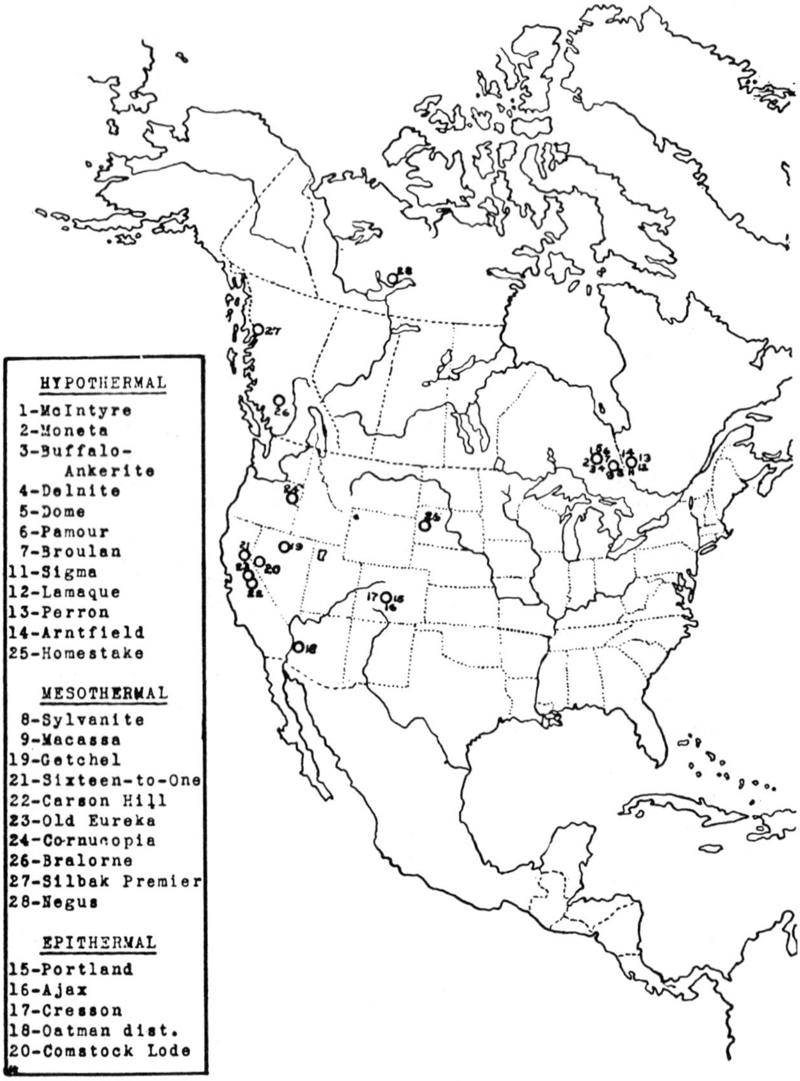

FIG. 1. Geographical Distribution of the Gold Deposits.

amined in thin and polished sections to obtain some idea of the environment of deposition, and to deduce the probable history of formation of each deposit. Comparison of the deposits thus analyzed disclosed some striking similarities. Certain significant features were found to be persistently characteristic of ores of very different type, and it became evident that the process of formation of each deposit conformed to the same fundamental pattern. This basic uniformity of process permits a generalized treatment of the subject of vein formation and justifies generalized conclusions regarding the deposition of gold.

The deposition of gold may be regarded as a subsidiary phase of the process of vein formation. The long and frequently complicated sequence of events which preceded the deposition of the metal will be sketched in a general way in order to consider this important subsidiary phase in its proper perspective. This can be done without neglecting any essential features because of a certain basic uniformity in the process of ore deposition.

GENERAL CONSIDERATION OF VEIN FORMATION.

Environment. Some of the deposits studied were formed at depths probably in excess of 20,000 feet, others, probably within a few hundred feet of the contemporary surface, whereas the majority fall somewhere between these extremes. The static pressures therefore may be assumed to have ranged from over 1,000 atmospheres to less than 100.

A study of the complex structural relations of many vein systems suggests, however, that static conditions rarely obtained. Great differential forces were at work producing complicated strain patterns. In a complex rock mass, the net result was a strongly, and erratically, anisotropic stress environment in which innumerable stress vectors, continually changing, produced a multiplicity of strain features in a confused, superimposed sequence along the composite structural zone. As a rule, such complicated stress environments are characteristic of vein structures formed at various depths. Among the deposits studied, two important exceptions to this rule, are the veins of the Pamour and Broulan mines in the eastern part of the Porcupine district, Ontario. These vein structures were developed at great depth, but a peculiar combination of circumstances produced relatively static stress conditions.

Temperature is the other important element in the environment of vein formation. Although it cannot be accurately determined for any one deposit, gradually accumulating evidence suggests an upper limit of about 573° C., the inversion point of quartz, and a lower limit which may not be much above 100° C. There is a tendency to assume that the greater the depth, the higher the temperature of vein formation. The validity of this generalization is open to question. Where local heat sources are involved, as they are in problems of ore genesis, the local temperature at a given point in the earth's crust must depend on a number of factors, such as the proximity of the heat source, the quantity of heat, and the relative rates of supply and dissipation. It is conceivable that a heat vehicle, such as a hydrothermal fluid, could produce as high local temperatures near the surface as at greater depth because of the low conductivity of rocks. However, we would expect the temperature to fall more rapidly near the surface. Temperature change during ore deposition will be considered later, but a generalization might be made. During a protracted period of ore deposition it is to be expected that dissipation of heat would finally exceed its accession, since hydrothermal activity is one of the closing phases of a dying magmatic epoch, and consequently that the temperature would gradually fall.

The Primary Ore Fluid. The long and frequently complicated train of events in vein formation was initiated when the "primary ore fluid" escaped from its magmatic source and ascended by way of structural channels into the overlying rocks. With the exception of certain substances such as iron and lime which could have been acquired in transit from rocks through which it passed, this juvenile fluid contained all the elements added to the rocks during ore deposition. In addition, it contained water and other substances which remained fluid and disappeared.

The concentration of the primary ore fluid is less evident than its elemental composition. The popular belief, that it was a dilute aqueous solution, may not be valid in all cases. It is true that almost any ore will show abundant evidence of the facile mobility of the ore fluid, but low viscosity is not necessarily synonomous with low concentration. Furthermore, the writer has found some evidence in the mineral relations of tourmaline and quartz in ores from Bourlamaque township, Quebec, which suggests that the fluid which deposited silica was of sufficient density to exert an appreciable buoyancy on the earlier tourmaline crystals. Although much more evidence is needed, the writer has concluded that the ore-forming fluid may have been rather highly concentrated. Lindgren [1] seriously considers this possibility, and Garrels,[2] in a recent paper dealing with the lead-zinc deposits of the Mississippi Valley, states his belief that high concentrations may be a common feature of hypogene solutions.

The Nature of the Process of Vein Formation. The most characteristic feature of all veins is that their constituent minerals were not formed simultaneously, but in a certain sequence. The same fundamental sequence was found in all the deposits studied. First, carbonates and pyrite were formed within the zones of alteration which border all hydrothermal deposits. During this phase the evidence is that only carbon dioxide and sulphur were provided by the primary ore fluid, the basic oxides and iron being furnished by the destruction of pre-existing minerals. Next, more carbonate and pyrite, and in some ores, arsenopyrite or pyrrhotite, were deposited in the earliest vein fractures. These and other fractures were then reopened, and great quantities of silica precipitated to form the predominant quartz of most veins. In the Porcupine district of Ontario and the Bourlamaque district of Quebec, tourmaline occurs in many veins. This interesting mineral invariably occupies a position in the sequence between pyrite and quartz. An alkalic variety of feldspar is found in many veins. Although this may be regarded as reconstituted rock feldspar, nevertheless it occupies a perfectly definite place in the mineral se-

1 Lindgren, W.: Mineral Deposits, 4th edit. P. 124.

2 Garrels, R. M.: The Mississippi Valley type lead-zinc deposits and the problem of mineral zoning. Econ. Geol., 36: 729–744, 1941.

quence. It was deposited before any quartz and slightly later than the earliest carbonate. Many vein contacts are selvedged with massed prisms of this feldspar, the inward-pointing terminations of which are embedded in quartz.

The great majority of veins, particularly those formed at intermediate or great depth, were not deposited as complete units. In the dynamic stress environment featured by a multitude of small, temporary and shifting strains, the vein structures were built piecemeal during a protracted period of local discontinuous dilations. The final size and shape of the vein structure is therefore the composite result of many small increments of mineral deposited where there were innumerable, small and repeatedly superimposed strains. Evidence of their complex construction may be found in nearly all veins. In simple veins, formed by axial reopening of the same vein fractures, the earliest minerals invariably occupy border positions and the later minerals, medial positions. In more complex veins, many textural and structural features, some of which are frequently rather obscure, disclose their composite nature. Replacement is another mechanism of vein formation, of minor importance in most of the veins studied, but prominent in a few. It involved close sheeting or brecciation of a particular zone gradually developing laterally, and accompanied by ingress of ore-forming fluid and progressive alteration of rock minerals. Rock feldspars were made over to a more alkalic variety, excess lime combined to form carbonate, and any original quartz merged with that of the ore-forming fluid.

At this point the primitive vein structures were practically solid bodies, similar in size and shape to those of today. The important difference was, of course, that as yet they contained no precipitated metallic gold. We may now consider the unique train of events which led finally to the enrichment of certain parts of these primitive mineral bodies, thus forming the ores.

CATACLASIS.

Although by the time the primitive vein structures were completed, the great deformational forces which shaped the larger

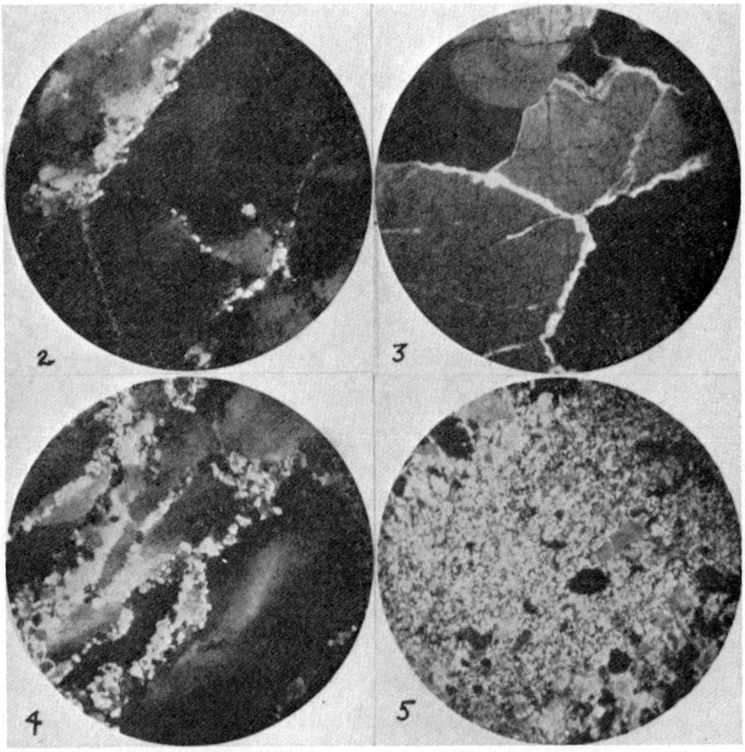

FIG. 2. The incipient stage of cataclastism, showing excessive optical strain and mottled extinction within the large quartz grains, and serrate grain boundaries. Tiny grains of clear "new" quartz are developing in chains along the grain contacts and starting to invade the interior of the original grains.

Buffalo-Ankerite Mine, western Porcupine district, Ontario. ×75, crossed nicols.

FIG. 3. The first hint of cataclasis. The vacuoles, at first random, have become aligned into two intersecting sets of vacuole planes which cross grain boundaries without deflection. The quartz exhibits no optical anomalies.

Eureka Mine, Mother Lode district, California. ×75, crossed nicols.

FIG. 4. A more advanced stage of cataclasis showing the clear fine-grained "new" quartz gradually destroying a large original crystal. There is a suggestion of the banding due to lateral variation in grain size, which will characterize the final state.

structures controlling vein formation had been largely dissipated, a similar, though more restricted, dynamic stress environment still persisted. Consequently, completed sections of the primitive vein structure, while still at somewhat elevated temperature, had to bear local differential stress. Under such conditions, quartz underwent a unique change designated as the "process of cataclasis." Goodspeed[3] recognized the nature of this change, and in his investigation of the ores of the Cornucopia Mine, Oregon, he classified the quartz either as "aclastic" or "cataclastic." The writer has found Goodspeed's classification of quartz applicable to the great majority of quartz veins formed at moderate to great depth. The process of cataclasis is most important in gold deposition.

Cataclasis is a progressive morphological change in vein quartz, and various stages in the development of cataclastic texture may be observed in almost any quartz ore. Original vein quartz usually contains myriads of minute, fluid-filled vacuoles, and the first hint of cataclasis is seen in an apparent rearrangement of these vacuoles. They appear to collect along two principal "vacuole planes." These planes are remarkably regular and are arranged in a definite pattern which extends by en echelon offsets across many grain boundaries. In the remarkable instance of a narrow, regular vein in the Sigma Mine, Bourlamaque district, Quebec, a regular pattern of vacuole planes persists for hundreds of feet. One set of planes is almost parallel and the other highly inclined to the plane of the vein. The arrangement would correspond to the planes of maximum shear in a theoretical strain-ellipsoid properly oriented with respect to either a tangential couple or a compressive force.

[3] Goodspeed, G. E.: Geology of the gold quartz veins of Cornucopia. A. I. M. E., Min. Tech., March, 1939.

McIntyre Mine, western Porcupine district, Ontario. ×75, crossed nicols.

FIG. 5. The final stage of cataclasis—a completely new texture with small, equidimensional grains of clear "new" quartz. Note the characteristic banding due to lateral variation in grain size.

Eureka Mine, Mother Lode district, California. ×75, crossed nicols.

The next stage is referred to as "incipient cataclastism." Grain boundaries become vague and minutely serrate, and chains and clusters of tiny clear grains of "new" quartz outline the original large turbid crystals. These original crystals lose their optical homogeneity and a striking mottled effect is observed at extinction. Gradually, the interior of the mottled crystals is invaded by chains and clusters of the "new" quartz; vacuole planes are destroyed and the vacuole fluid dispersed; the final result is a completely new texture composed of small, equidimensional grains of clear quartz. Figures 2–5 illustrate the typical development of cataclastic texture.

Cataclastism is not uniform throughout a vein, but is related to various local rolls and crenulations. The zones of most intense cataclastism in regular veins are arranged as parallel, en echelon streaks which lie almost parallel to the plane of the vein. As in the case of vacuole planes, it appears that the arrangement of these zones corresponds to the planes of maximum shear in a properly oriented strain-ellipsoid.

If differential stress is a cause of this unique change in quartz texture, we should find examples of differential strain; and these were abundant in many ores. Early, brittle pyrite crystals were crumbled and the fragments dragged in opposite directions; tourmaline laths, originally embedded in aclastic quartz, were broken into segments, which were slightly separated and offset and then solidly sealed in incipiently cataclastic quartz. Differential stress is therefore one of the prerequisites of cataclasis.

The other prerequisite is that the temperature be above a certain minimum value. In some ores it was found that the sequence of cataclastic strains—for they do occur in sequence and are frequently superimposed—finally gave place to a sequence of purely mechanical strains. The quartz was merely pulverized as with a hammer. After some practice it is easy to differentiate between a finely cataclastic texture and a brecciated texture. The only change of environment which could have occurred during this change in the type of strain was drop in temperature. Therefore, a certain minimum temperature is the other prime requisite of cataclasis. Although it cannot be defined too closely, it was probably between 200° and 300° C.

During cataclasis, the quartz is in a metastable condition and there is good evidence that some at least becomes transiently

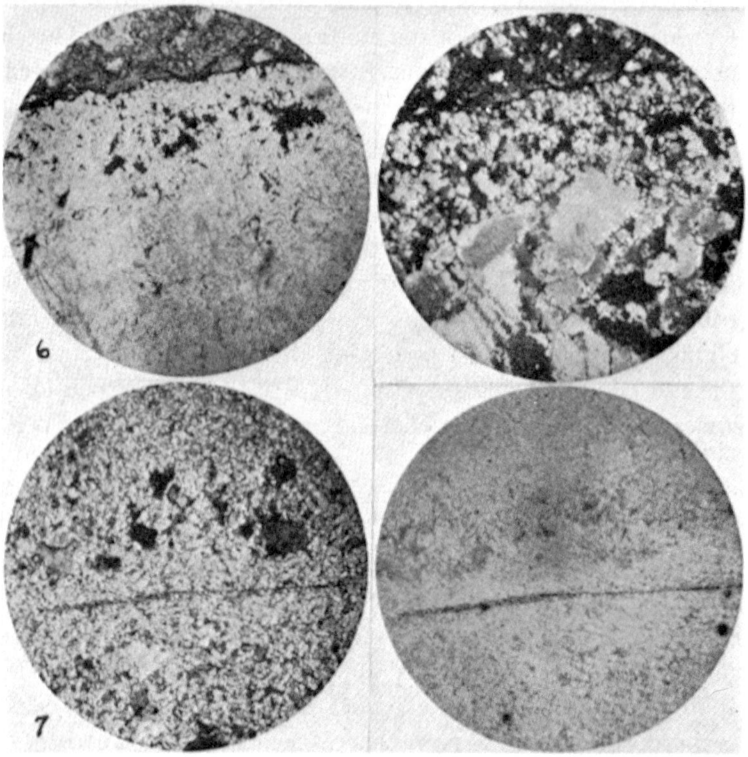

FIG. 6. Twin photomicrographs. Left, plane polarized light and right, crossed nicols. Minute arsenopyrite crystals in cataclastic quartz along a fragment contact. Note the distribution of the arsenopyrite crystals in relation to the intensity of cataclastism.

Delnite Mine, western Porcupine district, Ontario. ×75.

FIG. 7. Twin photomicrographs. Right, plane polarized light; left, crossed nicols. Minute arsenopyrite crystals forming a regular narrow band in quartz which is seen to be cataclastically banded, the arsenopyrite streak being restricted to the axis of the most finely cataclastic band. There is another similar band just above this field.

Eureka Mine, Mother Lode district, California. ×75.

mobile. For example, veinlets of clear quartz, which might be mistaken for separate later injections, are found to merge without break in both directions into zones of intense cataclastism. Again, brittle minerals such as galena, which were deposited later than quartz, may be fractured and cemented by clear quartz which is obviously the same as the adjacent highly cataclastic material. On the other hand, while the quartz exists in this metastable condition it still retains sufficient rigidity to maintain myriads of minute open seams which are the avenues of access of the later mineral-bearing fluid. Cataclasis might be broadly defined, therefore, as a process of micro-brecciation and simultaneous recrystallization.

Another feature of cataclasis is that, although the quartz has recrystallized, this has not resulted in decreased porosity of the vein material as a whole. On the contrary, cataclastic sections are very much more permeable than aclastic. Detailed examination shows the small "new" grains of cataclastic quartz packed together as in a loosely cemented sandstone, leaving myriads of minute inter-granular spaces—not interlocked like those of a quartzite.

DEVELOPMENT OF THE RESIDUAL ORE FLUID.

Among the variety of substances comprising the primary ore fluid, gold was a very insignificant constituent. However, as deposition proceeded, various non-auriferous constituents, carbon dioxide, sulphur, iron, arsenic, and silica were abstracted in sequence and fixed as vein minerals; consequently, the relative proportion of gold increased tremendously. Consideration of the structural and textural features of the primitive veins, prior to the deposition of gold, seems to supply an answer to the nature of this gold-rich fluid.

The veins of the Pamour and Broulan mines are unique in that no post-quartz strains [4] modified their original character. The vein quartz is still in the original form of coarse-grained subhedral crystals, unfractured and completely aclastic. Gold and associ-

[4] This refers only to post-quartz strains occurring during the period of ore deposition.

ated minerals occur interstitially among the subhedral quartz crystals as thin films along grain boundaries, and as larger plates molded upon crystal surfaces (Fig. 8). These veins attained their present dimensions with complete deposition of quartz; consequently, the fluid which contained the gold was confined to small interstices such as grain boundaries and small crystal-bounded vugs. There was no way for it to get into or out of these solid quartz bodies, and even limited migration along the veins was

FIG. 8. Gold, galena (partially etched), and sphalerite filling a vug between quartz crystal surfaces. Sphalerite appears to be a little earlier than gold and galena. Fractured and complete crystals of early pyrite are seen to the left. The subhedral quartz crystals have remained undisturbed since their formation.

Broulan Mine, eastern Porcupine district, Ontario. × 50, polished section.

almost prohibited by the scarcity of continuous passages. Similar reasoning applies to the more general type of vein. Among the deposits studied, the most careful examination failed to reveal any evidence of major post-quartz strains which could have been the means of access of large volumes of fluid. Cataclastic sections are typically local phenomena and the strains which caused them were of a local character. This is particularly well illustrated in the Sixteen-to-One Mine, Alleghany district, California, where the stoped areas of cataclastic ("live") quartz were found to be bounded on all sides by solid, aclastic ("dead") quartz. Consequently it is concluded that the total porosity of the typical primitive vein structure depended upon the intergranular spaces of aclastic sections and the more numerous interstices of cataclastic sections.

It may now be assumed that there is no mechanism [5] whereby a large volume of fluid could be moved through these relatively impermeable primitive vein structures against the tremendous frictional resistance of the myriads of minute, disconnected passages through which it would be obliged to pass. The alternative, that the volume of fluid must have been relatively small and of necessity rather highly concentrated, must therefore be true. This "residual ore fluid" contained relatively large quantities of gold and soda or potash, or both, and varying quantities of other substances, such as iron, arsenic, lead, zinc, copper, sulphur, and silica. In its mode of derivation, and its diverse composition, the residual ore fluid is analagous to the "residual magma" of the petrologists.

This conception of ore genesis is not altogether in harmony with currently popular views. From a theoretical standpoint, the derivation of this relatively concentrated residual ore fluid becomes difficult if the primary ore fluid is assumed to be a very dilute solution. With extraction of the common mineral constituents, the solution would become still more dilute, and derivation of the small concentrated residuum would then necessitate the separation of a large fluid fraction barren of gold and associated minerals. If, on the other hand, the original fluid were somewhat concentrated, a view which has much to commend it, derivation of the concentrated fluid residuum would be more easily understood.

Furthermore, this conception of the subtractive process whereby gold, present in the ore fluid since the beginning of ore deposition, finally becomes concentrated and appears as the solid metal late in the sequence, is inconsistent with the views of those who believe that gold is "introduced" into the hitherto barren veins from some

[5] Except by allowing an excessive length of time, of which there is no evidence.

extraneous source. The writer finds no support for such views in this study of the many ore samples examined. The exceptional occurrence of gold in the Pamour and Broulan mines, which has been mentioned, and the absence in other deposits of any possible channels of access, contradict a theory based on the later introduction of gold-bearing fluid. The distribution of various minerals and the timing of depositional events in gold deposits generally are too perfectly synchronized to be coincidental. Farmin[6] draws an apt illustration from the veins of the Grass Valley district, California. Here, dozens of veins traversing cubic miles of host rock, all have about the same tenor of post-quartz sulphides and gold. For these to have been introduced from an extraneous source would have required a synchronization of strains within specified zones in cubic miles of rock, and a nicety of timing which appear quite beyond the realm of possibility. The conclusion must be, therefore, that gold and other late minerals were present during the entire period of vein formation and, as persistent fluid constituents, became progressively concentrated in the shrinking volume of fluid, to appear as solid minerals only near the end of the entire sequence.

DEPOSITION OF GOLD.

The tenor of veins is by no means uniform. Many have an almost rhythmic alternation of mineralized and unmineralized sections, and of ore shoots separated by barren stretches. Even within ore shoots the gold and associated sulphides are apt to occur in streaks which tend to have an en echelon arrangement almost parallel to the plane of the vein. Such distribution is by no means accidental. It shows exactly the locus of permeation of the residual ore fluid. In ores such as those of the Pamour and Broulan mines, fluid permeation was widely diffused among many accidental interstices in the undisturbed vein material, and consequently the late sulphides and gold are similarly disseminated. In the more typical vein, the cataclastic sections, which are rela-

[6] Farmin, Rollin: Host-rock inflation by veins and dikes at Grass Valley, California. Econ. Geol., 36: 168, 1941.

tively more permeable than the intervening aclastic ones, were the loci to which the residual ore fluid diffused from all directions.

Within the cataclastic sections a subsidiary depositional sequence was initiated which was just as definite as that of the earlier stage of vein formation. The typical order of deposition is pyrite, arsenopyrite (if present), sphalerite, galena, and gold, with sericite[7] or carbonate (or both) continually depositing throughout the sequence.

The distribution of each mineral of this subsidiary phase was intimately related to the contemporary strain features. The earlier sulphides were deposited during one or more of the minor periods of cataclasis which were frequently superimposed. The fidelity with which these minerals were restricted to zones, or even lines of cataclasis, is remarkable, and it may be stated as a consistent rule that no mineral of this subsidiary phase of deposition ever occurs in solid, aclastic quartz. Furthermore, these minerals are restricted not only to the cataclastic zones, but to the most intensely cataclastic parts of such zones. This intimate relation is illustrated by the twin photomicrographs in Figures 6 and 7. The inference is that the most finely cataclastic quartz was also the most permeable to the residual ore fluid.

During this period of deposition, conditions had been gradually moderating, and in an environment of waning cataclasis swarms of tiny cracks appeared, forming parallel branching patterns through the cataclastic quartz. These were the 'trunk lines' of movement of the remaining residual fluid, now heavily charged with gold, usually a little galena, and, in one deposit, stibnite. From these trunk lines, the fluid diffused laterally into the minute interstices of cataclastic quartz or into tiny fractures in earlier brittle minerals, such as pyrite or feldspar, where the gold and the last traces of sulphide were deposited. Sericite, less commonly carbonate, is the invariable associate of gold, and it apparently continued to deposit after all the gold had crystallized. This is the last constituent of the residual ore fluid of which there is a

[7] White mica deposited during this period is referred to as "sericite" although it is suspected that the mineral contains an appreciable proportion of soda.

record. If any fluid remained it has disappeared in the course of time, leaving no clues to its identity.

THE MECHANISM AND TEMPERATURE OF GOLD DEPOSITION.

The larger mechanism by which gold became concentrated in the residual ore fluid has been discussed. The single and sufficient mechanism whereby this residual ore fluid was distributed appears to have been diffusion. The delicate control of differential permeability over the distribution of the gold-bearing fluid is well illustrated by the twin photomicrographs of Figures 11–13. In Figure 11, gold was deposited from a fluid which followed a serrate line of more intense cataclasis, and in Figure 13 the gold-bearing fluid, following one of the main trunk-line seams, diffused laterally into a zone of minutely cataclastic quartz where its gold precipitated as minute crystals. The faithful restriction of gold and contemporary stibnite crystals to cataclastic lines is illustrated by Figure 12.

Gold, the last metallic mineral to be deposited, is intimately related to minor strain features which post-dated the general period of cataclasis. For example, in the ores of the Cornucopia Mine gold occurs in very local fractures or zones of mechanical brecciation which traverse aclastic and cataclastic quartz alike (Fig. 9); a similar situation is found in the conglomerate ores of the Dome Mine. Moreover, in the peculiar ankerite veins of the latter mine, gold occurs in minutely serrate cracks which wander through the coarsely crystalline carbonate and which evidently originated by slight tensional adjustments within the vein structures (Fig. 10). Similar relations were observed in other ores. It appears reasonably certain, therefore, that a very low-stress or even non-stress environment is typical of the final stage in gold deposition.

The fact that gold was deposited where strains occurred subsequent to the period of cataclasis implies that the temperature of deposition was below that requisite for cataclasis. However, there is other evidence of a low temperature environment. Among the intimate mineral associates of gold a few provide some evidence of the temperature of gold deposition. Of these, realgar

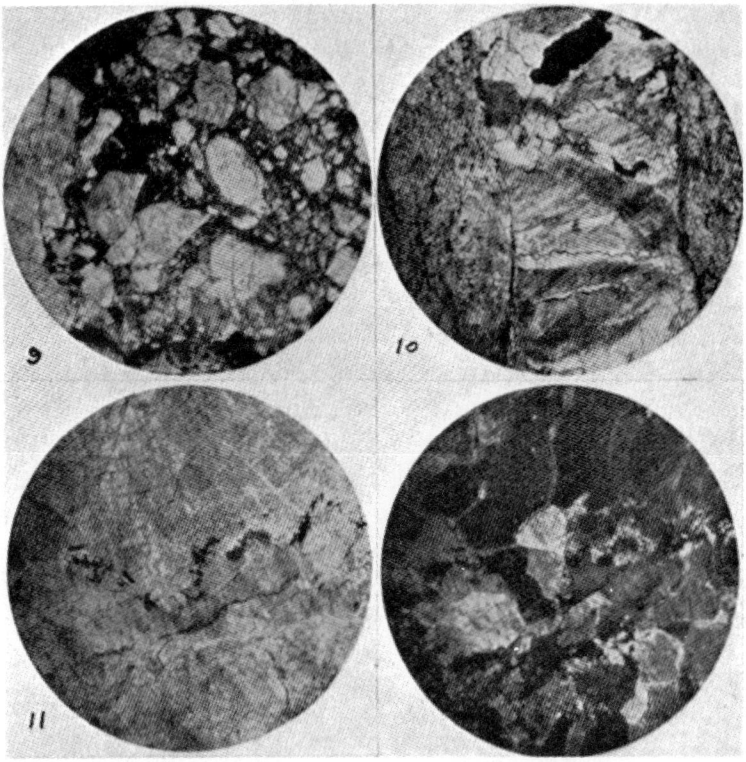

FIG. 9. A zone of mechanical brecciation which affected both aclastic and cataclastic quartz. The angular fragments of quartz are set in a matrix of sericite and carbonate. The typical occurrence of gold is as fine particles distributed along such zones. The black material is fractured early pyrite.

Cornucopia Mine, Oregon. ×35, plane polarized light.

FIG. 10. A carbonate veinlet showing the minutely jagged cracks which tend to follow crystal boundaries and cleavages. Gold occurs in minute particles here and there along these cracks. Black area is earlier pyrite.

Dome Mine, Porcupine district, Ontario. ×75, plane polarized light.

FIG. 11. Twin photomicrographs. Left, plane polarized light; right, crossed nicols. A styliolite-like sericite seam containing gold and several arsenopyrite prisms. Note that the seam faithfully follows cataclastic lines and grain contacts through quartz which exhibits incipient cataclastism. Although arsenopyrite and gold are both post-quartz minerals and

and stibnite are the most diagnostic. In the relatively shallow-seated ores of the Getchel Mine, Nevada, realgar is found intimately related to gold in the latest carbonate veinlets. Identical relations between gold and realgar are present in the deep-seated ores of the Homestake Mine, South Dakota. Since realgar is stable only at low temperatures, in the vicinity of 200° C., it is evident that the gold associated with it must have been deposited at the same moderate temperature. Stibnite, a typically low temperature mineral, is common to the epithermal ores of Cripple Creek, Colorado, and the relatively deep-seated ores of the Negus Mine, Northwest Territories where it is intimately related to gold. Warren [8] has recently discovered that native gold in many deposits contains detectable quantities of mercury. These data suggest that gold is deposited at a temperature which may not be above 200° C. and which may well be considerably below that temperature. Furthermore, it should be noted that this is the temperature of gold deposition in any deposit, whether it is classed as hypothermal, mesothermal, or epithermal.

The fact that the residual ore fluid was able to diffuse into almost submicroscopic interstices is ample proof of its extremely mobile nature. From the intimate mineral relations of gold it may be concluded that the residual ore fluid was a true solution, and that gold was not, as is supposed by some geologists, transported as colloidal particles. If such were the case, there should be a separation of constituents along the walls of the main trunk-line seams, the larger gold particles remaining at the walls and the fluid constituents diffusing into the adjacent minute interstices. This is not the case. Scattered through the finely cataclastic quartz which usually borders the trunk-line seams, we frequently find minute crystals of gold, galena, and in one deposit, stibnite, all

[8] Warren, H. V.: Personal communication, March, 1942.

---

appear to be closely associated, detailed examination showed that the arsenopyrite was deposited much earlier and that the fluid which precipitated gold merely followed the same permeable line.

Bralorne Mine, Bridge River district, British Columbia. ×75.

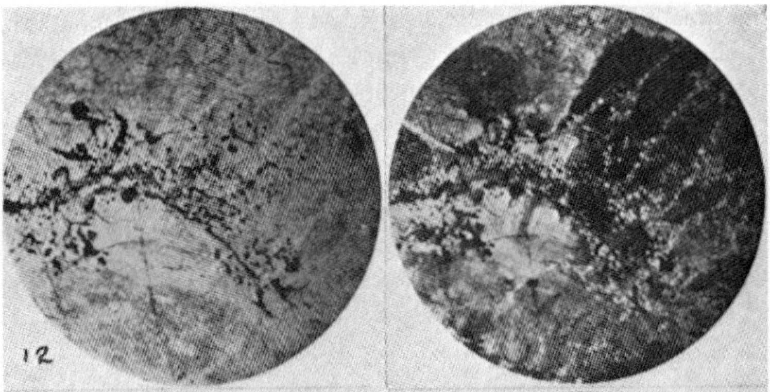

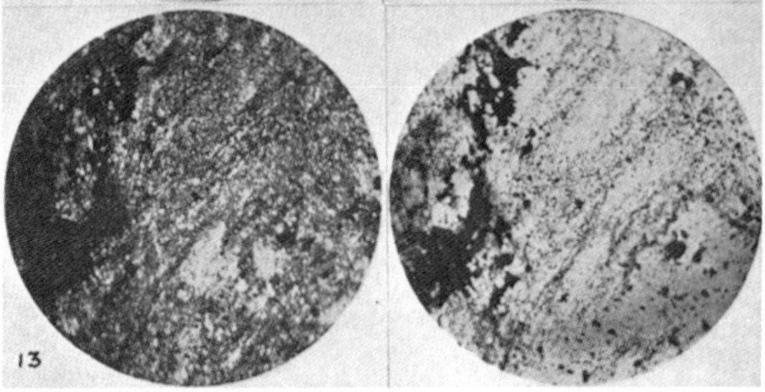

FIG. 12. Twin photomicrographs. Left, plane polarized light; right, crossed nicols. This is another illustration of the fidelity with which the residual ore fluid followed cataclastic lines which were lines of maximum permeability. This is the frayed termination of one of the characteristic "trunk line" seams. The black particles are gold, galena and stibnite. Many of the smaller ones are complete crystals.

Negus Mine, Northwest Territories. ×75.

FIG. 13. Twin photomicrographs. Right, plane polarized light; left, crossed nicols. A section through a gold-plated "trunk line" seam, showing the manner in which the gold-bearing fluid diffused laterally to deposit its gold as minute crystals in the interstices of a contact band of very finely cataclastic quartz. Just beyond the field the quartz rapidly becomes aclastic and there no gold occurs. Minute crystals of carbonate are scattered through the quartz, apparently being contemporary with the gold.

Negus Mine, Northwest Territories. ×75.

intimately and mutually associated. This implies that all constituents were transported in the same state; probably all were present as dissolved constituents of a residual ore solution.

CONCLUSIONS.

The presence and distribution of gold within veins is not a matter of chance, but is the logical result of a unique train of events occurring under certain favorable conditions. Gold, present in very minor proportion in the primitive hydrothermal fluid, by a rational process of subtractive crystallization of non-auriferous minerals, became progressively concentrated and finally comprised one of the principal metallic constituents of a very mobile, residual solution unlike the parent fluid. By virtue of its ability to diffuse readily, this residual fluid collected in any sections of the vein structure offering enhanced permeability. Cataclastic sections, having very superior permeability, were the favored areas to which the residual ore fluid diffused. It is a remarkable fact that where gold occurs in well-defined ore shoots, the quartz of such shoots is invariably cataclastic, and the limits of the ore mark the limits of cataclastic modification. A subsidiary sequence of mineral deposition occurred within these cataclastic sections. Gold was the last metallic mineral to crystallize, its closest sulphide associate being galena, and occasionally stibnite, realgar, and cinnabar.

Gold was deposited under uniform conditions of low and waning stress at a temperature possibly less than 200° C. Whatever the initial temperature of vein formation may have been, whatever the genetic classification of the deposit in which it occurs, gold remained a constituent of the residual ore fluid, and did not appear as the solid metal until a consistently low temperature had been attained.

ACKNOWLEDGMENTS.

The writer is deeply indebted to the officials at the many mines investigated for permitting him to visit their properties and obtain specimens for study, and to those who kindly sent him material for examination. The task of collecting and cataloguing material and arranging data was facilitated by Elizabeth U. White, to whom the writer expresses his thanks. He wishes also to express his thanks to Professors E. S. Moore and G. B. Langford for suggestions, and criticism of this paper.

17-2: RELATIONSHIPS OF MINOR STRUCTURES TO GOLD DEPOSITION IN CANADA

Frank Ebbutt

[*Editor's Note:* Plates I and II have been omitted.]

In this paper it is proposed to point out certain structural factors that have an important bearing on gold deposition. The subject will be treated more or less generally. Although the writer has had access to a great many of Canada's gold mines he does not wish to take the liberty of discussing details of any particular mine or mines; the aim will be to discuss controls that are evident in many places.

The production of gold in Canada has now become an important and very vital factor in the economy of the nation. In the period from 1858 to 1947, gold to the value of $2,998,538,965 has been produced, and the greater part of this amount has been gained during the last few decades. An industry that runs into such figures naturally warrants the most careful consideration of all the many factors that might lead to the discovery of additional deposits or the better appraisal of those now known. The factors that enter into gold mining today are many and most varied; they comprise science in many branches, economics of various kinds, and also many phases of politics. In Canada, gold is known to occur in a great number of places from the Atlantic to the Pacific and from the International Boundary to the Arctic, but in spite of its many occurrences the deposits are very much localized. There is an old and common saying that gold is where you find it, but there are many of us who keep asking ourselves why is the gold where we find it? Why is it so localized? Many before us have asked themselves these same questions and from time to time some have contributed ideas that have added to our knowledge, but still there is much to find out and we must continue to look for answers.

In Canada, gold is found in many varied circumstances as to type of host or wall-rock, type of quartz, kind of mineral association, attitude and size of deposit, setting with regard to intrusives, setting with regard to broad regional structure and intimate local structure, collectively a very wide diversity, yet there is one factor that is virtually always present but in the past has been frequently overlooked. That factor is late minor and highly localized structures within the deposit itself. This is the factor with which the writer proposes to deal, but before reaching that point it may be well to run quickly over some of the settings that the writer has observed over a period of years. For the sake of brevity, the rocks and minerals in which visible gold has been seen are tabulated and in each case listed roughly in order of frequency.

Types of Quartz in which Gold has been Observed

Porcelainous	Black
Milky	Granular
Grey	'Bull'
Blue	Jasper

Minerals in which Gold has been Observed

Pyrite	Sericite	Realgar
Arsenopyrite	Mariposite	Andalusite
Chalcopyrite	Epidote	Apatite
Sphalerite	Fuchsite	Cassiterite
Tourmaline	Chlorite	Cuprite
Galena	Garnet	Cobaltite
Calcite	Several tellurides	Niccolite
Scheelite	Rhodochrosite	Hematite
Ankerite	Muscovite	Hornblende
Magnetite	Biotite	Tetradymite
Stibnite	Dolomite	Stephanite
Cosalite	Enargite	Bismuthinite
Tetrahedrite	Barite	Albite
Bornite	Pyrrhotite	Marcasite
Jamesonite	Fluorite	Siderite
Molybdenite	Graphite	Pyroxene
Orthoclase	Talc	

Gold has also been observed in a great number of secondary minerals, but in these it is probably residual or secondary.

Rocks in which Gold has been Observed

Porphyry—several types	Sandstone
Slate—several types	Conglomerate
Greenstones	Flows—several types
Greywacke	Dykes—several types
Granite	

STRUCTURE

No doubt others could add to these lists, but for our present purpose they are certainly ample. If we consider these lists of gold hosts seriously it soon becomes evident that it would be folly to believe that all these various types of quartz, minerals, and rocks were precipitants of gold. It would seem to be equal folly to believe that the gold in these hosts came as a constituent part of such host. If this statement is clear and logical then we must ask ourselves once more, why is the gold where we find it. But there is another point that should be reviewed at this time—that is, the variation in the amount of gold from place to place. We all know that the grade of gold ores varies greatly from one mine to another or from one deposit to another. We also know that the content of gold varies greatly within the confines of any particular orebody or part thereof. Careful sampling and assaying will invariably demonstrate that the gold content varies from foot to foot in any direction, either along or across the strike or up or down the dip, plunge, or other feature. Such work nearly always shows that these variations have a very wide range, from waste to high grade. The same applies to the host quartz or any one of the other minerals listed. In many instances selected pyrite, arsenopyrite, or other mineral will carry abundant visible gold and/or microscopic gold, whereas pyrite or other minerals, evidently of the same type or generation as the auriferous piece, will be devoid of gold, yet this barren piece may be within less than inches of the high grade. It seems, in the face of such evidence, that we must exclude any particular precipitant or chemical affinity between the gold and/or any of the types of quartz and/or mineral or minerals listed. Again, if production records are considered closely, irrespective of the mine they are from or where the mine is, we quickly see that on almost any chosen vein there may be high-grade ore shoots, moderate-grade and non-commercial shoots, and completely barren parts. Frequently the ratio as to tonnage is greater for the barren material. There are many cases throughout this country where there are two veins nearby and almost identical in type and mineralogy, yet one is rich and the other poor or barren, regardless of whether the veins are believed to be epithermal, mesothermal, or hypothermal.

As the present symposium is on the structural relationships of ore deposits, let us look at our problem—why the gold is where we find it—from this structural point of view. The structural relationship of gold can be broadly covered by three items:

(1) The broad regional structure—such as a mountain range.

(2) The intimate local structure—superimposed on and related to, or in part controlled by, its predecessor (1). It is this structure that locally controls the emplacement of the deposit in its early stages.

(3) Important late, highly localized structures—which are only incipient in scale compared to (2) but definitely localized by the minor differences within the intimate local structure.

For the purposes of this discussion, items (1) and (2) will be only briefly reviewed, but some detailed attention will be given to item (3) and some typical examples illustrated.

The first topic—broad regional structure—will involve, in the west, the tectonic history of a great mountain assemblage, as for instance the Coast Range. A great area that has suffered uplift and all the complicated structural effects that accompanied such uplift, plus a great confusion of intrusives that are also closely related to the tectonics of this region. In the west there are many such ranges that can be explored and mapped because they are still very real and can easily be considered in three dimensions. In the east and north, such as, say, northern Quebec, Ontario, and Manitoba, again we are dealing with mountain ranges, but in these cases the mountains are so much older and have suffered such prolonged erosion that many people have difficulty in visualizing them as such, for the country is now an eroded and dissected peneplain. Only the bases or roots of these old ranges now exist, but we must learn more about what these ranges were like and what their limits were. They should be painstakingly sought out and named and indicated on our maps. When this is done it will be seen clearly that our mining camps have a very close relationship to these very old ranges.

The second item—intimate local structure—is the structure of a smaller area, such as a camp or even the confines of one property. Many of these have from time to time been mapped in considerable detail, but there are many cases where it is difficult to make good interpretations through lack of general knowledge pertaining to the larger setting—item (1), and naturally there is a very close relationship between (1) and (2). It is safe to say here that virtually all of our gold and base-metal deposits in Canada are in some way related to disturbed and/or faulted rocks. But all disturbed or faulted areas are by no means ore-bearing; there are a great many such areas that have no mineralization whatsoever. From this we see that those few structures that are mineralized must have become so because of some precise timing between local structural development and certain types of intrusive, from which the mineralizers were derived. We speak of 'early' and 'late' minerals when dealing with many types of ore; this involves early and late structural movements, and in many cases early and late intrusives.

This brings us to the third item—late highly localized structures—which are only incipient in scale compared with item (2) but are definitely localized by the minor differences within (2). From an economic point of view these

Photo by C. Cunningham

Fig. 1.—Typical face in high-grade ore, Empire vein, on 1151 east drift, Bralorne mine, Bridge River, B.C.

Note the ribboning of the vein and the free selvages showing minor movements subsequent to the deposition of the quartz.

small, late, superimposed structures are highly important. Many orebodies, irrespective of their location, show several generations of, say, quartz, pyrite, and/or other minerals, and in some rarer cases two generations of gold can be recognized, each slightly different from the other. Almost invariably, close inspection underground at the working face, followed by close study of material taken from such face, will clearly show that these several generations of quartz or other mineral, even gold, are each in turn related to some relatively minor structural factor that reopened the vein or other orebody at a critical time and allowed entrance of additional mineralizers. Further, as there are of necessity time intervals between these various shocks that bring about the minor reopenings, it is logical to believe the solutions available from the parent source would change in relation to these time intervals because conditions, such as pressure, temperature, etc., within the parent source would be changing as time went on.

If we now return to the list of types of quartz, minerals, and rocks in which gold is known to occur, we will see that in the greatest number of cases the gold occurs in the most brittle or harder of the common vein minerals, irrespective of their chemical characteristics. In other words, it is their physical characteristics or the way in which they respond or yield to small shocks, whether they fracture, shatter, or mash, that count. As an example, pyrite is frequently a host for gold, whereas pyrrhotite is only infrequently a host. Pyrite is a hard, brittle mineral and much more subject to fracturing than its very close relative pyrrhotite, which is prone to mash.

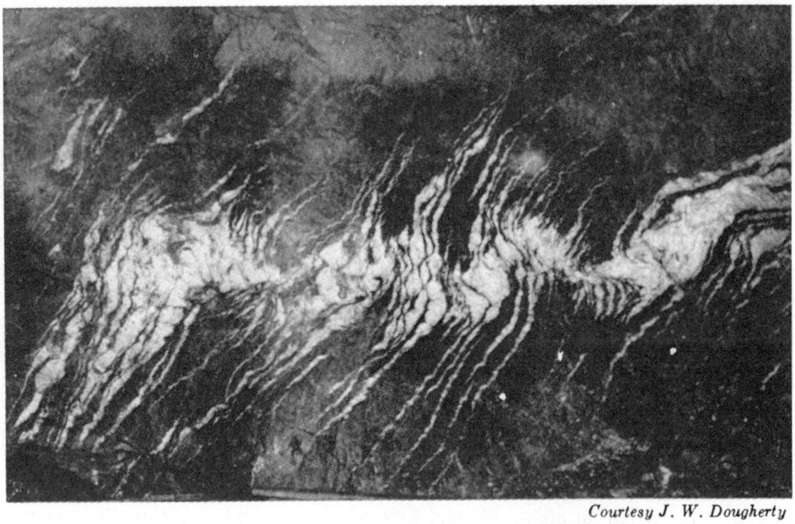

Courtesy J. W. Dougherty

Fig. 2.—Face in high-grade ore, Hollinger mine, Timmins, Ont. Quartz has been fractured by late shock.

It is these physical differences in relation to shock and the time of such shocks that determine the opportunities for gold deposition. It may be interesting here to point out how closely balanced the gold content frequently is to the type or strength of these late shocks, which bring about the minor structure within the orebody itself. In places where pyrite and sphalerite occur closely associated, if the shock is slight the pyrite is strong enough to withstand the shock without fracturing; the sphalerite, on the other hand, suffers and becomes the host for the gold. In such occurrences, however, close inspection will show that some pyrite crystals have become slightly loosened in their casts although the crystals have not actually fractured. In such cases, gold will commonly be found on the face of the loosened crystal or on the wall of the crystal cast. The same phenomenon is very true for quartz. In cases where a vein of quartz has withstood the late shock, but one wall has become freed and the wall-rock slightly fractured, gold will frequently be found. The gold is in the fractured wall-rock and on the loosened face of the quartz. There are countless different combinations and complications of this sort of thing, all in relation to the type of wall-rock, the type of quartz, and other minerals present, and their response or lack of response to these small late shocks. It is because of all these variables that it is seldom that any two gold-bearing orebodies look quite alike when given really close inspection. It is these countless differences that make the close inspection of any vein so interesting, and from which so much useful information can be gained.

If we go from one gold mine to another, or from one vein to another within the same mine, and note the grade of ore or the frequency of ore shoots, we find that in a great number of cases these ore shoots are related to changes in the strike and/or the dip of the vein or to sudden changes in the character of the wall-rock. In some cases these changes in attitude are very noticeable, in others it takes the closest kind of observation to detect them. Nevertheless, they are there, and it is at such points that minor shocks are more effective and, accordingly, that more gold is deposited.

To illustrate the effect of these late minor shocks on typical gold ores and the localization of the gold in late small structures, a suite of typical specimens from his own collection and those of others were selected by the writer and photographed by V. B. Meen. Most of these specimens are from Canadian occurrences, but a few were added from foreign sources, for comparison. Although, in this small suite, there are specimens from widely separated localities, all of them, nevertheless, present striking similarities. These similarities are not exceptions to the rule. They show how necessary late minor shocks are to prepare the barren quartz or other minerals for the gold deposition. The writer has many other specimens from many places and from different settings; some are from 6,000 or 7,000 feet below sea-level and some from 6,000 or 7,000 feet above, and from various continents and various latitudes. All show minor fracturing. In many cases this fracturing is unmistakable to the unaided eye, in some it can be seen only under a hand lens or microscope, but in every case the evidence is there. If it were necessary, and space permitted, many examples could be used to illustrate the point under discussion, but it is hoped that for this occasion the few submitted will be sufficient (see Figs. 1 and 2, and Plates I and II).

In conclusion, we can now list or discuss in question form a few of the important structural factors that have a very marked control or effect on the localization of gold. First, there is the initial fracture prior to the vein filling—is it straight, nearly so, or is it highly irregular? Did one wall move in some relation to the other prior to the vein filling, thus bringing about additional irregularities? Is there evidence of movement or movements after vein filling, what were the effects of these, and to which is the gold related? What are the wall-rocks—are both walls the same or is one more competent than the other? Are there dykes, and if so what effect does their presence have with regard to small, late shocks or movements? Are there strong or weak contacts nearby and what would be the effect of these? Is there some solid mass, perhaps an intrusive, that acts as a buttress and delays or localizes structural development? These are but a few of the many different situations that are met with, any one of which will necessitate the most careful and painstaking kind of detailed underground mapping, and the collection and study of type specimens in conjunction with sampling and close assaying of typical material. Mapping should be on a sufficiently large scale to show all minor structural details that can possibly be put on paper. At many of our Canadian gold mines this type of mapping is being done, and its quality in many cases compares most favourably with that done elsewhere in the world.

Acknowledgments

The writer wishes to record his indebtedness to V. B. Meen, Associate Director of the Royal Ontario Museum of Geology and Mineralogy, for the loan of several excellent specimens and for his kindness in photographing in colour these and the other specimens. The writer would also like to thank M. H. Frohberg, geologist for Macassa Mines, Limited, for the loan of some of his fine specimens.

CHAPTER

18

The Economics of Gold and Gold Mining

Nature herself makes it clear that the production of gold is laborious, the guarding of it difficult, the zest for it very great, and its use balanced between pleasure and pain.
Diodorus Siculus, first century B.C.

Gold is the most exquisite of all things . . . Whoever possesses gold can acquire all that he desires in this world. Truly, for gold can he gain entrance for his soul into paradise.
Christopher Columbus, 1500.

Gold mining and placering reach back into antiquity to a time at least more than 4,000 years before our era. During primitive times the winning of gold was a haphazard activity, nuggets of the native metal being plucked from some auriferous stream, the oxidation zones of gold or sulfide deposits, or from residuum near such deposits. Production was probably erratic and generally unorganized. By the time of the Pharaohs, however, organized gold mining in the Arabian and Egyptian shields appears to have been well established, and gold was seemingly relatively plentiful compared with silver, for in the Code of Menes, the first pharaoh, who reigned in Egypt about 3500 B.C., it was decreed that "one part of gold is equal to two and one-half parts of silver in value." Since that time up to the fall of the Roman Empire in the west (A.D. 476) gold appears to have been mined almost continuously mainly from placers but also from the oxidized zones of auriferous deposits in all organized societies throughout the world. There is evidence of ancient gold workings and placers in nearly all parts of the world, in Japan, China, India, USSR, Middle Asia, Turkey, the Mediterranean islands and lands, Africa, central Europe, Spain, France, Britain, Mexico, Central America, and South America. The early Minoan goldsmiths had a source of gold, for they produced some of the finest pieces of early gold work known to man. The Greeks mined gold extensively throughout their empire and appear to have prospected for it far and wide, judging from the legend of Jason and the Golden Fleece. The Romans, likewise, sought gold throughout their empire, obtaining it from placers and mines in Spain, France, Britain, Germany, central Europe, and the Middle East. The empires and national states that followed the fall of the Roman Empire sought and mined gold wherever it was discovered. In fact, historically, the cry "gold" has lured men across oceans and continents, over the highest mountain peaks, into the Arctic tundras, into scorching deserts and through impenetrable jungles. Its gleam prompted the expeditions of Jason of Thessaly, Cyrus and Darius of Persia,

Alexander of Greece, Caesar of Rome, the Vikings of Scandinavia, Vasco da Gama and Cabral of Portugal, Columbus of Genoa, Cortez and Pizzarro of Spain, Frobisher and Raleigh of England, and a host of others down through history. Inevitably the discovery and development of gold mines and placers has been followed by settlement and civilization as in California, Australia, New Zealand, Siberia, northern Canada, and many other countries and areas.

Nearly all of the gold mined by the ancients was wrought into idols, shrines, bowls, vases, flasks, cups, plates, and items of adornment. Later, about 700 B.C., gold and silver came into general usage for coins as a medium of exchange, certainly in all countries between the Indus and the Nile and probably on a much more widespread geographical basis. Today, gold maintains its use in artistic objects, jewelry, and coins and has found increasing employment in the industrial arts. Its traditional role as a hedge against currency holocausts continues.

HISTORY OF THE ECONOMICS OF GOLD TO THE END OF THE MIDDLE AGES

The worth (value, price) of gold has fluctuated through the centuries. At the time of Menes, the first pharaoh (c. 3500 B.C.), gold was valued at about 2.5 times that of silver, as just noted. Indeed, in early Sumerian, Egyptian, and Hittite times (3000-1500 B.C.) gold was worth much less than iron. By Philip II of Macedon's time (359-336 B.C.) gold and silver were generally valued in the ratio of 10:1. This value ratio did not change much through the centuries that followed, for we find that in England the ratio in 1464 was 11:1, according to the recorded mint price. Since then there has been considerable fluctuation in the value ratio as shown in Table 18-1. The precise reason for the fluctuation of the value ratio in more recent times is uncertain; it would seem to be related to the great increase in silver production as a result of the mining of argentiferous base metal sulfide bodies. This increase in the production of silver has consistently depressed its value (price) despite the fact that a great demand for the metal prevails for use in photography, electronics, and so on.

There is an interesting comment in the *Engineering and Mining Journal* (vol. 95, p. 1163, 1913) concerning the origin of the ratio of value between gold and silver that merits quotation.

> "I have endeavored to discover as to how and when the ratio of value between gold and silver commenced, and as to what was its origin," said Bedford McNeill, in his presidential address before the Institution of Mining and Metallurgy. "Sir David Barbour has called my attention to an interesting theory of Professor D'Arcy Thompson, 'that the origin of the ratio may have been astronomical, gold being associated with the sun, and silver with the moon; the period of the earth's revolution round the sun being 365¼ days, and that of the moon round the earth 27.32 days, it will be found that the ratio between these figures is very nearly 13⅓, and this figure 13⅓ was the ratio that was fixed between gold and silver during the Babylonian Empire (2000 B.C.) down to the Lydian Empire (500 B.C.)'. Professor D'Arcy Thompson has also pointed out that the Lydians made coins

Table 18-1. Value ratio of gold to silver through the ages*

Year	*Value Ratio*	*Remarks, Authority, etc.*†
Prior to 3500 B.C.	1 or less	No records
c. 3500	2.0	In Egypt. Code of Menes
708	13.5	In Assyria. Cuneiform inscriptions at Nineveh
500	13.0	In Persia. Herod III
400	12.0	In Greece. Plato
404-336	12.0	In Greece. Peloponnesian war to time of Alexander
300	10.0	In Greece. Fall in gold value due to influx of Alexander's spoil
207	13.7	In Rome
186	10.0	In Rome
58-49	8.9	In Rome. Fall in gold value due to influx of Caesar's spoil
A.D. 1-37	10.9	In Rome. Reigns of Augustus and Tiberius
54-68	11.8	In Rome. Reign of Nero
81-96	11.3	In Rome. Reign of Domitian
312	14.4	In Byzantium. Reign of Constantine
438	14.4	In Rome and Byzantium. Theodosian code
864	12.0	In Europe. Edictum Pistense
1344-1482	11.1	In England. Mint returns
1482-1492	12.2	In Europe
1492	11.0	In Europe. Discovery of America
1497	10.7	In Spain
1550	12.0	In Europe
1641	14.0	In Europe. Moran and John Locke on money
1690	16.0	In Europe. Sir Isaac Newton
1730	16.0	In Europe. Kelly's "Cambist"
1760	14.3	In London. Del Mar
1800	15.7	In Europe & U.S.A. *EMJ*
1850	15.7	In Europe & U.S.A. *EMJ*
1886	22.0	In Europe & U.S.A. Discovery of the Rand
1890	22.1	In Europe & U.S.A. *EMJ*
1900	34.4	In Europe & U.S.A. *EMJ*
1910	38.2	In Europe & U.S.A. *EMJ*
1915	38.2	In Europe & U.S.A. 1st Great War, *EMJ*
1920	20.8	In Europe & U.S.A. *EMJ*
1930	60.8	In Europe & U.S.A. *EMJ*
1935	45.5	In Europe & U.S.A. *EMJ*
1940	50.0	In Europe & U.S.A. 2nd Great War, *EMJ*
1950	41.2	In Europe & U.S.A. *EMJ*
1960	38.5	In Europe & U.S.A. *EMJ*
1970	21.1	In Europe & U.S.A. *EMJ*
1975	41.9	In Europe & U.S.A. *EMJ*
1983	46.2	In Europe & U.S.A. *EMJ*

*The figures given are calculated value ratios in the currency of the time and all should be read as follows: 2.5 indicates that at the stated time gold was two and one-half times the value of silver; 12 that it was twelve times, and so on.

†Sources of information include various documents in the British Museum and British Library, London; National Museum, Athens, Greece; Vatican Library, Rome; Jacob (1832) and Del Mar (1902). The values since 1800 were obtained mainly from the *Engineering and Mining Journal (EMJ)* annual volumes. No attempt was made to authenticate all of the values shown, and no adjustments have been made to compensate for differences between the various currencies of different countries. For a thorough discussion of the value ratios prior to 1902 see Del Mar (1902).

> of what was called 'electrum', and reckoned one electrum coin as being of the same value as 10 silver coins of the same weight. It was supposed that the Lydians were unable to separate the gold from the silver, but when we find by assay that the proportions in these old coins of gold to silver were approximately as 73 and 27, calculating gold as $13\frac{1}{3}$ times as valuable as silver, and we find that an electrum coin would be worth exactly 10 silver coins of the same weight, we cannot help but admit that the Lydian assayers must have been well versed in the science and practice of their art, and that such figures cannot be mere coincidences."

Since the earliest times gold has been hoarded by kings, states, popes, and individuals, the prime example being Croesus of the Mermnadae, last king of Lydia (560-546 B.C.). It is said that most of his fortune in gold came from the placers of the Pactolus, but it seems likely that his wealth had a more widespread origin, as the Lydians were renowned for their trading pursuits. How much gold was hoarded in ancient times is unknown, but the practice seems to have increased at the beginning of our era, so much so that during most of the Middle Ages both precious metals were scarce. Some of the scarcity was undoubtedly due to the exhaustion of most of the placers in Europe, the Middle East, and middle Asia and also to a general decline in mining, the economic limits (in terms of depth) of profitable ore having been reached in many districts mainly because of engineering problems (ventilation and flooding by groundwater). It is estimated that the total annual production of gold during the Middle Ages probably averaged only about 100,000 oz, compared with the roughly estimated annual production in the Roman Empire at the beginning of our era of 250,000 oz. The scarcity of gold during Medieval times was further exacerbated by the demand of the Orientals for payment in gold for their commodities, spices, silks, and so forth.

Geographical exploration brought an abrupt change at the end of the Middle Ages, the manifestations of which affect us to this day. With the rediscovery of America by the Spaniards in 1492 came the discovery of great stores of both gold and silver first in the islands of the West Indies, then in Mexico, and finally in Central and South America. At first, great quantities of gold were looted from the natives, but as this source declined, active placering and mining by slave labor provided a large and continuous influx of gold and silver to Spain. Despite an attempt by the Spanish government to keep the precious metals to itself, they soon filtered over the whole of Europe, increasing the supply of money and inflating prices. It is estimated that from 1492 to 1600 more than 10 million oz of gold came from the Spanish Americas, which was about 40% of the world production at that time. Marked increases in gold production from South America, particularly from the placers of Colombia, were registered during the period 1600-1800; for example, some 48 million oz were won during the eighteenth century, accounting for about 80% of the world production. Two further increases in the world gold production followed during the periods 1820-1880 and 1890-1920. The first marked the discoveries of the great placers and lode mining areas of Siberia, California, Australia, and New Zealand; the second followed the discoveries of placers and bedrock deposits in Alaska, Yukon, central Canada, and on the Witwatersrand of South Africa. A further factor in the second increase was the introduction of the cyanide process (1887)

Table 18-2. Estimate of annual world gold production (1982)

Country	*Production** in millions (10^6) of grams (metric tons)*
South Africa	664
USSR	270
Canada	62
China	60
U.S.A.	45
Brazil	45
Australia	27
Philippines	25
Papua-New Guinea	18
Chile	17
Colombia	13
Dominican Republic	13
Zimbabwe	13
Ghana	10
Mexico	7
Korea	7
Peru	5
Yugoslavia	4
Japan	3
Spain	3
India	2
Others	25
Total	1338

*Production estimated from publications of U.S. Bureau of Mines (Minerals Yearbooks); Mineral Policy Sector, Dept. Energy, Mines and Resources, Ottawa, Canada; and Consolidated Gold Fields, PLC., London, U.K.

for recovery of gold and silver from low-grade ores and ores containing microscopic particles of the two metals. Since 1920 there has been only one period, 1933-1939, when the production of gold increased significantly. During and for some thirty years since World War II a general decline in the production of gold in nearly all countries except South Africa and the Soviet Union has taken place. In South Africa the average annual production for the last five years has been about 670 metric tons; in the Soviet Union the annual production for the last five years has probably averaged 200 metric tons, although this figure is only a rough estimate, the true value being a closely guarded state secret. Today there is a modest increase in the production of gold in nearly all countries, stimulated by a rise in price for the metal.

The present world production of gold is shown in Table 18-2. About 50% of this production is derived from quartz-pebble conglomerate deposits, some 25% from nonlithified eluvial and alluvial placers, and the remainder from various vein and disseminated deposits.

The economics of ancient gold placer and bedrock mining is a difficult subject fraught with much uncertainty and conjecture. The first gold workings were undoubtedly stream placers, although there is some evidence in certain gold belts, especially in Egypt, Sudan (Nubia), and India (Kolar) that eluvial deposits, auriferous gossans, and the auriferous residuum of veins provided the bulk of the early gold. The stream and eluvial placers were worked with the simplest tools (stone hammers, antler picks, bone and wooden shovels) and the gold was won principally by using crude pans and baskets made of wood, split gourds (calabash), and so on. Perhaps a crude form of hushing (booming) was used in places to tear up the auriferous gravel and auriferous residuum of oxidized veins, and the gold may have been caught in primitive sluices lined with sheep's fleeces. With exhaustion of the placers the early miners took to the primary sources of gold, mainly in quartz veins and pyritized silicified zones. All of this bedrock gold mining, however, took place mainly above the water table in the zone of oxidation where the deposits were friable and easily grubbed out, trenched, or pitted along their strike with the crudest of tools (stone hammers, antler picks, bone and wooden shovels) and later with bronze and iron picks, chisels, wedges, and so on. Only rarely were adits, crude shafts, drifts, and stopes attempted and then only in the soft rocks of the zone of oxidation. The auriferous material (gossan, oxidized residuum) was broken up, then moiled in stone querns, sized in primitive basket sieves, and the gold separated by panning or by employing sloping wooden washeries or crude sluices.

The ancient Egyptians, Semites, Hittites, Lydians, Phoenicians, Etruscans, Indians, and Chinese used slaves, prisoners and captives of war, and convicts in their gold placers and mines, and hence the costs of mining were essentially those of minimum sustenance of the slave miners and relatively low-paid overseers. Materials for mining and metallurgical tools (stones, rock querns, antlers, bone, and wood) were in the nearest stream and forest and of relatively little import in the economics of mining. As far as I can ascertain from old dumps of moiled quartz, gossans, and oxidized residuum, obtained from veins where the gold was free, the limit (cut-off) of mining was exceptionally low, in many cases much less than 1 ppm (0.03 oz/ton). Where some of the gold was intimately tied up in sulfides or present as tellurides, the cut-off grade was much higher (15 ppm or higher), since the ancients knew little about roasting such ores. Similarly with the ancient placers, while the methods were crude, the washing (panning, sluicing) was most efficient, because one can now find remarkably little if any gold in the worked gravels (tailings).

The Greeks were particularly adept at prospecting for gold, and their methods of winning the metal from placers and bedrock mines were considerable improvements over those of the ancients. The Greeks improved the methods of placering by diverting streams and small rivers into canals and aquaducts for the transport of abundant water to the placer sites for the purpose of hushing (booming), sluicing and panning. Bedrock gold deposits were attacked with simple iron tools (hammers, chisels, picks, and mattocks), first by trenching and open-cut methods along the strike of the veins or silicified zones, and later by adits, shafts, drifts, and (underhand and overhand) stopes. The method of mining was essentially by chipping and wedging out the auriferous rock (generally quartz). Firesetting, followed by throwing cold water (or vinegar) on the hot rock, was also employed in places as related by Diodorus Siculus (First century B.C.) in his *Bibliotheca historica:* "The hardest gold-bearing earth is first burned with a hot fire, and when it is crumbled in this manner the miners continue the working of it by hand." Advance through the rock by firesetting and chipping was slow; some mining engineers consider that 12 cm a day was the maximum advance made along workings that averaged about 1 m high and 80 cm wide.

The auriferous ore from the mines was further processed in an *ergasterion* (mill) where it was crushed in stone mortars with iron pestles and then ground to a fine powder in stone mortars and querns or in banks of simple rotary (hourglass-type) stone mills. Stone sieves were employed to ensure uniformity of grain size. When the gold was free as in quartz, pans and sloping wooden washeries (tables) were employed directly, but when some of the gold was present in sulfides it is thought that the Greeks used a crude form of roasting prior to final concentration. Certain types of gold concentrates were mixed with a variety of substances and smelted (cupelled) in earthen (clay) pots as described by Agatharcides of Cnidus (c. 150 B.C.), the Greek historian and geographer, and quoted by Diodorus Siculus (op. cit.), Rickard (1932, p. 211), and Healy (1978, p. 154).

> Other skilled workmen take the concentrate (i.e. gold dust, sulphides, etc.) and put it by measure and weight into earthen pots. They mix with this a lump of lead according to the mass, lumps of salt, a little tin and barley husks. They put on a closely-fitting lid carefully smearing it with clay and heat it in a furnace for five days and nights continuously; then they allow the pots to cool and find no residual impurities in them; the gold they recover in a pure state with little wastage. This processing of gold is carried on round about the most distant boundaries of Egypt.

The Greeks probably learned about the art of amalgamation of gold with mercury from the Egyptians (Del Mar, 1902) but they do not seem to have employed the method to any extent in their gold *ergasterion.*

The Romans had a greater practical knowledge of auriferous placers and bedrock deposits than the Greeks, and were more efficient in prospecting for these deposits using indicators such as quartz pebbles for placers and *segullum* (gossans and red weathered residuum) for deposits in place. Likewise, the Romans were better mining engineers and advanced mining technology along more modern lines.

The Romans continued to work the placers not exhausted by the Greeks and Phoenicians in Thrace, Asia Minor, and Spain and exploited new ones in Spain, France, Great Britain, and central Europe. Hushing (booming) methods were improved and primitive forms of hydraulicking by diverting streams of water into constricted channels were developed. In some places holding tanks were used, and a crude form of monitor was employed. Sluices containing ulex (gorse) were used to catch the gold, and the first prototype of the "long ton" made its appearance. The boulder-riffle method of concentrating the gold seems to have been first introduced by the Romans in areas where wood was scarce; in this method boulders were arranged in such a way that as the water rushed along carrying the gold, it swirled around the boulders, depositing the nuggets and dust in the slack-water zones around and between the boulders. After an appropriate interval the sluicing water was turned off or diverted to permit removal of the boulders to allow the cleanup. Stretches of streams with natural riffles, such as slate and schist beds and folia oriented at right angles to the stream direction, also appear to have been employed in some Roman placer operations.

Open-pit mining was employed in the oxidized zones of deposits in many auriferous areas and undergound mining was developed along modern lines, using timbered access adits, drainage adits, shafts, drifts, and stopes. Some shafts were square and timbered and others circular and stone-lined. The problem of heavy flows of water, which had generally stymied the baling methods of the Greeks, was partly solved by the Romans with the introduction of water wheels, archimedean screws, and a type of force-pump. The ore-dressing and metallurgical methods of the Greeks for winning gold (panning, cupelling, roasting, smelting) were greatly improved by the Romans, who apparently utilized amalgamation extensively, and who were able to deal with some of the complex sulfide, arsenic, and telluride ores by smelting procedures. Harrison (1931) describes the great Roman gold mining works at Minas dos Mouros in Portugal, and states that their metallurgical practices (smelting) were capable of partly treating the high arsenical (5.9% As) pyritic ores containing about 15 ppm Au, leaving a tailing (slag and scoria) averaging about 9 ppm. Similar relative values have been found in the remains of the Roman smelting works elsewhere in central Europe, Greece, and Asia Minor.

The semi-socialist Greek city-states *(polis)* by 550 B.C. extended throughout the Mediterranean and Middle East, from eastern Iberia (Spain) at Saguntum to Phasis on the far eastern shore of the Black Sea, and by the time of the death of Alexander of Macedon (323 B.C.) their unified suzerainty extended southward to Egypt and eastward through Asia Minor, Media, Persia, Parthia, Carmania, Bactria, and Sogdiana to the valley of the Indus. Such a vast area from the Pillar of Hercules (Gibraltar) to Alexandria Opiana on the Indus was marked by a vast number of both placer and bedrock gold deposits that were mined extensively. Ownership of the mining rights of these deposits within the confines of the city-states seems to have rested principally with the governments of the states, which either worked the deposits directly or leased them to lessees who paid a straight forward yearly rent in some cases, but in others they may have paid a rudimentary royalty (10-20% of value of production) during their exploitation. In the intervening territories the mines were worked by the adjacent states or by private individuals; in the conquered lands (e.g., Egypt) the gold mines were worked directly by the (Ptolemaic) state, whereas in Persia it is thought that some gold and other mines were privately exploited. Nomadic gold washers seem to have roamed throughout Alexander's Empire, especially in Gandhara, Bactria, and Sogdiana (all in south central USSR, northern Afghanistan, and northern Pakistan); other nomads (Cimmerians, Scythians) in the region north of the Black Sea and in the southern Urals also seem to have washed gold extensively at this time, judging from the recent archeological finds of wrought gold articles in these regions.

The Greeks used slaves, captives of war, and convicts in their gold mining and metallurgical operations. The labor cost of mining and winning the gold was therefore minimal, as were also the costs of mining materials and equipment, which were rudimentary and easily accessible. The price of a mine slave at auction in Athens in the fourth century B.C. was 150-250 drachmas. One drachma was roughly equivalent to a day's wages for a freeman employed in public works during the fourth and fifth centuries B.C. (one drachma = 67.4 grains of silver = 0.14 oz = approx. US $1.40).

The cut-off grades in placers were exceedingly low; almost any stream or patch of eluvium showing gold was worked; similarly any bedrock deposit containing visible or recoverable free gold was attacked. Small placers and auriferous veins that today would not be considered in the least economic were exploited by the Greeks or by the peoples before them (Phoenicians) in many places. The return (profit) on effort expended on the Grecian gold mining operations is difficult to assess; judging from some classic writings the returns from some deposits were minimal, whereas others were enormous. Probably an average return of 15% on investment is a reasonable estimate for most deposits.

Two principal periods are recognized in Roman history—the Republic, beginning after the expulsion of the Etruscans in 509 B.C. and extending until 30 B.C. during which time the Italian city-states were subjugated, Carthage was reduced (Punic wars), and suzerainty was established over Spain, Macedonia, and Greece; and the Empire, beginning with the reign of Octavius (Augustus Caesar) in 30 B.C. and extending to the time of its fall in the west in A.D. 476. The eastern (Byzantine) part of the Empire survived until the late Middle Ages (1453, the fall of Constantinople to the Turks). At its height, during the reigns of Trajan, Hadrian, and Antoninus Pius (A.D. 98-160), the Roman Empire extended from Britain in the north, Spain in the west, to Mesopotamia in the east. It governed or controlled most of Britain, all of Iberia (Spain and Portugal), Gaul (France), Dacia (Romania), Dalmatia (Yugoslavia), Greece, Macedonia, Thrace, Asia Minor (Turkey), Assyria, Egypt, and the coastal regions of North Africa—countries and regions in which gold placers and mines taken over from the Phoenicians, Greeks, and other peoples had been exploited for almost ten centuries.

The expansive years of Roman gold mining spanned the last century of the Republic and the first two centuries of the Empire. This period of maximum Roman conquest and suzerainty coincided with the availability of abundant gold (and silver) specie that promoted a flourishing agriculture and internal commerce and an expansive external trade over the "Silk Route" to China and the "Spice Route" to India. During the fourth and fifth centuries a general "fatigue of the mind and spirit" set in: Certain of the Romans sought not for the common good but for individual aggrandizement, other Romans sought escape in mystical religions, the invading Germanic barbarians could be neither bought off nor fought off as previously, and the Empire in the west fell. Davies (1935) commented on this decline and fall.

> The most flourishing period of Roman mining was the late republic and early empire. New provinces were being explored, and after their gold-placers had been skimmed, the Romans undertook a systematic exploitation of their mines. It is not until the third century that we hear complaints of the failure of Rome's mineral resources. The decay of civic life in this period and the introduction of compulsory corporations caused a slackening in the output, and in the fourth century the emperors alternately applied persuasion and force to increase it. The archaeological evidence suggests that the western empire became bankrupt mainly owing to the failure of its mines, whereas the east survived because it reopened old workings, encouraged further prospecting, and appointed officers to supervise the mining districts. (p. 2)

Perhaps there is a lesson in all of this history for our modern civilization. Roman economics of mining gold were based on forced labor of slaves, convicts and Christians, and therefore not much different from the Greek practice. Some freedmen also appear to have participated in mining, according to some ancient tablets (Liversidge, 1968). Thus, the miner Memmius, son of Asclepuis, was contracted to work six months in a gold mine in Dacia (Romania) in the second century A.D. in return for seventy *denarii* (about U.S. $280) and his board. The introduction of more sophisticated mining and metallurgical machinery (e.g., archimedean screws, water wheels, smelting apparatus) by the Romans and the advent of deeper mining (necessitating timbering and control of water by drainage adits, etc.) raised the mining costs for bedrock mining considerably, compared with those of the Greeks. Similarly, the relatively large-scale mining of gold placers by the Romans, as along the Sil River in northwestern Spain, required extensive canals for conducting water to the crude hydraulic (monitor) operations, a feature that undoubtedly raised the cost of mining placer gold. There is no way of knowing what was the profit on Roman bedrock and placer operations. A 15% return on the capital expended by state-owned enterprises at the height of the mining boom during the early years of the Empire seems likely, but by the late and final years of the Empire it seems probable that most gold mining ventures were operated at cost and in some cases at severe losses, especially after the reign of Theodosius (395). In fact the actions of the Hierarchial Fisc, which exclusively controlled the operation of gold mines, seem to have been responsible for this state of affairs. Del Mar (1902) has commented on this particular situation.

> The reason why the Roman Fisc was unable to work the gold mines at a profit was due to its own policy. It fixed the ratio between silver and gold at 12 for 1, and exacted its tributes at this ratio. It was, therefore, cheaper for it to procure gold for silver in India, where the ratio was but 6 or 6½ for 1, than to delve for gold in Europe. The adoption of the Indian ratio by the Saracens rendered the commercial aspect of this policy profitable so long as the Arabian Empire survived in Egypt and Spain, both of which were gold-producing countries, and both willing to exchange their gold for the silver of the Romans. When the Hierarchy fell, many of the ancient gold mines of Europe were reopened. (p. 120)

The Roman state (emperor) appears to have held the mineral rights to gold deposits during both the Republic and Empire. Under the Republic, control of the gold mines was farmed out to *publicani* (farmer-generals of the Roman revenues and contracts usually of the order of *Equites*) who in turn contracted (leased) out the placers and bedrock deposits to *socii* (individuals, associations, or companies) who paid a rent amounting at times to half of the value of the gold extracted. During the early Empire the administration of the gold (royal metal) mines appears to have come under the emperor (and his surrogate the Hierarchial Fisc or Treasury), who contracted out the placers and mines directly or through a *procurator metallorum* (an Equestrian or freedman in control of a mining district), the *conductores* (Equites), or the *publicani.* In the later Empire the centralizing authority of the state (emperor) was modified to permit more activity in prospecting, development, and mining by *societates* (private capitalists and associations of capitalists)—a development that occurs down through mining history to the present day when the state because of its follies goes essentially bankrupt! This development is recorded in part in the *Lex metalli Vipascensis* (the Aljustrel Tablets of bronze found in 1876 and 1906 in the copper mining district of Vipasca, Portugal). Rickard (1932) and Davies (1935) have commented on the contents of these tablets in detail. Healy (1978) has summarized the mining laws and administration as related to Vipasca as follows:

> The most important evidence for Roman mine administration is contained in the *Lex metalli Vipascensis* of the time of Hadrian (A.D. 117-38) recorded on two bronze plaques discovered in the old slag heaps of the Aljustrel mines in Portugal.
>
> The procurator of mines, as in the case of Ulpius Aelianus, was frequently a freeman; he had full jurisdiction within the general disposition of the law. The ultimate ownership of mineral rights resided in the treasury and the Emperor issued regulations which defined appropriate taxes, the method of their collection, the legal conditions under which mining rights could be acquired and the technical obligations of those who exploited the rights. The structure of revenue collection was based on a prior determination of the value of the mining set. Any free inhabitant of the district could either start operations in a new site or take an abandoned shaft on which occupation rights had lapsed. To do so he had to affix a notice on the entrance, pay a fee to the official in charge and register his rights stating that he proposed to work the set. He was then allowed twenty-five days in which to collect the equipment and personnel to work the mine. Half the mineral won belonged to the procurator.
>
> The inscription contains regulations relating to the working of the mine including detailed safety measures to be observed, and outlines the provisions for joint ownership of occupation rights. Other clauses cover the prohibition of ore transport after dark, the theft of mineral ore, the maintenance of shafts, movement of boundary marks, deliberate damage to shafts, the restriction of trial shafts or workings within specified limits of drainage adits and, finally, the siting of spoil dumps. (p. 130)

The fall of the partly Christianized Roman Empire in the west ushered in what is known in western civilization as the medieval period or Middle Ages (476-1475). Interest in scientific problems persisted throughout the period, but there were no marked breakthroughs. In technology the major advance was the introduction of gunpowder which was to have a major impact on mining and on warfare. Concerning the latter the development of cannon made castles no longer impregnable and signaled the end of the feudal system.

Jacob (1832) remarks that during medieval times, from the sixth to the fifteenth centuries, "the precious metals were sought, not by exploring the bowels of the earth, but by the more summary process of conquest, tribute and plunder." Del Mar (1902) takes exception to this statement and goes on at some length to show that considerable mining of gold was pursued in the Middle Ages. More recent research tends to support his contention.

Certainly immediately after the fall of the Roman Empire in the west (476) and during the incursions by barbaric Germanic tribes the conditions were not opportune for gold mining or placering in western Europe. Similarly, the invasions and unrest that beset both the Byzantine and Islamic empires were not conducive to the stability required for gold mining. Nevertheless, there was much more mining and placering for gold and silver in the Middle Ages than is generally thought. The reason is of course that both precious metals were necessary for specie to conduct trade and for commercial expansion of the Byzantine, Islamic, and western European Empires, especially during the High Middle Ages.

Consider the Byzantine Empire first. In the early period when the empire controlled Egypt, the Aegean, Asia Minor, and the trade route through the Black Sea, the supply of gold was adequate, its source being the Balkans and Asia Minor with the principal path lying through Alexandria from Egypt (Nubia) and the lands bordering the Gulf of Aden and the Indian Ocean (Ethiopia, Somalia, Mozambique, Malagasy, etc.). After the fall of Egypt (Alexandria) in 641 to the Arabs the Black Sea trade route assumed prime importance, for a supply not only of gold, but also of corn and other staples formerly supplied mainly by Egypt. The gold reaching Constantinople along this route seems to have come principally from the lands north of the Black Sea under the suzerainty of the Khazars and Magyars where nomadic gold washers worked placers during the Middle Ages, and also from the placers of the Pontic goldfield on the southern coast of the Black Sea and from the Caucasus and Middle Asia. Other sources of gold to the Byzantine Empire were Romania, Thrace and Macedonia (Bulgaria), and the Aegean Islands. But the supply of gold from these areas was generally small and the resource soon depleted, a circumstance that promoted intensive prospecting in many of the Balkan areas, in Asia Minor, and the Aegean Islands. This effort helped only marginally as did also the Iconoclastic Movement (eighth and ninth centuries) that sought to tap the Church's accumulation of precious metals. Further problems of availability of precious metals and specie marked the early part of the thirteenth century, especially after the Fourth Crusade was diverted by the cupidity of Venetian traders into plundering and sacking Constantinople in 1204. The general scarcity of gold (and silver) continued in the Byzantine Empire after the sack of Constantinople until the fall of this great city to the Turks in 1453.

As the Moslem Empire grew from 632 to 732 it came into possession or control of gold mines and placers reaching from Spain to the Indus. In the eastern part of their empire gold (and silver) mines and placers were opened or reopened in Syria, Armenia, Arabia (Midian), Afghanistan, and in areas as far east as the Indus. The mines in Egypt and Nubia seem to have been in operation, and gold was obtained from many sources from countries bordering the Gulf of Aden and the Indian Ocean. Del Mar (1902) and numerous other writers claim that the cities such as Jibuti, Mombasa, Zanzibar, Mozambique, and Sofala were built on the African mainland and in Malagasy (Madagascar) as emporia for the prosecution of the gold trade. The source of the gold that supplied this trade in eastern Africa is uncertain, but was probably placers and eluvial deposits known to have been worked in the auriferous regions of Ethiopia, Somali, Kenya, Tanzania, Zimbabwe, Mozambique, and Malagasy (i.e., the ancient golden land of Punt; see also chap. 2).

The western part of the Islamic Empire provided gold (and much silver) from locations in Portugal, Spain, Morocco, and Mauritania. The trans-Saharan trade route through Timbuktu brought gold to Barbary (Fez, Tunis, Tripoli) from the placers and eluvials of Wangara, a fabled golden land, the source of its gold closely guarded as a secret for centuries. It has been estimated that more than a quarter of a million ounces of gold passed along the trans-Saharan caravan route to Barbary and thence to Europe during the fifteenth and sixteenth centuries. Many Moorish and later European expeditions were dispatched to discover Wangara, but with little result until the early part of the nineteenth century when it was realized that the fabled land was widespread, embracing the ancient kingdoms of Mali, Ghana, Songhai, and possibly Hausa. More specifically the Wangara of the ancients is equatable with the west African gold belts of Bambuk-Buri, Lobi-Ashanti, and northern Nigeria, drained respectively by the Senegal and its tributary the Falémé, the Volta, and the Niger rivers. An interesting account of Wangara and the gold trade of the Moors is by Bovill (1958). In Spain gold placers were opened or reopened by the Moors along the Lerida, a tributary of the Ebro, and at various sites washed by the Romans (Sil Valley, Esla Valley, Tagus River, Guadalquiver River, etc.). Gold mines, some formerly worked by the Romans, were reopened in many parts of Spain and Portugal, particularly in Baetica (Andalusia), Jaen, Bulache, Aroche, and in the mountains flanking the Tagus. The golden treasure won from the mines of the Iberian Peninsula and that obtained from western Africa ultimately ended up in the hoards of the Caliph and his fiefs, finally to be plundered by the Norsemen and the crusading Christians. According to Del Mar (1902) this metallic spoil, combined with that from the sack of Constantinople in 1204 by the Fourth Crusade and the plundering of Jewish money lenders provided the large sums of money to build the great gothic cathedrals of Europe (Westminister, Salisbury, York, Canterbury, Cologne, Milan, St. Peters).

Little is known about the economics of gold mining and placering in the Moslem Empire. Evidently infidels and natives were pressed into mining service by those who worked the mines and placers. Del Mar (1902) remarks that the mines in Spain were worked for the account of private individuals and not for that of the state or the Caliph.

During the years immediately before the fall of the Roman Empire in the west and during most of the early Middle Ages, gold mining and placering in central and western Europe almost ceased and were fitful when pursued, because no miners could safely operate for long because of barbarian incursions, incessant civil wars, and widespread brigandage. There are some notices in ecclesiastical and other records of gold mining by the Avars in the sixth and seventh centuries in the electrum mines at Kremnitz (Kremnica, in west central Slovakia) and at Vorrspatak (Rosia-Montana) in Transylvania (originally Dacia in the Roman Empire; now part of Romania). Similar eighth-century records reveal that the pagan Saxons, Avars, and Czechs reopened the silver (and some gold-silver) mines in the Harz Mountains, Bohemia, Croatia, the Banat (Romania-Yugoslavia), and in the metalliferous regions of the Carpathians, many of which had been prospected and worked by the Romans. In addition there was widespread gold washing when peaceful conditions permitted along many of the rivers in western and central Europe, notably on the Rhine (between Strasbourg and Philippsburg), the Rhone, the Garonne, the Danube, the Elbe, the Po, and the Tiber.

There is remarkably little information on the economics of gold (and silver) mining during the early Middle Ages. Trade was greatly restricted, in places nonexistent, and hence metallic specie was in little demand. The ownership of the few gold mines and placers seems to have belonged to those who could protect them by military force, and the records show that frequent exchanges of control took place. During the Merovingian period (482-639) the numerous regional Frankish kings and lords seem to have farmed out the deposits in their lands, but according to Rickard (1932) as soon as the mines proved profitable they were seized by the king as his royal patrimony, for the purpose of minting coinage. The bishops of the Church seem also to have had suzerainty over certain mining lands in their dioceses and to have been avaricious in demanding titles (royalties) on the metal produced. All labor in the gold mines and placers of the early Middle Ages seems to have been performed by serfs, slaves, and prisoners of war. The Avars (a nomadic tribe originally from central Asia) under Khagan (King) Bajan (565-602) exploited by slave and prisoner-of-war labor many of the electrum mines of central Europe (Kremnitz, Vorrspatak) under a system of fortified rings or enclosures. The Avars were also avid collectors of golden tribute. It is recorded that Bajan exacted from Byzantium an annual tribute of 120,000 gold pieces in addition to many golden gifts including a solid gold bed.

As the Merovingian period of medieval history in western Europe passed into the Carolingian period (679-887) trade increased, religious fervor was widespread, and the demand for metals increased—copper and tin in particular for making bronze church organ pipes and church bells, and iron for the manufacture of armor. Gold, however, was in extremely short supply; widespread minting of gold coins ceased almost completely in western Europe about 850, not to be continued again until about 1253 when the Florentine gold florin and other coins were struck in Florence, Genoa, and other Italian trade cities. Because the Carolingian kings and their successors, the Saxon emperors (911-1024), commanded no substantial gold reserves nor had access to any gold mining areas, they resorted to the development of a monetary system based on silver, which was more available in their lands, particularly in central Europe. Silver, at that time, was also a highly desired metal by the medieval Chinese and Indians and hence was acceptable as a medium of exchange for silks, spices, and other goods that flowed from China and from India to Europe.

One of the key events of the Middle Ages was the *drang nach osten,* the expansion of the Germanic peoples during the eighth to twelfth centuries into the region between the Rhine and the Elbe and thence to the Oder. Coincident with this expansion were widespread prospecting ventures mainly by Saxons, resulting in the discovery (or in some cases the rediscovery), development, and mining of the great silver-bearing deposits at Rammelshberg by Goslar (968), Freiberg (1170), Annaberg (1496), Schneeberg (1471), Marienberg (1520), Joachimsthal (Jachymov, 1576) and many others in the Erzgebirge of Bohemia and in Silesia. Saxon miners migrating eastward also opened or reopened many of the electrum and silver mines in the Carpathians and in Transylvania. Elsewhere in Europe silver mines were opened or reopened, often with the help of Saxon miners in Sardinia, in Serbia and Illyria, and in France and Britain. It was a silver age (eighth to sixteenth centuries) during which coin was struck almost entirely in silver. Our dollar owes its name to this period being derived from the name of the coin, the *Joachimsthaler* first struck in 1519 in Joachimsthal, later shortened to *thaler,* and corrupted to *dollar.*

While many forces were at work in the massive revival and expansion of metal mining in western and central Europe during the High and Late Middle Ages it appears that the ratio of the value between gold and silver was one of the principal economic factors. Del Mar (1902) comments on this in a succinct passage as follows:

> What had occurred to cause this revival? The Fall of the Sacred Empire in 1204. This event loosed the venerable but feeble grasp of the Basileus upon the prerogatives with which Caesar had invested his office, including those of mining, coinage, and the ratio of value between silver and gold. With the fall of the sacred ratio of 12 for 1 the independent kings raised the gold value of their silver coins and thus encouraged numerous silver mines to be opened or re-opened, which previously, and at the sacred ratio of 12 for 1, did not pay to work. (p. 123)

Summarizing the role of gold during the Middle Ages it can be said that the precious metal remained in considerable demand. The gold mines and placers were extensively worked by the Arabs in the regions under their control, and the Byzantine suzerainties and civilization in middle, southern, and eastern Asia also produced gold throughout the period. In most of western Europe (excluding Spain) gold mines and placers were largely exhausted and produced little of the metal, a circumstance that required recourse to silver for specie acceptable along the silk and spice trade routes to India and the Orient.

The great expansion of prospecting and development of mines for silver (and gold) in western Europe during the High and Late Middle Ages led to the emancipation of the Saxon miner from slavery and serfdom because such activity required individual initiative and curiosity. With time the Saxon miner became a free artisan sought for his expertise from Britain to Transylvania. Because the miner's expertise was necessary to operate mines at a profit, miners were granted special privileges during the Late Middle Ages, especially in Bohemia; these privileges included immunity from taxation, free brewing, and freedom from military service. Certain mining communities became free, self-governing towns, as the name of Freiberg, the famous medieval silver mining center amply attests. Rickard (1932) described this period of mining in the following passage.

> The emancipation of the miner, and his development as a free artisan, marks the heroic period of German mining. It coincided with an improvement in the technique of his art and a rapid expansion of the mining industry during the thirteenth, fourteenth, and fifteenth centuries. Aeneas Sylvius, writing in 1458, boasts of the mineral wealth of Germany, and says: "Gold-dust sparkles in the waters of the Rhine; there are rivers in Bohemia in which the Taborites find lumps of gold the size of peas." The finding of rich ore prompted rushes to the places of discovery, and led to the rapid building of towns; after the fashion of the modern mining 'boom', new communities appeared suddenly in mountain or desert like mushrooms over-night. It is recorded that when, in 1471, the rich mines of Schneeberg were started, a town sprang into existence as if by enchantment; and when new orebodies were uncovered at Joachimsthal, in 1576, the news attracted miners from afar, causing a rapid influx of population. Such stampedes, however, were under

> more definite control in those days than they have been in a later time on our American frontier, for the restraints of feudalism were still felt even during a mining excitement. (p. 553-554)

Toward the end of the Middle Ages many of the privileges enjoyed by the miners were revoked as capitalism developed. Again from Rickard (1932) we read respecting the merchant princes (e.g., the Fuggers of Augsburg) who became great investors in all types of mines:

> These merchant princes were operators of mines, and much of their wealth was drawn from the profitable production of the metals in Spain, Austria, and Germany. The successful mine-owner of today, enabled to acquire political power, finds in them a historic prototype. The interest taken by the Fuggers and other capitalists in the German mines, however, had the unfortunate effect of disintegrating the old miners' associations, and of contributing thereby to the subjection of the skilled workers, in consequence of which the German mining industry underwent a decline in the latter part of the sixteenth century, this result being accelerated probably by the stream of silver that was coming across the Atlantic from Mexico and Peru. (p. 567)

Agricola in his *De re Metallica* and various other records of the High and Late Middle Ages provides much information on the financing and operation of mines, only a resume of which is possible here. Details should be sought in Rickard (1932) and Nef (1952).

During the Middle Ages certain of the mineral rights belonged to the emperor(s), others to the lords, both lay and spiritual. Gold mines seem in general to have been operated under regalian control, although there were many variations of this control because certain emperors and princes let out gold prospects to lessees who in turn paid a royalty on the amount of gold produced. All mining operations during the first centuries of the High Middle Ages were essentially feudal enterprises, each prince, lord, or bishop working his own mine with the labor of his serfs. Later, the lords and bishops rented out their mine or mines as a form of fief whose owner in turn sublet the mining rights to miner lessees who paid royalties in the form of ore. With the passing of the feudal age, and especially because of the high costs of deeper mining, partnership agreements and associations (unions) among miners developed, and ultimately an early form of capitalism emerged, with all of its attendant problems described by Rickard (1932).

> The tributer, or lessee, also did not long survive. As the ore from the deeper workings proved refractory, it required more complicated methods of reduction, and a longer time in treatment, usually at the smelting-house that belonged to the landlord or his concessionaire. The delay in settlement, and the exactions made for smelting, put the tributers out of business, so that in the fifteenth century we find them mining on a piece-work basis, with allowances for the hardness of the rock and other things, the effect of which was to reduce them to the status of wage-earners. The chief lessee became a captain of

> industry, and hired men as he needed them. The mine associations, no longer connoting the personal labor of its members, began to include non-residents of the mine-cities, and even capitalists living at a distance, such as merchants in the great trading centres. In the fifteenth century companies were organized and stock was sold with a recklessness comparable to the vagaries of a modern mining boom, the result being a final cleavage between capital and labor. The guilds, which originally were fraternal organizations, created for charity and insurance, in due course became formidable fighting bodies, organized for purposes of industrial aggression and defence. Unsatisfactory conditions of employment induced the guilds to call a strike, and started contests that caused great loss to the mine-owners, until, during the Peasants' War, in 1525, these medieval unions were crushed by a combination of lords and capitalists. (p. 553)

Capital, defined as a producing surplus created by man, has existed probably since the advent of *Homo sapiens,* but in the general form with which we are now familiar capitalism had its beginnings during the early Crusades. Even before this time the Romans practiced a form of capitalism but it was greatly restricted, and it hardly touched mining or industry. The prevalence of slavery and a domestic (household) system of production, coupled with the general suspicion and dislike of the emperors for private business on a large scale, led to a situation in which fluid (negotiable) capital was amassed mainly in connection with public or semi-private enterprises, of which mining was one, as noted previously. When the Roman Empire in the west disintegrated, those forms of capitalistic enterprise that existed disappeared, leaving only landed property. Gradually during the High Middle Ages capital began to accumulate first in southern Europe in Italian cities (e.g., Genoa, Venice, Florence) controlling the trade routes to the east and later in the commercial and financial centers (e.g., Champagne, Bruges, Antwerp, and towns of the Hanseatic League) in Northern Europe. Most of this capital, however, was applied to commerce and finance and not to mining and industry. During the Late Middle Ages and the Renaissance large amounts of capital were accumulated by numerous family enterprises such as the Medici of Florence, who were financial agents for the papacy, the Fuggers of Augsburg, great merchants in spices, wool, and silk. The Fuggers also engaged in the financing of silver mines in the Tyrol, copper mines in Hungary, and the mercury mine at Almaden in Spain. It was such capitalistic family concerns that funded the early search and development of many of the silver and gold mining enterprises in central Europe. Later a crude form of the stock company made its appearance, principally to spread the risk of mining over a greater number of investing participants.

ECONOMICS OF GOLD DURING THE TRANSITION TO MODERN TIMES

At the close of the Middle Ages a general decline in both gold and silver mining occurred in Europe and western Asia and Africa (Egypt) as a result of many circumstances, among which were the rich new precious metal discoveries in America, long periods of instability resulting from civil and religious warfare in Europe, and a general conservative and moribund approach by both political and religious authorities to new engineering and scientific advances in prospecting and mining. One view by an economic historian of these and other causes of the decline in the mining of precious metals is given by Nef (1952) in a unique passage:

> At the end of the Middle Ages the rapid development of continental mining and metallurgy showed signs of waning. The discovery of ores extraordinarily rich in silver in South and Central America, and particularly the opening about 1546 of the famous mines of Potosi in Bolivia, dealt a heavy blow to the European silver-mining industry. Treasure from the new world could be delivered in Europe, even by the unwieldy Spanish galleons, more cheaply than the trained miners of Saxony, Bohemia, Tyrol, Hungary, and Silesia could dig and smelt their ores and ship their metal, with the help of the most skilful German, Hungarian, and Bohemian engineers and technical experts. While a few mining communities in Silesia, and at Freiberg and Goslar in Germany, continued to prosper after the middle of the sixteenth century, a slump had begun by that time in the output of silver and gold in most parts of Central Europe, and also in Alsace and Sweden.
>
> Fifty years later this slump in the production of precious metals had gone very far. On the eve of the Thirty Years' War (1618-48), which was to bring mining in Central Europe temporarily almost to a standstill, the annual output of silver was perhaps less than a third as great as it had been in the twenties and thirties of the sixteenth century. Even in Sweden, which unlike most of Central Europe prospered industrially during the hundred years following the Reformation, the production of silver in the best years of the mid-seventeenth century was hardly half what it had been in the fifteen-forties.
>
> At the close of the Middle Ages the progress of continental mining was bound up, almost as much as in the twelfth and thirteenth centuries, with the prosperity of mines rich in silver. The collapse of the market for European silver brought a reduction in the value of argentiferous copper and lead ores. Conditions proved almost equally unfavourable to other kinds of mining. Before the end of the sixteenth century, the rapid expansion in iron smelting and coal mining on the continent came to an end. Thus the signs of industrial revolution at the end of the Middle Ages proved deceptive. The remarkable growth in the output from mining and some other heavy industries lasted only during the interval between the Hundred Years War and the beginning of the religious wars in the mid-sixteenth century. It was two hundred years more before the rate of growth in industrial output again became as rapid on the continent generally as it had been during the late fifteenth and the early sixteenth centuries.
>
> What was lacking to bring about a development of mining and metallurgy that would lead directly to the wealthy industrial civilisation destined to dominate the whole of Western Europe in the

nineteenth century? Why was the new machinery for draining and ventilating mines at considerable depths, devised in connection with the mining of argentiferous copper ores, not taken over extensively in the mining of tin, lead, and above all coal on the continent, as it was in Great Britain in the late sixteenth and seventeenth centuries?

At the close of the Middle Ages warfare on the continent became more destructive and more damaging to heavy industry. The dissolution of the English monasteries and other ecclesiastical foundations played into the hands of private landlords and merchants eager to exploit mineral wealth. But in continental mining districts, the course taken by the religious struggle was different and the Church retained a greater portion of the landed property. Some churchmen possessed regalian rights by virtue of their political authority. Ecclesiastical foundations were less ready than lay landlords to invest large capitals in mines and metallurigcal plant. They were unwilling to lease their mines on as favourable terms as lay landlords. Again, the natural difficulties of carriage through mountainous country and the numerous tolls and taxes, which stood in the way of transporting heavy ores and coal for considerable distances, imposed handicaps upon the progress of mining on the continent. The great authority over the mines and the mining ventures, established at the end of the Middle Ages by so many continental rulers, discouraged private enterprise.

At the time of the Reformation political considerations frequently outweighed economic in the guidance of mining and metallurgy. While this helped European rulers to strengthen their authority over their subjects, it was on the whole unfavourable to the growth of industrial output, at least to the growth in the output of products like iron and coal, upon whose abundance the progress of modern industrial civilisation has been so largely based. It was only in the eighteenth and nineteenth centuries, after changes in the mining laws first of France and later of Germany made conditions more favourable to the initiative of private capitalists, that the output of mines again grew rapidly in either country.

In exploiting their silver resources, the Western European peoples were only following in the footsteps of their classical predecessors, who had ransacked the surface supplies of argentiferous ores in Spain and all along the shores of the Mediterranean. If the Western Europeans had turned aside from the supplies of iron and coal, as the classical peoples had done, they could hardly have created the industrial world of the late nineteenth century, which seemed to offer a foretaste of the millennium for those who measured happiness primarily in material terms.

It is perhaps beyond the scope of history to enquire what would have happened if there had been no America, and no regions in the North of Europe rich in mineral resources and rich also in industrious work-people. But it is certain that the progress of heavy industries in early modern times was not in those European countries which had been in the vanguard of civilised life during the Middle

Ages. The progress occurred in Sweden, Holland, and above all in Great Britain. All of these countries were protected for various reasons from the full force of the religious struggles and the actual battles. All established, partly with the help of these favours, traditions of constitutional government. Such conditions were helpful for the realisation of the dreams of Bacon and Descartes.

Some novel inventions and discoveries, indispensable for the eventual triumph of industrial civilization, had been made by the western peoples before the Reformation. If these inventions were actually to produce such a triumph, they had to be exploited relentlessly. Their development had to take a position of precedence in the minds of men. At the end of the Middle Ages conditions in continental Europe, where the inventions had been made, were not favourable to such precedence or to such relentless exploitation. Would the triumph of industrialism have occurred without a change of scene? An answer is impossible, but it is certain that the change of scene contributed to the triumph. (pp. 489-492)

ECONOMICS OF GOLD AND GOLD MINING DURING THE MODERN ERA

The modern era of gold mining can be said to have emerged at the beginning of the sixteenth century, that is with the age of the discovery, actually the rediscovery, by Europeans of the newly found lands of Africa, America, Asia, and later of Australia and New Zealand. During the whole of the Middle Ages it has been estimated that the total gold production for Europe, Africa, and the Middle East was little more than 100,000 oz annually or a total of about 3,000 tons (3×10^9 gms) during the period 500-1500. This situation was to change drastically after the rediscovery of America (1492), the general average production of the precious metal increasing more or less consistently to the present day, especially after the discovery of the great gold deposits of the Witwatersrand in South Africa in 1886.

Much has been written about the reasons for the great age of geographic exploration and colonization that characterized European history from the middle of the fifteenth to the close of the eighteenth centuries. Some of these reasons were technological and utilitarian; others were of a more idealistic nature. Among the latter may be mentioned curiosity and the spirit of adventure; the drive of empire building by various European states seeking prestige by adding new lands to their domains; and certain religious missionary motives, principally Christianization of the natives considered to be laboring under paganism. Laudable as these motives were, they remained always secondary to utilitarian drives, assisted by technological advances such as the great improvements in sailing ships, navigation aids, and geographical knowledge.

The utilitarian drives were mainly economic. There was, for instance, a strong economic impulse during the latter part of the fifteenth and early part of the sixteenth centuries to forge more and better trade contacts with the source of various Oriental commodities such as silk and spices, the latter in particular considered as necessary preservatives and condiments during an age when refrigeration was unknown. By venturing westward across the (Atlantic) ocean and southward around Africa the adventurers sought to break down the Italian monopoly over the direct (mainly Mediterranean) trade routes to the near East and to outflank the Turkish control of the trade routes onward to India, China, and other countries of the Orient. But this drive was secondary to another—the acquisition of precious metals, particularly gold, which had become relatively scarce during the last decades of the fifteenth century as a result of payment of costly Oriental goods, Papal taxes, and the exactions of Italian and other financiers. The commissions of all who sailed to discover foreign lands under the crowns of Portugal and Spain and later of Britain, France, Holland, and Germany contained sections requiring them to investigate the occurrence of precious metals and to acquire them, as it ultimately evolved, by any means fair or foul. That great quantities of precious metals were available in these foreign lands was well known. Marco Polo returning to Venice in 1295 from his business ventures to the Near and Far East recounted tales of people in Bhutan who had gold-plated teeth, a pyramid in Burma covered by an inch-thick gold plating, a tower filled with gold in the palace of the Caliph of Baghdad, and so on. The fabulous golden land of Punt (Ethiopia and the eastern seaboard of Africa) was known from antiquity, and the source of the gold that reached the Barbary Coast from the mythical land of Wangara (the Gold Coast and its hinterland) was sought for centuries.

Discussing the monetary conditions and political circumstances of the states of Europe at the times of the discovery of the newly found lands in America, Africa and Asia, Del Mar (1902) had the following statements.

> At this period the quantity of money in circulation, outside of the Moslem states, was extremely small; according to Mr. William Jacob, the vast acquisitions of the (Roman) Empire had disappeared chiefly through wear and tear, coupled with the lack of fresh supplies of the precious metals from the mines. Much of these metals had been taken as spoil by the Moslems and transported to their various empires in the Orient; much had been absorbed and sequestered by the temples and religious houses of the West; and much had also been hidden and lost in secret receptacles. It was estimated by Gregory King in 1685 that the whole stock of the precious metals, in coin and plate, in Europe, at the period of the Discovery of America in 1492 did not exceed £35,000,000 in value; and to this estimate Mr. Jacob, after the most careful researches, lent his full support. The population of Europe at that period could hardly have exceeded thirty millions; so that the quantity of coin and plate did not much exceed in value £1, or say $5 per capita. Of this amount it could hardly be supposed that more than one-half consisted of coins. The low level of prices at this period fully corroborates this view. Moreover, there was nothing to alleviate the scarcity of money; no means of accelerating its movement from hand to hand, and so of increasing its velocity or efficiency; no substitutes for coins; no negotiable instruments; no banks except those of the Italian republics; few or no good roads; no rapid means of communication; little peace or security; and no credit. Since the fall of the Roman empire every device by means of which this inadequate and always sinking Mea-

sure of Value could be enlarged had been tried, but in vain. The ratio of value between gold and silver in the coins had been altered by the kings of the western states with a frequency that almost defies belief. The coins had been repeatedly degraded and debased; clipping and counterfeiting were offences so common that notwithstanding the severest penalties, they were often committed by persons of the highest respectability, by prelates, by feudal noblemen and even by sovereign kings. The emission of leather moneys had been repeatedly attempted, but the general insecurity was too great and the condition of credit too low to admit of any extensive issues of this kind of money. Bills of exchange known to the East Indians as *hoondees* and familiar to the Greeks and Romans of the republican periods, had from the same cause almost entirely fallen into disuse. The cause was the low state of credit. The social state itself, so far as it depended upon that exchange of labour and its products which is impossible without the use of money, was upon the point of dissolution, when Columbus offered to the Crown of Castile his project for approaching the rich countries of the Orient by sailing westward.

What was the object of thus seeking Cathay and Japan? To discover them? They had long been discovered and were well known both to the Moslems, who had established subject states in the Orient, and to the Norsemen, who traded eastward with Tartary and India, and had even voyaged westward to the coasts of Labrador and Massachusetts. The Italians had long traded with the Orient through Alexandria and had even sent Marco Polo into China. No. The voyage of Columbus was not to discover Cathay, but to plunder it; to plunder it of those precious metals, to the use of which the Roman empire had committed all Europe and from the absence of which its various states were now suffering the throes of social decay and dissolution.

The terms which Columbus demanded and the Crown conceded in its contract with him, is a proof of this position. He demanded one eighth of all the profits of the voyage. To this the Crown consented, after making a better provision for itself, by requiring that in the first place one-fifth of all the treasure found or captured in the lands approached should be reserved for the king. The terms of this contract are given more fully in the author's "History of the Precious Metals," and therefore they need not be repeated here. From beginning to end it was essentially a business bargain; its object was not geographical discovery, but gold and silver; its aim was not the dissemination of the Christian religion, but the acquisition of plunder and especially that kind of plunder of which the Spanish states at that period stood in the sorest need.

Said the illustrious Von Humboldt: "America was discovered, not as has been so long falsely pretended, because Columbus predicted another Continent, but because he sought by the west a nearer way to the gold mines of Japan and the spice countries in the southeast of Asia." The expeditions of Cortes and Pizarro had precisely the same objects: to discover and acquire the precious metals, without permitting any considerations of religion or humanity to stand in the way of these objects.

Forty-five years after the Discovery of America the Crown of Spain made a contract with De Soto similar to that with Columbus. It will be instructive to examine its details. This document is dated Valladolid, April 20, 1537. It provides that De Soto shall be paid a salary of 1500 ducats (each, of the weight of about a half-sovereign or quarter-eagle of the present day,) and 100,000 maravedis for each one of three fortresses which he is to erect in the "Indies." To the alcade of the expedition it awards a salary of 200 gold pesos. De Soto may take with him free of duty (almojarifazgo) negro slaves to work the mines. All salaries except that of the alcade are to be paid from the proceeds of the enterprise, so that in case of its failure, there will be nothing to pay. Of gold obtained from mines, the king is to receive during the first year one-tenth, during the second year one-ninth, and so on until the proportion is increased to one-fifth; but of gold obtained by traffic or plunder, he is always to receive a fifth. De Soto shall not be required to pay any taxes. He shall have the entire disposal of the Indians. There shall be reserved 100,000 maravedis a year for a hospital for the Spaniards, which shall be free from taxes. No priests or attorneys shall accompany the expedition, except the alcade and such priests as may be appointed by the Crown. After the king's fifth is laid aside from the spoils of war, and the ransom of caciques, etc., then one-sixth shall go to to De Soto and the remainder divided among the men. In case of the death of a cacique, whether by murder, public execution, or disease, one-fifth of his property shall go to the king, then one-half of the remainder also to the king, leaving four-tenths to the expedition. Of treasure taken in battle or by traffic, one-fifth shall go the king; of treasure plundered from native temples, graves, houses or grounds, one-half to the king without discount, the remainder to the discoverer. Signed, CHARLES, The King.

Here is a charter to murder, torture and enslave human beings, to despoil temples and to desecrate graves. It is signed by the King of Spain who was also the Emperor of Germany; it is committed to a swash-buckler who by the most infamous means had made his fortune with Pizarro in Peru; it is as sordid a document as ever was penned; a disgrace to Spain, to Christianity, to civilization. It plainly and unequivocally lays bare the motive of this expedition. This was not to discover or explore North America, but to plunder it of gold and silver, to replenish the coffers of the king, to provide those blood-stained metals out of which man, in retrogressive periods, is obliged, through his own degeneracy and distrust of his fellow-men, to fabricate his Measure of Value. Said Sir Arthur Helps, the accomplished historian of the Spanish Conquest of America: "The blood-cemented walls of the Alcazar of Madrid might boast of being raised upon a complication of human suffering hitherto unparallelled in

> the annals of mankind . . . . Each ducat spent upon these palaces, was, at a moderate computation, freighted with ten human lives." Let us be still more moderate and say one human life to the ducat: even this was sufficiently atrocious. (pp. 136-138)

These are terrible words, but unfortunately true, and while they chastise and pillory the Spanish conquistadores they apply with equal force to the French, British, Dutch, Portuguese, and other adventurers of the period who were driven by the *auri sacra fames.* Truly it can be said that the period from the beginning of the sixteenth to eighteenth centuries was one of the blackest in the history of gold, being epitomized by the exhortation of King Ferdinand to the colonists of Hispaniola—"Get gold: humanely if you can; but at all hazards get gold."

A short history of the discovery of various gold belts in many parts of the world after the rediscovery of America by Columbus in 1492 is given in the Introduction. Here we shall deal only briefly with the impact that these discoveries have had upon the economy and course of civilization of the modern world.

Following the rediscovery of America by the Spaniards large amounts of both gold and silver, obtained first from the looting of the natives and then from the exploitation of new deposits in Mexico, Mesoamerica, and South America, flowed into Spain. Despite attempts by the Spanish government to keep the precious metals within the confines of the country, they soon filtered over the whole of Europe, increasing the supply of money and inflating prices, but generally contributing to a period of considerable economic growth. More than 10 million oz of gold came from the Spanish and Portuguese Americas during the period 1492-1600, an amount equal to about 40% of the world production at that time. Large increases in gold production from South America, particularly from the placers of the Chocó in Colombia and from Peru, were registered during the period 1600-1800; for example some 48 million oz were won during the eighteenth century, accounting for about 80% of the world production. Two further increases in the world gold production followed during the periods 1820-1880 and 1880-1925. The first marked the discoveries of the great placer and lode mining areas of Siberia (Yenisey River and Lena River Basins), California (Mother Lode system and associated placers), British Columbia (Cariboo), Australia (New South Wales, Victoria, Queensland), and New Zealand (Coromandel, Thames, Otago); the second followed the discoveries of placers and bedrock deposits in Alaska (Nome, Juneau), Yukon (Klondike), central Canada (Larder Lake, Porcupine, Rice Lake, Kirkland Lake, Noranda, and Red Lake), Australia (Kalgoorlie), and on the Witwatersrand of South Africa. Since 1920 there has been only one period, 1933-1939, when the production of gold increased significantly, mainly as a result of the Great Depression and an increase in the price of the metal. In the years following World War II a general decline in the production of gold in nearly all countries except South Africa and the Soviet Union has taken place.

Since 1971, following the recession of the fixed price of gold by the United States, and despite the consequent rise in the price of the metal, there has been little increase in the production of gold. In fact in some years a decrease has taken place. For instance, during 1972 the world production of the metal was approximately 1,400 metric tons; in 1975, 1,150 metric tons; in 1980, 1,145 metric tons; and at the time of writing (1984) about 1,400 metric tons.

The operation and financing of gold mines and placers since 1500 has varied widely. The Spaniards used slave labor (both native and imported) in the operation of their mines and placers in North and South America, as did also the Portuguese in Brazil, the French, British, and Dutch in South America, Africa, and Asia. Slavery in the true sense was generally abolished by 1850 by most European nations but persisted in the United States until 1865. Other forms of forced labor in mines, some producing gold, were employed in Nazi Germany during the period 1933-1945, and in USSR (the so-called corrective labor camps) from about 1920 to 1956. For instance, in 1932 the OGPU (United Government Political Administration) set up corrective labor camps (for convicts and political dissidents) for gold mining along the Kolyma River and elsewhere in the Siberian taiga, which were to persist under various administrations until 1956.

The financing of the early Spanish and Portuguese adventurers who sought Eldorado was essentially provided by monarchs who generally retained a royal fifth of the gold production from any source. (See the details of one contract in the reprint from Del Mar (1902) earlier in this section). Similar financial arrangements were followed by the French, British, Dutch, and other national heads of state. Later (seventeenth century), consortia of financiers (capitalists) funded gold prospecting and mining ventures. By the midnineteenth century joint stock companies, prospecting and development syndicates, and other forms of capitalistic groups provided the financial support for prospecting, development, and mining of gold deposits in capitalistic nations, and this form of financing has persisted in these nations to the present. In socialist and communist states financing has been entirely by the state through prospecting, development, and mining syndicates and trusts.

Discussion of the historical worth (price) of gold through the ages is a subject fraught with much difficulty because of two main problems: first, one of documentation and second, one of assessing former true values in terms of present day currencies. There are no adequate records of the price or worth of gold in pre-Classical times nor is there any way of determining the value of the metal because gold coins were not minted before 650 B.C. Judging from inscriptions, it would appear that gold was much less valuable than silver in the very early Mesopotamian and Egyptian civilizations. In the Code of Menes, the first pharaoh who reigned in Egypt about 3500 B.C., it was decreed that "one part of gold is equal to two and one-half parts of silver in value," giving a value ratio of gold to silver equivalent to 2.5. In Egypt and Mesopotamia and probably elsewhere in ancient centers of civilization gold rings fabricated to standard weights were used as money, but there is no way to evaluate these rings in terms of purchasing power for goods or labor. The gold ring form of currency prevailed in many old civilizations, in Celtic Ireland as long ago as 1600 B.C. and in ancient Britain until the Roman invasion and consolidation under Claudius (A.D. 43).

Herodotus, the Greek historian writing about 430 B.C., tells us that the Lydians, a people occupying a part of western Anatolia (Turkey) centered on the ancient city of Sardis (on the present Gediz River), were in addition to inventing the game of dice "the first people to know or to strike coins of gold and silver." This statement agrees with archaeological evidence unearthed at Sardis where small round gold (electrum) ingots, carrying manmade imprints were discovered in the present century. These gold pieces appear to be small selected nuggets of electrum of the same weight obtained from the nearby auriferous Pactolus stream renowned in the history of gold placers (see chap. 3). They are said to date from about 650 B.C., that is, during the reign of King Ardys (652-605 B.C.), the second of the Mermnadae dynasty. Later, during the reign of the Lydian king Alyattes (605-561 B.C.), father of Croesus, crude coins struck from gold ingots or nuggets bore the royal symbol of the head and forepart of a lion in relief. This coinage was continued during the reign of Croesus (561-546 B.C.), after which time it is apparent that both gold and silver coinage was in widespread usage in the region from the Nile to the Indus.

Alyattes' coins, known as *staters,* were 168 grains (0.35 troy oz) in weight and were variable in purity. Croesus' coins were, likewise, called staters, were also 168 grains, and were 98% gold, giving them a present value (at U.S. $400 an ounce)* of approximately U.S. $140. Their commodity value and worth in terms of wages during Croesus' time cannot be accurately determined because no records exist about these matters, except that one gold stater was worth ten of silver. The Lydian kingdom was conquered by the Persians under Cyrus (546), who was succeeded by Darius the Great (521-486 B.C.). The latter minted a great abundance of gold coins known as *darics* (from *dara=king*) at Sardis. These darics were variable in weight but contained about 130 grains (0.27 troy oz) of pure gold, giving them a present value of approximately U.S. $108. According to ancient records a soldier in Darius' army received one and a half gold darics (U.S. $162) a month. Philip of Macedon, Alexander the Great, and many of the Greek states minted staters containing about 130 grains pure gold; their commodity and wage values were only about 10% higher than that recorded for the Persian darics. The basic gold coin of the Roman republic and emperors was the aureus, equivalent to about 120 grains (0.25 troy oz) (average value U.S. $100), first struck during the time of Sulla and later continued by Julius Caesar and Augustus. Afterwards the aureus was severely debased by successive emperors either by reducing its weight or by alloying with silver and/or copper. A common soldier's pay at the time of Augustus was about 2 aureii per month, an increase of some 20% over that paid by Darius. During the second century of the Roman Empire free miners were paid about 12 denarii a month and their board for work in the gold mines of Dacia (Liversidge, 1968). A denarius at that time was worth about 1/25 aureus or about U.S. $4; hence a miner's pay per month (U.S. $48) was only about a quarter that of a common soldier.

During medieval times many gold coins were minted, first by the barbarians, followed by the *triens* or *tremissis* of the Merovingian Franks and the *solidus* of the Carolingian Franks. Byzantium and the Muslim Empire also struck a number of gold coins; the *solidus* (bezant) of Constantinople (originally introduced in A.D. 330 by Constantine the Great) equal in weight to 24 carats (about 70 grains), and the dinar (65.4 grains) of the muslim caliphates were standard. Each of these was worth about U.S. $57.

By the late thirteenth century Venice and Florence became the principle

*Estimated average price of gold over the period 1975-1984.

trading cities in Europe and minted their own gold coins; the Venetian ducat (53 grains) and the Florentine florin (55 grains) were standard and soon were imitated over most of Europe, particularly in Spain, France, the Germanic states, and England. The worth (price) of gold in the Middle Ages in commodity or labor value appears to have fluctuated only slightly since the days of the Roman Empire. An increase of ten % in commodity value appears probable, judging from the slight increase in the Au-Ag value ratios (11.1-12.2) (Table 18-1) and from the many statements in medieval literature that gold was relatively scarce during the period 1000-1500.

Following the rediscovery of America, as already mentioned, more than 10 million oz of new gold reached Europe from the Americas during the period 1500-1600. By comparison some 400 million oz of silver flowed into Europe from the same source. Because gold and silver at the time were more or less classed as commodities, they followed the general rule that an increase in supply was accompanied by a decrease in exchange value against other goods. This decrease, or in other words a rise in prices or inflation, tended to reduce the purchasing power of a given amount of gold, but by how much is problematical—perhaps about 20% in western Europe, although it may have been much more because according to some records the price of beef rose some 15% between 1500 and 1550 and the price of clothing and other goods rose in some cases as much as 50%. This change unbalanced the general economic structure of Europe and led to serious disturbances in the political structures of a number of countries.

During much of the sixteenth and seventeenth centuries there appear to have been some fluctuations in the exchange rate of gold (value in terms of commodities and labor), but the price seems to have reached a general equilibrium as judged by the Au-Ag value ratio. Certainly gold gained in value from about 1550 to 1700, as indicated by the increase in the Au-Ag value ratio from about 12 to 16 (Table 18-1). This increase was undoubtedly due to the great influx of silver that reached Europe from America during this period, a feature that greatly depressed the price of the white metal.

After 1700 the data on the price of gold is much more complete; during the first years of the eighteenth century the price of gold was about £4.17 per oz (U.S. $19.18), by 1800 it was £4.63 (U.S. $21.34), by 1814, £6.00 (U.S. $27.96), after which it declined to £4.63 (U.S. $21.29) in 1920, and remained in the range of £4.37 (U.S. $20.00) until 1930. If the price of gold is plotted for this period the curve shows only small fluctuations from a median value of about U.S. $20.00. The purchasing power of gold did not increase by much over this period whereas the price of commodities rose significantly (Govett and Harrowell, 1982). In 1934 the price of gold was fixed by the U.S. Treasury at U.S. $35.00, effectively devaluing the dollar by some 40%; it remained at this price with only minor fluctuations in some countries until 1968, when the price was raised to U.S. $38.00, again effectively devaluing the dollar. This devaluation resulted in a dramatic increase in the gold price, which reached U.S. $65.00 on the free market by the end of 1972. Thereafter, numerous international financial crises, "tinkering" with world currencies, uncertainty of political events in South Africa, inflation, and uncertainty concerning future world events fueled speculation and pushed the average price of gold to a high of U.S. $620.00 in January, 1980, despite U.S. Treasury and International Monetary Fund auctions of several millions of ounces of gold, which had temporarily depressed the price to about U.S. $140 in late 1975. Thereafter, the price stabilized at about U.S. $325.00 in 1985. The decline from the very high price in 1980 is attributed to numerous factors, among which may be mentioned the unprecedented interest rates in the United States (at times more than 18%), dishoarding of gold resulting in a glut of gold on the market, a decline of investor interest in gold, a strong U.S. dollar, lower gold purchases by Middle Eastern countries, increased sales of gold by the USSR, and decreased industrial demand for the precious metal.

Historically, numerous attempts have been made to establish a gold standard, usually defined as a monetary system in which the monetary unit is, or is maintained at the value of, a fixed weight of gold. One of the first moves in this direction was initiated by the Emperor Augustus when he established the aureus as the standard in the Roman Empire. The coin, however, was debased by later emperors and fell into disrepute by the fourth century. In the eastern Empire at Constantinople, however, the gold solidus or bezant, first struck by Constantine the Great in A.D. 330, lasted a long time. It was accepted without question as a medium of exchange from China to Britain and from the Baltic to Ethiopia, being superseded only by the florin of Florence and other Italian currencies at the midpoint of the thirteenth century. The Italian currencies dominated the trade routes of the then known world well into the first part of the sixteenth century.

It should not be assumed that these early forms of the gold standard excluded minting of other metals. On the contrary, silver and copper were in early use in ancient Greece and continued so during the Roman, Byzantine, Arab, and other empires. During the Late Middle Ages silver was the principal monetary metal and has continued more or less in this capacity in some countries to the present day. One can say, therefore, that bimetallism (usage of gold and silver as money) has been practiced almost from the invention of coins in the sixth century B.C.

In modern times what is commonly regarded as the gold standard (or its variant the gold-exchange standard) was put into operation in Great Britain in 1821, and in the 1870s the monometallic gold standard was adopted by Germany, France, Spain, Holland, and the Scandinavian countries. Numerous other countries, including Russia, the United States, and India, were to follow in the 1880s and 1890s, thus effectively instituting an international gold standard, whose lifetime, however, proved of short duration, being terminated by World War I when recourse to inconvertible paper money was introduced by most countries. Following this war there was a tendency to return to the gold standard, and in fact by 1928 it was well established in a number of countries. The Great Depression of the 1930s, which followed on the heels of the 1929 crash of the New York Stock Exchange, however, led to "tinkering" in the United States and other countries with the established (supposedly inviolable) money price of gold, thus putting finis to the original intent of the gold standard. In other words, most governments had concluded by the end of World War II that the belief that the gold (and gold money) supply was limited by natural forces and natural forces only, and hence independent of the foibles of politicians and monetary authorities, was abolished forever.

But apparently not entirely. It is generally recognized that what man did formerly in fixing the price of gold he could certainly do again. But what is the result when each nation fixes its own price of gold, and there is no world consensus on a monetary standard to regulate the flow of international trade? A famous Canadian economist and humorist, Stephen Leacock (1932), spoke about this aspect of world finance in a memorable address on gold.

> We have to have—I would not say 'stable prices'; you cannot have a world with all the prices stable; you can only have a world in which the price level moves slowly with a regular and intermittent oscillation. But you can have and must have a fixed exchange. You cannot have a foreign trade based on monetary speculation; you have got to have something absolutely stable for your dealings with foreign nations, and I tell you, gentlemen, I am convinced that there is nothing that will take the place of some absolutely metallic standard, some standard of things, and not of opinions. (p. 435)

Today (1984), as this book is being written, the world financial situation is in bad condition for reasons that are manifold and perplexing even to those who deal with such matters. It is obvious to anyone with only the slightest knowledge of economics that nations are living beyond their means, that there is no control on paper money, and that there is no standard by which to keep all politicians honest. There are some who vociferously advocate the gold standard or some modification thereof; others loudly proclaim the SDR (Special Drawing Rights) of the International Monetary Fund (actually paper gold); and still others are trudging down the trail of demonitizing gold. Whatever the outcome of these three movements, or others that may appear, it is obvious that discipline in international finance is essential, and I suspect that gold as a standard will ultimately enforce that discipline and thus avoid an international monetary holocaust.

USES OF GOLD

The historical uses of gold are of interest. From the dawn of civilization until about 650 B.C. the uses were restricted mainly to ornamentation, decoration, and the display of princely power. This span of time has often been called the decorative or ornamentative stage in the history of the precious metal. After 650 B.C. gold came into general use as money and circulated freely, with only minor restrictive periods, until about 1914. These years formed the monetary stage in the history of the metal, although considerable amounts continued to be used for ornamental purposes. After World War I the movement of gold and its use as money was greatly curtailed and eventually restricted in many countries, most of the metal, except that permitted for the manufacture of jewelery, going directly from the mints into the vaults of most countries. This period continues to the present day in many countries, although there has been a relaxation in some countries that permits their citizens to own and trade in the metal. Since 1950 there has been increased industrial use of gold—sufficient, it would seem, to suggest that we are entering the industrial age in the history of the metal.

Specifically, the uses of gold depend essentially on its traditional role as a monetary measure by governments and central banks in the settlement of international balances; on its intrinsic quality as the most beautiful of metals; and on its chemical inertness, great malleability and high electrical and thermal conductivity. In its international monetary role gold is utilized mainly in the form of high purity bars, tablets, or (more rarely) as coins with a specified gold content. In its other roles the metal is employed in the pure form or alloyed with other metals such as silver, copper, and the platinum metals.

It is difficult to state accurately the apportionment of the annual world production of gold among its various uses. There are several reasons for this difficulty, chief of which is the tendency for certain financial institutions and individuals to add to their hoards of the metal. Rarely is the magnitude of these hoards made public or is their annual increase or decrease known. The purchase of gold, for uses other than the manufacture of jewelery, industrial purposes, and so on, is prohibited by law in many countries, but despite this prohibition, hoarding is traditional in many Asiatic, African, South American, and European countries. It is estimated that about 20% of the annual world production of gold goes into the vaults of governments and central banks as a monetary reserve; some 20% probably passes into the hands of private and corporate hoarders; and the remaining 60%, some 800 metric tons, is used in fabricated articles. Of that 60%, some 80% is consumed in dentistry and the jewelery, coin, and medallion industry; the remainder is used in a great variety of electronics and other industries.

Gold is the royal and aristocratic ornamental metal par excellence with a seductive natural color, luster, and satinlike texture. Since ancient times this most beautiful metal has traditionally been used in articles of personal adornment, particularly in rings and a variety of costume jewelery. Considerable amounts of gold are also used in fountain pens, medallions, gold watches, and in gold fillings in dentistry. Some gold has also been used since antiquity in the fabrication of chalices, cups, plates, vases, shrines, and a great variety of other articles cherished for their intrinsic value. In all of these uses gold is generally alloyed with copper or silver, the fineness of the gold alloy being expressed in *carats*. Pure gold is 24-carat gold.

Pure gold or gold containing a low percentage of silver or copper can be hammered into extremely thin foil, about 0.000005 in. in thickness. In the foil or leaf form gold has been used for a great variety of decorative purposes on buildings, statues, and other articles since pre-Biblical times. Thin layers of gold welded to base metals such as nickel, copper, or brass can be rolled or drawn into complex forms without rupturing the thin gold layers. Material of this nature is called gold-filled plate and is used extensively in various types of jewelery, watch cases, spectacle rims, and so on.

The industrial uses of gold depend essentially on its softness, the ease with which it alloys with silver, copper or platinum metals, its great ductility and malleability, its great electrical and thermal conductivity, and its chemical inertness and hence corrosion-resistance to oxygen, sulfur, most chemical compounds, and nearly all of the single common acids. Most of the gold used in industry is consumed by the electronic and electrical engineering industries, the products ranging widely from coatings of vacuum tubes, specialized electrical contacts, electrical precision-drawn wire leads, and gold electroplate high-frequency conductors, to printed circuits in computers, radios, and television sets. Considerable amounts of gold are also used by industry in the fabrication of high-temperature brazing alloys, linings for specialized chemical and nuclear plants, platinum-gold alloy spinnerets for viscose rayon plants, infrared and thermal reflectors in aircraft and space vehicles, and heat shields for jet and rocket engines. Gold-coated window glass for buildings in hot climates and electrically heated windscreens having a thin transparent and conductive layer of gold are finding an increasing use in cars, aeroplanes, ships and locomotives throughout the world. Some gold is also consumed by the printing and furniture industries in the form of gold paints and by the ceramic industry in the form of organic "liquid golds" for application in pottery and glassware. On firing or other treatment the organic gold compounds are reduced, leaving a thin film of gold tightly bonded to the ceramicware or glass. Small amounts of gold are also used for coloring glass, and gold salts are used to a small extent in certain photographic and medical preparations, the latter in the form of the disodium aurothiomalate for the treatment of rheumatoid arthritis.

THE FUTURE OF GOLD

Gold plays two roles in today's world—one a commodity role and the other a monetary role. In the first its demand factor is keyed mainly to the manufacture of items of adornment (jewelery) and to its increasing usage in electronics and other industrial applications. As far as one can see, the world demand for gold in both these roles will increase in the long term but will be subject to economic fluctuations, some perhaps severe, in the short term.

Predictions of the demand factor in the monetary role of gold are hazardous to say the least—they have been so since the time of Menes, the first pharaoh. There are some who desire to see the complete demonitization of gold and to be rid of the "barbarous relic," to quote John Maynard Keynes (1930). But even Keynes could not quite bring himself to suggest abolition of "the whore of civilization," an allegorization bestowed on gold by some writers in the 1920s and 1930s. Nor have the Soviets followed Lenin's early admonition that gold be used to fabricate the furnishings of public lavatories. On the contrary, as I write, the Soviets are busy vigorously prospecting for gold throughout their empire and increasing their yearly production. Rather, they have followed Lenin's later advice that "the Soviet Union save its gold, sell it at the highest price, and buy goods with it at the lowest price—in other words when living among wolves, howl like the wolves." Nor, finally, do I observe any of the Western, Middle Eastern, or Far Eastern countries disposing of their gold at bargain prices; quite the reverse—they continue to husband and in many cases add to their hoards. From these factual indications I suspect that gold will retain its international monetary role unless Utopia should descend upon us suddenly, at which time we will have no need of an objective, tangible, enduring, weighable, readily transportable medium such as gold. Gold is the discipline that will ultimately curb the power of the printing press, all other opinions to the contrary.

The price of gold depends on many factors, some accurately definable, and some emotional and motivational and hence difficult to assess. Among the latter are the elements of fear and avarice. Under fear may be included particularly the degree of confidence that individuals place in their respective nation's paper notes, base metal coinage, or other inconvertible currency. When the confidence is high, purchase and hoarding of gold is low, depressing the price of the metal; when low, the reverse obtains. Influencing the confidence factor is the rate (rise or decline) of inflation and the cost of borrowing money (interest rate). When the latter is low, purchase of gold generally increases, thus stimulating an increase in the price of gold; when high, the reverse is the case. When inflation is high and increasing, gold is purchased as a hedge, a circumstance that generally results in an increase in the price of the metal; when inflation is low and decreasing, the reverse is true. When bankruptcy looms within an empire or nation, gold is hoarded and becomes scarce, as it did when the Roman Empire became bankrupt in the fifth century, when sundry empires in the Middle Ages went bankrupt, and when Germany approached bankruptcy in the 1930s. Political instability, both national and international, tends to raise the price of gold, as do also threats of civil disobedience, terrorism, wars (both civil and international), all of which may disrupt or curtail supplies of oil, metals, food, and commodity goods vital to the economy of importing nations.

Avarice is defined as an inordinate human desire of gaining and possessing wealth in any form; with respect to gold the motive is a form of greed or gain leading to speculation on the part of individuals, corporations, banks, and so on in the hope that the gold purchased will increase in price and hence confer a benefit on the purchaser. It is manifestly impossible to quantify all the factors of fear and avarice that bear on the price of gold. One can do no more than guess how the price of the metal will react to any particular national or international situation. In general, however, one can say that bad news tends to stimulate an increase in the price of gold, and good news, the opposite. The amount of the increase or decrease appears to depend directly upon the degree of solemnity of the bad news and pleasantry of the good.

Predictions of the demand factor in the future commodity (industrial) role of gold can be made with a fair degree of assurance. Since 1980 there has been about a 1% annual growth in the industrial usage of gold, and growth is likely to continue if the price of gold remains relatively stable in the range U.S. $400-$500 an ounce. With a higher gold price, substitute materials will become more economical. At present about 60% of the annual (newly mined) world production of gold (800 tons) is used in industry, including that employed in dentistry, jewelry, electronics, drugs, and so on.

Combining the speculations on the two demand factors bearing on gold (emotional and commodity) and the strong probability that a slow decline in the world production of gold will ensue as discussed in the abstract below, it appears to be a safe prediction that a relatively consistent upward trend in the price of gold will prevail, although one can expect abberations in the trend from time to time.

A forecast of the world production of gold in the next decades, say to the year 2000, depends on several factors, among which the most important are the cost of production of an ounce of gold under normal conditions and the availability (i.e., the discovery and development rate) of new deposits. The ratio between the world price and the cost of production of an ounce of gold (the profit margin) under normal mining conditions will probably remain relatively constant, at least in the short term. In the long term, prediction is precarious, principally because the cost of finding and developing new gold mines will increase and the mining of orebodies at great depths in known mines will certainly increase the costs.

The future availability of auriferous deposits, and hence the future world production of gold, depends on the rate of discovery and development of new deposits. Internationally this feature has been addressed by Pretorius (1981) in a classic analysis, of which the abstract follows.

> "*As South Africa goes, so the World goes*" has long applied to the production and supply of gold. Since South Africa has been the source of between 40 and 50 percent, at least, of all the gold that Man has ever mined in historical times, the truth of the dictum is readily apparent. However, this situation is changing, and, by the end of the 20th Century, the U.S.S.R. probably will have the most important influence on the total amount of gold mined.
>
> In 1979, of a World production of 1,380.6 metric tons of gold, South Africa contributed 703.3 tons (53 percent), the U.S.S.R. 376.5 tons (28 percent), Canada 49.1 tons (4 percent), the U.S.A. 28.3 (2 percent), and the rest of the World 182.7 tons (13 percent). Between 1970 and 1979, South Africa's production declined by 30 percent, Canada's by 34 percent, and the U.S.A.'s by 48 percent, but that of the U.S.S.R. increased by 28 percent. During this same period, World production fell by 15 percent, the nett effect of a decrease of 25 percent on the part of the Western World and an increase of 15 percent by the Communist countries. The total demand for gold between 1970 and 1979 rose by 40 percent, with a progressively greater excess of demand over supply from 1976 onwards.
>
> Sales of gold in 1979 amounted to 1,765 metric tons, 384 tons more than was actually produced. Private bullion purchases accounted for 26 percent, official coins 16 percent, and medallions 2 percent, so that 44 percent of the gold sold was employed as a hedge against inflation. Industrial demand absorbed 56 percent, with 42 percent being used for the manufacture of jewellery, 5 percent for dentistry, 5 percent for electronics, and 4 percent for other industrial purposes. The jewellery trade thus claims the largest share, by far, of gold coming on to the market. Sales of gold for the manufacture of jewellery are most susceptible to fluctuations in the price of the metal. Very marked decreases in demand characterised the elevated price of gold from 1973 to 1975, and the sharp upturn in price between 1978 and 1979 produced another conspicuous fall in the amount of gold purchased.
>
> Estimates of the trends of future gold production speculate that, by the year 2000, South Africa's production might have decreased to 50 percent of what it was in 1979. The assumption is made, in this calculation, that no new goldfields will be found in the Witwatersrand Basin, but that several new mines will come into operation within known fields. It is anticipated that production from Canada, the U.S.A., Brazil, and Australia will be enhanced. Despite this, total production from the Western World will have declined by 17 percent by 2000. The U.S.S.R.'s output will have increased by 35 percent, and a much greater amount will be mined in China, so that the overall figure for the Communist World will have improved by 56 percent.
>
> The nett effect on World production will be to increase the 2000

figure to only 5 percent above that for 1979. It is thought that World demand will have risen by 10 percent of the 1979 figure, after peaking in the early 1990s. The World shortfall will have increased by 30 percent between 1979 and 2000. Such a situation points to a continued positive upward trend in the price of gold over the closing two decades of the 20th Century.

A disturbing aspect of the above forecasts is that, before the end of the century, the U.S.S.R. will have supplanted South Africa as the World's leading producer of gold and that, within the first quarter of the 21st Century, output from the Communist countries will exceed that from the Western World. Thus, the U.S.S.R. will be in a position to manipulate, to a considerably greater degree, the availability of gold to the World market and, consequently, the price of the metal. The effects on Western economies could be serious. To offset the possibility of such a development, it is imperative that significant discoveries be made, in Western countries, of new Witwatersrand-type gold deposits. The decline in production cannot be countered by the enhanced exploitation of the type of mineralization which contributes to the major proportion of output from Canada, the U.S.A., Australia, and South America. Because of the time-lag between the commencement of exploration and the start of production, efforts will have to be intensified, in the immediate future, with respect to searching for new goldfields within the known Witwatersrand Basin and for further basins, containing the same type of gold mineralization, in Southern Africa and elsewhere, if the Western World is not to be relegated to a subordinate role in influencing the supply of, and demand for, gold in the next twenty years.

CONCLUSIONS

On the retrospect and prospect of the subject of gold I consider that my conclusions written in 1976 remain valid (Boyle, 1979, p. 492), and I repeat them here.

> As we look in retrospect at the last 5000 years of the history of gold we see that the noblest and most beautiful of metals has played a remarkable and sometimes dominant role in human experience and progress, first in an ornamental way, next in coinage, then as an international medium of exchange, and now as an indispensable element in industry. We see also that gold has elicited both good and evil works in man, as have also most other materials of this earth. On the good side of the coin, avarice for the metal has given us great discoveries both chemical and geographical; on the evil side the *auri sacra fames* has led to conquest, enslavement of nations, civil contention and the vilest treatment of men ever devised.
>
> As we look in prospect we perceive an industrial vista where gold will play an ever increasing role in the production of high speed computers, telecommunications, space vehicles, pharmaceuticals

and a thousand other artifacts of future civilizations. We perceive also that man will lose none of his fascination for the metal that he has long admired for its natural beauty and enduring qualities.

REFERENCES AND SELECTED BIBLIOGRAPHY

Altman, O., 1960. The role of gold in international liquidity, *Mines Mag.* **50** (10):39-42.

Bovill, E. W., 1958. *The Golden Trade of the Moors,* Oxford University Press, London, 281p.

Boyle, R. W., 1979. The geochemistry of gold and its deposits, *Canada Geol. Survey Bull.* 280, 584p.

Cousineau, E., and P. R. Richardson, 1979. *Gold: The World Industry and Canadian Corporate Strategy,* Centre for Resource Studies, Queen's University, Kingston, Ontario, 192p.

Crane, W. R., 1908. *Gold and Silver,* John Wiley & Sons, New York, 727p.

Davidson, I., and G. Weil, 1970. *The Gold War,* Secker & Warburg, London, 245p.

Davies, O., 1935. *Roman Mines in Europe,* Clarendon Press, Oxford, 291p.

Del Mar, A., 1902. *A History of the Precious Metals,* 2nd Ed., Cambridge Encyclopedia Company, New York, 478p.

Green, T., 1970. *The World of Gold,* Simon & Schuster, New York, 254p.

Green, T., 1973. *The World of Gold Today,* Arrow Books, London, 287p.

Govett, M. H., and M. R. Harrowell, 1982. *Gold: World Supply and Demand,* Australian Mineral Economics Pty., Sydney, Australia, 455p.

Harrison, F. A., 1931. Ancient mining activities in Portugal, *Mining Mag.* **45:**137-145.

Healy, J. F., 1978. Mining and Metallurgy in the Greek and Roman world, Thames & Hudson, London, 316p.

Hobson, B., 1971. *Historic Gold Coins of the World,* Doubleday & Co. Inc., Garden City, New York, 192p.

Jacob, W., 1832. *An Historical Inquiry into the Production and Consumption of the Precious Metals,* Carey & Lea, Philadelphia.

Kavanagh, P. M., 1968. Have 6,000 years of gold mining exhausted the world's gold reserves? *Canadian Inst. Min. Metall. Bull.* **61** (672):553-558.

Kavanagh, P. M., 1976. Gold reserves of the world; *Geoscience Can.* **3** (3):151-155.

Keynes, J. M., 1930. *A Treatise on Money,* 2 vols., Harcourt, Brace & Co., New York.

Leacock, S., 1932. If gold should cease to be "gold", *Canadian Inst. Min. Metall. Bull.* **25** (244):430-439.

Liversidge, J., 1968. *Britain in the Roman Empire,* Routledge & Kegan Paul, London, 526p.

Malcolmson, J. W., 1907. The history of gold and silver, *Eng. and Mining Jour.* **84:**1021-1023.

Morgan, E. V., 1965. *A History of Money,* Pelican Books, Inc., Baltimore, Md., 237p.

Nef, J. V., 1952. Mining and Metallurgy in Medieval Civilization, in *The Cambridge Economic History of Europe,* vol. 2, *Trade and Industry in the Middle Ages,* M. Postan and E. E. Rich, Eds., Cambridge University Press, Cambridge, pp. 429-492.

Paul, W., 1970. *Mining Lore,* Morris Printing Co., Portland, Oregon, 940p.

Potts, D. et al., 1981. Gold: Report of the fifth annual commodity meeting, *Inst. Min. Metall. Trans.* **90:**A122-A153.

Pretorius, D. A., 1981. Gold, Geld, Gilt: Future Supply and Demand, *Econ. Geol.* Res. Unit, Univ. Witwatersrand, Johannesburg, *Inf. Circ. No. 152,*15p.

Quiring, H., 1948. *Geschichte des Goldes,* F. Enke, Stuttgart, 318p.

Rickard, T. A., 1932. *Man and Metals,* 2 vols., McGraw-Hill, New York, 1068p.

Sutton, A. C., 1977. *The War on Gold,* Internat. Self-Counsel Press, North Vancouver, Canada, 238p.

Vilar, P., 1975. *A History of Gold and Money 1450-1920,* New Left Books, London, 360p.

Weil, G. L., and Davidson, I., 1970. *The gold war,* Secker & Warburg, London, 245p.

AUTHOR CITATION INDEX

SUBJECT INDEX